PAO-LU HSU
COLLECTED PAPERS

PAO-LU HSU

Pao-Lu Hsu
Collected Papers

Edited by Kai Lai Chung

With the Cooperation of
Ching-Shui Cheng and Tse-Pei Chiang

Springer-Verlag
New York Heidelberg Berlin

Pao-Lu Hsu
Department of Mathematics
Peking University
(1947-1970)
Peking, China

Editor
Kai Lai Chung
Department of Mathematics
Stanford University
Stanford, California 94305
U.S.A.

AMS Classification: 01A75, 62-XX

Library of Congress Cataloging in Publication Data
Hsu, Pao-lu, 1910–1970.
Pao-Lu Hsu collected papers.
1. Mathematics—Collected works. 2. Mathematical statistics—Collected works. I. Chung, Kai Lai, 1917– . II. Title.
QA3.H74 1983 510 82-19281

With 11 illustrations.

Printed and bound by Malloy Lithographing, Inc., Ann Arbor, MI.
Printed in the United States of America.

9 8 7 6 5 4 3 2 1

ISBN 0-387-90725-4 Springer-Verlag New York Heidelberg Berlin
ISBN 3-540-90725-4 Springer-Verlag Berlin Heidelberg New York

Preface

This volume contains all the published mathematical papers by the late Professor Pao-Lu Hsu. A volume of selected papers, including 19 items, six of which were in Chinese, was published by the Science Press in China in 1981. At the request of Beijing University, I took charge of the present, more complete compilation. My task was made easy by a previous list of Hsu's publications in the memorial article in the *Annals of Statistics*, Vol. 7, No. 3 (1979). The list has now been revised to contain 40 items instead of 37. Seven of the papers, Nos. 30, 31, 32, 33, 35, 38, and 40 on the new list, have been translated from the Chinese by Messrs. Sun, Cheng, Hsu, and Shih, as indicated on the first page of each article. This project was financed by Beijing University and accomplished with the goodwill and generosity of these colleagues, and the splendid cooperation of the staff of Springer-Verlag, with the personal blessing of Dr. H. Götze. These papers, as well as several others published after Hsu's return to China in 1947, are now available to international readers for the first time. In addition, Hsu's old friend, H. Robbins, contributed the hitherto unpublished portion of their joint paper No. 24; Hsu's colleagues Kiang and Tuan in Beijing rewrote their memorial tribute for this occasion. Last but not least, we are fortunate to obtain a rare photo of Hsu reproduced from his old library card (frontispiece). The photo of Hsu with M. S. Bartlett and H. Cramér (page vii) was taken by the late Mrs. Marta Cramér in Chapel Hill in December 1946, and was given to me by Professor Cramér. The reproduced letter in Hsu's handwriting dates from the same period. As a sequel to the last paragraph of the letter: Hsu did not get to the Princeton Bicentennial, because, as he wrote in a later letter, he did not have a black tie.

Commentary and discussion of Hsu's work in inference, multivariate analysis, and probability by Lehmann, Anderson, and myself are reproduced here. A few more comments by different people, but regrettably rather too few, have been appended to some of the papers. It should be noted that Nos. 37 and 38 were published under the name "Ban Cheng", meaning, literally, "class achieved", used by Hsu to indicate teamwork that he supervised. Paper No. 40 is Hsu's last written notes, though it is questionable whether he would have published it in this form. Indeed, in his gently self-mocking way Hsu would probably have disavowed some of his minor efforts. It seems fitting to include as an appendix the paper by Deemer and Olkin, which was based on Hsu's lectures, as a testimony to his fruitful teaching.

My editorial task is marred by one trivial but tiresome detail: the anglicization of Chinese names. With some trepidation I finally decided to let all extant forms of spelling coexist, sometimes side-by-side. Thus the same name may appear in different forms, with or without inversion or hyphen, in capital or small letters, and by the new or old transcription. A list of Chinese names in their original characters (unsimplified) is included (p. xii) for identification. I should also mention that no attempt was made to unify the terminology used in the various translations. The translator of No. 39 made some improvements in the text.

This volume will appear 12 years after Hsu's death. Although there was much concern over his well-being among his friends in the West, his demise was not confirmed until 1972. It took several more years before the tragic circumstances of his last days became known. It now seems that despite his fragile health, his untimely death at the age of 60 was at least partly attributable to the lack of care during a period of turmoil, persecution, and deprivation prevailing in China at that time. The account by Kiang and Tuan, and the brief comment by Hsu's pupil Zhang appended to the last item of this collection, testify to this sad revelation. If Professor Hsu were still with us today, he would certainly play a major role as a recognized leader in his fields of research, an indefatigable and inspiring teacher, and a good and charming person. His early passing was a great loss to China and to Science.

October 20, 1982 KAI LAI CHUNG

Hsu with M. S. Bartlett (left) and H. Cramér
in Chapel Hill, December 1946.

A Letter from Pao-Lu Hsu to Kai Lai Chung

May 12, Chapel Hill

Dear K.L.

I did not write because kind of busy. I seem to have obtained necessary and sufficient condition that

$$\mathop{\bigstar}_{\nu=1}^{n} V_{n\nu}(x) \to F(x) \qquad \text{(convolution)}$$

where $F(x)$ is the stable law whose ch. fcn. is $e^{-|t|^\alpha}$ $(0 \le \alpha \le 2)$. I believe that ~~this~~ in the domain of weak laws this is the best theorem yet. The method in this work also points to solution of most general problem of weak laws, namely when $F(x)$ is an infinitely divisible law. But further difficulty is great.

It will take me weeks to finish the work (say only the case $\alpha = 1$ has been handled!) & write it up. I am telling this to you for fear of possible clash, as you are hot getting probability results. It has been a miracle that we have never clashed. However, you seem to be reaching the summit of the "strong" while I pride myself in exploring the "weak", which people consider a dead field.

The problem in my hand is so natural that I am in constant worry of clashing with people. So, whenever you communicate with probability people, please spread the news.

Please tell me what will be going on at Princeton's Bicen— celebration around the tenth of June (sort of general occasion). Because I am a possible delegate to it, ~~and~~ I want to know as much as possible the nature of this occasion.

Sincerely,

P.L.

Contents

*Numbers in brackets refer to bibliography on pp. 2–4.

Permissions

Springer-Verlag would like to thank the original publishers of Pao-Lu Hsu's papers for granting permissions to reprint specific papers in this volume. The following list contains the credit lines for those articles [see bibliography, pp. 2–4] that are under copyright.

[1] Reprinted from *Bull. Amer. Math. Soc.* **41** 502–504. © 1935 by the American Mathematical Society.

[3] Reprinted from *Statist. Res. Mem.* **2**, 1–24. © 1938 by the University of London.

[4] Reprinted from *Statist. Res. Mem.* **2**, 91–104. © 1938 by the University of London.

[5] Reprinted from *Ann. Math. Statist.* **9**, 231–243. © 1938 by the Institute of Mathematical Statistics.

[6] Reprinted from *Ann. Eugenics* **9**, 250–258. © 1939 by Cambridge University Press.

[7] Reprinted from *Proc. Cambridge Philos. Soc.* **35**, 336–338. © 1939 by Cambridge University Press.

[9] Reprinted, with the permission of Kraus Reprint, a division of Kraus–Thomson Org. Ltd., from *Proc. Edinburgh Math. Soc.* **(2)6**, 185–189. © 1940 by the Edinburgh Mathematical Society.

[10] Reprinted from *Biometrika* **31**, 221–237. © 1940 by Biometrika Trust.

[11] Reprinted from *J. London Math. Soc.* **16**, 183–194. © 1941 by the London Mathematical Society.

[12] Reprinted from *Biometrika* **32**, 38–45. © 1941 by Biometrika Trust.

[13] Reprinted from *Biometrika* **32**, 62–69. © 1941 by Biometrika Trust.

[14] Reprinted from *Ann. Eugenics* **11**, 42–46. © 1941 by Cambridge University Press.

[15] Reprinted from *Ann. Eugenics* **11**, 39–41. © 1941 by Cambridge University Press.

[17] Reprinted from *Sankhyā* **6**, 253–254. © 1943 by the Indian Statistical Institute.

[18] Reprinted from *Ann. Math. Statist.* **16**, 1–29. © 1945 by the Institute of Mathematical Statistics.

[19] Reprinted from *Ann. Math. Statist.* **16**, 204–210. © 1945 by the Institute of Mathematical Statistics.

[20] Reprinted from *Ann. Math. Statist.* **16**, 278–286. © 1945 by the Institute of Mathematical Statistics.

[21] Reprinted from *Quart. J. Math. Oxford Ser.* **17**, 162–165. © 1946 by Oxford University Press.

[22] Reprinted from *C. R. Acad. Sci. Paris*, 467–469. © 1946 by L'Academie des Sciences, Paris.

[23] Reprinted from *Ann. Math. Statist.* **17**, 350–354. © 1946 by the Institute of Mathematical Statistics.

[25] Reprinted from Appendix III in *Limit Theorems of Sums of Independent Random Variables* (revised edition) by B. V. Gnedenko and A. N. Kolmogorov, translated by K. L. Chung. © 1968 by Addison-Wesley.

[26] Reprinted from *Proc. Berkeley Symp. Math. Statist. Probability*, 359–402. © 1949 by the University of California Press, Berkeley and Los Angeles.

[29] Reprinted, with the permission of Kraus Reprint, a division of Kraus–Thomson Org. Ltd., from *Proc. Edinburgh Math. Soc.* **(2)**, 37–44. © 1953 by the Edinburgh Mathematical Society.
Appendix reprinted from *Biometrika* **38**, 345–367. © 1951 by Biometrika Trust.

List of Chinese Names (surname first)

Bann Cheng	班成
Chen Jia-Ding	陳家鼎
Cheng Ching-Shui	鄭淸水
Chiang Tse-Pei	江澤培
Chung Kai-Lai	鍾開萊
Hsu Pao-Lu	許寶騄
Hsu Pei	徐佩
Kiang Tsai-Han	江澤涵
Shih Chung-Tuo	石仲拓
Sun Tze-Chien	孫自健
Tuan Hsio-Fu	段學復
Zhang Yao-Ting	張堯庭
Zheng Zhong-Guo	鄭忠國

The Annals of Statistics
1979, Vol. 7, No. 3, 467–470

PAO-LU HSU 1910–1970

BY T. W. ANDERSON, K. L. CHUNG AND E. L. LEHMANN
Stanford University and University of California, Berkeley

Pao-Lu Hsu was born in Peking in 1910 and obtained his B.Sc. degree from Tsing Hua University in 1933. He spent the years 1936–1940 at University College, London where he received a Ph.D. in 1938 and an D.Sc. in 1940. From London he returned to China to accept a professorship in the Department of Mathematics at Peking University.

The war years were difficult (letters from 1943–44 to Professor Neyman in Berkeley mention starvation), but Hsu continued his research. Efforts to bring him to the United States finally succeeded, and in 1945 he arrived just in time for the First Berkeley Symposium on Probability and Statistics. A semester of teaching at the University of California was followed by a semester at Columbia. This was the year in which Hotelling moved from Columbia to North Carolina, and he offered Hsu an associate professorship in the department he was creating there. Thus Hsu spent 1946–47 in Chapel Hill from where—in spite of many efforts to keep him in America—he returned to his professorship at Peking University.

Professor H. F. Tuan of Peking University informs us that Hsu died in his home on the campus of Peking University on December 18, 1970, from chronic tuberculosis. There was a memorial meeting in his honor.

Hsu was born into a mandarin family from the famed lake city of Hangchow, but was brought up in Peking. Perhaps because of this background he spoke the "common" (or "official") dialect with an interesting soft overtone. In appearance and carriage he was a Chinese scholar in the classic mold. His British education formed his taste in mathematics; he preferred the hard and concrete to the general and abstract. Setting even higher standards for his own work than for others, he would temper his critical sense with a gentle mocking humor. He could work feverishly on research for spells, but used to lament that life's diverse interests conflicted with a single-minded devotion to science. He enjoyed the company of men and women of the other culture and was versed in traditional Chinese letters. A particular hobby was chanting the musical drama of the Yuan dynasty with a small group of connoisseurs accompanied by ancient instruments. Partly owing to fragile health he never married but apparently came close to it. He returned to China in the summer of 1947, shortly before the Communist victory despite offers of positions in the U.S. by A. Wald and others. He was looking forward to being part of the emerging new society in his homeland. In his last years he was too ill to get around but continued to teach from his room. The Chinese scientific community

Received June 1978.
AMS 1970 *subject classifications.* Primary 01A70; secondary 00A15.
Key words and phrases. Obituary, bibliography.

gathered to pay homage to him at his funeral. Apart from his published papers and a few remarks given us by an old friend, we are unable to obtain further information about Hsu's life and work in the twenty some years he lived in Peking.

Hsu is affectionately remembered by many students and colleagues as a gentle, shy and modest man who was reticent about his personal life, but had a strong influence as a teacher and model of a scientist. Isadore Blumen, who was a student of Hsu during his year at Chapel Hill, writes: "It was Hsu's insistence on simplicity combined with depth of understanding, clarity without avoidance of difficulty, and above all a deep and obvious but unspoken commitment to the highest goals and standards of scholarship which attracted us to him". Ralph Bradley remembers Hsu's lectures as "models for the future", and Herbert Robbins says "he was unforgettable, and I fear irreplaceable". While Hsu may not have been the formal supervisor of any Ph.D. students he strongly influenced a number, among them Kai-Lai Chung, Sen-Ming Len and Shou-ren Wang before he came to the West, and Isadore Blumen, Albert Bowker, Erich Lehmann and Ingram Olkin in the U.S.

Hsu's statistical work was concerned primarily with inference in univariate and multivariate linear models and with the associated distribution theory, both exact and asymptotic. A more detailed discussion of this and his probabilistic work follows.

We gratefully acknowledge the assistance of the following persons, who provided us with their recollections of Hsu, with information about his life, and with suggestions for and corrections of our account: I. Blumen, R. Bradley, S. S. Chern, F. N. David, C. Eisenhart, J. Neyman, I. Olkin, E. S. Pearson, H. Robbins, S. Stigler, H. F. Tuan.

Pao-Lu Hsu: Publications*

⟨1935⟩

[1] On the limit of a sequence of point sets. *Bull. Amer. Math. Soc.* 41, 502–504.

[2] A note on the indices and numbers of nondegenerate critical points of biharmonic functions (with Kiang Tsai-han). *Science Quarterly, Nat. Univ. Peking* (3) 5, 389–392.

⟨1938⟩

[3] Contribution to the theory of "Student's" t-test as applied to the problem of two samples. *Statist. Res. Mem.* 2, 1–24.

[4] On the best unbiased quadratic estimate of the variance. *Statist. Res. Mem.* 2, 91–104.

[5] Notes on Hotelling's generalized T. *Ann. Math. Statist.* 9, 231–243.

⟨1939⟩

[6] On the distribution of roots of certain determinantal equations. *Ann. Eugenics* 9, 250–258.

[7] A new proof of the joint product moment distribution. *Proc. Cambridge Philos. Soc.* 35, 336–338.

⟨1940⟩

[8] On n-fold iterated limits. *J. Chinese Math. Soc.* 2, 40–63.

[9] An algebraic derivation of the distribution of rectangular coordinates. *Proc. Edinburgh Math. Soc.* (2) 6, 185–189.

[10] On generalized analysis of variance. *Biometrika* 31, 221–237.

* *Editor's note*: This is a revised list. The original list contains 37 items.

⟨1941⟩

[11] On the limiting distribution of roots of a determinantal equation. *J. London Math. Soc.* 16, 183–194.
[12] On the limiting distribution of the canonical correlations. *Biometrika* 32, 38–45.
[13] Analysis of variance from the power function standpoint. *Biometrika* 32, 62–69.
[14] Canonical reduction of the general regression problem. *Ann Eugenics* 11, 42–46.
[15] On the problem of rank and the limiting distribution of Fisher's test function. *Ann. Eugenics* 11, 39–41.

⟨1942⟩

[16] The limiting distribution of a general class of statistics. *Acad. Sinica Science Record* 1, 37–41.

⟨1943⟩

[17] Some simple facts about the separation of degrees of freedom in factorial experiments. *Sankhyā* 6, 253–254.

⟨1945⟩

[18] The approximate distributions of the mean and variance of a sample of independent variables. *Ann. Math. Statist.* 16, 1–29.
[19] On the approximate distribution of ratios. *Ann. Math. Statist.* 16, 204–210.
[20] On the power functions of the e^2-test and the T^2-test. *Ann. Math. Statist.* 16, 278–286.

⟨1946⟩

[21] On a factorization of pseudo-orthogonal matrices. *Quart. J. Math. Oxford Ser.* 17, 162–165.
[22] Sur un théorème de probabilités dénombrables (with Kai Lai Chung). *C. R. Acad. Sci. Paris*, 467–469.
[23] On the asymptotic distributions of certain statistics used in testing the independence between successive observations from a normal population. *Ann. Math. Statist.* 17, 350–354.

⟨1947⟩

[24] Complete convergence and the law of large numbers (with H. Robbins). *Proc. Nat. Acad. Sci. U.S.A.* 33, 25–31.

⟨1948⟩

[25] A general weak limit theorem for independent distributions. (Appendix III in *Limit Theorems of Sums of Independent Random Variables* by B. V. Gnedenko and A. N. Kolmogorov, translated by K. L. Chung, revised edition, Addison-Wesley 1968).

⟨1949⟩

[26] The limiting distribution of functions of sample means and application to testing hypotheses. *Proc. Berkeley Symp. Math. Statist. Probability*, 359–402. Univ. of California press, Berkeley and Los Angeles.

⟨1951⟩

[27] Absolute moments and characteristic function. *J. Chinese Math. Soc.* (*New Series*) 1, 257–280. (In English with abstract in Chinese.)
[28] A lemma on the coefficient of reduction of a sum of independent variates. *Acad. Sinica Science Record* 4, 197–200.

⟨1953⟩

[29] On symmetric, orthogonal, and skew-symmetric matrices. *Proc. Edinburgh Math. Soc.* (2), 37–44.

⟨1954⟩

[30] On characteristic functions which coincide in a neighborhood of zero. *Acta Math. Sinica* (Succeeding *J. Chinese Math. Soc.* (*New Series*)) 4, 21–32. (In Chinese with abstract in English.)

⟨1955⟩

[31] On a kind of transformations of matrices. *Acta Math. Sinica* 5, 333–346. (In Chinese with abstract in English.)

[32] On a kind of transformations of matrix pairs. *Acta Sci. Natur. Univ. Pekinensis* 1, 1–16. (In Chinese with abstract in English.)

⟨1957⟩

[33] Simultaneous transformation of a hermitian matrix and a symmetric or a skew-symmetric matrix. *Acta Sci. Natur. Univ. Pekinesis* 3, 167–210.

⟨1958⟩

[34] The absolute continuity of the distribution functions in the class *L*. *Acta Sci. Natur. Univ. Pekinesis* 4, 145–150. (In Chinese with abstract in English.)

[35] The differentiability of the probability transition function of a purely discontinuous stationary Markov process on the euclidean space. *Acta Sci. Natur. Univ. Pekinesis* 4, 257–270. (In Chinese with abstract in English.)

⟨1964⟩

[36] An association scheme $M_3(6)$ which is not a L_3-scheme. *Acta Math. Sinica* 14, 177–178. (In Chinese.)

[37] The limiting distributions of order statistics. *Acta Math. Sinica* 14, 694–714. (Team work in Chinese.)

[38] Partially balanced incomplete block designs. *Prog. Math.* 7, 240–281. (Team work in Chinese.)

⟨1965⟩

[39] On the coincidence property of stochastic matrices (with Cheng Jia-ding and Zheng Zhong-guo). *Acta Sci. Natur. Univ. Pekinesis* 1 (1980), 21–47.

⟨1970⟩

[40] BIB matrices, simple codes and orthogonal codes. (Last lecture notes.)

DEPARTMENT OF STATISTICS
STANFORD UNIVERSITY
STANFORD, CALIFORNIA 94305

DEPARTMENT OF MATHEMATICS
STANFORD UNIVERSITY
STANFORD, CALIFORNIA 94305

DEPARTMENT OF STATISTICS
UNIVERSITY OF CALIFORNIA
BERKELEY, CALIFORNIA 94720

The Annals of Statistics
1979, Vol. 7, No. 3, 471–473

HSU'S WORK ON INFERENCE

BY E. L. LEHMANN

University of California, Berkeley

Hsu spent four years (1936-1940) at University College, London, where E. S. Pearson had recently succeeded his father in the chair of statistics and where, during the first two years, Neyman was a Reader in the Statistics Department. During this period Hsu wrote a remarkable series of papers on statistical inference which show the strong influence of the Neyman-Pearson point of view.

In 1938 Hsu's first two statistical papers appeared in Vol. II of the Neyman-Pearson edited *Statistical Research Memoirs*. The first of these [2] is concerned with what today is called the Behrens-Fisher problem. If X_i and $Y_j (i = 1, \cdots, m; j = 1, \ldots, n)$ denote samples from normal distributions $N(\xi, \sigma^2)$ and $N(\eta, \tau^2)$, Hsu considers the class of statistics $u = (\bar{Y} - \bar{X})^2/(A_1 S_x^2 + A_2 S_Y^2)$ where $S_X^2 = \Sigma(X_i - \bar{X})^2$ and $S_Y^2 = \Sigma(Y_j - \bar{Y})^2$. This reduces to u_1, Student's t, for $A_1 = A_2 = N/mn(N - 2)$ where $N = m + n$ and to the Behrens-Fisher statistic u_2 for $A_1 = 1/m(m - 1)$, $A_2 = 1/n(n - 1)$.

Hsu finds a series expansion for the density of u, and utilizes this to study the power function of the rejection regions $u \geqslant C$ in terms of the parameters $\theta = \tau^2/\sigma^2$ and $\lambda = (\eta - \xi)^2/(\frac{\sigma^2}{m} + \frac{\tau^2}{n})$. It is an exact (not asymptotic) analysis, described by Scheffé (1970) as "a model of mathematical rigor". In the process, he obtains stochastic bounds for u_2 which were later taken up independently and generalized by Hájek, Lawton and others (cf. Eaton and Olshen (1972)). Hsu's main conclusion, obtained by a combination of his analytical study with some numerical work, is that for $\lambda = 0$ and varying θ neither u_1 nor u_2 control the rejection probability at all well (except when $m = n$) although of the two, u_2 is less sensitive to variation of θ.

In the second paper [3], Hsu treats the question of optimal estimators of the variance σ^2 in the Gauss-Markov model. In the spirit of the Gauss-Markov theorem, he considers estimators Q which are (a) quadratic and (b) unbiased. In addition he imposes the restriction (c) that the variance of Q be independent of the unknown means. (This is a forerunner of the condition he imposed in [12] for the power function of analysis of variance tests).

Hsu then obtains a necessary and sufficient condition for the usual unbiased estimate S^2 of σ^2 to have uniformly minimum variance within this class of estimators. He illustrates the condition on a number of examples and, in particular, shows that S^2 has the desired property in the one-sample case. The problem was

Received June 1978.

AMS 1970 *subject classifications.* Primary 01A70; secondary 62F05, 62F10.

Key words and phrases. Obituary, hypothesis testing, estimation.

taken up again by Rao (1952) who replaced (c) by the condition that Q be definite, while Seely (1971) instead considered an equivalent invariance condition. Hsu's paper may be regarded as the point of origin of the recent large literature on best quadratic estimates of variances and variance components.

The remaining papers in this series devoted to small-sample inference are concerned with the problem of testing univariate and multivariate linear hypotheses, particularly with power properties of the tests of these hypotheses. In the first of these papers [4], Hsu obtains the power of Hotelling's T^2-test, showing that the nonnull distribution of T^2 is that of the ratio of a noncentral χ^2 to an independent central χ^2 variable. He also proves that the test is in a certain sense locally most powerful. In addition, he points out some new applications of T^2.

From a study of T^2 it was a natural step to a multivariate extension of the general linear (univariate) hypothesis discussed by Kolodzieczyk (1935). Generalizing some examples considered earlier in the literature, Hsu in [9] formulates the general multivariate linear hypothesis in its canonical form. He obtains the nonnull distribution of the likelihood ratio test statistic when the covariance matrix is known. For the case of unknown covariance matrix, if θ_i denote the nonvanishing roots of the associated determinantal equation, Hsu considers the test statistics $W = \Pi(1 - \theta_i)$ and $V = \Sigma\theta_i/(1 - \theta_i)$, and proves that their asymptotic power agrees as the sample size tends to infinity. In a succeeding paper [13], Hsu shows how the general multivariate regression problem can be reduced to the earlier canonical form so that the theory of [9] becomes applicable.

Perhaps the most important paper in this series is [12], where Hsu obtains the first optimum property for the likelihood ratio test of the univariate linear hypothesis, in fact essentially the first nonlocal optimum property for any hypothesis specifying the value of more than one parameter. It had been shown by Kolodzieczyk that a uniformly most powerful similar test does not exist for linear hypotheses with more than one constraint. If λ denotes the noncentrality parameter, which is the only parameter on which the nonnull distribution of the likelihood ratio test for this problem depends, Hsu proves that this test is uniformly most powerful among all tests with power function depending only on λ. It was later realized that this condition is equivalent to invariance of the power function under the natural class of transformations. The resulting optimum property is essentially also equivalent to the fact that the test is uniformly most powerful invariant.

This paper started two lines of development. On the one hand, Hsu's formulation was applied to multivariate problems (Hotelling's T^2 and the multiple correlation coefficient) by his student Simaika (1941) and modifications of the formulation were considered by Wald (1942), Hsu (1945) [19], Wolfowitz (1949), Lehmann (1959) and others. On the other hand, in the course of this paper Hsu provided a new method for obtaining all similar tests. At Hsu's suggestion this method was applied to other problems by Simaika (1941) and Lehmann (1947), and was formalized by means of the concept of completeness by Lehmann and Scheffé (1950).

REFERENCES

EATON, MORRIS L. and OLSHEN, RICHARD A. (1972). Random quotients and the Behrens-Fisher problem. *Ann. Math. Statist.* **43** 1852–1860.

HSU, P. L. (1945). On the power functions for the E^2-test and the T^2-test. *Ann. Math. Statist.* **16** 278–286.

KOLODZIECZYK, S. (1935). On an important class of statistical hypotheses. *Biometrika* **21** 161-190.

LEHMANN, E. L. (1947). On optimum tests of composite hypotheses with one constraint. *Ann. Math. Statist.* **18** 473–494.

LEHMANN, E. L. (1959). Optimum invariant tests. *Ann. Math. Statist.* **30** 881–884.

LEHMANN, E. L. and SCHEFFÉ, HENRY (1950). Completeness, similar regions and unbiased estimation, *Sankhyā* **10** 305–340.

RAO, C. RADHAKRISHNA (1952). Some theorems on minimum variance estimation. *Sankhyā* **12** 27–44.

SCHEFFÉ, HENRY (1970). Practical solutions of the Behrens-Fisher problem. *J. Amer. Statist. Assoc.* **65** 1501–1508.

SEELY, JUSTUS (1971). Quadratic subspaces and completeness. *Ann. Math. Statist.* **42** 710–721.

SIMAIKA, J. B. (1941). On an optimum property of two important statistical tests. *Biometrika* **32** 70–80.

WALD, ABRAHAM (1942). On the power function of the analysis of variance test. *Ann. Math. Statist.* **13** 434–439.

WOLFOWITZ, J. (1949). The power of the classical tests associated with the normal distribution. *Ann. Math. Statist.* **20** 540–551.

DEPARTMENT OF STATISTICS
UNIVERSITY OF CALIFORNIA
BERKELEY, CALIFORNIA 94720

The Annals of Statistics
1979, Vol. 7, No. 3, 474–478

HSU'S WORK IN MULTIVARIATE ANALYSIS

BY T. W. ANDERSON
Stanford University

From 1938 to 1945 Hsu published papers in the forefront of the development of the mathematical theory of multivariate analysis. It can be presumed that he was influenced by his proximity to R. A. Fisher, who was also at University College, London. After 1945 he lectured on multivariate analysis at Columbia University and the University of North Carolina, where he trained students who pursued research in this area. As a highly-trained mathematician Hsu promoted the use of matrix theory in statistical theory as well as proving new theorems about matrices.

A crucial element of multivariate theory is the distribution of the sample covariance matrix **S**. When the p-component vectors are independently distributed each according to $N(\mathbf{0}, \mathbf{\Sigma})$, $(N-1)\mathbf{S} = \mathbf{A} = \Sigma_{\alpha=1}^{N}(\mathbf{X}_\alpha - \overline{\mathbf{X}})(\mathbf{X}_\alpha - \overline{\mathbf{X}})'$ has the so-called Wishart distribution with density (for **A** positive definite) of

$$w(\mathbf{A}|\mathbf{\Sigma}, n) = K(\mathbf{\Sigma}, n)|\mathbf{A}|^{\frac{1}{2}(n-p-1)}e^{-\frac{1}{2}\operatorname{tr}\mathbf{\Sigma}^{-1}\mathbf{A}}, \tag{1}$$

where $K(\mathbf{\Sigma}, n)$ is a constant depending on $\mathbf{\Sigma}$, its order p, and the "degrees of freedom" $n = N - 1$. For $p = 2$ the density of a_{11}, a_{22}, and $r_{12} = a_{12}a_{11}^{-\frac{1}{2}}a_{22}^{-\frac{1}{2}}$ was obtained by Fisher (1915) in his famous paper, which marked the beginning of rigorously derived exact small-sample distribution theory. Wishart's paper of 1928 derived the density (1) by use of a geometric argument, which was, roughly speaking, a generalization of that of Fisher. Since Wishart's publication many alternative proofs have been given. That of Hsu [5], based on algebra and analysis, is particularly elegant. To derive the density for p and n Hsu assumed it for $p - 1$ and $n - 1$. In addition to the matrix with this density there is needed a $(p - 1)$-component normal vector and an n-component normal vector. With a little algebraic manipulation only the analytic derivation of the χ_n^2-distribution for the norm squared of the n-component vector is needed to complete the proof.

Mahalanobis, Bose and Roy (1937) approached the distribution of **A** by writing $\mathbf{A} = \mathbf{TT}'$, where **T** is lower triangular ($t_{ij} = 0, i < j$), and derived the distribution of the $p(p + 1)/2$ elements of T, called "rectangular coordinates", from the distribution of $\mathbf{X}_1, \cdots, \mathbf{X}_N$. The (nonzero) elements of **T** were shown to be independent when $\mathbf{\Sigma} = \mathbf{I}$. Each off-diagonal element of **T** has the standard univariate normal distribution, and the ith (nonnegative) diagonal element has a χ-distribution with $n - i + 1$ degrees of freedom. Hsu's derivation of the distribution of rectangular coordinates in [8] is algebraic and analytic instead of the more geometric method of Mahalanobis, Bose and Roy. It is in the same spirit as the derivation of the Wishart distribution in [5].

Received June 1978.
AMS 1970 *subject classifications.* Primary 01A70; secondary 62H10.
Key words and phrases. Obituary, multivariate analysis.

The next most basic distributions to be derived in multivariate analysis were those of the roots of certain determinantal equations. That this was the "natural" development is signified by the fact that such distributions were discovered independently and almost simultaneously by Hsu [6], Fisher (1939), Girshick (1939), Mood (1951) and Roy (1939). (For Girshick and Mood the research was intended for doctoral dissertations, but having been "beat out" by Fisher and Hsu they turned to other topics. Bose (1977) has some interesting remarks on the independence of the work of Fisher and Roy.) For **A** positive semidefinite and **B** positive definite let $\theta_1 \geqslant \theta_2 \geqslant \cdots \geqslant \theta_p \geqslant 0$ be the roots of

$$|\mathbf{A} - \theta(\mathbf{A} + \mathbf{B})| = 0. \tag{2}$$

If **A** and **B** are independently distributed according to $W(\boldsymbol{\Sigma}, m)$ and $W(\boldsymbol{\Sigma}, n)$, respectively, with $m \geqslant p$ and $n \geqslant p$, the density of $\theta_1, \cdots, \theta_p$ is a constant times

$$\Pi_{i=1}^{p}\theta_i^{\frac{1}{2}(m-p-1)}\Pi_{i=1}^{p}(1-\theta_i)^{\frac{1}{2}(n-p-1)}\Pi_{i=1}^{p}\Pi_{j=1+1}^{p}(\theta_i - \theta_j). \tag{3}$$

If the roots of $|\mathbf{A} - \phi\mathbf{B}| = 0$ are $\phi_1 \geqslant \cdots \geqslant \phi_p \geqslant 0$, then $\phi_i = \theta_i(1-\theta_i)^{-1}$, $i = 1, \cdots, p$. Hsu made the transformation $\mathbf{A} = \mathbf{W}\mathbf{D}_\phi\mathbf{W}'$, $\mathbf{B} = \mathbf{W}\mathbf{W}'$, where $\mathbf{D}_\phi$ is a diagonal matrix with $\phi_1, \cdots, \phi_p$ as diagonal elements. The transformation is one-to-one if $w_{1j} > 0$ (with probability 1). The derivation involves (i) substituting the above expressions into the density of **A** and **B**, namely, $w(\mathbf{A}|\boldsymbol{\Sigma}, m)w(\mathbf{B}|\boldsymbol{\Sigma}, n)$, (ii) multiplying by the Jacobian of the transformation, (iii) integrating out the elements of **W** to obtain the marginal density of $\phi_1, \cdots, \phi_p$, and (iv) converting to $\theta_1, \cdots, \theta_p$. The difficult part of this procedure is the calculation of the Jacobian. In fact, in this paper Hsu states the Jacobian for arbitrary p but proves it only for $p = 3$. (The 12×12 matrix of partial derivatives is written explicitly and evaluated.)

A tractable method of evaluating the Jacobian was subsequently developed by Hsu and was presented in his lectures at the University of North Carolina. Unfortunately Hsu never wrote up his lecture notes, but several students recorded them. Deemer and Olkin (1951) elaborated the methods and results and made them available (with the approval of Hsu, then in China). The case of $m < p$ was also treated in [6]. The roots of $|\mathbf{A} - \lambda\boldsymbol{\Sigma}| = 0$ have the density which is a constant times

$$\Pi_{i=1}^{p}\lambda_i^{\frac{1}{2}(m-p-1)}e^{\frac{1}{2}\Sigma_{i=1}^{p}\lambda_i}\,\Pi_{i=1}^{p}\Pi_{j=1+1}^{p}(\lambda_i - \lambda_j) \tag{4}$$

for $\lambda_1 \geqslant \lambda_2 \geqslant \cdots \geqslant \lambda_p > 0$. The exposition in Chapter 13 of Anderson (1958) is based on the papers of Hsu and Deemer and Olkin. The constants in the densities are not given here in order to contain the exposition, but they are, in fact, important (in obtaining moments of likelihood ratio test criteria, for instance) and require careful treatment.

In obtaining the distribution of the roots it is crucial that the Wishart distributions be central. Suppose $\mathscr{E}\mathbf{X}_{\alpha t} = \boldsymbol{\mu}_t$, $\alpha = 1, \cdots, N_t$, $t = 1, \cdots, k$. Let $\mathbf{A} = \Sigma_{t=1}^{k}N_t(\overline{\mathbf{X}}_t - \overline{\mathbf{X}})(\overline{\mathbf{X}}_t - \overline{\mathbf{X}})'$ and $\mathbf{B} = \Sigma_{t=1}^{k}\Sigma_{\alpha=1}^{N_t}(\mathbf{X}_{\alpha t} - \overline{\mathbf{X}}_t)(\mathbf{X}_{\alpha t} - \overline{\mathbf{X}}_t)'$. Then **A** and **B** are independent, $\mathbf{B}$ has a central Wishart distribution and $\mathbf{A}$ has the noncentral Wishart distribution depending on $\mathbf{\Psi} = N^{-1}\Sigma_{\alpha=1}^{k} N_t(\boldsymbol{\mu}_t - \bar{\boldsymbol{\mu}})(\boldsymbol{\mu}_t - \bar{\boldsymbol{\mu}})'$ as well as on $\mathbf{\Sigma}$ and $k - 1$. [The noncentral Wishart distributions for ranks of $\mathbf{\Psi}$ being one or two were obtained independently by Anderson and Girshick (1944)—again, intended as dissertation topics]. In [10] Hsu treated the roots of $|\mathbf{A} - \phi\mathbf{B}| = 0$ when $\mathbf{B}$ has a central Wishart distribution and $\mathbf{A}$ has a noncentral Wishart distribution. The distribution of these random roots depends on $\boldsymbol{\mu}_1, \cdots, \boldsymbol{\mu}_k$, and $\mathbf{\Sigma}$ only through the roots of

$$|\mathbf{\Psi} - \lambda\mathbf{\Sigma}| = 0. \tag{5}$$

The exact distribution of $\phi_1, \cdots, \phi_p$ when $\mathbf{\Psi} \neq \mathbf{0}$ is very complicated. Hsu treated the asymptotic problem as $N_t \to \infty$ such that N_t/N_s, $t, s = 1, \cdots, k$, and $\mathbf{\Psi}$ remain fixed. Then the elements of $(1/N^{\frac{1}{2}})(\mathbf{A} - \mathbf{\Sigma} - \mathbf{\Psi})$ and $(1/N^{\frac{1}{2}})(\mathbf{B} - \mathbf{\Sigma})$ have a joint limiting normal distribution. If the roots of (5) are different and positive, then the properly normalized elements of $\mathbf{D}_\phi$ or $\mathbf{D}_\theta$ and $\mathbf{W}$, have a joint asymptotic normal distribution.

Hsu treated the more general situation where the roots of (5) can have arbitrary multiplicity and some may be 0. Let the different nonzero roots of (5) be $\lambda_1 > \cdots > \lambda_\nu > 0$ with multiplicities $m_1, \cdots, m_\nu$, respectively; then the root 0 has multiplicity $p - \Sigma_{j=1}^{\nu} m_j$. Let

$$\zeta_i = N^{\frac{1}{2}}(\phi_i - \lambda_h)/\left(2\lambda_h^2 + 4\lambda_h\right)^{\frac{1}{2}}, \tag{6}$$

$$i = m_1 + \cdots + m_{h-1} + 1, \cdots, m_1 + \cdots + m_h, \quad h = 1, \cdots, \nu,$$

$$\zeta_i = N\phi_i, \qquad i = m_1 + \cdots + m_\nu + 1, \cdots, p. \tag{7}$$

Then the $\nu + 1$ sets of normalized roots are independent in the limiting distribution. The density for a set corresponding to $\lambda_h > 0$ is a constant times

$$e^{-\frac{1}{2}\Sigma_{i=1}^{m} x_i^2} \prod_{i=1}^{m} \prod_{j=i+1}^{m} (x_i - x_j) \tag{8}$$

for $x_1 > \cdots > x_m > 0$ $(m = m_h)$. The density for the set corresponding to the zero root is a constant times (4) where p is replaced by the number of zero roots of (5) and n is replaced by the sum of $k - 1$ and the number of zero roots of (5) (if $k - 1 > p$). As we shall see, the provision for zero roots is essential to treat problems of rank; the inclusion of multiple positive roots is for mathematical generality. The treatment of such generality required great ingenuity and mathematical technique. The corresponding treatment of $\mathbf{D}_\phi$ (or $\mathbf{D}_\theta$) and simultaneously $\mathbf{W}$ was done by Anderson (1951b).

If $\mathbf{A} = \Sigma_{\alpha=1}^{N}(\mathbf{X}_\alpha - \bar{\mathbf{X}})(\mathbf{X}_\alpha - \bar{\mathbf{X}})'$ and $\mathbf{\Sigma}$ are partitioned into p_1 and p_2 rows and columns,

$$\mathbf{A} = \begin{bmatrix} \mathbf{A}_{11} & \mathbf{A}_{12} \\ \mathbf{A}_{21} & \mathbf{A}_{22} \end{bmatrix}, \qquad \mathbf{\Sigma} = \begin{bmatrix} \mathbf{\Sigma}_{11} & \mathbf{\Sigma}_{12} \\ \mathbf{\Sigma}_{21} & \mathbf{\Sigma}_{22} \end{bmatrix}, \tag{9}$$

the sample and population canonical correlations are the roots of

$$(10) \qquad \begin{vmatrix} -\lambda \mathbf{A}_{11} & \mathbf{A}_{12} \\ \mathbf{A}_{21} & -\lambda \mathbf{A}_{22} \end{vmatrix} = 0, \qquad \begin{vmatrix} -\lambda \boldsymbol{\Sigma}_{11} & \boldsymbol{\Sigma}_{12} \\ \boldsymbol{\Sigma}_{21} & -\lambda \boldsymbol{\Sigma}_{22} \end{vmatrix} = 0,$$

respectively. In [11] Hsu finds the asymptotic distribution of the normalized sample roots for arbitrary multiplicities of the canonical correlations including 0. This paper parallels [10].

In [9] Hsu considers Wilks' likelihood ratio criterion, equivalent to $W = |\mathbf{B}|/|\mathbf{A} + \mathbf{B}|$, and Lawley's trace criterion $V = \operatorname{tr} \mathbf{AB}^{-1}$ for testing the null hypothesis that $\boldsymbol{\mu}_1 = \cdots = \boldsymbol{\mu}_k$. He showed that $NV + N \log W \to 0$ in probability if $N\boldsymbol{\Psi}$ as defined above has a limit. In effect this implies that the tests behave similarly when the alternatives approach the null hypothesis at the rate of the reciprocal of the square root of the sample sizes. An implication of the reduction to canonical form [13] is that this conclusion holds for the general linear hypothesis. The problem considered in [14] can be stated in the above terms. That $\boldsymbol{\mu}_1, \cdots, \boldsymbol{\mu}_k$ lie on an l-dimension hyperplane in the p-space is equivalent to $\boldsymbol{\Psi}$ being of rank l or that the number of 0 roots of (5) is $p - l$. To test this hypothesis Fisher (1938) proposed the test criterion $\Sigma_{j=l+1}^{p}\phi_j$ (Anderson (1951a) showed that the likelihood ratio test is based on $\Pi_{j=l+1}^{p}(1 + \phi_i)$.) Hsu deduced from [11] that N times Fisher's criterion has a limiting χ^2-distribution with $(p - l)(k - 1 - l)$ degrees of freedom.

Asymptotic distributions of ratios. In [18] and [22] Hsu applied Cramér's (1937) theory of asymptotic expansions and Berry's (1941) theorem to the distributions of ratios. Let $\overline{Y}_m = \Sigma_{i=1}^{m} Y_i/m$ and $\overline{X}_n = \Sigma_{i=1}^{n} X_i/n$, where the Y_i's and X_i's are independent, $Y_1, \cdots, Y_m$ are independently, identically distributed random variables with finite positive absolute moment of some specified order and $X_1, \ldots, X_n$ are independently, identically distributed positive random variables with finite positive moment of the same order. The probability distribution of $\overline{Y}/\overline{X}$ can be written

$$(11) \qquad \Pr\{\overline{Y}/\overline{X} \leqslant z\} = \Pr\{-z\overline{X} + \overline{Y} \leqslant 0\}.$$

Each of the distribution functions of $\overline{X}$ and $\overline{Y}$ suitably normalized can be expanded with the Berry bound on the error. Then by means of the convolution of these two distributions Hsu found an expansion of (11) with a bound on the error.

A serial correlation which is used to test independence of $X_1, \cdots, X_N$ normally distributed with the same means and variances can be written as $T = Q/S$, where $Q = \Sigma_{i,j=1}^{N} a_{ij}(X_i - \overline{X})(X_j - \overline{X})$ and $S = \Sigma_{i=1}^{N}(X_i - \overline{X})^2$. By an orthogonal transformation of $X_1, \cdots, X_N$ to $Y_1, \cdots, Y_N$ with a change of scale, one can write $Q = \Sigma_{i=1}^{N-1}\lambda_i Y_i^2$ and $S = \Sigma_{i-1}^{N-1} Y_i^2$ where under the hypothesis of independence each Y_i has the distribution $N(0, 1)$, $i = 1, \cdots, N-1$. Then

$$(12) \qquad \Pr\{T \leqslant z\} = \Pr\{\Sigma_{i=1}^{N-1}(\lambda_i - z)Y_i^2 \leqslant 0\}.$$

If a_{ij} depends on N, then $\lambda_1, \ldots, \lambda_{N-1}$ depend on N. Under certain conditions on these sequences of sets of roots the distribution (12) has an asymptotic expansion as $N \to \infty$.

In [24], which is a further development of the earlier [15], Hsu considers the limiting distribution of a function $f(\bar{\mathbf{u}}_1, \cdots, \bar{\mathbf{u}}_k)$, where $\bar{\mathbf{u}}, \cdots, \bar{\mathbf{u}}_k$ are independent vector sample means, as the respective sample sizes tend to infinity. On the basis of the central limit theorem applied to the means, and the Taylor expansion of $f(\cdot)$, Hsu obtains in the limit a normal distribution or the distribution of a weighted sum of squares of normal variables (if the variances of the linear terms tend to zero). Hsu applies his general results to obtain the asymptotic distributions of many test criteria, particularly in multivariate analysis.

REFERENCES

ANDERSON, T. W. (1951a). Estimating linear restrictions on regression coefficients for multivariate normal distributions. *Ann. Math. Statist.* **22** 327–351.

ANDERSON, T. W. (1951b). The asymptotic distribution of certain characteristic roots and vectors. *Proc. Second Berkeley Symp. Math. Statist. Probability*. Univ. California Press, Berkeley and Los Angeles, 103–130.

ANDERSON, T. W. (1958). *An Introduction to Multivariate Statistical Analysis*. John Wiley and Sons, New York.

ANDERSON, T. W. AND GIRSHICK, M. A. (1944). Some extensions of the Wishart distribution. *Ann. Math. Statist.* **15** 345–357.

BERRY, A. C. (1941). The accuracy of the Gaussian approximation to the sum of independent variates. *Trans. Amer. Math. Soc.* **49** 122–136.

BOSE, R. C. (1977). Early history of multivariate statistical analysis. *J. Mult. Anal.* **4** 3–22.

CRAMÉR, H. (1937). *Random Variables and Probability Distributions*. Chap. 7. Cambridge Univ. Press.

DEEMER, WALTER L. and OLKIN, INGRAM (1951). The Jacobians of certain matrix transformations useful in multivariate analysis. Based on lectures of P. L. Hsu at the University of North Carolina, 1947. *Biometrika* **38** 345–367.

FISHER, R. A. (1915). Frequency distribution of the values of the correlation coefficient in samples from an indefinitely large population. *Biometrika* **10** 507–521.

FISHER, R. A. (1938). The statistical utilization of multiple measurements. *Ann. Eugen.* **8** 376–386.

FISHER, R. A. (1939). The sampling distribution of some statistics obtained from non-linear equations. *Ann. Eugen.* **9** 238–249.

GIRSHICK, M. A. (1939). On the sampling theory of roots of determinantal equations. *Ann. Math. Statist.* **10** 203–224.

LAWLEY, D. N. (1938). A generalization of Fisher's z-test. *Biometrika* **30** 180–187.

MAHALANOBIS, P. C., BOSE, R. C. and ROY, S. N. (1937). Normalisation of statistical variates and the use of rectangular co-ordinates in the theory of sampling distributions. *Sankhyā* **3** 1–40.

MOOD, A. M. (1951). On the distribution of the characteristic roots of normal second-moment matrices. *Ann. Math. Statist.* **22** 266–273.

ROY, S. N. (1939). p-statistics or some generalisations in analysis of variance appropriate to multivariate problems. *Sankyā* **4** 381–396.

WISHART, JOHN (1928). The generalized product moment distribution in samples from a normal multivariate population. *Biometrika* **20** 32–52.

DEPARTMENT OF STATISTICS
STANFORD UNIVERSITY
STANFORD, CALIFORNIA 94305

The Annals of Statistics
1979, Vol. 7, No. 3, 479–483

HSU'S WORK IN PROBABILITY

BY K. L. CHUNG
Stanford University

Hsu returned from England to Kunming in the middle of the Sino-Japanese war, apparently in 1940. He gave a long course beginning with the theory of measure and integration, through probability theory, and leading to mathematical statistics. For the first part he based his lectures on Carathéodory's *Vorlesungen über Reelle Funktionen*, for the second on Cramér's *Random Variables and Probability Distributions*. He was a polished and vivacious lecturer with complete notes written out beforehand in a notebook he carried. He enjoyed making a fine point in class. For instance, when he was doing the inversion formula for characteristic functions, he took delight in the fact that one could integrate over a single point by the Lebesgue-Stieltjes integral. He had a tattered copy of Cramér's book with many marginal notes and used to say that it was written in a more difficult way than necessary but contained all that was essential for probability. He was a true virtuoso in the method of characteristic functions. His papers [17], [23], [25], [26], [31] and [35] all showed his fascination for, as well as mastery of, this precious tool. Mathematical literature was hard to come by in Kunming during those years, and some of the old volumes shipped from Peking (then called Peiping) were kept in caves to preserve them from air raids. (We did not actually live in caves, but frequently had to run to them for hours at a stretch under raids or alerts.) Among the books so kept was, for example, Kolmogorov's *Grundbegriffe der Wahrscheinlichkeitsrechnung*. At my request Hsu had this book extracted from the caves, but I remember his saying of it: "This is another kind of mathematics". He was not, of course, a probabilist *per se* at a time when such an appellation hardly existed anywhere in the world; by education and inclination he was more fond of problem-solving than generalities. However he never tried to dissuade others from following their own bent and getting interested in those other kinds of things. Indeed, if he could be drawn into a new topic, he would work at it in earnest and quickly produce something of value. For instance, when Borel's work on what he called "denumerable probabilities" attracted my attention, Hsu took an interest also, which led to the paper [21]. This was an early approach to what became known as "recurrent events" under Feller's development. Papers [23] and [30] were probably written under similar circumstances.

But, of course, Hsu's major contribution to probability and its applications flowed from his consummate skill in manipulating characteristic functions. Let me describe and comment on these papers in their chronological order.

Received June 1978.
AMS 1970 *subject classifications.* Primary 01A70; secondary 60E05.
Key words and phrases. Obituary, probability theory.

Paper [17] is, without doubt, one of Hsu's most important works. He wrote it shortly after A. C. Berry obtained the correct order of magnitude of the error estimate in the classical central limit theorem. Namely, if $\{\xi_n, n \geqslant 1\}$ are independent and identically distributed random variables having distribution $P(x)$ with mean 0 and variance 1, and finite third absolute moment β_3, then for all x the distribution

$$F(x) = \Pr\left\{ \frac{1}{n^{\frac{1}{2}}} \Sigma_{r=1}^{n} \xi_r \leqslant x \right\}$$

differs from the standard normal distribution by less than $A\beta_3/n^{\frac{1}{2}}$, where A is an absolute constant. (This result was also proved by Esseen, but Esseen's paper was published later and not known to Hsu at the time.) Hsu extended Berry's method to give a simpler proof of Cramér's theorem on the asymptotic expansion of $F(x)$ when $P(x)$ is nonsingular. But much deeper and farther reaching are the corresponding results when the sample mean

$$\bar{\xi} = \frac{1}{n} \Sigma_{r=1}^{n} \xi_r$$

is replaced by the sample variance

$$\eta = \frac{1}{n} \Sigma_{r=1}^{n} (\xi_r - \bar{\xi})^2.$$

In Hsu's own words: "so much for the known results for the approximate distribution of $\bar{\xi}$. By a purely formal operational method Cornish and Fisher obtain terms of successive approximation to the distribution of any random variable X with the help of semiinvariants. It is hardly necessary to emphasize the importance of turning Cornish and Fisher's formal result (asymptotic expansion without appraisal of the remainder) into a mathematical theorem of asymptotic expansion which gives the order of magnitude of the remainder. In this paper we achieve this for the simplest function next to $\bar{\xi}$, viz. η." The basic idea is as follows. Let

$$X = \frac{1}{n^{\frac{1}{2}}} \Sigma_{r=1}^{n} \frac{\xi_r^2 - 1}{(\alpha_4 - 1)^{\frac{1}{2}}}, \qquad Y = \frac{1}{n^{\frac{1}{2}}} \Sigma_{r=1}^{n} \xi_r,$$

where α_4 is the fourth moment of $P(x)$. These are related to η by

$$\left(\frac{n}{\alpha_4 - 1} \right)^{\frac{1}{2}} (\eta - 1) = X - \frac{1}{((\alpha_4 - 1)n)^{\frac{1}{2}}} Y^2$$

and the crux of the method is to approximate the distribution of the random vector (X, Y) through its characteristic function. Even a cursory perusal of the proof would make it obvious that this is a task of a higher degree of complexity and not just a routine analogue of the results of Berry and Cramér, since X and Y are highly correlated. To achieve the goal Hsu literally added a new dimension to the method of approximation used before him. He was still teaching the course mentioned above when he obtained the result, and was obviously elated. I recall his saying that he made several previous attempts at the problem and the breakthrough came when he realized that he must attack the bivariate approximation rather than evading the difficulties by shortcuts. Hsu's method is applicable to many other functions commonly used in statistics such as sample moments of higher order and Student's statistic. Some of these were carried out by his students. Although Hsu's paper was cited by Bikjalis in his work on normal approximations, a recent book on the subject failed to mention it. In the course of writing this article I asked the authors about this glaring omission and received the regretful reply that they were not aware of Hsu's paper! This is a serious omission and suggests that the analytic power generated by Hsu's method has yet to be fully explored.

In paper [23] the main result is that for a sequence of independent and identically distributed random variables with zero mean and finite variance, we have for every $\varepsilon > 0$:

$$\Sigma_n P\left\{\frac{1}{n}|X_1 + \cdots + X_n| > \varepsilon\right\} < \infty.$$

This is an interesting strengthening of the classical strong law of large numbers in the direction of the Borel-Cantelli lemma. The idea of such a result is probably due to Robbins, but the method of proof is vintage Hsu. The problem is reduced, after a convoluted Fourier inversion, to the estimation of several pieces of an integral involving the characteristic function—a remarkable tour de force. Erdös (1949) later sharpened the result and showed that the conditions on the moments are necessary as well as sufficient. (According to Robbins, their original manuscript also contained a proof of the necessity of the conditions which was not published as being too complicated.) Erdös used a combinatorial type of argument. Baum and Katz extended the result to moments of fractional and higher order.

Paper [35] forms Appendix III in the translation of Gnedenko-Kolmogorov (1968) and its story may be worth telling here. Hsu had always been impressed with the necessary and sufficient conditions for the general form of the central limit theorem due to Feller, if only for the adroit use of characteristic functions there. In a letter dated May 12, 1947, from Chapel Hill, he told me that he had just solved the same problem for all symmetric stable laws. He wrote: "I believe that in a domain of the weak laws this is the best theorem yet. The method in this work also points to solution of the most general problem of weak law, namely [when the limit is] an infinitely divisible law. But further difficulty is great. . . . The problem in my hand is so natural that I am in constant worry of clashing with people. So, whenever you communicate with probability people, please spread the news." A latter dated May 26, 1947, announced the final result to me. It gives the necessary and sufficient conditions under which the row sums of a triangular array of infinitesimal random variables, independent in each row, converge in distribution to a given infinitely divisible distribution. His conditions differ somewhat from those in Gnedenko (1944) in that the "two tails" are combined. The proof is direct and makes use of the kernel $(\sin t/t)^3$. The complete manuscript containing this result and its self-contained proof is paper [35]. It was given to me shortly before Hsu's departure for China in the summer of 1947. (I have erroneously placed the year to be 1946 in the preface to the translation cited above.) Despite his premonition, Hsu did not learn of Gnedenko's 1944 paper until later. In the only letter he wrote me after his return to China, dated March 19, 1950, he acknowledged Gnedenko's priority and urged me in most emphatic terms to keep the only copy of his paper for him. He had clearly staked his own strength on solving this challenging problem, the culmination of a series of papers by such names as Lévy, Khintchine, Kolmogorov, Marcinkiewicz (whom Hsu met), Feller and Gnedenko. He toiled hard and triumphed quickly. Does it matter that another striver has climbed the same peak before? Hsu's method is direct and "starts from nothing". The last three words give a free rendering of a more elusive Chinese saying which Hsu used on another occasion, meaning minimal reliance on precedents. He had always considered this nonreliance as a virtue of mathematical work, and it is apparent in all his papers where use of previous results for intermediate steps are shunned so far as possible. The new edition of the translation of Gnedenko-Kolmogorov (1968) appeared in the middle of the cultural revolution in China; I do not know if Hsu ever saw this paper in print.

Papers [25] and [26] were reviewed in some detail by me in the *Mathematical Reviews*, Vol. 17 (1956), page 274. Since this is easily available, I need not repeat the contents here. Suffice it to say that they are again products of his expertise with characteristic functions. Paper [31] falls in the same category but apparently was never reviewed. In it it is proved that every distribution in the class L is absolutely continuous. So far as I know this was a new result. It was rediscovered by Fisz and Varadarajan in 1963 (*Z. Wahrscheinlichkeitstheorie und Verw. Gebiete*, Vol. 1).

Paper [30] stands out in Hsu's work in probability as it has nothing to do with a characteristic function. Like [31], it is published in Chinese with an English abstract, appeared in the same issue of the journal, and apparently was never reviewed. It deals with the differentiability properties of the transition probability function of a homogeneous Markov process whose state space is Euclidean and whose sample paths are purely jumping. Such processes were studied by D. G. Kendall in 1955, who analysed the differentiation of the transition probability function $P(t, x, E)$ at $t = 0$. In the case of Markov chains, where the state space is discrete, the term-by-term differentiability of the Chapman-Kolmogorov equations:

$$P'(s + t) = P'(s)P(t), \qquad s > 0,$$
$$P'(s + t) = P(s)P'(t), \qquad t > 0,$$

was proved by Austin analytically, and by myself probabilistically (for references to D. G. Kendall and Austin see Chung (1967).) Hsu extended these equations to the Euclidean case, "using more elementary methods and obtaining more precise results". For the first equation above he actually derived an integral representation of $P(t, x, E)$ in terms of the post-exit process, thus generalizing my approach but by purely analytic means. The intimate relation between the analytic and stochastic structures is implicit in his work, and can easily be made explicit.

I did not learn of Hsu's death until my arrival in Peking in June 1972. Hsiao-fu Tuan told me that Hsu spent his last years re-doing a lot of classical combinatorial analysis in his own way and left copious notes. So far the contents of these notes remain unknown.

REFERENCES

CHUNG, K. L. (1967). *Markov Chains With Stationary Transition Probabilities, 2nd ed.* Springer-Verlag.

ERDÖS, P. (1949). On a theorem of Hsu and Robbins. *Ann. Math. Statist.* **20** 286–291.

GNEDENKO, B. V. (1944). Limit theorems for sums of independent random variables. *Uspehi Mat. Nauk.* **10** 115–165. English translation: Translation No. 45, *Amer. Math. Soc.*, New York, 1951.

GNEDENKO, B. V. and KOLMOGOROV, A. N. (1968). *Limit Distribution of Sums of Independent Random Variables*, rev. ed. (translated by K. L. Chung, with Appendices by J. L. Doob and P. L. Hsu). Addison-Wesley.

DEPARTMENT OF MATHEMATICS
STANFORD UNIVERSITY
STANFORD, CALIFORNIA 94305

In Memory of Professor Pao-Lu Hsu

JIANG ZE-HAN (TSAI-HAN KIANG)   DUAN XUE-FU (HSIO-FU TUAN)

Professor Xu Bao-lu (Pao-Lu Hsu) was born in Beijing on September 1, 1910, though his forefathers were natives of Hangzhou, Zhejiang Province. A well-known mathematician, he was a full professor of rank one at Peking University, and concurrently a member of the Division of Mathematics and Physics of Academia Sinica. Professor Xu was also a member of the Fourth National Committee of the Chinese People's Political Consultative Conference. His old illness was aggravated in the late sixties and caused his untimely death on December 18, 1970. He died a bachelor, having never married.

Before enrolling in Tsinghua University in 1930 to learn mathematics, he had studied chemistry for two years at Yangjing University (1928–1930). At Tsinghua he got his Bachelor of Science degree, and went on to the then Peking University for two years' work as assistant in the Mathematics Department. In 1936 Mr. Xu passed the examination and went to Britain on public funds for further study as a graduate student at London University, while at the same time doing studies at Cambridge. In the third year in Britain he emerged as a lecturer at London University, working while studying. He earned his Ph.D. in 1938 and Sc.D. in 1940. That same year, while the War of Resistance against Japan was raging in China, he undauntedly came back to his motherland to be professor at the then Peking University, and gave lectures at the Southwestern Associated University at Kunming, Yunnan. In 1945 he went to the United States of America by invitation and lectured as visiting professor at the University of California at Berkeley, Columbia University, and the University of North Carolina. In October 1947 he resolutely returned to Beijing, and thereafter was engaged in teaching mathematics in Peking University for more than 20 years until he breathed his last.

He was truly a patriotic scientist. It was in 1947 that he returned from the United States of America, where he could have lived much better and had far better facilities for work; at that time he still held valid letters of appointment from some American universities. Upon the liberation of Beijing he sent telegrams to his colleagues back in the States informing them of his joy at the liberation. Although plagued by poor health, he threw his great enthusiasm and energy into his work. In 1950, having recovered somewhat from his long illness, he began lectures on "functions of real variables", took on graduate students, and conducted a seminar on "Fourier transforms and probability". Even his spare time was devoted to helping his comrades with their Russian. The heavy work brought about a recurrence of his illness in the summer of 1951, and he was hospitalized. Concerned about his health, the authorities repeatedly suggested that he go abroad to recuperate. All such suggestions met with his polite refusal; he insisted on continuing his work here. Disregarding his poor health he helped correct the translations from the Russian of "A Brief Course of Mathematical Analysis" by Khintchin, and "A Course in Differential Equations" by Stepanov. From 1956 on his health steadily worsened. Having difficulty in moving about, he worked at home. With a blackboard hanging on the wall of his room, he gave lectures to upper-class students, graduate students, and young teachers; he was in charge of seminars and other academic activities. By the early 1960s his health had deteriorated to such an extent that he could stand in front of the blackboard for only a few minutes before he had to sit down and rest. Nevertheless his teaching did not cease at that time, nor did his work cease under the extremely difficult conditions in the later years. It was only a month before his death that his manuscript on the relationship between experimental design and algebraic coding theory was at last completed, being his final legacy. Found beside his bed the day after his death were piles of manuscripts which serve as a testimony of the superhuman fortitude with which he exerted himself over a period of more than 20 years in his devotion to the development of science in his motherland and to the bringing up of the younger generation of mathematicians.

Mr. Xu was the first Chinese scientist to do research and reach advanced world levels in the field of the theory of probability and mathematical statistics, making a great contribution to the development of his discipline in our country. Back in the early fifties he began to enroll graduate students to study this subject, and presided over seminars on it. In 1956, under Premier Zhou En-lai's guidance, China's first program for the development of science and research was drawn up, which laid full emphasis on the theory of probability and mathematical statistics as an important branch of mathematics, thus bestowing on it an unprecedented momentum. The same year the Ministry of Education arranged for more than 40 teachers and students from various universities and institutes to assemble in Peking University to do research under Mr. Xu. This measure was highly instrumental in filling in what was virtually a blank field in China. The same year Professor Xu was put in charge of China's first teaching and research group on probability and statistics at Peking University. His efforts led to plans and the curriculum for training specialized students. The teaching material for many of the specialized courses was based mainly on his lecture notes. In the ten years from the forming of this teaching and research group until the disastrous cultural revolution, eight classes specializing in probability and statistics completed their studies. Of these, five classes were under his personal direction for discussions and the completion of their theses. In addition he conducted numerous seminars on limit theorems, Markov processes, multivariate analysis, experimental design, order statistics, statistics of stochastic processes, decision theory, combinatorial mathematics, etc. Those present at the seminars included undergraduate and graduate students and teachers of Peking University, researchers from the Mathematics Institute of the Chinese Academy of Science and Technology, and teachers from other universities who had come to Peking University for advanced studies. Papers on the results of these seminars have already been published. As a prominent specialist on probability and mathematical statistics, Professor Xu put forward many ideas and proposals for the future development of this branch of science in our country. He suggested the establishment of a statistics laboratory in the Mathematics Department. In 1963, after consultations with others, he proposed publishing a periodical on probability and statistics; if there were financial difficulties, he would pay part of the expenses himself. It is indisputable, that his ideas would have been of great significance for the discovery and training of personnel, and facilitating the rapid introduction of probability and mathematical statistics into production where they could find widespread use. Unfortunately, for a variety of reasons, his ideas failed to become reality.

Always strict as regarded his work, he was extremely conscientious in his preparations and lucid in his lecturing. This serious attitude towards work can be seen most clearly in his lecture notes written after the liberation: "Abstract Theory of Integration," "Theory of Matrices," and "Point Set Toplogy," which are of lasting value for reference in our teaching. In his later years, his energies focused on young teachers and upper-class students, giving them systematic and detailed guidance. To those whose English was not up to par, he taught English. For those whose foundation in mathematics was somewhat shaky, he carefully corrected their homework assignments. Thus, under his guidance all students made much headway, whatever their original level. One of his favorite methods of training was holding seminars, at which he often added interesting remarks giving background information on the history and present state of the topic under discussion, and pointing out difficulties. The young people derived great benefit from these. It was his custom, after giving a report on a special topic, to rewrite the material, introducing new methods and directions for further study in depth, and then give the report again, thus leading young people directly into scientific research. He particularly encouraged young people to be bold in expressing their own opinions at the seminars, to ask about what they did not understand, to put forward any new ideas they had on the subject, even if not yet mature. He was frank in his criticism and mocking of bad tendencies. He laid emphasis on a down-to-earth style of work for young people, and boldness in doing scientific research, often giving admonition against having fancy ambitions with little hard work and ending up accomplishing nothing. The accomplishments of scientific research and the resulting theses by young people never failed to bring him the greatest pleasure. Besides going over the manuscripts carefully, down to the last punctuation mark, he did much to help the young person make further progress by deeper discussions. He also showed great interest in the popularizing of mathematics, as when he caught sight of some titles in the *Mathematics Bulletin*. He spent great energy in training the younger generation of mathematicians. The younger generation is truly fortunate to have such a commendable example of integrity to follow in their studies.

Professor Xu Bao-lu is widely recognized as one of China's best mathematicians. He had a good command of such foreign languages as English, French, German, and Russian, and was also well versed in the ancient Chinese classics. After liberation, despite poor health, a heavy teaching load and various kinds of interference, his research work never stopped and thirteen of his papers were published during this period. His outstanding achievements in mathematics, especially in the theory of probability and statistics, won him the respect of foreign mathematicians. In his memory the *U.S. Annals of Statistics* in 1979 asked some noted scholars and former colleagues and students of Mr. Xu's to write articles about his life and work; his contributions to the theory of probability and mathematical statistics were highly praised.

It is ten years since Mr. Xu left us. Still it is imperative for us to emulate Mr. Xu's selfless spirit of devotion to the motherland and to science, to shoulder sincerely and conscientiously the task of educating the younger generation, and to scale dauntlessly the heights of science in our efforts to raise rapidly the scientific and cultural level of our country.

September 24, 1980

Reprinted from
Bull. Amer. Math. Soc.
41, 502–504 (1935).

ON THE LIMIT OF A SEQUENCE OF POINT SETS

BY HSÜ PAO-LU

A variable point P_n is said to approach the point P as its limit if to an arbitrary positive ϵ there corresponds an m such that

$$\overline{P_n P} < \epsilon, \qquad (n > m).$$

In other words, P is to have the property that every neighborhood of it contains almost all* the points P_n.

In attempting to generalize this definition to a sequence of point sets $M_1, M_2, \cdots$, one is naturally led to begin with a definition of the neighborhood of a set and then write down (Definition A_0) the last sentence of the last paragraph, replacing the letter P by M.

DEFINITION. *By the ϵ-neighborhood of a set M is meant the set of all points which have a distance $<\epsilon$ from some point of M. We shall denote it by $(\epsilon)_M$.*

DEFINITION A_0. *A point set M is called a limit of the sequence of sets $M_1, M_2, \cdots$, if every neighborhood of it contains almost all the sets M_i as partial sets.*

But the above definition is far from being useful, because the limit would then not be unique. In the first place, if the set M is a limit in the sense of Definition A_0, and if M has a cluster point C, then the set $M-C$ has also the property of being a limit of the sequence. Secondly every set containing M as a partial set is a fortiori a limit.

The first difficulty is overcome by requiring M to be closed, and the second difficulty is met by adding still another condition (γ):

DEFINITION A. *A set M is said to be the limit of the sequence of sets $M_1, M_2, \cdots$, if it has the following properties:*

(α) *M is closed.*

* Thereby is meant that at most a finite number of the points P_i can lie outside the neighborhood.

(β) *For an arbitrary* $\epsilon>0$, $(\epsilon)_M \supset M_i$* *for almost all indices i.*

(γ) *For an arbitrary* $\epsilon>0$, $(\epsilon)_{M_i} \supset M$ *for almost all indices i.*

Observe that in the case of a sequence of points, (α) is fulfilled, (β) and (γ) become equivalent, and the definition reduces to the old one.

The following are immediate consequences of the definition:

(1) If a sequence of sets has a limit, the limit is unique.

(2) If a sequence $\mathfrak{S}$ of point sets has the limit M, every partial sequence of $\mathfrak{S}$ has the same limit M.

Further results hereby obtained consist of two fundamental criteria for the existence of a limit, when we restrict the sets of the sequence to lying in the same finite region of space. Given a sequence $\mathfrak{S}$ of sets $M_1, M_2, \cdots$, an L-point of $\mathfrak{S}$ shall be defined as a point which is the limit of a sequence of points $P_1, P_2, \cdots$, where each P_i belongs to the set M_i.

THEOREM A. *Let*

$$\mathfrak{S}: \qquad M_1, M_2, \cdots$$

be a sequence of point sets such that all the M_i*'s lie in the same finite region of space. Then* $\mathfrak{S}$ *has a limit when and only when, whatever partial sequence* $\mathfrak{S}_1$ *be selected from* $\mathfrak{S}$*, the set of L-points of* $\mathfrak{S}_1$ *coincides with the set of L-points of* $\mathfrak{S}$*. The limit of* M_i *is then the set of L-points of* $\mathfrak{S}$.

THEOREM B. *A necessary and sufficient condition for the sequence of sets* M_i*, lying in the same finite region of space, to have a limit is that, to an arbitrary positive* ϵ*, there corresponds an* M_m *such that*

(β') $(\epsilon)_{M_m} \supset M_i$ *for almost all indices i,*

(γ') $(\epsilon)_{M_i} \supset M_m$ *for almost all indices i.*

EXAMPLE 1. If each M_i is closed, M_1 is bounded, and $M_i \supset M_{i+1}$ for all i's, then a limit M exists and is equal to the set of points common to all the M_i's.

EXAMPLE 2. If $M_{i+1} \supset M_i$ and all the M_i's lie in the same finite region of space, then a limit M exists and is equal to the closed cover† of the set of points which belong to one of the M_i's.

* Read: "the ϵ-neighborhood of M contains M_i as a partial set."

† The closed cover of a set is the sum of the set and its first derived set.

In the case where each M_i is a point, the meaning of Theorem A is obvious, while Theorem B leads directly to the fundamental criterion for the variable point P_n to approach a limit, namely, to an arbitrary $\epsilon > 0$ there corresponds an m such that $\overline{P_n P_{n'}} < \epsilon$, provided that $n, n' < m$.

But what do these theorems tell us when the set M_i corresponds to a point function?

To be exact, consider a sequence of functions $f_1(P), f_2(P), \cdots$, defined in the same bounded set N of $(n-1)$-dimensional space, and converging toward a limiting function $f(P)$, in each point P of N. Moreover, let the functions $f_i(P)$, $f(P)$ be bounded; that is, $|f_i(P)| < G$, $|f(P)| < G$, where G is the same number for all the functions. To each f_i corresponds then a bounded set M_i, formed of the points $(x_1, x_2, \cdots, x_{n-1}, x_n)$, where $P:(x_1, x_2, \cdots, x_{n-1})$ is a point of N and $x_n = f_i(P)$. Furthermore, all the sets M_i lie in the same finite region of space. Let M be the set corresponding to f as M_i to f_i. The following result is immediate. *If f_i is uniformly convergent, M_i has the closed cover of M as its limit.* But the converse is not true. For example, let N be the interval $(0, 1)$, and

$$f_i = \begin{cases} \epsilon_i & \text{when } 0 \leqq x \leqq \eta_i, \\ 1 - \epsilon_i & \text{when } \eta_i < x \leqq 1, \end{cases}$$

where $\epsilon_i > 0$, $\eta_i > 0$, $\epsilon_i > \epsilon_{i+1}$, $\eta_i < \eta_{i+1}$, $\lim_{n=\infty} \epsilon_n = 0$, $\lim_{n=\infty} \eta_n = 1/2$. Here f_i converges non-uniformly, while M_i has the closed cover of M as its limit.

Nevertheless it is true that *if N is closed and f is continuous, then f_i converges uniformly when M_i approaches M as its limit.* Thus under appropriate restrictions Theorem B is equivalent to the condition for uniform convergence, namely, f_i converges uniformly when and only when, to an arbitrary $\epsilon > 0$, there corresponds an m, independent of P, such that

$$| f_n(P) - f_{n'}(P) | < \epsilon, \qquad (m < n, n').$$

Under the same restrictions Theorem A may be translated as follows. *The function f_i is uniformly convergent when and only when, for every sequence of points $P_1, P_2, \cdots$ of N with the limit P, the sequence of numbers $f_1(P_1), f_2(P_2), \cdots$ has the limit $f(P)$.*

NATIONAL UNIVERSITY OF PEKING, PEIPING, CHINA

A Note on the Indices and Numbers of Nondegenerate Critical Points of Biharmonic Functions

BY KIANG TSAI-HAN (江澤涵) AND HSÜ PAO-LU (許寶騄)

1. A real function $h(x, y)$ of $2n$ real variables $(x, y) = (x_1, y_1; x_2, y_2; \dots; x_n, y_n)$ in an open region R in the $2n$-dimensional Euclidian space is called a *biharmonic function*, when in R it is continuous together with its first and second partial derivatives and its second partial derivatives satisfy the following relations:

$$\frac{\partial^2 h}{\partial x_r \partial x_s} + \frac{\partial^2 h}{\partial y_r \partial y_s} = 0, \quad \frac{\partial^2 h}{\partial x_r \partial y_s} - \frac{\partial^2 h}{\partial x_s \partial y_r} = 0, \quad r, s = 1, 2, \cdots, n.$$

Consider a real quadratic form $q(x, y)$ in the $2n$ variables (x, y) and satisfying the conditions for being biharmonic. Let us set

$$\frac{\partial^2 q}{\partial x_r \partial x_s} = 2a_{rs}, \quad \frac{\partial^2 q}{\partial x_r \partial y_s} = 2b_{rs},$$

a_{rs} and b_{rs} being all real constants. Then the matrix of the quadratic form $q(x, y)$ may be written as $\|A_{rs}\|$, $r, s = 1, 2, \cdots, n$, where

$$A_{rs} = A_{sr} = \begin{Vmatrix} a_{rs} & b_{rs} \\ b_{rs} & -a_{rs} \end{Vmatrix}.$$

Let us call a $2n$-rowed square matrix of such type a *biharmonic matrix*.

A $2n$-rowed square matrix $\|u_{ij}\|$, $i, j = 1, 2, \cdots, 2n$, is biharmonic, when and only when

$$(1) \qquad u_{ij} = u_{ji}, \quad u_{\mu_i \mu_j} = (-1)^{\mu_i + \mu_j + 1} u_{\nu_i \nu_j},$$

where $(\mu_k, \nu_k) = (2m-1, 2m)$ or $(2m, 2m-1)$; $k = i, j$ and $m = 1, 2, \cdots, n$. To show this, let us write our matrix in the form $\|B_{rs}\|$, $r, s = 1, 2, \cdots, n$, and let $(\mu_k, \nu_k) = (2m-1, 2m)$ or $(2m, 2m-1)$. Then $u_{\mu_i \mu_j}$ and $u_{\nu_i \nu_j}$ are elements belonging to a certain B_{rs}. They are both along the principal diagonal or both not according as $\mu_i + \mu_j$ is even or odd. Hence the second relations in (1) are necessary and sufficient for B_{rs} to be of the form A_{rs}. Hence (1) is necessary and sufficient for $\|u_{ij}\|$ to be biharmonic.

2. Let $U=\|u_{ij}\|$ be the matrix of a biharmonic quadratic form $q(x, y)$. Let M_p be the p-rowed principal minor of U whose diagonal consists of the elements $u_{\mu_i \mu_i}$, $i=1, 2,\cdots, p$. We write

$$M_p=|u_{\mu_i \mu_j}|, \quad i, j=1, 2,\cdots, p.$$

To M_k we make correspond the p-rowed principal minor of U:

$$M_p^*=|u_{\nu_i \nu_j}|, \quad i, j=1, 2,\cdots, p,$$

where $(\mu_k, \nu_k)=(2m-1, 2m)$ or $(2m, 2m-1)$, $k=1, 2,\cdots, p$. Let $v_{ij}=u_{\mu_i \mu_j}$ and $v_{ij}^*=u_{\nu_i \nu_j}$. To an arbitrary term of M_p:

$$w=\pm v_{1\eta_1} v_{2\eta_2} \cdots v_{p\eta_p}$$

there corresponds the term of M_p^*:

$$w^*=\pm v^*_{1\eta_1} v^*_{2\eta_2} \cdots v^*_{p\eta_p}.$$

From (1), $w^*=(-1)^p w$, and therefore $M_p^*=(-1)^p M_p$. Hence *when p is any positive odd number less than $2n$, the p-rowed principal minors of U divide themselves into pairs M_p, M_p^* such that $M_p+M_p^*=0$, or the sum of all the p-rowed principal minors is zero.*

From this result, we obtain the following expression for the characteristic function $f(\lambda)=|U-\lambda I|$:

$$f(\lambda)=\lambda^{2n}+S_2 \lambda^{n-2}+S_4 \lambda^{n-4}+\cdots+S_{2n-2} \lambda^2+S_{2n},$$

where S_i is the sum of all the i-rowed principal minors of U. The roots of $f(\lambda)=0$, being all real,[1] divide themselves into pairs $\pm \lambda_i$, $i,=1, 2,\cdots, n$, where $\lambda_i \geqq 0$. From a well-known theorem[2] in algebra, we have the following:

Theorem 1. *A real biharmonic quadratic form in the $2n$ variables $(x_1, y_1; x_2, y_2;\cdots; x_n, y_n)$ can be reduced by a real orthogonal transformation into the following form:*

$$\lambda_1 (x_1^2-y_1^2)+\lambda_2 (x_2^2-y_2^2)+\cdots+\lambda_n (x_n^2-y_n^2),$$

where $\lambda_i \geqq 0$, $i=1, 2,\cdots, n$.

1 M. Bôcher, *Higher Algebra*, Th. 1, p. 170

2 Bôcher, *loc. cit.*, Th. 2, p. 171.

Corollary 1. *The rank and the index*[1] *of a real biharmonic quadratic form in* $2n$ *variables are always* $2r$ *and* r *respectively, where* r *is a definite integer such that* $0 \leqq r \leqq n$.

When the rank is $2n$, *the discriminant is positive or negative according as* n *is even or odd.*

3. The preceding theorem finds an application in the study of 'critical points" of certain biharmonic functions. Let $h(x, y)$ be a biharmonic function defined as above. A point (x^0, y^0) in R is called a *critical point* of $h(x, y)$ when it is a common zero point of all the first partial derivatives of $h(x, y)$. A critical point of $h(x, y)$ is said to be *nondegenerate* when the Hessian of $h(x, y)$ does not vanish at the point. Hence in the Taylor's expansion of $h(x, y)$ at a nondegenerate critical point (x^0, y^0), the quadratic terms in $(x-x^0, y-y^0)$ constitute a nonsingular biharmonic quadratic form. The index of this quadratic form is called the *index of the nondegenerate critical point*[2]. From the preceding theorem we infer immediately the following:

Corollary 2. *A nondegenerate critical point of a real biharmonic function has always the index* n.

Suppose in R a number of $(2n-1)$-dimensional closed orientable manifolds B bound an open connected subregion D of R, and the points on a manifold in B neighboring any particular point (x^0, y^0) on the manifold satisfy a relation of the form $F(x, y)=0$, where $F(x, y)$ is a real single-valued function of (x, y) of class C^3 neighboring (x^0, y^0), at least one of whose first partial derivatives does not vanish at (x^0, y^0). Suppose there exists a biharmonic function $h(x, y)$ in R whose critical points in D are all nondegenerate. Suppose further that at each point of each boundary manifold the directional derivative h_N of h along the normal to the manifold in the sense leading from points in D to points

---

1 L. E. Dickson, *Modern Algebraic Theories*, p. 71.

2 The definitions concerning critical points are as given by M. Morse. See (I) Morse, "Relations between the critical points of a real function of n independent variables," *Trans. Amer. Math. Soc.*, Vol. 27 (1925), pp. 345-396; or (II) Morse and Schaack, "The critical point theory under general boundary conditions," *Annals of Math.*, Vol. 35 (1934), pp. 545-571.

not in D does not vanish. The function $h(x, y)$ will be called *an admissible biharmonic function in the admissible region D.*[1]

Let us denote by $B' [B'']$ the totality of boundary manifolds at all of whose points $h_N > 0$ $[< 0]$ and call $B' [B'']$ the positive [negative] boundary of D with respect to $h(x, y)$. Let R_i, $i=0, 1, \cdots, 2n$, and R'_j $[R''_j]$, $j=0, 1, \cdots, 2n-1$, be the ith and jth connectivity numbers or Betti numbers[2] of $D+B$ and B' $[B'']$ respectively. Let M be the total number of critical points of $h(x, y)$ in D. Then, as one of the authors derived a theorem[3] from Theorem 6, Morse I, p. 390, we can derive from this theorem of Morse and our Corollary 2 the following relations:

$$
(2)\qquad \begin{cases} R_i - R''_i = 0, \quad i=0, 1, \cdots, n-2, n+1, \cdots, 2n, \\ R''_{n-1} - R_{n-1} \geqq 0, \\ R_n - R''_n \geqq 0, \\ M = (R_n - R_{n-1}) - (R''_n - R''_{n-1}) \geqq 0. \end{cases}
$$

Now the function $H(x, y) \equiv -h(x, y)$ is also an admissible biharmonic function in the admissible region D and the critical points of the functions $H(x, y)$ and $h(x, y)$ are equal in number and identical in position in D, and the positive boundary B' of D with respect to $h(x, y)$ is the negative boundary of D with respect to $H(x, y)$. Thus the relations (2) hold when R''_j are replaced by R'_j. From (2) and these derived relations, we obtain the following:

---

1 See Morse II, *loc. cit.*, the boundary conditions β, pp. 561-562.

2 The connectivity numbers or Betti numbers are as defined by J. W. Alexander, "Combinatorial analysis situs," *Trans. Amer. Math. Soc.*, Vol. 28 (1926), pp. 301-329, or by S. Lefschetz, *Topology* (1930). That $D+B$, B', and B'' are complexes has been shown by S. S. Cairns, "Triangulation of regular loci," *Annals of Math.*, Vol. 35 (1934), pp. 579-587.

3 T. H. Kiang, "On the critical points of nondegenerate Newtonian potentials." *Amer. Jour. of Math.*, Vol. 54 (1932), Th. B, pp. 93-97. See also A. B. Brown, "Relations between critical points of a real analytic function of N independent variables," *Amer. Jour. of Math.*, Vol. 52 (1930), Th. 2, p. 252.

$$R_i = R_i' = R_i'', \ i=0,\ 1,\cdots,\ n-2,\ n+1,\cdots,\ 2n.$$

(3) $$R_{n-1} \leqq R_{n-1}',\ R_{n-1} \leqq R_{n-1}'',$$

$$R_n \geqq R_n',\ R_n \geqq R_n''$$

and

$$M=(R_n-R_{n-1})-(R_n'-R_{n-1}')=(R_n-R_{n-1})-(R_n''-R_{n-1}'')\geqq 0.$$

From Poincaré's duality theorem[1],

$$R_n'-R_{n-1}'=R_n''-R_{n-1}''=0,$$

and therefore

(4) $$M=R_n-R_{n-1}\geqq 0.$$

Theorem 2. *If there exists an admissible biharmonic fnnction $h(x, y)$ of $2n$ real variables in an admissible open region D in the $2n$-dimensional Euclidian space, then there exist the relations* (3) *and* (4) *among the number M of the critical points of $h(x, y)$ in D, the ith connectivity numbers R_i of the closure of D, and the jth connectivity numbers R_i' and R_j'' of the positive boundary and negative boundary of D respectively.*

Let us note that $R_0=1$ for connected D. Hence, for $n>1$, $R_0'=R_0''=1$, or each of the positive and negative boundaries B' and B'' consists of a single $(2n-1)$-dimensional manifold. For $n=1$, the function $u(x, y)$ is both biharmonic and harmonic and the relation (4) reduces to $M=R_1-R_0$, which has been obtained under different conditions on $h(x, y)$ and D by one of the authors in a previous paper[2].

May 1935.

---

1 Lefschetz, *Topology*, p. 140

2 T. H. Kiang, "Critical points of harmonic functions and Green's functions in plane regions," *This Quarterly*, Vol. 5 (1932), pp. 115-123.

Reprinted from
Statist. Res. Mem.
2, 1–24 (1938).

CONTRIBUTION TO THE THEORY OF "STUDENT'S" t-TEST AS APPLIED TO THE PROBLEM OF TWO SAMPLES

BY P. L. HSU

I. INTRODUCTION

Consider two unconnected normal populations, Π_1 and Π_2, with means ξ_1, ξ_2 and variances σ_1^2, σ_2^2 respectively, and suppose that a sample is drawn from each of the populations. The general question whether the populations are alike or not presents itself in three different phases which we may describe as follows in the terminology commonly employed in testing statistical hypotheses:

(A) to test the hypothesis H_1, that $\xi_1 = \xi_2$ and $\sigma_1^2 = \sigma_2^2$, against the set of all the alternative hypotheses which specify nothing except that $\xi_1 \neq \xi_2$ or $\sigma_1^2 \neq \sigma_2^2$ or both;

(B) to test the hypothesis H_2, that $\xi_1 = \xi_2$ while the alternatives specify nothing except that $\xi_1 \neq \xi_2$;

(C) in connexion with Π_1 and Π_2 it is assumed as given that σ_1^2 and σ_2^2 have the same (though unknown) value, to test the hypothesis H_3, that $\xi_1 = \xi_2$, against the set of alternatives that $\xi_1 \neq \xi_2$.

It is to be noticed that if either H_1 or H_3 is true, then the populations Π_1 and Π_2 will be identical, while if H_2 is the hypothesis under test, we are only interested in whether or not the means are the same, neglecting altogether any difference that may exist between the population variances.*

R. A. Fisher (1925) was the first to prove that whenever the two normal populations are identical a certain statistic calculated from the two samples will follow exactly the t-distribution of "Student". The square of this t we shall later denote by u_1 and consider it in detail. Thus, without discriminating the hypotheses H_1 and H_3, Fisher introduced the t-test as the criterion for the identity of Π_1 and Π_2. As is well known, the t-test used for this purpose is an exact one in the sense that the distribution of t is entirely known under the assumption that H_1 is true. Later on R. Sato (1937), using the method of treating composite hypotheses due to Neyman (1935), obtained some general results which involve the fact that, if the hypothesis to be tested is H_3, then the t-test is the uniformly most powerful of all the unbiassed exact tests that can possibly be constructed. It follows that the t-test completely answers the question (C).

* Such a question might arise, for example, if we wished to know whether one variety of sugar beet had on the average a higher sugar content than a second, differences in the variability among individual plants being immaterial.

Fisher (1934, p. 122) expressed the opinion that for the hypothesis H_1 the same t-test should be used, while Neyman and Pearson (1930) suggested that the λ-test might be used. The properties of λ have been studied by P. V. Sukhatme (1935), who has also prepared a table of 5 % significance level of λ. These are the attempts to answer the question (A).

As to the testing of the hypothesis H_2, criteria have been suggested by Fisher (1936) and M. S. Bartlett (cf. Welch, 1938), the former based his criterion on fiducial arguments and the latter devised some tests that have the advantage of being exact.

The purpose of this paper is to give a thorough survey of the possibilities that certain test criteria, namely, u_1 ($=t^2$) and u_2 (which is closely related to u_1), may be advantageously used for H_1 or H_2. The whole discussion is, of course, dependent on a knowledge of the distributions of u_1 and u_2, and thence of their power functions,* in terms of the parameters ξ_i and σ_i. Unfortunately most of the results obtained below are of a rather negative character. The only positive conclusion arrived at is that, in the case where the sample sizes are the same, the test u_1 ($=u_2$ here) may be safely used in testing the hypothesis H_2. The same problem has been taken up by B. L. Welch (1938); the results obtained by him by an approximate method are in full accordance with the general results described below.

II. The distribution of a statistic u

(1) We have assumed that two independent normal populations have means ξ_1, ξ_2 and variances σ_1^2, σ_2^2 respectively. Let samples of n_j ($j=1,2$) individuals be drawn, giving rise to $\bar{x}_j$, the sample means, and $\Sigma_j = \Sigma\,(x-\bar{x}_j)^2$, the total variations about the means within the samples, for $j=1,2$. Denote by

$$\delta = \xi_1-\xi_2, \qquad \theta = \sigma_1^2/\sigma_2^2, \tag{1}$$

$$\sigma^2 = \frac{\sigma_1^2}{n_1}+\frac{\sigma_2^2}{n_2}, \qquad \lambda = \frac{\delta^2}{2\sigma^2}. \tag{2}$$

Thus H_1 is the hypothesis that $\lambda = 0$ and $\theta = 1$, H_2 is the hypothesis that $\lambda = 0$, while the hypothesis H_3 states that $\lambda = 0$ under the *a priori* assumption that $\theta = 1$.

Instead of considering t we shall always deal with its square. Accordingly we put

$$u = (\bar{x}_1-\bar{x}_2)^2/(A_1\Sigma_1+A_2\Sigma_2), \tag{3}$$

where A_1 and A_2 are some known positive constants. Define also B_1 and B_2 by

$$B_j\sigma^2 = A_j\sigma_j^2 \quad (j=1,2). \tag{4}$$

Later on we shall identify u with some particular tests by giving special values to A_1 and A_2.

* For a definition of the power function of a test and a discussion of some of its properties see Neyman and Pearson (1936, 1938).

A method of finding the distribution of the ratio of two independent random variates is due to H. Cramér (1937, p. 46). If $z = X_1/X_2$ and if X_1 and X_2 are independent, then under certain restrictions the elementary probability law of z is given by

$$p(z) = \frac{1}{2\pi i}\int_{-\infty}^{\infty} \phi_1(t)\,\phi_2'(-tz)\,dt, \tag{5}$$

where $\phi_j(t)$ is the characteristic function of X_j $(j=1,2)$ and $\phi_2'(-tz)$ denotes the function

$$\frac{d}{dy}\phi_2(y)\bigg|_{y=-tz}$$

We shall use the formula (5) to get the distribution of u, without, however, stopping to prove the legitimacy of doing so; the reader can easily satisfy himself in this respect.

The numerator and denominator of u have respectively the characteristic functions

$$\phi_1(t) = e^{-\lambda}e^{\lambda/(1-2\sigma^2 it)}(1-2\sigma^2 it)^{-\frac{1}{2}}, \tag{6}$$

$$\begin{aligned}\phi_2(t) &= (1-2\sigma_1^2 A_1 it)^{-\frac{1}{2}(n_1-1)}(1-2\sigma_2^2 A_2 it)^{-\frac{1}{2}(n_2-1)}\\ &= (1-2B_1\sigma^2 it)^{-\frac{1}{2}(n_1-1)}(1-2B_2\sigma^2 it)^{-\frac{1}{2}(n_2-1)}.\end{aligned} \tag{7}$$

Hence

$$\begin{aligned}\phi_2'(-tu) = \sigma^2 i[&B_1(n_1-1)(1+2B_1\sigma^2 itu)^{-\frac{1}{2}(n_1+1)}(1+2B_2\sigma^2 itu)^{-\frac{1}{2}(n_2-1)}\\ &+B_2(n_2-1)(1+2B_1\sigma^2 itu)^{-\frac{1}{2}(n_1-1)}(1+2B_2\sigma^2 itu)^{-\frac{1}{2}(n_2+1)}].\end{aligned} \tag{8}$$

Substituting (6) and (8) into (5) and making the transformation $2\sigma^2 t = \tau$ in the resulting integral, we get the elementary probability law of u:

$$p(u) = B_1(n_1-1)J_1 + B_2(n_2-1)J_2, \tag{9}$$

where

$$J_1 = \frac{e^{-\lambda}}{4\pi}\int_{-\infty}^{\infty} e^{\lambda/(1-i\tau)}(1-i\tau)^{-\frac{1}{2}}(1+B_1 i\tau u)^{-\frac{1}{2}(n_1+1)}(1+B_2 i\tau u)^{-\frac{1}{2}(n_2-1)}d\tau, \tag{10}$$

$$J_2 = \frac{e^{-\lambda}}{4\pi}\int_{-\infty}^{\infty} e^{\lambda/(1-i\tau)}(1-i\tau)^{-\frac{1}{2}}(1+B_1 i\tau u)^{-\frac{1}{2}(n_1-1)}(1+B_2 i\tau u)^{-\frac{1}{2}(n_2+1)}d\tau. \tag{11}$$

(2) *The expansion of* $p(u)$ *into an infinite series.* If X_1 and X_2 are distributed independently as χ^2's with f_1 and f_2 degrees of freedom respectively, then it is well known that their ratio $z = X_1/X_2$ will have the elementary probability law

$$p(z) = \frac{1}{B(\frac{1}{2}f_1, \frac{1}{2}f_2)} z^{\frac{1}{2}f_1-1}(1+z)^{-\frac{1}{2}(f_1+f_2)} \quad \text{for } z>0, \tag{12}$$

while $p(z) = 0$ for $z<0$. On the other hand, we may apply (5) to find $p(z)$. Thus, if $\phi_j(t)$ is the characteristic function of X_j for $j=1,2$, then

$$\phi_1(t) = (1-2it)^{-\frac{1}{2}f_1}, \tag{13}$$

$$\phi_2'(-tz) = f_2 i(1+2itz)^{-\frac{1}{2}f_2}. \tag{14}$$

I-2

Substituting (13) and (14) into (5) and putting $2t = \tau$ in the resulting integral, we get

$$p(z) = \frac{f_2}{4\pi}\int_{-\infty}^{\infty}(1-i\tau)^{-\frac{1}{2}f_1}(1+i\tau z)^{-\frac{1}{2}f_2-1}d\tau. \tag{15}$$

The right-hand sides of (12) and (15) must be identical; we have therefore

$$\frac{f_2}{4\pi}\int_{-\infty}^{\infty}(1-i\tau)^{-\frac{1}{2}f_1}(1+i\tau z)^{-\frac{1}{2}f_2-1}d\tau = \frac{1}{B(\frac{1}{2}f_1, \frac{1}{2}f_2)}z^{\frac{1}{2}f_1-1}(1+z)^{-\frac{1}{2}(f_1+f_2)} \quad \text{for } z \geqslant 0$$
$$= 0 \quad \text{for } z < 0, \tag{16}$$

identically in z, f_1 and f_2, at least when f_1 and f_2 are positive integers.

With the help of (16) we can evaluate the integral, say,

$$\Psi(z, f_2) = \frac{f_2 e^{-\lambda}}{4\pi}\int_{-\infty}^{\infty}e^{\lambda/(1-i\tau)}(1-i\tau)^{-\frac{1}{2}}(1+i\tau z)^{-\frac{1}{2}f_2-1}d\tau, \tag{17}$$

where f_2 is some positive integer. In fact, the integrand of (17) is $\sum_{k=0}^{\infty}\psi_k(\tau)$, where

$$\psi_k(\tau) = \frac{\lambda^k}{k!}(1-i\tau)^{-k-\frac{1}{2}}(1+i\tau z)^{-\frac{1}{2}f_2-1}. \tag{18}$$

We have

$$\left|\sum_{k=0}^{n}\psi_k(\tau)\right| \leqslant \sum_{k=0}^{\infty}|\psi_k(\tau)| = \sum_{k=0}^{\infty}\frac{\lambda^k}{k!}(1+\tau^2)^{-\frac{1}{2}k-\frac{1}{4}}(1+\tau^2z^2)^{-\frac{1}{4}f_2-\frac{1}{2}}$$
$$= e^{\lambda/\sqrt{(1+\tau^2)}}(1+\tau^2)^{-\frac{1}{4}}(1+\tau^2z^2)^{-\frac{1}{4}f_2-\frac{1}{2}} \leqslant e^{\lambda}(1+\tau^2)^{-\frac{1}{4}}(1+\tau^2z^2)^{-\frac{1}{4}f_2-\frac{1}{2}}. \tag{19}$$

The last written function being summable over $(-\infty, \infty)$ whenever $z \neq 0$, the series $\sum_{k=0}^{\infty}\psi_k(\tau)$ may be integrated term-by-term;* thus we have

$$\Psi(z, f_2) = \frac{f_2 e^{-\lambda}}{4\pi}\sum_{k=0}^{\infty}\frac{\lambda^k}{k!}\int_{-\infty}^{\infty}(1-i\tau)^{-k-\frac{1}{2}}(1+i\tau z)^{-\frac{1}{2}f_2-1}d\tau. \tag{20}$$

Applying (16) to (20) with $f_1 = 2k+1$, we get

$$\Psi(z, f_2) = e^{-\lambda}\sum_{k=0}^{\infty}\frac{\lambda^k}{k!}\frac{1}{B(k+\frac{1}{2}, \frac{1}{2}f_2)}z^{k-\frac{1}{2}}(1+z)^{-\frac{1}{2}(f_2+1)-k} \quad \text{for } z > 0$$
$$= 0 \quad \text{for } z < 0, \tag{21}$$

identically in z and f_2, at least when f_2 is a positive integer.

We are now in the position to evaluate J_1 and J_2. Suppose $0 < B_1 \leqslant B_2$, then from the identity

$$1 + B_2 i\tau u = \frac{B_2}{B_1}(1 + B_1 i\tau u)\left(1 - \frac{1 - B_1/B_2}{1 + B_1 i\tau u}\right) \tag{22}$$

we have

$$(1 + B_2 i\tau u)^{-\frac{1}{2}(n_2-1)} = \left(\frac{B_1}{B_2}\right)^{\frac{1}{2}(n_2-1)}(1 + B_1 i\tau u)^{-\frac{1}{2}(n_2-1)}$$
$$\times \sum_{h=0}^{\infty}\frac{\Gamma(\frac{1}{2}(n_2-1)+h)}{\Gamma(\frac{1}{2}(n_2-1))\,h!}\left(1 - \frac{B_1}{B_2}\right)^h(1 + B_1 i\tau u)^{-h}. \tag{23}$$

* By the Lebesgue theorem of term-by-term integration; cf. e.g. Kestelman (1937, p. 137, Theorem 214).

Substituting (23) into (10) we get

$$J_1 = \frac{e^{-\lambda}}{4\pi}\left(\frac{B_1}{B_2}\right)^{\frac{1}{2}(n_2-1)}\int_{-\infty}^{\infty}\sum_{h=0}^{\infty}\phi_h(\tau)\,d\tau, \tag{24}$$

where

$$\phi_h(\tau) = \frac{\Gamma(\frac{1}{2}(n_2-1)+h)}{\Gamma(\frac{1}{2}(n_2-1))\,h!}\left(1-\frac{B_1}{B_2}\right)^h e^{\lambda/(1-i\tau)}(1-i\tau)^{-\frac{1}{2}}(1+B_1 i\tau u)^{-\frac{1}{2}N-h}, \tag{25}$$

$$N = n_1+n_2. \tag{26}$$

In order to show that we can reverse the order of integration and summation in (24), we have

$$\begin{aligned}\left|\sum_{h=0}^{n}\phi_h(\tau)\right| \leqslant \sum_{h=0}^{\infty}|\phi_h(\tau)| &= e^{\lambda/(1+\tau^2)}(1+\tau^2)^{-\frac{1}{4}}(1+B_1^2\tau^2u^2)^{-\frac{1}{4}N} \\ &\quad\times\sum_{h=0}^{\infty}\frac{\Gamma(\frac{1}{2}(n_2-1)+h)}{\Gamma(\frac{1}{2}(n_2-1))\,h!}\left(1-\frac{B_1}{B_2}\right)^h(1+B_1^2\tau^2u^2)^{-\frac{1}{2}h} \\ &\leqslant e^{\lambda}(1+\tau^2)^{-\frac{1}{4}}(1+B_1^2\tau^2u^2)^{-\frac{1}{4}N}\left[1-\frac{1-B_1/B_2}{1+B_1^2\tau^2u^2}\right]^{-\frac{1}{2}(n_2-1)} \\ &\leqslant e^{\lambda}\left(\frac{B_1}{B_2}\right)^{-\frac{1}{2}(n_2-1)}(1+\tau^2)^{-\frac{1}{4}}(1+B_1^2\tau^2u^2)^{-\frac{1}{4}N}.\end{aligned}$$

Provided $u \neq 0$, the last written function is summable over $(-\infty, \infty)$; therefore*

$$\begin{aligned}J_1 &= \frac{e^{-\lambda}}{4\pi}\left(\frac{B_1}{B_2}\right)^{\frac{1}{2}(n_2-1)}\sum_{h=0}^{\infty}\int_{-\infty}^{\infty}\phi_h(\tau)\,d\tau \\ &= \frac{e^{-\lambda}}{4\pi}\left(\frac{B_1}{B_2}\right)^{\frac{1}{2}(n_2-1)}\sum_{h=0}^{\infty}\frac{\Gamma(\frac{1}{2}(n_2-1)+h)}{\Gamma(\frac{1}{2}(n_2-1))\,h!}\left(1-\frac{B_1}{B_2}\right)^h \\ &\qquad\times\int_{-\infty}^{\infty}e^{\lambda/(1-i\tau)}(1-i\tau)^{-\frac{1}{2}}(1+B_1 i\tau u)^{-\frac{1}{2}N-h}\,d\tau.\end{aligned} \tag{27}$$

Comparing (20) with (27) we get immediately

$$J_1 = \left(\frac{B_1}{B_2}\right)^{\frac{1}{2}(n_2-1)}\sum_{h=0}^{\infty}\frac{\Gamma(\frac{1}{2}(n_2-1)+h)}{\Gamma(\frac{1}{2}(n_2-1))\,h!}\left(1-\frac{B_1}{B_2}\right)^h\frac{1}{N+2h-2}\Psi(B_1u, N+2h-2). \tag{28}$$

Similarly we have

$$J_2 = \left(\frac{B_1}{B_2}\right)^{\frac{1}{2}(n_2+1)}\sum_{h=0}^{\infty}\frac{\Gamma(\frac{1}{2}(n_2+1)+h)}{\Gamma(\frac{1}{2}(n_2+1))\,h!}\left(1-\frac{B_1}{B_2}\right)^h\frac{1}{N+2h-2}\Psi(B_1u, N+2h-2). \tag{29}$$

From (9), (28) and (29) it follows after some reduction that

$$p(u) = B_1^{\frac{1}{2}(n_2+1)}B_2^{-\frac{1}{2}(n_2-1)}\sum_{h=0}^{\infty}C_h\left(1-\frac{B_1}{B_2}\right)^h\Psi(B_1u, N+2h-2), \tag{30}$$

wherein we have put

$$C_h = \frac{\Gamma(\frac{1}{2}(n_2-1)+h)}{\Gamma(\frac{1}{2}(n_2-1))\,h!}. \tag{31}$$

* See footnote on p. 4.

Formula (30) is true only when $B_1 \leqslant B_2$. In the case $B_2 \leqslant B_1$ we have only to interchange the indices 1 and 2. Thus

$$p(u) = B_1^{-\frac{1}{2}(n_1-1)} B_2^{\frac{1}{2}(n_1+1)} \sum_{h=0}^{\infty} C'_h \left(1 - \frac{B_2}{B_1}\right)^h \Psi(B_2 u, N+2h-2) \tag{32}$$

for $B_2 \leqslant B_1$, where

$$C'_h = \frac{\Gamma(\frac{1}{2}(n_1-1)+h)}{\Gamma(\frac{1}{2}(n_1-1))\, h!}. \tag{33}$$

With the functions Ψ given by (21) the right-hand side of (30) is a repeated series of positive terms. It is therefore identical with the double series

$$p(u) = \left(\frac{B_1}{B_2}\right)^{\frac{1}{2}(n_2-1)} e^{-\lambda} \sum_{h,k=0}^{\infty} C_h \left(1 - \frac{B_1}{B_2}\right)^h \frac{\lambda^k}{k!} p_{hk}(u) \quad \text{for } B_1 \leqslant B_2, \tag{34}$$

where

$$p_{hk}(u) = \frac{B_1^{k+\frac{1}{2}} u^{k-\frac{1}{2}}}{B(k+\frac{1}{2}, \frac{1}{2}(N-2)+h)} (1+B_1 u)^{-\frac{1}{2}(N-1)-h-k}. \tag{35}$$

Similarly we have

$$p(u) = \left(\frac{B_2}{B_1}\right)^{\frac{1}{2}(n_1-1)} e^{-\lambda} \sum_{h,k=0}^{\infty} C'_h \left(1 - \frac{B_2}{B_1}\right)^h \frac{\lambda^k}{k!} p'_{hk}(u) \quad \text{for } B_2 \leqslant B_1, \tag{36}$$

where

$$p'_{hk}(u) = \frac{B_2^{k+\frac{1}{2}} u^{k-\frac{1}{2}}}{B(k+\frac{1}{2}, \frac{1}{2}(N-2)+h)} (1+B_2 u)^{-\frac{1}{2}(N-1)-h-k}. \tag{37}$$

If, as is permissible, we sum the double series (34) first with respect to h and then with respect to k, we shall also obtain

$$p(u) = \left(\frac{B_1}{B_2}\right)^{\frac{1}{2}(n_2-1)} e^{-\lambda} \sum_{k=0}^{\infty} \frac{\lambda^k}{k!} \frac{1}{B(k+\frac{1}{2}, \frac{1}{2}(N-2))} B_1^{k+\frac{1}{2}} u^{k-\frac{1}{2}} (1+B_1 u)^{-\frac{1}{2}(N-1)-k}$$
$$\times F\left(\tfrac{1}{2}(n_2-1), \tfrac{1}{2}(N-1)+k, \tfrac{1}{2}(N-2), \frac{1-B_1/B_2}{1+B_1 u}\right), \tag{38}$$

and a similar formula from (36).

Let us indicate rapidly a few particular cases of $p(u)$.

Case (i) $\lambda = 0$: Denoting the corresponding distribution by $p_0(u)$ we have

$$p_0(u) = \left(\frac{B_1}{B_2}\right)^{\frac{1}{2}(n_2-1)} \sum_{h=0}^{\infty} C_h \left(1 - \frac{B_1}{B_2}\right)^h p_h(u), \tag{39}$$

where

$$p_h(u) = \frac{B_1^{\frac{1}{2}} u^{-\frac{1}{2}}}{B(\frac{1}{2}, \frac{1}{2}(N-2)+h)} (1+B_1 u)^{-\frac{1}{2}(N-1)-h}. \tag{40}$$

Case (ii) $B_1 = B_2 = B$: Using the notation $p(u \mid B)$ in this connexion we have

$$p(u \mid B) = e^{-\lambda} \sum_{k=0}^{\infty} \frac{\lambda^k}{k!} \bar{p}_k(u), \tag{41}$$

where

$$\bar{p}_k(u) = \frac{B^{k+\frac{1}{2}} u^{k-\frac{1}{2}}}{B(k+\frac{1}{2}, \frac{1}{2}(N-2))} (1+Bu)^{-\frac{1}{2}(N-1)-k}. \tag{42}$$

Case (iii) $\lambda = 0$ and $B_1 = B_2 = B$:

$$p_0(u \mid B) = \frac{B^{\frac{1}{2}} u^{-\frac{1}{2}}}{B(\frac{1}{2}, \frac{1}{2}(N-2))} (1+Bu)^{-\frac{1}{2}(N-1)}. \tag{43}$$

Case (iv) $B_1 = 0$: Putting $B_1 = 0$ in (9) and (11) and remembering (21) we get

$$p(u \mid B_1 = 0) = e^{-\lambda} \sum_{k=0}^{\infty} \frac{\lambda^k}{k!} \frac{1}{B(k+\frac{1}{2}, \frac{1}{2}(n_2-1))} B_2^{k+\frac{1}{2}} u^{k-\frac{1}{2}} (1+B_2 u)^{-\frac{1}{2}n_2 - k}. \tag{44}$$

(44) is also the limiting form of (38) as B_1 tends to the limit zero.

Case (v) $\lambda = 0$ and $B_1 = 0$:

$$p_0(u \mid B_1 = 0) = \frac{1}{B(\frac{1}{2}, \frac{1}{2}(n_2-1))} B_2^{\frac{1}{2}} u^{-\frac{1}{2}} (1+B_2 u)^{-\frac{1}{2}n_2}. \tag{45}$$

(3) *Finite expressions of* $p_0(u)$. If both n_1 and n_2 are odd, then the integrands of (10) and (11) are uniform functions of τ in the upper half plane having only poles at the points $\tau = i/B_1 u$ and $i/B_2 u$. Accordingly the integrals in (10) and (11) can be evaluated by the calculus of residues, resulting in finite analytic formulae for J_1 and J_2. In the case $\lambda = 0$ we shall show how to get the finite expression for $p_0(u)$ without, however, resorting to the method of residues.

Putting $\lambda = 0$ and $n_j = 2m_j + 1$ $(j = 1, 2)$ in (10) and (11) we get

$$J_1 = \frac{1}{4\pi} \int_{-\infty}^{\infty} (1 - i\tau)^{-\frac{1}{2}} (1 + B_1 i\tau u)^{-(m_1+1)} (1 + B_2 i\tau u)^{-m_2} d\tau, \tag{46}$$

$$J_2 = \frac{1}{4\pi} \int_{-\infty}^{\infty} (1 - i\tau)^{-\frac{1}{2}} (1 + B_1 i\tau u)^{-m_1} (1 + B_2 i\tau u)^{-(m_2+1)} d\tau. \tag{47}$$

If we denote by

$$I_{r_1 r_2} = \frac{1}{4\pi} \int_{-\infty}^{\infty} (1 - i\tau)^{-\frac{1}{2}} (1 + B_1 i\tau u)^{-r_1} (1 + B_2 i\tau u)^{-r_2} d\tau, \tag{48}$$

then from (9), (46) and (47) we have

$$p_0(u) = 2m_1 B_1 I_{m_1+1, m_2} + 2m_2 B_2 I_{m_1, m_2+1}. \tag{49}$$

From the identity (22) follows easily the recurrence formula

$$I_{r_1 r_2} = \frac{1}{\Delta} (B_1 I_{r_1, r_2-1} - B_2 I_{r_1-1, r_2}), \tag{50}$$

where

$$\Delta = B_1 - B_2 \tag{51}$$

provided $\Delta \neq 0$. On the other hand, from (16) we have, on putting $f_1 = 1$ and $f_2 = 2(r-1)$,

$$\left.\begin{aligned} I_{r0} &= \frac{1}{2(r-1)\, B(\frac{1}{2}, r-1)} (B_1 u)^{-\frac{1}{2}} (1 + B_1 u)^{\frac{1}{2}-r}, \\ I_{0r} &= \frac{1}{2(r-1)\, B(\frac{1}{2}, r-1)} (B_2 u)^{-\frac{1}{2}} (1 + B_2 u)^{\frac{1}{2}-r}. \end{aligned}\right\} \tag{52}$$

With the help of (49), (50) and (52) we can compute $p_0(u)$ by repeated steps.

Below we give, for $n_1 = 9$ and $n_2 = 5$, the scheme of computation for $p_0(u)$:

$$p_0(u) = 8B_1 I_{52} + 4B_2 I_{43} \tag{i}$$

$$= \frac{1}{\Delta}(8B_1^2 I_{51} - 8B_1 B_2 I_{42} + 4B_1 B_2 I_{42} - 4B_2^2 I_{33}) \tag{ii}$$

$$= \frac{1}{\Delta}(8B_1^2 I_{51} - 4B_1 B_2 I_{42} - 4B_2^2 I_{33}) \tag{iii}$$

$$= \frac{1}{\Delta^2}(\underline{8B_1^3 I_{50}} - 8B_1^2 B_2 I_{41} - 4B_1^2 B_2 I_{41} + 4B_1 B_2^2 I_{32} - 4B_1 B_2^2 I_{32} + 4B_2^3 I_{23}) \tag{iv}$$

$$= \frac{1}{\Delta^2}(-12B_1^2 B_2 I_{41} + 4B_2^3 I_{23}) \tag{v}$$

$$= \frac{1}{\Delta^3}(\underline{-12B_1^3 B_2 I_{40}} + 12B_1^2 B_2^2 I_{31} + 4B_1 B_2^3 I_{22} - 4B_2^4 I_{13}) \tag{vi}$$

$$= \frac{1}{\Delta^4}(\underline{12B_1^3 B_2^2 I_{30}} - 12B_1^2 B_2^3 I_{21} + 4B_1^2 B_2^3 I_{21} - 4B_1 B_2^4 I_{12} - 4B_1 B_2^4 I_{12} + \underline{4B_2^5 I_{03}}) \tag{vii}$$

$$= \frac{1}{\Delta^4}(-8B_1^2 B_2^3 I_{21} - 8B_1 B_2^4 I_{12}) \tag{viii}$$

$$= \frac{1}{\Delta^5}(\underline{-8B_1^3 B_2^3 I_{20}} + 8B_1^2 B_2^4 I_{11} - 8B_1^2 B_2^4 I_{11} + \underline{8B_1 B_2^5 I_{02}}). \tag{ix}$$

Explanation. To obtain the lines (ii), (iv), (vi), (vii) and (ix) apply formula (50) to the lines immediately preceding; the lines (iii), (v) and (viii) are obtained by collecting similar terms in the lines immediately preceding. An underlined term cannot be reduced any further and has therefore not been copied in the subsequent lines, but is, of course, a term of the final expression of $p_0(u)$.

Thus, collecting all the underlined terms, we have, for $n_1 = 9$ and $n_2 = 5$,

$$p_0(u) = 8\frac{B_1^3}{\Delta^2} I_{50} - 12\frac{B_1^3 B_2}{\Delta^3} I_{40} + 12\frac{B_1^3 B_2^2}{\Delta^4} I_{30} - 8\frac{B_1^3 B_2^3}{\Delta^5} I_{20} + 8\frac{B_1 B_2^5}{\Delta^5} I_{02} + 4\frac{B_2^5}{\Delta^4} I_{03}. \tag{53}$$

The formula (53) is true only when $B_1 \neq B_2$. If B_2 is held fixed and B_1 allowed to approach B_2 as its limit, the right-hand side of (53) will, of course, tend to the limit given in (43), namely

$$\frac{B_2^{\frac{1}{2}} u^{-\frac{1}{2}}}{B(\frac{1}{2}, 6)}(1 + B_2 u)^{-\frac{13}{2}},$$

but in a rather complicated manner. Even the way in which the integral of (53), taken between the limits 0 and ∞, becomes unity is intricate enough. Thus

$$\int_0^\infty p_0(u)\,du = \frac{B_1^2}{\Delta^2} - \frac{2B_1^2 B_2}{\Delta^3} + \frac{3B_1^2 B_2^2}{\Delta^4} - \frac{4B_1^2 B_2^3}{\Delta^5} + \frac{4B_1 B_2^4}{\Delta^5} + \frac{B_2^4}{\Delta^4}$$

and
$$-\frac{4B_1^2B_2^3}{\Delta^5}+\frac{4B_1B_2^4}{\Delta^5}=-\frac{4B_1B_2^3}{\Delta^4},$$

$$-\frac{4B_1B_2^3}{\Delta^4}+\frac{3B_1^2B_2^2}{\Delta^4}+\frac{B_2^4}{\Delta^4}=\frac{3B_1B_2^2}{\Delta^3}-\frac{B_2^3}{\Delta^3},$$

$$\frac{3B_1B_2^2}{\Delta^3}-\frac{B_2^3}{\Delta^3}-\frac{2B_1^2B_2}{\Delta^3}=-\frac{2B_1B_2}{\Delta^2}+\frac{B_2^2}{\Delta^3},$$

$$-\frac{2B_1B_2}{\Delta^2}+\frac{B_2^2}{\Delta^2}+\frac{B_1^2}{\Delta^2}=\frac{(B_1-B_2)^2}{\Delta^2}=1.$$

In the manner described above we have computed $p_0(u)$ for several particular values of n_1 and n_2 The results are given below, the expressions I_{r0} and I_{0r} being defined in (52).

$n_1 = 5$, $n_2 = 15$:

$$p_0(u)=\frac{B_1^8}{\Delta^7}\left(4I_{30}-14\frac{B_2}{\Delta}I_{20}\right)+\frac{B_2^3}{\Delta^2}\sum_{r=2}^{8}(9-r)(2r-2)\left(\frac{B_1}{\Delta}\right)^{8-r}I_{0r}; \tag{54}$$

$n_1 = 3$, $n_2 = 5$:

$$p_0(u)=2\frac{B_1^3}{\Delta^2}I_{20}-\frac{B_2^2}{\Delta}\left(2\frac{B_1}{\Delta}I_{02}+4I_{03}\right); \tag{55}$$

$n_1 = n_2 = 3$:

$$p_0(u)=\frac{2}{\Delta}(B_1^2I_{20}-B_2^2I_{02}); \tag{56}$$

$n_1 = n_2 = 5$:

$$p_0(u)=\frac{4}{\Delta^2}(B_1^3I_{30}+B_2^3I_{03})-\frac{4B_1B_2}{\Delta^3}(B_1^2I_{20}-B_2^2I_{02}); \tag{57}$$

$n_1 = n_2 = 7$:

$$p_0(u)=\frac{6}{\Delta^3}(B_1^4I_{40}-B_2^4I_{04})-\frac{12B_1B_2}{\Delta^4}(B_1^3I_{30}+B_2^3I_{03})+\frac{12B_1^2B_2^2}{\Delta^5}(B_1^2I_{20}-B_2^2I_{02}); \tag{58}$$

$n_1 = n_2 = 9$:

$$p_0(u)=\frac{8}{\Delta^4}(B_1^5I_{50}+B_2^5I_{05})-\frac{24B_1B_2}{\Delta^5}(B_1^4I_{40}+B_2^4I_{04})$$
$$+\frac{40B_1^2B_2^2}{\Delta^6}(B_1^3I_{30}-B_2^3I_{03})-\frac{40B_1^3B_2^3}{\Delta^7}(B_1^2I_{20}-B_2^2I_{02}); \tag{59}$$

$n_1 = n_2 = 11$:

$$p_0(u)=\frac{10}{\Delta^5}(B_1^6I_{60}-B_2^6I_{06})-\frac{40B_1B_2}{\Delta^6}(B_1^5I_{50}+B_2^5I_{05})+\frac{90B_1^2B_2^2}{\Delta^7}(B_1^4I_{40}-B_2^4I_{04})$$
$$-\frac{140B_1^3B_2^3}{\Delta^8}(B_1^3I_{30}+B_2^3I_{03})+\frac{140B_1^4B_2^4}{\Delta^9}(B_1^2I_{20}-B_2^2I_{02}). \tag{60}$$

III. The power function of u

(1) From the definitions (1), (2) and (4) of the quantities θ, σ^2 and the B_i we have

$$B_1 = \frac{n_1 n_2 A_1 \theta}{n_1 + n_2 \theta}, \quad B_2 = \frac{n_1 n_2 A_2}{n_1 + n_2 \theta}, \tag{61}$$

so that

$$B_1/B_2 = \rho\theta, \quad \text{where} \quad \rho = A_1/A_2. \tag{62}$$

If u_0 is any positive constant, we shall write

$$\beta(\lambda, \theta) = \int_{u_0}^{\infty} p(u)\, du, \tag{63}$$

because the integral on the right-hand side is a function of λ and θ when the expressions (61) are substituted for the B_i.

The function $\beta(\lambda, \theta)$ represents the probability that the statistic u calculated from the samples should have a value not less than u_0 when the population characters λ and θ have given values. In testing the various hypotheses considered in section I, if u be used as the criterion so that a hypothesis is rejected if $u > u_0$, then $\beta(\lambda, \theta)$ will be the power function connected with the u-test.*

It is our purpose to study two particular forms of u, namely, (i) the statistic u_1, obtained when

$$\left.\begin{aligned} A_1 = A_2 = N/(N-2)\, n_1 n_2, \quad B_1 = N\theta/(N-2)\,(n_1 + n_2\theta), \\ B_2 = N/(N-2)\,(n_1 + n_2\theta), \quad B_1/B_2 = \theta \end{aligned}\right\} \tag{64}$$

and (ii) the statistic u_2, obtained when

$$A_j = 1/n_j(n_j - 1) \quad (j = 1, 2). \tag{65}$$

It is seen that u_1 is the square of "Student's" t, while u_2 has as its denominator an estimate of the variance of $\bar{x}_1 - \bar{x}_2$ when the equality of the population variances σ_1^2 and σ_2^2 is not assumed.

We shall write

$$\beta_j(\lambda, \theta) = \int_{u_j^0}^{\infty} p(u_j)\, du_j \quad (j = 1, 2), \tag{66}$$

and, for short,

$$\beta_j(\theta) = \beta_j(0, \theta) \quad (j = 1, 2), \tag{67}$$

where the u_j^0 are some positive constants. $\beta_j(\lambda, \theta)$ is the power function connected with the test criterion u_j, for $j = 1, 2$.

For the sake of illustration let us compute $\beta_1(\theta)$ from the distribution (54) with $n_1 = 5$ and $n_2 = 15$. From the formulae (52) for I_{r0} and I_{0r} we have

$$\left.\begin{aligned} \int_{u_1^0}^{\infty} I_{r0}\, du &= \frac{1}{2(r-1)\, B_1} I_{\mu_1}(r-1, \tfrac{1}{2}), \\ \int_{u_1^0}^{\infty} I_{0r}\, du &= \frac{1}{2(r-1)\, B_2} I_{\mu_2}(r-1, \tfrac{1}{2}), \end{aligned}\right\} \tag{68}$$

* See footnote on p. 2.

where $$\mu_j \doteq (1+B_j u_1^0)^{-1} \quad (j=1,2), \tag{69}$$

and the I_{μ_j} are the ordinary incomplete Beta functions whose standard notation is $I_x(p,q)$ and whose values have been tabled (K. Pearson, 1934).

From (54) and (68) we have

$$\beta_1(\theta) = \left(\frac{B_1}{\Delta}\right)^7 \left(I_{\mu_1}(2,\tfrac{1}{2}) - 7\frac{B_2}{\Delta} I_{\mu_1}(1,\tfrac{1}{2})\right) + \left(\frac{B_2}{\Delta}\right)^2 \sum_{r=2}^{8} (9-r)\left(\frac{B_1}{\Delta}\right)^{8-r} I_{\mu_2}(r-1,\tfrac{1}{2}). \tag{70}$$

In this particular case we have also

$$B_1 = 20\theta/18(5+15\theta), \quad B_2 = 20/18(5+15\theta), \tag{71}$$

$$B_1\Delta^{-1} = \theta(\theta-1)^{-1}, \quad B_2\Delta^{-1} = (\theta-1)^{-1}. \tag{72}$$

Therefore, substituting (72) into (70),

$$\beta_1(\theta) = \left(\frac{\theta}{\theta-1}\right)^7 \left(I_{\mu_1}(2,\tfrac{1}{2}) - \frac{7}{\theta-1} I_{\mu_1}(1,\tfrac{1}{2})\right) + \frac{1}{(\theta-1)^2} \sum_{r=2}^{8} (9-r)\left(\frac{\theta}{\theta-1}\right)^{8-r} I_{\mu_2}(r-1,\tfrac{1}{2}), \tag{73}$$

where the μ_j should be expressed in terms of θ by substituting (71) into (69). Finally, let us take $u_1^0 = (2{\cdot}101)^2$, the 5 % level for t^2 with $N-2 = 18$ degrees of freedom.

In the same way $\beta_1(\theta)$ was computed for two more particular cases. For $n_1 = 3$ and $n_2 = 5$ we have

$$\beta_1(\theta) = \left(\frac{\theta}{\theta-1}\right)^2 I_{\mu_1}(1,\tfrac{1}{2}) - \frac{1}{\theta-1}\left(\frac{\theta}{\theta-1} I_{\mu_2}(1,\tfrac{1}{2}) + I_{\mu_2}(2,\tfrac{1}{2})\right) \tag{74}$$

with $$B_1 = 8\theta/6(3+5\theta), \quad B_2 = 8/6(3+5\theta), \quad u_1^0 = (2{\cdot}447)^2; \tag{75}$$

for $n_1 = n_2 = 7$ we have

$$\beta_1(\theta) = \frac{1}{(\theta-1)^3}(\theta^3 I_{\mu_1}(3,\tfrac{1}{2}) - I_{\mu_2}(3,\tfrac{1}{2})) - \frac{3\theta}{(\theta-1)^4}(\theta^2 I_{\mu_1}(2,\tfrac{1}{2}) + I_{\mu_2}(2,\tfrac{1}{2}))$$
$$+ \frac{6\theta^2}{(\theta-1)^2}(\theta I_{\mu_1}(1,\tfrac{1}{2}) - I_{\mu_2}(1,\tfrac{1}{2})) \tag{76}$$

with $$B_1 = \theta/6(1+\theta), \quad B_2 = 1/6(1+\theta), \quad u_1^0 = (2{\cdot}179)^2. \tag{77}$$

The numerical values for u_1^0 in (75) and (77) are again the 5 % levels for t^2 with $N-2$ degrees of freedom.

By means of (73), (74) and (76) we have calculated $\beta_1(\theta)$ for various values of θ. The results are given below in Table I. When $n_1 = 5$ and $n_2 = 15$, $\beta_1(\theta)$ increases with θ throughout the whole range $(0,\infty)$; when $n_1 = 3$ and $n_2 = 5$, $\beta_1(\theta)$ begins by decreasing as θ increases from 0, passes through a minimum near the point $\theta = 0{\cdot}05$ and then keeps increasing as θ increases further towards ∞; when $n_1 = n_2 = 7$, $\beta_1(\theta)$ is smallest at $\theta = 1$, increases steadily as θ departs from unity in both directions, and there is the symmetric relation $\beta_1(\theta) = \beta_1(\theta^{-1})$.

TABLE I. *Values of $\beta_1(\theta)$*

θ	0	0·02	0·05	0·10	0·20	0·50	1	2	5	10	20	50	∞
$n_1=5$, $n_2=15$	0·0024	0·0028	0·0032	0·0049	0·084	0·025	0·05	0·098	0·178	0·230	0·266	0·286	0·317
$n_1=3$, $n_2=5$	0·031	0·030+	0·030−	0·030	0·031	0·038	0·05	0·072	0·103	0·145	0·172	0·194	0·216
$n_1=7$, $n_2=7$	0·072	0·070	0·067	0·063	0·058	0·051	0·05	0·051	0·058	0·063	0·067	0·070	0·072

(2) Let us now go back to the general formulae (34) and (36). Since the terms of the double series are all positive functions, there is no question as to the validity of the term-by-term integration.* Hence from (34) and (36) we get respectively the formulae (78) and (79):

$$\beta(\lambda,\theta)=\int_{u_0}^{\infty}p(u)\,du=\left(\frac{B_1}{B_2}\right)^{\frac{1}{2}(n_2-1)}e^{-\lambda}\sum_{h,k=0}^{\infty}C_h\left(1-\frac{B_1}{B_2}\right)^h\frac{\lambda^k}{k!}P_{hk},\quad(B_1\leqslant B_2),\tag{78}$$

$$\beta(\lambda,\theta)=\int_{u_0}^{\infty}p(u)\,du=\left(\frac{B_2}{B_1}\right)^{\frac{1}{2}(n_1-1)}e^{-\lambda}\sum_{h,k=0}^{\infty}C'_h\left(1-\frac{B_2}{B_1}\right)^h\frac{\lambda^k}{k!}P'_{hk},\quad(B_2\leqslant B_1),\tag{79}$$

where

$$P_{hk}=\int_{u_0}^{\infty}p_{hk}(u)\,du=I_{\mu_1}(\tfrac{1}{2}(N-2)+h,\,k+\tfrac{1}{2}),\tag{80}$$

$$P'_{hk}=\int_{u_0}^{\infty}p'_{hk}(u)\,du=I_{\mu_2}(\tfrac{1}{2}(N-2)+h,\,k+\tfrac{1}{2}),\tag{81}$$

$$\mu_j=(1+B_ju_0)^{-1}\quad(j=1,2),\tag{82}$$

and the expressions (61) for the B_j are to be substituted into (78), (79) and (82) in order to make explicit the dependence of $\beta(\lambda,\theta)$ on θ.

Since $P_{hk}\leqslant 1$ for all h and k, the double series in (78), regarded as a power series in λ, is dominated by

$$\sum_{k=0}^{\infty}\left\{\sum_{h=0}^{\infty}C_h\left(1-\frac{B_1}{B_2}\right)^h\right\}\frac{\lambda^k}{k!}=\left(\frac{B_1}{B_2}\right)^{-\frac{1}{2}(n_2-1)}\sum_{k=0}^{\infty}\frac{\lambda^k}{k!},$$

so that its radius of convergence is ∞. We can therefore differentiate (78) term by term with respect to λ; we get

$$\begin{aligned}\frac{\partial}{\partial\lambda}\beta(\lambda,\theta)&=-\left(\frac{B_1}{B_2}\right)^{\frac{1}{2}(n_2-1)}e^{-\lambda}\sum_{h,k=0}^{\infty}C_h\left(1-\frac{B_1}{B_2}\right)^h\frac{\lambda^k}{k!}P_{hk}\\&\quad+\left(\frac{B_1}{B_2}\right)^{\frac{1}{2}(n_2-1)}e^{-\lambda}\sum_{h,k=0}^{\infty}C_h\left(1-\frac{B_1}{B_2}\right)^h\frac{\lambda^{k-1}}{(k-1)!}P_{hk}\\&=\left(\frac{B_1}{B_2}\right)^{\frac{1}{2}(n_2-1)}e^{-\lambda}\sum_{h,k=0}^{\infty}C_h\left(1-\frac{B_1}{B_2}\right)^h\frac{\lambda^k}{k!}(P_{h,k+1}-P_{hk})\quad(B_1\leqslant B_2).\end{aligned}\tag{83}$$

* Cf. Kestelman (1937, p. 123, Theorem 191). The double series Σ_{hk} of positive terms being indistinguishable from the repeated series $\Sigma_h\Sigma_k$, we have only to argue twice with simple series.

Similarly

$$\frac{\partial}{\partial\lambda}\beta(\lambda,\theta) = \left(\frac{B_2}{B_1}\right)^{\frac{1}{2}(n_1-1)} e^{-\lambda} \sum_{h,k=0}^{\infty} C'_h\left(1-\frac{B_2}{B_1}\right)^h \frac{\lambda^k}{k!}(P'_{h,k+1}-P'_{hk}) \quad (B_2 \leqslant B_1). \quad (84)$$

If the smaller of B_1 and B_2 is positive, we have

$$P_{h,k+1}-P_{hk} = \frac{1}{(k+\frac{1}{2})\,B(\frac{1}{2}(N-2)+h,k+\frac{1}{2})}\mu_1^{\frac{1}{2}(N-2)+h}(1-\mu_1)^{k+\frac{1}{2}} > 0$$

and similarly $P'_{h,k+1}-P'_{hk} > 0$, and consequently the right-hand sides of (83) and (84) are always positive. If, say, $B_1 = 0$ (i.e. $\theta=0$), then from the distribution (44) of u it is easy to get $\beta(\lambda,0)$ and thence to prove that its first derivative is everywhere positive. We conclude therefore that *for any n_1, n_2 and fixed θ the power of u is increasing with λ.*

Let us now express $\beta(\lambda,\theta)$ explicitly in terms of θ. This consists of substituting (62) in (78) and (79) and results in the formulae

$$\beta(\lambda,\theta) = (\rho\theta)^{\frac{1}{2}(n_2-1)} e^{-\lambda} \sum_{h,k=0}^{\infty} C_h(1-\rho\theta)^h \frac{\lambda^k}{k!} P_{hk} \quad (\theta \leqslant \rho^{-1}), \quad (85)$$

$$\beta(\lambda,\theta) = (\rho^{-1}\theta^{-1})^{\frac{1}{2}(n_1-1)} e^{-\lambda} \sum_{h,k=0}^{\infty} C'_h(1-\rho^{-1}\theta^{-1})^h \frac{\lambda^k}{k!} P'_{hk} \quad (\theta \geqslant \rho^{-1}), \quad (86)$$

where P_{hk} and P'_{hk} are expressed in terms of θ by means of (80), (81), (82) and (61).

We may observe that in order to obtain (86) from (85) we (i) interchange n_1 and n_2, (ii) replace θ by θ^{-1} and ρ by ρ^{-1} and (iii) replace each P_{hk} by P'_{hk}. Assume now that the A_j depend on the n_j in such a way that the former are interchanged whenever the latter are. Thus in the cases $u = u_1$ with $A_1 = A_2$ and $u = u_2$ with $A_j = 1/n_j(n_j-1)$ the A_j have the desired property. If such is the case, then all the three steps described above are effected simply by interchanging n_1 and n_2 and replacing θ by θ^{-1}. For, if n_1 and n_2 are interchanged, so are A_1 and A_2, and consequently $\rho = A_1/A_2$ becomes its reciprocal; if further θ is replaced by θ^{-1}, then B_1 and B_2 are interchanged and therefore P_{hk} becomes P'_{hk}. We have thus proved that the relation

$$\beta(\lambda,\theta,n_1,n_2) = \beta(\lambda,\theta^{-1},n_2,n_1), \quad (87)$$

holds both for u_1 and for u_2, the n_j being entered as arguments in (87) to make the relation clear. We may henceforth consider (85) only; (86) is dispensed with by virtue of (87).

It is now a question of differentiating (85) with respect to θ. Let us first differentiate term by term and then examine for uniform convergence. We have

$$\frac{dP_{hk}}{d\theta} = \frac{1}{B(\frac{1}{2}(N-2)+h,k+\frac{1}{2})}\mu_1^{\frac{1}{2}(N-2)+h-1}(1-\mu_1)^{k-\frac{1}{2}}\frac{d\mu_1}{d\theta}, \quad (88)$$

$$\frac{d\mu_1}{d\theta} = -\frac{u_0}{(1+B_1u_0)^2}\frac{dB_1}{d\theta} = -\frac{B_1u_0}{(1+B_1u_0)^2}\frac{d}{d\theta}\log B_1 = -\nu\left(\frac{1}{1+B_1u_0}-\frac{1}{(1+B_1u_0)^2}\right)$$

$$= -\nu\mu_1(1-\mu_1), \quad (89)$$

where $$\nu = \frac{d}{d\theta}\log B_1 = \frac{n_2}{\theta(n_1+n_2\theta)}. \tag{90}$$

We have therefore $$\frac{dP_{hk}}{d\theta} = -\nu(\tfrac{1}{2}(N-2)+h)\,\eta_{hk}, \tag{91}$$

where $$\eta_{hk} = \frac{1}{(\frac{1}{2}(N-2)+h)\,B(\frac{1}{2}(N-2)+h,\,k+\frac{1}{2})}\,\mu_1^{\frac{1}{2}(N-2)+h}(1-\mu_1)^{k+\frac{1}{2}}. \tag{92}$$

Thus the term-by-term integration of the double series

$$\sum_{h,k=0}^{\infty} C_h \frac{\lambda^k}{k!}\,\theta^{\frac{1}{2}(n_2-1)}(1-\rho\theta)^h\,P_{hk} \tag{93}$$

results in the series

$$\sum_{h,k=0}^{\infty} C_h \frac{\lambda^k}{k!}\big[\{\tfrac{1}{2}(n_2-1)\,\theta^{\frac{1}{2}(n_2-3)}(1-\rho\theta)^h - \rho h\theta^{\frac{1}{2}(n_2-1)}(1-\rho\theta)^{h-1}\}\,P_{hk}$$
$$-\theta^{\frac{1}{2}(n_2-1)}(1-\rho\theta)^h\,\nu\{\tfrac{1}{2}(N-2)+h\}\,\eta_{hk}\big]. \tag{94}$$

It is not difficult to verify the relation $\eta_{hk} = P_{hk}-P_{h+1,k}$, which shows that $0 \leqslant \eta_{hk} \leqslant 1$. Since we also have $0 < \nu\theta \leqslant 1$ from (90), it is seen that the double series (94) is dominated by

$$\theta^{\frac{1}{2}(n_2-3)}\sum_{h,k=0}^{\infty} C_h \frac{\lambda^k}{k!}\,[\tfrac{1}{2}(n_2-1)\,(1-\rho\theta)^h+\rho\theta h(1-\rho\theta)^{h-1}+\{\tfrac{1}{2}(N-2)+h\}\,(1-\rho\theta)^h]$$
$$=\theta^{\frac{1}{2}(n_2-3)}\sum_{h,k=0}^{\infty} C_h \frac{\lambda^k}{k!}\,[\{\tfrac{1}{2}(n_2-1)+\tfrac{1}{2}(N-2)\}\,(1-\rho\theta)^h+(1+\rho\theta)\,h(1-\rho\theta)^{h-1}],$$

which is uniformly convergent in any interval $0<\theta_1\leqslant\theta\leqslant\rho^{-1}$. Therefore the series (94) is absolutely uniformly convergent in any interval $0<\theta_1\leqslant\theta\leqslant\rho^{-1}$, and consequently (i) (94) represents the derivative of (85) with respect to θ whenever $0<\theta<\rho^{-1}$, (ii) (94) is also the sum of the repeated series when we first sum with respect to h and then with respect to k.

As a consequence of (i) and (ii) we can write

$$\frac{\partial}{\partial\theta}\beta(\lambda,\theta) = \rho^{\frac{1}{2}(n_2-1)}e^{-\lambda}\theta^{\frac{1}{2}(n_2-3)}\sum_{k=0}^{\infty}\frac{\lambda^k}{k!}\,g_k \quad (\theta<\rho^{-1}), \tag{95}$$

where $$g_k = \sum_{h=0}^{\infty} C_h[\tfrac{1}{2}(n_2-1)\,(1-\rho\theta)^h-\rho h\theta(1-\rho\theta)^{h-1}]\,P_{hk}$$
$$-\nu\theta\sum_{h=0}^{\infty} C_h\{\tfrac{1}{2}(N-2)+h\}\,(1-\rho\theta)^h\,\eta_{hk}. \tag{96}$$

The first series on the right-hand side of (96) may be written as

$$\sum_{h=0}^{\infty} D_h(P_{hk}-P_{h+1,k}) = \sum_{h=0}^{\infty} D_h\,\eta_{hk}, \tag{97}$$

where $$D_h = \sum_{j=0}^{h} C_j[\tfrac{1}{2}(n_2-1)\,(1-\rho\theta)^j-\rho j\theta(1-\rho\theta)^{j-1}]$$
$$=\tfrac{1}{2}(n_2-1)\sum_{j=0}^{h} C_j(1-\rho\theta)^j-\rho\theta\sum_{j=0}^{h-1}(j+1)\,C_{j+1}(1-\rho\theta)^j. \tag{98}$$

From the definition (31) of C_j we have

$$(j+1)\,C_{j+1} = \{\tfrac{1}{2}(n_2-1)+j\}\,C_j,$$

so that
$$\tfrac{1}{2}(n_2-1)\,C_j = (j+1)\,C_{j+1}-jC_j,$$

whence, substituting into (98),

$$D_h = \tfrac{1}{2}(n_2-1)\,C_h(1-\rho\theta)^h + \sum_{j=0}^{h-1}\{(j+1)\,C_{j+1}(1-\rho\theta)^{j+1}-jC_j(1-\rho\theta)^j\}$$

$$= \{\tfrac{1}{2}(n_2-1)+h\}\,C_h(1-\rho\theta)^h. \tag{99}$$

Thus we obtain, from (96), (97) and (99),

$$g_k = \sum_{h=0}^{\infty}\{\tfrac{1}{2}(n_2-1)+h-\nu\theta(\tfrac{1}{2}(N-2)+h)\}\,C_h(1-\rho\theta)^h\,\eta_{hk},$$

and finally, replacing ν by its expression (90),

$$g_k = \frac{1}{n_1+n_2\theta}\sum_{h=0}^{\infty}\{\tfrac{1}{2}n_2(n_2-1)\,\theta-\tfrac{1}{2}n_1(n_1-1)+hn_2\theta\}\,C_h(1-\rho\theta)^h\,\eta_{hk}. \tag{100}$$

(3) *Study of* $\beta(0,\theta)$. If in particular $\lambda=0$, then it follows from (95) and (100) that

$$\frac{d}{d\theta}\beta(0,\theta) = \beta'(0,\theta) = \rho^{\frac{1}{2}(n_2-1)}\theta^{\frac{1}{2}(n_2-3)}(n_1+n_2\theta)^{-1}g \quad (\theta<\rho^{-1}), \tag{101}$$

where
$$g = \sum_{h=0}^{\infty}\{\tfrac{1}{2}n_2(n_2-1)\,\theta-\tfrac{1}{2}n_1(n_1-1)+hn_2\theta\}\,C_h(1-\rho\theta)^h\,\eta_h, \tag{102}$$

$$\eta_h = \frac{1}{\{\frac{1}{2}(N-2)+h\}\,B\{\frac{1}{2}(N-2)+h,\frac{1}{2}\}}\mu_1^{\frac{1}{2}(N-2)+h}(1-\mu_1)^{\frac{1}{2}}. \tag{103}$$

It is easy to verify that η_h decreases as h increases. The terms of the series (102) may be negative at the beginning, but they must become and remain positive as h increases. If the $(l+1)$st term is the first positive one, then we have obviously

$$g<\eta_l\sum_{h=0}^{\infty}\{\tfrac{1}{2}n_2(n_2-1)\,\theta-\tfrac{1}{2}n_1(n_1-1)+hn_2\theta\}\,C_h(1-\rho\theta)^h$$

$$=\eta_l[\{\tfrac{1}{2}n_2(n_2-1)\,\theta-\tfrac{1}{2}n_1(n_1-1)\}\,(\rho\theta)^{-\frac{1}{2}(n_2-1)}+\tfrac{1}{2}n_2(n_2-1)\,\theta(\rho\theta)^{-\frac{1}{2}(n_2+1)}\,(1-\rho\theta)]$$

$$=\eta_l\rho^{-\frac{1}{2}(n_2+1)}\theta^{-\frac{1}{2}(n_2-1)}\{\tfrac{1}{2}n_2(n_2-1)-\tfrac{1}{2}n_1(n_1-1)\,\rho\}. \tag{104}$$

It follows from (101) and (104) that

$$\beta'(0,\theta)<\tfrac{1}{2}(\rho\theta)^{-1}\,(n_1+n_2\theta)^{-1}\,\eta_l\{n_2(n_2-1)-n_1(n_1-1)\,\rho\} \quad (\theta<\rho^{-1}). \tag{105}$$

If $\theta>\rho^{-1}$, we make use of the relation (87) and obtain similarly

$$\frac{d}{d(\theta^{-1})}\beta(0,\theta)<\tfrac{1}{2}\rho\theta(n_2+n_1\theta^{-1})^{-1}\,\eta'_{l'}\{n_1(n_1-1)-n_2(n_2-1)\,\rho^{-1}\} \quad (\theta>\rho^{-1}),$$

whence
$$\beta'(0,\theta)>\tfrac{1}{2}(n_1+n_2\theta)^{-1}\,\eta'_{l'}\{n_2(n_2-1)-n_1(n_1-1)\,\rho\} \quad (\theta>\rho^{-1}), \tag{106}$$

where $\eta'_{l'}$ is again some positive quantity.

Take first the particular case $u = u_2$ so that $\rho = n_2(n_2-1)/n_1(n_1-1)$. The right-hand sides of (105) and (106) both vanish and we have

$$\frac{d}{d\theta}\beta_2(\theta) \lesseqgtr 0 \quad \text{according as} \quad \theta \lesseqgtr \rho^{-1}.$$

Hence $\beta_2(\theta)$ has the smallest value at $\theta = \rho^{-1}$ and increases as θ departs from ρ^{-1} in either direction. $\beta_2(\theta)$ is smallest at $\theta = 1$ if and only if $\rho = 1$, i.e. $n_1 = n_2$.

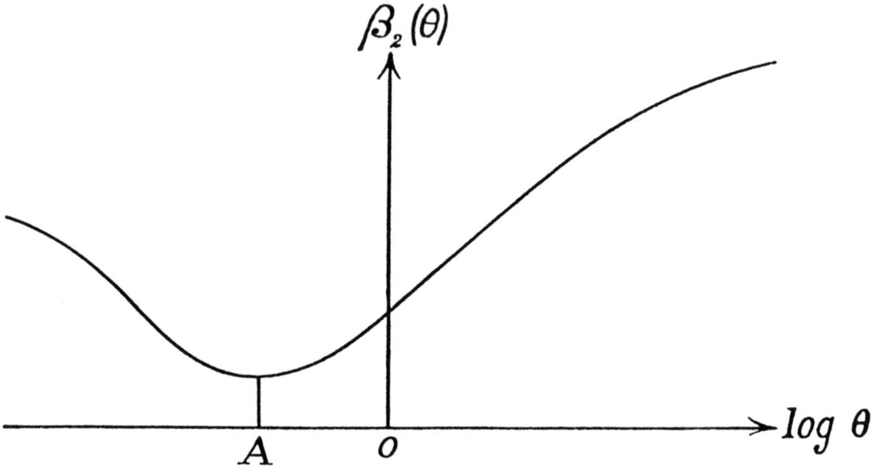

Fig. 1.

Suppose $n_1 < n_2$; then the curve of $\beta_2(\theta)$ takes the form of Fig. 1. But the infinite branch of the curve as $\log\theta \to -\infty$ may or may not rise as high as $\beta_2(1)$. Thus, for example, using (73) and (74) we obtain the following numerical results:

$$\text{(i)}\ n_1 = 5,\ n_2 = 15,\ u_2^0 = 5{\cdot}2;\quad \beta_2(1) = 0{\cdot}050,\ \beta_2(0) = 0{\cdot}039 < \beta_2(1),$$

$$\text{(ii)}\ n_1 = 3,\ n_2 = 5,\ u_2^0 = 6{\cdot}6;\quad \beta_2(1) = 0{\cdot}050,\ \beta_2(0) = 0{\cdot}063 > \beta_2(1).$$

In the case $n_1 > n_2$ we have only to reflect the curve of Fig. 1 in the vertical axis.

Next let $u = u_1$ so that $\rho = 1$. If $n_1 = n_2$, then u_1 coincides with u_2 and the shape of the curve of $\beta_1(\theta)$ is known. It has a minimum at $\theta = 1$ and rises steadily as θ increases from 1 to ∞ or diminishes from 1 to 0.

Suppose now $n_1 < n_2$; then from (106) with $\rho = 1$ we see that

$$\frac{d}{d\theta}\beta_1(\theta) = \beta_1'(\theta) > 0 \quad (\theta > 1).$$

Hence $\beta_1(\theta)$ is decreasing to $\beta_1(1)$ as θ decreases from ∞ to 1. If $\theta < 1$, then $\beta_1'(\theta)$ is given by (101) and (102) wherein $\rho = 1$. We notice that, if θ lies between $n_1(n_1-1)/n_2(n_2-1)$ and 1, then every term of (102) is positive and therefore $\beta_1'(\theta) > 0$. Hence, *as θ further decreases from unity $\beta_1(\theta)$ continues to diminish until θ reaches some value $\bar{\theta}$ which is certainly less than* $n_1(n_1-1)/n_2(n_2-1)$. As to what will happen if θ diminishes further from θ we have yet no general knowledge, but the numerical examples given in Table I (p. 12) show that sometimes (e.g. when $n_1 = 5$, $n_2 = 15$) $\beta_1(\theta)$ is constantly decreasing as θ diminishes to 0 and sometimes (e.g. when $n_1 = 3$, $n_2 = 5$) $\beta_1(\theta)$ passes through a minimum and then rises again as θ becomes very small.

In order to know a little more about $\beta_1(\theta)$ let us find the limit of $\beta_1'(\theta)$ as $\theta \to 0$. From (101), (102) and (103) we find

$$\beta_1'(\theta) = (n_1+n_2\theta)^{-1}\theta^{\frac{1}{2}(n_2-3)}\mu_1^{\frac{1}{2}(N-2)}(1-\mu_1)^{\frac{1}{2}}f(\theta), \tag{107}$$

where

$$f(\theta) = \sum_{h=0}^{\infty}\{\tfrac{1}{2}n_2(n_2-1)\theta - \tfrac{1}{2}n_1(n_1-1)+hn_2\theta\}\frac{1}{(\frac{1}{2}(N-2)+h)\,B(\frac{1}{2}(N-2)+h,\frac{1}{2})}C_hx^h, \tag{108}$$

$$x = \mu_1(1-\theta). \tag{109}$$

Now

$$\sum_{h=0}^{\infty}\frac{C_hx^h}{(\frac{1}{2}(N-2)+h)\,B(\frac{1}{2}(N-2)+h,\frac{1}{2})} = \frac{\Gamma(\frac{1}{2}(N-1))}{\Gamma(\frac{1}{2}N)\,\Gamma(\frac{1}{2})}F(\tfrac{1}{2}(n_2-1),\tfrac{1}{2}(N-1),\tfrac{1}{2}N,x), \tag{110}$$

$$\sum_{h=0}^{\infty}\frac{hC_hx^h}{(\frac{1}{2}(N-2)+h)\,B(\frac{1}{2}(N-2)+h,\frac{1}{2})} = \tfrac{1}{2}(n_2-1)\,x\frac{\Gamma(\frac{1}{2}(N+1))}{\Gamma(\frac{1}{2}(N+2))\,\Gamma(\frac{1}{2})}F(\tfrac{1}{2}(n_2+1),\tfrac{1}{2}(N+1),\tfrac{1}{2}(N+2),x). \tag{111}$$

Using the identity

$$F(\alpha,\beta,\gamma,x) = (1-x)^{\gamma-\alpha-\beta}F(\gamma-\alpha,\gamma-\beta,\gamma,x), \tag{112}$$

and remembering that

$$\begin{aligned}1-x = 1-\mu_1+\mu_1\theta &= B_1u_1^0(1+B_1u_1^0)^{-1}+\theta(1+B_1u_1^0)^{-1}\\ &= \theta(1+B_2u_1^0)\,(1+B_1u_1^0)^{-1},\end{aligned} \tag{113}$$

we have

$$F(\tfrac{1}{2}(n_2-1),\tfrac{1}{2}(N-1),\tfrac{1}{2}N,x) = \theta^{-\frac{1}{2}(n_2-2)}\left(\frac{1+B_2u_1^0}{1+B_1u_1^0}\right)^{-\frac{1}{2}(n_2-2)}F(\tfrac{1}{2}(n_1+1),\tfrac{1}{2},\tfrac{1}{2}N,x), \tag{114}$$

$$F(\tfrac{1}{2}(n_2+1),\tfrac{1}{2}(N+1),\tfrac{1}{2}(N+2),x) = \theta^{-\frac{1}{2}n_2}\left(\frac{1+B_2u_1^0}{1+B_1u_1^0}\right)^{-\frac{1}{2}n_2}F(\tfrac{1}{2}(n_1+1),\tfrac{1}{2},\tfrac{1}{2}(N+2),x). \tag{115}$$

From all the relations (107)–(115) and the relation

$$1-\mu_1 = \theta B_2u_1^0(1+B_2u_1^0)^{-1},$$

it follows that

$$\begin{aligned}\beta_1'(\theta) = \frac{1}{2}\frac{\mu_1^{\frac{1}{2}(N-2)}}{n_1+n_2\theta}\left(\frac{B_2u_1^0}{1+B_1u_1^0}\right)^{\frac{1}{2}}\Bigg[&\{n_2(n_2-1)\,\theta-n_1(n_1-1)\}\\ &\times\frac{\Gamma(\frac{1}{2}(N-1))}{\Gamma(\frac{1}{2}N)\,\Gamma(\frac{1}{2})}\left(\frac{1+B_2u_1^0}{1+B_1u_1^0}\right)^{-\frac{1}{2}(n_2-2)}F(\tfrac{1}{2}(n_1+1),\tfrac{1}{2},\tfrac{1}{2}N,x)+n_2(n_2-1)\\ &\times\frac{\Gamma(\frac{1}{2}(N+1))}{\Gamma(\frac{1}{2}(N+2))\,\Gamma(\frac{1}{2})}\left(\frac{1+B_2u_1^0}{1+B_1u_1^0}\right)^{-\frac{1}{2}n_2}F(\tfrac{1}{2}(n_1+1),\tfrac{1}{2},\tfrac{1}{2}(N+2),x)\Bigg].\end{aligned} \tag{116}$$

Now put $\theta = 0$ in the right-hand side of (116); we have

$$B_1 = 0, \quad B_2 = N/(N-2)n_1, \quad \mu_1 = 1, \quad x = 1,$$

$$F(\tfrac{1}{2}(n_1+1), \tfrac{1}{2}, \tfrac{1}{2}N, 1) = \frac{\Gamma(\frac{1}{2}N)\,\Gamma(\frac{1}{2}(n_2-2))}{\Gamma(\frac{1}{2}(N-1))\,\Gamma(\frac{1}{2}(n_2-1))},$$

$$F(\tfrac{1}{2}(n_1+1), \tfrac{1}{2}, \tfrac{1}{2}(N+2), 1) = \frac{\Gamma(\frac{1}{2}(N+2))\,\Gamma(\frac{1}{2}n_2)}{\Gamma(\frac{1}{2}(N+1))\,\Gamma(\frac{1}{2}(n_2+1))}.$$

It results, after necessary reductions, that

$$\lim_{\theta\to 0} \beta_1'(\theta) = \frac{(B_2 u_1^0)^{\frac{1}{2}}\,(1+B_2 u_1^0)^{-\frac{1}{2}n_2}}{n_1(n_2-2)\,B(\frac{1}{2}(n_2-1), \frac{1}{2})}\{n_2(n_2-2) - n_1(n_1-1)(1+B_2 u_1^0)\}. \quad (117)$$

If $n_1 < n_2$ and u_1^0 is taken to be the 5 % level of t^2 corresponding to $N-2$ degrees of freedom, then by actual trials it is found that the right-hand side of (117) is negative when and only when $n_2 - n_1 \leqslant 2$. In other words, the curve of $\beta_1(\theta)$ is rising as θ diminishes from some small value to 0 when and only when $n_2 - n_1$ is 1 or 2. The particular examples given in Table I for (*a*) $n_1 = 5$, $n_2 = 15$ and (*b*) $n_1 = 3$, $n_2 = 5$ both verify this general result.

A graph of $\beta_1(\theta)$ for $\beta_1(1) = 0{\cdot}05$ and $n_2 - n_1 > 2$ is shown in Fig. 2*a*, wherein A is a point such that $OA = \log\{n_1(n_1-1)/n_2(n_2-1)\}$. We are able to sketch such a curve since we have proved that $\beta_1(\theta)$ is steadily decreasing as θ diminishes from ∞ to some value which is less than $n_1(n_1-1)/n_2(n_2-1)$ and also as θ decreases from some small value towards 0. There is a dotted part of the curve, indicating that its validity has not been established in a general manner. But the numerical results of Table I for $n_1 = 5$ and $n_2 = 15$ show that sometimes the dotted branch represents the correct behaviour of $\beta_1(\theta)$. A curve of $\beta_1(\theta)$ for $n_1 = n_2$ is also shown in Fig. 2*a*, and in this case the whole curve is proved to be valid.

There remain the possibilities that $n_2 - n_1$ is 1 or 2. Two curves corresponding to these cases and having $\beta_1(1) = 0{\cdot}05$ are shown together in Fig. 2*b*. The points A and A' represent the positions $\log\theta = \log\{n_1(n_1-1)/n_2(n_2-1)\}$ for $n_1 - n_2 = 2$ and 1 respectively. Here the infinite branches of the curves are rising as $\log\theta \to -\infty$, a fact that we have just established. But the value of $\beta_1(0)$ is greater or less than $\beta_1(1) = 0{\cdot}05$ according as $n_2 - n_1 = 1$ or 2. This fact, made conspicuous in Fig. 2*b*, may be established by actual trials on using the formula

$$\beta_1(0) = I_{\mu_2}(\tfrac{1}{2}(n_2-1), \tfrac{1}{2}), \quad \text{where} \quad \mu_2 = \left(1 + \frac{Nu_1^0}{(N-2)n_1}\right)^{-1}, \quad (118)$$

which is a consequence of the distribution (45). In Fig. 2*b* the dotted parts of the curves again signify that we are in doubt. It may be mentioned, however, that the examples $n_1 = 3$ and $n_2 = 5$ given in Table I conforms with the dotted branch of the curve for $n_2 - n_1 = 2$.

If $n_2 < n_1$, the curves of $\beta_1(\theta)$ are simply the reflexions of those drawn in Fig. 2 in the vertical axis.

Thus, if $n_1 \leqslant n_2$, we can have $\beta_1(0) > \beta_1(1) = 0{\cdot}05$ when and only when $n_2 - n_1 = 0$ or 1. Using (118) we have calculated the values of $\beta_1(0)$ both when $n_2 = n_1 + 1$ and when $n_2 = n_1$ for various values of n_1. The results are given in Table II.

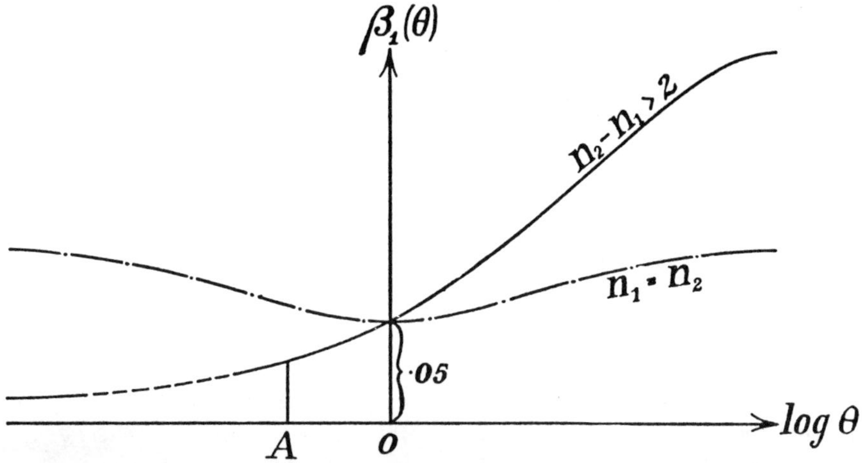

Fig. 2a.

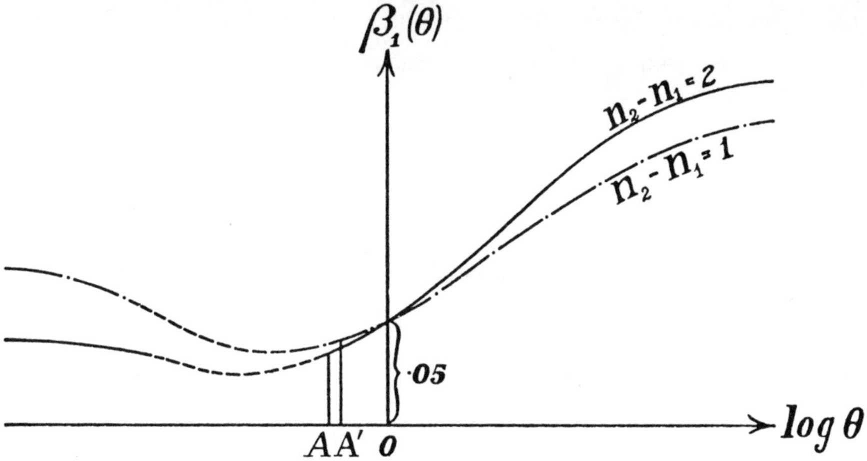

Fig. 2b.

TABLE II. *Values of* $\beta_1(0)$

n_1	2	3	4	5	6	7	8	9	10	∞
$n_2 = n_1$	0·146	0·109	0·0918	0·0824	0·0764	0·0722	0·0692	0·0668	0·0650	0·050
$n_2 = n_1 + 1$	0·0545	0·0558	0·0551	0·0545	0·0538	0·0534	0·0530	0·0527	0·0525	0·050

(4) Some knowledge of the functions $\beta_j(\theta) = \beta_j(0, \theta)$ being now available, it is our purpose to find in the plane of λ and θ the region formed of points (λ, θ) at which $\beta_j(\lambda, \theta) < \beta_j(0, 1)$, while always assuming that $n_1 < n_2$. We have shown that

2-2

$\beta_j(\lambda,\theta)$ increases with λ and that $\beta_j(0,\theta)$ increases with θ for all $\theta>1$. Hence, whenever $\lambda>0$ and $\theta>1$, we will have $\beta_j(\lambda,\theta)>\beta_j(0,\theta)>\beta_j(0,1)$. We must therefore consider only the values of θ which are less than unity.

Consider the function $\beta_1(\lambda,\theta)$ first. For any fixed λ and any $\theta<1$ it follows from (95) and (100) on putting $\rho=1$ that

$$\frac{\partial}{\partial\theta}\beta_1(\lambda,\theta)=e^{-\lambda}\theta^{\frac{1}{2}(n_2-3)}\sum_{k=0}^{\infty}\frac{\lambda^k}{k!}g_k, \tag{119}$$

where

$$g_k=(n_1+n_2\theta)^{-1}\sum_{h=0}^{\infty}\{\tfrac{1}{2}n_2(n_2-1)\theta-\tfrac{1}{2}n_1(n_1-1)+hn_2\theta\}C_h(1-\theta)^h\eta_{hk}. \tag{120}$$

If θ lies between $n_1(n_1-1)/n_2(n_2-1)$ and 1, then g_k is positive for every k and therefore

$$\frac{\partial}{\partial\theta}\beta_1(\lambda,\theta)>0 \quad \text{for} \quad n_1(n_1-1)/n_2(n_2-1)\leqslant\theta<1.$$

Thus, for any fixed λ, the curve of $\beta_1(\lambda,\theta)$ as θ diminishes from unity behaves exactly as those shown in Fig. 2, i.e. $\beta_1(\lambda,\theta)$ is decreasing as θ decreases from unity towards some value which is less than $n_1(n_1-1)/n_2(n_2-1)$. Let us examine, as before, the behaviour of $\beta_1(\lambda,\theta)$ as θ approaches 0. If, just as before, we express each g_k in terms of hypergeometric series and apply the formula (112), we shall get

$$\lim_{\theta\to 0}\frac{\partial}{\partial\theta}\beta_1(\lambda,\theta)=\frac{e^{-\lambda}(B_2u_1^0)^{\frac{1}{2}}(1+B_1u_0)^{-\frac{1}{2}n_2}}{2n_1\Gamma(\frac{1}{2}(n_2-1))}\sum_k\frac{\lambda^k}{k!}\frac{\Gamma(\frac{1}{2}(n_2-2)+k)}{\Gamma(k+\frac{1}{2})}$$
$$\times\{n_2(n_2-2+2k)-n_1(n_1-1)(1+B_2u_1^0)\}. \tag{121}$$

Let us now take u_1^0 to be the 5 % level of t^2 corresponding to $N-2$ degrees of freedom and consider separately the following three possibilities:

(1) $n_2-n_1>2$. In this case we have shown before that the quantity

$$n_2(n_2-2+2k)-n_1(n_1-1)(1+B_2u_1^0)$$

is positive for $k=0$. It is therefore positive for all k and we get

$$\lim_{\theta\to 0}\frac{\partial}{\partial\theta}\beta_1(\lambda,\theta)>0.$$

Hence the infinite branch of the curve of $\beta_1(\lambda,\theta)$, for any given λ, has again the same form as that shown in Fig. 2*a* for $n_2-n_1>2$.

To every $\theta<1$ there corresponds a value $\lambda=\lambda_\theta$ with the property that

$$\beta_1(\lambda,\theta)\lesseqqgtr 0{\cdot}05 \quad \text{according as} \quad \lambda\lesseqqgtr\lambda_\theta.$$

This is an immediate consequence of the facts that $\beta_1(0,\theta)<0{\cdot}05$ and $\beta_1(\lambda,\theta)$ is increasing with λ. We shall show that λ_θ will increase as θ decreases provided θ does not lie within the range for which the curve in Fig. 2*a* is dotted. In fact, if $\theta_2<\theta_1$, then $\beta_1(\lambda_{\theta_2},\theta_2)=\beta_1(\lambda_{\theta_1},\theta_1)>\beta_1(\lambda_{\theta_1},\theta_2)$ because $\beta_1(\lambda_{\theta_1},\theta)$ is decreasing with θ. Hence $\lambda_{\theta_2}>\lambda_{\theta_1}$ because $\beta_1(\lambda,\theta_2)$ is increasing with λ.

Thus we may sketch the graph of λ_θ against $\log\theta$ as in Fig. 3a, wherein the dotted part corresponds to the dotted part of the curve in Fig. 2a.

(2) $n_2 = n_1+2$. Now λ being fixed, we cannot say that $\beta_1(\lambda, \theta)$ is rising as $\theta\to 0$, although $\beta_1(0,\theta)$ is so, because we can find some positive λ so that the right-hand side of (121) is positive. The behaviour of λ_θ for very small values of θ is therefore unknown. But the curve for $n_2 = n_1+2$ in Fig. 2b suggests that the graph of λ_θ against $\log\theta$ might be that shown in Fig. 3b, wherein the solid branch is certainly valid.

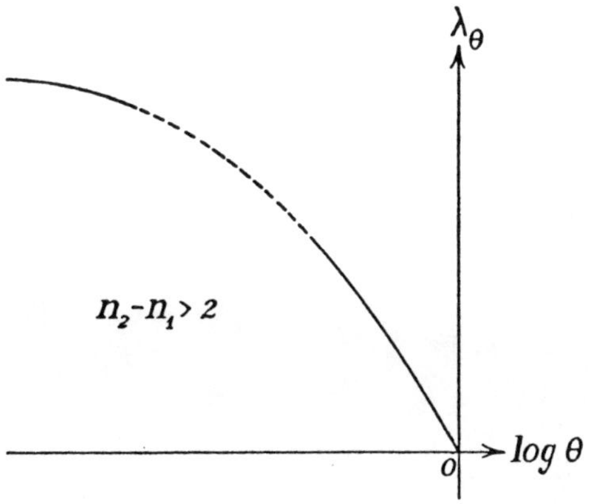

Fig. 3a.

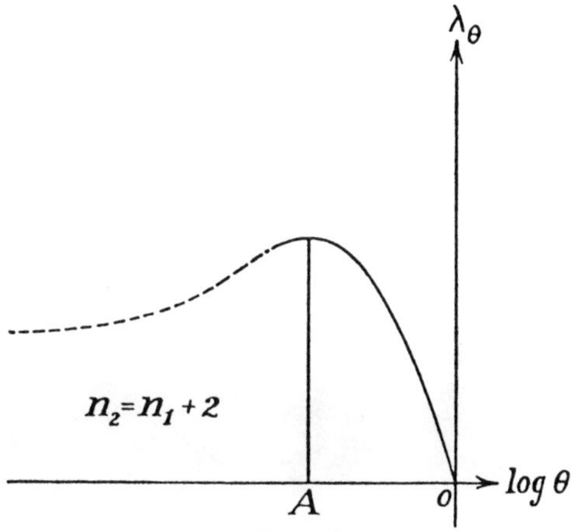

Fig. 3b.

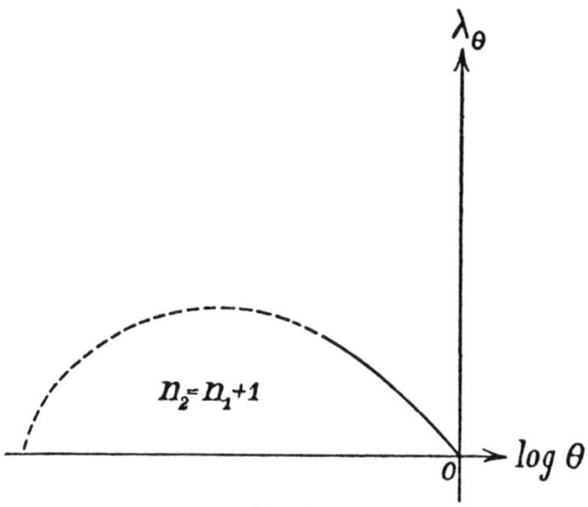

Fig. 3c.

(3) $n_2 = n_1+1$. In this case the range of $\log\theta$ for which $\beta_1(\theta) < 0{\cdot}05$ is limited, since $\beta_1(0) > 0{\cdot}05$. Hence the graph of $\lambda = \lambda_\theta$ takes some form as shown in Fig. 3c.

In the case $n_2-n_1 > 1$ let us denote by λ_0 the quantity $\lim\lambda_\theta$ as $\theta\to 0$. It is the solution of the equation

$$\beta_1(\lambda, 0) = 0{\cdot}05,$$

or, as may be seen from (44),

$$e^{-\lambda}\sum_{k=0}^{\infty}\frac{\lambda^k}{k!}I_{\mu_2}(\tfrac{1}{2}(n_2-1), k+\tfrac{1}{2}) = 0{\cdot}05, \quad \text{where } \mu_2 = (1+B_2u_1^0)^{-1},\ B_2 = N/(N-2)n_1. \tag{122}$$

We have computed λ_0 from (122) for some particular values of n_1 and n_2. The results are

TABLE III. *Values of λ_0*

n_1	5	10	15	20	30	90	∞
$n_2=n_1+5$	1·35	0·53	0·33	0·21	0·09	0·01	0
$n_2=n_1+10$	3·02	1·35	0·85	0·66	0·45	0·08	0

Also, for $n_1 = 30$ and $n_2 = 60$ we have $\lambda_0 = 1{\cdot}41$. Thus, for example, if

$$\sigma_1^2 = 1, \quad \sigma_2^2 = 200, \quad |\delta| = 2{\cdot}6, \quad n_1 = 30, \quad n_2 = 60,$$

then, since $\theta = 0{\cdot}005$ as is nearly zero, λ_0 is nearly 1·4. But actually

$$\lambda = \frac{2{\cdot}6^2}{2(\frac{1}{30} + \frac{200}{60})} = 1{\cdot}004.$$

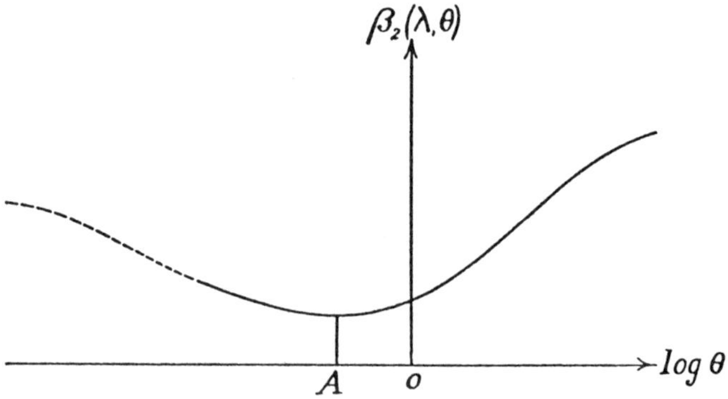

Fig. 4.

Therefore it is likely that $\lambda < \lambda_0$, so that, using the u_1-test, the chance of detecting the falsehood of the hypothesis that both $\delta = 0$ and $\sigma_1^2 = \sigma_2^2$ is less than 0·05. This chance decreases as σ_2^2 increases and has actually the limit, as $\sigma_2^2 \to \infty$,

$$I_{0{\cdot}88}(29{\cdot}5, 0{\cdot}5) = 0{\cdot}006.$$

Next consider the case $u = u_2$, where $\rho = n_2(n_2-1)/n_1(n_1-1)$. It is seen from (95) and (100) that, for any given λ,

$$\begin{aligned} \frac{\partial}{\partial\theta}\beta_2(\lambda,\theta) &> 0 \quad \text{for} \quad \theta > n_1(n_1-1)/n_2(n_2-1) \\ &= 0 \quad \text{for} \quad \theta = n_1(n_1-1)/n_2(n_2-1). \end{aligned}$$

Thus, whatever be the fixed λ, the variation of $\beta_2(\lambda, \theta)$ with θ is always represented by the curve in Fig. 4, with a dotted branch indicating doubt.

If we define the function λ_θ by the relation

$$\beta_2(\lambda, \theta) \gtreqless \beta_2(0, 1) \quad \text{according as} \quad \lambda \gtreqless \lambda_\theta,$$

then we may sketch the curves of λ_θ as shown in Fig. 5.

According as $\beta_2(0)$ is greater or less than $\beta_2(1)$, the curve $\lambda = \lambda_\theta$ is represented by Fig. 5*a* or Fig. 5*b*.

IV. CONCLUSION, THE QUESTION OF BIAS

We have started considering three hypotheses in connexion with the two sample problem and some test based on the statistic u of equation (3). The merit or demerit of u as a test criterion for these hypotheses can only be decided from a knowledge of the power function $\beta(\lambda, \theta)$, which expresses the probability that the test will detect the falsehood of the hypothesis tested when an alternative hypothesis is true.

(A) If the hypothesis to be tested is $H_1\{\lambda = 0, \theta = 1\}$, then a point (λ, θ) with $\lambda > 0$ and $\theta \neq 1$ may be called a point of bias if $\beta(\lambda, \theta) < \beta(0, 1)$. When we are

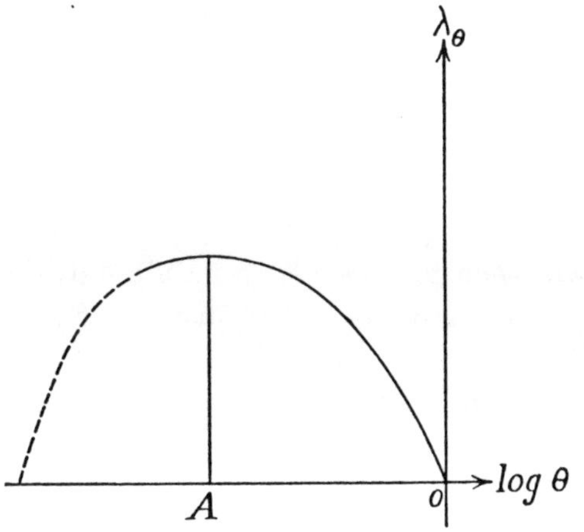

Fig. 5a.

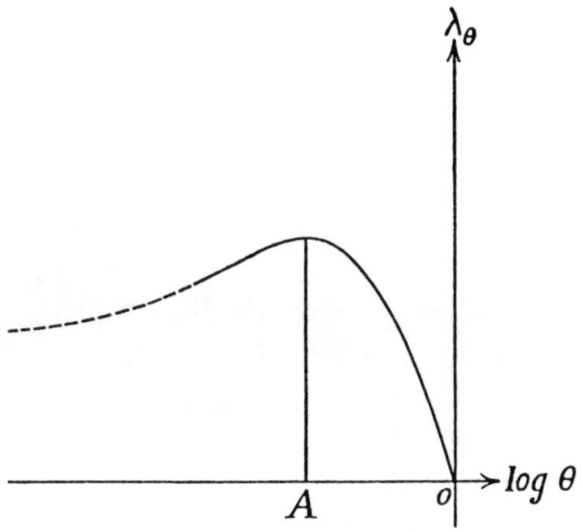

Fig. 5b.

sampling from populations represented by such points and using, say, the 5 % significance level for u, we shall be less likely to reject the hypothesis H_1 than when it is true. This is clearly a most unsatisfactory result. If $n_1 < n_2$, the domain consisting of the points of bias is shown in Fig. 3 in the case of u_1 and in Fig. 5 in the case of u_2, wherein all the points under the curve $\lambda = \lambda_\theta$ are points of bias.* The seriousness of the bias has been illustrated in several cases by numerical results. If u_2 is used instead of u_1, the bias is likely to occur in more limited domains and to be less serious, but u_2 is also likely to be less sensitive to the variation of θ. In the case $n_1 = n_2$ the test u_1 $(= u_2)$ becomes unbiassed, but is extremely insensitive to the variation of θ so long as each sample contains more than two or three individuals. This fact is illustrated numerically in Table II (p. 19). As $n_1 = n_2$ becomes large, the test tends to be entirely indifferent to the variation of θ. It follows that both u_1 and u_2 are to be ruled out as inadequate for testing the hypothesis H_1.

(B) If $H_2\{\lambda = 0\}$ is the hypothesis to be tested, then both u_1 and u_2 are unbiassed tests in the sense that both $\beta_1(\lambda, \theta)$ and $\beta_2(\lambda, \theta)$ are smallest at $\lambda = 0$ for any fixed

* If $n_1 > n_2$, we obtain the domains of bias through the reflexion of Figs. 3 and 5 in the vertical axis.

θ. The chief difficulty with regard to these tests is, of course, that the sampling distributions of u_1 and u_2 are not independent of θ, so that they do not enable us to control the risk of the first kind error, i.e. of rejecting H_2 when it is true. Thus, if $n_1 = 5$ and $n_2 = 15$ (see Table I, p. 12), the chance of committing an error of the first kind, when u_1 is used as the criterion and when we fix this chance as 0·05 for $\theta = 1$, will vary with the true value of θ and may be anything between 0·0024 and 0·317. The test u_2 has the advantage over u_1 because it is likely to be more insensitive to the variation of θ. For example, if $n_1 = 5$, $n_2 = 15$, and if $\beta_2(\theta)$ is taken to be 0·050 at $\theta = 1$, then we have found on p. 16 that $u_2^0 = 5{\cdot}2$. The minimum of $\beta_2(\theta)$ occurs at $\theta = n_1(n_1-1)/n_2(n_2-1) = 0{\cdot}095$ and may be calculated from (85) by putting $\lambda = 0$ and $\rho\theta = 1$; we find the result is 0·035. The maximum of $\beta_2(\theta)$ is the limit as $\theta \to \infty$, or, what is the same thing, the limit of $\beta_2(\theta)$ as $\theta \to 0$ if we take $n_1 = 15$ and $n_2 = 5$. This is found to be 0·085. Thus if we choose the significance level so that $\beta_2(1) = 0{\cdot}050$, then $0{\cdot}035 \leqslant \beta_2(\theta) \leqslant 0{\cdot}085$ for all θ. This may be compared with the range (0·0024, 0·317) of $\beta_1(\theta)$ when $\beta_1(1) = 0{\cdot}05$. This advantage of u_2 over u_1 has also been pointed out by Welch (1938).

If $n_1 = n_2 = n$ and if n is not very small, then u_1 $(=u_2)$ is so indifferent (see Table II) to the variation of θ that we may safely use it to test H_2 just as if θ were known to be unity. In fact, if, say, $n = 10$, then, in using the u_1-test for H_2 and taking the significance level as the 5 % level of t^2, we can say that (i) the test is unbiassed, (ii) whenever H_2 is true, the chance of rejecting it is between 0·05 and 0·065, and (iii) if $\theta = 1$, then there is no other unbiassed test, adjusted to give the same chance 0·05 of the first kind error, which is more able to detect a departure of λ from 0.

REFERENCES

CRAMÉR, H. (1937). *Random Variables and Probability Distributions*. Camb. Univ. Press.

FISHER, R. A. (1925). "Application of 'Student's' distribution." *Metron*, **5**, 90.

FISHER, R. A. (1934). *Statistical Methods for Research Workers*, 5th ed. Edinburgh: Oliver and Boyd.

FISHER, R. A. (1936). "The fiducial argument in statistical inference." *Ann. Eugen., Camb.*, **6**, 391.

KESTELMAN, H. (1937). *Modern Theories of Integration*. Oxford Univ. Press.

NEYMAN, J. (1935). "Sur la vérification des hypothèses statistiques composées." *Bull. Soc. Math. France*, **63**, 246.

NEYMAN, J. and PEARSON, E. S. (1930). "On the problem of two samples." *Bull. int. Acad. Cracovie*, A, p. 73.

NEYMAN, J. and PEARSON, E. S. (1936). "Contributions to the theory of testing statistical hypotheses. I. Unbiassed critical regions of type A and type A_1." *Statist. Res. Mem.* **1**, 1.

NEYMAN, J. and PEARSON, E. S. (1938). "Contributions to the theory of testing statistical hypotheses. II. Certain theorems on unbiassed critical regions of type A. III. Unbiassed tests of simple statistical hypotheses specifying the values of more than one unknown parameter." *Statist. Res. Mem.* **2**, 25, below.

PEARSON, KARL (1934). *Tables of the Incomplete Beta-Function*. Biometrika publication.

SATO, R. (1937). Thesis for Ph.D. degree in the University of London, not yet published.

SUKHATME, P. V. (1935). "A contribution to the problem of two samples." *Proc. Ind. Acad. Sci.* A, **2**, 584.

WELCH, B. L. (1938). "The significance of the difference between two means when the population variances are unequal." *Biometrika*, **29**, 350.

Reprinted from
Statist. Res. Mem.
2, 91–104 (1938).

ON THE BEST UNBIASSED QUADRATIC ESTIMATE OF THE VARIANCE

By P. L. HSU

1. The population set-up we are here concerned with is the one that forms the basis of the Markoff theorem* of linear estimation. By virtue of this theorem we are able to construct linear forms in the observational data which provide estimates of certain unknown parameters which have no bias and whose variances have the smallest possible values. Part of the Markoff theorem, which we shall mention later, also states that a certain quadratic form in the observational data serves as an unbiassed estimate of the unknown variance σ^2. While this is so, there still remains the problem of how to estimate the quantity σ^2 by means of a quadratic form with the minimum variance. It is our purpose to solve this problem; the method of reaching the solution is also on the lines followed by Markoff, namely, the method of undetermined coefficients (Markoff. 1913).

In connexion with the use of matrices (see for example Turnbull & Aitken, 1932), let us explain once for all the notation we adopt. By capital italic letters, as A, we denote matrices which have more than one row and at least as many columns as rows. The elements of such a matrix are denoted by corresponding small letters with double indices, e.g. a_{ij}, the first index standing for the order number of the rows and the second index for that of the columns. Matrices with one row only (also known as row vectors) will be denoted by small letters, e.g. x; the elements are denoted by the same letter with a single index. If A and x are any matrices, then A' and x' denote respectively the matrices obtained from A and x by interchanging the rows and the columns. Thus x' is a matrix of one column only or a column vector. A diagonal matrix will, for example, be denoted by D_τ, indicating that the ith diagonal element is τ_i and the non-diagonal elements are all zeros. The unit matrix and the zero matrix are denoted by I and O respectively.

Assume, exactly as Markoff did, that the independent random variates $x_1, x_2, \ldots, x_n$ have a common unknown variance† σ^2, that their population means $\xi_1, \xi_2, \ldots, \xi_n$ are linear combinations of $(s < n)$ unknown parameters $p_1, p_2, \ldots, p_s$ with known constant coefficients. Thus

$$\xi = pA, \tag{1}$$

where A, the $s \times n$-matrix of the coefficients a_{ij}, is of rank s.

* For an extension and proof of the theorem see David & Neyman (1938). Section 1 of my paper should be read in conjunction with their contribution. Still more general results have been obtained by Aitken (1935).

† Of course Markoff originally assumed that the variance of x_i is σ^2/P_i, with known P_i. Here we put each P_i equal to unity. There is no loss of generality in doing this, because the variates $x_i\sqrt{P_i}$ have the same variance σ^2.

We shall define the best unbiassed quadratic estimate of σ^2 as the quadratic form

$$Q \equiv Q(x_1, x_2, \ldots, x_n), \tag{2}$$

which satisfies the following conditions:

(i) $\mathscr{E}(Q) \equiv \sigma^2$ for any values of the p_i,

(ii) the variance of Q, σ_Q^2, is independent of the values of the p_i,

(iii) if Q_1 be any other quadratic form having the properties (i) and (ii), then $\sigma_Q^2 \leqq \sigma_{Q_1}^2$.

Part (β)* of the Markoff theorem states that the quadratic form

$$Q_0 = \frac{1}{n-s} S_0 \tag{3}$$

satisfies condition (i). It also satisfies (ii), as will be seen in the sequel; and a necessary and sufficient condition will be found in order that it should satisfy (iii). For this purpose we require to know explicitly the matrix of Q_0.

Now the computation of S_0 involves the following steps: first, in the equations

$$\sum_{j=1}^{n} (x_j - \xi_j)\, a_{ij} = 0 \quad (i = 1, 2, \ldots, s), \tag{4}$$

substitute for the ξ_j their values from (1) and solve for the unknowns p_i; secondly, substitute these values of the p_i in (1) to get the ξ_j; and lastly, substitute the resulting values of the ξ_j in the following equation to get S_0,

$$S_0 = \sum_{j=1}^{n} (x_j - \xi_j)^2. \tag{5}$$

Equations (4) are clearly equivalent to

$$(x - \xi)\, A' = O,$$

i.e.

$$\xi A' = xA'. \tag{6}$$

Substituting (1) into (6) we have

$$pAA' = xA',$$

whence, solving for p,

$$p = xA'(AA')^{-1}; \tag{7}$$

the inverse matrix exists because AA' is a non-singular matrix. Substituting (7) again into (1) we get the required values of the ξ_i:

$$\xi = xB, \tag{8}$$

whence

$$x - \xi = xM, \tag{9}$$

where B and M denote the matrices

$$B = A'(AA')^{-1}A, \quad M = I - B. \tag{10}$$

* David & Neyman (1938, p. 107 below).

In order to substitute (9) into (5), we notice that the matrix multiplication $(x-\xi)(x-\xi)'$ results in a matrix of a single element which is S_0. From (9) we have

$$(x-\xi)(x-\xi)' = xMM'x' = xM^2x', \tag{11}$$

since M is symmetric. But from (10) we have

$$B^2 = A'(AA')^{-1}AA'(AA')^{-1}A = A'(AA')^{-1}A = B,$$

and consequently

$$M^2 = (I-B)^2 = I-2B+B = I-B = M. \tag{12}$$

From (11) and (12) we get

$$(x-\xi)(x-\xi)' = xMx', \tag{13}$$

showing that S_0 is the single element of the matrix product on the right-hand side of (13). In other words, the quadratic form S_0 has M as its matrix; *the quadratic form Q_0 has the matrix $M/(n-s)$.*

2. We shall now start with a quadratic form Q having an arbitrary matrix Λ, i.e.

$$Q = \Sigma\lambda_{ij}x_ix_j \quad \text{with} \quad \lambda_{ji}=\lambda_{ij}, \tag{14}$$

where, as a rule, the symbol Σ means summation extending over all indices from 1 to n. Our purpose is to determine Λ so that Q should satisfy the conditions (i), (ii) and (iii).

Since
$$\mathscr{E}(Q) = \sigma^2\Sigma\lambda_{ii} + \Sigma\lambda_{ij}\xi_i\xi_j, \tag{15}$$

condition (i) requires that
$$\Sigma\lambda_{ii} = 1, \tag{16}$$

and that
$$\Sigma\lambda_{ij}\xi_i\xi_j \equiv 0 \tag{17}$$

identically in the p_i. Assuming that (16) and (17) are satisfied by the λ_{ij}, we have

$$\sigma_Q^2 = \mathscr{E}(Q^2) - \sigma^4, \tag{18}$$

and therefore condition (ii) requires that $\mathscr{E}(Q^2)$ be independent of the p_i. In order to calculate $\mathscr{E}(Q^2)$, write

$$y_i = x_i - \xi_i \quad (i=1,2,\ldots,n), \tag{19}$$

so that
$$\mathscr{E}(y_i) = 0, \quad \mathscr{E}(y_i^2) = \sigma^2, \quad \mathscr{E}(y_i^4) = \beta_i\sigma^4 \quad (i=1,2,\ldots,n), \tag{20}$$

where β_i denotes the quantity

$$\beta_i = \sigma^{-4}\mathscr{E}(x_i-\xi_i)^4 \quad (i=1,2,\ldots,n). \tag{21}$$

We have then
$$Q^2 = \Sigma\lambda_{ij}\lambda_{kl}(y_i+\xi_i)(y_j+\xi_j)(y_k+\xi_k)(y_l+\xi_l). \tag{22}$$

Owing to the identity (17) and the first equation (20) the aggregate of those terms of Q^2, which are independent of the y_i, is

$$\Sigma\lambda_{ij}\lambda_{kl}\xi_i\xi_j\xi_k\xi_l = (\Sigma\lambda_{ij}\xi_i\xi_j)^2 \equiv 0,$$

and the terms involving the first degree of the y_i all have zero expectation. Hence, if Π_r denotes the aggregate of terms of Q^2 constituting the homogeneous polynomial of rth degree in the y_i, then

$$\mathscr{E}(Q^2) = \mathscr{E}(\Pi_4)+\mathscr{E}(\Pi_3)+\mathscr{E}(\Pi_2). \tag{23}$$

Since Π_4 is independent of the p_i we must have

$$\mathscr{E}(\Pi_3)+\mathscr{E}(\Pi_2) \equiv 0, \tag{24}$$

and consequently
$$\mathscr{E}(\Pi_3) \equiv 0, \tag{25}$$

$$\mathscr{E}(\Pi_2) \equiv 0, \tag{26}$$

because each Π_r is a homogeneous polynomial of degree $4-r$ in the p_i.

The full expression for Π_2 is

$$\Pi_2 = \Sigma\lambda_{ij}\lambda_{kl}(y_iy_j\xi_k\xi_l+y_iy_k\xi_j\xi_l+y_iy_l\xi_j\xi_k+y_jy_k\xi_i\xi_l+y_jy_l\xi_i\xi_k+y_ky_l\xi_i\xi_j), \tag{27}$$

which, by symmetry, is reduced to

$$\Pi_2 = 2\Sigma\lambda_{ij}\lambda_{kl}y_iy_j\xi_k\xi_l+4\Sigma\lambda_{ij}\lambda_{kl}y_iy_k\xi_j\xi_l, \tag{28}$$

or
$$\Pi_2 = 2(\Sigma\lambda_{ij}y_iy_j)(\Sigma\lambda_{ij}\xi_i\xi_j)+4(\Sigma\lambda_{ij}y_i\xi_j)^2. \tag{29}$$

The first product term of (29) identically vanishes by (17); there remains the necessity that

$$\mathscr{E}(\Pi_2) = 4\mathscr{E}(\Sigma\lambda_{ij}y_i\xi_j)^2 \equiv 0. \tag{30}$$

The identity (30) holds good when and only when

$$\Sigma\lambda_{ij}y_i\xi_j \equiv 0 \tag{31}$$

for all values of the y_i, and this requires that

$$\sum_{j=1}^{n}\lambda_{ij}\xi_j \equiv 0 \quad (i = 1, 2, \ldots, n), \tag{32}$$

identically in the p_i.

Return now to the condition (25); the full expression of Π_3 is

$$\Pi_3 = \Sigma\lambda_{ij}\lambda_{kl}(y_jy_ky_l\xi_i+y_ky_ly_i\xi_j+y_ly_iy_j\xi_k+y_iy_jy_k\xi_l), \tag{33}$$

which, by symmetry, is reduced to

$$\Pi_3 = 4\Sigma\lambda_{ij}\lambda_{kl}y_ky_ly_i\xi_j = 4(\Sigma\lambda_{ij}y_i\xi_j)(\Sigma\lambda_{ij}y_iy_j), \tag{34}$$

and is identically zero provided the relation (31) holds good. Moreover, it is evident that the relations (32) imply (17). *Hence the conditions* (16) *and* (32) *are necessary and sufficient in order that* Q *should have the properties* (i) *and* (ii). Assuming these conditions satisfied and remembering (18), we can now calculate the variance of Q from the formula

$$\sigma_Q^2 = \mathscr{E}(\Pi_4)-\sigma^4. \tag{35}$$

The full expression of Π_4 is

$$\Pi_4 = \Sigma\lambda_{ij}\lambda_{kl}y_iy_jy_ky_l, \tag{36}$$

and, owing to the first equation (20), the only aggregate of terms having non-zero expectations is

$$\Pi_4' = \Sigma\lambda_{ii}^2y_i^4+2\sum_{i\neq j}\lambda_{ij}^2y_i^2y_j^2+\sum_{i\neq j}\lambda_{ii}\lambda_{jj}y_i^2y_j^2. \tag{37}$$

Hence, making use of the last two equations (20),

$$\mathscr{E}(\Pi_4) = \mathscr{E}(\Pi_4') = \sigma^4 \sum \beta_i \lambda_{ii}^2 + 2\sigma^4 \sum_{i \neq j} \lambda_{ij}^2 + \sigma^4 \sum_{i \neq j} \lambda_{ii}\lambda_{jj}$$

$$= \sigma^4 \Sigma(\beta_i - 3)\lambda_{ii}^2 + 2\sigma^4 \Sigma \lambda_{ij}^2 + \sigma^4(\Sigma\lambda_{ii})^2. \tag{38}$$

Using (16) to reduce the last term of (38) and substituting in (35), we finally obtain

$$\sigma_Q^2 = \sigma^4[\Sigma(\beta_i - 3)\lambda_{ii}^2 + 2\Sigma\lambda_{ij}^2]. \tag{39}$$

We shall denote the expression inside the bracket of (39) by F, thus

$$\sigma_Q^2 = F\sigma^4, \tag{40}$$

$$F = \Sigma(\beta_i - 3)\lambda_{ii}^2 + 2\Sigma\lambda_{ij}^2, \tag{41}$$

and our task is to minimize F with respect to the λ_{ij} under the conditions (16) and (32). Before going any further we must get a clearer form for the equations (32). Evidently these equations are equivalent to

$$\xi\Lambda \equiv 0. \tag{42}$$

Substituting for the ξ_i their values in terms of the p_i from (1) we have

$$pA\Lambda \equiv 0 \tag{43}$$

for any values of the p_i. Therefore (32) are equivalent to the condition

$$A\Lambda = 0 \tag{44}$$

to be satisfied by the coefficients λ_{ij}.

3. We have a quadratic form, F, in the λ_{ij}, these latter being connected by the linear relations (16) and (44), and we have to determine the λ_{ij} so that the value of F be minimum. We must, however, observe that the relations (44) are not necessarily independent. Here is an example showing dependence. If $n = 4$ and

$$A = \left\| \begin{matrix} 1 & 1 & 0 & 0 \\ 0 & 0 & 1 & 1 \end{matrix} \right\|,$$

then (44) will include the following four relations, every three of which imply the fourth:

$$\lambda_{13} + \lambda_{23} = 0, \quad \lambda_{14} + \lambda_{24} = 0, \quad \lambda_{13} + \lambda_{14} = 0, \quad \lambda_{23} + \lambda_{24} = 0.$$

Despite this observation it is still legitimate to go through the usual procedure of undetermined multipliers. The only difficulty is that, since we cannot strike out those relations in (44) which are dependent on the others, we shall have introduced too many parameters which we shall fail to eliminate in the usual manner. The result will be that we cannot solve the problem at once, but shall obtain instead a necessary condition on the matrix A.

Denote by Θ the $s \times n$-matrix

$$\Theta = A\Lambda; \tag{45}$$

then condition (44) requires that Θ be the zero matrix:

$$\theta_{ij} = 0 \quad (i = 1, 2, \ldots, s; \; j = 1, 2, \ldots, n). \tag{46}$$

In order to minimize F under the conditions (16) and (46), introduce the multipliers u and v_{ij} and write

$$F_1 = \Sigma(\beta_i - 3)\lambda_{ii}^2 + 2\Sigma\lambda_{ij}^2 - 2u\Sigma\lambda_{ii} - 8\sum_{i=1}^{s}\sum_{j=1}^{n} v_{ij}\theta_{ij}. \tag{47}$$

The results of setting $\dfrac{\partial F_1}{\partial \lambda_{ij}} = 0$ are

$$(\beta_j - 1)\lambda_{jj} = u + 4\sum_{k=1}^{s} v_{kj}a_{kj} \quad (j = 1, 2, \ldots, n), \tag{48}$$

$$\lambda_{ij} = \sum_{k=1}^{s} v_{ki}a_{kj} + \sum_{k=1}^{s} v_{kj}a_{ki} \quad (i \neq j;\ i, j = 1, 2, \ldots, n). \tag{49}$$

But the first sum on the right-hand side of (49) is the element in the ith row and jth column of the matrix $V'A$, and the second sum is the corresponding element of the matrix $A'V$. Hence, for $i \neq j$, λ_{ij} is the element in the ith row and jth column of the matrix

$$V'A + A'V. \tag{50}$$

We can therefore express Λ in the form

$$\Lambda = V'A + A'V + D_\tau. \tag{51}$$

On multiplying the two sides of (51) by the matrix M both on the left and on the right, we get for the left-hand side the matrix $M\Lambda M$, which is equal to Λ itself on account of the definition (10) of M and the condition (44) on Λ. For the right-hand side we have

$$MV'AM + MA'VM + MD_\tau M. \tag{52}$$

The first two terms of (52) are zero matrices because

$$AM = A - AA'(AA')^{-1}A = A - A = O$$

and similarly $MA' = O$. We have therefore reduced (51) to the simple equation

$$\Lambda = MD_\tau M. \tag{53}$$

Thus, in order that the numbers λ_{ij} satisfy the conditions (16) and (44) and minimize the expression F, the matrix Λ must assume the form (53). Condition (44) is automatically satisfied if Λ has such a form, because $AM = O$. There remains the condition (16), which now constitutes a single linear relation connecting up the diagonal elements τ_i of D_τ.

We may notice in passing that equations (48) have not been used. Let us only observe that if $\beta_i = 3$ for each i, then (48) give

$$\lambda_{jj} = \tfrac{1}{2}u + 2\sum_{k=1}^{s} v_{kj}a_{kj} \quad (j = 1, 2, \ldots, n). \tag{54}$$

But $2\sum_{k=1}^{s} v_{kj}a_{kj}$ is precisely the jth diagonal element of the matrix (50), whose every non-diagonal element is equal to that of Λ. Hence

$$\Lambda = V'A + A'V + \frac{u}{2}I; \tag{55}$$

consequently $D_\tau = \frac{1}{2}uI$ and (53) gives

$$\Lambda = \tfrac{1}{2}uM^2 = \tfrac{1}{2}uM, \tag{56}$$

because $M^2 = M$ (cf. (12)). Finally, condition (16) will give $\frac{1}{2}u = (n-s)^{-1}$ and we see that Λ coincides with the matrix of Q_0. But this result will be included later in a more general one.

4. Equation (53) does not solve our problem; it is now necessary to express F in term of the new parameters τ_i and determine the values of the τ_i to minimize F under the condition (16). By actually performing the multiplication in (53) we get

$$\lambda_{ij} = \sum_{k=1}^{n} m_{ik} m_{jk} \tau_k \quad (i, j = 1, 2, \ldots, n), \tag{57}$$

whence
$$\lambda_{ii}^2 = \sum_{k,l=1}^{n} m_{ik}^2 m_{jl}^2 \tau_k \tau_l \quad (i = 1, 2, \ldots, n), \tag{58}$$

and
$$\sum \lambda_{ij}^2 = \sum m_{ik} m_{jk} m_{il} m_{jl} \tau_k \tau_l = \sum_{k,l} \left(\sum_i m_{ik} m_{il} \right)^2 \tau_k \tau_l. \tag{59}$$

The quantity $\sum_i m_{ik} m_{il}$, being the element in the kth row and lth column of M^2, is simply equal to m_{kl}, because $M^2 = M$ by (12). Hence we have

$$\Sigma \lambda_{ij}^2 = \Sigma m_{kl}^2 \tau_k \tau_l. \tag{60}$$

Substituting (58) and (60) in (41) we have

$$F = \Sigma \mu_{kl} \tau_k \tau_l, \tag{61}$$

where
$$\mu_{kl} = \sum_i (\beta_i - 3) m_{ik}^2 m_{il}^2 + 2m_{kl}^2 \quad (k, l = 1, 2, \ldots, n). \tag{62}$$

Again, from (57) we have

$$\Sigma \lambda_{ii} = \sum_k \left(\sum_i m_{ik}^2 \right) \tau_k = \Sigma m_{kk} \tau_k, \tag{63}$$

because $\sum_i m_{ik}^2$ is the kth diagonal element of M^2, and is consequently equal to the kth diagonal element, m_{kk}, of M. Hence equation (16), expressed in terms of the τ_i, is

$$\Sigma m_{kk} \tau_k = 1. \tag{64}$$

We have therefore proved the following:

THEOREM 1. *In order to get the best unbiassed quadratic estimate Q of σ^2, find the numbers τ_k to minimize the quadratic form* (61) *under the condition* (64); *out of these τ_k we can then construct Q with the matrix Λ given by* (53).

The remaining discussion will be devoted to the condition under which the best estimate Q should coincide with Q_0. For this purpose we shall assume that:

$$\beta_i > 1 \quad (i = 1, 2, \ldots, n). \tag{65}$$

If (61) is a positive definite form, then the usual process of minimizing F through introducing an undetermined multiplier will consist in (i) solving the equations

$$\sum_{l=1}^{n} \mu_{kl}\tau_l = \nu m_{kk} \quad (k=1,2,\ldots,n), \tag{66}$$

for the τ_l, obtaining, say

$$\tau_k = b_k \nu \quad (k=1,2,\ldots,n), \tag{67}$$

and (ii) substituting (67) in (64) to get the value of ν. This resulting ν is then the minimum value of F and the corresponding values of the τ_k in (67) are those which minimize F.

In our present case F is a non-negative form, because from its definition (41) it is never negative. Furthermore $F>0$ if (64) is true, because, granted the condition (65), the vanishing of F requires that each $\lambda_{ij}=0$ and is therefore incompatible with the condition (16). We have, nevertheless, the difficulty that F is not necessarily a positive definite form, i.e. the determinant $|\mu_{kl}|$ is not necessarily non-vanishing. For example, if $n=3$ and

$$A = \| 0, \quad 1, \quad 1 \|,$$

then

$$M = \begin{Vmatrix} 1 & 0 & 0 \\ 0 & \frac{1}{2} & -\frac{1}{2} \\ 0 & -\frac{1}{2} & \frac{1}{2} \end{Vmatrix}$$

and consequently we shall get

$$|\mu_{kl}| = \begin{vmatrix} \beta_1 - 1 & 0 & 0 \\ 0 & \frac{1}{16}(\beta_2+\beta_3+2) & \frac{1}{16}(\beta_2+\beta_3+2) \\ 0 & \frac{1}{16}(\beta_2+\beta_3+2) & \frac{1}{16}(\beta_2+\beta_3+2) \end{vmatrix} = 0.$$

In cases like this the solution of equations (66) will cease to have a definite meaning. We shall, however, justify these equations by the following lemma, proved below in an Appendix:

LEMMA 1. *If F is a non-negative form and if $F=0$ is incompatible with* (64), *then*

(i) *the equations* (66) *are consistent whatever be the value of ν;*

(ii) *F has a positive minimum, say ν_1, under the condition* (64);

(iii) *the values of the τ_k which satisfy* (64) *and for which $F=\nu_1$ must also satisfy* (66) *with $\nu=\nu_1$;*

(iv) *conversely, if $(\tau_1,\ldots,\tau_n,\nu)$ be any solution of* (64) *and* (66), *then ν can have one value only, which is the minimum value of F, and F assumes the value ν at $(\tau_1,\ldots,\tau_n)$.*

We have shown at the end of § 1 that the matrix of the quadratic form Q_0 is

$$A_0 = \frac{1}{n-s} M. \tag{68}$$

Remembering that $M^2 = M$, we see that the matrix A in (53) coincides with A_0 if $D_\tau = \frac{1}{n-s} I$, i.e. if

$$\tau_k = \frac{1}{n-s} \quad (k = 1, 2, \ldots, n). \tag{69}$$

The value, ν_0, of F calculated from A_0 is obtained by substituting (69) into (61). Since the τ_k of (69) satisfy (64) because $\Sigma m_{kk} = n-s$, we conclude, in accordance with the result (iii) of the lemma, that if ν_0 is the minimum value of F then the τ_k of (69) must satisfy the equations (66). Conversely, if they do, then, as an immediate consequence of the result (iv) of the lemma, A_0 will lead to the minimum value of F.

We have proved thus far that if and only if the equations (66) are satisfied by the values (69) will Q_0 be *an* estimate with *the* minimum variance. But in the original problem of minimizing F defined by (41) under the conditions (16) and (44), the quadratic form F is positive definite if the condition (65) is taken into account. Hence there exists one and only one set of the λ_{ij} which give F its minimum value. Therefore we can state the following lemma:

LEMMA 2. *The necessary and sufficient condition that Q_0 be the best unbiassed quadratic estimate of σ^2 is that the equations* (66) *should admit the solution* (69).

Substituting (69) into (66) we get

$$\sum_l \mu_{kl} = (n-s)\,\nu_0 m_{kk} \quad (k = 1, 2, \ldots, n). \tag{70}$$

From the definition (62) of the μ_{kl} we have

$$\sum_l \mu_{kl} = \sum_i (\beta_i - 3)\, m_{ik}^2 \sum_l m_{il}^2 + 2 \sum_l m_{kl}^2 = \sum_i (\beta_i - 3)\, m_{ii} m_{ik}^2 + 2m_{kk} \tag{71}$$

because $\sum_l m_{il}^2 = m_{ii}$ $(i = 1, 2, \ldots, n)$ as a consequence of $M^2 = M$. From (70) and (71) we get

$$\sum_i (\beta_i - 3)\, m_{ii} m_{ik}^2 = ((n-s)\,\nu_0 - 2)\, m_{kk} \quad (k = 1, 2, \ldots, n). \tag{72}$$

Summing up the equations (72) for $k = 1, 2, \ldots, n$ and remembering that

$$\sum_k m_{ik}^2 = m_{ii}, \quad \Sigma m_{kk} = n-s,$$

we have

$$\Sigma(\beta_i - 3)\, m_{ii}^2 = (n-s)^2 \nu_0 - 2(n-s)$$

whence

$$\nu_0 = \frac{1}{(n-s)^2} \Sigma(\beta_i - 3)\, m_{ii}^2 + \frac{2}{n-s} \tag{73}$$

and, on substituting in (72),

$$(n-s) \sum_{i=1}^{n} (\beta_i - 3)\, m_{ii} m_{ik}^2 = m_{kk} \Sigma(\beta_i - 3)\, m_{ii}^2 \quad (k = 1, 2, \ldots, n). \tag{74}$$

7-2

We have therefore proved the following theorem:

THEOREM 2. *In order that Q_0 be the best unbiassed quadratic estimate of σ^2, it is necessary and sufficient that the numbers m_{ij} and β_i should satisfy the relations* (74). *The variance of Q_0 is then*

$$\sigma^2_{Q_0} = \nu_0 \sigma^4, \tag{75}$$

where ν_0 is defined by (73).

In the particular case where $\beta_i = \beta$ for all i, equations (74) become

$$(n-s)(\beta-3)\sum_i m_{ii} m_{ik}^2 = (\beta-3) m_{kk} \sum m_{ii}^2 \quad (k=1,2,\ldots,n), \tag{76}$$

and hold good for any m_{ij} if $\beta = 3$, or for any β if

$$(n-s)\sum_i m_{ii} m_{ik}^2 = m_{kk} \sum m_{ii}^2 \quad (k=1,2,\ldots,n). \tag{77}$$

A particular case in which (77) is true is where the m_{ii} are the same for all i. Since $\Sigma m_{ii} = n-s$, the common value must be $(n-s)/n$. Substituting this value for each m_{ii} in (77) the equality is seen to hold if we remember that $\sum_i m_{ik}^2 = m_{kk}$.

5. *Examples.* In this section we shall give three illustrative examples. Before doing this let us recall briefly what has been done in the previous sections.

Denoting by Q_0 the quadratic form (3) which the Markoff theorem states to be an unbiassed estimate of σ^2, we consider the matrix M of $(n-s)\,Q_0$ and its $\frac{1}{2}n(n+1)$ elements m_{ij}. We denote further by β_i the quantities defined by equation (21); as it stands β_i is the quantity $\gamma_{4i} : \gamma_{2i}^2$, commonly known as "the β_2 of x_i", where γ_{ki} denotes the kth moment about the mean of the variate x_i. As a very special result, Q_0 will be the best unbiassed quadratic estimate of σ^2 if the β_i are the same and the m_{ii} are the same for all i. The condition both necessary and sufficient in order that Q_0 be the best unbiassed quadratic estimate of σ^2 is that the numbers m_{ij} and β_i should satisfy the equations (74). If this condition is not fulfilled, then Theorem 1 gives the method of finding the best unbiassed quadratic estimate of σ^2.

Example 1. Case of one sample, where

$$\beta_i = \beta, \quad \xi_i = p \quad (i=1,2,\ldots,n). \tag{78}$$

We have

$$Q_0 = \frac{1}{n-1}\sum_{i=1}^{n}(x_i-\bar{x})^2, \quad \bar{x} = \frac{1}{n}\sum_{i=1}^{n} x_i, \tag{79}$$

$$m_{ii} = \frac{n-1}{n} \quad (i=1,2,\ldots,n). \tag{80}$$

Hence, whatever be the population, the best unbiassed quadratic estimate of σ^2 will be Q_0.

Example 2. An observation being denoted by x_{ti} $(t=1,\ldots,m;\ i=1,\ldots,n)$, let the β_2 of all x_{ti} be the same and let

$$\mathscr{E}(x_{ti}) = A + B_t + C_i, \tag{81}$$

where A, B_t, C_i are unknown parameters satisfying the relations $\sum_t B_t = \sum_i C_i = 0$. We have, as is well known,

$$Q_0 = \frac{1}{(m-1)(n-1)} \sum_{t,i} (x_{ti} - x_{t.} - x_{.i} + x_{..})^2, \tag{82}$$

where
$$x_{t.} = \frac{1}{n} \sum_{i=1}^{n} x_{ti}, \quad x_{.i} = \frac{1}{m} \sum_{t=1}^{m} x_{ti}, \quad x_{..} = \frac{1}{mn} \sum_{t,i} x_{ti}. \tag{83}$$

The coefficient of x_{ti}^2 in Q_0 is the same for all t and i; therefore Q_0 is the best unbiassed quadratic estimate of σ^2.

In other words if, as in the case of a randomized block experiment (for m treatments in n blocks), we regard as appropriate the set-up

$$x_{ti} = A + B_t + C_i + u_{ti}, \tag{84}$$

then Q_0 is the best unbiassed quadratic estimate of σ^2 provided that the residuals u_{ti} are independent and satisfy the following conditions, namely,

$$(a)\ \mathscr{E}(u_{ti}) = 0, \quad (b)\ \mathscr{E}(u_{ti}^2) = \sigma^2, \quad (c)\ \beta_2(u_{ti}) = \beta_2(u), \tag{85}$$

i.e. the first, second and fourth moments of u have common values for all t and i. The variates u_{ti} need not be normally distributed, in particular $\beta_2(u)$ need not equal 3.

Example 3. Case of k samples. Let the ith observation in the tth sample be x_{ti} $(i = 1, \ldots, n_t;\ t = 1, \ldots, k)$ and let

$$\mathscr{E}(x_{ti}) = \xi_t, \quad \beta_2(x_{ti}) = \beta_t \tag{86}$$

for all t and i. Write $N = \Sigma n_t$.

We have
$$(N-k)\, Q_0 = \sum_{t,i} (x_{ti} - x_{t.})^2, \quad x_{t.} = \frac{1}{n_t} \sum_{i=1}^{n_t} x_{ti}. \tag{87}$$

If we arrange the x_{ti} in order so that

$$x_i = x_{1i} \ (i = 1, \ldots, n_1), \quad x_{n_1+1} = x_{21}, \text{ etc.}, \tag{88}$$

we shall then have

$$M = \begin{Vmatrix} M_1 & 0 & \ldots & 0 \\ 0 & M_2 & \ldots & 0 \\ \ldots & \ldots & \ldots & \ldots \\ 0 & 0 & \ldots & M_k \end{Vmatrix}, \tag{89}$$

where each M_t denotes the matrix

$$M_t = \begin{Vmatrix} \frac{n_t - 1}{n_t} & -\frac{1}{n_t} & \ldots & -\frac{1}{n_t} \\ -\frac{1}{n_t} & \frac{n_t - 1}{n_t} & \ldots & -\frac{1}{n_t} \\ \ldots & \ldots & \ldots & \ldots \\ -\frac{1}{n_t} & -\frac{1}{n_t} & \ldots & \frac{n_t - 1}{n_t} \end{Vmatrix} \tag{90}$$

From the explicit form of M it is easy to reduce the equations (74) to the following:

$$(N-k)\frac{n_t-1}{n_t}(\beta_t-3) = \sum_{t=1}^{k}\frac{(n_t-1)^2}{n_t}(\beta_t-3) \quad (t=1,\ldots,k). \tag{91}$$

In other words, Q_0 is the best unbiassed quadratic estimate of σ^2 if and only if the quantities $(\beta_t-3)(n_t-1)/n_t$ are the same for all t. In order to save space we shall not give the details of how to obtain the best estimate Q. The result is

$$Q = \nu\sum_{t,i}\frac{1}{\alpha_t+2}(x_{ti}-x_{t.})^2, \quad \sigma_Q^2 = \nu\sigma^4, \tag{92}$$

where
$$\nu^{-1} = \sum_t\frac{n_t-1}{\alpha_t+2}, \quad \alpha_t = \frac{n_t-1}{n_t}(\beta_t-3). \tag{93}$$

On the other hand, we have
$$\sigma_{Q_0}^2 = \nu_0\sigma^4,$$

where
$$\nu_0 = \frac{1}{(N-k)^2}\sum_t(n_t-1)(\alpha_t+2). \tag{94}$$

Let us calculate the quantity $L = (\sigma_{Q_0}^2-\sigma_Q^2)/\sigma_Q^2$ as a measure of the gain in accuracy that would follow from using Q instead of Q_0. We obtain

$$L = \frac{1}{(N-k)^2}\sum_{s,t=1}^{k}\frac{(n_s-1)(n_t-1)(\alpha_s-\alpha_t)^2}{(\alpha_s+2)(\alpha_t+2)}. \tag{95}$$

If in particular all the β_t have the same value, let us denote this common value by β_2 in order to conform with the usual moment notation; we have then

$$L = \frac{(\beta_2-3)^2}{(N-k)^2}\sum_{s,t=1}^{k}\frac{(n_s-1)(n_t-1)(n_s-n_t)^2}{n_s n_t\{(\beta_2-1)(n_s-1)+2\}\{(\beta_2-1)(n_t-1)+2\}}. \tag{96}$$

It will be seen that in this particular case L will vanish if and only if either the n_t are the same or else $\beta_2 = 3$. If the n_t are not the same, we have the following inequalities:

$$L < L_1 = \frac{1}{(N-k)^2}\sum_{s,t=1}^{k}\frac{(n_s-n_t)^2}{n_s n_t} \quad \text{if } \beta_2 \geqq 2, \tag{97}$$

$$L \leqq L_2 = \frac{1}{(N-k)^2}\sum_{s,t=1}^{k}\frac{(n_s-1)(n_t-1)(n_s-n_t)^2}{n_s n_t} \quad \text{if } 1 \leqq \beta_2 \lesseqqgtr 2. \tag{98}$$

L_1 and L_2 are also respectively the limiting values of L as $\beta_2 \to \infty$ and $\beta_2 \to 1$.

The numerical examples in the case of three samples as given below furnish us with some idea of the order of magnitude of L. It will be noticed that only when β_2 has very unusual values, i.e. when it is nearly unity and when the sample sizes differ considerably, will L not be negligibly small.

(i) $n_1 = 20$, $n_2 = 15$, $n_3 = 5$:

β_2	1·00	1·05	1·20	1·80	2·00	4·00	6·00	9·00	∞
L	0·196	0·116	0·040	0·0033	0·0016	0·00025	0·00086	0·0011	0·0027

(ii) $n_1 = 10$, $n_2 = 9$, $n_3 = 8$:

β_2	1	1·05	1·2	1·8	2	4	6	9	∞
L	0·0082	0·0054	0·0021	0·00017	0·000083	0·000012	0·000042	0·000069	0·00013

(iii) $n_1 = 10$, $n_2 = 6$, $n_3 = 2$:

β_2	1	1·05	1·2	1·8	2	4	6	9	∞
L	0·211	0·161	0·079	0·011	0·0060	0·0013	0·0053	0·0094	0·021

REFERENCES

Aitken, A. C. (1935). *Proc. Roy. Soc.* Edinburgh, **55**, 42.

David, F. N., & Neyman, J. (1938). *Statist. Res. Mem.* **2**, 105.

Markoff, A. A. (1913). *Calculus of Probability* (Russian). St Petersburg.

Turnbull, H. W. & Aitken, A. C. (1932). *An Introduction to the Theory of Canonical Matrices.* London and Glasgow.

APPENDIX

Here we give the proof of Lemma 1 on p. 98. With a different notation we restate the lemma as follows:

Lemma 1. *If*
$$F = \sum_{i,j=1}^{n} c_{ij}x_i x_j = xCx' \quad (c_{ij}=c_{ji}) \tag{1}$$
is a non-negative quadratic form and if $F = 0$ *is incompatible with the linear relation*
$$bx' = 1, \tag{2}$$
then, (i) *the system of linear equations*
$$xC = \nu b \tag{3}$$
is consistent whatever be the number ν; (ii) F *has a positive minimum, say* ν_1, *under the condition* (2); (iii) *if* $(x_1, x_2, \ldots, x_n)$ *be any system of values which satisfy* (2) *and for which* $F = \nu_1$, *then* $(x_1, \ldots, x_n)$ *must also satisfy* (3) *with* ν *replaced by* ν_1; (iv) *conversely, if* $(x_1, \ldots, x_n, \nu)$ *be any solution of* (2) *and* (3), *then* ν *can have one value only, namely the minimum value* ν_1 *of* F, *and* $(x_1, \ldots, x_n)$ *give the value* ν_1 *for* F.

Proof. Let C be of rank ρ, so that
$$C = \Gamma' D_\alpha \Gamma, \tag{4}$$
where Γ is a $\rho \times n$-matrix such that
$$\Gamma\Gamma' = I. \tag{5}$$
Let Δ be the $(n-\rho) \times n$-matrix forming with Γ the $n \times n$-orthogonal matrix, i.e.
$$\Delta\Delta' = I, \quad \Delta\Gamma' = O, \quad \Gamma'\Gamma + \Delta'\Delta = I. \tag{6}$$

If we perform the orthogonal transformation

$$\left.\begin{aligned} y &= \|y_1, \ldots, y_\rho\| = x\Gamma', \\ z &= \|z_1, \ldots, z_{n-\rho}\| = x\Delta', \end{aligned}\right\} \tag{7}$$

or

$$x = y\Gamma + z\Delta, \tag{8}$$

then (1) and (2) will take the following forms:

$$F = yD_\alpha y' = \sum_{i=1}^{\rho} \alpha_i y_i^2, \quad \alpha_i > 0 \quad (i = 1, \ldots, \rho), \tag{9}$$

$$y\Gamma b' + z\Delta b' = 1. \tag{10}$$

Since we have assumed that $F = 0$ is incompatible with (10), we must have

$$\Delta b' = O, \tag{11}$$

and consequently (10) is reduced to

$$y\Gamma b' = 1. \tag{12}$$

The ordinary process of minimizing (9) under the condition (12) leads to the equation

$$yD_\alpha = \nu_1 b\Gamma', \tag{13}$$

where ν_1 is the minimum value of F. ν_1 is obtained by substituting (13) into (12); its value is therefore positive. This proves (ii).

All the matrices x which satisfy (2) and which minimize (1) are contained in (8), with y fixed by (13). We can verify that (3) is satisfied by any such x. In fact, we get from (4), (8) and (13) that

$$xC = x\Gamma' D_\alpha \Gamma = yD_\alpha \Gamma = \nu_1 b\Gamma'\Gamma. \tag{14}$$

From (11) we have further

$$O = \nu_1 b\Delta'\Delta. \tag{15}$$

Combining (14) and (15) we obtain, on taking into account the last equation (6),

$$xC = \nu_1 b(\Gamma'\Gamma + \Delta'\Delta) = \nu_1 b, \tag{16}$$

which proves (iii).

If ν is any number, equation (3) may be written as

$$x\Gamma' D_\alpha \Gamma = \nu b\Gamma'\Gamma, \tag{17}$$

whence, multiplying both sides with Γ' on the right,

$$x\Gamma' D_\alpha = \nu b\Gamma'. \tag{18}$$

This last equation may be regarded as a system of ρ linear equations in the ρ unknowns y_i defined by the first equation (7). Since the determinant $|D_\alpha|$ of this system is non-vanishing we have the proof of (i).

Finally, let $(x_1, \ldots, x_n, \nu)$ be any solution of (2) and (3). If we perform the transformation (8) on this set of x_i we shall regain the equations (13) and (12) for ν. Hence $\nu = \nu_1$, and for the present set of x_i the function F takes on the value $xCx' = \nu_1 bx' = \nu_1$. This proves (iv) and thereby the whole lemma.

Reprinted from
T. Ann. Math. Statist.
9, 231–243 (1938).

NOTES ON HOTELLING'S GENERALIZED T

BY P. L. HSU

1. Frequency Distribution When the Hypothesis Tested is Not True

a. THE PROBLEM. Let the simultaneous elementary probability law of the $k(f+1)$ variables z_i and z'_{ir} $(i = 1, 2, \cdots, k; r = 1, 2, \cdots, f)$ be

$$(1)\quad p(z, z') = (\sqrt{2\pi})^{-k(f+1)} | C |^{\frac{1}{2}(f+1)} \exp\left[-\frac{1}{2} \sum_{i,j=1}^{k} c_{ij} \{ (z_i - \zeta_i)(z_j - \zeta_j) + v'_{ij} \} \right],$$

where

$$v'_{ij} = \sum_{r=1}^{f} z'_{ir} z'_{jr} \qquad (i, j = 1, 2, \cdots, k)$$

C stands for the matrix $\| c_{ij} \|$ and $| C |$, the corresponding determinant. It is required to find the elementary probability law of the statistic

$$T = | V' |^{-1} \sum_{i,j=1}^{k} V'_{ij} z_i z_j,$$

where $| V' | = | v'_{ij} |$ and V'_{ij} denotes the cofactor of the element v'_{ij} in the matrix $\| v'_{ij} \|$.

The quantity fT is a generalization of "Student's" t considered by Hotelling [1]*. It is an appropriate criterion to test the hypothesis, say H_0, that the ζ_i in the parent population as given by (1) all vanish. The distribution of T when the hypothesis H_0 is true has already been obtained by Hotelling. But our knowledge of the test is hardly complete unless we know also the distribution of T when the ζ_i do not all vanish. Indeed, only such a knowledge can enable us to control the risk of error of the second kind, i.e. of failure to detect that H_0 is untrue [3, 4].

b. THE SOLUTION. Let H be a $k \times k$ non-singular matrix such that $H'CH = I$, the unit matrix, where H' denotes the transposed matrix of H. Let the sets of variables $(z_1, z_2, \cdots, z_k)$ and $(z'_{1r}, z'_{2r}, \cdots, z'_{kr})(r = 1, 2, \cdots f)$ be subject to the same collineation by means of H, so that

$$\| z_1, z_2, \cdots, z_k \| = \| t_1, t_2, \cdots, t_k \| \cdot H'$$

$$\| z'_{1r}, z'_{2r}, \cdots, z'_{kr} \| = \| t'_{1r}, t'_{2r}, \cdots, t'_{kr} \| \cdot H' \quad (r = 1, 2, \cdots, f)$$

where the t_i and t'_{ir} are the new variables. Let further the quantities τ_i be defined by

* References are given at the end of the paper.

(2) $$\|\zeta_1, \zeta_2, \cdots, \zeta_k\| = \|\tau_1, \tau_2, \cdots, \tau_k\| \cdot H'.$$

Then, as is easy to verify, the simultaneous distribution of the new variables will be given by

(3) $$p_1(t, t') = (\sqrt{2\pi})^{-k(f+1)} \exp\left[-\frac{1}{2}\sum_{i=1}^{k}\{(t_i - \tau_i)^2 + u'_{ii}\}\right],$$

while the statistic T, as a function of the t's, retains the original form:

(4) $$T = |U|^{-1}\sum_{i,j=1}^{k} U'_{ij}t_it_j$$

where

$$u'_{ij} = \sum_{r=1}^{f} t'_{ir}t'_{jr} \qquad (i, j = 1, 2, \cdots, k),$$

$|U'| = |u'_{ij}|$, and U'_{ij} is the cofactor of the element u'_{ij} in the matrix $\|u'_{ij}\|$. By virtue of (2) we have the following relation between the old and new parametric constants:

(5) $$\sum_{i=1}^{k} \tau_i^2 = \sum_{i,j=1}^{k} c_{ij}\zeta_i\zeta_j.$$

Our problem is thus reduced to finding the derived distribution of T defined by (4) from the parent population given by (3).

We solve this problem by obtaining an expression for the Laplace integral $E(e^{-\beta T})$, i.e. the mathematical expectation of $e^{-\beta T}$ for real non-negative β. A few words are perhaps needed to explain the fact that the Laplace transform of an elementary probability law determines the latter uniquely except on a null set of points. If $f(x)$ is an elementary probability law which vanishes for all negative x and if

$$g(\beta) = \int_0^\infty e^{-\beta x}f(x)\,dx \quad \text{for} \quad \beta \geq 0,$$

then, letting c be any fixed positive constant, we have

$$g(c - \beta) = \int_0^\infty e^{\beta x}e^{-cx}f(x)\,dx$$

for all $\beta \leq c$. We get therefore

$$m_h = \int_0^\infty x^h e^{-cx}f(x)\,dx = \frac{d^h}{d\beta^h}\,g(c - \beta)\Big|_{\beta=0}, \quad (h = 0, 1, 2, \cdots)$$

the definite integral being obviously finite for all $h \geq 0$. Now a sufficient condition that the set of numbers m_h determines the function $e^{-cx}f(x)$ uniquely, with the exception of a null set at most, is that the latter multiplied by $e^{k\sqrt{x}}$ be summable $(0, \infty)$ for some positive k (cf. [6], p. 320). Since this condition is trivially satisfied by the function $e^{-cx}f(x)$, this function, and therefore $f(x)$ itself, must be uniquely determined by the m_h. In other words, $f(x)$ is uniquely determined by its Laplace transform $g(\beta)$. We now proceed to find the Laplace integral $E(e^{-\beta T})$.

Introduce the function

$$g(t, t', \theta, \alpha) = (\sqrt{2\pi})^k |U'|^{\frac{1}{2}} \exp\left[-\frac{1}{2}\left\{\sum_{i,j=1}^{k} u'_{ij}\theta_i\theta_j + 2i\alpha \sum_{i=1}^{k} t_i\theta_i\right\}\right]$$

and write

$$F(t, t', \theta, \alpha) = p_1(t, t')g(t, t', \theta, \alpha),$$

where all the arguments take real values only. For any functions $\varphi(\theta)$ and $\psi(t, t')$ let us write

$$\int \varphi(\theta)\, d\theta = \int_{-\infty}^{\infty} \cdots \int_{-\infty}^{\infty} \varphi(\theta)\, d\theta_1 \cdots d\theta_k$$

$$\int \psi(t, t')\, d(t, t') = \int_{-\infty}^{\infty} \cdots \int_{-\infty}^{\infty} \psi(t, t')\, dt_1 \cdots dt_k\, dt'_{11} \cdots dt'_{kf}.$$

We have

$$\int d(t, t') \int |F(t, t', \theta, \alpha)|\, d\theta = \int p_1(t, t')\, d(t, t') \int g(t, t', \theta, 0)\, d\theta = 1$$

whence we know that

$$\int d(t, t') \int F\, d\theta = \int d\theta \int F\, d(t, t') \tag{6}$$

On the right-hand side of (6) we find

$$\int p_1(t, t')\, d(t, t') \int g(t, t', \theta, \alpha)\, d\theta = \int e^{-\frac{1}{2}\alpha^2 T} p_1(t, t')\, d(t, t') = E(e^{-\frac{1}{2}\alpha^2 T})$$

while for the integral on the right-hand side of (6) we have

$$\int F\, d(t, t')$$

$$= (\sqrt{2\pi})^{-k(f+2)} \exp\left(-\frac{1}{2}\sum_{i=1}^{k} \tau_i^2\right) \int \exp\left[-\frac{1}{2}\sum_{i=1}^{k} \{t_i^2 + 2(i\alpha\theta_i - \tau_i\}\right] dt \tag{7}$$

$$\times \int |U'|^{\frac{1}{2}} \exp\left[-\frac{1}{2}\sum_{i,j=1}^{k} (\theta_i\theta_j + \delta_{ij})\, u'_{ij}\right] dt',$$

where we mean by the δ_{ij} the quantities

$$\left.\begin{aligned} \delta_{ij} &= 0 \qquad \text{for } i \neq j \\ \delta_{ii} &= 1 \end{aligned}\right\} (i, j = 1, 2, \cdots, k)$$

In the equation (7) the integral with respect to the t_i is immediately written down as

$$(\sqrt{2\pi})^k \exp\left[\frac{1}{2}\sum_{i=1}^{k}(\tau_i - i\alpha\theta_i)^2\right]$$

As to the integral with respect to the t'_{ir}, we may evaluate it by the method by which Wilks [7] evaluated the moments of the generalized variance. The result is

$$2^{\frac{1}{2}k}(\sqrt{2\pi})^{kf} \mid \theta_i\theta_j + \delta_{ij} \mid^{-\frac{1}{2}(f+1)} \frac{\Gamma\frac{1}{2}(f+1)}{\Gamma(\frac{1}{2}(f+1-k))}$$

Making the substitution into (7) we get, after necessary reductions,

$$\int F\, d(t, t') = \frac{(\sqrt{\pi})^{-k}\Gamma(\frac{1}{2}(f+1))}{\Gamma(\frac{1}{2}(f+1-k))} \mid \theta_i\theta_j + \delta_{ij} \mid^{-\frac{1}{2}(f+1)}$$

$$\times \exp\left[-\sum_{i=1}^{k}\{\tfrac{1}{2}\alpha^2\theta_i^2 + i\alpha\tau_i\theta_i\}\right]$$

whence, noticing that $\mid \theta_i\theta_j + \delta_{ij} \mid = 1 + \sum_{i=1}^{k}\theta_i^2$

$$E(e^{-\frac{1}{2}\alpha^2 T}) = \frac{(\sqrt{\pi})^{-k}\Gamma(\frac{1}{2}(f+1))}{\Gamma(\frac{1}{2}(f+1-k))} \int \left(1 + \sum_{i=1}^{k}\theta_i^2\right)^{-\frac{1}{2}(f+1)} \tag{8}$$

$$\exp\left[-\sum_{i=1}^{k}\{\tfrac{1}{2}\alpha^2\theta_i^2 + i\alpha\tau_i\theta_i\}\right] d\theta$$

Equation (8) gives the Laplace transform of the elementary probability law, $p(T)$, of T. There is no essential difficulty in getting $p(T)$ by inversion directly from (8). Nevertheless, it may be of interest to get $p(T)$ indirectly by identifying the right-hand side of (8) with the Laplace transform of another elementary probability law which is otherwise *known*. For this purpose consider the simultaneous elementary probability law

$$p(x, y) = (\sqrt{2\pi})^{-(f_1+f_2)} \exp\left[-\frac{1}{2}\sum_{i=1}^{f_1}(x_i - \xi_i)^2 - \frac{1}{2}\sum_{j=1}^{f_2} y_j^2\right]$$

and let us find the derived distribution of the statistic

$$L = \sum_{i=1}^{f_1} x_i^2 \Big/ \sum_{j=1}^{f_2} y_j^2$$

As before, we introduce the function

$$g(x, y, \theta, \alpha) = (\sqrt{2\pi})^{-f_1}\left(\sum_{j=1}^{f_2} y_j^2\right)^{\frac{1}{2}f_1} \exp\left[-\frac{1}{2}\left(\sum_{j=1}^{f_2} y_j^2 \sum_{i=1}^{f_1}\theta_i^2 + 2i\alpha\sum_{i=1}^{f_1} x_i\theta_i\right)\right]$$

write

$$F(x, y, \theta, \alpha) = p(x, y)g(x, y, \theta, \alpha)$$

and ascertain that

$$\int d(x, y) \int F\, d\theta = \int d\theta \int F\, d(x, y) \tag{9}$$

On the left-hand side of (9) we find

$$\int e^{-\frac{1}{2}\alpha^2 L} p(x, y)\, d(x, y) = E(e^{-\frac{1}{2}\alpha^2 L})$$

while for the integral on the right-hand side of (9), we have

$$\begin{aligned}\int F\, d(x, y) &= (\sqrt{2\pi})^{-(2f_1+f_2)} \exp\left(-\frac{1}{2}\sum_{i=1}^{f_1} \xi_i^2\right) \\ &\quad \times \int \exp\left[-\frac{1}{2}\sum_{i=1}^{f_1} \{x_i^2 + 2(i\alpha\theta_i - \xi_i)x_i\}\right] dx \\ &\quad \times \int \left(\sum_{j=1}^{f_2} y_j^2\right)^{\frac{1}{2}f_1} \exp\left[-\frac{1}{2}\left(1 + \sum_{i=1}^{f_1} \theta_i^2\right)\sum_{j=1}^{f_2} y_j^2\right] dy \\ &= \frac{(\sqrt{\pi})^{-f_1}\Gamma(\frac{1}{2}(f_1 + f_2))}{\Gamma(\frac{1}{2}f_2)} \left(1 + \sum_{i=1}^{f_1} \theta_i^2\right)^{-\frac{1}{2}(f_1+f_2)} \\ &\quad \exp\left[-\frac{1}{2}\sum_{i=1}^{f_1} (\alpha^2\theta_i^2 + 2i\alpha\xi_i\theta_i)\right]\end{aligned}$$

Writing

$$f_1 = k, \qquad f_2 = f + 1 - k \tag{10}$$

we get finally

$$\begin{aligned}E(e^{-\frac{1}{2}\alpha^2 L}) &= \frac{(\sqrt{\pi})^{-k}\Gamma(\frac{1}{2}(f + 1))}{\Gamma(\frac{1}{2}(f + 1 - k))} \int \left(1 + \sum_{i=1}^{k} \theta_i^2\right)^{-\frac{1}{2}(f+1)} \\ &\quad \exp\left[-\frac{1}{2}\sum_{i=1}^{k} (\alpha^2\theta_i^2 + 2i\alpha\xi_i\theta_i)\right]\end{aligned} \tag{11}$$

From the identity of (8) and (11) we conclude that T is distributed exactly the same as L with the appropriate "degrees of freedom" f_1 and f_2 given by (10). But the elementary probability law of L has already been derived by P. C. Tang [5]. Using his result we immediately write down the elementary probability law of T:

$$p(T) = e^{-\lambda} \sum_{h=0}^{\infty} \frac{\lambda^h}{h!} \frac{1}{B(\frac{1}{2}f_1 + h, \frac{1}{2}f_2)} T^{\frac{1}{2}f_1+h-1}(1 + T)^{-\frac{1}{2}(f_1+f_2)-h} \tag{12}$$

where f_1 and f_2 are given by (10) and

$$\lambda = \frac{1}{2}\sum_{i=1}^{k} \tau_i^2 = \frac{1}{2}\sum_{i,j=1}^{k} c_{ij}\zeta_i\zeta_j \tag{13}$$

in accordance with (5). The tables of probability integrals prepared by Tang can, of course, be used to suit our purpose.

2. An Optimum Property of the T-Test. To any reader familiar with the Neyman-Pearson theory of testing statistical hypotheses [3, 4], the theorem stated below may be of considerable interest.

Denote by W the $k(f + 1)$-dimensional space of the z_i and z'_{ir} and let w be any region in W which may possibly serve as a critical region for the rejection of the hypothesis H_0. Let us speak of a critical region w as belonging to the class D if w satisfies the following condition:

$$\int_w p(z, z')d(z, z') = \epsilon + \frac{\alpha}{2} \sum_{i,j=1}^{k} c_{ij}\zeta_i\zeta_j + R \tag{14}$$

where $\epsilon < 1$ is a positive constant independent of the ζ_i, c_{ij} and the region w, α is a constant depending on w only, but not on the ζ_i or c_{ij}, and where R for any given set of values of the c_{ij} is an infinitesimal of at least the third order as all the ζ_i tend to zero.

THEOREM. *Of all the regions belonging to the class D, the particular region which gives the largest possible value to the coefficient α in the equation (14) is the region defined by $T \geq T_\epsilon$, where T_ϵ is a constant so determined that the probability, when all ξ_i vanish, of the observed T being not less than T_ϵ is exactly ϵ.*

The significance of the theorem is clear. Every critical region belonging to the class D serves as an unbiased exact test of the hypothesis H_0, ϵ being the preassigned chance of rejecting H_0 if it is true. Further, as is seen from (14), as the ζ_i start to depart from zero, the increased chance of rejecting H_0 due to its falsehood is approximately proportional to the quantity $\Sigma c_{ij}\zeta_i\zeta_j$. The coefficient α therefore measures the power of the critical region w to detect the falsehood of H_0, at least when the departure of the ζ_i from zero is small. Our theorem asserts that in this particular sense the T-test is the most powerful of its kind.

The method of proof is very much the same as that by which Neyman and Pearson proved some of their general theorems concerning unbiased tests. However, as the present theorem has not yet been contained in their more general results, we shall give it a full proof without referring, save in one occasion, to these authors.

PROOF. Write

$$v'_{ij} + z_iz_j = s_{ij} \qquad (i, j = 1, 2, \cdots, k)$$

$$p_0(z, z') = (\sqrt{2\pi})^{-k(f+1)}C^{\frac{1}{2}(f+1)} \exp\left[-\frac{1}{2}\sum_{i,j=1}^{k} c_{ij}s_{ij}\right] \tag{15}$$

and denote by $p_0(s)$ the simultaneous elementary probability law of the variables s_{ij} derived from (15). Let W_1 be the domain of all possible positions of the point $(s_{11}, s_{12}, \cdots, s_{kk})$ in the $\frac{1}{2}k(k + 1)$-dimensional space.

We know, although we omit the proof of it, that there is no elementary probability law of the variables s_{ij} other than $p_0(s)$ which has the same moments of all orders as those derived from $p_0(z, z')$. It then follows that if $g(s)$ be any summable function of the s_{ij} and if

(16) $$\int_{W_1} \left(\prod_{i,j=1}^{k} s_{ij}^{r_{ij}} \right) g(s) p_0(s)\, ds = 0$$

for all positive integers r_{ij} or zero, then we must have $g(s) \equiv 0$ except perhaps on a null set of points.

It follows therefore that the identity

(17) $$\int_{W_1} g(s) p_0(s)\, ds \equiv 0$$

implies the identity $g(s) \equiv 0$ provided $g(s)$ does not involve the parameters c_{ij}. For, substituting for $p_0(s)$ its expression as given by Wishart [8] we shall have

(18) $$K \int_{W_1} g(s) p_0(s)\, ds \equiv \int_{W_1} g(s) \mid S \mid^{\frac{1}{2}(f-k)} \exp\left[-\frac{1}{2} \sum_{i,j=1}^{k} c_{ij} s_{ij} \right] ds \equiv 0$$

where $\mid S \mid = \mid s_{ij} \mid$ and K is some constant. Differentiating (18) successively with respect to the c_{ij} and dividing the results by K, we shall regain the equations (16). Hence it follows that $g(s) \equiv 0$.

This being established, let w be any region belonging to D and rewrite the equation (14), so that

(19) $$(\sqrt{2\pi})^{-k(f+1)} C^{\frac{1}{2}(f+1)} \int_{w} \exp\left[-\frac{1}{2} \sum_{i,j=1}^{k} c_{ij} \{ (z_i - \zeta_i)(z_j - \zeta_j) + v'_{ij} \} \right] d(z, z') = \epsilon + \frac{\alpha}{2} \sum_{i,j=1}^{k} c_{ij} \zeta_i \zeta_j + R$$

Setting all the ζ_i to zero in both sides of (19), we have

(20) $$\int_{w} p_0(z, z')\, d(z, z') \equiv \epsilon$$

identically in the c_{ij}. Differentiating (19) once with respect to ζ_i and afterwards setting all the ζ_i to zero, we easily get

(21) $$\int_{w} z_i p_0(z, z')\, d(z, z') \equiv 0 \qquad (i = 1, 2, \cdots k)$$

for all possible values of the c_{ij}.

Finally, differentiating (19) with respect to ζ_i and then to ζ_j and putting all $\zeta_i = 0$ in the result we obtain

$$\int_{w} \left\{ \left(\sum_{h=1}^{k} c_{ih} z_h \right) \left(\sum_{h=1}^{k} c_{jh} z_h \right) - c_{ij} \right\} p_0(z, z')\, d(z, z') \equiv \alpha c_{ij} \qquad (i, j = 1, 2, \cdots k)$$

whence, renumbering (20)

(22) $$\sum_{h,l=1}^{k} c_{ih} c_{jl} q_{hl} \equiv \beta c_{ij} \qquad (i, j = 1, 2, \cdots, k)$$

in which we denote by $\beta = \alpha + \epsilon$ and

$$q_{hl} = \int_w z_h z_l p_0(z, z')\, d(z, z') \qquad (h, l = 1, 2, \cdots, k)$$

If we denote by Q the matrix of order k formed of the elements q_{hl}, we see that (22) may be written as

$$CQC \equiv \beta C,$$

whence, since C has its inverse matrix, C^{-1},

$$Q \equiv \beta C^{-1}$$

i.e.,

(23) $$q_{ij} \equiv \beta c_{ij}^{(-1)} \qquad (i, j = 1, 2, \cdots, k)$$

where $c_{ij}^{(-1)}$ denotes the element in the matrix C^{-1} which corresponds to the element c_{ij} in the matrix C.

Conditions (20), (21) and (23) are necessary for the region w to belong to the class D. They are evidently also sufficient.

Let us evaluate the integrals in (20), (21) and the q_{ij} by first evaluating the surface integrals on any surface, say $G(s)$, on which all the s_{ij} have constant values, and then integrating the results with respect to the s_{ij} over a region, say w_1, of the s_{ij} contained in W_1. Thus we may write (20), (21) and (23) in the form

(24) $$\int_{w_1} f(s) p_0(s)\, ds \equiv \epsilon, \qquad \int_{w_1} g_i(s) p_0(s)\, ds \equiv 0, \qquad \int_{w_1} \varphi_{ij}(s) p_0(s) \equiv \beta c_{ij}^{(-1)},$$
$$(i, j = 1, 2, \cdots, k),$$

where

$$f(s) = \frac{1}{p_0(s)} \int_{G(s)} p_0(z, z')\, dG(s)$$

$$g_i(s) = \frac{1}{p_0(s)} \int_{G(s)} z_i p_0(z, z')\, dG(s)$$

$$\varphi_{ij}(s) = \frac{1}{p_0(s)} \int_{G(s)} z_i z_j p_0(z, z')\, dG(s)$$

It is readily verified that the function $p_0(z, z')/p_0(s)$ is free from the parameters c_{ij}, and consequently so are the functions $f(s)$, $g_i(s)$, $\varphi_{ij}(s)$. Besides, we can extend the definition of these functions in the whole domain W_1 by assigning them the value zero outside of the region w_1. Doing this we can now write the equations (24) as

(25) $$\int_{W_1} (f(s) - \epsilon) p_0(s)\, ds \equiv 0, \qquad \int_{W_1} g_i(s) p_0(s)\, ds \equiv 0,$$
$$\int_{W_1} [\varphi_{ij}(s) - \gamma s_{ij}]\, p_0(s)\, ds \equiv 0 \qquad (i, j = 1, 2, \cdots, k)$$

where $\gamma = \dfrac{1}{f+1} \beta$.

Now all the equations (25) are of the form (17); consequently, according to the already established result and remembering the definitions of the functions $f(s)$, $g_i(s)$ and $\varphi_{ij}(s)$, we must have

$$\int_{G(s)} p_0(z, z')\, dG(s) = \epsilon p_0(s) \tag{26}$$

$$\int_{G(s)} z_i p_0(z, z')\, dG(s) = 0 \tag{27}$$

$$\int_{G(s)} z_i z_j p_0(z, z')\, dG(s) = \gamma s_{ij} p_0(s) \tag{28}$$

in the whole domain W_1.

Hence the most general region belonging to the class D is constructed as follows. On any surface $s_{ij} = \text{const.}$ $(i, j = 1, 2, \cdots k)$ we take an areal region such that it satisfies the equations (26)–(28); we then allow the s_{ij} to vary in the whole domain W_1. Equations (28) may now be replaced by

$$\int_{G(s)} \left(\frac{z_1^2}{s_{11}} - \frac{z_i z_j}{s_{ij}}\right) p_0(z, z') = 0, \qquad (i, j = 1, 2, \cdots, k) \tag{28'}$$

Let us call w_0 the region defined by $T \geq T_\epsilon$. Since w_0 belongs to the class D (cf. (12)), its cross section, say $G_0(s)$, by any surface $s_{ij} = \text{const.}$ $(i, j = 1, 2, \cdots, k)$ must satisfy the equations (26), (27) and (28'). Since $\gamma = \frac{1}{f+1}(\alpha + \epsilon)$, all we have to prove now is that among all the areal regions $G(s)$ satisfying the equations (26), (27) and (28') it is the region $G_0(s)$ that gives the largest possible value to $\gamma p_0(s)$. Now

$$\gamma p_0(s) = \int_{G(s)} \frac{z_1^2}{s_{11}} p_0(z, z')\, d(z, z') \tag{29}$$

and, according to a Lemma of Neyman and Pearson, [3, p. 10] the right-hand side of (29) will attain its maximum value if $G(s)$ is defined by an inequality of the form

$$\frac{z_1^2}{s_{11}} \geq \sum_{i,j=1}^{k} a_{ij}\left(\frac{z_1^2}{s_{11}} - \frac{z_i z_j}{s_{ij}}\right) + \sum_{i=1}^{k} b_i z_i + c \tag{30}$$

where the a_{ij}, b_i and c are constants so determined as to enable the region $G(s)$ to satisfy the equations (26)–(28). We shall show presently that the region $G_0(s)$ is defined by such an inequality.

The inequality $T \geq T_\epsilon$ may be written as

$$\frac{|v'_{ij}|}{|v'_{ij} + z_i z_j|} \leq \frac{1}{1 + T_\epsilon}$$

f.e.

$$\frac{|s_{ij} - z_i z_j|}{|s_{ij}|} \leq \frac{1}{1 + T_\epsilon},$$

or

$$\sum_{i,j=1}^{k} s_{ij}^{(-1)} z_i z_j \geq \frac{T_\epsilon}{1 + T_\epsilon} \tag{31}$$

where $s_{ij}^{(-1)}$ denotes the (i, j)th element in the inverse matrix of $\| s_{ij} \|$. The region $G_0(s)$ is therefore defined by the same inequality (31) in which we regard the s_{ij} as constants.

If we put

$$a_{ij} = \frac{1}{k} s_{ij} s_{ij}^{(-1)}, \qquad b_i = 0, \qquad c = \frac{1}{k} \frac{1}{1 + T_\epsilon} \qquad (i, j = 1, 2, \cdots, k)$$

in (30) we can easily reduce the inequality (30) into (31).

The proof is now complete.

3. Note on Applications of T. It is already known that the T-test may be used for the following purposes (a) and (b):

(a) Given a k-variate normal surface

$$p(x) = (\sqrt{2\pi})^{-k} C^{\frac{1}{2}} \exp\left[-\frac{1}{2} \sum_{i,j=1}^{k} c_{ij}(x_i - \xi_i)(x_j - \xi_j) \right]$$

with the unknown ξ_i and c_{ij}. n observations

$$(x_{1l}, x_{2l}, \cdots, x_{kl}), \qquad (l = 1, 2, \cdots, n)$$

having been made, it is required to test the hypothesis that the ξ_i have the particular values ξ_i^0 for $i = 1, 2, \cdots, k$.

Here we use the T-test with

$$\left.\begin{aligned} z_i &= \sqrt{n}(\bar{x}_i - \xi_i^0), \qquad v'_{ij} = \sum_{l=1}^{m} (x_{il} - \bar{x}_i)(\bar{x}_{jl} - \bar{x}_j) \\ \zeta_i &= \sqrt{n}\,(\xi_i - \xi_i^0), \qquad f = n - 1 \end{aligned}\right\} (i, j = 1, 2, \cdots, k)$$

where

$$\bar{x}_i = \frac{1}{n} \sum_{l=1}^{n} x_{il}$$

(b) Given two k-variate normal surfaces

$$p_1(x) = (\sqrt{2\pi})^{-k} C^{\frac{1}{2}} \exp\left(-\frac{1}{2} \sum_{i,j=1}^{k} c_{ij}(x_i - \xi_i)(x_j - \xi_j) \right)$$

$$p_2(x) = (\sqrt{2\pi})^{-k} C^{\frac{1}{2}} \exp\left(-\frac{1}{2} \sum_{i,j=1}^{k} c_{ij}(y_i - \eta_i)(y_j - \eta_j) \right)$$

where the c_{ij} are common to the two surfaces while all the ξ_i, ξ_j, c_{ij} are unknown. Samples of n_1 and n_2 having been drawn respectively from the two populations, to test the hypothesis that $\xi_i = \eta_i$ for all i.

Let the samples be

$$(x_{1l}, x_{2l}, \cdots, x_{kl}), \qquad (l = 1, 2, \cdots, n_1)$$

and

$$(y_{1h}, y_{2h}, \cdots, y_{kh}), \qquad (h = 1, 2, \cdots, n_2)$$

Let

$$\bar{x}_i = \frac{1}{n_1}\sum_{i=1}^{n_1} x_{il}, \qquad \bar{y}_i = \frac{1}{n_2}\sum_{h=1}^{n_2} y_{ih} \qquad (i = 1, 2, \cdots, k)$$

We use the T-test with

$$z_i = \sqrt{\frac{n_1 n_2}{n_1 + n_2}}(\bar{x}_i - \bar{y}_i), \qquad v'_{ij} = \sum_{l=1}^{n_1}(x_{il} - \bar{x}_i)(x_{jl} - \bar{x}_j) + \sum_{h=1}^{n_2}(y_{ih} - \bar{y}_i)(y_{ih} - \bar{y}_j)$$

$$\zeta_i = \sqrt{\frac{n_1 n_2}{n_1 + n_2}}(\xi_i - \eta_i), \qquad f = n_1 + n_2 - 2$$

$$(i, j = 1, 2, \cdots, k)$$

A third application of T, which appears to be novel, is the following:

(c) Given a $(k + 1)$-variate normal surface

$$p(x) = (\sqrt{2\pi})^{-(k+1)} D^{\frac{1}{2}} \exp\left[-\frac{1}{2}\sum_{i,j=1}^{k+1} d_{ij}(x_i - \xi_i)(x_j - \xi_j)\right], \qquad D = |d_{ij}|,$$

where the ξ_i and d_{ij} are all unknown. n observations

$$(x_{1l}, x_{2l}, \cdots, x_{k+1,l}) \qquad (l = 1, 2, \cdots, n)$$

having been made, to test the hypothesis that all the ξ_i are equal.

If we put

$$y_i = x_i - x_{k+1} \qquad (i = 1, 2, \cdots, k),$$

then we have a k-variate normal surface for the variables y_i.

$$p(y) = (\sqrt{2\pi})^{-k} C^{\frac{1}{2}} \exp\left[-\frac{1}{2}\sum_{i,j=1}^{k} c_{ij}(y_i - \eta_i)\right]$$

where $\eta_i = \xi_i - \xi_{k+1}$ $(i = 1, 2, \cdots, k)$. Thus the problem is reduced to testing the hypothesis that $\eta_i = 0$ for $i = 1, 2, \cdots, k$ and therefore belongs to the type (a). Write

$$y_{il} = x_{il} - x_{k+1,l} \qquad (i = 1, 2, \cdots, k;\ l = 1, 2, \cdots, n)$$

and

$$\bar{y}_i = \frac{1}{2}\sum_{l=1}^{n} y_{il}, \qquad (i = 1, 2, \cdots, k).$$

We use the T-test with

$$\left.\begin{aligned} z_i &= \sqrt{n}\,\bar{y}_i, \qquad v'_{ij} = \sum_{l=1}^{n} (y_{il} - \bar{y})(y_{jl} - \bar{y}_j) \\ \zeta_i &= \sqrt{n}\,\eta_i, \qquad f = n + 1 \end{aligned}\right\} (i, j = 1, 2, \cdots, k)$$

Although there are no simple expressions for the c_{ij}, there is one for the parameter $\Sigma\, c_{ij}\eta_i\eta_j$, on which alone the distribution of T depends. We have indeed

$$\sum_{i,j=1}^{k} c_{ij}\eta_i\eta_j = \frac{1}{D} \begin{vmatrix} \sigma_{11} & \cdots & \sigma_{1,\,k+1} & \xi_1 & 1 \\ \cdots & \cdots & \cdots & \cdots & \cdots \\ \sigma_{k+1,1} & \cdots & \sigma_{k+1,\,k+1} & \xi_{k+1} & 1 \\ \xi_1 & \cdots & \xi_{k+1} & 0 & 0 \\ 1 & \cdots & 1 & 0 & 0 \end{vmatrix}$$

where

$$D = \begin{vmatrix} \sigma_{11} & \cdots & \sigma_{1,\,k+1} & 1 \\ \cdots & \cdots & \cdots & \cdots \\ \sigma_{k+1,1} & \cdots & \sigma_{k+1,\,k+1} & 1 \\ 1 & \cdots & 1 & 0 \end{vmatrix}$$

where σ_{ij} is the covariance between x_i and x_j.

Expressing T in terms of the original variables x, we have

$$T = -\frac{1}{D'} \begin{vmatrix} s_{11}\ s_{12} & \cdots & s_{1,\,k+1} & \bar{x}_1 & 1 \\ \cdots & \cdots & \cdots & \cdots & \cdots \\ s_{k+1,1} & \cdots & s_{k+1,\,k+1} & \bar{x}_{k+1} & 1 \\ \bar{x}_1 & \cdots & \bar{x}_{k+1} & 0 & 0 \\ 1 & \cdots & 1 & 0 & 0 \end{vmatrix}$$

where

$$D' = \begin{vmatrix} s_{11} & \cdots & s_{1,\,k+1} & 1 \\ \cdots & \cdots & \cdots & \cdots \\ s_{k+1,1} & \cdots & s_{k+1,\,k+1} & 1 \\ 1 & \cdots & 1 & 0 \end{vmatrix}$$

and where

$$\bar{x}_i = \frac{1}{n}\sum_{l=1}^{n} x_{il}, \qquad s_{ij} = \frac{1}{n}\sum_{l=1}^{n} (x_{il} - \bar{x}_i)(x_{jl} - \bar{x}_j), \qquad (i, j = 1, 2, \cdots, k+1)$$

Therefore T is independent of which variable has been taken as the $(k + 1)$st.

UNIVERSITY COLLEGE, LONDON.

REFERENCES

[1] H. HOTELLING, *Ann. Math. Statist.*, Vol. 2 (1931) pp. 359–378.

[2] J. Neyman, *Bull. Soc. Math. France*, Vol. 63 (1935) pp. 246–266.
[3] J. Neyman and E. S. Pearson, *Statist. Res. Mem.*, Vol. 1 (1936) pp. 1–37.
[4] J. Neyman and E. S. Pearson, *Statist. Res. Mem.*, Vol. 2 (1938).
[5] P. C. Tang, *Statist. Res. Mem.*, Vol. 2 (1938).
[6] E. C. Titchmarsh, *Introduction to the Theory of Fourier Integrals*. Oxford Univ. Press (1937).
[7] S. S. Wilks, *Biometrika*, Vol. 24 (1932) pp. 471–494.
[8] J. Wishart, *Biometrika*, Vol. 20A (1928) pp. 32–52.

Reprinted from
Ann. Eugenics
9, 250–258 (1939).

ON THE DISTRIBUTION OF ROOTS OF CERTAIN DETERMINANTAL EQUATIONS

By P. L. HSU

THE extension (Fisher, 1938) of Fisher's discriminant analysis to more than two multivariate samples has directed attention to the problem of the exact distribution of the roots of a certain type of determinantal equation, which are required for various significance tests. While Fisher was solving this problem he submitted it to me for its interest in relation to matrix algebra. The purpose of the present paper is to give a complete demonstration of the analytic solution, including the case in which the number of variates, p, exceeds one of the sample numbers n_1.

Consider a set of $p(n_1+n_2)$ random variables

$$y_{ir},\, z_{it} \quad (i = 1, 2, \ldots, p;\; r = 1, 2, \ldots, n_1;\; t = 1, 2, \ldots, n_2),$$

following the distribution law

$$\text{const.}\ \exp\left[-\tfrac{1}{2}\sum_{i,j=1}^{p} \alpha_{ij}(a_{ij}+b_{ij})\right]\prod dy\,dz, \qquad \ldots\ldots(1)$$

where

$$a_{ij} = \sum_{r=1}^{n_1} y_{ir}y_{jr}, \quad b_{ij} = \sum_{t=1}^{n_2} z_{it}z_{jt}.$$

We shall assume that $n_2 \geqslant p$, so that the matrix $\|b_{ij}\|$ is almost always positively definite. Thus the determinantal equation

$$|\,a_{ij}-\theta(a_{ij}+b_{ij})\,| = 0 \qquad \ldots\ldots(2)$$

is known to possess exactly p or n_1 (whichever is smaller) real, not identically vanishing, roots, each lying between 0 and 1. Denoting them by $\theta_1, \theta_2, \ldots$, in the order of descending magnitude, we shall now establish the simultaneous distribution of these θ's.

THEOREM 1. *The simultaneous distribution of the roots of* (2) *is given by*

$$CP\left\{\prod_{i=1}^{p}\theta_i\right\}^{\frac{1}{2}(n_1-p-1)}\left\{\prod_{i=1}^{p}(1-\theta_i)\right\}^{\frac{1}{2}(n_2-p-1)}\prod_{i=1}^{p} d\theta_i,\ \textit{if}\ p\leqslant n_1, \qquad \ldots\ldots(3)$$

and

$$C_1P_1\left\{\prod_{i=1}^{n_1}\theta_i\right\}^{\frac{1}{2}(p-n_1-1)}\left\{\prod_{i=1}^{n_1}(1-\theta_i)\right\}^{\frac{1}{2}(n_2-p-1)}\prod_{i=1}^{n_1} d\theta_i,\ \textit{if}\ n_1\leqslant p, \qquad \ldots\ldots(4)$$

where

$$C = \pi^{\frac{1}{2}p}\prod_{i=1}^{p}\frac{\Gamma\frac{1}{2}(n_1+n_2-i+1)}{\Gamma\frac{1}{2}(n_1-i+1)\,\Gamma\frac{1}{2}(n_2-i+1)\,\Gamma\frac{1}{2}(p-i+1)},$$

$$C_1 = \pi^{\frac{1}{2}n_1}\prod_{i=1}^{n_1}\frac{\Gamma\frac{1}{2}(n_1+n_2-i+1)}{\Gamma\frac{1}{2}(n_1-i+1)\,\Gamma\frac{1}{2}(n_1+n_2-p-i+1)\,\Gamma\frac{1}{2}(p-i+1)},$$

and

$$P = \prod_{i=1}^{p}\prod_{j=i+1}^{p}(\theta_i-\theta_j), \quad P_1 = \prod_{i=1}^{n_1}\prod_{j=i+1}^{n_1}(\theta_i-\theta_j).$$

I. *Proof of* (3), case $n_1 \geqslant p$.

Instead of the θ_i, consider new variables ϕ_i, where

$$\phi_i = \theta_i(1-\theta_i)^{-1},$$

so that $\phi_1, \phi_2, \ldots, \phi_p$ are the roots, in descending order, of the equation

$$| a_{ij} - \phi b_{ij} | = 0. \qquad \ldots\ldots(5)$$

Obviously we can set, in (1), $\alpha_{ii} = 1$ and $\alpha_{ij} = 0$ for $i \neq j$. Hence, instead of (1) we may regard the parent population distribution as

$$\{\sqrt{(2\pi)}\}^{-p(n_1+n_2)} \exp\left[-\tfrac{1}{2}\sum_{i=1}^{p}(a_{ii}+b_{ii})\right] \Pi\, dy\, dz. \qquad \ldots\ldots(6)$$

Since $n_1 \geqslant p$ we can write down the distribution law (Wishart & Bartlett, 1933) of the a_{ij} and b_{ij}:

$$C_2 \,| a_{ij} |^{\frac{1}{2}(n_1-p-1)} \,| b_{ij} |^{\frac{1}{2}(n_2-p-1)} \exp\left[-\tfrac{1}{2}\sum_{i=1}^{p}(a_{ii}+b_{ii})\right] \Pi\, da\, db, \qquad \ldots\ldots(7)$$

where
$$C_2 = 2^{-\frac{1}{2}p(n_1+n_2)} \pi^{-\frac{1}{4}p(p-1)} \left\{ \prod_{i=1}^{p} \Gamma\tfrac{1}{2}(n_1-i+1)\, \Gamma\tfrac{1}{2}(n_2-i+1) \right\}^{-1}$$

In the space of the a_{ij} and b_{ij} we shall consider only those points for which the following conditions are satisfied:

(i) that the matrix $\|b_{ij}\|$ is positively definite,

(ii) that equation (5) has no multiple roots.

The remaining points in the space will form a set whose measure is zero. Consider now the transformation

$$\|a_{ij}\| = W D_\phi W', \quad \|b_{ij}\| = W W', \qquad \ldots\ldots(8)$$

where

$$W = \left\| \begin{matrix} w_{11} & w_{12} & \cdots & w_{1p} \\ w_{21} & w_{22} & \cdots & w_{2p} \\ \cdots & \cdots & \cdots & \cdots \\ w_{p1} & w_{p2} & \cdots & w_{pp} \end{matrix} \right\|$$

is a matrix of real elements, W' denotes the transposed matrix of W, and D_ϕ is the diagonal matrix

$$D_\phi = \left\| \begin{matrix} \phi_1 & & & \\ & \phi_2 & & \\ & & \cdots & \\ & & & \phi_p \end{matrix} \right\|$$

Since the ϕ_i are all distinct and arranged in fixed order of descending magnitude, it can easily be shown that the matrix W is uniquely determined by the a_{ij} and b_{ij} *except only that any entire column of W may change its sign.* Hence, if we impose the condition that the first row of W may assume positive values only, the equations (8) will establish a (1, 1)-

correspondence between the quantities a_{ij}, b_{ij} and the quantities ϕ_i, w_{ij}. The domain of the new variables will accordingly be specified as follows:

$$\infty > \phi_1 > \phi_2 > \ldots > \phi_p \geqslant 0,$$
$$0 \leqslant w_{1j} < \infty \quad (j = 1, 2, \ldots, p),$$
$$-\infty < w_{ij} < \infty \quad (i = 2, \ldots, p;\ j = 1, 2, \ldots, p).$$

We may now proceed to compute the Jacobian, say Δ, of the transformation (8).

We shall establish the formula

$$\Delta = \pm 2^p \,|\, W \,|^{p+2} P', \qquad \ldots\ldots(9)$$

where
$$P' = \prod_{i=1}^{p} \prod_{j=i+1}^{p} (\phi_i - \phi_j),$$

for the case $p = 3$. It is seen that the method of proof will hold good for any p, but owing to cumbersome notation we restrict ourselves to the particular case.

For $p = 3$ let us arrange the variables so that

$$\Delta = \frac{\partial(a_{11}, a_{12}, a_{13}, a_{22}, a_{23}, a_{33}, b_{11}, b_{12}, b_{13}, b_{22}, b_{23}, b_{33})}{\partial(\phi_1, \phi_2, \phi_3, w_{11}, w_{12}, w_{13}, w_{21}, w_{22}, w_{23}, w_{31}, w_{32}, w_{33})}.$$

Then

$$\Delta = \begin{vmatrix}
w_{11}^2 & w_{12}^2 & w_{13}^2 & 2\phi_1 w_{11} & 2\phi_2 w_{12} & 2\phi_3 w_{13} & . & . & . & . & . & . \\
w_{11}w_{21} & w_{12}w_{22} & w_{13}w_{23} & \phi_1 w_{21} & \phi_2 w_{22} & \phi_3 w_{23} & \phi_1 w_{11} & \phi_2 w_{12} & \phi_3 w_{13} & . & . & . \\
w_{11}w_{31} & w_{12}w_{32} & w_{13}w_{33} & \phi_1 w_{31} & \phi_2 w_{32} & \phi_3 w_{33} & . & . & . & \phi_1 w_{11} & \phi_2 w_{12} & \phi_3 w_{13} \\
w_{21}^2 & w_{22}^2 & w_{23}^2 & . & . & . & 2\phi_1 w_{21} & 2\phi_2 w_{22} & 2\phi_3 w_{23} & . & . & . \\
w_{21}w_{31} & w_{22}w_{32} & w_{23}w_{33} & . & . & . & \phi_1 w_{31} & \phi_2 w_{32} & \phi_3 w_{33} & \phi_1 w_{21} & \phi_2 w_{22} & \phi_3 w_{23} \\
w_{31}^2 & w_{32}^2 & w_{33}^2 & . & . & . & . & . & . & 2\phi_1 w_{31} & 2\phi_2 w_{32} & 2\phi_3 w_{33} \\
. & . & . & 2w_{11} & 2w_{12} & 2w_{13} & . & . & . & . & . & . \\
. & . & . & w_{21} & w_{22} & w_{23} & w_{11} & w_{12} & w_{13} & . & . & . \\
. & . & . & w_{31} & w_{32} & w_{33} & . & . & . & w_{11} & w_{12} & w_{13} \\
. & . & . & . & . & . & 2w_{21} & 2w_{22} & 2w_{23} & . & . & . \\
. & . & . & . & . & . & w_{31} & w_{32} & w_{33} & w_{21} & w_{22} & w_{23} \\
. & . & . & . & . & . & . & . & . & 2w_{31} & 2w_{32} & 2w_{33}
\end{vmatrix} \qquad \ldots\ldots(10)$$

Denote by W_{ij} the co-factor of w_{ij}. In order to evaluate the determinant Δ, multiply it by another determinant, Δ_1, according to the column-by-column law:

$$\Delta_1 = \begin{vmatrix}
W_{11}^2 & W_{12}^2 & W_{13}^2 & . & . & . & . & . & . & . & . & . \\
2W_{11}W_{21} & 2W_{12}W_{22} & 2W_{13}W_{23} & . & . & . & . & . & . & . & . & . \\
2W_{11}W_{31} & 2W_{12}W_{32} & 2W_{13}W_{33} & . & . & . & . & . & . & . & . & . \\
W_{21}^2 & W_{22}^2 & W_{23}^2 & 1 & . & . & . & . & . & . & . & . \\
2W_{21}W_{31} & 2W_{22}W_{32} & 2W_{23}W_{33} & . & 1 & . & . & . & . & . & . & . \\
W_{31}^2 & W_{32}^2 & W_{33}^2 & . & . & 1 & . & . & . & . & . & . \\
-\phi_1 W_{11}^2 & -\phi_2 W_{12}^2 & -\phi_3 W_{13}^2 & . & . & . & 1 & . & . & . & . & . \\
-2\phi_1 W_{11}W_{21} & -2\phi_2 W_{12}W_{22} & -2\phi_3 W_{13}W_{23} & . & . & . & . & 1 & . & . & . & . \\
-2\phi_1 W_{11}W_{31} & -2\phi_2 W_{12}W_{32} & -2\phi_3 W_{13}W_{33} & . & . & . & . & . & 1 & . & . & . \\
-\phi_1 W_{21}^2 & -\phi_2 W_{22}^2 & -\phi_3 W_{23}^2 & . & . & . & . & . & . & 1 & . & . \\
-2\phi_1 W_{21}W_{31} & -2\phi_2 W_{22}W_{32} & -2\phi_3 W_{23}W_{33} & . & . & . & . & . & . & . & 1 & . \\
-\phi_1 W_{31}^2 & -\phi_2 W_{32}^2 & -\phi_3 W_{33}^2 & . & . & . & . & . & . & . & . & 1
\end{vmatrix}$$

The result of this multiplication is the determinant

$$\Delta_1 \Delta = \begin{vmatrix} |W|^2 & \cdot & \cdot & \cdot & \cdot & \cdot & \cdot & \cdot & \cdot & \cdot & \cdot & \cdot \\ \cdot & |W|^2 & \cdot & \cdot & \cdot & \cdot & \cdot & \cdot & \cdot & \cdot & \cdot & \cdot \\ \cdot & \cdot & |W|^2 & \cdot & \cdot & \cdot & \cdot & \cdot & \cdot & \cdot & \cdot & \cdot \\ w_{21}^2 & w_{22}^2 & w_{23}^2 & \cdot & \cdot & \cdot & 2\phi_1 w_{21} & 2\phi_2 w_{22} & 2\phi_3 w_{23} & \cdot & \cdot & \cdot \\ w_{21}w_{31} & w_{22}w_{32} & w_{23}w_{33} & \cdot & \cdot & \cdot & \phi_1 w_{31} & \phi_2 w_{32} & \phi_3 w_{33} & \phi_1 w_{21} & \phi_2 w_{22} & \phi_3 w_{23} \\ w_{31}^2 & w_{32}^2 & w_{33}^2 & \cdot & \cdot & \cdot & \cdot & \cdot & \cdot & 2\phi_1 w_{31} & 2\phi_2 w_{32} & 2\phi_3 w_{33} \\ \cdot & \cdot & \cdot & 2w_{11} & 2w_{12} & 2w_{13} & \cdot & \cdot & \cdot & \cdot & \cdot & \cdot \\ \cdot & \cdot & \cdot & w_{21} & w_{22} & w_{23} & w_{11} & w_{12} & w_{13} & \cdot & \cdot & \cdot \\ \cdot & \cdot & \cdot & w_{31} & w_{32} & w_{33} & \cdot & \cdot & \cdot & w_{11} & w_{12} & w_{13} \\ \cdot & \cdot & \cdot & \cdot & \cdot & \cdot & 2w_{21} & 2w_{22} & 2w_{23} & \cdot & \cdot & \cdot \\ \cdot & \cdot & \cdot & \cdot & \cdot & \cdot & w_{31} & w_{32} & w_{33} & w_{21} & w_{22} & w_{23} \\ \cdot & \cdot & \cdot & \cdot & \cdot & \cdot & \cdot & \cdot & \cdot & 2w_{31} & 2w_{32} & 2w_{33} \end{vmatrix}$$

Δ_1 has the value $2^2 W_{11} W_{12} W_{13} | W |^2$. The product determinant can be reduced in order with the help of Laplace's theorem, and cancelling out the common factors $2^2 | W |^2$, we get the equation

$$W_{11} W_{12} W_{13} \Delta = 2^3 | W |^5 \begin{vmatrix} \phi_1 w_{21} & \phi_2 w_{22} & \phi_3 w_{23} & \cdot & \cdot & \cdot \\ \phi_1 w_{31} & \phi_2 w_{32} & \phi_3 w_{33} & \phi_1 w_{21} & \phi_2 w_{22} & \phi_3 w_{23} \\ \cdot & \cdot & \cdot & \phi_1 w_{31} & \phi_2 w_{32} & \phi_3 w_{33} \\ w_{21} & w_{22} & w_{23} & \cdot & \cdot & \cdot \\ w_{31} & w_{32} & w_{33} & w_{21} & w_{22} & w_{23} \\ \cdot & \cdot & \cdot & w_{31} & w_{32} & w_{33} \end{vmatrix}. \qquad \ldots\ldots(11)$$

Now (11) is an identity of which both sides are polynomials in the w_{ij}. Hence the right-hand side of (11) must be identically divisible by the polynomial $W_{11} W_{12} W_{13}$. But the determinant $| W |$ is an irreducible polynomial of its elements.* Therefore the other determinant in the right-hand side of (11) is identically divisible by $W_{11} W_{12} W_{13}$. Further, the quotient is of degree zero in the w_{ij} because the dividend and the divisor have the same degree. We conclude, therefore, that

$$\Delta = k | W |^5, \qquad \ldots\ldots(12)$$

where k is some quantity depending on the ϕ_i only.

We may determine the value of k from the reduced equation (11). But, keeping the general case in mind, we should go back to the original expression (10) of Δ. If we put therein $w_{ii} = 1$ and $w_{ij} = 0$ for $i \neq j$, we get the value $\pm 2^p P'$. Owing to the identity (12) this is the value of k. The formula (9) is thus established.

By direct substitution into (6) and multiplication by the Jacobian, we get the joint distribution of the new variables, ϕ_i and w_{ij}:

$$2^p C_2 | W'W |^{\frac{1}{2}(n_1+n_2-p)} \left\{ \prod_{i=1}^{p} \phi_i \right\}^{\frac{1}{2}(n_1-p-1)} P' \exp\left[-\tfrac{1}{2} \sum_{i,j=1}^{p} (1+\phi_i) w_{ij}^2 \right] \Pi \, d\phi \, dw, \qquad \ldots\ldots(13)$$

wherein we use $| W'W |^{\frac{1}{2}} = \sqrt{| W |^2}$ instead of $| W |$ because only the absolute value of $| W |$ matters.

It remains to integrate with respect to all the w_{ij} to get the required distribution of the ϕ_i. Instead of the restriction $0 \leqslant w_{1i} (i = 1, 2, \ldots, p)$ we may integrate with respect to all the

* Cf. Bôcher (1929), p. 176, Theorem 1, also p. 213, Theorem 5 and p. 216, Exercise 1.

w_{ij} from $-\infty$ to ∞, and divide the result by 2^p. Using Wilks's method (1932) of evaluating the moments of the generalized variance we easily obtain

$$\int |W'W|^{\frac{1}{2}(n_1+n_2-p)} \exp\left[-\tfrac{1}{2}\sum_{i,j=1}^{p}(1+\phi_i)w_{ij}^2\right]\Pi\, dw = C_3\left\{\prod_{i=1}^{p}(1+\phi_i)\right\}^{-\frac{1}{2}(n_1+n_2)} \quad \text{......(14)}$$

where
$$C_3 = 2^{\frac{1}{2}p(n_1+n_2)}\pi^{\frac{1}{2}p^2}\prod_{i=1}^{p}\frac{\Gamma\frac{1}{2}(n_1+n_2-i+1)}{\Gamma\frac{1}{2}(p-i+1)}.$$

Substituting (14) into (13) and dividing the result by 2^p we get the distribution of the ϕ_i:

$$CP'\left\{\prod_{i=1}^{p}\phi_i\right\}^{\frac{1}{2}(n_1-p-1)}\left\{\prod_{i=1}^{p}(1+\phi_i)\right\}^{-\frac{1}{2}(n_1+n_2)}\prod_{i=1}^{p}d\phi_i, \quad \text{......(15)}$$

because $C_2C_3 = C$.

The transformation
$$\phi_i = \theta_i(1-\theta_i)^{-1}$$

constitutes the last step for the establishment of (3).

II. *Proof of* (4), case $n_1 < p$.

For this case the previous method of proof cannot be used, for now there is no joint probability distribution for the set of variables a_{ij}. Starting with the distribution (6) and writing

$$c_{ij} = a_{ij} + b_{ij},$$

we have, by a straightforward application of Wishart & Bartlett's distribution, the joint probability law of the c_{ij} and the y_{ir}:

$$C_4 |c_{ij}-a_{ij}|^{\frac{1}{2}(n_2-p-1)}\exp\left[-\tfrac{1}{2}\sum_{i=1}^{p}c_{ii}\right]\Pi\, dc\, dy,$$

where
$$C_4 = 2^{-\frac{1}{2}p(n_1+n_2)}\pi^{-\frac{1}{2}pn_1-\frac{1}{4}p(p-1)}\left\{\prod_{i=1}^{p}\Gamma\tfrac{1}{2}(n_2-i+1)\right\}^{-1}.$$

Since $\|c_{ij}\|$ is a positively definite matrix, it can be expressed as a matrix product TT', where T has all its elements real. Writing

$$Y = \begin{Vmatrix} y_{11} & y_{12} & \cdots & y_{1n_1} \\ y_{21} & y_{22} & \cdots & y_{2n_1} \\ \cdots & \cdots & \cdots & \cdots \\ y_{p1} & y_{p2} & \cdots & y_{pn_1} \end{Vmatrix},$$

it follows that
$$\|a_{ij}\| = YY'.$$

We now introduce a new set of variables, namely the elements of the matrix

$$U = \begin{Vmatrix} u_{11} & u_{12} & \cdots & u_{1n_1} \\ u_{21} & u_{22} & \cdots & u_{2n_1} \\ \cdots & \cdots & \cdots & \cdots \\ u_{p1} & u_{p2} & \cdots & u_{pn_1} \end{Vmatrix}$$

by means of the transformation $Y = TU$. The Jacobian of this transformation being $|T|^{n_1} = \pm |c_{ij}|^{\frac{1}{2}n_1}$, the joint distribution law of the c_{ij} and u_{ir} is

$$C_4 | I - UU' |^{\frac{1}{2}(n_2-p-1)} | c_{ij} |^{\frac{1}{2}(n_1+n_2-p-1)} \exp\left[-\tfrac{1}{2}\sum_{i=1}^{p} c_{ii}\right] \prod dc\,du. \qquad \text{......(16)}$$

The roots of (2) are precisely the latent roots of the matrix UU', i.e. the roots of the equation

$$| UU' - \theta I | = 0. \qquad \text{......(17)}$$

Therefore from (16) we may integrate with respect to all the c_{ij}. After the integrations have been performed, what remains is the distribution law of the u_{ir}:

$$C_5 | I - UU' |^{\frac{1}{2}(n_2-p-1)} \prod du, \qquad \text{......(18)}$$

where
$$C_5 = \pi^{-\frac{1}{2}pn_1} \prod_{i=1}^{p} \frac{\Gamma\frac{1}{2}(n_1+n_2-i+1)}{\Gamma\frac{1}{2}(n_2-i+1)}.$$

The distribution of the non-vanishing roots of (17) is obtained by the following argument. In the case where $n_1 \geq p$ the distribution of $\theta_1, \theta_2, \ldots, \theta_p$, which would be derived from (17) and (18), is already known to be given by the formula (3). Now suppose that $n_1 < p$. Then, as is easily shown, the non-vanishing roots of (17) are precisely the roots, all non-vanishing, of the equation

$$| U'U - \theta I | = 0. \qquad \text{......(19)}$$

Further, it may be seen that (18) is identically the same as

$$C_5 | I - U'U |^{\frac{1}{2}(n_2-p-1)} \prod du. \qquad \text{......(20)}$$

Now if we write
$$p' = n_1, \quad n_1' = p, \quad n_2' = n_1 + n_2 - p, \qquad \text{......(21)}$$

$$V = \left\| \begin{matrix} v_{11} & v_{12} & \cdots & v_{1n_1'} \\ v_{21} & v_{22} & \cdots & v_{2n_1'} \\ \cdots & \cdots & \cdots & \cdots \\ v_{p'1} & v_{p'2} & \cdots & v_{p'n_1'} \end{matrix} \right\| = U',$$

then the equation (19) will become

$$| VV' - \theta I | = 0, \qquad \text{......(22)}$$

and the expression (20) will take the form

$$C_5' | I - VV' |^{\frac{1}{2}(n_2'-p'-1)} \prod dv, \qquad \text{......(23)}$$

where
$$C_5' = \pi^{-\frac{1}{2}p'n_1'} \prod_{i=1}^{p'} \frac{\Gamma\frac{1}{2}(n_1'+n_2'-i+1)}{\Gamma\frac{1}{2}(n_2'-i+1)} = C_5.$$

Since now we have $n_1' > p'$ the distribution of the roots of (22), as derived from (23), must be exactly the formula (3), provided that the integers p, n_1, and n_2 therein are replaced respectively by p', n_1' and n_2'. If we make such changes in (3) and remember (21), we obtain the formula (4).

The proof of Theorem 1 is now complete.

Theorem 2. *If the* $\frac{1}{2}p(p+1)$ *variables* $s_{ij}(i \leqslant j = 1, 2, \ldots, p)$ *have such a domain of existence that the symmetric matrix* $\| s_{ij} \|$ *is always non-singular, and if they are so distributed that their joint probability density function depends only on the latent roots, say* $\lambda_1, \lambda_2, \ldots, \lambda_p$, *arranged in the order of descending magnitude, of* $\| s_{ij} \|$, *i.e. if*

$$df = g(\lambda_1, \lambda_2, \ldots, \lambda_p) \prod ds_{ij},$$

then the joint distribution law of the λ_i *is the following:*

$$\pi^{\frac{1}{2}p(p+1)} \left\{ \prod_{i=1}^{p} \Gamma \tfrac{1}{2}(p-i+1) \right\}^{-1} \left\{ \prod_{i=1}^{p} \prod_{j=i+1}^{p} (\lambda_i - \lambda_j) \right\} g(\lambda_1, \ldots, \lambda_p) \prod d\lambda. \qquad \ldots\ldots(24)$$

Proof. It is a familiar argument that the general formula (24) will follow if we can find a particular example for g for which (24) holds true. This is because the multiplier of g in (24) is entirely independent of the function g itself. The required example is found in the proof of Theorem 1.

Let us take the case $n_1 \geqslant p$ in the distribution law (18). We have seen that the roots of (17) follow the distribution (3). Writing

$$s_{ij} = \sum_{r=1}^{n_1} u_{ir} u_{jr},$$

we get, as a consequence of Wishart's formula, the following distribution for the s_{ij}:

$$C_6 \mid s_{ij} \mid^{\frac{1}{2}(n_1-p-1)} \mid \delta_{ij} - s_{ij} \mid^{\frac{1}{2}(n_2-p-1)} \prod ds, \qquad \ldots\ldots(25)$$

where $\delta_{ii} = 1$ and $\delta_{ij} = 0$ for $i \neq j$, and

$$C_6 = \pi^{-\frac{1}{2}p(p-1)} \prod_{i=1}^{p} \frac{\Gamma \frac{1}{2}(n_1 + n_2 - i + 1)}{\Gamma \frac{1}{2}(n_1 - i + 1) \Gamma \frac{1}{2}(n_2 - i + 1)}.$$

Now the probability density function appearing in (25) depends on the latent roots of $\| s_{ij} \|$ only. It is in fact equal to

$$C_6 \left\{ \prod_{i=1}^{p} \lambda_i \right\}^{\frac{1}{2}(n_1-p-1)} \left\{ \prod_{i=1}^{p} (1-\lambda_i) \right\}^{\frac{1}{2}(n_2-p-1)} \qquad \ldots\ldots(26)$$

On the other hand, the λ_i follow the distribution (3), with the replacement of θ by λ. In other words, the probability density function of the λ_i is (26) multiplied by the expression which coincides with the multiplier of g in (24). Thus Theorem 2 is proved.

As an application of Theorem 2 let us take the following example. Suppose that the pn random variables y_{ir} $(i = 1, 2, \ldots, p;\ r = 1, 2, \ldots, n)$ follow the distribution

$$\text{const.} \exp\left[-\tfrac{1}{2} \sum_{i,j=1}^{p} \alpha_{ij} s_{ij} \right] \prod dy,$$

where

$$s_{ij} = \sum_{r=1}^{p} y_{ir} y_{jr}.$$

Let $\lambda_1, \lambda_2, \ldots$, in descending order of magnitude, be the not identically vanishing roots of the equation

$$|s_{ij} - \lambda\alpha^{ij}| = 0,$$

where the α^{ij} are the elements of the reciprocal matrix of $\|\alpha_{ij}\|$. In order to obtain the distribution of the λ_i, it is obvious that we may set $\alpha_{ii} = 1$ and $\alpha_{ij} = 0$ for $i \neq j$. Hence we may regard the parent population distribution as

$$(\sqrt{2\pi})^{-pn} \exp[-\tfrac{1}{2}\sum_{i=1}^{p} s_{ii}] \prod dy, \qquad \ldots\ldots(27)$$

and find the distribution of the latent roots $\lambda_1, \lambda_2, \ldots$, of $\|s_{ij}\|$. Now we have, from (27), supposing $n \geqslant p$, the following distribution for the s_{ij}:

$$C_7 |s_{ij}|^{\frac{1}{2}(n-p-1)} \exp\left[-\tfrac{1}{2}\sum_{i=1}^{p} s_{ii}\right] \prod ds = C_7 \left\{\prod_{i=1}^{p} \lambda_i\right\}^{\frac{1}{2}(n-p-1)} \exp\left[-\tfrac{1}{2}\sum_{i=1}^{p} \lambda_i\right] \prod ds,$$

where
$$C_7 = 2^{-\frac{1}{2}pn} \pi^{-\frac{1}{4}p(p-1)} \left\{\prod_{i=1}^{p} \Gamma \tfrac{1}{2}(n-i+1)\right\}^{-1}.$$

The probability density function of the $\|s_{ij}\|$ being a function of the λ_i only, we get, by virtue of Theorem 2, the joint distribution of the λ_i when $n \geqslant p$:

$$2^{-\frac{1}{2}pn} \pi^{\frac{1}{2}p} \left\{\prod_{i=1}^{p} \Gamma\tfrac{1}{2}(n-i+1)\, \Gamma\tfrac{1}{2}(p-i+1)\right\}^{-1} \left\{\prod_{i=1}^{p} \prod_{j=i+1}^{p} (\lambda_i - \lambda_j)\right\}$$
$$\times \left\{\prod_{i=1}^{p} \lambda_i\right\}^{\frac{1}{2}(n-p-1)} \exp\left[-\tfrac{1}{2}\sum_{i=1}^{p} \lambda_i\right] \prod_{i=1}^{p} d\lambda_i. \qquad \ldots\ldots(28)$$

In the same manner we get the distribution, when $n \leqslant p$, of $\lambda_1, \lambda_2, \ldots, \lambda_n$ which is the above formula with the letters p and n interchanged. It can be verified that (28) is the limiting form of (15) when we set $\phi_i = n_2^{-1}\lambda_i$ and allow n_2 to approach infinity.

The connexion of the θ's with the problem of several multivariate normal samples is familiar through Fisher's work. We shall proceed to show that the quantities known as the canonical correlations between two sets of random variables, as introduced by Hotelling (1936), are distributed like the square roots of the θ's.

Consider a system of $p+q$ random variables, x_i, and a sample of size n, x_{il} ($i = 1, 2, \ldots, p+q$; $l = 1, 2, \ldots, n$). Write

$$\bar{x}_i = \frac{1}{n}\sum_{l=1}^{n} x_{il}, \quad s_{ij} = \sum_{l=1}^{n} (x_{il} - \bar{x}_i)(x_{jl} - \bar{x}_j), \quad (i, j = 1, 2, \ldots, p+q),$$

$$a'_{ij} = -\frac{\begin{vmatrix} 0 & s_{i,p+1} & \cdots & s_{i,p+q} \\ s_{p+1,j} & s_{p+1,p+1} & \cdots & s_{p+1,p+q} \\ \cdots & \cdots & \cdots & \cdots \\ s_{p+q,j} & s_{p+q,p+1} & \cdots & s_{p+q,p+q} \end{vmatrix}}{\begin{vmatrix} s_{p+1,p+1} & \cdots & s_{p+1,p+q} \\ \cdots & \cdots & \cdots \\ s_{p+q,p+1} & \cdots & s_{p+q,p+q} \end{vmatrix}}, \quad b'_{ij} = \frac{\begin{vmatrix} s_{ij} & s_{i,p+1} & \cdots & s_{i,p+q} \\ s_{p+1,j} & s_{p+1,p+1} & \cdots & s_{p+1,p+q} \\ \cdots & \cdots & \cdots & \cdots \\ s_{p+q,j} & s_{p+q,p+1} & \cdots & s_{p+q,p+q} \end{vmatrix}}{\begin{vmatrix} s_{p+1,p+1} & \cdots & s_{p+1,p+q} \\ \cdots & \cdots & \cdots \\ s_{p+q,p+1} & \cdots & s_{p+q,p+q} \end{vmatrix}}$$

$$(i, j = 1, 2, \ldots, p).$$

Then the square roots of the not identically vanishing roots of the equation

$$|a'_{ij} - \theta(a'_{ij} + b'_{ij})| = 0$$

are Hotelling's canonical correlations between the two sets of variables $(x_1, x_2, \ldots, x_p)$ and $(x_{p+1}, x_{p+2}, \ldots, x_{p+q})$.

It is seen that the quantities a'_{ij} and b'_{ij} are symmetric bi-linear forms (quadratic forms if $i = j$) in the variables $x_{i1}, x_{i2}, \ldots, x_{in}$ and $x_{j1}, x_{j2}, \ldots, x_{jn}$, with coefficients depending on the values of the second set of variables. It may be shown* that if a certain orthogonal transformation, depending on the values of the second set of variables, is applied separately to each of the systems $(x_{i1}, x_{i2}, \ldots, x_{in})$ for $i = 1, 2, \ldots, p$, the bi-linear forms a'_{ij} and b'_{ij} can be simultaneously transformed into a_{ij} and b_{ij}, where

$$a_{ij} = \sum_{r=1}^{n_1} y_{ir} y_{jr}, \quad b_{ij} = \sum_{t=1}^{n_2} y_{i,n_1+t} y_{j,n_1+t} \quad (i,j = 1, 2, \ldots, p),$$

and

$$n_1 = q, \quad n_2 = n - q - 1.$$

Assume now that the set of variables $(x_1, x_2, \ldots, x_p)$ is normally distributed, and that the set of variables $(x_{p+1}, x_{p+2}, \ldots, x_{p+q})$ is distributed independently of the former set, but otherwise in any manner whatsoever† (provided that the matrix $\|s_{p+i,p+j}\|$ is almost always positively definite). Then the joint distribution law of the y_{ir} $(i = 1, 2, \ldots, p;$ $r = 1, 2, \ldots, n_1 + n_2)$ is precisely (1). We have therefore proved the following theorem.

THEOREM 3. *Let $(x_1, x_2, \ldots, x_p)$ and $(x_{p+1}, x_{p+2}, \ldots, x_{p+q})$ be two mutually independent sets of variables of which the first set is normally distributed. Let $\theta_1, \theta_2, \ldots$ be the squares of the canonical correlations between the two sets, arranged in the descending order of magnitude. Then the joint distribution of the θ_i is given by*

$$K\left\{\prod_{i=1}^{p} \prod_{j=i+1}^{p} (\theta_i - \theta_j)\right\} \left\{\prod_{i=1}^{p} \theta_i\right\}^{\frac{1}{2}(q-p-1)} \left\{\prod_{i=1}^{p} (1-\theta_i)\right\}^{\frac{1}{2}(n-p-q-2)} \prod_{i=1}^{p} d\theta_i$$

if $p \leqslant q$, where

$$K = \pi^{\frac{1}{2}p} \prod_{i=1}^{p} \frac{\Gamma\frac{1}{2}(n-i)}{\Gamma\frac{1}{2}(n-q-i)\,\Gamma\frac{1}{2}(p-i+1)\,\Gamma\frac{1}{2}(q-i+1)}.$$

If $q \leqslant p$, the distribution is represented by the above formula with the letters p and q interchanged.

This is a generalization of Hotelling's result for $p = q = 2$.

REFERENCES

M. BÔCHER (1929). *Introduction to Higher Algebra*. New York.

W. G. COCHRAN (1934). "The distribution of quadratic forms in a normal system, with applications to the analysis of covariance." *Proc. Camb. Phil. Soc.* **30**, 178–91.

R. A. FISHER (1938). "The statistical utilization of multiple measurements." *Ann. Eugen., Lond.*, **8**, 376–86.

H. HOTELLING (1936). "Relations between two sets of variates." *Biometrika*, **28**, 321–77.

S. S. WILKS (1932). "Certain generalizations in the analysis of variance." *Biometrika*, **24**, 471–94.

J. WISHART & M. S. BARTLETT (1933). "The generalized product moment distribution in a normal system." *Proc. Camb. Phil. Soc.* **29**, 260–70.

* E.g. by means of the theorem of Cochran (1934).

† Here we make a generalization of Hotelling's work as he assumes that both sets of variables are normally distributed.

Reprinted from
Proc. Cambridge Philos. Soc.
35, 336–339 (1939).

A NEW PROOF OF THE JOINT PRODUCT MOMENT DISTRIBUTION

BY P. L. HSU

Communicated by E. S. PEARSON

Received 8 December 1938

The joint distribution of the variances and covariances calculated from a normal sample has been derived by Wishart (**1**) geometrically and again by Wishart and Bartlett (**2**) through the use of the characteristic function and inversion*. I now state their result in the following theorem.

THEOREM. *If the probability density function of the km variables* x_{ir} *is*

$$p(x) = \frac{|c_{ij}|^{\frac{1}{2}m}}{(2\pi)^{\frac{1}{2}km}} \exp\left[-\tfrac{1}{2}\sum_{i,j=1}^{k} c_{ij}s_{ij}\right], \tag{1}$$

where

$$s_{ij} = \sum_{r=1}^{m} x_{ir}x_{jr} \quad (i, j = 1, 2, \ldots, k), \tag{2}$$

then the $\frac{1}{2}k(k+1)$ *variables* s_{ij} *have the probability density function*

$$V_k(s_{ij}, m) \equiv \frac{|c_{ij}|^{\frac{1}{2}m} |s_{ij}|^{\frac{1}{2}(m-k-1)}}{2^{\frac{1}{2}km}\pi^{\frac{1}{4}k(k-1)} \prod_{i=1}^{k} \Gamma\{\frac{1}{2}(m-i+1)\}} \exp\left[-\tfrac{1}{2}\sum_{i,j=1}^{k} c_{ij}s_{ij}\right] \tag{3}$$

at all points where the matrix $\| s_{ij} \|$ *is positive definite; elsewhere the density is zero.*

* See also Ingham (**3**).

A new proof is given below, the method being that of mathematical induction.

Proof. In what follows, probability density functions are always represented by the letter p; the arguments of p indicate the variables concerned.

When $k = 1$, the theorem is true, since (3) represents the χ^2-distribution with m degrees of freedom. Let us assume that the theorem is true for $k-1$ variables, and then establish its validity for k variables. The proof by induction is then complete.

Let us perform the transformation

$$y_{it} = \sum_{r=1}^{m} a_{tr} x_{ir} \quad (t = 1, 2, \ldots, m-1), \quad y_{im} = s_{kk}^{-\frac{1}{2}} \sum_{r=1}^{m} x_{ir} x_{kr} \tag{4}$$

for $i = 1, 2, \ldots, k-1$, where the a_{tr} are so chosen as to make (4) an *orthogonal* transformation of the variables $(x_{i1}, x_{i2}, \ldots, x_{im})$ to the new variables $(y_{i1}, y_{i2}, \ldots, y_{im})$ for each fixed i $(1 \leqslant i \leqslant k-1)$. This gives

$$p(y_{ir}, x_{kr}) = \frac{|c_{ij}|^{\frac{1}{2}m}}{(2\pi)^{\frac{1}{2}km}} \exp\left[-\tfrac{1}{2}\sum_{i,j=1}^{k-1} c_{ij} s'_{ij}\right]$$

$$\times \exp\left[-\frac{1}{2}\left(c_{kk}s_{kk} + 2\sqrt{s_{kk}}\sum_{i=1}^{k-1} c_{ik} y_{im} + \sum_{i,j=1}^{k-1} c_{ij} y_{im} y_{jm}\right)\right] \tag{5}$$

and

$$s_{ij} = s'_{ij} + y_{im} y_{jm}, \quad s_{ik} = \sqrt{s_{kk}}\, y_{im} \quad (i, j = 1, 2, \ldots, k-1), \tag{6}$$

where

$$s'_{ij} = \sum_{r=1}^{m-1} y_{ir} y_{jr} \quad (i, j = 1, 2, \ldots, k-1). \tag{7}$$

By our assumption, the s'_{ij} are jointly distributed with the density $V_{k-1}(s'_{ij}, m-1)$. Also, for the variables $x_{k1}, x_{k2}, \ldots, x_{km}$ we can introduce polar coordinates, namely, s_{kk} and $m-1$ angles, and get rid of the angles by integration. We then obtain

$$p(s'_{ij}, y_{im}, s_{kk}) = \frac{|c_{ij}|^{\frac{1}{2}m} s_{kk}^{\frac{1}{2}(m-2)} |s'_{ij}|^{\frac{1}{2}(m-k-1)}}{2^{\frac{1}{2}km}\pi^{\frac{1}{4}k(k-1)} \prod_{i=1}^{k} \Gamma\{\frac{1}{2}(m-i+1)\}}$$

$$\times \exp\left[-\frac{1}{2}\left\{\sum_{i,j=1}^{k-1} c_{ij}(s'_{ij} + y_{im} y_{jm}) + 2\sqrt{s_{kk}}\sum_{i=1}^{k-1} c_{ik} y_{im} + c_{kk} s_{kk}\right\}\right], \tag{8}$$

wherever $\|s'_{ij}\|$ is a positive definite matrix, and we obtain zero otherwise.

Now we can introduce the final set of variables s_{ij} by the transformation

$$s'_{ij} = s_{ij} - \frac{s_{ik} s_{jk}}{s_{kk}} \quad y_{im} = \frac{s_{ik}}{\sqrt{s_{kk}}}, \quad (i, j = 1, 2, \ldots, k-1). \tag{9}$$

The Jacobian of this transformation is $s_{kk}^{-\frac{1}{2}(k-1)}$. We obtain

$$p(s_{ij}) = \frac{|c_{ij}|^{\frac{1}{2}m} s_{kk}^{\frac{1}{2}(m-k-1)} |S_1|^{\frac{1}{2}(m-k-1)}}{2^{\frac{1}{2}km}\pi^{\frac{1}{4}k(k-1)} \prod_{i=1}^{k} \Gamma\{\frac{1}{2}(m-i+1)\}} \exp\left[-\tfrac{1}{2}\sum_{i,j=1}^{k} c_{ij}s_{ij}\right], \tag{10}$$

where $|S_1|$ is the $(k-1)$-rowed determinant whose elements are $s_{ij}-s_{ik}s_{jk}/s_{kk}$.

It is easy to verify that (i) $s_{kk}|S_1| = |s_{ij}|$ and (ii) S_1 is positive definite if and only if $||s_{ij}||$ is positive definite. Hence the right-hand side of (10) coincides with $V_k(s_{ij}, m)$, and $p(s_{ij})$ vanishes whenever $||s_{ij}||$ is not positive definite. This completes the proof.

REFERENCES

(1) WISHART, J. *Biometrika*, 20A (1928), 32–52.
(2) WISHART, J. and BARTLETT, M. S. *Proc. Cambridge Phil. Soc.* 29 (1933), 260–70.
(3) INGHAM, A. E. *Proc. Cambridge Phil. Soc.* 29 (1933), 271–6.

UNIVERSITY COLLEGE
LONDON

Reprinted from
J. Chinese Math. Soc. **2** (1940), 40–63.

ON n-FOLD ITERATED LIMITS

BY PAO-LU HSU*

The question of two-fold repeated limits in connection with a function of two real variables has been treated by Pringsheim[1], London[2], Hobson[3], Townsend[4], and Osgood[5]. According to the definition given by the first three writers, the repeated limit $\lim_{x=a}\lim_{y=b} f(x, y)$ may exist without the existence of $\lim_{y=b} f(x, y)$. We shall, however, use the simpler definition, so that for the existence of a certain iterated limit all the iterated limits of lower order must exist. Assuming the existence of the two limits $\lim_{x=a}\lim_{y=b} f(x, y)$ and $\lim_{y=b}\lim_{x=a} f(x, y)$, Townsend[4] has given a necessary and sufficient condition for their equality. It is not difficult to generalize his theorem to the case of n variables (cf. Theorem 1 below). We prefer, however, to subject our functions to a condition (Condition A below) which was first given by Osgood[6] and which leads at once to the existence of all the n-fold iterated limits and the equality between some of them.

In the second part of the paper the condition of uniform convergence is introduced, and it is shown how a single such condition helps to eliminate the assumption of the existence of a great number of the intermediate iterated limits.

Notation. The function $s(x_1, \ldots, x_n)$ is a function of n real variables, defined in the points $(x_1, \ldots, x_n)$, where each x_i belongs to a point set M_i, $i = 1, \ldots, n$. Each M_i has the upper bound ∞. Those limits of the function are considered for which some or all of the variables are allowed to become positively infinite. The results apply equally to the case in which M_i is any set with a cluster point a_i and the limits in question are those for which x_i approaches a_i as its limit. They remain true also when each x_i stands for a point instead of a real variable. The following notation will be used:

$$\lim_{x_i=\infty} s = [i]s, \qquad \lim_{x_j=\infty} \{[i]s\} = [j]\{[i]s\} = [j,i]s, \quad \text{etc.}$$

Hence $[l, \ldots, j, i]s$ is a function in which $x_i, x_j, \ldots, x_l$ do not appear. Sometimes it may be necessary to write out explicitly the arguments of the functions $[l, \ldots, i]s$,

---

*Research fellow of the China Foundation for the Promotion of Education and Culture

[1] Pringsheim, *Math. Annalen*, **53** (1900), 289–321.

[2] London, *ibid.*, 322–370.

[3] Hobson, *Functions of a Real Variable*, Vol. 1, p. 405; Vol. 2, p. 383.

[4] Townsend, *Functions of Real Variables*, p. 91.

[5,6] A theorem on double limits, sometimes known as the Moore–Osgood theorem, was used for the first time by Osgood in his lectures in the year 1905.

as, e.g.,

$$[k]s(x_1, \cdots, x_{k-1}, x_{k+1}, \cdots, x_n).$$

In case all the x's become infinite, the limit becomes a single number. We will then drop the s and write the limit as

$$[m, \cdots, j, i].$$

I

Theorem 1. *Let the following n-fold iterated limits exist:*

(i) $$[1, \cdots, n],$$

(ii) $$[1, \cdots, k-1, n, k, \cdots, n-1],$$

where k is a fixed positive integer $< n$. A necessary and sufficient condition for them to be equal is that

$$|s - [n]s| < \varepsilon, \tag{1}$$

$$\lambda_0 \leqslant x_1, \qquad \lambda_1 \leqslant x_2, \cdots, \qquad \lambda_{1, \cdots, k-1} \leqslant x_n,$$

$$\lambda_{1, \cdots, k-1, n} \leqslant x_k, \cdots, \qquad \lambda_{1, \cdots, k-1, n, k, \cdots, n-2} \leqslant x_{n-1},$$[7]

where, for a given ε, λ_0 is a constant, λ_1 depends on x_1, $\lambda_{1,2}$ depends on x_1, x_2, etc.

Proof. Let us take $k > 1$. The reader can easily supply the proof for the case $k = 1$. To prove the sufficiency of the condition, write

$$[1, \cdots, n] = [1, \cdots, k-1, n, k, \cdots, n-1] = S.$$

Now $|s - [n]s|$ is less than or equal to the sum of the following $2n - 1$ terms:

$A_{n-1} = |s - [n-1]s|$, $A_{n-2} = |[n-1]s - [n-2, n-1]s|, \cdots,$
$A_k = |[k+1, \cdots, n-1]s - [k, \cdots, n-1]s|,$
$E = |[k, \cdots, n-1]s - [n, k, \cdots, n-1]s|,$
$A_{k-1} = |[n, k, \cdots, n-1]s - [k-1, n, k, \cdots, n-1]s|, \cdots,$
$A_2 = |[3, \cdots, k-1, n, k, \cdots, n-1]s - [2, \cdots, k-1, n, k, \cdots, n-1]s|,$
$A_1 = |[2, \cdots, k-1, n, k, \cdots, n-1]s - S|;$
$B_1 = |S - [2, \cdots, n]s|,$
$B_2 = |[2, \cdots, n]s - [3, \cdots, n]s|, \cdots,$ $B_{k-1} = |[k-1, \cdots, n]s - [k, \cdots, n]s|,$
$B_k = |[k, \cdots, n]s - [k+1, \cdots, n]s|, \cdots,$
$B_{n-2} = |[n-2, n-1, n]s - [n-1, n]s|$, $B_{n-1} = |[n-1, n]s - [n]s|,$

$$|s - [n]s| \leqslant \sum_{r=1}^{n-1} (A_r + B_r) + E. \tag{2}$$

[7]In case $k = 1$ the series of relations read:

$$\lambda_0 \leqslant x_n, \qquad \lambda_n \leqslant x_1, \qquad \lambda_{n,1} \leqslant x_2, \cdots, \qquad \lambda_{n,1, \cdots, n-1} \leqslant x_{n-1}.$$

But now

$$\begin{cases} A_1 < \varepsilon, & B_1 < \varepsilon, & \lambda_0 \leqslant x_1 \\ A_2 < \varepsilon, & B_2 < \varepsilon, & \lambda_1 \leqslant x_2 \\ \cdots\cdots & \cdots\cdots & \cdots\cdots \\ A_{k-1} < \varepsilon, & B_{k-1} < \varepsilon, & \lambda_{1, \cdots, k-2} \leqslant x_{k-1} \\ E < \varepsilon, & & \lambda_{1, \cdots, k-1} \leqslant x_n \\ A_k < \varepsilon, & B_k < \varepsilon, & \lambda_{1, \cdots, k-1, n} \leqslant x_k \\ \cdots\cdots & \cdots\cdots & \cdots\cdots \\ A_{n-2} < \varepsilon, & B_{n-2} < \varepsilon, & \lambda_{1, \cdots, k-1, n, k, \cdots, n-3} \leqslant x_{n-2} \\ A_{n-1} < \varepsilon, & B_{n-1} < \varepsilon, & \lambda_{1, \cdots, k-1, n, k, \cdots, n-2} \leqslant x_{n-1} \end{cases}$$

Hence

$$|s - [n]s| < (2n-1)\varepsilon, \quad \lambda_0 \leqslant x_1, \quad \lambda_1 \leqslant x_2, \cdots, \lambda_{1, \cdots, k-1, n, k, \cdots, n-2} \leqslant x_{n-1},$$

as was to be proved.

To prove the sufficiency of Condition (1) write

$$A = |[1, \cdots, k-1, n, k, \cdots, n-1] - [2, \cdots, n]s|$$

and notice that A is less than or equal to the sum of the following $2n-1$ terms:

$$\begin{aligned} &C_1 = |[1, \cdots, k-1, n, k, \cdots, n-1] - [2, \cdots, k-1, n, k, \cdots, n-1]s, |, \\ &C_2 = |[2, \cdots, k-1, n, k, \cdots, n-1]s - [3, \cdots, k-1, n, k, \cdots, n-1]s|, \cdots, \\ &C_{k-1} = |[k-1, n, k, \cdots, n-1]s - |[n, k, \cdots, n-1]s|, \\ &F = |[n, k, \cdots, n-1]s - [k, \cdots, n-1]s|, \\ &C_k = |[k, \cdots, n-1]s - [k+1, \cdots, n-1]s|, \cdots, C_{n-1} = |[n-1]s - s|; \\ &D_n = |s - [n]s|, \\ &D_{n-1} = |[n]s - [n-1, n]s|, \cdots, D_k = |[k+1, \cdots, n]s - [k, \cdots, n]s|, \\ &D_{k+1} = |[k, \cdots, n]s - [k-1, \cdots, n]s|, \cdots, D_2 = |[3, \cdots, n]s - [2, \cdots, n]s|, \end{aligned}$$

$$A \leqslant C_1 + D_n + \sum_{r=2}^{n-1} (C_r + D_r) + F. \tag{3}$$

Then

$$C_1 < \varepsilon, \qquad \mu_1 \leqslant x_1.$$

Let h_1 be $\geqslant \lambda_0, \mu_1$. Take an arbitrary value $x_1^0 \geqslant h_1$ for x_1 and hold it fast. Substituting in (3) we find:

$$A < \varepsilon + D_n + \sum_{r=2}^{n-1} (C_r + D_r) + F, \qquad x_1 = x_1^0. \tag{4}$$

But now C_2 and D_2 are functions of x_2 alone and

$$C_2 < \varepsilon, \qquad D_2 < \varepsilon, \qquad \mu_2 \leqslant x_2.$$

Let $x_2^0 \geqslant \lambda_1, \mu_2$ (λ_1 now becoming a constant) be a fixed value of x_2. Substituting in (4) we find:

$$A < 3\varepsilon + D_n + \sum_{r=3}^{n-1} (C_r + D_r) + F, \qquad x_1 = x_1^0, \quad x_2 = x_2^0.$$

In this manner we dispose successively of $C_3 + D_3$, $C_4 + D_4$, etc., until we come to the inequality:

$$A < (2k-3)\varepsilon + D_n + \sum_{r=k}^{n-1} (C_r + D_r) + F, \qquad x_i = x_i^0, \qquad i = 1, \cdots, k-1. \quad (5)$$

But now F is a function of x_n alone and

$$F < \varepsilon, \qquad \mu_n \leqslant x_n .$$

Let $x_n^0 \geqslant \mu_n, \lambda_{1,\cdots,k-1}$ be a fixed value of x_n. Substituting in (5) we find:

$$A < 2(k-1)\varepsilon + D_n + \sum_{r=k}^{n-1} (C_r + D_r), \qquad x_i = x_i^0, \qquad i = 1, \cdots, k-1, n. \quad (6)$$

$C_k + D_k$ is disposed of as above by letting $x_k = x_k^0$, and so on. Thus finally:

$$A < 2(n-1)\varepsilon + D_n, \qquad x_i = x_i^0, \qquad i = 1, \cdots, n. \quad (7)$$

But at each step x_i^0 has been chosen in accordance with Condition (1). Hence

$$D_n < \varepsilon, \qquad x_i = x_i^0, \qquad i = 1, \cdots, n,$$

i.e.,

$$A < (2n-1)\varepsilon, \quad x_1 = x_1^0, \quad \text{or} \quad A < (2n-1)\varepsilon, \qquad h_1 \leqslant x_1 . \quad (8)$$

This shows that

$$\lim_{x_1=\infty} \{[2, \cdots, n]s\} = [1, \cdots, k-1, n, k, \cdots, n-1],$$

and this is the relation that was to be proved.

Remark. It is evident that the theorem is true when the two limits under consideration are replaced by any two of the following form:

(i) $$[\eta_1, \cdots, \eta_n]$$

(ii) $$[\eta_1, \cdots, \eta_{k-1}, \eta_n, \eta_{k+1}, \cdots, \eta_{n-1}]$$

provided in Condition (1) i is replaced by η_i for each i.

The following corollary is the case of Theorem 1 when $n = 2$:

Corollary. *If both limits*

$$\lim_{x_1=\infty} \lim_{x_2=\infty} s(x_1, x_2), \qquad \lim_{x_2=\infty} \lim_{x_1=\infty} s(x_1, x_2)$$

exist, then they are equal when and only when

$$\left| \lim_{x_2=\infty} s(x_1, x_2) - s(x_1, x_2) \right| < \varepsilon, \qquad \lambda_1 \leqslant x_2, \qquad \lambda_2 \leqslant x_1 .$$

This is Townsend's theorem cited above.

Condition A. The following condition imposed on the function s will be designated as Condition A *with regard to* x_n:

$$|s(x_1, \cdots, x_{n-1}, x_n) - s(x_1, \cdots, x_{n-1}, x_n')| < \varepsilon,$$
$$\lambda \leqslant x_i, x_n', \qquad i = 1, \cdots, n,^8 \tag{A}$$

where λ depends only on the ε chosen.

When the limit $[n]s$ exists, let $x_n' = \infty$ in (A), so that

$$|s - [n]s| \leqslant \varepsilon, \qquad \lambda \leqslant x_i, \qquad i = 1, \cdots, n.$$

Hence Condition A includes Condition 1 of Theorem 1 provided $[n]s$ exists.
Thus from the proof of sufficiency in Theorem 1 follows immediately:

Theorem 2. *If* $s(x_1, \cdots, x_n)$ *satisfies Condition A with regard to* x_n *and the following limits exist*:

(i) $$[1, \cdots, k-1, n, k, \cdots, n-1],$$

(ii) $$[2, \cdots, n]s,$$

then the limit $[1, \cdots, n]$ *exists and is equal to the limit* (i).

Theorem 3. *If* $s(x_1, \cdots, x_n)$ *satisfies Condition A with regard to* x_n *and all the* $(n-1)$*-fold iterated limits exist, then all the* n*-fold iterated limits exist and two of them are equal, provided the order of appearance of* $1, \cdots, n-1$ *is the same in their symbolic representation.*

Proof. All that is necessary to prove is the existence of all the n-fold iterated limits, because the other part of the conclusion follows from Theorem 2 (cf. Remark on Theorem 1).

The existence proof is given by mathematical induction. The theorem is certainly true for $n = 1$. Suppose it is true for a given $n = 1$.

In (A), put $x_h = \infty$, $1 \leqslant h \leqslant n-1$:

$$|[h]s(x_1, \cdots, x_{h-1}, x_{h+1}, \cdots, x_n) - [h]s(x_1, \cdots, x_{h-1}, x_{h+1}, \cdots, x_{n-1}, x_n')| \leqslant \varepsilon,$$
$$\lambda \leqslant x_i, x_n', \qquad i = 1, \cdots, h-1, h+1, \cdots, n.$$

Further, all the $(n-2)$-fold iterated limits of $[h]s$ exist. By hypothesis all the $(n-1)$-fold limits of $[h]s$ exist, i.e., all the n-fold limits of s of the form $[*, \cdots, *, h]$ exist. But also all the $(n-1)$-fold limits $[*, \cdots, *, n]s$ exist. Hence all the n-fold limits of the form $[*, \cdots, *, n]$ exist by Theorem 2. The proof is now complete.

---

[8] Hence Condition A means the same thing as

$$\lim_{(x_1, \cdots, x_n, x_n') = (\infty, \cdots, \infty, \infty)} \{s(x_1, \cdots, x_{n-1}, x_n) - s(x_1, \cdots, x_{n-1}, x_n')\} = 0.$$

Theorem 4. *If $s(x_1, \cdots, x_n)$ satisfies Condition A with regard to $x_n, x_{n-1}, \cdots, x_{n-p+1}$ $(1 \leqslant p \leqslant n-1)$ and if all the $(n-1)$-fold iterated limits exist, then two n-fold limits are equal provided the order of appearance of $1, \cdots, n-p$ is the same in their symbolic representation.*

Proof. To prove the theorem by induction, notice that for $p = 1$ it reduces to Theorem 3. Suppose the theorem true for a given $p < n - 1$. Add the further assumption that s satisfies Condition A with regard to x_{n-p}. What we want to prove is that under these hypotheses two n-fold limits are equal provided the order of appearance of $1, \cdots, n-p-1$ is the same in their symbolic representation.

Let L and K be two such limits. In L and K let $1, \cdots, n-p$ appear respectively in the order $(\eta_1, \cdots, \eta_{l-1}, n-p, \eta_l, \cdots, \eta_{n-p+1})$ and $(\eta_1, \cdots, \eta_{k-1}, n-p, \eta_k, \cdots, \eta_{n-p-1})$. By hypothesis we have:

$$\begin{cases} L = \left[\eta_1, \cdots, \eta_{l-1}, n-p, \eta_1, \cdots, \eta_{n-p-1}, n-p+1, \cdots, n\right], \\ K = \left[\eta_1, \cdots, \eta_{n-1}, n-p, \eta_k, \cdots, \eta_{n-p-1}, n-p+1, \cdots, n\right]. \end{cases}$$

Since s satisfies Condition A with regard to x_{n-p}, it follows from Theorem 3 that the two limits in the right-hand member of the above equations are equal. Hence $L = K$, and the proof is complete.

Remark. Since there are $(n-p)!$ permutations of the digits $1, \cdots, n-p$, the theorem asserts that all the $n!$ n-fold iterated limits fall into $(n-p)!$ groups, with $n(n-1) \cdots (n-p+1)$ members in each, such that all the members in the same group are equal. In case $p = n-1$, all the n-fold iterated limits have the same value.

But when $p < n - 1$, will the limits in two distinct groups be in general equal? The answer is in the negative. We can, in fact, form functions, which satisfy the conditions of Theorem 4 and such that among the $n!$ n-fold iterated limits there are $(n-p)!$ distinct ones. To show this we first prove:

Theorem 5. *Given an arbitrary set of $n!$ real numbers $\alpha_{ij, \cdots, k}$ where $(i, j, \cdots, k)$ ranges over all the permutations of $1, \cdots, n$, there exists a function $s(x_1, \cdots, x_n)$, defined in the points*

$$1 \leqslant x_i, \qquad i = 1, \cdots, n,$$

such that all the n-fold iterated limits exist and each limit

$$\left[i, j, \cdots, k\right] = \alpha_{i,j, \cdots, k}.$$

Proof. The theorem is trivial for $n = 1$, because the function $s(x_1) = \alpha_1$ will do. Assume the theorem true for a given $n - 1$. Consider all the numbers α whose last index is h. By hypothesis they form the $(n-1)!$ $(n-1)$-fold iterated limits of a certain function

$$s_h(x_1, \cdots, x_{h-1}, x_{h+1}, \cdots, x_n), \qquad 1 \leqslant x_i, \\ i = 1, \cdots, h-1, h+1, \cdots, n.$$

Having determined the n functions

$$s_1(x_2, \cdots, x_n), s_2(x_1, x_3, \cdots, x_n), \cdots, s_n(x_1, \cdots, x_{n-1}),$$

we form the function

$$s = \frac{1}{x_1 + \cdots + x_n} \sum_{k=1}^{n} x_k s_k(x_1, \cdots, x_{k-1}, x_{k+1}, \cdots, x_n).$$

This is a function such as we set out to obtain, for

$$[h]s = s_h(x_1, \cdots, x_{h-1}, x_{h+1}, \cdots, x_n), \qquad h = 1, \cdots, n,$$

and therefore

$$[i, j, \cdots, h] = \alpha_{i,j}, \cdots, h, \qquad h = 1, \cdots, n.$$

The proof is now complete.

Theorem 6. *There exists a function $s(x_1, \cdots, x_n)$ satisfying all the conditions of Theorem 4 and such that two n-fold limits are not equal if the order of appearance of $1, \cdots, n-p$ is not the same in their symbolic representation.*

Proof. Determine a function $f(x_1, \cdots, x_{n-p})$ for which all the $(n-p)!$-fold iterated limits exist and are distinct. This is possible by Theorem 5. Then the function

$$s(x_1, \cdots, x_n) = f(x_1, \cdots, x_{n-p}) + \sum_{i=n-p+1}^{n} \varphi_i(x_i),$$

where $\lim_{x=\infty} \varphi_i(x)$ exists, $i = n-p+1, \cdots, n$, is a case in point, for evidently all the conditions of Theorem 4 are fulfilled. Let $\lim_{x=\infty} \varphi_i(x) = \Phi_i$, $i = n - \mathrm{p} + 1$, $\cdots, n$. Then

$$[\eta_1, \cdots, \eta_{n-p}, n-p+1, \cdots, n] = [\eta_1, \cdots, \eta_{n-p}] + \sum_{i=n-p+1}^{n} \Phi_i,$$

where $[\eta_1, \cdots, \eta_{n-p}]$ is the corresponding $(n-p)$-fold limit of f. Hence the n-fold limits of s are different for different permutations $(\eta_1, \cdots, \eta_{n-p})$.

When s satisfies Condition A with regard to x_n and $[n]s$ exists, then the question of whether s satisfies Condition A with regard to x_{n-1} is equivalent to the same question relative to the function $[n]s$. Also, in general:

Theorem 7. *If the limit $[n-p+2, \cdots, n]s$ exists, then the following sets of conditions are equivalent.*

$$\begin{cases} (A_1) & |s(x_1, \cdots, x_n) - s(x_1, \cdots, x_n') < \varepsilon, \quad \lambda \leqslant x_i, x_n', \quad i = 1, \cdots, n, \\ (A_2) & |s(x_1, \cdots, x_n) - s(x_1, \cdots, x_{n-1}', x_n)| < \varepsilon, \quad \lambda \leqslant x_i, x_{n-1}', \quad i = 1, \cdots, n, \\ \cdots & \cdots \\ (A_p) & |s(x_1, \cdots, x_n) - s(x_1, \cdots, x_{n-p+1}', \cdots, x_n)| < \varepsilon, \quad \lambda \leqslant x_i, x_{n-p+1}', \\ & \qquad i = 1, \cdots, n; \end{cases}$$

$$(B_1) \quad |s(x_1, \cdots, x_n) - s(x_1, \cdots, x_n')| < \varepsilon, \qquad \lambda \leqslant x_i, x_n', \qquad i = 1, \cdots, n,$$

$$(B_2) \quad |[n]s(x_1, \cdots, x_{n-1}) - [n]s(x_1, \cdots, x_{n-1}')| < \varepsilon, \quad \lambda \leqslant x_i, x_{n-1}', \quad i = 1, \cdots, n-1,$$

$$\cdots\cdots$$

$$(B_p) \quad |[n-p+2, \cdots, n]s(x_1, \cdots, x_{n-p+1}) - [n-p+2, \cdots, n]s(x_1, \cdots, x_{n-p+1}')| < \varepsilon, \quad \lambda \leqslant x_i, x_{n-p+1}', \quad i = 1, \cdots, n-p+1.$$

Proof. From

$$(A_k) \qquad |s(x_1, \cdots, x_n) - s(x_1, \cdots, x_{n-k+1}', \cdots, x_n)| < \varepsilon, \quad \lambda \leqslant x_i, x_{n-k+1}', \qquad i = 1, \cdots, n,$$

it follows immediately that

$$|[n-k+2, \cdots, n]s(x_1, \cdots, x_{n-k+1}) - [n-k+2, \cdots, n]s(x_1, \cdots, x_{n-k+1}')| \leqslant \varepsilon,$$
$$\lambda \leqslant x_i, x_{n-k+1}', \qquad i = 1, \cdots, n-k+1,$$

and this relation implies (B_k). Conversely, (A_k) follows from $(B_1), \cdots, (B_k)$. For, in (B_j), let $x_{n-j+1} = \infty, j = 1, \cdots, k-1$. Hence

$$\left.\begin{array}{r} |[n]s - s| \leqslant \varepsilon \\ |[n-1,n]s - [n]s| \leqslant \varepsilon \\ \cdots\cdots \\ |[n-k+2, \cdots, n]s - [n-k+3, \cdots, n]s| \leqslant \varepsilon \end{array}\right\} \lambda \leqslant x_i, \qquad i = 1, \cdots, n.$$

On adding up all the above inequalities, there results the relation:

$$|s - [n-k+2, \cdots, n]s| \leqslant (k+1)\varepsilon, \qquad \lambda \leqslant x_i, \qquad i = 1, \cdots, n.$$

Hence

$$
\begin{aligned}
&\left|s(x_1, \cdots, x_n) - s(x_1, \cdots, x'_{n-k+1}, \cdots, x_n)\right. \\
&\quad \leqslant \left|s - [n-k+2, \cdots, n]s\right| \\
&\qquad + \left|[n-k+2, \cdots, n]s(x_1, \cdots, x_{n-k+1})\right. \\
&\qquad\quad \left. - [n-k+2, \cdots, n]s(x_1, \cdots, x_{n-k+1})\right| \\
&\qquad + \left|[n-k+2, \cdots, n]s(x_1, \cdots, x'_{n-k+1})\right. \\
&\qquad\quad \left. - s(x_1, \cdots, x'_{n-k+1}, \cdots, x_n)\right| \\
&\quad \leqslant (k-1)\varepsilon + \varepsilon + (k-1)\varepsilon = (2k-1)\varepsilon, \\
&\lambda_1 \leqslant x_i, x'_{n-k+1}, \qquad i = 1, \cdots, n,
\end{aligned}
$$

and this relation is equivalent to (A).

On combining Theorem 7 with Theorem 4 the following theorem is obtained.

Theorem 8. *If $s(x_1, \cdots, x_n)$ satisfies Conditions $(B_1), \cdots, (B_k)$ and all the $(n-1)$-fold iterated limits exist, then all the n-fold iterated limits exist, and two limits are equal provided the order of appearance of $1, \cdots, n-p$ is the same in their symbolic representation.*

II

Condition U. If s satisfies the condition:

$$(U) \qquad \left|s(x_1, \cdots, x_n) - s(x_1, \cdots, x_{n-1}, x'_n)\right| < \varepsilon, \qquad h \leqslant x_n, x'_n,$$

where h depends only on ε, then s is said to satisfy *Condition U with regard to x_n*.

Lemma. *If s satisfies Condition U with regard to x_n and if the limit $[k]s$ exists, where $k < n$, then the limits $[n]s$, $[k,n]s$, $[n,k]s$ exist, and the last two are equal.*

Proof. From Condition U follows immediately the existence of $[n]s$ and the relation:

$$\left|s[n]s\right| \leqslant \varepsilon, \qquad h \leqslant x_n.$$

In (U) let $x_k = \infty$. Then

$$
\begin{aligned}
&\left|[k]s(x_1, \cdots, x_{k-1}, x_{k+1}, \cdots, x_n)\right. \\
&\quad \left. - [k]s(x_1, \cdots, x_{k-1}, x_{k+1}, \cdots, x_{n-1}, x'_n)\right| \leqslant \varepsilon, \qquad h \leqslant x_n, x'_n.
\end{aligned}
$$

Hence the limit $[n,k]s$ exists and

$$\left|[n,k]s - [k]s\right| \leqslant \varepsilon, \qquad k \leqslant x_n.$$

Again:

$$\left|[n,k]s - [n]s\right| \leqslant \left|[n,k]s - [k]s\right| + \left|[k]s - s\right| + \left|s - [n]s\right|.$$

If we put $x_n = h$ in the right-hand member, we see that the first and last terms are $\leqslant \varepsilon$ while the second member can be made $< \varepsilon$ by taking x large enough. Hence

$$\lim_{x_k=\infty} \{[n]s\} = [n,k]s,$$

as was to be proved.

Theorem 9. *If s satisfies Condition U with regard to x_n and the limit $[\eta_1, \cdots, \eta_l]s$ exists, where $l < n$ and $\eta_i < n$, $i = 1, \cdots, l$, then all the limits:*

$$[\eta_1, \cdots, \eta_{k-1}, n, \eta_k, \cdots, \eta_l]s, \qquad k = 1, \cdots, l+1,$$

exist and are equal.

Proof. The theorem is true, by the preceding Lemma, for $l = 1$. Suppose it is true for a given $l-1$. Consider the function $[\eta_l]s$. Then by hypothesis

$$[\eta_1, \cdots, \eta_{k-1}, n, \eta_k, \cdots, \eta_{l-1}]\{[\eta_l]s\}, \qquad k = 1, \cdots, l,$$

exist and are equal. Hence all the following limits:

$$[\eta_1, \cdots, \eta_{k-1}, n, \eta_k, \cdots, \eta_l]s, \qquad k = 1, \cdots, l,$$

exist and are equal. Furthermore from the Lemma,

$$[\eta_l, n]s \text{ exists, and } [\eta_l, n]s = [n, \eta_l]s.$$

Hence finally:

$$[\eta_1, \cdots, \eta_l, n]s \text{ exists, and } [\eta_1, \cdots, \eta_l, n]s = [\eta_1, \cdots, \eta_{l-1}, n, \eta_l]s.$$

Thus the theorem is also true for l, and the proof is complete.

Corollary 1. *If s satisfies Condition U with regard to x_n and each of the following l-fold iterated limits exists:*

$$[\eta_1, \cdots, \eta_l]s, \qquad \eta_i < n, \qquad i = 1, \cdots, l,$$

where $l = 1, \cdots, n-1$, then all the m-fold iterated limits exist, $m = 1, \cdots, n$, and

$$[\eta_1, \cdots, \eta_{k-1}, n, \eta_k, \cdots, \eta_{m-1}]s, \qquad k = 1, \cdots, m,$$

are the same function, where $m = 1, \cdots, n$.

Corollary 2. *If s satisfies Condition U with regard to each x_i, then each l-fold iterated limit exists, where $l = 1, \cdots, n$, and two l-fold limits are equal provided the same variables have become infinite.*

Besides the foregoing limits it is possible furthermore to consider those iterated limits in which some or all the variables become infinite simultaneously. Introduce the notation:

$$\lim_{(x_1, \cdots, x_k)=(\infty, \cdots, \infty)} \{[j, \cdots, l]s\} = [(i, \cdots, k), j, \cdots, l]s,$$

the meaning of

$$[(i, \cdots, k)]s, \quad [\eta_1, \cdots, \eta_{k-1}, (\eta_k, \cdots, \eta_{h-1}), \eta_h, \cdots, \eta_l]s, \quad \text{etc.},$$

being now obvious.

Theorem 10. *Under the hypothesis of Corollary* 2 *under Theorem* 9, *the digits in the symbolic representation of an iterated limit can be bracketed in any way whatever.*

Thus, e.g.,

$$[1,2,3,4,5,6,7,8]s = [1,(2,3,4),5,(6,7),8]s = [(1,2),3,4,(5,6),7,8]s.$$

Proof. It is obviously sufficient to prove the following equality:

$$[(\eta_1, \cdots, \eta_{k-1}), \eta_k, \cdots, \eta_l]s - [\eta_1, \cdots, \eta_l]s.$$

For this purpose write down Condition U with regard to x_i:

$$|s(x_1, \cdots, x_n) - s(x_1, \cdots, x_i', \cdots, x_n)| < \varepsilon, \qquad h \leqslant x_i, x_i'$$

and let successively $x_k = \infty, \cdots, x_j = \infty, x_i = \infty$. Then

$$|[i, j, \cdots, k]s - [j, \cdots, k]s| \leqslant \varepsilon, \qquad h \leqslant x_i.$$

Hence

$$|[\eta_1, \cdots, \eta_l]s - [\eta_2, \cdots, \eta_l]s| \leqslant \varepsilon, \qquad h \leqslant x_{\eta_1},$$

$$\cdots \quad \cdots \quad \cdots \quad \cdots$$

$$|[\eta_{k-1}, \cdots, \eta_l]s - [\eta_k, \cdots, \eta_l]s| \leqslant \varepsilon, \qquad h \leqslant x_{\eta_{k-1}}.$$

On adding the above inequalities it appears that

$$|[\eta_1, \cdots, \eta_l]s - [\eta_k, \cdots, \eta_l]s| \leqslant (k-1)\varepsilon, \qquad h \leqslant x_i, \quad i = \eta_1, \cdots, \eta_{k-1}.$$

Hence

$$[\eta_1, \cdots, \eta_l]s = [(\eta_1, \cdots, \eta_{k-1}), \eta_k, \cdots, \eta_l]s,$$

as was to be proved.

Reprinted from
Proc. Edinburgh Math. Soc.
(2)6, 185–189 (1940).

An algebraic derivation of the distribution of rectangular coordinates

By P. L. Hsu.

Communicated by A. C. Aitken.

(*Received* 20*th April*, 1940. *Read* 3*rd May*, 1940.)

Let x_{ir} $(i = 1, \ldots, q;\ r = 1, \ldots, m;\ q \leqq m)$ be random variables which have an elementary probability law $p(x_{11}, \ldots, x_{qm})$. Let

$$s_{ij} = \sum_{r=1}^{m} x_{ir} x_{jr}.$$

The fundamental assumption is that $p(x_{11}, \ldots, x_{qm})$ is explicitly a function of the set of s_{ij} alone, so that

$$p(x_{11}, \ldots, x_{qm}) = f(s_{11}, s_{12}, \ldots, s_{qq}). \tag{1}$$

The $\frac{1}{2}q(q+1)$ functions t_{ij} $(i \leqq j)$, defined by the equation

$$S = TT', \tag{2}$$

where

$$S = \begin{vmatrix} s_{11} & \ldots & s_{1q} \\ \ldots & \ldots & \ldots \\ s_{q1} & \ldots & s_{qq} \end{vmatrix}, \quad T = \begin{vmatrix} t_{11} & 0 & \ldots & 0 \\ t_{12} & t_{22} & \ldots & 0 \\ \ldots & \ldots & \ldots & \ldots \\ t_{1q} & t_{2q} & \ldots & t_{qq} \end{vmatrix}, \quad t_{ii} \geqq 0,$$

and T' is the transposed matrix of T, are a generalisation of the rectangular coordinates of multivariate normal samples defined and studied by Mahalanobis and others in a joint paper[1].

We have, directly from (2),

$$s_{ij} = t_{1i} t_{1j} + t_{2i} t_{2j} + \ldots + t_{ii} t_{ij} \tag{3}$$

To express the t_{ij} in terms of the s_{ij}, we notice from (2) that, for $i \leqq j$,

$$\begin{Vmatrix} s_{11} & \ldots & s_{1,i-1} & s_{1j} \\ \ldots & \ldots & \ldots & \ldots \\ s_{i-1,1} & \ldots & s_{i-1,i-1} & s_{i-1,j} \\ s_{i1} & \ldots & s_{i,i-1} & s_{ij} \end{Vmatrix} =$$

$$\begin{Vmatrix} t_{11} & 0 & \ldots & 0 & 0 \\ \ldots & \ldots & \ldots & \ldots & \ldots \\ t_{1,i-1} & t_{2,i-1} & \ldots & t_{i-1,i-1} & 0 \\ t_{1i} & t_{2i} & \ldots & t_{i-1,i} & t_{ii} \end{Vmatrix} \begin{vmatrix} t_{11} & \ldots & t_{1,i-1} & t_{1j} \\ 0 & \ldots & t_{2,i-1} & t_{2j} \\ \ldots & \ldots & \ldots & \ldots \\ 0 & \ldots & t_{i-1,i-1} & t_{i-1,j} \\ 0 & \ldots & 0 & t_{ij} \end{vmatrix}$$

---

[1] Mahalanobis, Bose and Roy, *Sankhya*, 3 (1937), 1-40. This paper will be referred to as (*M*).

whence, taking determinants,

$$\begin{vmatrix} s_{11} & \cdots & s_{1,i-1} & s_{1j} \\ \cdots & \cdots & \cdots & \cdots \\ s_{i-1,1} & \cdots & s_{i-1,i-1} & s_{i-1,j} \\ s_{i1} & \cdots & s_{i,i-1} & s_{ij} \end{vmatrix} = t_{11}^2 \cdots t_{i-1,i-1}^2\, t_{ii}\, t_{ij}. \tag{4}$$

Setting $i = j$ in (4) we have

$$\begin{vmatrix} s_{11} & \cdots & s_{1i} \\ \cdots & \cdots & \cdots \\ s_{i1} & \cdots & s_{ii} \end{vmatrix} = t_{11}^2 \cdots t_{ii}^2, \tag{5}$$

whence

$$t_{ii} = \begin{vmatrix} s_{11} & \cdots & s_{1i} \\ \cdots & \cdots & \cdots \\ s_{i1} & \cdots & s_{ii} \end{vmatrix}^{\frac{1}{2}} \begin{vmatrix} s_{11} & \cdots & s_{1,i-1} \\ \cdots & \cdots & \cdots \\ s_{i-1,1} & \cdots & s_{i-1,i-1} \end{vmatrix}^{-\frac{1}{2}}. \tag{6}$$

Dividing both sides of (4) by the corresponding sides of (5) and using (6) we obtain

$$t_{ij} = \begin{vmatrix} s_{11} & \cdots & s_{1,i-1} & s_{1j} \\ \cdots & \cdots & \cdots & \cdots \\ s_{i-1,1} & \cdots & s_{i-1,i-1} & s_{i-1,j} \\ s_{i1} & \cdots & s_{i,i-1} & s_{ij} \end{vmatrix} \begin{vmatrix} s_{11} & \cdots & s_{1i} \\ \cdots & \cdots & \cdots \\ s_{i1} & \cdots & s_{ii} \end{vmatrix}^{-\frac{1}{2}} \begin{vmatrix} s_{11} & \cdots & s_{1,i-1} \\ \cdots & \cdots & \cdots \\ s_{i-1,1} & \cdots & s_{i-1,i-1} \end{vmatrix}^{-\frac{1}{2}}. \tag{7}$$

(3) and (7) are the relations connecting the rectangular coordinates t_{ij} and the product moments s_{ij} and have been derived in (M).

In the case where the elementary probability law $p(x_{11}, \ldots, x_{qm})$ is a normal one, a geometrical demonstration has been given in (M) of the distribution of the t_{ij}. Here we shall give a purely algebraic derivation of the distribution, assuming that (1) holds true.

THEOREM. *If* (1) *is true, the elementary probability law of the rectangular coordinates t_{ij} defined by* (2) *is*

$$2^q \pi^{\frac{1}{2}mq - \frac{1}{4}q(q-1)} \{\prod_{i=1}^{q} \Gamma(\tfrac{1}{2}(m - i + 1))\}^{-1} (\prod_{i=1}^{q} t_{ii}^{m-i}) f(s_{11}, s_{12}, \ldots, s_{qq}), \tag{8}$$

where the arguments of f are to be regarded as the functions (3) *of the t_{ij}.*

Proof. By virtue of (1) we may express the elementary probability as

$$f(s_{11}, s_{12}, \ldots, s_{qq})\, dx_{11} \cdots dx_{qm}.$$

Let us write more fully as follows:

$$f\begin{bmatrix} s_{11} & & & \\ s_{12} & s_{22} & & \\ \cdots & \cdots & \cdots & \\ s_{1q} & s_{2q} & \cdots & s_{qq} \end{bmatrix} dx_{11} \ldots dx_{qm}. \tag{9}$$

Let the sets of variables $(x_{21}, \ldots, x_{2m}), \ldots, (x_{q1}, \ldots, x_{qm})$ be subjected to the same linear transformation below whose coefficients are functions of $x_{11}, \ldots, x_{1m}$:

$$\left.\begin{aligned} y_{i1} &= \Big(\sum_{r=1}^{m} x_{1r}^2\Big)^{-\frac{1}{2}} \sum_{r=1}^{m} x_{1r}\, x_{ir} \\ y_{i2} &= \sum_{r=1}^{m} c_{2r}\, x_{ir} \\ &\cdots\cdots\cdots\cdots \\ y_{im} &= \sum_{r=1}^{m} c_{mr}\, x_{ir} \end{aligned}\right\}, \quad (i = 2, \ldots, q), \tag{10}$$

where the c's are so determined as to make (10) an orthogonal transformation. As the Jacobian is 1, we get the result

$$f\begin{bmatrix} \sum_{r=1}^{m} x_{1r}^2 & & & \\ y_{21}\Big(\sum_{r=1}^{m} x_{1r}^2\Big)^{\frac{1}{2}} & \sum_{r=1}^{m} y_{2r}^2 & & \\ \cdots & \cdots & \cdots & \\ y_{q1}\Big(\sum_{r=1}^{m} x_{1r}^2\Big)^{\frac{1}{2}} & \sum_{r=1}^{m} y_{2r}\, y_{qr} & \cdots & \sum_{r=1}^{m} y_{qr}^2 \end{bmatrix} dx_{11} \ldots dx_{1m}\, dy_{21} \ldots dy_{qm}.$$

If for $x_{11}, \ldots, x_{1m}$ we substitute spherical coordinates, viz. the radius vector, t_{11}, and $m-1$ angles, $\theta_{12}, \ldots, \theta_{1m}$, we obtain

$$t_{11}^{m-1}\Big\{\prod_{a=3}^{m} (\cos\theta_{1a})^{a-2}\Big\} f\begin{bmatrix} t_{11}^2 & & & \\ t_{11}\, y_{21} & \sum_{r=1}^{m} y_{2r}^2 & & \\ \cdots & \cdots & \cdots & \\ t_{11}\, y_{q1} & \sum_{r=1}^{m} y_{2r}\, y_{qr} & \cdots & \sum_{r=1}^{m} y_{qr}^2 \end{bmatrix} dt_{11}\, d\theta_{12} \ldots d\theta_{1m}\, dy_{21} \ldots dy_{qm}.$$

Writing t_{1i} for y_{i1} $(i = 2, \ldots q)$ and

$$s'_{ij} = \sum_{r=2}^{m} y_{ir}\, y_{jr} \quad (i, j = 2, \ldots, q),$$

we get

$$t_{11}^{m-1}\{\prod_{a=3}^{m}(\cos\theta_{1a})^{a-2}\}\times$$

$$f\begin{bmatrix} t_{11}^2 & & & \\ t_{11}t_{12} & t_{12}^2+s'_{22} & & \\ \cdots & \cdots & \cdots & \cdots \\ t_{11}t_{1q} & t_{12}t_{1q}+s'_{2q} & \cdots & t_{1q}^2+s'_{qq}\end{bmatrix} dt_{11}\,..\,dt_{1q}\,d\theta_{12}\,..\,d\theta_{1m}\,dy_{22}\,..\,dy_{qm}. \quad (11)$$

It is seen that, as far as the y-variables are concerned, we have the same situation that only the product moments s'_{ij} figure in the elementary probability law. Hence the same procedure which carries (9) to (11) may be repeated. In doing so we introduce the following variables: $t_{22}, t_{23}, \ldots, t_{2q}$; $\theta_{23}, \ldots, \theta_{2q}$; $z_{33}, z_{34}, \ldots, z_{qm}$, to replace the y's, and write down the elementary probability:

$$t_{11}^{m-1}t_{22}^{m-2}\{\prod_{a=3}^{m}(\cos\theta_{1a})^{a-2}\}\{\prod_{a=4}^{m}(\cos\theta_{2a})^{a-3}\}\times$$

$$f\begin{bmatrix} t_{11}^2 & & & & \\ t_{11}t_{12} & t_{12}^2+t_{22}^2 & & & \\ t_{11}t_{13} & t_{12}t_{13}+t_{22}t_{23} & t_{13}^2+t_{23}^2+s''_{33} & & \\ \cdots & \cdots & \cdots & \cdots & \cdots \\ t_{11}t_{1q} & t_{12}t_{1q}+t_{22}t_{2q} & t_{13}t_{1q}+t_{23}t_{2q}+s''_{3q} & \cdots & t_{1q}^2+t_{2q}^2+s''_{qq}\end{bmatrix}\times$$

$$dt_{11}\ldots dt_{1q}\,dt_{22}\ldots dt_{2q}\,d\theta_{12}\ldots d\theta_{1m}\,d\theta_{23}\ldots d\theta_{2m}\,dz_{33}\ldots dz_{qm},$$

where

$$s''_{ij}=\sum_{r=3}^{m}z_{ir}z_{jr}\quad(i,j=3,\ldots,q).$$

Proceeding in this manner we finally obtain

$$(\prod_{i=1}^{q}t_{ii}^{m-i})\{\prod_{i=1}^{q}\prod_{a=i+2}^{m}(\cos\theta_{ia})^{a-i-1}\}\times$$

$$f\begin{bmatrix} t_{11}^2 & & & \\ t_{11}t_{12} & t_{12}^2+t_{22}^2 & & \\ \cdots & \cdots & \cdots & \\ t_{11}t_{1q} & t_{12}t_{1q}+t_{22}t_{2q} & \cdots & t_{1q}^2+t_{2q}^2+\cdots+t_{qq}^2\end{bmatrix}\times$$

$$d\theta_{12}\ldots d\theta_{q+1,m}\,dt_{11}\,dt_{12}\ldots dt_{qq}. \quad (12)$$

Now the functions t_{ij} introduced in the proof are precisely the rectangular co-ordinates defined by (2) or (3). For, in each step of transformation from (9) to (12) the change in the arguments of f is effected by direct substitution. Equating the arguments of f in (9) and (12) we get

$$s_{ij} = t_{1i}t_{1j} + t_{2i}t_{2j} + \ldots\ldots + t_{ii}t_{ij},$$

which gives (3).

If from (12) the θ's are integrated over the following domain:

$$-\pi \leqq \theta_{ia} \leqq \pi \qquad (a = i+1;\ i = 1, \ldots\ldots, q),$$
$$-\tfrac{1}{2}\pi \leqq \theta_{ia} \leqq \tfrac{1}{2}\pi \quad (a = i+2, \ldots\ldots, m;\ i = 1, \ldots\ldots, q),$$

the result is the elementary probability law (8). The proof is therefore complete.

One more step leads to the distribution of the product moments s_{ij}, as is done in (M). The transformation is given by (7) and the reciprocal transformation by (3), which has the Jacobian

$$2^q\, t_{11}^q\, t_{22}^{q-1} \ldots\ldots t_{qq}.$$

Dividing (8) by this Jacobian and then carrying out the substitution we get the following elementary probability law of the s_{ij}:

$$\pi^{\frac{1}{2}mq-\frac{1}{4}q(q-1)}\{\prod_{i=1}^{q}\Gamma(\tfrac{1}{2}(m-i+1))\}^{-1}\begin{vmatrix} s_{11} & \ldots\ldots & s_{1q} \\ \ldots & \ldots\ldots & \ldots \\ s_{q1} & \ldots\ldots & s_{qq}\end{vmatrix}^{\frac{1}{2}(m-q+1)} f(s_{11}, s_{12}, \ldots, s_{qq}). \quad (13)$$

The multiplier of $f(s_{11}, s_{12}, \ldots\ldots, s_{qq})$ in (13) is well known[1].

The result (12) brings out the following important fact: If $x_{11}, x_{12}, \ldots\ldots, x_{qm}$ are all the observational data and if only the rectangular coordinates t_{ij} or the product moments s_{ij} are utilised for statistical purposes, the part of the observational data thus thrown away may be regarded as a set of angles which are distributed independently of the t_{ij} or s_{ij} and whose elementary probability law does not involve any of the unknown parameters that may figure in the elementary probability law of the x's.

---

[1] *Cf.* Wishart and Bartlett, *Proc. Camb. Phil. Soc.*, 29 (1933), 271-6.

153, CHESTERTON ROAD,
CAMRRIDGE.

Reprinted from
Biometrika
31, 221–237 (1940).

ON GENERALIZED ANALYSIS OF VARIANCE. (I)

BY P. L. HSU

1. *The Wilks-Lawley hypothesis.* Suppose that, as a result of some random sampling, we are in possession of $p(n_1+n)$ quantities, where $n \geqslant p$, calculated from the observational data. Calling these y_{ir} and $z_{ir'}$ ($i=1,2,\ldots,p$; $r=1,2,\ldots,n_1$; $r'=1,2,\ldots,n$), we assume that in repeated sampling they follow the probability distribution

$$\text{Const.} \times \exp\left\{-\frac{1}{2}\sum_{i,j=1}^{p}\alpha_{ij}\sum_{r=1}^{n_1}(y_{ir}-\eta_{ir})(y_{jr}-\eta_{jr})-\frac{1}{2}\sum_{i,j=1}^{p}\alpha_{ij}\sum_{r=1}^{n}z_{ir}z_{jr}\right\}\Pi\, dy\, dz, \quad (1)$$

and that we have no previous knowledge of the values of the η_{ir}. Our problem is to test the hypothesis H_0:

$$\eta_{ir}=0 \quad \text{for} \quad i=1,2,\ldots,p;\ r=1,2,\ldots,n_1. \qquad (H_0)$$

For $p=1$ the hypothesis is that to which every linear hypothesis* may be reduced, and the test amounts to that ordinarily employed in the analysis of variance. For $n_1=1$ the problem calls for Hotelling's generalized "Student's" test.†

It should be emphasized that the y_{ir}, z_{ir} and η_{ir} will not usually correspond to the original physical observations and physical constants belonging to the sampled populations, but are so derived from them that (1) is true and that H_0 is equivalent to some hypothesis regarding population constants that we want to test. In another paper we shall deal with a general class of "linear hypothesis" on multivariate normal means, showing how all of them can be reduced by a rotation of the sample space to the "canonical" H_0. Here we content ourselves with the following example.

Example. Case of k samples. $x_{i\nu t}$ ($i=1,2,\ldots,p$; $\nu=1,2,\ldots,k$; $t=1,2,\ldots,m_\nu$) are k samples drawn respectively from k p-variate normal populations all of which have the same set of variances and covariances. Let the population means be $\xi_{i\nu}$ ($i=1,2,\ldots,p$; $\nu=1,2,\ldots,k$). Let

$$\sum_{\nu=1}^{k}m_\nu = M, \quad \bar{\xi}_i = \frac{1}{M}\sum_{\nu=1}^{k}m_\nu\xi_{i\nu},$$

$$\bar{x}_{i\nu} = \frac{1}{m_\nu}\sum_{t=1}^{m_\nu}x_{i\nu t}, \quad \bar{x}_i = \frac{1}{M}\sum_{\nu=1}^{k}m_\nu\bar{x}_{i\nu},$$

$$s_{ij\nu} = \sum_{t=1}^{m_\nu}(x_{i\nu t}-\bar{x}_{i\nu})(x_{j\nu t}-\bar{x}_{j\nu}),$$

$$(i,j=1,2,\ldots,p;\ \nu=1,2,\ldots,k).$$

* Kolodziezczyk (1935), Tang (1938).

† Hotelling (1931), Hsu (1938), Bose and Roy (1938)

Suppose that we want to test the hypothesis:

$$\xi_{i1} = \xi_{i2} = \ldots = \xi_{ik} \quad \text{for} \quad i = 1, 2, \ldots, p.$$

Then it is known that if we call y_{ir} and z_{ir} certain linear functions of the $x_{i\nu t}$ and η_{ir} certain linear functions of the $\xi_{i\nu}$, (1) will hold true and the hypothesis under test is equivalent to H_0. Thus reducing the problem, we have

$$n_1 = k-1, \quad n = M-k,$$

$$\sum_{r=1}^{n_1} y_{ir}y_{jr} = \sum_{\nu=1}^{k} m_\nu(\bar{x}_{i\nu}-\bar{x}_i)(\bar{x}_{j\nu}-\bar{x}_j), \quad \sum_{r=1}^{n} z_{ir}z_{jr} = \sum_{\nu=1}^{k} s_{ij\nu},$$

$$\sum_{r=1}^{n_1} \eta_{ir}\eta_{jr} = \sum_{\nu=1}^{k} m_\nu(\xi_{i\nu}-\bar{\xi}_i)(\xi_{j\nu}-\bar{\xi}_j),$$

$$(i, j = 1, 2, \ldots, p).$$

In particular, if, $k = 2$ then $n_1 = 1$, and we may drop the second index of y and η. We have then

$$y_i y_j = \frac{m_1 m_2}{m_1+m_2}(\bar{x}_{i1}-\bar{x}_{i2})(\bar{x}_{j1}-\bar{x}_{j2}),$$

$$\sum_{r=1}^{n} z_{ir}z_{jr} = \sum_{t=1}^{m_1}(x_{i1t}-\bar{x}_{i1})(x_{j1t}-\bar{x}_{j1}) + \sum_{t=1}^{m_2}(x_{i2t}-\bar{x}_{i2})(x_{j2t}-\bar{x}_{j2}),$$

$$\eta_i\eta_j = \frac{m_1 m_2}{m_1+m_2}(\xi_{i1}-\xi_{i2})(\xi_{j1}-\xi_{j2}).$$

$$(i, j = 1, 2, \ldots, p).$$

The hypothesis in the above example has been studied by Wilks (1932), while an attempt to test the general hypothesis H_0 was made by Lawley (1938). Both assumed no prior knowledge of the values of the α_{ij}. We propose to call H_0 the Wilks-Lawley hypothesis.

2. *Case where the α_{ij} are known, generalizations of Mahalanobis's distance.* From now on we shall write

$$a_{ij} = \sum_{r=1}^{n_1} y_{ir}y_{jr}, \quad b_{ij} = \sum_{r=1}^{n} z_{ir}z_{jr},$$

$$\psi_{ij} = \sum_{r=1}^{n_1} \eta_{ir}\eta_{jr}, \quad \Psi = \sum_{i,j=1}^{p} \alpha_{ij}\psi_{ij},$$

$$(i, j = 1, 2, \ldots, p).$$

In order to test H_0 when the α_{ij} are known, we calculate the likelihood ratio* and get

$$-2\log(\text{likelihood ratio}) = \sum_{i,j=1}^{p} \alpha_{ij}a_{ij} = S, \text{ say}$$

and decide to reject H_0 if the value of S exceeds some fixed constant, chosen so as to fix at some desired level the risk of rejecting H_0 when it is true.

* For the definition and interpretation of the term see Neyman & Pearson (1928).

THEOREM 1. *For arbitrary values of the η_{ir} the distribution of S is that of the sum of pn_1 non-central squares:**

$$2^{-\frac{1}{2}pn_1} S^{\frac{1}{2}pn_1-1} e^{-\frac{1}{2}(S+\Psi)} \left\{ \sum_{h=0}^{\infty} \frac{\Psi^h S^h}{4^h h!\,\Gamma(h+\frac{1}{2}pn_1)} \right\} dS. \tag{2}$$

In particular, if H_0 is true, then $\Psi = 0$ and S follows the χ^2 distribution with pn_1 degrees of freedom.

Proof. Let the sets of variables $(y_{1r}, y_{2r}, \ldots, y_{pr})$, for $r = 1, 2, \ldots, n_1$, be subject to the same linear transformation such that, calling the new variables $(u_{1r}, u_{2r}, \ldots, u_{pr})$,

$$\sum_{i,j=1}^{p} \alpha_{ij} y_{ir} y_{jr} = \sum_{i=1}^{p} u_{ir}^2, \quad (r = 1, 2, \ldots, n_1).$$

This is possible because the matrix $\|\alpha_{ij}\|$ is positive definite. Let μ_{ir} be the same linear function of the η's as u_{ir} is of the y's. Then

$$S = \sum_{i=1}^{p} \sum_{r=1}^{n_1} u_{ir}^2 \tag{3}$$

and the u_{ir} follow the distribution

$$(2\pi)^{-\frac{1}{2}pn_1} \exp\left\{ -\frac{1}{2} \sum_{i=1}^{p} \sum_{r=1}^{n_1} (u_{ir} - \mu_{ir})^2 \right\} \Pi\, du. \tag{4}$$

From (3) and (4) follows the result (2),† as we have

$$\sum_{i=1}^{p} \sum_{r=1}^{n_1} \mu_{ir}^2 = \sum_{i,j=1}^{p} \alpha_{ij} \sum_{r=1}^{n_1} \eta_{ir} \eta_{jr} = \Psi.$$

From (2) we get
$$\mathscr{E}(S) = \Psi + pn_1, \tag{5}$$
$$\mathscr{E}(S^2) = \Psi^2 + 2(pn_1+2)\Psi + pn_1(pn_1+2),$$
whence
$$\sigma^2(S) = 4\Psi + 2pn_1.$$

If these general results are applied to the above example of k samples, we have

$$S = \sum_{i,j=1}^{p} \alpha_{ij} \sum_{\nu=1}^{k} m_\nu (\bar{x}_{i\nu} - \bar{x}_i)(\bar{x}_{j\nu} - \bar{x}_j) = S_k,$$

$$\Psi = \sum_{i,j=1}^{p} \alpha_{ij} \sum_{\nu=1}^{k} m_\nu (\xi_{i\nu} - \bar{\xi}_i)(\xi_{j\nu} - \bar{\xi}_j) = \Psi_k,$$

say, and the distribution

$$2^{-\frac{1}{2}p(k-1)} S_k^{\frac{1}{2}p(k-1)-1} e^{-\frac{1}{2}(S_k+\Psi_k)} \left\{ \sum_{h=0}^{\infty} \frac{\Psi_k^h S_k^h}{4^h h!\,\Gamma(h+\frac{1}{2}p(k-1))} \right\} dS_k. \tag{6}$$

For $k = 2$ we have

$$S_2 = \frac{m_1 m_2}{m_1+m_2} \sum_{i,j=1}^{p} \alpha_{ij} (\bar{x}_{i1} - \bar{x}_{i2})(\bar{x}_{j1} - \bar{x}_{j2}),$$

$$\Psi_2 = \frac{m_1 m_2}{m_1+m_2} \sum_{i,j=1}^{p} \alpha_{ij} (\xi_{i1} - \xi_{i2})(\xi_{j1} - \xi_{j2}),$$

* Fisher (1928), p. 669. † Fisher (1928), Tang (1938), pp. 138–9.

15-2

and the distribution

$$2^{-\frac{1}{2}p}S_2^{\frac{1}{2}p-1}e^{-\frac{1}{2}(S_2+\Psi_2)}\left\{\sum_{h=0}^{\infty}\frac{\Psi_2^h S_2^h}{4^h h!\Gamma(h+\frac{1}{2}p)}\right\}dS_2. \tag{7}$$

The quantity Ψ_2 is equal to

$$\frac{pm_1m_2}{m_1+m_2}\Delta,$$

where Δ is Mahalanobis's "distance"* between two populations having the same variances and covariances weighted according to m_1 and m_2. The statistic

$$\frac{m_1+m_2}{pm_1m_2}S_2-\frac{1}{m_1}-\frac{1}{m_2},$$

which, according to (5), is an unbiassed estimate of Δ, is called the D^2 statistic and the distribution (7) has already been obtained.† These concepts and formulae are completely generalized here by S_k, Ψ_k and the formulae (5) and (6) for the case $k>2$.

3. *Tests suggested by Wilks and Lawley: Generalized* E^2, $1-E^2$ *and* $E^2(1-E^2)^{-1}$. From now on we shall assume complete ignorance of the values of the α_{ij}. In the case $p=1$ the following functions are well known:

$$E^2=\sum_{r=1}^{n_1}y_r^2\Big/\left(\sum_{r=1}^{n_1}y_r^2+\sum_{r=1}^{n}z_r^2\right),$$

$$1-E^2=\sum_{r=1}^{n}z_r^2\Big/\left(\sum_{r=1}^{n_1}y_r^2+\sum_{r=1}^{n}z_r^2\right),$$

$$E^2/(1-E^2)=\sum_{r=1}^{n_1}y_r^2\Big/\sum_{r=1}^{n}z_r^2.$$

Wilks, while studying the k-sample case,‡ suggested the functions U and W respectively as generalizations of E^2 and $1-E^2$. In our general case these are defined as

$$U=\frac{|a_{ij}|}{|a_{ij}+b_{ij}|}\quad\text{if } n_1\geqslant p, \tag{8}$$

and

$$W=\frac{|b_{ij}|}{|a_{ij}+b_{ij}|}. \tag{9}$$

Both are ratios of two determinants. The above definition for U based on the idea of a generalized E^2 breaks down if $n_1<p$, as then $|a_{ij}|$ vanishes identically. This difficulty can be met if we define U as the product of the non-identically vanishing roots of the determinantal equation $|a_{ij}-\theta(a_{ij}+b_{ij})|=0$. If $n_1\geqslant p$, this definition evidently coincides with (8).

* Mahalanobis (1936). † Bose (1936). ‡ Wilks (1932).

As a generalization of $E^2(1-E^2)^{-1}$ we take, instead of UW^{-1}, the function V suggested by Lawley* and defined as

$$V = \sum_{i,j=1}^{p} b^{ij} a_{ij}, \tag{10}$$

where b^{ij} is the element (i, j) of the reciprocal matrix $\|b_{ij}\|^{-1}$

If we denote by $\theta_1, \theta_2, \ldots, \theta_{l_1}$, where l_1 is the smaller of the integers p and n_1, the non-identically vanishing roots of the equation

$$|a_{ij} - \theta(a_{ij} + b_{ij})| = 0;$$

then, according to (9) and (10) and the new definition of V, we have

$$U = \prod_{i=1}^{l_1} \theta_i, \tag{11}$$

$$W = \prod_{i=1}^{l_1} (1-\theta_i), \tag{12}$$

$$V = \sum_{i=1}^{l_1} \frac{\theta_i}{1-\theta_i}. \tag{13}$$

W and V are test functions for H_0 suggested respectively by Wilks and Lawley. If we use W (or V), we reject H_0 if W is smaller (or V is greater) than some prescribed constant. In the following sections we shall study the distributions of U, V and W in repeated sampling.

4. *Case when H_0 is true.* If H_0 is true, then, putting all the $\eta_{ir} = 0$ in (1), we get the parent distribution

$$\text{const.} \times \exp\left\{-\frac{1}{2}\sum_{i,j=1}^{p} \alpha_{ij}(a_{ij}+b_{ij})\right\} \Pi\, dy\, dz. \tag{14}$$

The following theorem has been proved elsewhere.†

THEOREM 2. *Let l_1 be the smaller, and l_2 the larger, of the integers p and n_1. Let $\theta_1, \theta_2, \ldots, \theta_{l_1}$ be the non-identically vanishing roots of the determinantal equation*

$$|a_{ij} - \theta(a_{ij} + b_{ij})| = 0, \tag{15}$$

arranged in the order of descending magnitude: $1 \geqslant \theta_1 \geqslant \theta_2 \geqslant \ldots \geqslant \theta_{l_1} \geqslant 0$. Then the simultaneous distribution of the θ's, as derived from (14), *is*

$$\pi^{\frac{1}{2}l_1}\left\{\prod_{i=1}^{l_1} \frac{\Gamma\frac{1}{2}(n-p+l_2+i)}{\Gamma\frac{1}{2}(l_2-l_1+i)\,\Gamma\frac{1}{2}(n-p+i)\,\Gamma\frac{1}{2}i}\right\}$$
$$\times\left\{\prod_{i=1}^{l_1}\theta_i\right\}^{\frac{1}{2}(l_2-l_1-1)}\left\{\prod_{i=1}^{l_1}(1-\theta_i)\right\}^{\frac{1}{2}(n-p-1)}\left\{\prod_{i=1}^{l_1}\prod_{j=i+1}^{l_1}(\theta_i-\theta_j)\right\}\left\{\prod_{i=1}^{l_1} d\theta_i\right\}. \tag{16}$$

Therefore the distributions of U, W and V are those of the functions in (11), (12) and (13), where the θ's follow the distribution (16).

* Lawley (1938). † Hsu (1939), Fisher (1939).

Take the particular case $l_1 = 2$. We have then

$$U = \theta_1\theta_2, \quad W = (1-\theta_1)(1-\theta_2), \quad V = \frac{\theta_1}{1-\theta_1} + \frac{\theta_2}{1-\theta_2},$$

whence
$$U^{\frac{1}{2}} + W^{\frac{1}{2}} \leqslant 1, \quad V = W^{-1}(1-U-W),$$

and the distribution (16) becomes

$$\frac{\Gamma(n-p+l_2+1)}{4\Gamma(l_2-1)\Gamma(n-p+1)}(\theta_1\theta_2)^{\frac{1}{2}(l_2-3)}\{(1-\theta_1)(1-\theta_2)\}^{\frac{1}{2}(n-p-1)}(\theta_1-\theta_2)\,d\theta_1 d\theta_2. \quad (17)$$

From (17) we easily obtain the joint distribution of U and W:

$$\frac{\Gamma(n-p+l_2+1)}{4\Gamma(l_2-1)\Gamma(n-p+1)} U^{\frac{1}{2}(l-3)}W^{\frac{1}{2}(n-p-1)}\,dU\,dW.$$

Integrating with respect to W ranging from 0 to $(1-U^{\frac{1}{2}})^2$, we get the distribution of U:

$$\frac{1}{2B(l_2-1, n-p+2)} U^{\frac{1}{2}(l_2-3)}(1-U^{\frac{1}{2}})^{n-p+1}\,dU. \quad (18)$$

Similarly, the distribution of W is

$$\frac{1}{2B(l_2, n-p+1)} W^{\frac{1}{2}(n-p-1)}(1-W^{\frac{1}{2}})^{l_2-1}\,dW. \quad (19)$$

Again, from (18) we find the distribution of V:

$$\frac{\Gamma(n-p+l_2+1)}{4\Gamma(l_2-1)\Gamma(n-p+1)}(1+V)^{-\frac{1}{2}(n-p+3)}\left\{\int_{\frac{4(1+V)}{(2+V)^2}}^{1} y^{\frac{1}{2}(n-p+1)}(1-y)^{\frac{1}{2}(l_2-3)}\,dy\right\}dV. \quad (20)$$

Now in the particular case considered where $l_1 = 2$ we have either (i) $p = 2$, $l_2 = n_1$ if $p \leqslant n_1$, or (ii) $n_1 = 2$, $l_2 = p$ if $n_1 \leqslant p$. Hence, from (18), (19) and (20) we get the following three pairs of distributions:

$$\frac{1}{2B(n_1-1, n)} U^{\frac{1}{2}(n_1-3)}(1-U^{\frac{1}{2}})^{n-1}\,dU, \quad 2 = p \leqslant n_1,$$

$$\frac{1}{2B(p-1, n-p+2)} U^{\frac{1}{2}(p-3)}(1-U^{\frac{1}{2}})^{n-p+1}\,dU, \quad 2 = n_1 \leqslant p,$$

$$\frac{1}{2B(n_1, n-1)} W^{\frac{1}{2}(n-3)}(1-W^{\frac{1}{2}})^{n_1-1}\,dW, \quad 2 = p \leqslant n_1,$$

$$\frac{1}{2B(p, n-p+1)} W^{\frac{1}{2}(n-p-1)}(1-W^{\frac{1}{2}})^{p-1}\,dW, \quad 2 = n_1 \leqslant p,$$

$$\frac{\Gamma(n_1+n-1)}{4\Gamma(n_1-1)\Gamma(n-1)}(1+V)^{-\frac{1}{2}(n+1)}\,dV\int_{4(1+V)(2+V)^{-2}}^{1} y^{\frac{1}{2}(n-1)}(1-y)^{\frac{1}{2}(n_1-3)}\,dy, \quad 2 = p \leqslant n_1,$$

$$\frac{\Gamma(n+1)}{4\Gamma(p-1)\Gamma(n-p+1)}(1+V)^{-\frac{1}{2}(n-p+3)}\,dV\int_{4(1+V)(2+V)^{-2}}^{1} y^{\frac{1}{2}(n-p+1)}(1-y)^{\frac{1}{2}(p-3)}\,dy, \quad 2 = n_1 \leqslant p.$$

The general expression for the moments of U and W can easily be deduced from (16). We have, from (16),

$$C(l_2, n)\int f(\theta; l_2, n)\prod_{i=1}^{l_1} d\theta_i = 1,$$

where
$$C(l_2, n) = \pi^{\frac{1}{2}l_1}\prod_{i=1}^{l_1}\frac{\Gamma\frac{1}{2}(n-p+l_2+i)}{\Gamma\frac{1}{2}(l_2-l_1+i)\,\Gamma\frac{1}{2}(n-p+i)\,\Gamma\frac{1}{2}i},$$

$$f(\theta; l_2, n) = \left\{\prod_{i=1}^{l_1}\theta_i\right\}^{\frac{1}{2}(l_2-l_1-1)}\left\{\prod_{i=1}^{l_1}(1-\theta_i)\right\}^{\frac{1}{2}(n-p-1)}\left\{\prod_{i=1}^{l_1}\prod_{j=i+1}^{l_1}(\theta_i-\theta_j)\right\}.$$

Hence
$$\int f(\theta; l_2, n)\prod_{i=1}^{l_1} d\theta_i = \frac{1}{C(l_2, n)},$$

whence
$$\mathscr{E}(U^{q_1}W^{q_2}) = C(l_2, n)\int f(\theta; l_2+q_1, n+q_2)\prod_{i=1}^{l_1} d\theta_i = \frac{C(l_2, n)}{C(l_2+2q_1, n+2q_2)}$$
$$= \prod_{i=1}^{l_1}\frac{\Gamma\frac{1}{2}(n-p+l_2+i)\,\Gamma\frac{1}{2}(l_2-l_1+2q_1+i)\,\Gamma\frac{1}{2}(n+2q_2-p+i)}{\Gamma\frac{1}{2}(n-p+l_2+2q_1+2q_2+i)\,\Gamma\frac{1}{2}(l_2-l_1+i)\,\Gamma\frac{1}{2}(n-p+i)}.$$

The case of k samples (see p. 221) of two variables ($p=2$) has been studied at length by Pearson & Wilks (1933). The hypothesis H_0 and the functions U and W are called in their paper H_3, U_2 and L_2 respectively. It is found there that the test based on U_2 (i.e. U) cannot be regarded as an adequate test of H_0. We shall show in the next section that H_0 is true if and only if two population constants, called λ_1 and λ_2, both vanish. The disadvantage of U referred to by Pearson and Wilks, may be expressed by saying that it is unlikely to be able to detect the falsehood of H_0 if only one of λ_1 and λ_2 vanishes.

5. *Simplification of the parent distribution function.* The matrix $\|\alpha_{ij}\|$ is positive definite; hence it can be expressed as $\mathbf{CC}'$,* where $\mathbf{C}$ is some non-singular real matrix. Write

$$\mathbf{Y} = \begin{Vmatrix} y_{11} & y_{12} & \cdots & y_{1n_1} \\ y_{21} & y_{22} & \cdots & y_{2n_1} \\ \cdots & \cdots & \cdots & \cdots \\ y_{p1} & y_{p2} & \cdots & y_{pn_1}\end{Vmatrix}, \quad \mathbf{Z} = \begin{Vmatrix} z_{11} & z_{12} & \cdots & z_{1n} \\ z_{21} & z_{22} & \cdots & z_{2n} \\ \cdots & \cdots & \cdots & \cdots \\ z_{p1} & z_{p2} & \cdots & z_{pn}\end{Vmatrix}, \quad \mathbf{G} = \begin{Vmatrix} \eta_{11} & \eta_{12} & \cdots & \eta_{1n_1} \\ \eta_{21} & \eta_{22} & \cdots & \eta_{2n_1} \\ \cdots & \cdots & \cdots & \cdots \\ \eta_{p1} & \eta_{p2} & \cdots & \eta_{pn_1}\end{Vmatrix},$$

$$\|\sigma_{ij}\| = \|\alpha_{ij}\|^{-1} = (\mathbf{C}')^{-1}\mathbf{C}^{-1}.$$

THEOREM 3. *Let* $\mathbf{G}$ *be of rank* l *(hence* $l\leqslant p$, $l\leqslant n_1$ *and* l *is the rank of* $GG' = \|\psi_{ij}\|$*), and let* $\lambda_1, \lambda_2, \ldots, \lambda_l$ *be the non-vanishing roots of the determinantal equation* $|\psi_{ij}-\lambda\sigma_{ij}| = 0$. *Let* $\theta_1, \theta_2, \ldots, \theta_{l_1}$ *be the non-identically vanishing roots of the determinantal equation* $|a_{ij}-\theta(a_{ij}+b_{ij})| = 0$. *Then the joint distribution of the* θ*'s, as derived from* (1), *depends upon the parameters* $\lambda_1, \lambda_2, \ldots, \lambda_l$ *alone in such a way that instead of* (1) *we may regard the parent distribution as*

$$(2\pi)^{-\frac{1}{2}p(n_1+n)}\exp(-\tfrac{1}{2}\Psi)\exp\left\{-\frac{1}{2}\sum_{i=1}^{p}a_{ii}-\frac{1}{2}\sum_{i=1}^{p}b_{ii}+\sum_{i=1}^{l}\sqrt{\lambda_i}\,y_{ii}\right\}\Pi\,dy\,dz. \quad (21)$$

* The accent denotes the transposed matrix.

Hence, in studying the distribution of any functions of the θ's, such as U, V and W, we may replace (1) by (21).

LEMMA 1 (Hotelling).* *Let* $\mathbf{A}$ *and* $\mathbf{B}$ *be any positive definite matrices of orders* m *and* n *respectively, and* $\mathbf{C}$ *be any real matrix of order* $m \times n$ *and rank* l. *There exist two non-singular real matrices,* $\mathbf{N}_1$ *and* $\mathbf{N}_2$, *such that*

$$\mathbf{N}_1\mathbf{A}\mathbf{N}_1' = \mathbf{I},\dagger \quad \mathbf{N}_2\mathbf{B}\mathbf{N}_2' = \mathbf{I}, \quad \mathbf{N}_1\mathbf{C}\mathbf{N}_2' = \mathbf{D},$$

where

$$\mathbf{D} = \begin{Vmatrix} \mathbf{D}_{\sqrt{\lambda}} & \mathbf{O} \\ \mathbf{O} & \mathbf{O} \end{Vmatrix},\dagger \qquad \mathbf{D}_{\sqrt{\lambda}} = \begin{Vmatrix} \sqrt{\lambda_1} & & & \\ & \sqrt{\lambda_2} & & \\ & & \cdots & \\ & & & \sqrt{\lambda_l} \end{Vmatrix},$$

and the λ_i *are the non-vanishing roots of the determinantal equation*

$$|\mathbf{C}\mathbf{B}^{-1}\mathbf{C}' - \lambda\mathbf{A}| = 0.$$

Allowing both $\mathbf{A}$ and $\mathbf{B}$ to be unit matrices in Lemma 1 we get the following:

Corollary. Let $\mathbf{C}$ *be any real matrix of rank* l. *There exist two real orthogonal matrices* $\mathbf{\Gamma}_1$ *and* $\mathbf{\Gamma}_2$ *such that*

$$\mathbf{\Gamma}_1\mathbf{C}\mathbf{\Gamma}_2' = \mathbf{D}$$

where

$$\mathbf{D} = \begin{Vmatrix} \mathbf{D}_{\sqrt{\lambda}} & \mathbf{O} \\ \mathbf{O} & \mathbf{O} \end{Vmatrix}, \qquad \mathbf{D}_{\sqrt{\lambda}} = \begin{Vmatrix} \sqrt{\lambda_1} & & & \\ & \sqrt{\lambda_2} & & \\ & & \cdots & \\ & & & \sqrt{\lambda_l} \end{Vmatrix},$$

and the λ_i *are the non-vanishing latent roots of the matrix* $\mathbf{C}\mathbf{C}'$.

Proof of Theorem 3. We write $\operatorname{tr}\mathbf{A}$ for the sum of the diagonal elements of any square matrix $\mathbf{A}$. It is easily verified that $\operatorname{tr}(\mathbf{AB}) = \operatorname{tr}(\mathbf{BA})$ whenever $\mathbf{AB}$ is a square matrix.

The expression inside the bracket in (1) may be written as

$$\begin{aligned} &-\tfrac{1}{2}\Psi - \tfrac{1}{2}\operatorname{tr}(\mathbf{C}\mathbf{C}'\mathbf{Y}\mathbf{Y}') - \tfrac{1}{2}\operatorname{tr}(\mathbf{C}\mathbf{C}'\mathbf{Z}\mathbf{Z}') + \operatorname{tr}(\mathbf{C}\mathbf{C}'\mathbf{G}\mathbf{Y}') \\ &\quad = -\tfrac{1}{2}\Psi - \tfrac{1}{2}\operatorname{tr}(\mathbf{C}'\mathbf{Y}\mathbf{Y}'\mathbf{C}) - \tfrac{1}{2}\operatorname{tr}(\mathbf{C}'\mathbf{Z}\mathbf{Z}'\mathbf{C}) + \operatorname{tr}(\mathbf{C}'\mathbf{G}\mathbf{Y}'\mathbf{C}). \qquad (22) \end{aligned}$$

The λ_i, as defined in Theorem 3, are the non-vanishing latent roots of the equation $|\mathbf{G}\mathbf{G}' - \lambda(\mathbf{C}')^{-1}\mathbf{C}^{-1}| = 0$. On pre- and post-multiplying by $|\mathbf{C}'|$ and $|\mathbf{C}|$ respectively this becomes $|\mathbf{C}'\mathbf{G}\mathbf{G}'\mathbf{C} - \lambda\mathbf{I}| = 0$. Hence the λ_i are the non-vanishing latent roots of the matrix $(\mathbf{C}'\mathbf{G})(\mathbf{C}'\mathbf{G})'$. Therefore, by the corollary to Lemma 1, there exist two real orthogonal matrices, $\mathbf{\Gamma}_1$ and $\mathbf{\Gamma}_2$, such that

$$\mathbf{\Gamma}_1\mathbf{C}'\mathbf{G}\mathbf{\Gamma}_2 = \mathbf{D} = \begin{Vmatrix} \mathbf{D}_{\sqrt{\lambda}} & \mathbf{O} \\ \mathbf{O} & \mathbf{O} \end{Vmatrix}. \qquad (23)$$

* Hotelling (1936), pp. 326–30.

† $\mathbf{I}$ and $\mathbf{O}$ stand for a unit matrix and a zero matrix respectively.

The transformation of variables

$$\mathbf{Y} = (\mathbf{C}')^{-1}\mathbf{\Gamma}_1'\mathbf{U}\mathbf{\Gamma}_2', \quad \mathbf{Z} = (\mathbf{C}')^{-1}\mathbf{\Gamma}_1'\mathbf{V}, \tag{24}$$

where $$\mathbf{U} = \begin{Vmatrix} u_{11} & u_{12} & \cdots & u_{1n_1} \\ u_{21} & u_{22} & \cdots & u_{2n_1} \\ \cdots & \cdots & \cdots & \cdots \\ u_{p1} & u_{p2} & \cdots & u_{pn_1} \end{Vmatrix}, \quad \mathbf{V} = \begin{Vmatrix} v_{11} & v_{12} & \cdots & v_{1n} \\ v_{21} & v_{22} & \cdots & v_{2n} \\ \cdots & \cdots & \cdots & \cdots \\ v_{p1} & v_{p2} & \cdots & v_{pn} \end{Vmatrix}$$

are the matrices of the new variables, has a constant Jacobian and leaves the θ's invariant, because the equation (15) becomes

$$|a'_{ij} - \theta(a'_{ij} + b'_{ij})| = 0, \tag{15'}$$

where $$a'_{ij} = \sum_{r=1}^{n_1} u_{ir}u_{jr}, \quad b'_{ij} = \sum_{r=1}^{n} v_{ir}v_{jr}, \quad (i, j = 1, 2, \ldots, p).$$

To show this, we have, remembering the orthogonality of $\mathbf{\Gamma}_2$,

$$\|a_{ij}\| = \mathbf{YY}' = (\mathbf{C}')^{-1}\mathbf{\Gamma}_1'\mathbf{UU}'\mathbf{\Gamma}_1\mathbf{C}^{-1},$$
$$\|b_{ij}\| = \mathbf{ZZ}' = (\mathbf{C}')^{-1}\mathbf{\Gamma}_1'\mathbf{VV}'\mathbf{\Gamma}_1\mathbf{C}^{-1}.$$

Hence the equation (15) is transformed into

$$|\mathbf{UU}' - \theta(\mathbf{UU}' + \mathbf{VV}')| = 0,$$

which is another way of writing (15′).

On the other hand, substituting (24) into (22) and remembering (23) we get the expression

$$-\tfrac{1}{2}\Psi - \tfrac{1}{2}\operatorname{tr}(\mathbf{UU}') - \tfrac{1}{2}\operatorname{tr}(\mathbf{VV}') + \operatorname{tr}(\mathbf{DU}')$$
$$= -\tfrac{1}{2}\Psi - \frac{1}{2}\sum_{i=1}^{p} a'_{ii} - \frac{1}{2}\sum_{i=1}^{p} b'_{ii} + \sum_{i=1}^{l} \sqrt{\lambda_i}\, u_{ii}.$$

Replacing again the letters u and v by y and z respectively, we obtain the result.

It may be noticed that

$$\Psi = \operatorname{tr}(\mathbf{CC}'\mathbf{GG}') = \operatorname{tr}(\mathbf{C}'\mathbf{GG}'\mathbf{C}) = \sum_{i=1}^{l} \lambda_i. \tag{25}$$

Remark. For the case of k samples (see Example on p. 221) Fisher (1938) has considered the problem of testing for the colinearity or coplanarity of the k populations. It can be proved that the hypothesis of colinearity (or coplanarity) states that all except one (or two) of the λ's vanish.

6. *Behaviour of V when H_0 is not necessarily true.* The Laplace transform of a probability density function $p(x)$ vanishing identically for $x < 0$, viz. the integral

$$\int_0^\infty e^{-\alpha x} p(x)\, dx,$$

has the following property.

LEMMA 2. *Let $p_n(x)$, $(n=1,2,3,\ldots)$, and $p(x)$ be probability density functions vanishing identically for $x<0$, and such that*

$$\lim_{n\to\infty}\int_0^\infty e^{-\alpha x}p_n(x)\,dx=\int_0^\infty e^{-\alpha x}p(x)\,dx \tag{26}$$

for every $\alpha>0$. Then $$\lim_{n\to\infty}\int_0^x p_n(u)\,du=\int_0^x p(u)\,du.$$

Proof. The function $f_n(x)=p(x)-p_n(x)$ is summable in $(0,\infty)$ and, by (26),

$$\lim_{n\to\infty}\int_0^\infty e^{-\alpha x}f_n(x)\,dx=0 \quad \text{for every } \alpha>0. \tag{27}$$

The function $$g_n(z)=\int_0^\infty e^{-zx}f_n(x)\,dx$$

is an analytic function of z, regular at every z with a positive real part and uniformly bounded. By (27) $g_n(z)$ tends to a limit whenever z is real and positive. Hence by Vitali's theorem of convergence* $g_n(z)$ tends to a limit uniformly in the half-plane on the right of the imaginary axis. This limit is an analytic function regular at every z with a positive real part and vanishes whenever z is real and positive; therefore it vanishes identically. Hence

$$\lim_{n\to\infty}\int_0^\infty e^{-\alpha x+itx}p_n(x)\,dx=\int_0^\infty e^{-\alpha x+itx}p(x)\,dx \tag{28}$$

for all $\alpha>0$ and real t. Let

$$L_n(\alpha)=\int_0^\infty e^{-\alpha x}p_n(x)\,dx,\quad L(\alpha)=\int_0^\infty e^{-\alpha x}p(x)\,dx,$$

so that $$\lim_{n\to\infty}L_n(\alpha)=L(\alpha). \tag{29}$$

It follows from (28) and (29) that

$$\lim_{n\to\infty}\frac{1}{L_n(\alpha)}\int_0^\infty e^{-\alpha u+itu}p_n(u)\,du=\frac{1}{L(\alpha)}\int_0^\infty e^{-\alpha u+itu}p(u)\,du,$$

whence, by a well-known property of the characteristic function,

$$\lim_{n\to\infty}\frac{1}{L_n(\alpha)}\int_0^x e^{-\alpha u}p_n(u)\,du=\frac{1}{L(\alpha)}\int_0^x e^{-\alpha u}p(u)\,du,$$

whence $$\lim_{n\to\infty}\int_0^x e^{-\alpha u}p_n(u)\,du=\int_0^x e^{-\alpha u}p(u)\,du,$$

whence $$\lim_{\alpha\to+0}\lim_{n\to\infty}\int_0^x e^{-\alpha u}p_n(u)\,du=\int_0^x p(u)\,du.$$

The order of the above repeated limit can be interchanged because

$$\int_0^x e^{-\alpha u}p_n(u)\,du\to\int_0^x p_n(u)\,du$$

uniformly in n as $\alpha\to+0$. This completes the proof.

* Titchmarsh (1932), p. 168.

We shall now derive an expression for the Laplace integral $\mathscr{E}(\exp(-\alpha V))$. To simplify the notation we shall use a single letter, e.g. y to denote a set of variables, e.g. all the y_{ir}, when they figure as arguments of a function. We shall write dy for $\Pi\, dy$ and a single integration sign for multiple integrals with respect to the variables from $-\infty$ to ∞. We denote the integral $\mathscr{E}(\exp(-\alpha V))$ by $L(\alpha)$.

The following equation can be verified by direct integration, on referring to (10) as the definition of V:

$$\exp(-\tfrac{1}{2}\alpha^2 V) = \int f(y, z, \alpha, t)\, dt,$$

where

$$f(y,z,\alpha,t) = (2\pi)^{-\frac{1}{2}pn_1} |\, b_{ij}\,|^{\frac{1}{2}n_1} \exp\left\{-\frac{1}{2}\sum_{i,j=1}^{p} b_{ij}s_{ij} + i\alpha \sum_{i=1}^{p}\sum_{r=1}^{n_1} y_{ir}t_{ir}\right\},$$

$$s_{ij} = \sum_{r=1}^{n_1} t_{ir}t_{jr} \quad (i, j = 1, 2, \ldots, p),$$

and where the letter i, when not figuring as an index, stands for $\sqrt{-1}$. Denoting by $p(y, z)$ the coefficient of $\Pi\, dy\, dz$ in (21), we have

$$L(\tfrac{1}{2}\alpha^2) = \int p(y,z)\, dy\, dz \int f(y,z,\alpha,t)\, dt = \int dt \int p(y,z) f(y,z,\alpha,t)\, dy\, dz, \quad (30)$$

as the change of the order of the integration is obviously legitimate for every real α.

By direct substitution it results

$$\int p(y,z) f(y,z,\alpha,t)\, dy\, dz = (2\pi)^{-\frac{1}{2}p(2n_1+n)} \exp(-\tfrac{1}{2}\Psi) f_1(\alpha,t) f_2(\alpha,t), \quad (31)$$

where
$$f_1(\alpha,t) = \int \exp\left\{-\frac{1}{2}\sum_{i=1}^{p} a_{ii} + \sum_{i=1}^{l} \sqrt{\lambda_i}\, y_{ii} + i\alpha \sum_{i=1}^{p}\sum_{r=1}^{n_1} t_{ir}y_{ir}\right\} dy, \quad (32)$$

$$f_2(\alpha,t) = \int |\, b_{ij}\,|^{\frac{1}{2}n_1} \exp\left\{-\frac{1}{2}\sum_{i=1}^{p} b_{ii} - \frac{1}{2}\sum_{i,j=1}^{p} s_{ij}b_{ij}\right\} dz. \quad (33)$$

The integral (32) is readily evaluated and runs

$$f_1(\alpha,t) = (2\pi)^{\frac{1}{2}pn_1} \exp(\tfrac{1}{2}\Psi) \exp\left\{-\tfrac{1}{2}\alpha^2 \sum_{i=1}^{p}\sum_{r=1}^{n_1} t_{ir}^2 - i\alpha \sum_{i=1}^{l} \sqrt{\lambda_i}\, t_{ii}\right\}. \quad (34)$$

The integral (33) can be evaluated by using Wilks's formula for the moments of the generalized variance.* The result is

$$f_2(\alpha,t) = 2^{\frac{1}{2}pn_1}(2\pi)^{\frac{1}{2}pn} K\, |\, \delta_{ij} + s_{ij}\,|^{-\frac{1}{2}(n_1+n)}, \quad (35)$$

where $\delta_{ij} = 1$ if $i = j$ and 0 otherwise, and where

$$K = \prod_{i=1}^{p} \frac{\Gamma\frac{1}{2}(n_1+n-i+1)}{\Gamma\frac{1}{2}(n-i+1)}.$$

* Wilks (1932).

From (34), (35), (31) and (30) we obtain

$$L(\tfrac{1}{2}\alpha^2) = \pi^{-\frac{1}{2}pn_1} K \int |\delta_{ij}+s_{ij}|^{-\frac{1}{2}(n_1+n)} \exp\left\{-\tfrac{1}{2}\alpha^2 \sum_{i=1}^{p} s_{ii} - i\alpha \sum_{i=1}^{l} \sqrt{\lambda_i}\, t_{ii}\right\} dt,$$

whence

$$L(\alpha) = \pi^{-\frac{1}{2}pn_1} K \int |\delta_{ij}+s_{ij}|^{-\frac{1}{2}(n_1+n)} \exp\left\{-\alpha \sum_{i=1}^{p} s_{ii} - i\sqrt{(2\alpha)} \sum_{i=1}^{l} \sqrt{\lambda_i}\, t_{ii}\right\} dt. \quad (36)$$

It may be observed that the right-hand side of (36) involves no sum of n terms; hence it is particularly useful when we wish to make n approach infinity.

We may calculate the moments of V by means of (36). Thus

$$\begin{aligned} \mathscr{E}(V) &= \pi^{-\frac{1}{2}pn_1} K \int |\delta_{ij}+s_{ij}|^{-\frac{1}{2}(n_1+n)} \left\{\sum_{i=1}^{p} s_{ii} + \left(\sum_{i=1}^{l} \sqrt{\lambda_i}\, t_{ii}\right)^2\right\} dt \\ &= \frac{1}{n-p-1}(\Psi + pn_1), \end{aligned} \quad (37)$$

$$\begin{aligned} \mathscr{E}(V^2) &= \pi^{-\frac{1}{2}pn_1} K \int |\delta_{ij}+s_{ij}|^{-\frac{1}{2}(n_1+n)} \left\{\left(\sum_{i=1}^{p} s_{ii}\right)^2 + 2\left(\sum_{i=1}^{p} s_{ii}\right)\left(\sum_{i=1}^{l} \sqrt{\lambda_i}\, t_{ii}\right)^2 \right. \\ &\qquad \left. + \frac{1}{3}\left(\sum_{i=1}^{l} \sqrt{\lambda_i}\, t_{ii}\right)^4\right\} dt \\ &= \frac{\Psi^2}{(n-p-1)(n-p-3)} - \frac{4\sum_{i \neq j} \lambda_i \lambda_j}{(n-p)(n-p-1)(n-p-3)} \\ &\quad + 2\Psi\left\{\frac{pn_1+2}{(n-p-1)(n-p-3)} - \frac{2(p-1)(n_1-1)}{(n-p)(n-p-1)(n-p-3)}\right\} \\ &\quad + \frac{pn_1(pn_1+2)}{(n-p-1)(n-p-3)} - \frac{2(p-1)(n_1-1)}{(n-p)(n-p-1)(n-p-3)}. \end{aligned} \quad (38)$$

Remark. Consider again the case of k samples (see Example on p. 221) and write V_k and Ψ_k correspondingly. It is seen from (37) that the statistic

$$(M-k-p-1)V_k - p(k-1)$$

is an unbiassed estimate of Ψ_k. We may regard V_k as a generalization of V_2, which is the "Studentized" D^2 statistic* but for a constant factor. V_2 is, of course, also identical with Hotelling's generalized "Student's" ratio except for a constant factor.

We shall next study the behaviour of V as $n \to \infty$. As n is now allowed to vary, we shall attach to various letters the index n. In particular we write Ψ_n for Ψ to emphasize the fact that the value of Ψ depends also on n, a fact which is not brought out by the formal definition of Ψ. This is because the η_{ir}, as we explained on p. 221, are themselves linear functions of the original population constants with coefficients depending upon n.

* Bose and Roy (1938).

It is in general true that, except when H_0 is true, in which case $\Psi_n \equiv 0$, Ψ_n is either $O(1)$ or $O(n)$ as $n \to \infty$. That both are possible is seen in the Example on p. 221 for $k = 2$; Ψ_n is $O(n)$ or $O(1)$ according as $m_1 = m_2 \to \infty$ or one of m_1, m_2 remains fixed while the other tends to infinity.

THEOREM 4. *If $\Psi_n = O(n)$, then the mean of V is $\frac{1}{n}\Psi_n + O\left(\frac{1}{n}\right)$ and the variance of V is $O\left(\frac{1}{n}\right)$. Hence the random variable $V - \frac{1}{n}\Psi_n \to 0$ in probability.*

This is an immediate consequence of (37) and (38).

If $\Psi_n = O(1)$, and if Ψ_n tends to a limit as $n \to \infty$, the limiting distribution of nV will be that of S (cf. (2)) with Ψ replaced by its limit. This is our next theorem. We assume now that

$$\lim_{n\to\infty} \Psi_n = \Psi_0. \tag{39}$$

The case $\Psi_n \equiv 0$ (i.e. H_0 is true) is covered by (39) on putting $\Psi_0 = 0$. But (39) may also be regarded as covering the case $\Psi_n = O(n)$ in the sense that the repeated sampling is not made from the same population, but from populations whose distribution constants vary with n in such a way that (39) holds true. Thus, for example, in the case of two samples with $m_1 = m_2$ (see p. 222) we have

$$\Psi_n = \tfrac{1}{2}m\Phi = \tfrac{1}{4}(n+2)\Phi,$$

where
$$\Phi = \sum_{i,j=1}^{p} \alpha_{ij}(\xi_{i1}-\xi_{i2})(\xi_{j1}-\xi_{j2}).$$

Here we have $\Psi_n = O(n)$ if the population constant $\Phi > 0$. Ψ_n will be $O(1)$ if we consider Φ as decreasing to the order n^{-1} as n increases indefinitely. This idea of regarding population constants as varying with the sample size can be found in the works of Fisher* and Neyman.†

THEOREM 5. *If Ψ_n tends to a limit Ψ_0 as $n \to \infty$, then*

$$\lim_{n\to\infty} Pr\{nV \leqslant x\} = \int_0^x p(x)\,dx, \tag{40}$$

where
$$p(x) = 2^{-\frac{1}{2}pn_1} x^{\frac{1}{2}pn_1-1} e^{-\frac{1}{2}(x+\Psi_0)} \sum_{h=0}^{\infty} \frac{\Psi_0^h x^h}{4^h h!\,\Gamma(h+\frac{1}{2}pn_1)}. \tag{41}$$

In particular, if $\Psi_0 = 0$ (i.e. H_0 is true), the limiting distribution of nV is that of χ^2 with pn_1 degrees of freedom.

Proof. Write $L_1^{(n)}(\alpha)$ for the Laplace transform of the probability density function of nV, so that

$$L_1^{(n)}(\alpha) = \mathscr{E}(\exp(-\alpha nV)) = L(n\alpha).$$

* Fisher (1928), p. 663.

† Neyman (1937), pp. 169–70, (1938), pp. 70–1.

If we replace α by $n\alpha$ in (36) and subsequently make the change of variables $t_{ir} = n^{-\frac{1}{2}}\tau_{ir}$, we get

$$L_1^{(n)}(\alpha) = \pi^{-\frac{1}{2}pn_1} n^{-\frac{1}{2}pn_1} K_n \int \left| \delta_{ij} + \frac{s'_{ij}}{n} \right|^{-\frac{1}{2}(n_1+n)} \exp\left\{ -\alpha \sum_{i=1}^{p} s'_{ii} - i\sqrt{(2\alpha)} \sum_{i=1}^{l} \sqrt{\lambda_i}\, \tau_{ii} \right\} d\tau, \tag{42}$$

where K_n is now written for K and

$$s'_{ij} = \sum_{r=1}^{n_1} \tau_{ir}\tau_{jr}, \quad (i, j = 1, 2, \ldots, p).$$

We shall now find the limit of $L_1^{(n)}(\alpha)$ as $n \to \infty$. We have, using Stirling's formula,

$$\lim_{n\to\infty} n^{-\frac{1}{2}pn_1} K_n = 2^{-\frac{1}{2}pn_1}. \tag{43}$$

For every system of fixed values of the τ_{ir} we have

$$\left| \delta_{ij} + \frac{s'_{ij}}{n} \right| = 1 + \frac{1}{n} \sum_{i=1}^{p} s'_{ii} + O\left(\frac{1}{n^2}\right),$$

whence

$$\lim_{n\to\infty} \left| \delta_{ij} + \frac{s'_{ij}}{n} \right|^{\frac{1}{2}(n_1+n)} = \exp\left(-\frac{1}{2} \sum_{i=1}^{p} s'_{ii} \right). \tag{44}$$

Calling I_n the integral in (42), we have

$$\begin{aligned} I_n = \int \left\{ \left| \delta_{ij} + \frac{s'_{ij}}{n} \right|^{-\frac{1}{2}(n_1+n)} - \exp\left(-\tfrac{1}{2}\textstyle\sum s'_{ii}\right) \right\} \exp\left\{ -\alpha \sum_{i=1}^{p} s'_{ii} - i\sqrt{(2\alpha)} \sum_{i=1}^{l} \sqrt{\lambda_i}\, \tau_{ii} \right\} d\tau \\ + \int \exp\left\{ -\left(\tfrac{1}{2} + \alpha\right) \sum_{i=1}^{p} s_{ii} - i\sqrt{(2\alpha)} \sum_{i=1}^{l} \sqrt{\lambda_i}\, \tau_{ii} \right\} d\tau \\ = A_n + B_n. \end{aligned} \tag{45}$$

The absolute value of the integrand of A_n is less than

$$2\exp\left(-\alpha \sum_{i=1}^{p} s'_{ii} \right).$$

Hence by (44) and dominated convergence, $A_n \to 0$ as $n \to \infty$. The value of B_n is easily found to be

$$B_n = (2\pi)^{\frac{1}{2}pn_1} (1 + 2\alpha)^{-\frac{1}{2}pn_1} \exp\left(-\frac{\alpha \Psi_n}{1 + 2\alpha} \right).$$

From (42), (43) and (45) we obtain the result

$$\lim_{n\to\infty} L_1^{(n)}(\alpha) = (1 + 2\alpha)^{-\frac{1}{2}pn_1} \exp\left(-\frac{\alpha \Psi_0}{1 + 2\alpha} \right). \tag{46}$$

If $p(x)$ is defined as in (41), then the integral

$$\int_0^\infty e^{-\alpha x} p(x)\, dx$$

is identically equal to the right-hand side of (46). Theorem 5 is thus proved on remembering Lemma 2.

7. *Behaviour of W when H_0 is not necessarily true and n is large.* We shall establish the following theorem.

THEOREM 6. *If $\Psi_n = O(1)$, the statistics nV and $-n\log W$ tend to be certainly identical in the sense that*

$$\lim_{n\to\infty}(nV + n\log W) = 0 \textit{ in probability.} \tag{47}$$

Corollary. If Ψ_n tends to a limit Ψ_0 as $n\to\infty$, $-n\log W$ has the same limiting distribution (40) *as nV as $n\to\infty$.*

Remark. Besides the Corollary there is another significance to be attached to Theorem 6. Since the test functions V and W are not functionally related (except when $p = 1$ or $n_1 = 1$), there is the question of choosing one of them to be consistently used in carrying out the actual tests. This can only be decided by a comparison of their power* to detect the falsehood of H_0 when the η_{ir} do not all vanish. While this may be a difficult problem for small samples, Theorem 6 appears to have answered the question for large samples. In fact, if n is large, it is almost certain that the values of nV and $-n\log W$ calculated from the sample will differ very little. That $-n\log W$ and nV tend to have the same power function is a consequence of the Corollary.

Proof of Theorem 6. For every θ such that $0\leqslant\theta\leqslant 1$ we have

$$0\leqslant \frac{\theta}{1-\theta}+\log(1-\theta)\leqslant \frac{1}{2}\left(\frac{\theta}{1-\theta}\right)^2.$$

Hence, remembering (12) and (13),

$$0\leqslant V+\log W\leqslant \frac{1}{2}\sum_{i=1}^{l_1}\left(\frac{\theta_i}{1-\theta_i}\right)^2\leqslant \frac{1}{2}\left(\sum_{i=1}^{l_1}\frac{\theta_i}{1-\theta_i}\right)^2 = \tfrac{1}{2}V^2,$$

whence, for every $\eta>0$,

$$\begin{aligned}Pr\{|\,nV+n\log W\,|>\eta\} &= Pr\{nV+n\log W>\eta\}\leqslant Pr\left\{\frac{n}{2}V^2>\eta\right\}\\ &\leqslant \frac{1}{\eta}\mathscr{E}\left(\frac{n}{2}V^2\right) = O\left(\frac{1}{n}\right)\end{aligned}$$

because of (38). This establishes (47).

Proof of the Corollary. Let

$$F_n(x) = Pr\{nV\leqslant x\},\quad G_n(x) = Pr\{-n\log W\leqslant x\}.$$

Then for every x and $\eta>0$ we have†

$$0\leqslant G_n(x)-F_n(x)\leqslant F_n(x+\eta)-F_n(x-\eta)+Pr\{nV+n\log W\geqslant\eta\}.$$

* Neyman & Pearson (1936, 1938).

† Fréchet (1937), p. 164.

Let $F(x)$ be the limit of $F_n(x)$ as $n \to \infty$, which is known to exist by virtue of Theorem 6. We have then

$$0 \leqslant G_n(x) - F_n(x) \leqslant | F_n(x+\eta) - F(x+\eta) | + | F_n(x-\eta) - F(x-\eta) | \\ + F(x+\eta) - F(x-\eta) + Pr\{nV + n \log W \geqslant \eta\}. \qquad (48)$$

Given any $\epsilon > 0$, choose and fix an $\eta > 0$ so small that $F(x+\eta) - F(x-\eta) < \epsilon$. By (47) the rest of the terms in the right-hand side of (48) is smaller than ϵ for all sufficiently large n. Hence $G_n(x) - F_n(x) \to 0$, i.e. $G_n(x) \to F(x)$, which completes the proof.

Summary

The Wilks-Lawley hypothesis concerning population means of multivariate normal populations is put in the canonical form H_0 (§ 1), and, assuming the population variances and covariances known, the test function S is derived together with its exact distribution (§ 2). In the case where the population variances and covariances are unknown, two possible test functions, denoted by V and W are considered (§§ 3 and 4), and their distributions in certain special cases are given. In § 5 it is shown that the sample space can be so transformed that all the variables are independently distributed and that only a minimum number of unknown parameters remain. These parameters are the roots of a certain determinantal equation; the hypothesis H_0, and the hypotheses of colinearity and coplanarity of populations, all specify the values zero for all or some of these parameters. Returning to the test functions V and W, it is shown that as the sample size n increases indefinitely the two functions nV and $-n \log W$ tend to be certainly identical ((47), § 7), and both of them tend to have the same distribution function as S ((40), § 6).

References

Bose, R. C. (1936). "On the exact distribution of the D^2-statistic." *Sankhya*, **2**, 143–54.

Bose, R. C. & Roy, S. N. (1938). "The distribution of the 'Studentized' D^2-statistic." *Proceedings of the first session of the Indian Statistical Conference, Calcutta*, pp. 19–38.

Fisher, R. A. (1928). "The general sampling distribution of the multiple correlation coefficient." *Proc. Roy. Soc.* **A**, **121**, 653–73.

—— (1938). "The statistical utilization of multiple measurements." *Ann. Eugen., Lond.*, **8**, 376–86.

—— (1939). "The sampling distribution of some statistics obtained from non-linear equations." *Ann. Eugen., Lond.*, **9**, 238.

Fréchet, M. (1937). *Généralités sur les probabilités. Variables aléatoires.* Paris.

Hotelling, H. (1931). "The generalization of Student's ratio." *Ann. Math. Statist.* **2**, 359–78.

—— (1936). "Relations between two sets of variates." *Biometrika*, **28**, 321–77.

Hsu, P. L. (1938). "Notes on Hotelling's generalized T." *Ann. Math. Statist.* **9**, 231–43.

—— (1939). "On the distribution of roots of certain determinantal equations." *Ann. Eugen., Lond.*, **9**, 250.

Kolodziezczyk, St (1935). "On an important class of statistical hypotheses." *Biometrika*, **27**, 161–90.

Lawley, D. N. (1938). "A generalization of Fisher's z-test." *Biometrika*, **30**, 180–8.

Mahalanobis, P. C. (1936). "On the generalized distance in statistics." *Proc. Nat. Inst Sci. India*, **2**, 49–55.

Neyman, J. (1937). "Smooth test for goodness of fit." *Skand. AktuarTidskr.* pp. 149–99.

—— (1938). "Tests of statistical hypotheses which are unbiassed in the limit." *Ann. Math. Statist.* **9**, 69–86.

Neyman, J. & Pearson, E. S. (1928). "On the use and interpretation of certain test criteria for purposes of statistical inference." *Biometrika*, **20** A, 175–240.

—— —— (1936, 1938). "Contributions to the theory of testing statistical hypotheses. I." *Statist. Res. Mem.* **1**, 1–37. "Contributions to the theory of testing statistical hypotheses. II." *Statist. Res. Mem.* **2**, 25–57.

Pearson, E. S. & Wilks, S. S. (1933). "Methods of statistical analysis appropriate for k samples of two variables." *Biometrika*, **25**, 353–78.

Tang, P. C. (1938). "The power function of the analysis of variance tests." *Statist. Res. Mem.* **2**, 126–49.

Titchmarsh, E. C. (1932). *The Theory of Functions.* Oxford University Press.

Wilks, S. S. (1932). "Certain generalization in the analysis of variance." *Biometrika*, **24**, 471–94.

Reprinted from
J. London Math. Soc.
16, 183–194 (1941).

ON THE LIMITING DISTRIBUTION OF ROOTS OF A DETERMINANTAL EQUATION

P. L. Hsu*.

1. *Preliminary.* Consider k p-variate populations and the matrix

$$\left\| \begin{matrix} \xi_{11} \; \xi_{21} \cdots \xi_{p1} \; 1 \\ \xi_{12} \; \xi_{22} \cdots \xi_{p2} \; 1 \\ \cdots \quad \cdots \quad \cdots \\ \xi_{1k} \; \xi_{2k} \cdots \xi_{pk} \; 1 \end{matrix} \right\|, \tag{1}$$

where ξ_{it} is the mean of the i-th variate of the t-th population. Modern research in multivariate statistics has culminated in R. A. Fisher's discriminant analysis†, in connection with which a primary problem is to test, by means of samples, whether the rank of (1) is equal to some hypothetical value. The geometrical meaning of the rank is obvious: if the rank is 1, the k centroids $(\xi_{1t}, \ldots, \xi_{pt})$ are coïncident; if it is 2, the centroids are collinear but not coincident, and so on.

We shall assume that the populations are normal ones with the same set of second moments about the means which we shall denote by

$$\sigma_{ij} = \sigma_{ji} \quad (i, j = 1, \ldots, p).$$

Suppose that k samples, of sizes $m_1, \ldots, m_k$, are drawn respectively from the k populations. If we put

$$\xi_i = \frac{1}{N} \sum_{t=1}^{k} m_t \xi_{it}, \quad \psi_{ij} = \frac{1}{N} \sum_{t=1}^{k} m_t (\xi_{it} - \xi_i)(\xi_{jt} - \xi_j), \quad N = \sum_{t=1}^{k} m_t, \tag{2}$$

it may be readily seen that the rank of the matrix $\|\psi_{ij}\|$ is *one* less than that of (1). The rank of $\|\psi_{ij}\|$ is in turn equal to the number of positive roots of the determinantal equation

$$|\psi_{ij} - \lambda \sigma_{ij}| = 0, \tag{3}$$

---

* Received 24 March, 1940; received 9 May, 1940.
† Fisher (1938, 1939).

because clearly the matrix $\|\psi_{ij}\|$ is positive, though possibly indefinite, and the matrix $\|\sigma_{ij}\|$ is positive definite. The problem of rank is thus equivalent to the problem of testing whether (3) has a root zero of a given multiplicity, and may therefore be regarded as a special case of a wider class of problems, namely the problems of testing hypotheses concerning the true values of the roots of (3) and of estimating these roots.

Consider again the k samples. Denoting by $x_{1t}, \ldots, x_{pt}$ the means of the t-th sample, by $x_1, \ldots, x_p$ the grand sample means, by s_{ijt} $(i, j = 1, \ldots, p)$ the second moments about the means of the t-th sample, and writing

$$\bar{a}_{ij} = \sum_{t=1}^{k} m_t(x_{it}-x_i)(x_{jt}-x_j), \quad \bar{b}_{ij} = \sum_{t=1}^{k} m_t s_{ijt},$$

$$l_1 = \min(p, k-1), \quad l_2 = \max(p, k-1),$$

we may readily see that the matrix $\|\bar{a}_{ij}\|$ is positive and of rank l_1,* and that, provided that $N-k \geqslant p$, the matrix $\|\bar{b}_{ij}\|$ is positive definite*. Hence the determinantal equation in ϕ

$$|\bar{a}_{ij} - \phi\bar{b}_{ij}| = 0 \tag{4}$$

has a root zero of multiplicity $p-l_1$ and l_1 real positive roots.

The l_1 roots of (4) which are not zero play an important part in discriminant analysis†. Their distribution depends solely on the roots of (3) (see § 3, first paragraph) and their exact distribution is known‡ in the case where all the roots of (3) vanish. In this paper we obtain the limiting distribution for the most general case, as the sample sizes become infinite while their ratios remain constant.

2. *A lemma on limiting distributions.*

Lemma 1. *Let Q_n $(n = 1, 2, \ldots)$ be a random point with a finite number, independent of n, of coordinates, and let its domain of existence be the Borel*

* More exactly, the probability that the rank of $\|\bar{a}_{ij}\|$ is less than l_1, or that $\|\bar{b}_{ij}\|$ is singular, is zero.

† It may be shown that these roots are what Fisher termed the "components" of discrimination in his paper of 1938.

‡ Hsu (1939). For the case $l_1 = p$, the same distribution is also obtained by Fisher (1939) and Roy (1939).

set E_n. Let

$$E_n \subset E_{n+1} \quad (n = 1, 2, \ldots) \tag{5}$$

and put

$$\lim_{n\to\infty} E_n = E. \tag{6}$$

Let the probability function of Q_n approach a limiting probability function, which is continuous, as $n \to \infty$.

Let $f_n(P)$ $(n = 1, 2, \ldots)$ be a real point function defined and Borel measurable in E_n. Let

$$\lim_{n\to\infty} f_n(P) = 0 \text{ throughout } E.^* \tag{7}$$

Then the random variable $f_n(Q_n) \to 0$ i.pr.† as $n \to \infty$.

Proof. Let $\eta > 0$ be arbitrarily chosen, and let p_n be the probability that $|f_n(Q_n)| \leqslant \eta$. It is required to prove that $p_n \to 1$ as $n \to \infty$.

Let W_n be the probability function of Q_n and let W be its limit. By (6), $W(E) = 1$. Writing

$$G_n = S\{|f_n(P)| \leqslant \eta;\ P \in E_m\},\ddagger$$

we have, by definition,

$$p_n = W_n(G_n). \tag{8}$$

Let a positive integer m be arbitrarily chosen and fixed. Then, by (5),

$$G_n \supset S\{|f_n(P)| \leqslant \eta;\ P \in E_m\} \quad \text{for} \quad n \geqslant m. \tag{9}$$

On the set E_m we define a function $g(P)$ which is *the smallest integer such that $n \geqslant m$ and $n \geqslant g(P)$ implies that $|f_n(P)| \leqslant \eta$.* $g(P)$ exists because of (7), and is obviously unique. Clearly $g(P) = 1$ or $g(P) \geqslant m+1$. $g(P)$ is Borel measurable since it assumes only the values $1, m+1, m+2, \ldots,$ and since

$$S\{g(P) = 1\} = \prod_{n=m}^{\infty} S\{|f_n(P)| \leqslant \eta;\ P \in E_m\}$$

---

* If P' is any point of E, $f_n(P')$ is defined for all sufficiently large n. Hence the meaning of (7) is clear.

† "i.pr." stands for "in probability."

‡ $S\{\ldots\}$ denotes the set of points having the property stated in the brackets.

and

$$S\{g(P)=n_0\}=S\{|f_{n_0-1}(P)|>\eta\}\prod_{n=n_0}^{\infty}S\{|f_n(P)|\leqslant\eta\}$$

$$(n_0=m+1,\ m+2,\ \ldots).$$

From (9) and the definition of $g(P)$ it follows that

$$G_n\supset S\{g(P)\leqslant n;\ P\in E_m\}$$

for $n\geqslant m$, whence

$$G_n\supset S\{g(P)\leqslant r;\ P\in E_m\}$$

for $n\geqslant r\geqslant m$. Let the last set be H_{mr}; since this is a Borel set, we have, by (8),

$$p_n\geqslant W_n(H_{mr}). \tag{10}$$

Since the limiting probability function W is continuous, the relation $W_n(S)\to W(S)$ is true for every Borel set S. Hence (10) gives

$$\varliminf_{n\to\infty} p_n\geqslant W(H_{mr}).$$

Hence

$$1\geqslant\varliminf_{n\to\infty} p_n\geqslant\lim_{m\to\infty}\lim_{r\to\infty}W(H_{mr})=\lim_{m\to\infty}W(E_m)=W(E)=1,$$

and the lemma is proved.

3. *The limiting distribution of the roots of* (4).

As mentioned at the beginning, the underlying assumption is that all the populations are normal and that their sets of second moments about the means are the same. It is known* that if $\|\psi_{ij}\|$ is of rank r and if $\lambda'_1, \ldots, \lambda'_r$ are the positive roots of (3), then the distribution of the roots of (4) is the same as the distribution, derived from the density

$$(2\pi)^{-\frac{1}{2}(n_1+n)}\exp\left\{-\tfrac{1}{2}\left(\sum_{i=1}^{p}\sum_{\beta=1}^{n_1}y_{i\beta}^2+\sum_{i=1}^{p}\sum_{\gamma=1}^{n}z_{i\gamma}^2-2\sqrt{N}\sum_{i=1}^{r}\sqrt{\lambda_i'}\,y_{ii}+N\sum_{i=1}^{r}\lambda_i'\right)\right\}, \tag{11}$$

of the roots of the equation

$$|a_{ij}-\phi b_{ij}|=0, \tag{12}$$

* Hsu (1940), 227, Theorem 3. Notice that the λ_i there are the $N\lambda_i'$ of this paper.

where
$$n_1 = k-1, \quad n = N-k,$$
$$a_{ij} = \sum_{\beta=1}^{n_1} y_{i\beta} y_{j\beta}, \quad b_{ij} = \sum_{\gamma=1}^{n} z_{i\gamma} z_{j\gamma}. \tag{13}$$

Let the sample sizes be
$$m_t = mq_t \quad (t = 1, \ldots, k),$$
and write $q = \Sigma q_t$, so that $N = mq$. We notice first that, by their definition (2), the ψ_{ij}, and therefore the roots λ', remain constant as m varies. Let the distinct roots be $\lambda_1, \ldots, \lambda_\nu$, of multiplicities $\mu_1, \ldots, \mu_\nu$ respectively. Put
$$a_0 = 0, \quad a_1 = \mu_1, \quad a_h = \mu_1 + \ldots + \mu_h \quad (h = 1, \ldots, \nu), \quad r = a_\nu,$$
and arrange the roots in order of descending magnitude
$$\lambda'_{a_{h-1}+1} = \ldots = \lambda'_{a_h} = \lambda_h \quad (h = 1, \ldots, \nu), \tag{14}$$
$$\lambda_1 > \ldots > \lambda_\nu > 0.$$

We now substitute $y_{ii} + \sqrt{(N\lambda_i')}$ for y_{ii} $(i = 1, \ldots, r)$ in (11) and modify the elements of (12) accordingly. The results are the new parent distribution represented by the density
$$(2\pi)^{-\frac{1}{2}(n_1+n)} \exp\left\{-\tfrac{1}{2}\left(\sum_{i=1}^{p} \sum_{\beta=1}^{n_1} y_{i\beta}^2 + \sum_{i=1}^{p} \sum_{\gamma=1}^{n} z_{i\gamma}^2\right)\right\} \tag{15}$$
and a new equation in ϕ, say $D(\phi) = 0$; the first element, for instance, of $D(\phi)$ is $a_{11} + 2\sqrt{(N\lambda_1')}\, y_{11} + N\lambda_1' - \phi b_{11}$. In $D(\phi)$ we put
$$b_{ii} = N + \sqrt{(2N)}\, u_{ii}, \quad b_{ij} = \sqrt{(N)}\, u_{ij} \quad (i \neq j), \tag{16}$$
$$v = N^{-\frac{1}{2}},$$
and divide each element by N. Taking (14) into account, we get the equation
$$|v^2 \mathbf{A} + v\mathbf{C} - v\phi\mathbf{U} + \mathbf{D}| = 0, \tag{17}$$
where
$$\mathbf{A} = \begin{Vmatrix} a_{11} & \ldots & a_{1p} \\ \ldots & \ldots & \ldots \\ a_{1p} & \ldots & a_{pp} \end{Vmatrix}, \qquad \mathbf{U} = \begin{Vmatrix} \sqrt{(2)}\, u_{11} & \ldots & u_{1p} \\ \ldots & \ldots & \ldots \\ u_{1p} & \ldots & \sqrt{(2)}\, u_{pp} \end{Vmatrix},$$
$$\mathbf{D} = \begin{Vmatrix} (\lambda_1 - \phi)\,\mathbf{I}_{\mu_1} & & & \\ & \ddots & & \\ & & (\lambda_\nu - \phi)\,\mathbf{I}_{\mu_\nu} & \\ & & & -\phi \mathbf{I}_{p-r} \end{Vmatrix}$$

is a diagonal matrix, $\mathbf{I}_\mu$ standing for the μ-rowed unit matrix*, and

$$\mathbf{C} = \begin{Vmatrix} \mathbf{C}_{11} \dots \mathbf{C}_{1\nu} \; \mathbf{E}_1' \\ \dots \quad \dots \quad \dots \\ \mathbf{C}_{1\nu}' \dots \mathbf{C}_{\nu\nu}' \; \mathbf{E}_\nu' \\ \mathbf{E}_1 \; \dots \mathbf{E}_\nu \; \mathbf{O} \end{Vmatrix}.$$

Here the accent denotes the transposed matrix. $\mathbf{O}$ denotes a zero matrix,

$$\mathbf{C}_{hh} = \sqrt{(\lambda_h)} \| y_{ij} + y_{ji} \| \quad (i, j = a_{h-1}+1, \dots, a_h),$$

$$\mathbf{E}_h = \sqrt{(\lambda_h)} \| y_{ij} \| \quad (i = r+1, \dots, p; \; j = a_{h-1}+1, \dots, a_h).$$

The elements of $\mathbf{C}_{hg}$ for $h \neq g$, which are independent of ϕ and which do not involve v explicitly, are not written, since they are not used later. Our problem is now reduced to finding the limit as $m \to \infty$ of the distribution of the roots of (17) derived from (15).

LEMMA 2. *The limiting distribution of the u_{ij} in* (16) *is that of $\frac{1}{2}p(p+1)$ mutually independent normal variates with the same mean* 0 *and standard deviation* 1.

Proof. If $v_{ii} = (2n)^{-\frac{1}{2}}(b_{ii} - n)$ and $v_{ij} = n^{-\frac{1}{2}} b_{ij}$ $(i \neq j)$, then it follows immediately from (15) and the well-known central limit theorem† that the v_{ij} have the limiting distribution of Lemma 2. But $N \sim n$ and $u_{ij} = v_{ij} + \Delta_{ij}$, where

$$\Delta_{ii} = \frac{k}{\sqrt{(2N)}} + \left\{ \sqrt{\left(\frac{n}{N}\right)} - 1 \right\} v_{ii},$$

$$\Delta_{ij} = \left\{ \sqrt{\left(\frac{n}{N}\right)} - 1 \right\} v_{ij} \quad (i \neq j).$$

Each $\Delta_{ii} \to 0$ i.pr. as a trivial case of Lemma 1. Hence‡ the u_{ij} have the same limiting distribution.

---

* We also write $\mathbf{I}$ for a unit matrix of unspecified order.

† *Cf.* Cramér (1937), 113–114.

‡ If $y_n^{(1)}, \dots, y_n^{(\mu)}$ have a limiting distribution and if $z_n^{(i)} \to 0$ i.pr. $(i = 1, \dots, \mu)$ as $n \to \infty$, then $y_n^{(1)} + z_n^{(1)}, \dots, y_n^{(\mu)} + z_n^{(\mu)}$ have the same distribution. This proposition is proved by Doob (1935) for the case $\mu = 1$; the extension of the proof to the case $\mu > 1$ is so easy that it is left to the reader.

We now come to the main lemma.

LEMMA 3. *Let $\phi_1, \ldots, \phi_{l_1}$ be the roots, other than the zero root of multiplicity $p-l_1$, of (4), arranged in descending order of magnitude. Let*

$$\phi_i = \lambda_h + N^{-\frac{1}{2}}(2\lambda_h^2+4\lambda_h)^{\frac{1}{2}}\zeta_i \quad (i = a_{h-1}+1, \ldots, a_h;\ h = 1, \ldots, \nu), \tag{18}$$

$$\phi_i = N^{-1}\zeta_i \quad (i = r+1, \ldots, l_1). \tag{19}$$

Then, as $m \to \infty$, the sets $(\zeta_1, \ldots, \zeta_{a_1}), \ldots, (\zeta_{a_{\nu-1}+1}, \ldots, \zeta_r)$ and $(\zeta_{r+1}, \ldots, \zeta_{l_1})$ tend to be distributed independently of one another; for $h = 1, \ldots, \nu$ the limiting distribution of $\zeta_{a_{h-1}+1}, \ldots, \zeta_{a_h}$ is the same as the distribution of the latent roots, in descending order, of the μ_h-rowed symmetric matrix $\|w_{ij}\|$ derived from the density

$$2^{-\frac{1}{2}\mu_h}\pi^{-\frac{1}{4}\mu_h(\mu_h+1)}\exp\{-\tfrac{1}{2}(w_{11}^2+\ldots+w_{\mu_h,\mu_h}^2+2w_{12}^2+\ldots+2w_{\mu_h-1,\mu_h}^2)\}; \tag{20}$$

and the limiting distribution of $\zeta_{r+1}, \ldots, \zeta_{l_1}$ is the same as the distribution of the latent roots, in descending order, of the symmetric matrix $\|d_{ij}\|$ whose elements are

$$d_{ij} = \sum_{s=1}^{n_1-r} x_{is}x_{js} \quad (i, j = 1, \ldots, p-r), \tag{21}$$

where the x's are mutually independent normal variates with the same mean 0 and standard deviation 1.

Proof. We regard the ϕ's as the roots of (17) and the parent distribution as given by the density (15). We adopt the following method of arranging the roots of (17) in a definite order, say $\phi_1', \ldots, \phi_{l_1}'$, not necessarily the descending order of magnitude. In (17) we regard the u's and y's as fixed and consider the variation of the roots with v. The roots are classified into $\nu+1$ sets, containing $\mu_1, \ldots, \mu_\nu, l_1-r$ members respectively; those of the h-th set are $\lambda_h+o(1)$ $(h = 1, \ldots, \nu)$, and those of the last set are $o(1)$, for small v. Let ϕ' be any of the roots of the first set. Putting $\phi' = \lambda_1+v\eta$, substituting in (17), cancelling the common factor v from the first μ_1 rows, and finally making $v \to 0$, we get the equation

$$|\mathbf{C}_{11}-\lambda_1\mathbf{U}_{11}-\eta\mathbf{I}| = 0, \tag{22}$$

where $\mathbf{U}_{11}$ is the μ_1-rowed sub-matrix in the upper left corner of $\mathbf{U}$.

Let the roots of (22), arranged in descending order, be $\bar{\eta}_1, \ldots, \bar{\eta}_{\mu_1}$. Then the roots of the first set are $\lambda_1+v\eta_i+o(v)$. If we write

$$\phi_i' = \lambda_1+v\bar{\eta}_i+o(v) \quad (i = 1, \ldots, \mu_1)$$

then *the roots* $\phi_1', \ldots, \phi'_{\mu_1}$ *are completely specified.* For we may neglect the possibility that (22) has multiple roots, since the probability that this is so is zero. In the same manner we arrange the roots of the first ν sets in a definite order on putting

$$\phi_i' = \lambda_h + v\bar{\eta}_i + o(v) \quad (i = a_{h-1}+1, \ldots, a_h;\ h = 1, \ldots, \nu), \tag{23}$$

where $\bar{\eta}_{a_{h-1}+1}, \ldots, \eta_{a_h}$ are the roots, in descending order, of the equation

$$|\mathbf{C}_{hh} - \lambda_h \mathbf{U}_{hh} - \eta \mathbf{I}| = 0, \tag{24}$$

in which $\mathbf{U}_{hh}$ stands for the μ_h-rowed sub-matrix of $\mathbf{U}$ occupying the same position as $\mathbf{C}_{hh}$ in $\mathbf{C}$.

Consider now the roots of the last set, and let ϕ' be any one of them. Putting $\phi' = v^2\zeta$, substituting this in (17), cancelling the common factor v from the last $p-r$ rows and columns, and finally letting $v \to 0$, we obtain the equation

$$\begin{vmatrix} \lambda_1 \mathbf{I}_{\mu_1} & \cdots & \mathbf{O} & \mathbf{E}_1' \\ \cdots & \cdots & \cdots & \cdots \\ \mathbf{O} & \cdots & \lambda_\nu \mathbf{I}_{\mu_\nu} & \mathbf{E}_\nu' \\ \mathbf{E}_1 & \cdots & \mathbf{E}_\nu & \bar{\mathbf{A}} - \zeta \mathbf{I} \end{vmatrix} = 0,$$

where $\bar{\mathbf{A}}$ is the $(p-r)$-rowed sub-matrix in the lower right corner of $\mathbf{A}$. This equation may be written as

$$\left| \bar{\mathbf{A}} - \frac{1}{\lambda_1} \mathbf{E}_1 \mathbf{E}_1' - \cdots - \frac{1}{\lambda_\nu} \mathbf{E}_\nu \mathbf{E}_\nu' - \zeta \mathbf{I} \right| = 0,$$

or, on recalling the elements of $\bar{\mathbf{A}}$ and $\mathbf{E}_h$,

$$|e_{ij} - \zeta \delta_{ij}| = 0, \tag{25}$$

where $\delta_{ii} = 1$, $\delta_{ij} = 0$ $(i \neq j)$ and

$$e_{ij} = \sum_{\beta=1}^{n_1-r} y_{i\beta} y_{j\beta} \quad (i, j = r+1, \ldots, p). \tag{26}$$

The rank of $\|d_{ij}\|$ being $l_1 - r$, the equation (25) has necessarily a zero root of multiplicity $p - l_1$. Let the other roots, in descending order, be $\bar{\zeta}_{r+1}, \ldots, \bar{\zeta}_{l_1}$. Then, by the same reasoning as before, the equations

$$\phi'_i = v^2 \bar{\zeta}_i + o(v^2) \quad (i = r+1, \ldots, l_1) \tag{27}$$

completely specify each individual root which belongs to the last set.

To summarize the results, each member of the roots $\phi_1', \ldots, \phi_{l_1}'$ of (17) is specified, in value as well as in order of appearance, by means of (23) and (27), where the η's and ζ's are defined by means of (24) and (25).

In this way each individual ϕ_i' is a well-defined function of the u's and y's, and so $\phi_1', \ldots, \phi_{l_1}'$ have a joint distribution. If we again write $\phi_1' = \lambda_1 + N^{-\frac{1}{2}}\eta_1'$, this time to define η_1', *not* ϕ_1', then η_1' is the root which is $\bar{\eta}_1 + o(1)$, for all fixed u's and y's and for large N, of the equation in η which we obtain on putting $\phi = \lambda_1 + N^{-\frac{1}{2}}\eta$ in (17) and cancelling the common factor $N^{-\frac{1}{2}}$ from the first μ_1 rows. We now prove that

$$\eta_1' - \bar{\eta}_1 \to 0 \text{ i.pr.}$$

by applying Lemma 1, letting Q_N be the random point whose coordinates are $y_{i\beta}$ and u_{ij} $(i, j = 1, \ldots, p;\ \beta = 1, \ldots, n_1)$, and f_N the function $\eta_1' - \bar{\eta}_1$. The only condition which we must prove to be satisfied is that $E_N \subset E_{N+1}$, where E_N is the domain of joint existence of the $y_{i\beta}$ and u_{ij}. From the definition (16) of the u_{ij} it is clear that E_N is the set of points $(y_{11}, \ldots, y_{pn_1}, u_{11}, \ldots, u_{p-1,p})$ such that the matrix $\|N\mathbf{I} + \sqrt{N}\,\mathbf{U}\|$ is positive definite, i.e. that the inequality

$$\sqrt{N} \sum_{i=1}^{p} x_i^2 + \sqrt{2} \sum_{i=1}^{p} u_{ii} x_i^2 + 2 \sum_{i<j} u_{ij} x_i x_j > 0 \tag{28}$$

holds for all real numbers $x_1, \ldots, x_p$ not all zero. If, for a given system of values of the x's and u's, (28) is true for N, then evidently it is true for $N+1$. Hence the condition $E_N \subset E_{N+1}$ is satisfied, and we have established that $\eta'_1 - \bar{\eta}_1 \to 0$ i.pr.

Similarly, if we define η_i' and ζ_i' by

$$\phi_i' = \lambda_h + N^{-\frac{1}{2}}\eta_i' \qquad (i = a_{h-1}+1, \ldots, a_h;\ h = 1, \ldots, \nu),$$

$$\phi_i' = N^{-1}\zeta_i' \qquad (i = r+1, \ldots, l_1),$$

then $\eta_i' - \bar{\eta}_i \to 0$ i.pr. and $\zeta_i' - \bar{\zeta}_i \to 0$ i.pr. Hence* η_i', ζ_i' have the same limiting distribution as $\bar{\eta}_i, \bar{\zeta}_i$.

We now define η_i and ζ_i by putting

$$\phi_i = \lambda_h + N^{-\frac{1}{2}}\eta_i \quad (i = a_{h-1}+1, \ldots, a_h;\ h = 1, \ldots, \nu), \tag{29}$$

$$\phi_i = N^{-1}\zeta_i \qquad (i = r+1, \ldots, l_1), \tag{30}$$

where $\phi_1, \ldots, \phi_{l_1}$ is the rearrangement of $\phi_1', \ldots, \phi_{l_1}'$ in their descending order of magnitude. Now $\phi_1 = \phi_1'$ if and only if $\phi_1' > \phi_i'$ for $i > 1$.

* See footnote on p. 188.

Hence*

$$\Pr\{\eta_1-\eta_1' \neq 0\} \leqslant \sum_{i=2}^{l_1} \Pr\{\phi_1' < \phi_1'\}.$$

If ϕ_i' belongs to the first set, say $\phi_i' = \lambda_h + N^{-\frac{1}{2}}\eta_i'$, then

$$\Pr\{\phi_1' < \phi_i'\} = \Pr\{\eta_1' - \eta_i' < 0\} \to \Pr\{\bar{\eta}_1 - \bar{\eta}_i < 0\} = 0;$$

and similarly $\Pr\{\phi_1' < \phi_i'\} \to 0$ for each i. Hence $\Pr\{\eta_1 - \bar{\eta}_1 \neq 0\} \to 0$ and *à fortiori* $\eta_1 - \eta_1' \to 0$ i.pr. Similarly $\eta_i - \eta_i' \to 0$ i.pr. and $\zeta_i - \zeta_i' \to 0$ i.pr. for each i. Hence† η_i, ζ_i have the same limiting distribution as η_i', ζ_i', i.e., the same distribution as $\bar{\eta}_i$, $\bar{\zeta}_i$.

To complete the proof, we have only to observe that, by (15) and Lemma 2, the limiting distribution of η_i, $\bar{\zeta}_i$ is the same as the distribution of the roots of (24) and (25), in which all the variables $y_{i\beta}$ and u_{ij} are independently normally distributed with the same mean 0 and standard deviation 1. Since the e_{ij} in (26) are identical in form with the d_{ij} in (21), the assertion about the limiting distribution of $\zeta_{r+1}, \ldots, \zeta_{l_1}$ is established.

Take now the equation (24) for $h = 1$, namely

$$|\mathbf{C}_{11} - \lambda_1 \mathbf{U}_{11} - \eta \mathbf{I}| = 0. \tag{31}$$

If all the u's and y's are independent normal variates with mean 0 and standard deviation 1, then the diagonal elements of $\mathbf{C}_{11} - \lambda_1 \mathbf{U}_{11}$, which are $2\sqrt{\lambda_1}\, y_{ii} - \sqrt{2}\,\lambda_1 u_{ii}$, are normal variates with mean 0 and standard deviation $(2\lambda_1^2 + 4\lambda_1)^{\frac{1}{2}}$; the non-diagonal elements, which are

$$\sqrt{\lambda_1}(y_{ij} + y_{ji}) - \lambda_1 u_{ij},$$

are normal variates with mean 0 and standard deviation $(\lambda_1^2 + 2\lambda_1)^{\frac{1}{2}}$. Hence we may write (31) as

$$|(2\lambda_1^2 + 4\lambda_1)^{\frac{1}{2}}\, \mathbf{W} - \eta \mathbf{I}| = 0,$$

where $\mathbf{W} = \|w_{ij}\|$ is a symmetric matrix whose elements are distributed with the density (20). The substitution $\eta = (2\lambda_1^2 + 4\lambda_1)^{\frac{1}{2}}\zeta$ completes the proof of the assertion about the limiting distribution of $\zeta_1, \ldots, \zeta_{a_1}$. The same proof holds good for all the other ζ's.

Finally, the statement that the different sets of ζ's tend to be independently distributed follows from the fact that no two of the equations (24) and (25) involve a common u or y. The proof is therefore complete.

* Pr{...} denotes the probability of the event in {...}.

† See footnote on p. 188.

THEOREM. *Let*

$$\zeta_i = \sqrt{N}(2\lambda_h^2+4\lambda_h)^{-\frac{1}{2}}(\phi_i-\lambda_h) \quad (i = a_{h-1}, \ldots, a_h;\ h = 1, \ldots, \nu),$$

$$\zeta_i = N\phi_i \quad (i = r+1, \ldots, l_1).$$

Then the limiting distribution of $\zeta_1, \ldots, \zeta_{l_1}$, *as* $m \to \infty$, *is represented by the density*

$$D(\zeta_1, \ldots, \zeta_{a_1}) D(\zeta_{a_\nu+1}, \ldots, \zeta_{a_2}) \ldots D(\zeta_{a_{\nu-1}+1}, \ldots, \zeta_r) D_1(\zeta_{r+1}, \ldots, \zeta_{l_1}), \tag{32}$$

where

$$D(x_1, \ldots, x_\mu) = 2^{-\frac{1}{2}\mu}\left(\prod_{i=1}^{\mu}\Gamma(\tfrac{1}{2}i)\right)^{-1}\left\{\prod_{i=1}^{\mu}\prod_{j=i+1}^{\mu}(x_i-x_j)\right\}\exp\left(-\tfrac{1}{2}\sum_{i=1}^{\mu}x_i^2\right), \tag{33}$$

$$(\infty > x_1 \geqslant \ldots \geqslant x_\mu > -\infty),$$

$$D_1(\zeta_{r+1}, \ldots, \zeta_{l_1}) = 2^{-\frac{1}{2}(p-r)(n_1-r)}\pi^{\frac{1}{2}(l_1-r)}\left\{\prod_{i=1}^{l_1-r}\Gamma(\tfrac{1}{2}l_2-\tfrac{1}{2}r-\tfrac{1}{2}i+\tfrac{1}{2})\Gamma(\tfrac{1}{2}i)\right\}^{-1}$$

$$\times\left\{\prod_{i=r+1}^{l_1}\prod_{j=i+1}^{l_1}(\zeta_i-\zeta_j)\right\}\left\{\prod_{i=r+1}^{l_1}\zeta_i\right\}^{\frac{1}{2}(l_2-l_1-1)}\exp\left(-\tfrac{1}{2}\sum_{i=r+1}^{l_1}\zeta_i\right), \tag{34}$$

$$(\infty > \zeta_{r+1} \geqslant \ldots \geqslant \zeta_l \geqslant 0).$$

Proof. As a consequence of Lemma 3 we have only to deduce the actual limiting distributions of the ζ's as indicated in that lemma.

The density (34) is known to represent the latent roots of the matrix whose elements d_{ij} are of the form (21). In order to prove (33), we derive the distribution of the latent roots of $\|w_{ij}\|$ from the density (20). Now (20) may be written as

$$2^{-\frac{1}{2}\mu_h}\pi^{-\frac{1}{4}\mu_h(\mu_h+1)}\exp\left(-\tfrac{1}{2}\sum_{i=a_{h-1}+1}^{a_h}\zeta_i^2\right), \tag{35}$$

a function of the latent roots only. Therefore* the distribution of the latent roots has the density equal to the product of (34) and

$$\pi^{\frac{1}{4}\mu_h(\mu_h+1)}\left(\prod_{i=1}^{\mu_h}\Gamma(\tfrac{1}{2}i)\right)^{-1}\left\{\prod_{i=a_{h-1}+1}^{a_h}\prod_{j=i+1}^{a_h}(\zeta_i-\zeta_j)\right\},$$

which verifies (33).

Finally (32) is true because the limiting distribution is such that different sets of ζ's are mutually independent. The theorem is therefore proved.

---

* Hsu (1939), 256, 257.

References.

Cramér, H., *Random variables and probability distributions* (Cambridge University Press, 1937.)

Doob, J. L., "The limiting distributions of certain statistics", *Ann. Math. Stat.* 6 (1935), 160–9.

Fisher, R. A., "The statistical utilization of multiple measurements", *Ann. Eugen.*, 8 (1938), 376–86.

Fisher, R. A., "The sampling distribution of some statistics obtained from non-linear equations", *Ann. Eugen.*, 9 (1939), 238–49.

Hsu, P. L., "On the distribution of roots of certain determinantal equations", *Ann. Eugen.*, 9 (1939), 250–8.

Hsu, P. L., "On generalized analysis of variance", *Biometrika*, 31 (1940), 221–37.

Roy, S. N., "p-statistics or some generalizations in analysis of variance appropriate to multivariate problems", *Sankhya*, 4 (1939), 381–96.

Reprinted from
Biometrika
32, 38–45 (1941).

ON THE LIMITING DISTRIBUTION OF THE CANONICAL CORRELATIONS

BY P. L. HSU

1. The purpose of this paper is to deduce the limiting distribution of Hotelling's canonical correlations* under the most general assumption on the population canonical correlations. The result is stated in the theorem at the end of this paper.

The method employed here is essentially the same as that by which we derived (Hsu, 1940) the limiting distribution of Fisher's discriminating components. In what follows, steps in the derivation are given while strictly rigorous reasoning is left out. The latter may be found in the author's 1940 paper.

The parent distribution is represented by the density

$$\text{const.}\exp\left\{-\tfrac{1}{2}\left(\sum_{i,j=1}^{p}\alpha_{ij}a_{ij}+2\sum_{i=1}^{p}\sum_{g=1}^{q}\beta_{ig}b_{ig}+\sum_{g,h=1}^{q}\gamma_{gh}c_{gh}\right)\right\},\quad (p\leqslant q), \qquad \ldots\ldots(1)$$

where

$$a_{ij}=\sum_{t=1}^{n}x_{it}x_{jt},\quad b_{ig}=\sum_{t=1}^{n}x_{it}y_{gt},\quad c_{gh}=\sum_{t=1}^{n}y_{gt}y_{ht}. \qquad \ldots\ldots(2)$$

By virtue of Hotelling's reduction, the matrix of variances and covariances is taken to be

$$\left\|\begin{matrix}\alpha_{11} & \cdots & \alpha_{1p} & \beta_{11} & \cdots & \beta_{1q}\\ \cdots & \cdots & \cdots & \cdots & \cdots & \cdots\\ \alpha_{p1} & \cdots & \alpha_{pp} & \beta_{p1} & \cdots & \beta_{pq}\\ \beta_{11} & \cdots & \beta_{p1} & \gamma_{11} & \cdots & \gamma_{1q}\\ \cdots & \cdots & \cdots & \cdots & \cdots & \cdots\\ \beta_{1q} & \cdots & \beta_{pq} & \gamma_{q1} & \cdots & \gamma_{qq}\end{matrix}\right\|^{-1}=\left\|\begin{matrix}1 & \cdots & 0 & \rho_1' & \cdots & 0 & \cdots & 0\\ \cdots & \cdots & \cdots & \cdots & \cdots & \cdots & \cdots & \cdots\\ 0 & \cdots & 1 & 0 & \cdots & \rho_p' & \cdots & 0\\ \rho_1' & \cdots & 0 & 1 & \cdots & 0 & \cdots & 0\\ \cdots & \cdots & \cdots & \cdots & \cdots & \cdots & \cdots & \cdots\\ 0 & \cdots & \rho_p' & 0 & \cdots & 1 & \cdots & 0\\ \cdots & \cdots & \cdots & \cdots & \cdots & \cdots & \cdots & \cdots\\ 0 & \cdots & 0 & 0 & \cdots & 0 & \cdots & 1\end{matrix}\right\|, \qquad \ldots\ldots(3)$$

where $\rho_1', \ldots, \rho_p'$ are the population canonical correlations. The sample canonical correlations, $r_1, \ldots, r_p$, are the positive roots of the equation†

$$\left|\begin{matrix}ra_{11} & \cdots & ra_{1p} & b_{11} & \cdots & b_{1q}\\ \cdots & \cdots & \cdots & \cdots & \cdots & \cdots\\ ra_{p1} & \cdots & ra_{pp} & b_{p1} & \cdots & b_{pq}\\ b_{11} & \cdots & b_{p1} & rc_{11} & \cdots & rc_{1q}\\ \cdots & \cdots & \cdots & \cdots & \cdots & \cdots\\ b_{1q} & \cdots & b_{pq} & rc_{q1} & \cdots & rc_{qq}\end{matrix}\right|=0. \qquad \ldots\ldots(4)$$

* Hotelling (1936). For further work on the distribution of the canonical correlations, see Madow (1938), Girschick (1939) and Hsu (1939).

† We use r in (4) instead of $-r$ as in Hotelling's original definition because it is known that the non-vanishing roots of (4) form pairs each of which have the same absolute value but opposite signs.

We set

$$\left.\begin{aligned} \rho'_1 = \dots = \rho'_{\mu_1} = \rho_1, \\ \rho'_{\mu_1+1} = \dots = \rho'_{\mu_1+\mu_2} = \rho_2, \\ \dots\dots\dots\dots\dots\dots\dots\dots\dots\dots \\ \rho'_{\mu_1+\dots+\mu_{\nu-1}+1} = \dots = \rho'_s = \rho_\nu, \end{aligned}\right\} \quad \dots\dots(5)$$

$$\rho'_{s+1} = \dots = \rho'_p = 0, \quad \dots\dots(6)$$

$$\rho_1 > \rho_2 > \dots > \rho_\nu > 0, \quad \dots\dots(7)$$

$$r_1 \geqslant r_2 \geqslant \dots \geqslant r_p, \quad \dots\dots(8)$$

and proceed to find the limiting distribution of $r_1, \dots, r_p$ as $n \to \infty$.

2. LEMMA 1. *We have the identity*

$$\begin{vmatrix} \mathbf{P} & \mathbf{Q} \\ \mathbf{R} & \mathbf{S} \end{vmatrix} = |\mathbf{S}| . |\mathbf{P} - \mathbf{Q}\mathbf{S}^{-1}\mathbf{R}|. \quad \dots\dots(9)$$

This results from the identity

$$\left\| \begin{matrix} \mathbf{P} & \mathbf{Q} \\ \mathbf{R} & \mathbf{S} \end{matrix} \right\| . \left\| \begin{matrix} \mathbf{I} & \mathbf{O} \\ -\mathbf{S}^{-1}\mathbf{R} & \mathbf{I} \end{matrix} \right\| = \left\| \begin{matrix} \mathbf{P} - \mathbf{Q}\mathbf{S}^{-1}\mathbf{R} & \mathbf{Q} \\ \mathbf{O} & \mathbf{S} \end{matrix} \right\|$$

on taking determinants on both sides.

LEMMA 2. *Let*

$$\left.\begin{aligned} a_{ii} &= n + \sqrt{n}\, u_{ii}, & a_{ij} &= \sqrt{n}\, u_{ij} \quad (i \neq j), \\ b_{ii} &= n\rho'_i + \sqrt{n}\, v_{ii}, & b_{ig} &= \sqrt{n}\, v_{ig} \quad (i \neq g), \\ c_{gg} &= n + \sqrt{n}\, w_{gg}, & c_{gh} &= \sqrt{n}\, w_{gh} \quad (g \neq h), \end{aligned}\right\} \quad \dots\dots(10)$$

$$(i, j = 1, \dots, p;\ g, h = 1, \dots, q).$$

The distribution of the u's, v's and w's approach that of $\frac{1}{2}(p+q)(p+q+1)$ normal variates whose means are zero and whose second moments are specified in the following statements:

(i) *any v or w which has at least one suffix number $> p$ is uncorrelated with all the others*;

(ii) *any member of one of the sets (u_{ii}, v_{ii}, w_{ii}), $(i = 1, \dots, p)$, $(u_{ij}, v_{ij}, v_{ji}, w_{ij})$, $(i, j = 1, \dots, p;\ i \neq j)$ is uncorrelated with all the members of all the other sets*;

(iii) *for $i, j = 1, \dots, p$ we have*

$$\left.\begin{aligned} \mathscr{E}(u_{ii}^2) &= \mathscr{E}(w_{ii}^2) = 2, \\ \mathscr{E}(v_{ii}^2) &= 1 + \rho_i'^2, \\ \mathscr{E}(u_{ii} w_{ii}) &= 2\rho_i'^2, \\ \mathscr{E}(u_{ii} v_{ii}) &= \mathscr{E}(w_{ii} v_{ii}) = 2\rho'_i, \\ \mathscr{E}(u_{ij}^2) &= \mathscr{E}(v_{ij}^2) = \mathscr{E}(w_{ij}^2) = 1 \quad (i \neq j), \\ \mathscr{E}(u_{ij} w_{ij}) &= \mathscr{E}(v_{ij} v_{ji}) = \rho'_i \rho'_j \quad (i \neq j), \\ \mathscr{E}(u_{ij} v_{ij}) &= \mathscr{E}(w_{ij} v_{ij}) = \rho'_j \quad (i \neq j). \end{aligned}\right\} \quad \dots\dots(11)$$

This follows immediately from the well-known central limit theorem.* The second moments may easily be computed by virtue of (3).

COROLLARY. *Under the assumption* (6) *any* u, v *or* w *which has at least one suffix number* $> s$ *is uncorrelated with all the others, and* $\mathscr{E}(v_{ii}^2) = 1$ *for* $i = s+1, \ldots, p$.

This results from (11) on putting $\rho_i' = 0$ for $i = s+1, \ldots, p$.

3. We may now find the limiting distribution of $r_{s+1}, \ldots, r_p$, the $p-s$ smallest correlations.

We substitute (10) in (4) and then divide each element by n. There results the equation

$$\begin{vmatrix}
r\left(1+\frac{u_{11}}{\sqrt{n}}\right) & \cdots & \frac{ru_{1s}}{\sqrt{n}} & \frac{ru_{1,s+1}}{\sqrt{n}} & \cdots & \frac{ru_{1p}}{\sqrt{n}} & \rho_1'+\frac{v_{11}}{\sqrt{n}} & \cdots & \frac{v_{1s}}{\sqrt{n}} & \frac{v_{1,s+1}}{\sqrt{n}} & \cdots & \frac{v_{1q}}{\sqrt{n}} \\
\hdotsfor{12} \\
\frac{ru_{s1}}{\sqrt{n}} & \cdots & r\left(1+\frac{u_{ss}}{\sqrt{n}}\right) & \frac{ru_{s,s+1}}{\sqrt{n}} & \cdots & \frac{ru_{sp}}{\sqrt{n}} & \frac{v_{s1}}{\sqrt{n}} & \cdots & \rho_s'+\frac{v_{ss}}{\sqrt{n}} & \frac{v_{s,s+1}}{\sqrt{n}} & \cdots & \frac{v_{sq}}{\sqrt{n}} \\
\frac{ru_{s+1,1}}{\sqrt{n}} & \cdots & \frac{ru_{s+1,s}}{\sqrt{n}} & r\left(1+\frac{u_{s+1,s+1}}{\sqrt{n}}\right) & \cdots & \frac{ru_{s+1,p}}{\sqrt{n}} & \frac{v_{s+1,1}}{\sqrt{n}} & \cdots & \frac{v_{s+1,s}}{\sqrt{n}} & \frac{v_{s+1,s+1}}{\sqrt{n}} & \cdots & \frac{v_{s+1,q}}{\sqrt{n}} \\
\hdotsfor{12} \\
\frac{ru_{p1}}{\sqrt{n}} & \cdots & \frac{ru_{ps}}{\sqrt{n}} & \frac{ru_{p,s+1}}{\sqrt{n}} & \cdots & r\left(1+\frac{u_{pp}}{\sqrt{n}}\right) & \frac{v_{p1}}{\sqrt{n}} & \cdots & \frac{v_{ps}}{\sqrt{n}} & \frac{v_{p,s+1}}{\sqrt{n}} & \cdots & \frac{v_{pq}}{\sqrt{n}} \\
\rho_1'+\frac{v_{11}}{\sqrt{n}} & \cdots & \frac{v_{s1}}{\sqrt{n}} & \frac{v_{s+1,1}}{\sqrt{n}} & \cdots & \frac{v_{p1}}{\sqrt{n}} & r\left(1+\frac{w_{11}}{\sqrt{n}}\right) & \cdots & \frac{rw_{1s}}{\sqrt{n}} & \frac{rw_{1,s+1}}{\sqrt{n}} & \cdots & \frac{rw_{1q}}{\sqrt{n}} \\
\hdotsfor{12} \\
\frac{v_{1s}}{\sqrt{n}} & \cdots & \rho_s'+\frac{v_{ss}}{\sqrt{n}} & \frac{v_{s+1,s}}{\sqrt{n}} & \cdots & \frac{v_{ps}}{\sqrt{n}} & \frac{rw_{s1}}{\sqrt{n}} & \cdots & r\left(1+\frac{w_{ss}}{\sqrt{n}}\right) & \frac{rw_{s,s+1}}{\sqrt{n}} & \cdots & \frac{rw_{sq}}{\sqrt{n}} \\
\frac{v_{1,s+1}}{\sqrt{n}} & \cdots & \frac{v_{s,s+1}}{\sqrt{n}} & \frac{v_{s+1,s+1}}{\sqrt{n}} & \cdots & \frac{v_{p,s+1}}{\sqrt{n}} & \frac{rw_{s+1,1}}{\sqrt{n}} & \cdots & \frac{rw_{s+1,s}}{\sqrt{n}} & r\left(1+\frac{w_{s+1,s+1}}{\sqrt{n}}\right) & \cdots & \frac{rw_{s+1,q}}{\sqrt{n}} \\
\hdotsfor{12} \\
\frac{v_{1q}}{\sqrt{n}} & \cdots & \frac{v_{sq}}{\sqrt{n}} & \frac{v_{s+1,q}}{\sqrt{n}} & \cdots & \frac{v_{pq}}{\sqrt{n}} & \frac{rw_{q1}}{\sqrt{n}} & \cdots & \frac{rw_{qs}}{\sqrt{n}} & \frac{rw_{q,s+1}}{\sqrt{n}} & \cdots & r\left(1+\frac{w_{qq}}{\sqrt{n}}\right)
\end{vmatrix} = 0. \qquad \ldots\ldots(12)$$

The equation (12) has $p-s$ roots which are $o(1)$ for large n. To evaluate these, we substitute $n^{-\frac{1}{2}}\eta$ for r in (12), delete the common factor $n^{-\frac{1}{2}}$ from the rows

$$s+1, s+2, \ldots, p, p+s+1, p+s+2, \ldots, p+q,$$

* Cf. Cramér (1937), pp. 113–14.

and then let $n \to \infty$. There results the equation

$$\begin{vmatrix} 0 & \cdots & 0 & 0 & \cdots & 0 & \rho_1' & \cdots & 0 & 0 & \cdots & 0 \\ \cdots & & & & & & & & & & & \cdots \\ 0 & \cdots & 0 & 0 & \cdots & 0 & 0 & \cdots & \rho_s' & 0 & \cdots & 0 \\ 0 & \cdots & 0 & -\eta & \cdots & 0 & v_{s+1,1} & \cdots & v_{s+1,s} & v_{s+1,s+1} & \cdots & v_{s+1,q} \\ \cdots & & & & & & & & & & & \cdots \\ 0 & \cdots & 0 & 0 & \cdots & -\eta & v_{p1} & \cdots & v_{ps} & v_{p,s+1} & \cdots & v_{pq} \\ \rho_1' & \cdots & 0 & 0 & \cdots & 0 & 0 & \cdots & 0 & 0 & \cdots & 0 \\ \cdots & & & & & & & & & & & \cdots \\ 0 & \cdots & \rho_s' & 0 & \cdots & 0 & 0 & \cdots & 0 & 0 & \cdots & 0 \\ v_{1,s+1} & \cdots & v_{s,s+1} & v_{s+1,s+1} & \cdots & v_{p,s+1} & 0 & \cdots & 0 & -\eta & \cdots & 0 \\ \cdots & & & & & & & & & & & \cdots \\ v_{1q} & \cdots & v_{sq} & v_{s+1,q} & \cdots & v_{pq} & 0 & \cdots & 0 & 0 & \cdots & -\eta \end{vmatrix} = 0,$$

i.e.

$$\begin{vmatrix} -\eta & \cdots & 0 & v_{s+1,s+1} & \cdots & v_{s+1,q} \\ \cdots & & & & & \cdots \\ 0 & \cdots & -\eta & v_{p,s+1} & \cdots & v_{pq} \\ v_{s+1,s+1} & \cdots & v_{p,s+1} & -\eta & \cdots & 0 \\ \cdots & & & & & \cdots \\ v_{s+1,q} & \cdots & v_{pq} & 0 & \cdots & -\eta \end{vmatrix} = 0. \qquad \text{......(13)}$$

By (9) the left-hand side of (13) is equal to

$$(-\eta)^{q-p} \begin{vmatrix} d_{s+1,s+1} - \eta^2 & \cdots & d_{s+1,q} \\ \cdots & & \cdots \\ d_{q,s+1} & \cdots & d_{qq} - \eta^2 \end{vmatrix}, \qquad \text{......(14)}$$

where

$$d_{ij} = \sum_{g=s+1}^{q} v_{ig} v_{jg} \quad (i, j = s+1, \ldots, p). \qquad \text{......(15)}$$

Let $\zeta_{s+1}', \ldots, \zeta_p'$, in descending order of magnitude, be the latent roots of the matrix $\| d_{ij} \|$. Then the $p-s$ roots of (12) which are $o(1)$ for large n are

$$n^{-\frac{1}{2}} \zeta_i'^{\frac{1}{2}} + o(n^{-\frac{1}{2}}) \quad (i = s+1, \ldots, p).$$

If we define $\zeta_{s+1}, \ldots, \zeta_p$ by putting

$$r_i = n^{-\frac{1}{2}} \zeta_i^{\frac{1}{2}} \quad (i = s+1, \ldots, p), \qquad \text{......(16)}$$

then the ζ's have the same limiting distribution as the ζ''s. Hence the limiting distribution of the ζ_i may be derived as the distribution of the latent roots of $\| d_{ij} \|$, in which the v's are regarded as having a distribution which is the limiting distribution described in Lemma 2. By virtue of the Corollary this is the distribution of $(p-s)(q-s)$ mutually independent normal variates with zero mean and unit standard deviation. Therefore* the limiting distribution of the $\zeta_i = nr_i^2$ has the density function

$$2^{-\frac{1}{2}(p-s)(q-s)}\pi^{\frac{1}{2}(p-s)}\left\{\prod_{i=1}^{p-s}\Gamma\tfrac{1}{2}(q-s-i+1)\,\Gamma\tfrac{1}{2}i\right\}^{-1}\left\{\prod_{i=s+1}^{p}\prod_{j=i+1}^{p}(\zeta_i-\zeta_j)\right\}$$
$$\times\left(\prod_{i=s+1}^{p}\zeta_i\right)^{\frac{1}{2}(q-p-1)}\exp\left(-\tfrac{1}{2}\sum_{i=s+1}^{p}\zeta_i\right), \quad \text{......(17)}$$

$$\infty > \zeta_{s+1} \geqslant \ldots \geqslant \zeta_p \geqslant 0.$$

The transformation $\zeta_i = \eta_i^2$ gives the following density for the limiting distribution of the $\eta_i = n^{\frac{1}{2}} r_i$ $(i = s+1, \ldots, p)$:

$$f_1(\eta_{s+1}, \ldots, \eta_p) = 2^{p-s-\frac{1}{2}(p-s)(q-s)}\pi^{\frac{1}{2}(p-s)}\left\{\prod_{i=1}^{p-s}\Gamma\tfrac{1}{2}(q-s-i+1)\,\Gamma\tfrac{1}{2}i\right\}^{-1}$$
$$\times\left\{\prod_{i=s+1}^{p}\prod_{j=i+1}^{p}(\eta_i^2-\eta_j^2)\right\}\left(\prod_{i=s+1}^{p}\eta_i\right)^{q-p}\exp\left(-\tfrac{1}{2}\sum_{i=s+1}^{p}\eta_i^2\right), \quad \text{......(18)}$$

$$\infty > \eta_{s+1} \geqslant \ldots \geqslant \eta_p \geqslant 0.$$

4. We now proceed to find the limiting distribution of $r_1, \ldots, r_s$. By virtue of (9) we may write the left-hand side of (4) as

$$r^{q-p}\,|\,\mathbf{C}\,|\,.\,|\,r^2\mathbf{A} - \mathbf{B}\mathbf{C}^{-1}\mathbf{B}'\,|,$$

where

$$\mathbf{A} = \begin{Vmatrix} a_{11} & \cdots & a_{1p} \\ \cdots & \cdots & \cdots \\ a_{p1} & \cdots & a_{pp} \end{Vmatrix}, \quad \mathbf{B} = \begin{Vmatrix} b_{11} & \cdots & b_{1q} \\ \cdots & \cdots & \cdots \\ b_{p1} & \cdots & b_{pq} \end{Vmatrix}, \quad \mathbf{C} = \begin{Vmatrix} c_{11} & \cdots & c_{1q} \\ \cdots & \cdots & \cdots \\ c_{q1} & \cdots & c_{qq} \end{Vmatrix}. \quad \text{......(19)}$$

Hence, if we set $\qquad \theta_i = r_i^2 \quad (i = 1, \ldots, p), \qquad$ ......(20)

the θ_i are the roots of the equation

$$|\,\mathbf{B}\mathbf{C}^{-1}\mathbf{B}' - \theta\mathbf{A}\,| = 0. \quad \text{......(21)}$$

Substituting (10) in (21) and dividing each element by n, we get

$$|\,(\mathbf{\Delta} + n^{-\frac{1}{2}}\mathbf{V})(\mathbf{I} + n^{-\frac{1}{2}}\mathbf{W})^{-1}(\mathbf{\Delta}' + n^{-\frac{1}{2}}\mathbf{V}') - \theta(\mathbf{I} + n^{-\frac{1}{2}}\mathbf{U})\,| = 0, \quad \text{......(22)}$$

where

$$\mathbf{U} = \begin{Vmatrix} u_{11} & \cdots & u_{1p} \\ \cdots & \cdots & \cdots \\ u_{p1} & \cdots & u_{pp} \end{Vmatrix}, \quad \mathbf{V} = \begin{Vmatrix} v_{11} & \cdots & v_{1q} \\ \cdots & \cdots & \cdots \\ v_{p1} & \cdots & v_{pq} \end{Vmatrix}, \quad \mathbf{W} = \begin{Vmatrix} w_{11} & \cdots & w_{1q} \\ \cdots & \cdots & \cdots \\ w_{q1} & \cdots & w_{qq} \end{Vmatrix}, \quad \text{......(23)}$$

$$\mathbf{\Delta} = \begin{Vmatrix} \rho_1' & \cdots & 0 & 0 & \cdots & 0 \\ \cdots & \cdots & \cdots & \cdots & \cdots & \cdots \\ 0 & \cdots & \rho_p' & 0 & \cdots & 0 \end{Vmatrix}. \quad \text{......(24)}$$

* Hsu (1939), pp. 256–7.

Neglecting higher powers of $n^{-\frac{1}{2}}$ we write $\mathbf{I}-n^{-\frac{1}{2}}\mathbf{W}$ for $(\mathbf{I}+n^{-\frac{1}{2}}\mathbf{W})^{-1}$ and carry out the matrix multiplication to the term with $n^{-\frac{1}{2}}$ as a factor. There results the equation

$$|\mathbf{\Delta\Delta}'+n^{-\frac{1}{2}}(\mathbf{V\Delta}'+\mathbf{\Delta V}'-\mathbf{\Delta W\Delta}')-\theta(\mathbf{I}+n^{-\frac{1}{2}}\mathbf{U})| = 0, \qquad \text{......(25)}$$

i.e.

$$\begin{vmatrix} \rho_1'^2-\theta+n^{-\frac{1}{2}}(2\rho_1'v_{11}-\rho_1'^2w_{11}-\theta u_{11}) & n^{-\frac{1}{2}}(\rho_1'v_{21}+\rho_2'v_{12}-\rho_1'\rho_2'w_{12}-\theta u_{12}) & \cdots & n^{-\frac{1}{2}}(\rho_1'v_{p1}+\rho_p'v_{1p}-\rho_1'\rho_p'w_{1p}-\theta u_{1p}) \\ n^{-\frac{1}{2}}(\rho_1'v_{21}+\rho_2'v_{12}-\rho_1'\rho_2'w_{12}-\theta u_{12}) & \rho_2'^2-\theta+n^{-\frac{1}{2}}(2\rho_2'v_{22}-\rho_2'^2w_{22}-\theta u_{22}) & \cdots & n^{-\frac{1}{2}}(\rho_2'v_{p2}+\rho_p'v_{2p}-\rho_2'\rho_p'w_{2p}-\theta u_{2p}) \\ \cdots & \cdots & \cdots & \cdots \\ n^{-\frac{1}{2}}(\rho_1'v_{p1}+\rho_p'v_{1p}-\rho_1'\rho_p'w_{1p}-\theta u_{1p}) & n^{-\frac{1}{2}}(\rho_2'v_{p2}+\rho_p'v_{2p}-\rho_2'\rho_p'w_{2p}-\theta u_{2p}) & \cdots & \rho_p'^2-\theta+n^{-\frac{1}{2}}(2\rho_p'v_{pp}-\rho_p'^2w_{pp}-\theta u_{pp}) \end{vmatrix}$$

$$= 0. \qquad \text{......(26)}$$

On account of (5) there are μ_1 roots of (26) which are $\rho_1^2+o(1)$ for large n. To evaluate these we substitute $\rho_1^2+n^{-\frac{1}{2}}\rho_1\zeta$ for θ in (26). Since the first μ_1 of the ρ''s are equal to ρ_1, there will be a common factor $n^{-\frac{1}{2}}$ in each of the first μ_1 rows. After deleting this factor and then letting $n\to\infty$, we get the equation

$$\begin{vmatrix} 2v_{11}-\rho_1(u_{11}+w_{11})-\zeta & v_{12}+v_{21}-\rho_1(u_{12}+w_{12}) & \cdots & v_{1\mu_1}+v_{\mu_1 1}-\rho_1(u_{1\mu_1}+w_{1\mu_1}) \\ v_{12}+v_{21}-\rho_1(u_{12}+w_{12}) & 2v_{22}-\rho_1(u_{22}+w_{22})-\zeta & \cdots & v_{2\mu_1}+v_{\mu_1 2}-\rho_1(u_{2\mu_1}+w_{2\mu_1}) \\ \cdots & \cdots & \cdots & \cdots \\ v_{1\mu_1}+v_{\mu_1 1}-\rho_1(u_{1\mu_1}+w_{1\mu_1}) & v_{2\mu_1}+v_{\mu_1 2}-\rho_1(u_{2\mu_1}+w_{2\mu_1}) & \cdots & 2v_{\mu_1\mu_1}-\rho_1(u_{\mu_1\mu_1}+w_{\mu_1\mu_1})-\zeta \end{vmatrix}$$

$$= 0, \qquad \text{......(27)}$$

i.e.

$$\begin{vmatrix} z_{11}-\zeta & \cdots & z_{1\mu_1} \\ \cdots & \cdots & \cdots \\ z_{\mu_1 1} & \cdots & z_{\mu_1\mu_1}-\zeta \end{vmatrix} = 0, \qquad \text{......(28)}$$

where

$$z_{ij} = v_{ij}+v_{ji}-\rho_1(u_{ij}+w_{ij}) \quad (i,j=1,\ldots,\mu_1). \qquad \text{......(29)}$$

Let $\zeta_1',\ldots,\zeta_{\mu_1}'$ be the roots, in descending order of magnitude, of (28). Then the μ_1 roots of (26) which are $\rho_1^2+o(1)$ for large n are

$$\rho_1^2+n^{-\frac{1}{2}}\rho_1\zeta_i'+o(n^{-\frac{1}{2}}), \quad (i=1,\ldots,\mu_1).$$

If we define $\zeta_1,\ldots,\zeta_{\mu_1}$ by putting

$$\theta_i = \rho_1^2+n^{-\frac{1}{2}}\rho_1\zeta_i \quad (i=1,\ldots,\mu_1), \qquad \text{......(30)}$$

then the ζ's have the same limiting distribution as the ζ''s. Hence the limiting distribution of the ζ_i may be derived as the distribution of the latent roots of the matrix $\| z_{ij} \|$, in which the u's, v's and w's are regarded as having a distribution which is the limiting distribution described in Lemma 2.

Now, if all the u's, v's and w's are normal variates with zero mean, so are the z's. Owing to the fact (ii) of Lemma 2 all the z's are uncorrelated. Using the formulae (11) to calculate the variances of the z's, we easily obtain

$$\mathscr{E}(z_{ii}^2) = 4(1-\rho_1^2)^2, \quad \mathscr{E}(z_{ij}^2) = 2(1-\rho_1^2)^2 \quad (i \neq j), \qquad \ldots\ldots(31)$$
$$(i, j = 1, \ldots, \mu_1).$$

Hence
$$z_{ii} = 2(1-\rho_1^2)\, t_{ii}, \quad z_{ij} = \sqrt{2}\,(1-\rho_1^2)\, t_{ij} \quad (i \neq j), \qquad \ldots\ldots(32)$$
$$(i, j = 1, \ldots, \mu_1),$$

where the t's are mutually independent normal variates with zero mean and unit standard deviation.

Setting $\zeta = 2(1-\rho_1^2)\,\eta$ in (28), we get, by (32),

$$\begin{vmatrix} t_{11}-\eta & 2^{-\frac{1}{2}}t_{12} & \ldots & 2^{-\frac{1}{2}}t_{1\mu_1} \\ 2^{-\frac{1}{2}}t_{21} & t_{22}-\eta & \ldots & 2^{-\frac{1}{2}}t_{2\mu_1} \\ \ldots & \ldots & \ldots & \ldots \\ 2^{-\frac{1}{2}}t_{\mu_1 1} & 2^{-\frac{1}{2}}t_{\mu_1 2} & \ldots & t_{\mu_1\mu_1}-\eta \end{vmatrix} = 0. \qquad \ldots\ldots(33)$$

Let $\eta'_1, \ldots, \eta'_{\mu_1}$ be the roots of (33) in descending order of magnitude.

The density function

$$(2\pi)^{-\frac{1}{4}\mu_1(\mu_1+1)} \exp\{-\tfrac{1}{2}(t_{11}^2 + \ldots + t_{\mu_1\mu_1}^2 + t_{12}^2 + \ldots + t_{\mu_1-1,\mu_1}^2)\} \qquad \ldots\ldots(34)$$

is equal to
$$(2\pi)^{-\frac{1}{4}\mu_1(\mu_1+1)} \exp\left(-\tfrac{1}{2}\sum_{i=1}^{\mu_1} \eta_i'^2\right), \qquad \ldots\ldots(35)$$

which is a function of the latent roots only. Hence* the distribution of the latent roots has the density

$$f(\eta'_1, \ldots, \eta'_{\mu_1}) = 2^{-\frac{1}{4}\mu_1(\mu_1+1)} \left(\prod_{i=1}^{\mu_1} \Gamma\tfrac{1}{2}i\right)^{-1} \left\{\prod_{i=1}^{\mu_1} \prod_{j=i+1}^{\mu_1} (\eta'_i - \eta'_j)\right\} \exp\left(-\tfrac{1}{2}\sum_{i=1}^{\mu_1} \eta_i'^2\right), \ldots\ldots(36)$$
$$\infty > \eta'_1 \geqslant \ldots \geqslant \eta'_{\mu_1} > -\infty.$$

The density function (36) represents the limiting distribution of $\eta'_1, \ldots, \eta'_{\mu_1}$, where

$$\theta_i = \rho_1^2 + 2n^{-\frac{1}{2}}\rho_1(1-\rho_1^2)\,\eta'_i \quad (i = 1, \ldots, \mu_1). \qquad \ldots\ldots(37)$$

Hence
$$r_i = \theta_i^{\frac{1}{2}} = \rho_1 + n^{-\frac{1}{2}}(1-\rho_1^2)\,\eta'_i + o(n^{-\frac{1}{2}}) \quad (i = 1, \ldots, \mu_1).$$

If we define $\eta_1, \ldots, \eta_{\mu_1}$ by

$$r_i = \rho_1 + n^{-\frac{1}{2}}(1-\rho_1^2)\,\eta_i \quad (i = 1, \ldots, \mu_1), \qquad \ldots\ldots(38)$$

then the $\eta_i = n^{\frac{1}{2}}(1-\rho_1^2)^{-1}(r_i - \rho_1)$ have the same limiting distribution as the η'_i. Hence the limiting distribution of $\eta_1, \ldots, \eta_{\mu_1}$ has the density $f(\eta_1, \ldots, \eta_{\mu_1})$.

In exactly the same manner we may prove that for $k = 1, \ldots, \nu$ the

$$\eta_i = n^{\frac{1}{2}}(1-\rho_k^2)^{-1}(r_i - \rho_k) \quad (i = \mu_1 + \ldots + \mu_{k-1} + 1, \ldots, \mu_1 + \ldots + \mu_k)$$

* Hsu (1939), p. 256, Theorem 2.

have the limiting distribution represented by the density

$$f(\eta_{\mu_1+\ldots+\mu_{k-1}+1}, \ldots, \eta_{\mu_1+\ldots+\mu_k}),$$

where, in general,

$$f(x_1, \ldots, x_m) = 2^{-\frac{1}{4}m(m+1)}\left(\prod_{i=1}^{m} \Gamma\tfrac{1}{2}i\right)^{-1}\left\{\prod_{i=1}^{m}\prod_{j=i+1}^{m}(x_i - x_j)\right\}\exp\left(-\tfrac{1}{2}\sum_{i=1}^{m} x_i^2\right), \quad \ldots\ldots(39)$$

$$\infty > x_1 \geqslant \ldots \geqslant x_m > -\infty.$$

Furthermore, the sets $(\eta_1, \ldots, \eta_{\mu_1})$, $(\eta_{\mu_1+1}, \ldots, \eta_{\mu_1+\mu_2})$, $\ldots$, $(\eta_{\mu_1+\ldots+\mu_{\nu-1}+1}, \ldots, \eta_s)$ are such that the equations corresponding to (27) for two different sets involve only mutually uncorrelated u's, v's and w's owing to (ii) of Lemma 2. Therefore the limiting distribution must be such that these sets are independent of one another. Also, recalling (14) and (i) of Lemma 2, it is seen that the limiting distribution of $\eta_1, \ldots, \eta_p$ must be such that the sets $(\eta_1, \ldots, \eta_{\mu_1})$, $(\eta_{\mu_1+1}, \ldots, \eta_{\mu_1+\mu_2})$, $\ldots$, $(\eta_{\mu_1+\ldots+\mu_{\nu-1}+1}, \ldots, \eta_s)$ and $(\eta_{s+1}, \ldots, \eta_p)$ are independent of one another.

In conclusion we sum up the results in the following theorem:

THEOREM. *Let the population canonical correlations be $\rho'_1, \ldots, \rho'_p$, where*

$$\rho'_1 = \ldots = \rho'_{\mu_1} = \rho_1,$$
$$\rho'_{\mu_1+1} = \ldots = \rho'_{\mu_1+\mu_2} = \rho_2,$$
$$\ldots\ldots\ldots\ldots\ldots\ldots\ldots\ldots\ldots\ldots$$
$$\rho'_{\mu_1+\ldots+\mu_{\nu-1}+1} = \ldots = \rho'_s = \rho_\nu,$$
$$\rho'_{s+1} = \ldots = \rho'_p = 0,$$
$$\rho_1 > \rho_2 > \ldots > \rho_\nu > 0.$$

Let the sample canonical correlations be $r_1, \ldots, r_p$, where

$$r_1 \geqslant r_2 \geqslant \ldots \geqslant r_p.$$

Let

$$\eta_i = n^{\frac{1}{2}}(1 - \rho_i'^2)^{-1}(r_i - \rho'_i) \quad (i = 1, \ldots, p).$$

Then the limiting distribution of $\eta_1, \ldots, \eta_p$ is represented by the density function

$$f(\eta_1, \ldots, \eta_{\mu_1}) f(\eta_{\mu_1+1}, \ldots, \eta_{\mu_1+\mu_2}) \ldots f(\eta_{\mu_1+\ldots+\mu_{\nu-1}+1}, \ldots, \eta_s) f_1(\eta_{s+1}, \ldots, \eta_p),$$

where the functions f and f_1 are given by (39) *and* (18) *respectively.*

REFERENCES

CRAMÉR, H. (1937). *Random Variables and Probability Distributions.* Camb. Univ. Press.

GIRSCHICK, M. A. (1939). "On the sampling theory of roots of determinantal equations." *Ann. Math. Statist.* **10**, 205.

HOTELLING, H. (1936). "Relations between two sets of variates." *Biometrika*, **28**, 321–77

HSU, P. L. (1939). "On the distribution of roots of certain determinantal equations." *Ann. Eugen., Lond.*, **9**, 250–8.

—— (1940). "On the limiting distribution of roots of a determinantal equation." *Proc. Lond. Math. Soc.* (at press).

MADOW, W. G. (1938). "Contributions to the theory of multi-variate statistical analysis." *Trans. Amer. Math. Soc.* **44**, 454.

Reprinted from
Biometrika
32, 62–69 (1941).

ANALYSIS OF VARIANCE FROM THE POWER FUNCTION STANDPOINT

BY P. L. HSU

A FRESH study on the classical analysis of variance tests in the light of the Neyman-Pearson theory was started by Kołodziejczyk (1935), who formulated the class of linear hypotheses for which these tests may be employed. As a linear hypothesis is defined relative to the set of admissible hypotheses, the study of the E^2-test (by which we denote any test falling under the usual methods of analysis of variance) may be made with reference to its power function. P. C. Tang (1938) showed how the power function was related to R. A. Fisher's (C) distribution (Fisher, 1928) and so was able to appraise the chance of detecting the falsehood of a linear hypothesis using the E^2-test. The great theoretical value of the power function lies, however, in its use in comparing the relative merits of alternative tests of the same hypothesis. In this paper we shall prove a theorem (p. 63) which asserts that out of a certain class of tests the E^2-test is uniformly most powerful.

In his paper Tang has used an orthogonal transformation in the sample space which enabled the general linear hypothesis to be reduced to the following simple form: Given the elementary probability law

$$p(y_1, \ldots, y_m, z_1, \ldots, z_n) = \{\sqrt{(2\pi)}\,\sigma\}^{-(m+n)} \exp\left\{-\frac{1}{2\sigma^2}\left(\sum_{i=1}^{m} (y_i - \eta_i)^2 + \sum_{i=1}^{n} z_i^2\right)\right\}, \quad (1)$$

where all real values of $\eta_1, \ldots, \eta_m$ and all positive values of σ are admissible, the hypothesis is that n_1 ($\leqslant m$) of the η's have the true value 0:

$$\eta_1 = \eta_2 = \ldots = \eta_{n_1} = 0. \quad (2)$$

We call the above hypothesis H_0.

We shall set

$$E^2 = \sum_{i=1}^{n_1} y_i^2 \Big/ \left(\sum_{i=1}^{n_1} y_i^2 + \sum_{i=1}^{n} z_i^2\right), \quad (3)$$

and call w_0 (of size ϵ) the critical region for the rejection of H_0 defined by the inequality

$$E^2 \geqslant E_\epsilon^2, \quad (4)$$

where E_ϵ^2 is a constant so determined that the probability that (4) is true, given that (2) is true, equals ϵ.

The power function of w_0 as given by Tang can be written

$$e^{-\lambda} \sum_{h=0}^{\infty} \{h!\, B(\tfrac{1}{2}n_1 + h, \tfrac{1}{2}n)\}^{-1} \lambda^h \int_{E_\epsilon^2}^{1} (E^2)^{\frac{1}{2}n_1+h-1} (1-E^2)^{\frac{1}{2}n-1}\, d(E^2), \quad (5)$$

where $$\lambda = \frac{1}{2\sigma^2} \sum_{i=1}^{n_1} \eta_i^2. \tag{6}$$

An outstanding feature of the power function (5) is that it depends on the single parameter λ. Our problem is, does there exist another critical region of size ϵ whose power function depends on λ alone and which is more powerful than w_0 for certain values of λ? The answer is contained in the following theorem and is in the negative.

THEOREM. *Suppose that the critical region w satisfies the following conditions:*

(*a*) *w is of size ϵ,*

(*b*) *the power function of w depends on the single parameter λ.*

Let $\beta(\lambda)$ be the power function of w and $\beta_0(\lambda)$ be the power function (5) *of w_0. Then*

$$\beta(\lambda) \leqslant \beta_0(\lambda) \tag{7}$$

for all positive values of λ.

Proof. In the place of $z_1, \ldots, z_n$ we substitute spherical co-ordinates, viz. the radius vector $r = (\Sigma z^2)^{\frac{1}{2}}$ and $n-1$ angles, $\theta_1, \ldots, \theta_{n-1}$. We deduce from (1) that

$$p(y_1, \ldots, y_m, \theta_1, \ldots, \theta_{n-1}, r) = p(y_1, \ldots, y_{n_1})\, p(y_{n_1+1}, \ldots, y_m)\, p(\theta_1, \ldots, \theta_{n-1})\, p(r), \tag{8}$$

where $$p(y_1, \ldots, y_{n_1}) = \{\sqrt{(2\pi)}\,\sigma\}^{-n_1} \exp\left\{-\tfrac{1}{2} \sum_{i=1}^{n_1} (y_i - \eta_i)^2\right\}, \tag{9}$$

$$p(y_{n_1+1}, \ldots, y_m) = \{\sqrt{(2\pi)}\,\sigma\}^{-(m-n_1)} \exp\left\{-\tfrac{1}{2} \sum_{i=n_1+1}^{m} (y_i - \eta_i)^2\right\}, \tag{10}$$

$$p(r) = 2^{-\frac{1}{2}(n-2)} \sigma^{-n} (\Gamma \tfrac{1}{2} n)^{-1} r^{n-1} \exp\left(-\frac{1}{2\sigma^2} r^2\right), \tag{11}$$

and $p(\theta_1, \ldots, \theta_{n-1})$ is the well-known product of cosines which involves none of the parameters $\eta_1, \ldots, \eta_m$ and σ.

We now make the following successive transformations:

$$r = s^{\frac{1}{2}}, \quad s = t - \sum_{i=1}^{n_1} y_i^2, \quad y_i = t^{\frac{1}{2}} u_i \quad (i = 1, \ldots, n_1), \tag{12}$$

and also write $$\gamma_i = \sigma^{-1} \eta_i \quad (i = 1, \ldots, n_1). \tag{13}$$

It follows that

$$p(y_{n_1+1}, \ldots, y_m, \theta_1, \ldots, \theta_{n-1}, u_1, \ldots, u_{n_1}, t) = p(y_{n_1+1}, \ldots, y_m)\, p(\theta_1, \ldots, \theta_{n-1})\, p(u_1, \ldots, u_{n_1}, t), \tag{14}$$

where

$$p(u_1, \ldots, u_{n_1}, t) = (\sqrt{2}\,\sigma)^{-(n+n_1)} \pi^{-\frac{1}{2}n} (\Gamma \tfrac{1}{2} n)^{-1} e^{-\lambda} t^{\frac{1}{2}(n+n_1-2)} \exp\left(-\frac{t}{2\sigma^2}\right) \times \left(1 - \sum_{i=1}^{n_1} u_i^2\right)^{\frac{1}{2}(n-2)} \exp\left(\frac{\sqrt{t}}{\sigma} \sum_{i=1}^{n_1} \gamma_i u_i\right). \tag{15}$$

From now on we shall write y for the set of variables $y_{n_1+1}, \ldots, y_m$ and dy for $dy_{n_1+1}, \ldots, dy_m$, and use similar abbreviations θ, u, $d\theta$ and du.

Suppose now that the critical region w satisfies the conditions (a) and (b). Let $\Gamma(y, \theta, u, t)$ be the characteristic function of w, i.e. $\Gamma(y, \theta, u, t) = 1$ or 0 according as the sample point falls within w or not. Let W be the sample space. Then the power function of w is

$$\beta(\lambda) = \int_W \Gamma(y, \theta, u, t)\, p(y, \theta, u, t)\, dy\, d\theta\, du\, dt, \tag{16}$$

whence,

$$(\sqrt{2}\,\sigma)^{-(n+n_1)} \int_W \Gamma(y, \theta, u, t)\, p(y)\, p(\theta)\, t^{\frac{1}{2}(n+n_1-2)} \times \exp\left(-\frac{t}{2\sigma^2}\right)\left(1 - \sum_{i=1}^{n_1} u_i^2\right)^{\frac{1}{2}(n-2)} dy\, d\theta\, du\, dt = \pi^{\frac{1}{2}n_1}\Gamma(\tfrac{1}{2}n)\,\epsilon, \tag{17}$$

$$(\sqrt{2}\,\sigma)^{-(n+n_1)} \int_W \Gamma(y, \theta, u, t)\, p(y)\, p(\theta)\, t^{\frac{1}{2}(n+n_1-2)} \times \exp\left(-\frac{t}{2\sigma^2}\right)\left(1 - \sum_{i=1}^{n_1} u_i^2\right)^{\frac{1}{2}(n-2)} \exp\left(\frac{\sqrt{t}}{\sigma}\sum_{i=1}^{n_1} \gamma_i u_i\right) dy\, d\theta\, du\, dt$$
$$= \pi^{\frac{1}{2}n_1}\Gamma(\tfrac{1}{2}n)\, e^{\lambda}\beta(\lambda) = F(\lambda) = F\left(\sum_{i=1}^{n_1}\gamma_i^2\right), \text{ say.} \tag{18}$$

Let W_1 be the sample space of θ, u and t, and put

$$(\sqrt{2}\,\sigma)^{-(n+n_1)} \int_{W_1} \Gamma(y, \theta, u, t)\, p(\theta)\, t^{\frac{1}{2}(n+n_1-2)} \times \exp\left(-\frac{t}{2\sigma^2}\right)\left(1 - \sum_{i=1}^{n_1} u_i^2\right)^{\frac{1}{2}(n-2)} \exp\left(\frac{\sqrt{t}}{\sigma}\sum_{i=1}^{n_1} \gamma_i u_i\right) d\theta\, du\, dt$$
$$- F(\lambda) = \phi(y, \gamma, \sigma). \tag{19}$$

Then, by (18),

$$\int_{-\infty}^{\infty} \cdots \int_{-\infty}^{\infty} \phi(y)\, p(y)\, dy = 0, \tag{20}$$

i.e.

$$\int_{-\infty}^{\infty} \cdots \int_{-\infty}^{\infty} \phi(y_{n_1+1}, \ldots, y_m, \gamma_1, \ldots, \gamma_{n_1}, \sigma) \exp\left(-\frac{1}{2\sigma^2}\sum_{i=n_1+1}^{m} y_i^2\right) \times \exp\left(\sum_{i=n_1+1}^{m} \alpha_i y_i\right) dy_{n_1+1} \ldots dy_m = 0, \tag{21}$$

on writing α_i for $(2\sigma^2)^{-1}\eta_i$ $(i = n_1 + 1, \ldots, m)$.

Equation (20) must hold true for all real values of the α's. Hence it follows from the well-known theorem on Laplace transformation* that

$$\phi(y, \gamma, \sigma) \exp\left(-\frac{1}{2\sigma^2}\sum_{i=n_1+1}^{m} y_i^2\right) = 0,$$

* Cf. Doetsch (1937), p. 35, Theorem 1. Though the theorem referred to is stated for the case where the number of y's is one, it may easily be extended to the case of more than one y by induction.

i.e. $\phi(y,\gamma,\sigma)=0$, whence, by (19),

$$(\sqrt{2}\,\sigma)^{-(n+n_1)}\int_{W_1}\Gamma(y,\theta,u,t)\,p(\theta)\,t^{\frac{1}{2}(n+n_1-2)}$$
$$\times\exp\left(-\frac{t}{2\sigma^2}\right)\left(1-\sum_{i=1}^{n_1}u_i^2\right)^{\frac{1}{2}(n-2)}\exp\left(\frac{\sqrt{t}}{\sigma}\sum_{i=1}^{n_1}\gamma_iu_i\right)d\theta\,du\,dt=F\left(\sum_{i=1}^{n_1}\gamma_i^2\right). \quad (22)$$

In particular, from $\phi(y,0,\sigma)=0$ and (17) we have

$$(\sqrt{2}\,\sigma)^{-(n+n_1)}\int_{W_1}\Gamma(y,\theta,u,t)\,p(\theta)\,t^{\frac{1}{2}(n+n_1-2)}$$
$$\times\exp\left(-\frac{t}{2\sigma^2}\right)\left(1-\sum_{i=1}^{n_1}u_i^2\right)^{\frac{1}{2}(n-2)}d\theta\,du\,dt=\pi^{\frac{1}{2}n_1}\Gamma(\tfrac{1}{2}n)\,\epsilon. \quad (23)$$

Letting W_2 be the sample space of θ and u, we get respectively from (23) and (22) that

$$(\sqrt{2}\,\sigma)^{-(n+n_1)}\int_0^\infty t^{\frac{1}{2}(n+n_1-2)}$$
$$\times\exp\left(-\frac{t}{2\sigma^2}\right)dt\int_{W_2}\Gamma(y,\theta,u,t)\,p(\theta)\left(1-\sum_{i=1}^{n_1}u_i^2\right)^{\frac{1}{2}(n-2)}d\theta\,du=\pi^{\frac{1}{2}n_1}\Gamma(\tfrac{1}{2}n)\,\epsilon, \quad (24)$$

$$(\sqrt{2}\,\sigma)^{-(n+n_1)}\int_0^\infty t^{\frac{1}{2}(n+n_1-2)}\exp\left(-\frac{t}{2\sigma^2}\right)dt\int_{W_2}\Gamma(y,\theta,u,t)\,p(\theta)\left(1-\sum_{i=1}^{n_1}u_i^2\right)^{\frac{1}{2}(n-2)}$$
$$\times\exp\left(\frac{\sqrt{t}}{\sigma}\sum_{i=1}^{n_1}\gamma_iu_i\right)d\theta\,du=F\left(\sum_{i=1}^{n_1}\gamma_i^2\right). \quad (25)$$

Hence, on developing the left-hand side of (25) into power series in the γ's, we must have

$$\int_0^\infty t^{\frac{1}{2}(n+n_1-2+h)}\exp\left(-\frac{t}{2\sigma^2}\right)dt$$
$$\times\int_{W_2}\Gamma(y,\theta,u,t)\,p(\theta)\left(1-\sum_{i=1}^{n_1}u_i^2\right)^{\frac{1}{2}(n-2)}\left(\sum_{i=1}^{n_1}\gamma_iu_i\right)^h d\theta\,du=0 \text{ for odd } h, \quad (26)$$

$$2^{-\frac{1}{2}(n+n_1)}\sigma^{-(n+n_1+2h)}\int_0^\infty t^{\frac{1}{2}(n+n_1-2)+h}$$
$$\times\exp\left(-\frac{t}{2\sigma^2}\right)dt\int_{W_2}\Gamma(y,\theta,u,t)\,p(\theta)\left(1-\sum_{i=1}^{n_1}u_i^2\right)^{\frac{1}{2}(n-2)}\left(\sum_{i=1}^{n_1}\gamma_iu_i\right)^{2h}d\theta\,du$$
$$=a_h\left(\sum_{i=1}^{n}\gamma_i^2\right)^h \quad (h=1,2,3,\ldots). \quad (27)$$

Further, equations (24) and (27) may be written as

$$\int_0^\infty t^{\frac{1}{2}(n+n_1-2)}\exp\left(-\frac{t}{2\sigma^2}\right)$$
$$\times\left[\int_{W_2}\Gamma(y,\theta,u,t)\,p(\theta)\left(1-\sum_{i=1}^{n_1}u_i^2\right)^{\frac{1}{2}(n-2)}d\theta\,du-\frac{\pi^{\frac{1}{2}n_1}\Gamma(\frac{1}{2}n)}{\Gamma\frac{1}{2}(n+n_1)}\epsilon\right]dt=0, \quad (28)$$

$$\int_0^\infty t^{\frac{1}{2}(n+n_1-2)}\exp\left(-\frac{t}{2\sigma^2}\right)\left[\int_{W_2}\Gamma(y,\theta,u,t)\,p(\theta)\left(1-\sum_{i=1}^{n_1}u_i^2\right)^{\frac{1}{2}(n-2)}\right.$$
$$\left.\times\left(\sum_{i=1}^{n_1}\gamma_i u_i\right)^{2h}d\theta\,du-\frac{a_h\left(\sum_{i=1}^{n_1}\gamma_i^2\right)^h}{2^h\,\Gamma\{\frac{1}{2}(n+n_1)+h\}}\right]dt=0\quad(h=1,2,3,\ldots).\tag{29}$$

Equations (28), (26) and (29) must hold true for all positive values of σ. Hence, by the theorem of Laplace transformation,* the functions within the square brackets in (28) and (29) and the inner integral in (26) must vanish identically:

$$\int_{W_2}\Gamma(y,\theta,u,t)\,p(\theta)\left(1-\sum_{i=1}^{n_1}u_i^2\right)^{\frac{1}{2}(n-2)}d\theta\,du=\frac{\pi^{\frac{1}{2}n_1}\,\Gamma(\frac{1}{2}n)}{\Gamma\frac{1}{2}(n+n_1)}\,\epsilon,\tag{30}$$

$$\int_{W_2}\Gamma(y,\theta,u,t)\,p(\theta)\left(1-\sum_{i=1}^{n_1}u_i^2\right)^{\frac{1}{2}(n-2)}\left(\sum_{i=1}^{n_1}\gamma_i u_i\right)^h d\theta\,du=0\text{ for odd }h,\tag{31}$$

$$\int_{W_2}\Gamma(y,\theta,u,t)\,p(\theta)\left(1-\sum_{i=1}^{n_1}u_i^2\right)^{\frac{1}{2}(n-2)}\left(\sum_{i=1}^{n_1}\gamma_i u_i\right)^{2h}d\theta\,du$$
$$=\frac{a_h\left(\sum_{i=1}^{n_1}\gamma_i^2\right)^h}{2^h\,\Gamma\{\frac{1}{2}(n+n_1)+h\}}\quad(h=1,2,3,\ldots).\tag{32}$$

From (31) and (32) we infer that

$$\int_{W_2}\Gamma(y,\theta,u,t)\,p(\theta)\left(1-\sum_{i=1}^{n_1}u_i^2\right)^{\frac{1}{2}(n-2)}\exp\left(\sum_{i=1}^{n_1}\gamma_i u_i\right)d\theta\,du=G\left(\sum_{i=1}^{n_1}\gamma_i^2\right).\tag{33}$$

Now for any given values of y and t the integral $\int_{W_2}\Gamma(y,\theta,u,t)\,f(\theta,u)\,d\theta\,du$ equals $\int_{w_2}f(\theta,u)\,d\theta\,du$, where w_2 is the set of points in the sample space of θ and u for which $\Gamma(y,\theta,u,t)=1$. Hence (30) and (33) are equivalent to

$$\int_{w_2}p(\theta)\left(1-\sum_{i=1}^{n_1}u_i^2\right)^{\frac{1}{2}(n-2)}d\theta\,du=\frac{\pi^{\frac{1}{2}n_1}\,\Gamma(\frac{1}{2}n)}{\Gamma\frac{1}{2}(n+n_1)}\,\epsilon,\tag{34}$$

$$\int_{w_2}p(\theta)\left(1-\sum_{i=1}^{n_1}u_i^2\right)^{\frac{1}{2}(n-2)}\exp\left(\sum_{i=1}^{n_1}\gamma_i u_i\right)d\theta\,du=G\left(\sum_{i=1}^{n_1}\gamma_i^2\right).\tag{35}$$

The conditions (34) and (35) are necessary and sufficient that the critical region w should have the properties (a) and (b).

On the other hand, from (12) we have

$$E^2=\sum_{i=1}^{n_1}u_i^2;\tag{36}$$

hence w_0 is the region defined by the inequality

$$\sum_{i=1}^{n_1}u_i^2\geqslant E_\epsilon^2.\tag{37}$$

* Cf. footnote, p. 64.

Since w_0 is of size ϵ, we must get the same right-hand side of (34) when in the left-hand side we substitute w_0 for w_2. Hence

$$\int_{w_2} p(\theta)\left(1-\sum_{i=1}^{n_1} u_i^2\right)^{\frac{1}{2}(n-2)} d\theta\,du = \int_{w_0} p(\theta)\left(1-\sum_{i=1}^{n_1} u_i^2\right)^{\frac{1}{2}(n-2)} d\theta\,du. \tag{38}$$

Let
$$\int_{w_0} p(\theta)\left(1-\sum_{i=1}^{n_1} u_i^2\right)^{\frac{1}{2}(n-2)} \exp\left(\sum_{i=1}^{n_1}\gamma_i u_i\right) d\theta\,du = G_0\left(\sum_{i=1}^{n_1}\gamma_i^2\right). \tag{39}$$

With the help of (37) and (38) we may now appeal to the lemma proved in the Appendix and conclude that

$$G(\lambda) \leqslant G_0(\lambda), \tag{40}$$

whence, replacing γ_i by $\sigma^{-1}\sqrt{t}\,\gamma_i$ in the integrals in (35) and (39),

$$\int_{w_2} p(\theta)\left(1-\sum_{i=1}^{n_1} u_i^2\right)^{\frac{1}{2}(n-2)} \exp\left(\frac{\sqrt{t}}{\sigma}\sum_{i=1}^{n_1}\gamma_i u_i\right) d\theta\,du$$
$$\leqslant \int_{w_0} p(\theta)\left(1-\sum_{i=1}^{n_1} u_i^2\right)^{\frac{1}{2}(n-2)} \exp\left(\frac{\sqrt{t}}{\sigma}\sum_{i=1}^{n_1}\gamma_i u_i\right) d\theta\,du. \tag{41}$$

If we multiply both sides of (41) by

$$\frac{e^{-\lambda}}{(\sqrt{2}\,\sigma)^{n+n_1}\pi^{\frac{1}{2}n_1}\Gamma(\frac{1}{2}n)}\,p(y)\,t^{\frac{1}{2}(n+n_1-2)}\exp\left(-\frac{t}{2\sigma^2}\right),$$

and integrate over the sample space of y and t, we get, in accordance with (18),

$$\beta(\lambda) \leqslant \beta_0(\lambda).$$

Hence the theorem is proved.

APPENDIX

LEMMA. *Let $g(x)\geqslant 0$ be defined for $x\geqslant 0$ and vanish for $x>1$, such that $g(v_1^2+\ldots+v_n^2)$ is summable. Let $f(w_1,\ldots,w_m)\geqslant 0$ be summable. In the product space of the v's and w's let R be a region such that*

$$\int_R f(w_1,\ldots,w_m)\,g(v_1^2+\ldots+v_n^2)\exp(\gamma_1 v_1+\ldots+\gamma_n v_n)\,dv\,dw = G(\gamma_1^2+\ldots+\gamma_n^2). \tag{1}$$

Let w_0 be the region defined by the inequality

$$v_1^2+\ldots+v_n^2 \geqslant k. \tag{2}$$

Let $\int_{R_0} f(w_1,\ldots,w_m)\,g(v_1^2+\ldots+v_n^2)\exp(\gamma_1 v_1+\ldots+\gamma_n v_n)\,dv\,dw = G_0(\gamma_1^2+\ldots+\gamma_n^2).$ (3)*

Finally, let

$$\int_R f(w_1,\ldots,w_m)\,g(v_1^2+\ldots+v_n^2)\,dv\,dw = \int_{R_0} f(w_1,\ldots,w_m)\,g(v_1^2+\ldots+v_n^2)\,dv\,dw. \tag{4}$$

Then
$$G(x)\leqslant G_0(x) \tag{5}$$
for all positive x.

* Notice that (3) is not a separate condition on R_0, but is implied by (2).

5-2

Proof. In (1) we set $\gamma_1 = x$, $\gamma_2 = \ldots = \gamma_n = 0$, and get

$$G(x^2) = \int_R f(w_1, \ldots, w_m)\, g(v_1^2 + \ldots + v_n^2) \exp(xv_1)\, dv\, dw.$$

This and the conditions on f and g imply that G is continuous.

Multiplying both sides of (1) by $\exp(-\Sigma\gamma_1^2)$ and integrating over the region $0 \leqslant a \leqslant \Sigma\gamma_i^2 \leqslant b$, we get

$$K\int_a^b x^{\frac{1}{2}(n-2)} e^{-x} G(x)\, dx$$
$$= \int_R f(w_1, \ldots, w_m)\, g(\Sigma v_i^2)\, dv\, dw \int_{a \leqslant \Sigma\gamma_i^2 \leqslant b} \exp(-\Sigma\gamma_i^2 + \Sigma\gamma_i v_i)\, d\gamma, \quad (6)$$

where K is some numerical constant. Applying a rotation in the space of the γ's to the inner integral in the right-hand side of (6), we obtain

$$K\int_a^b x^{\frac{1}{2}(n-2)} e^{-x} G(x)\, dx$$
$$= \int_R f(w_1, \ldots, w_m)\, g(\Sigma v_i^2)\, dv\, dw \int_{a \leqslant \Sigma x_i^2 \leqslant b} \exp\{-\Sigma x_i^2 + (\Sigma v_i^2)^{\frac{1}{2}} x_1\}\, dx$$
$$= \sum_{h=0}^{\infty} c_h I_h(R),$$

where
$$I_h(R) = \frac{1}{(2h)!}\int_R f(w_1, \ldots, w_m)\, g(\Sigma v_i^2)^h\, dv\, dw,$$
$$c_h = \int_{a \leqslant \Sigma x_i^2 \leqslant b} x_1^{2h} \exp(-\Sigma x_i^2)\, dx.$$

Similarly, we have

$$K\int_a^b x^{\frac{1}{2}(n-2)} e^{-x} G_0(x)\, dx = \sum_{h=0}^{\infty} c_h I_h(R_0).$$

An appeal to a general lemma of Neyman and Pearson,* on remembering (4), leads to the inequality

$$I_h(R) \leqslant I_h(R_0).$$

Hence
$$\int_a^b x^{\frac{1}{2}(n-2)} e^{-x}\{G(x) - G_0(x)\}\, dx \leqslant 0.$$

Since a and b are arbitrary and since the integrand is a continuous function, the latter must be $\leqslant 0$. Hence $G(x) \leqslant G_0(x)$.

* Neyman and Pearson (1936), p. 11.

REFERENCES

Doetsch, G. (1937). *Theorie und Anwendung der Laplace-Transformation.* Berlin: Julius Springer.

Fisher, R. A. (1928). "The general sampling distribution of the multiple correlation coefficient." *Proc. Roy. Soc.* A, **121**, 654–73.

Kołodziejczyk, St (1935). "On an important class of statistical hypotheses." *Biometrika*, **27**, 161–90.

Neyman, J. & Pearson, E. S. (1936). "Contributions to the theory of testing statistical hypotheses." *Statist. Res. Mem.* **1**, 1–37.

Tang, P. C. (1938). "The power function of the analysis of variance tests." *Statist. Res. Mem.* **2**, 126–49.

Reprinted from
Ann. Eugenics
11, 42–46 (1941).

CANONICAL REDUCTION OF THE GENERAL REGRESSION PROBLEM

By P. L. HSU

1. The hypothesis H

Given that relative distribution of x_{ir} $(i = 1, \ldots, p;\ r = 1, \ldots, N)$ given

$$w_{\mu r}\ (\mu = 1, \ldots, q;\ r = 1, \ldots, N),$$

where $N \geqslant p+q$, is

$$(2\pi)^{-\frac{1}{2}pN}\,|\,\alpha_{ij}\,|^{\frac{1}{2}N}\exp\left\{-\tfrac{1}{2}\sum_{i,j}\alpha_{ij}\sum_{r}(x_{ir}-\beta_{i1}w_{1r}-\ldots-\beta_{iq}w_{qr})(x_{jr}-\beta_{j1}w_{1r}-\ldots-\beta_{jq}w_{qr})\right\}$$
$$\times\prod_{i,r}dx_{ir}, \tag{1}$$

and that the matrix $||\,w_{\mu r}\,||$ has always the rank q, we call H the hypothesis that

$$\beta_{is} = 0 \quad (i = 1, \ldots, p;\ s = 1, \ldots, n_1). \tag{H}$$

Throughout this paper, the ranges of the indices, unless explicitly stated otherwise, shall be

$$i, j = 1, \ldots, p, \quad \mu, \nu = 1, \ldots, q,$$
$$s = 1, \ldots, n_1, \quad t = 1, \ldots, n.$$

J. F. Daly (1940, p. 21) has formulated the same hypothesis H, describing (1) as the distribution of samples of N individuals. However, by merely avoiding such a description we arrive at a generalization which, though formally trivial, is not unessential. In fact, for a fixed μ the statement that $w_{\mu 1}, \ldots, w_{\mu N}$ are a sample, i.e. observations made on some random variate w_μ, covers the special case where w_μ is the constant 1, so that $(w_{\mu 1}, \ldots, w_{\mu N}) = (1, \ldots, 1)$; but it does not cover the case where $(w_{\mu 1}, \ldots, w_{\mu N})$ is a constant point, e.g. $(0, \ldots, 0, 1, \ldots, 1)$, not necessarily of the form $(1, \ldots, 1)$. This latter situation is allowed in our formulation. If all the points $(w_{\mu 1}, \ldots, w_{\mu N})$ are constant points, (1) represents simply the distribution of the x_{ir}.

As an important special case we may take $N = N_1 + \ldots + N_q$ and

$$\left.\begin{aligned} w_{\mu r} &= 1 \quad \text{when} \quad N_1+\ldots+N_{\mu-1}+1 \leqslant r \leqslant N_1+\ldots+N_\mu \\ w_{\mu r} &= 0 \quad \text{otherwise} \end{aligned}\right\} \quad \text{for } \mu = 1, \ldots, q-1$$
$$w_{qr} = 1 \quad (r = 1, \ldots, N).$$

The hypothesis H, on taking $n_1 = q-1$, is then the hypothesis that q p-variate normal populations, known to have the same set of variances and covariances, have also a common set of means.

When $p = 1$, H coincides with the linear hypothesis studied in full by St Kołodziejczyk (1935) and P. C. Tang (1938).

In a previous paper (Hsu, 1940) the author has considered the 'Wilks-Lawley hypothesis' H_0 which runs as follows: given the parent distribution

$$(2\pi)^{-\frac{1}{2}p(n+n_1)}\,|\alpha_{ij}|^{\frac{1}{2}(n+n_1)}\exp\left\{-\tfrac{1}{2}\sum_{i,j}\alpha_{ij}\sum_s(y_{is}-\eta_{is})(y_{js}-\eta_{js})-\tfrac{1}{2}\sum_{i,j}\alpha_{ij}\sum_t z_{it}z_{jt}\right\}\prod_{i,j,s,t}dy_{is}dz_{jt}, \tag{2}$$

H_0 is the hypothesis that $\quad \eta_{is}=0 \quad (i=1,\ldots,p;\ s=1,\ldots,n_1).$ $\hfill (H_0)$

In the next section we shall show that H may be reduced to H_0 by means of a transformation of variables. Such a reduction has more than a formal interest. In fact, there remains unsolved the problem of how to construct the most powerful test, if any in one sense or another, for H; hence it is desirable that the problem of testing H be reduced to its simplest form, namely testing H_0.

2. Reduction of H to H_0

Let $$\mathbf{X}=||x_{ir}||,\quad \mathbf{W}=||w_{\mu r}||.$$

We write $\mathbf{A}'$ for the transposed of any matrix $\mathbf{A}$. It has been assumed that $\mathbf{W}$ is of rank q; hence $\mathbf{WW}'$ is a positive definite matrix. Now every q-rowed positive definite matrix is expressible as $\mathbf{EE}'$, where $\mathbf{E}$ is a non-singular real matrix of the triangular form:*

$$\mathbf{E}=\begin{Vmatrix} e_{11} & e_{12} & \cdots & e_{1q}\\ 0 & e_{22} & \cdots & e_{2q}\\ \cdots & \cdots & \cdots & \cdots\\ 0 & 0 & \cdots & e_{qq}\end{Vmatrix},$$

where $$e_{\mu\mu}>0\quad(\mu=1,\ldots,q).$$

Hence $$\mathbf{WW}'=\mathbf{EE}'.$$

It follows that if we write $$\mathbf{\Gamma}_1=\mathbf{E}^{-1}\mathbf{W},$$

then $$\mathbf{\Gamma}_1\mathbf{\Gamma}_1'=\mathbf{I}.$$

The real matrix $\mathbf{\Gamma}_1$ satisfying the condition $\mathbf{\Gamma}_1\mathbf{\Gamma}_1'=\mathbf{I}$ may be taken as the first q rows of a real orthogonal matrix of order N. Hence we can construct a real orthogonal matrix $\mathbf{\Gamma}$ such that $$\mathbf{\Gamma}'=||\mathbf{\Gamma}_1',\mathbf{\Gamma}_2'||.$$

This being done, let us introduce new variables $y_{i\mu}$ and z_{it} $(i=1,\ldots,p;\ \mu=1,\ldots,q;\ t=1,\ldots,n)$, where $n=N-q$, in the place of the x_{ir} by means of the transformation

$$\mathbf{Y}\equiv||y_{i\mu}||=\mathbf{X\Gamma}_1',\quad \mathbf{Z}\equiv||z_{it}||=\mathbf{X\Gamma}_2',$$

i.e. $||\mathbf{Y},\mathbf{Z}||=\mathbf{X\Gamma}'$. Owing to the orthogonality of $\mathbf{\Gamma}$ we have for the relative distribution of the $y_{i\mu}$ and z_{it} given the $w_{\mu r}$ the same normal distribution as (1) only with a different set of regression functions. Let us write $\epsilon(\mathbf{X}\,|\,\mathbf{W})$ for the matrix of the regression functions of the x_{ir} on the $w_{\mu r}$, and similarly $\epsilon(\mathbf{Y}\,|\,\mathbf{W})$ and $\epsilon(\mathbf{Z}\,|\,\mathbf{W})$. Then

$$\epsilon(\mathbf{X}\,|\,\mathbf{W})=\mathbf{BW},\quad\text{where}\quad \mathbf{B}=||\beta_{i\mu}||.$$

* Without appealing to classical theorems on matrices we may directly prove this statement by actually determining the elements of $\mathbf{E}$, from last column backwards, so that $\mathbf{EE}'$ is equal to a given positive definite matrix.

Hence
$$\epsilon(\mathbf{Y} \mid \mathbf{W}) = \mathbf{BW\Gamma}_1' = \mathbf{BE\Gamma}_1\mathbf{\Gamma}_1' = \mathbf{BE},$$
$$\epsilon(\mathbf{Z} \mid \mathbf{W}) = \mathbf{BW\Gamma}_2' = \mathbf{BE\Gamma}_1\mathbf{\Gamma}_2' = \mathbf{0}.$$

Therefore we obtain

$$(2\pi)^{-\frac{1}{2}pN} \mid \alpha_{ij} \mid^{\frac{1}{2}N} \exp\left\{-\tfrac{1}{2}\sum_{i,j}\alpha_{ij}\sum_{\mu}(y_{i\mu}-\eta_{i\mu})(y_{j\mu}-\eta_{j\mu})-\tfrac{1}{2}\sum_{i,j}\alpha_{ij}\sum_{t}z_{it}z_{jt}\right\}\prod_{i,j,\mu,t}dy_{i\mu}dz_{jt}, \tag{3}$$

where the $\eta_{i\mu}$ are the elements of **BE**, viz.

$$\eta_{i\mu} = \beta_{i1}e_{1\mu} + \ldots + \beta_{i\mu}e_{\mu\mu} \quad (i = 1, \ldots, p;\ \mu = 1, \ldots, q).$$

For each $\mu > n_1$, the regression functions $\eta_{i\mu}$ $(i = 1, \ldots, p)$ involve the terms $\beta_{i\mu}e_{\mu\mu}$, where $e_{\mu\mu} > 0$ and $\beta_{i\mu}$ is an unknown parameter about which the hypothesis H asserts nothing whatsoever. We are therefore justified in disregarding the factor of (3) which represents the relative distribution of the $y_{i\mu}$ for $i = 1, \ldots, p$ and $\mu = n_1+1, \ldots, q$. After deleting this factor there remains finally the distribution (2):

$$(2\pi)^{-\frac{1}{2}p(n+n_1)} \mid \alpha_{ij} \mid^{\frac{1}{2}(n+n_1)} \exp\left\{-\tfrac{1}{2}\sum_{i,j}\alpha_{ij}\sum_{s}(y_{is}-\eta_{is})(y_{js}-\eta_{js})-\tfrac{1}{2}\sum_{i,j}\alpha_{ij}\sum_{t}z_{it}z_{jt}\right\}\prod_{i,j,s,t}dy_{is}dy_{jt},$$

where
$$\eta_{is} = \beta_{i1}e_{1s} + \ldots + \beta_{is}e_{ss} \quad (i = 1, \ldots, p;\ s = 1, \ldots, n_1). \tag{4}$$

If all the $w_{\mu r}$ are constants, then the e's are constants and so are the η's. Owing to (4) and the fact that $e_{ss} > 0$ the hypothesis H is true if and only if $\eta_{is} = 0$ for all i and s; this establishes the equivalence of H and H_0. If some of the $w_{\mu r}$ are not constants, then the e's, being functions of the $w_{\mu r}$, are in general not all constants. In this case we have to refer to the formula (4) and understand by H_0 the hypothesis that $\beta_{is} = 0$ $(i = 1, \ldots, p;\ s = 1, \ldots, n_1)$.

3. Relations between old and new variates

We write
$$a_{ij} = \sum_s y_{is}y_{js}, \quad b_{ij} = \sum_t z_{it}z_{jt}, \quad c_{ij} = a_{ij}+b_{ij}, \quad \psi_{ij} = \sum_s \eta_{is}\eta_{js}$$

and shall establish the fundamental formulae (17), (18) and (19) which express the product sums a_{ij} and c_{ij} as functions of the x_{ir} and $w_{\mu r}$, and the ψ_{ij} as functions of the β_{is} and $w_{\mu r}$.

Let us write the matrices $\mathbf{\Gamma}_1$, **E**, **W**, **Y** and **B** in the partitioned form:

$$\mathbf{\Gamma}_1 = \begin{Vmatrix}\mathbf{\Gamma}_{11}\\ \mathbf{\Gamma}_{12}\end{Vmatrix}, \quad \mathbf{W} = \begin{Vmatrix}\mathbf{W}_1\\ \mathbf{W}_2\end{Vmatrix}, \quad \mathbf{Y} = \|\mathbf{Y}_1, \mathbf{Y}_2\|, \quad \mathbf{B} = \|\mathbf{B}_1, \mathbf{B}_2\|, \quad \mathbf{E} = \begin{Vmatrix}\mathbf{E}_{11} & \mathbf{E}_{12}\\ \mathbf{0} & \mathbf{E}_{22}\end{Vmatrix},$$

where $\mathbf{\Gamma}_{11}$ and $\mathbf{W}_1$ have n_1 rows, $\mathbf{Y}_1$ and $\mathbf{B}_1$ have n_1 columns, and $\mathbf{E}_{11}$ has n_1 rows and n_1 columns. Then

$$\| a_{ij} \| = \mathbf{Y}_1\mathbf{Y}_1' = \mathbf{X\Gamma}_{11}'\mathbf{\Gamma}_{11}\mathbf{X}', \tag{5}$$
$$\| b_{ij} \| = \mathbf{Y}_1\mathbf{Y}_1' + \mathbf{ZZ}' = \mathbf{X}(\mathbf{\Gamma}_{11}'\mathbf{\Gamma}_{11}+\mathbf{\Gamma}_2'\mathbf{\Gamma}_2)\mathbf{X}' = \mathbf{X}(\mathbf{I}-\mathbf{\Gamma}_{12}'\mathbf{\Gamma}_{12})\mathbf{X}', \tag{6}$$
$$\| \psi_{ij} \| = \mathbf{B}_1\mathbf{E}_{11}\mathbf{E}_{11}'\mathbf{B}_1'. \tag{7}$$

We have
$$\mathbf{E}^{-1} = \begin{Vmatrix}\mathbf{E}_{11}^{-1} & -\mathbf{E}_{11}^{-1}\mathbf{E}_{12}\mathbf{E}_{22}^{-1}\\ \mathbf{0} & \mathbf{E}_{22}^{-1}\end{Vmatrix}.$$

The equations $\boldsymbol{\Gamma}_1 = \mathbf{E}^{-1}\mathbf{W}$ and $\mathbf{E}\mathbf{E}' = \mathbf{W}\mathbf{W}'$ give

$$\boldsymbol{\Gamma}_{11} = \mathbf{E}_{11}^{-1}(\mathbf{W}_1 - \mathbf{E}_{12}\mathbf{E}_{22}^{-1}\mathbf{W}_2), \tag{8}$$

$$\boldsymbol{\Gamma}_{12} = \mathbf{E}_{22}^{-1}\mathbf{W}_2, \tag{9}$$

$$\mathbf{E}_{11}\mathbf{E}_{11}' + \mathbf{E}_{12}\mathbf{E}_{12}' = \mathbf{W}_1\mathbf{W}_1', \tag{10}$$

$$\mathbf{E}_{12}\mathbf{E}_{22}' = \mathbf{W}_1\mathbf{W}_2', \tag{11}$$

$$\mathbf{E}_{22}\mathbf{E}_{22}' = \mathbf{W}_2\mathbf{W}_2'. \tag{12}$$

From (11) we get $\mathbf{E}_{12} = \mathbf{W}_1\mathbf{W}_2'\mathbf{E}_{22}'^{-1}$, whence

$$\mathbf{E}_{12}\mathbf{E}_{12}' = \mathbf{W}_1\mathbf{W}_2'(\mathbf{E}_{22}\mathbf{E}_{22}')^{-1}\mathbf{W}_2\mathbf{W}_1' = \mathbf{W}_1\mathbf{W}_2'(\mathbf{W}_2\mathbf{W}_2')^{-1}\mathbf{W}_2\mathbf{W}_1'$$

by (12). Writing

$$\mathbf{M} = \mathbf{I} - \mathbf{W}_2'(\mathbf{W}_2\mathbf{W}_2')^{-1}\mathbf{W}_2, \tag{13}$$

we have $\mathbf{E}_{12}\mathbf{E}_{12}' = \mathbf{W}_1\mathbf{W}_1' - \mathbf{W}_1\mathbf{M}\mathbf{W}_1'$, whence, by (10)

$$\mathbf{E}_{11}\mathbf{E}_{11}' = \mathbf{W}_1\mathbf{M}\mathbf{W}_1'. \tag{14}$$

Substituting $\mathbf{W}_1\mathbf{W}_2'\mathbf{E}_{22}'^{-1}$ for $\mathbf{E}_{12}$ in (8) we get

$$\boldsymbol{\Gamma}_{11} = \mathbf{E}_{11}^{-1}\mathbf{W}_1(\mathbf{I} - \mathbf{W}_2'(\mathbf{E}_{22}\mathbf{E}_{22}')^{-1}\mathbf{W}_2) = \mathbf{E}_{11}^{-1}\mathbf{W}_1(\mathbf{I} - \mathbf{W}_2(\mathbf{W}_2\mathbf{W}_2')^{-1}\mathbf{W}_2) = \mathbf{E}_{11}^{-1}\mathbf{W}_1\mathbf{M},$$

whence

$$\boldsymbol{\Gamma}_{11}'\boldsymbol{\Gamma}_{11} = \mathbf{M}\mathbf{W}_1'(\mathbf{E}_{11}\mathbf{E}_{11}')^{-1}\mathbf{W}_1\mathbf{M},$$

and so, by (14),

$$\boldsymbol{\Gamma}_{11}'\boldsymbol{\Gamma}_{11} = \mathbf{M}\mathbf{W}_1'(\mathbf{W}_1\mathbf{M}\mathbf{W}_1')^{-1}\mathbf{W}_1\mathbf{M}. \tag{15}$$

Again, from (9) we have

$$\boldsymbol{\Gamma}_{12}'\boldsymbol{\Gamma}_{12} = \mathbf{W}_2'(\mathbf{E}_{22}\mathbf{E}_{22}')^{-1}\mathbf{W}_2 = \mathbf{W}_2(\mathbf{W}_2\mathbf{W}_2')^{-1}\mathbf{W}_2,$$

and so

$$\boldsymbol{\Gamma}_{12}'\boldsymbol{\Gamma}_{12} = \mathbf{I} - \mathbf{M}. \tag{16}$$

Substituting (15), (16) and (14) respectively in (5), (6) and (7) we obtain

$$||a_{ij}|| = \mathbf{X}\mathbf{M}\mathbf{W}_1'(\mathbf{W}_1\mathbf{M}\mathbf{W}_1')^{-1}\mathbf{W}_1\mathbf{M}\mathbf{X}', \tag{17}$$

$$||c_{ij}|| = \mathbf{X}\mathbf{M}\mathbf{X}', \tag{18}$$

$$||\psi_{ij}|| = \mathbf{B}_1\mathbf{W}_1\mathbf{M}\mathbf{W}_1'\mathbf{B}_1'. \tag{19}$$

4. Generalized canonical correlations

The statistics fundamental in the study of problems arising out of the distribution (1) are the roots of the determinantal equation

$$|a_{ij} - \phi b_{ij}| = 0. \tag{20}$$

From $N \geqslant p+q$, i.e. $n \geqslant p$, it follows that $||b_{ij}|| = \mathbf{Z}\mathbf{Z}'$ is positive definite, and consequently (20) has all roots real, non-negative and finite. Substituting $\theta(1-\theta)^{-1}$ for ϕ (whence $\theta = \phi(1+\phi)^{-1}$) in (20) we get the equation

$$|a_{ij} - \theta c_{ij}| = 0, \tag{21}$$

whose roots are all real, non-negative and < 1. The rank of $||a_{ij}|| = \mathbf{Y}_1\mathbf{Y}_1'$ is l_1, where

$$l_1 = \mathrm{Min}(p, n_1).$$

Hence (21) has a root zero of multiplicity $p-l$. The remaining roots, say $\theta_1, \theta_2, \ldots, \theta_{l_1}$, are random variates whose joint distribution, when H_0 (i.e. H) is true, has already been obtained (Fisher, 1939; Hsu, 1939).

Let us obtain a determinantal equation which is equivalent to (21) but which has a different form, a form similar to the equation satisfied by Hotelling's canonical correlations (Hotelling, 1936). For this purpose let us recall the following identity.

$$\begin{vmatrix} \mathbf{P} & \mathbf{Q} \\ \mathbf{R} & \mathbf{S} \end{vmatrix} = |\mathbf{S}| \, . \, |\mathbf{P}-\mathbf{Q}\mathbf{S}^{-1}\mathbf{R}|, \tag{22}$$

which results from the identity

$$\begin{Vmatrix} \mathbf{P} & \mathbf{Q} \\ \mathbf{R} & \mathbf{S} \end{Vmatrix} . \begin{Vmatrix} \mathbf{I} & \mathbf{O} \\ -\mathbf{S}^{-1}\mathbf{R} & \mathbf{I} \end{Vmatrix} = \begin{Vmatrix} \mathbf{P}-\mathbf{Q}\mathbf{S}^{-1}\mathbf{R} & \mathbf{Q} \\ \mathbf{O} & \mathbf{S} \end{Vmatrix}$$

on taking determinants.

Consider now the following equation in r:

$$\begin{vmatrix} -r\mathbf{X}\mathbf{M}\mathbf{X}' & \mathbf{X}\mathbf{M}\mathbf{W}_1' \\ \mathbf{W}_1\mathbf{M}\mathbf{X}' & -r\mathbf{W}_1\mathbf{M}\mathbf{W}_1' \end{vmatrix} = 0. \tag{23}$$

Using (22) we may write the left-hand side of (23) as

$$(-r)^{n_1} \, |\mathbf{W}_1\mathbf{M}\mathbf{W}_1'| \, . \, |r^{-1}\mathbf{X}\mathbf{M}\mathbf{W}_1'(\mathbf{W}_1\mathbf{M}\mathbf{W}_1')^{-1}\mathbf{W}_1\mathbf{M}\mathbf{X}' - r\mathbf{X}\mathbf{M}\mathbf{X}'|,$$

which is, by virtue of (17) and (18), equal to

$$(-1)^{n_1} r^{n_1-p} \, |\mathbf{W}_1\mathbf{M}\mathbf{W}_1'| \, . \, |a_{ij}-r^2c_{ij}|.$$

Hence the non-vanishing roots of (23) are the non-vanishing roots of

$$|a_{ij}-r^2c_{ij}| = 0.$$

In other words, the non-vanishing roots of (23) are $\pm r_1, \ldots, \pm r_{l_1}$ and their squares are precisely the above defined statistics $\theta_1, \ldots, \theta_{l_1}$.

It can easily be verified that, on taking $\mathbf{W}_2 = \| 1, \ldots, 1 \|$, (23) is precisely the equation through which Hotelling defined the canonical correlations between the sets of variates $(x_1, \ldots, x_p)$ and $(w_1, \ldots, w_{q-1})$. Hence the square roots of $\theta_1, \ldots, \theta_{l_1}$ furnish a generalization of the canonical correlations.

5. Summary

The general regression problem, formulated as the problem of testing the hypothesis H (§ 1), is reduced to that of testing H_0 (§ 2) by means of a transformation of variables. The product moments of the new variables introduced are expressed as functions of the original variables by means of the matrix equations (17) and (18). The statistics defined in § 4 are shown to be adequate generalization of Hotelling's canonical correlations.

References

J. F. Daly (1940). 'On the unbiassed character of likelihood-ratio tests for independence in normal systems.' *Ann. Math. Stat.* 2, 1–32.

R. A. Fisher (1939). 'The sampling distribution of some statistics obtained from non-linear equations.' *Ann. Eugen., Lond.*, 9, 238–49.

H. Hotelling (1936). 'Relations between two sets of variates.' *Biometrika*, 2, 321–77.

P. L. Hsu (1939). 'On the distribution of roots of certain determinantal equations.' *Ann. Eugen., Lond.*, 9, 250–8.

—— (1940). 'On generalized analysis of variance.' *Biometrika*, 31, 221–37.

St Kolodziejczyk (1935). 'On an important class of statistical hypotheses.' *Biometrika*, 27, 161–90.

P. C. Tang (1938). 'The power function of the analysis of variance tests.' *Statist. Res. Mem.* 2, 126–49.

Reprinted from
Ann. Eugenics
11, 39–41 (1941).

ON THE PROBLEM OF RANK AND THE LIMITING DISTRIBUTION OF FISHER'S TEST FUNCTION

By P. L. HSU

We formulate the problem of rank as follows: the distribution of $x_{ir}(i = 1, \ldots, p; r = 1, \ldots, N)$ relative to given values of $w_{\mu r}(\mu = 1, \ldots, q; r = 1, \ldots, N)$ is

$$(2\pi)^{-\frac{1}{2}pN} \mid \alpha_{ij} \mid^{\frac{1}{2}N} \exp\left\{-\frac{1}{2}\sum_{i,j=1}^{p} \alpha_{ij} \sum_{r=1}^{N} (x_{ir}-\beta_{i1}w_{1r}-\ldots-\beta_{iq}w_{qr})(x_{jr}-\beta_{j1}w_{1r}-\ldots-\beta_{jq}w_{qr})\right\} \times \Pi dx_{ir}, \tag{1}$$

where $N \geqslant p+q$ and where the matrix $\| w_{\mu r} \|$ has always the rank q; the problem is to test whether the matrix

$$\mathbf{B}_1 = \begin{Vmatrix} \beta_{11} \cdots \beta_{1n_1} \\ \cdots\cdots\cdots\cdots \\ \beta_{p1} \cdots \beta_{pn_1} \end{Vmatrix}, \quad (n_1 \leqslant q) \tag{2}$$

has some hypothetical rank, say l.

We write

$$\mathbf{X} = \begin{Vmatrix} x_{11} \cdots x_{1N} \\ \cdots\cdots\cdots\cdots \\ x_{p1} \cdots x_{pN} \end{Vmatrix}, \quad \mathbf{W}_1 = \begin{Vmatrix} w_{11} \cdots w_{1N} \\ \cdots\cdots\cdots\cdots \\ w_{n_11} \cdots w_{n_1N} \end{Vmatrix}, \quad \mathbf{W}_2 = \begin{Vmatrix} w_{n_1+1,1} \cdots w_{n_1+1,N} \\ \cdots\cdots\cdots\cdots \\ w_{q1} \cdots w_{qN} \end{Vmatrix}, \tag{3}$$

$$\mathbf{M} = \mathbf{I}-\mathbf{W}_2'(\mathbf{W}_2\mathbf{W}_2')^{-1}\mathbf{W}_2. \tag{4}$$

It has been shown elsewhere* that, starting from the distribution (1), new variables y_{is} $(i = 1, \ldots, p; s = 1, \ldots, n_1)$ and z_{it} $(i = 1, \ldots, p; t = 1, \ldots, n)$, where $n = N-q \geqslant p$, may be introduced so that we have the distribution (relative to given values of the w's)

$$(2\pi)^{-\frac{1}{2}p(n+n_1)} \mid \alpha_{ij} \mid^{\frac{1}{2}(n+n_1)} \exp\left\{-\frac{1}{2}\sum_{i,j=1}^{p} \alpha_{ij} \sum_{s=1}^{n_1} (y_{is}-\eta_{is})(y_{js}-\eta_{js})-\frac{1}{2}\sum_{i,j=1}^{p} \alpha_{ij} \sum_{t=1}^{n} z_{it}z_{jt}\right\} \times \Pi d\eta_{is}dz_{jt}, \tag{5}$$

and that, writing

$$a_{ij} = \sum_{s=1}^{n_1} y_{is}y_{js}, \quad b_{ij} = \sum_{t=1}^{n} z_{it}z_{jt}, \tag{6}$$

$$\psi_{ij} = \sum_{s=1}^{n_1} \eta_{is}\eta_{js}, \tag{7}$$

we have the relations

$$\| a_{ij} \| = \mathbf{XMW}_1'(\mathbf{W}_1\mathbf{MW}_1')^{-1}\mathbf{W}_1\mathbf{MX}', \tag{8}$$

$$\| a_{ij}+b_{ij} \| = \mathbf{XMX}', \tag{9}$$

$$\| \psi_{ij} \| = \mathbf{B}_1\mathbf{W}_1\mathbf{MW}_1'\mathbf{B}_1'. \tag{10}$$

* Hsu (1940*c*).

It has also been shown that, if
$$l_1 = \mathrm{Min}\,(p, n_1) \tag{11}$$
and if $\theta_1, \ldots, \theta_{l_1}$ are the non-vanishing roots of the determinantal equation
$$|a_{ij} - \theta(a_{ij} + b_{ij})| = 0, \tag{12}$$
then $J\theta_1, \ldots, J\theta_{l_1}$ are the positive roots of the equation
$$\begin{vmatrix} -r\mathbf{XMX}' & \mathbf{XMW}_1' \\ \mathbf{W}_1\mathbf{MX}' & -r\mathbf{W}_1\mathbf{MW}_1' \end{vmatrix} = 0. \tag{13}$$

An important special case of (1), in which the w's are constants and assume appropriate values of 0 and 1, is the distribution of several samples taken independently from normal populations assumed to have the same set of variances and covariances. R. A. Fisher, when developing his theory of discriminant analysis of several samples,* has encountered the problem of rank. He put forward the following test function
$$\Phi_l = \phi_{l+1} + \ldots + \phi_{l_1} \tag{14}$$
where
$$\phi_i = \theta_i/(1-\theta_i) \quad (i = 1, \ldots, l_1) \tag{15}$$
and
$$\phi_1 \geqslant \phi_2 \geqslant \ldots \geqslant \phi_{l_1}, \tag{16}$$
so that $\phi_{l+1}, \ldots, \phi_{l_1}$, are the $l - l_1$ smallest non-vanishing roots of the determinantal equation
$$|a_{ij} - \phi b_{ij}| = 0. \tag{17}$$

In the following theorem we establish the limiting distribution of $n\Phi_l$ as $N \to \infty$, thus verifying the approximate distribution suggested by Fisher. It is important, however, to notice that the result holds true not only for the case of several samples, but also when the distribution of the w's satisfies a very general condition.

Theorem. Let $\mathbf{B}_1$ *be of rank* l, *so that the determinantal equation*
$$|\psi_{ij} - n\lambda\sigma_{ij}| = 0 \tag{18}$$
has zero as root of multiplicity $p-l$, *where the* σ_{ij} *are the variances and covariances of the distribution* (1). *Let the other roots of* (18) *be* $\lambda_1, \ldots, \lambda_l$. *Suppose that, as* $N \to \infty$ (i.e. $n = N - q \to \infty$) *the distribution of* $\lambda_1, \ldots, \lambda_l$ *approaches a limiting distribution which is such that for each* $i = 1, \ldots, l$ *the probability that* $\lambda_i = 0$ *is zero. Then the limiting distribution of* $n\Phi_l$ *is the* χ^2*-distribution with* $(p-l)(n_1-l)$ *degrees of freedom.*

Proof. It is known† that, in order to derive the distribution of $\phi_1, \ldots, \phi_{l_1}$ from the parent distribution (5), the latter may be replaced by the simpler form
$$(2\pi)^{-\frac{1}{2}p(n+n_1)} \exp\left\{-\frac{1}{2}\sum_{i=1}^{p}\sum_{s=1}^{n_1} y_{is}^2 - \frac{1}{2}\sum_{i=1}^{p}\sum_{t=1}^{n} z_{it}^2 + \sum_{i=1}^{l} J(n\lambda_i)\, y_{ii} - \tfrac{1}{2}n\sum_{i=1}^{l}\lambda_i\right\} \Pi dy dz. \tag{19}$$

* Fisher (1938).

† Hsu (1940*a*, p. 227, Theorem 3).

In another recent paper (1940*b*, Lemma 3) the author has proved that, under the assumption that the λ's are positive constants, the limiting distribution of $n\phi_{l+1}, \ldots, n\phi_{l_1}$ derived from (19), is that of the latent roots of the symmetric matrix $||d_{ij}||$ whose elements are

$$d_{ij} = \sum_{s=1}^{n_1-l} v_{is}v_{js} \quad (i, j = 1, \ldots, p-l), \tag{20}$$

where all the v's are mutually independent normal variates with mean 0 and standard deviation 1. This result holds true for our present more general case in which the λ's are random variates satisfying the condition of the Theorem, the proof being applicable without modification. Hence the limiting distribution of $n\Phi_l$ is that of $\sum_{i=1}^{p-l} d_{ii} = \sum_{i=1}^{p-l}\sum_{s=1}^{n_1-l} v_{is}^2$ which gives the χ^2-distribution with $(p-l)(n_1-l)$ degrees of freedom.

REFERENCES

R. A. Fisher (1938). 'The statistical utilization of multiple measurements.' *Ann. Eugen., Lond.*, **8**, 376–86.
P. L. Hsu (1940*a*). 'On generalized analysis of variance.' *Biometrika*, **31**, 221–37.
—— (1940*b*). 'On the limiting distribution of roots of a determinantal equation.' *Proc. Lond. Math. Soc.*
—— (1940*c*). 'Canonical reduction of the general regression problem.' *Ann. Eugen., Lond.*, **11**, 42–46.

THE LIMITING DISTRIBUTION OF A GENERAL CLASS OF STATISTICS[1)]

A great many statistics occurring in common practice are functions, with a high degree of regularity, of means:

$$f(\bar{u}_1, \bar{u}_2, \cdots, \bar{u}_m),$$

where $\bar{u}_i = n^{-1} \sum_{r=1}^{n} u_{ir}$ and where every pair u_{ir}, u_{js} $(r \neq s)$ are statistically independent. In this paper we solve the problem of finding the limiting distributions of all the statistics of such a nature. We shall give a reformulation of the problem in the language of pure theory of probability and state three theorems which constitute the solution.

THE PROBLEM. *Consider a set of random variables*

$$\begin{array}{l} X_{11}, X_{12}, \cdots, X_{1n}, \cdots \\ X_{21}, X_{22}, \cdots, X_{2n}, \cdots \\ \cdots\cdots\cdots\cdots\cdots\cdots \\ X_{m1}, X_{m2}, \cdots, X_{mn}, \cdots \end{array} \tag{1}$$

satisfying the following conditions:
(a) *for all i, j and $r \neq s$ the pair X_{ir} and X_{js} are mutually independent*; (b) *the simultaneous distribution of $X_{1r}, X_{2r}, \ldots, X_{mr}$ is independent of r*; (c) each *X_{ir} has a finite non-vanishing second moment about its mean.*
Let $f(x_1, x_2, \ldots, x_m)$ be a real function satisfying certain conditions of regularity. It is required to find the limiting distribution for $n \to \infty$ of

$$T^{(n)} = f(\bar{X}_1^{(n)}, \bar{X}_2^{(n)}, \cdots, \bar{X}_m^{(n)}). \tag{2}$$

where

$$\bar{X}_i^{(n)} = \frac{1}{n} \sum_{r=1}^{n} X_{ir} \quad (i = 1, \cdots, m), \tag{3}$$

We now state the theorems which constitute the solution:

THEOREM 1. *Let the random variables X_{ir} satisfy the conditions* (a), (b), (c) *and let*

$$\xi_i = \varepsilon(X_{ir}), \quad \eta_{ij} = \varepsilon(X_{ir} - \xi_i)(X_{jr} - \xi_j) \quad (i, j = 1, \cdots, m) \tag{4}$$

Let $f(x_1, \ldots, x_m)$ be a real function of m variables defined in the whole m-dimensional space and possessing everywhere continuous second derivatives with respect to all its arguments. Let

$$f_0 = f(\xi_1, \cdots, \xi_m), \; f_i = \frac{\partial}{\partial x_i} f(x_1, \cdots, x_m)|_{(x)=(\xi)} \quad (i = 1, \cdots, m), \tag{5}$$

---

1) 原载: *Acad. Sinica Science Record*, 1(1942), 37—41.

and let the quantity

$$\sigma^2 = \sum_{i,j=1}^{m} \eta_{ij} f_i f_j \tag{6}$$

be positive (it may vanish, but cannot be negative), Let

$$T^{(n)} = f(\bar{X}_1^{(n)}, \cdots, \bar{X}_m^{(n)}), \tag{7}$$

Then, for all real x, we have

$$\lim_{n\to\infty} P_r\{T^{(n)} \leqslant f_0 + n^{-\frac{1}{2}}\sigma x\} = (2\pi)^{-\frac{1}{2}} \int_{-\infty}^{x} e^{-\frac{1}{2}y^2} dy. \tag{8}$$

THEOREM 2. *Let the conditions in* THEOREM 1 *be subjected to the following modifications: (i) $f(x_1, \ldots, x_m)$ possesses continuous third derivatives with respect to all its arguments; (ii) the quantity σ^2 in (6) is equal to zero; (iii) denoting by*

$$f_{ij} = \frac{\partial^2}{\partial x_i \partial x_j} f(x_1, \cdots, x_m)|_{(x)=(\xi)} \quad (i, j = 1, \cdots, m) \tag{9}$$

the symmetric matrix $||f_{ij}||$ is definite (non-negative or non-positive).

Let

$$U_i^{(n)} = n^{\frac{1}{2}}(\bar{X}_i^{(n)} - \xi_i) \quad (i = 1, \cdots, m). \tag{10}$$

Then the limiting distribution of $n(T^{(n)} - f_0)$ is the same as the limiting distribution of $\frac{1}{2} \sum_{ij=1}^{n} f_{ij} U_i^{(n)} U_j^{(n)}$.

THEOREM 3. *Let the symmetric matrix $||\eta_{ij}||$ be of rank $\rho > 0$ and have the positive latent roots $\lambda_1, \lambda_2, \cdots, \lambda_\rho$. Let*

$$Q^{(n)} = \sum_{i=1}^{m} (U_i^{(n)})^2. \tag{11}$$

Then for all real x, we have

$$\lim_{n\to\infty} Pr\{Q^{(n)} \leqslant x\} \tag{12}$$

$$= \frac{(\lambda_1 \lambda_2, \cdots, \lambda_\rho)^{\frac{1}{2}}}{(2\pi)^{\frac{1}{2}\rho}} \int_{y_1^2+y_2^2+\cdots+y_\rho^2<x} e^{-\frac{1}{2}(\lambda_1 y_1^2+\cdots+\lambda_\rho y_\rho^2)} dy_1 dy_2 \cdots dy_\rho$$

If further the η_{ij} satisfy the following equations:

$$\sum_{l=1}^{m} \eta_{il} \eta_{jl} = \eta_{ij} \quad (i = 1, \cdots, m), \tag{13}$$

then we have

$$\lim_{n\to\infty} Pr\{Q^{(n)} \leqslant x\} = \{2^{\frac{1}{2}\rho}\Gamma(\tfrac{1}{2}\rho)\}^{-1} \int_0^x y^{\frac{1}{2}\rho-1} e^{-\frac{1}{2}y} dy, \tag{14}$$

the familiar Chi-square distribution with ρ degrees of freedom.

EXPLANATION. (*a*) The upshot of the first two theorems is that one can effect a Tayler expansion of functions of random variables just as one does for functions of ordinary variables, regarding the remainder as an infinitesimal of higher order "in probability." Thus Theorem 1 amounts to the expansion

$$T^{(n)} = f(\bar{X}_1^{(n)}, \cdots, \bar{X}_m^{(n)}) = f_0 + n^{-\frac{1}{2}} \sum_{i=1}^{m} f_i U_i^{(n)} + R^{(n)}, \tag{15}$$

where $n^{\frac{1}{2}} R^{(n)}$ tends to zero in probability as $n \to \infty$: Theorem 2 amounts to the expansion

$$T^{(n)} = f_0 + \frac{1}{2n}\sum_{i=1}^{m} f_{ij}U_i^{(n)}U_j^{(n)} + S^{(n)}. \tag{16}$$

where $nS^{(n)}$ tends to zero is probability as $n \to \infty$ and where the term $O(n^{-\frac{3}{2}})$is omitted because on the assumption that $\sigma^2 = 0$ this term has unit probability of being zero. Evidently we can deal with the case where both the terms $O(n^{-\frac{1}{2}})$ and $O(n^{-1})$ have the unit probability of being zero by investigating the limiting distribution of the term $O(n^{-\frac{3}{2}})$. But in practical statistics one has not encountered a case where the expansion beyond the term $O(n^{-1})$ is necessary.

(β) The condition in Theorem 2 that the matrix $||f_{ij}||$ should be definite is invariably satisfied by statistics that occur in practice.

(γ) Theorem 2 is, at first sight, incomplete in two respects. First, it involves a tacit assumption, namely that the limiting distribution of $\frac{1}{2}\sum f_{ij}U_i^{(n)}U_j^{(n)}$ exists; secondly, it does not give an explicit formula for this limiting distribution, even if the existence of the latter be granted. Both these difficulties are, in fact, answered in Theorem 3. For, as the matrix $||f_{ij}||$ is definite, the quadratic form is expressible as $\pm\sum_{i=1}^{\mu}(U_i'^{(n)})^2$ where each $U_i'^{(n)}$ is a linear homogeneous function of $U_1^{(n)}, \cdots, U_m^{(n)}$ with real coefficients depending only on the f_{ij}. If we write

$$U_i'^{(n)} = \sum_{j=1}^{\mu} b_{ij}U_j^{(n)} \quad (i = 1, \cdots, \mu), \tag{17}$$

we have

$$U_i'^{(n)} = n^{\frac{1}{2}}(\bar{X}'^{(n)} - \xi_i') \quad (i = 1, \cdots, \mu) \tag{18}$$

where

$$\bar{X}_i' = \frac{1}{n}\sum_{r=1}^{n} X_{ir}', \quad X_{ir}' = \sum_{j=1}^{\mu} b_{ij}X_{jr} \quad \text{and} \tag{19}$$

$$\xi_i' = \sum_{j=1}^{\mu} b_{ij}\xi_j \quad (i = 1, \cdots, \mu).$$

Evidently the random variables X_{ir} satisfy the same conditions (a) and (b) as the X_{ir}, but not necessarily (c), i.e. for certain values of i the variables $X_{i1}, X_{i2}, \ldots$ may have the unit probability of being equal to X_i. Hence for each i either $U_i'^{(n)}$ has the unit probability of being zero or it has exactly the same structure as $U_i^{(n)}$. Therefore either $\sum(U_i'^{(n)})^2$ has the unit probability of being zero or it has exactly the same structure as the $Q^{(n)}$ in (11), and consequently possesses a limiting distribution in the form (12) as claimed by Theorem 3. Thus not only does the validity of Theorem 3 imply the existence of the limiting distribution of $\sum f_{ij}U_i^{(n)}U_j^{(n)}$ as required by Theorem 2, but the reduction to a sum of squares and formula (12) together furnish the way of finding the actual limiting distribution in question.

Reprinted from
Sankhyā
6, 253–254 (1943).

SOME SIMPLE FACTS ABOUT THE SEPARATION OF DEGREES OF FREEDOM IN FACTORIAL EXPERIMENTS

By P. L. HSU, Ph.D., D.Sc.
National University of Peking, Kumming, China.

This subject is treated in *The Design and Analysis of Factorial Experiments*, by F. Yates, 1937 (Imperial Bureau of Soil Science—Technical Communication No. 35). Some results not given in that booklet will appear in this note. It is believed that the derivation is sufficiently simple to justify its total omission.

To each positive integer k let there correspond a real k-rowed orthogonal matrix, Δ_k, whose elements in the first row shall each be $k^{-\frac{1}{2}}$. The following is a well-known possible Δ_k :

$$\begin{Vmatrix} 1 & 1 & 1 & \cdots & 1 & 1 \\ 1 & -1 & 0 & \cdots & 0 & 0 \\ 1 & 1 & -2 & \cdots & 0 & 0 \\ \cdots & \cdots & \cdots & \cdots & \cdots & \cdots \\ 1 & 1 & 1 & \cdots & 1 & -(k-1) \end{Vmatrix}$$

with a normalizing factor for each row. For example, the third row should be divided by $\sqrt{1+1+4} = \sqrt{6}$. Besides this there are other forms of Δ_k which may prove more useful. Examples of Δ_3, Δ_4, Δ_5, Δ_6 and Δ_7 are given below with a normalizing factor for each row.

$$\Delta_3 : \begin{Vmatrix} 1 & 1 & 1 \\ -1 & 0 & 1 \\ 1 & -2 & 1 \end{Vmatrix}; \qquad \Delta_4 : \begin{Vmatrix} 1 & 1 & 1 & 1 \\ -1 & -1 & 1 & 1 \\ -1 & 1 & -1 & 1 \\ 1 & -1 & -1 & 1 \end{Vmatrix};$$

$$\Delta_5 : \begin{Vmatrix} 1 & 1 & 1 & 1 & 1 \\ -1 & -1 & 0 & 1 & 1 \\ -1 & 1 & 0 & -1 & 1 \\ 1 & -1 & 0 & -1 & 1 \\ 1 & 1 & -4 & 1 & 1 \end{Vmatrix}, \qquad \begin{Vmatrix} 1 & 1 & 1 & 1 & 1 \\ -1 & 0 & 0 & 0 & 1 \\ 0 & -1 & 0 & 1 & 0 \\ 1 & 0 & -2 & 0 & 1 \\ 2 & -3 & 2 & -3 & 2 \end{Vmatrix}$$

$$\Delta_6 : \begin{Vmatrix} 1 & 1 & 1 & 1 & 1 & 1 \\ -1 & 0 & 1 & -1 & 0 & 1 \\ 1 & -2 & 1 & 1 & -2 & 1 \\ -1 & -1 & -1 & 1 & 1 & 1 \\ 1 & 0 & -1 & -1 & 0 & 1 \\ -1 & 2 & -1 & 1 & -2 & 1 \end{Vmatrix}, \qquad \begin{Vmatrix} 1 & 1 & 1 & 1 & 1 & 1 \\ -1 & 0 & 0 & 0 & 0 & 1 \\ 0 & -1 & 0 & 0 & 1 & 0 \\ 0 & 0 & -1 & 1 & 0 & 0 \\ 1 & -1 & 0 & 0 & -1 & 1 \\ 1 & 1 & -2 & -2 & 1 & 1 \end{Vmatrix}$$

$$\Delta_7 : \begin{Vmatrix} 1 & 1 & 1 & 1 & 1 & 1 & 1 \\ -1 & 0 & 1 & 0 & -1 & 0 & 1 \\ 1 & -2 & 1 & 0 & 1 & -2 & 1 \\ -1 & -1 & -1 & 0 & 1 & 1 & 1 \\ 1 & 0 & -1 & 0 & 1 & 0 & -1 \\ -1 & 2 & -1 & 0 & 1 & -2 & 1 \\ 1 & 1 & 1 & -6 & 1 & 1 & 1 \end{Vmatrix}, \quad \begin{Vmatrix} 1 & 1 & 1 & 1 & 1 & 1 & 1 \\ -1 & 0 & 0 & 0 & 0 & 0 & 1 \\ 0 & -1 & 0 & 0 & 0 & 1 & 0 \\ 0 & 0 & -1 & 0 & 1 & 0 & 0 \\ 1 & 0 & 0 & -2 & 0 & 0 & 1 \\ 2 & 0 & -3 & 2 & -3 & 0 & 2 \\ 2 & -5 & 2 & 2 & 2 & -5 & 2 \end{Vmatrix}$$

Suppose that the experiment consists of two factors, at k and l levels respectively. It is here more convenient to use 1 for the lowest level instead of the customary 0. The yield totals will therefore be denoted by $x_{\alpha\beta}(\alpha = 1 \ldots, k\,;\, \beta = 1,\ldots, l)$. Let

$$\| y_{\alpha\beta} \| = \Delta_k \| x_{\alpha\beta} \| \Delta'_l$$

the accent denoting the transposed matrix; then y^2_1 is the correction for mean, $\sum\limits_{\alpha}' y^2_{\alpha 1}$ and $\sum\limits_{\beta}' y^2_{1\beta}$ are respectively the sums of squares for the main effects, and $\sum\limits_{\alpha,\beta}' y^2_{\alpha\beta}$ is the interaction sum of squares. Thus all the $kl - 1$ degrees of freedom are separated into single degrees of freedom. (Σ' denotes summation with respect to indices each of which runs from 2 onwards).

The passage from $\| x_{\alpha\beta} \|$ to $\| y_{\alpha\beta} \|$ will be called the $(k\,;\,l)$ operation.

For the $k \times l \times m$ design there are m matrices $\|x_{\alpha\beta\gamma}\|$, wherein γ is held fast and α, β stand for running indices. Let each be subjected to the $(k;\, l)$ operation and let the resulting matrices be $\|y_{\alpha\beta\gamma}\|$. Then the following analysis is valid except for numerical multiptiers.

Correction for mean $y^2_{11\cdot}$

Main effects $\sum\limits_{\alpha}' y^2_{\alpha 1 \cdot}$, $\sum\limits_{\beta}' y^2_{1\beta\cdot}$ and $\sum\limits_{\gamma}(y_{11\gamma} - y_{11\cdot})^2$

Two factor interactions $\sum\limits_{\alpha,\beta}' y^2_{\alpha\beta\cdot}$, $\sum\limits_{\alpha}' \sum\limits_{\gamma}(y_{\alpha 1\gamma} - y_{\alpha 1\cdot})^2$ and $\sum\limits_{\beta}' \sum\limits_{\gamma}(y_{1\beta\gamma} - y_{1\beta\cdot})^2$

Three factor interaction $\sum\limits_{\alpha,\beta}' \sum\limits_{\gamma}(y_{\alpha\beta\gamma} - y_{\alpha\beta\cdot})^2$

For the $k \times l \times m \times n$ design, construct for each fixed (γ, δ) the matrix $\| x_{\alpha\beta\gamma\delta} \|$; there are mn such matrices. Let each be subjected to the $(k\,;\,l)$-operation and let the resulting matrices be $\| y_{\alpha\beta\gamma\delta} \|$. Next, rearrange the y's so that α, β are held fast and γ, δ are the running indices of the matrices $\| y_{\alpha\beta\gamma\delta} \|$; there are kl such matrices. Let each be subjected to the $(m\,;\,n)$-operation and let the resulting matrices be $\| u_{\alpha\beta\gamma\delta} \|$. Then the following analysis is valid.

Correction for mean u^2_{1111}

Main effects $\sum\limits_{\alpha}' u^2_{\alpha 111}$, $\sum\limits_{\beta}' u^2_{1\beta 11}$, $\sum\limits_{\gamma}' u^2_{11\gamma 1}$ and $\sum\limits_{\delta}' u^2_{111\delta}$

Two factor interactions $\sum\limits_{\alpha,\beta}' u^2_{\alpha\beta 11}$, $\sum\limits_{\alpha,\gamma}' u^2_{\alpha 1\gamma 1}$, etc.

Three-factor interactions $\sum\limits_{\alpha,\beta,\gamma,\delta}' u^2_{\alpha\beta\gamma 1}$, etc.

Four-factor interaction $\sum\limits_{\alpha,\beta,\gamma,\delta}' u^2_{\alpha\beta\gamma\delta}$

This method can be extended to any number of factors.

Paper received : 24 November, 1942

Reprinted from
Ann. Math. Statist.
16, 1–29 (1945).

THE APPROXIMATE DISTRIBUTIONS OF THE MEAN AND VARIANCE OF A SAMPLE OF INDEPENDENT VARIABLES

BY P. L. HSU

The National University of Peking

1. Introduction. In this paper we shall study the mean and variance of a large number, n (a sample of size n) of mutually independent random variables:

$$\xi_1, \xi_2, \cdots, \xi_n, \tag{1}$$

having the same probability distribution represented by a (cumulative) distribution function $P(x)$. The rth moment, absolute moment, and semi-invariant of $P(x)$ are denoted by α_r, β_r, and γ_r respectively. It is assumed that for a certain integer $k \geq 3$, $\beta_k < \infty$ and that $\alpha_2 > 0$. Hence there is no loss of generality in assuming that

$$\alpha_1 = 0, \qquad \alpha_2 = 1. \tag{2}$$

The characteristic function corresponding to $P(x)$ is denoted by $p(t)$.

We put

$$\bar{\xi} = \frac{1}{n}\sum_{r=1}^{n} \xi_r, \qquad \eta = \frac{1}{n}\sum_{r=1}^{n} (\xi_r - \bar{\xi})^2 \tag{3}$$

$$F(x) = Pr\{\sqrt{n}\,\bar{\xi} \leq x\}, \qquad G(x) = Pr\left\{\frac{\sqrt{n}(\eta - 1)}{\sqrt{\alpha_4 - 1}} \leq x\right\}. \tag{4}$$

The definition of $G(x)$ implies that $\alpha_4 < \infty$ and $\alpha_4 - 1 > 0$. The case $\alpha_4 - 1 = 0$ provides an easy degenerated case which will be treated separately (section 4).

Cramér's theorem of asymptotic expansion[1] reads as follows:

THEOREM 1. *If $P(x)$ is non-singular and if $\beta_k < \infty$ for some integer $k \geq 3$, then*

$$F(x) = \Phi(x) + \Psi(x) + R(x) \tag{5}$$

where

$$\Phi(x) = \frac{1}{\sqrt{2\pi}} \int_{-\infty}^{x} e^{-\frac{1}{2}y^2}\, dy. \tag{6}$$

$\Psi(x)$ is a certain linear combination of successive derivatives $\Phi^{(3)}(x), \cdots, \Phi^{(3(k-3))}(x)$ with each coefficient of the form $n^{-\frac{1}{2}\nu}$ times a quantity depending only on $k, \alpha_3, \cdots, \alpha_{k-1}$ $(1 \leq \nu \leq k - 3)$ and

$$| R(x) | \leq Q/n^{\frac{1}{2}(k-2)} \tag{7}$$

where Q is a constant depending only on k and $P(x)$.

---

[1] H. CRAMÉR: *Random Variables and Probability Distributions* (1937), Ch. 7. This book will be referred to as (C).

In particular, putting $k = 3$ we get that $|F(x) - \Phi(x)| \leq Qn^{-\frac{1}{2}}$ provided $P(x)$ is non-singular and $\beta_3 < \infty$. If the condition of non-singularity of $P(x)$ be removed, then Liapounoff's theorem[2] furnishes the weaker result: $|F(x) - \Phi(x)| \leq A\beta_3 n^{-\frac{1}{2}} \log n$ where A is a numerical constant.

Very recently Berry[3] succeeded in removing the factor $\log n$ from Liapounoff's theorem under no other condition than that $\beta_3 < \infty$. We state here Berry's theorem:

THEOREM 2. *If $\beta_3 < \infty$, then*

$$|F(x) - \Phi(x)| \leq \frac{A\beta_3}{\sqrt{n}} \tag{8}$$

where A is a numerical constant.

An essential step in the proof of these results is the selection of a weighting function $w(x)$ and the appraisal of the integral

$$\int_{-\infty}^{\infty} w(u)\{F(u + x) - \Phi(u + x) - \Psi(u + x)\}\, du \tag{9}$$

($\Psi \equiv 0$ when $k = 3$). In his book[1] Cramér proves Theorem 1 by taking $w(u) = \frac{1}{\Gamma(\omega)}(-u)^{\omega-1}$ when $u < 0$ and $w(u) = 0$ when

$$u \geq 0 \quad (0 < \omega < 1) \tag{10}$$

and proves Liapounoff's theorem by taking

$$w(u) = \frac{1}{\sqrt{2\pi}\epsilon} e^{-u^2/2\epsilon^2}. \tag{11}$$

On the other hand, Berry uses the following weighting function in his proof of Theorem 2:

$$w(u) = \frac{1 - \cos Tu}{u^2}. \tag{12}$$

The unfortunate selection of the function (11) accounts for the presence of the factor $\log n$ in Liapounoff's theorem.

Now Cramér's proof of Theorem 1, based on the integral (9) with $w(u)$ defined in (10), makes use of a result on that integral due to M. Riesz. A more elementary proof than this can be devised. In fact, one has only to use, with Berry, the function (12) and to adopt his elementary appraisal[4] of the integral

---

[2] (C), Ch. 7.

[3] A. C. BERRY: "The accuracy of the Gaussian approximation to the sum of independent variates." *Trans. Amer. Math. Soc.*, Vol. 49 (1941), pp. 122–136. This paper will be referred to as (B).

[4] Berry proves the inequality (in our notation):

$$\left|\int_{-\infty}^{\infty} \frac{1 - \cos Tx}{x^2}\{F(x + a) - \Phi(x + a)\}\, dx\right| \leq \int_0^T \frac{(T - t)\,|f(t) - e^{-\frac{1}{2}t^2}|}{t}\, dt$$

(9) in order to obtain the proof of Theorem 1. One of our purposes is therefore to give an elementary proof of Theorem 1, without reference to the above-mentioned result due to M. Riesz. Section 2 is devoted to this work.

We ought to add that Cramér's theorem and Berry's theorem correspond to Theorems 1 and 2 for the case in which the random variables (1) do not follow the same distribution. The proof given in Section 2 is adaptable to these more general theorems when subjected to appropriate modifications; the assumption of a common distribution function for (1) is only made for the sake of convenience.

So much for the known results for the approximate distribution of $\bar{\xi}$. By a purely formal operational method Cornish and Fisher[5] obtain terms of successive approximation to the distribution function of any random variable X with the help of its semi-invariants. It is hardly necessary to emphasize the importance of turning Cornish and Fisher's formal result (asymptotic expansion without appraisal of the remainder) into a mathematical theorem of asymptotic expansion which gives the order of magnitude of the remainder. In this paper we achieve this for the simplest function of (1) next to $\bar{\xi}$, viz. the η in (3). We do not seek to remove the assumption of a common distribution for (1), as there will be no practical significance (e.g. in statistics) of η if the variables (1) do not have the same probability distribution. Section 3 is devoted to the proof of the following theorems:

THEOREM 3. *If* $\alpha_6 < \infty$ *and* $\alpha_4 - 1 - \alpha_3^2 \neq 0$ *(it cannot be negative), then*

$$(13)\qquad | G(x) - \Phi(x) | \leq \frac{A}{\sqrt{n}} \left(\frac{\alpha_6}{\alpha_4 - 1 - \alpha_3^2} \right)^{3/2}$$

where A *is a numerical constant.*

THEOREM 4. *Let* $P(x)$ *be non-singular and let* $\alpha_{2k} < \infty$ *for some integer* $k > 3$. *Then*

$$(14)\qquad G(x) = \Phi(x) + \chi(x) + R_1(x),$$

where $\Phi(x)$ *is the function* (6), $\chi(x)$ *is a linear combination of the derivatives* $\Phi'(x)$, $\cdots, \Phi^{(3(k-3))}(x)$ *with each coefficient of the form* $n^{-\frac{1}{2}\nu}$ *times a quantity depending only on* k *and* $\alpha_3, \alpha_4, \cdots, \alpha_{2k-2}$, *and*

(B), p. 128. The "appraisal" mentioned here refers to (50) which is contained in B, p. 128. But Berry's appraisal of the integral in the right-hand side of the above inequality is in default. He writes

$$\frac{\epsilon}{6} \int_0^{c/\epsilon} \left(\frac{1.1}{\epsilon} - t \right) t^2 e^{-\frac{1}{2}t^2}\, dt = \frac{1.1}{6} \sqrt{\frac{\pi}{2}} - \frac{\epsilon}{3} - \frac{1}{6} \int_{c/\epsilon}^{\infty} \left\{ (1.1 - c)t^2 + c - \frac{2c}{t^2} \right\} e^{-t^2/2}\, dt$$

(B, p. 132, line 3) whilst the last integral ought to be

$$\int_{c/\epsilon}^{\infty} \{(1.1 - c)t^2 + c - 2\epsilon t\} e^{-t^2/2}\, dt.$$

[5] E. A. Cornish and R. A. Fisher: "Moments and cumulants in the specification of distributions." (Revue de l'Institut International de Statistique (1937), pp. 1–14.)

$$|R_1(x)| \leq \frac{Q_k}{n^{\frac{1}{2}(k-2)}} \quad \text{if } k = 4, 5 \text{ or } 6 \tag{15}$$

$$|R_1(x)| \leq \frac{Q'_k}{n^{k(k-1)/(2k+3)}} \quad \text{if } k \geq 7 \tag{16}$$

where Q_k and Q'_k are constants depending only on k and $P(x)$.

It may be noticed that Theorem 3 is a "Berryian" theorem about $G(x)$, its characteristic feature being the absence of any condition on the distribution function except the two on its moments, and that Theorem 4 is a "Cramerian" theorem about $G(x)$, the characteristic feature being the assumption of non-singularity of $P(x)$ besides that $\alpha_{2k} < \infty$.

In proving these theorems we have devised a method which is applicable to getting similar results about functions other than η, such as functions commonly used in applied statistics: the higher moments about the means, the moment ratios (e.g. K. Pearson's b_1 and b_2), the covariance, the coefficient of correlation, and "Student's" t-statistic. Works on such functions are being done by my university colleagues, and the results will be published shortly.

If ξ is any of the random variables (1), then

$$0 \leq \epsilon\{a(\xi^2 - 1) + b\xi\} = a^2(\alpha_4 - 1) + 2ab\alpha_3 + b^2$$

for all real (a, b). Hence $\alpha_4 - 1 - \alpha_3^2 \geq 0$, and $\alpha_4 - 1 - \alpha_3^2 = 0$ means that there is unit probability that ξ assumes exactly two values. This easily degenerated case is first eliminated in Theorem 3 by the assumption $\alpha_4 - 1 - \alpha_3^2 \neq 0$ and then considered in section 4. In Theorem 4 the condition $\alpha_4 - 1 - \alpha_3^2 \neq 0$ is implied since ξ cannot be a random variable of the nature just described owing to the non-singularity of $P(x)$.

2. Lemmas. Throughout this paper A, B, C, etc. will denote positive numerical constants; A_k, B_k (A_{km}, B_{km}), etc., will denote positive constants depending only on some integer k (integers k and m), and Q_k (Q_{km}) will denote a positive constant depending only on k (k and m) and the distribution function $P(x)$. ϑ, Θ, Θ_k, (Θ_{km}), Λ_k (Λ_{km}) will denote respectively quantities such that $|\vartheta| \leq 1$, $|\Theta| \leq A$, $|\Theta_k| \leq A_k$ ($|\Theta_{km}| \leq A_{km}$), $|\Lambda_k| \leq Q_k$ ($|\Lambda_{km}| \leq Q_{km}$). These symbols do not necessarily stand for the same quantity at each occurrence. Thus $2\vartheta = \Theta$, $k\Theta_k = \Theta_k$ etc. In particular any positive functions of $k, \alpha_3, \cdots, \alpha_k$ is a Q_k.

1.1. Cramér obtains the asymptotic expansion of the characteristic function of the distribution of $\sqrt{n}\xi$, viz. $\epsilon(e^{it\sqrt{n}\xi})$, when (1) do not have the same distribution, valid for $|t| \leq Q_k n^{1/6}$. Since we assume a common distribution for (1), so that the characteristic function is $\left\{p\left(\frac{t}{\sqrt{n}}\right)\right\}^n$, we are able to derive an asymptotic expansion valid for $|t| \leq Q_k\sqrt{n}$. The extension to $\left\{p\left(\frac{t_1}{\sqrt{n}},\right.\right.$

$\cdots, \frac{t_m}{\sqrt{n}}\Big)\Big\}^n$ presents no difficulty. This is done in the following three lemmas, of which Lemma 3 contains the final result.

LEMMA 1.

$$\log p(t) = \sum_{r=2}^{k-1} \frac{\gamma_r (it)^r}{r!} + \Theta_k \beta_k |t|^k, \quad \text{for } |t| \leq \beta_k^{1/k}. \tag{17}$$

PROOF: Since $p(t) = 1 + \sum_{r=1}^{k-1} \frac{\alpha_r (it)^r}{r!} + \frac{\vartheta \beta_k |t|^k}{k!} = 1 + q(t)$ say, we have, for $\beta_k^{1/k} |t| \leq 1$,

$$q(t) \leq \sum_{r=2}^{k} \frac{\beta_r |t|^r}{r!} \leq \sum_{r=2}^{k} \frac{(\beta_k^{1/k} |t|)^r}{r!} < \sum_{r=2}^{\infty} \frac{1}{r!} = e - 2 < \frac{3}{4}.$$

Hence

$$\log p(t) = \sum_{1 \leq j \leq [\frac{1}{2}(k-1)]} (-1)^{j+1} \frac{\{q(t)\}^j}{j} + \Theta |q(t)|^{[\frac{1}{2}(k+1)]} \tag{18}$$

For $1 \leq j \leq [\frac{1}{2}(k-1)]$ let us expand each $(-1)^{j+1} j^{-1} \{q(t)\}^j$ to get a polynomial $q_j(t)$ of degree $k - 1$ and a remainder $r_j(t)$. In doing this we regard $q(t)$ formally as a polynomial of degree k in t. For this polynomial we have the majorating relation

$$q(t) \ll e^{\beta_k^{1/k} |t|},$$

whence

$$\frac{(-1)^j}{j} \{q(t)\}^j \ll e^{j \beta_k^{1/k} |t|},$$

which gives

$$|r_j(t)| \leq \sum_{r=k}^{\infty} \frac{j^r \beta_k^{r/k} |t|^r}{r!} \leq j^k \beta_k |t|^k e^{j \beta_k^{1/k} |t|} \leq j^k e^j \beta_k |t|^k \leq A_k \beta_k |t|^k. \tag{19}$$

Similarly,

$$|q(t)|^{[\frac{1}{2}(k+1)]} \leq A_k \beta_k |t|^k. \tag{20}$$

From (18), (19), (20) we obtain

$$\log p(t) = \sum_{1 \leq j \leq [\frac{1}{2}(k-1)]} q_j(t) + \Theta_k \beta_k |t|^k. \tag{21}$$

Since the sum in (21) must equal the sum in (17), the Lemma is proved.

LEMMA 2. *Let* $(\zeta_1, \zeta_2, \cdots, \zeta_m)$ *be a random point with* $\epsilon(\zeta_i) = 0$ *and* $\epsilon(|\zeta_i|^k) = \beta_{ki} < \infty$ *for some integer* $k \geq 3$ $(i = 1, \cdots, m)$. *Let* $p(t_1, \cdots, t_m)$ *be the characteristic function. Then for* $|t_i| \leq m^{-2+1/k} \beta_{ki}^{-1/k} \sqrt{n}$ $(i = 1, \cdots, m)$ *we have*

$$n \log p\left(\frac{t_1}{\sqrt{n}}, \cdots, \frac{t_m}{\sqrt{n}}\right) = \sum_{r=2}^{k-1} \frac{i^r U_r}{r!\, n^{\frac{1}{2}(r-2)}} + \frac{\Theta_k V_k}{n^{\frac{1}{2}(k-2)}} \tag{22}$$

where U_r and V_r are the rth *semi-invariant and the absolute moment respectively of* $\Sigma t_i\zeta_i$

PROOF: If $|t_i| \le m^{-2+1/k}\beta_{ki}^{-1/k}\sqrt{n}$, then $V_k^{1/k} \le m^{(k-1)/k}(\Sigma\beta_{ki}|t_i|^k)^{1/k} \le m^{(k-1)/k}(\Sigma\beta_{ki}^{1/k}|t_i|) \le \sqrt{n}$. Since $p\left(\frac{t_1}{\sqrt{n}}, \cdots, \frac{t_m}{\sqrt{n}}\right)$ is the value at $t = \frac{1}{\sqrt{n}}$ of the characteristic function of $\Sigma t_i\zeta_i$, it follows from Lemma 1 that for $\sqrt{n} \ge V_k^{1/k}$ we have (22).

LEMMA 3. *Let $(\zeta_1, \cdots, \zeta_m)$ be a random point with $\epsilon(\zeta_i) = 0$, $\epsilon(\zeta_i^2) = 1$ and $\epsilon(|\zeta_i|^k) = \beta_{ki} < \infty$ for some integer $k \ge 3$. Let $\rho_{ij} = \epsilon(\zeta_i\zeta_j)(\rho_{ii} = 1; i, j = 1, \cdots, m)$ and the matrix $\|\rho_{ij}\|$ be positive definite. Let*

$$\Delta = \text{det.}\,|\rho_{ij}|, \qquad \varphi(t_1, \cdots, t_m) = e^{-\frac{1}{2}\sum_{i,j=1}^{m}\rho_{ij}t_it_j} \tag{23}$$

Let $p(t_1, \cdots, t_m)$ be the characteristic function. Then there exists a B_{km} such that for $|t_i| \le \frac{B_{km}\Delta\sqrt{n}}{\beta_{ki}^{3/k}}$ $(i = 1, \cdots, m)$ we have

$$\begin{aligned}\left\{p\left(\frac{t_1}{\sqrt{n}}, \cdots, \frac{t_m}{\sqrt{n}}\right)\right\}^n &= \varphi(t_1, \cdots, t_m)\{1 + \psi(it_1, \cdots, it_m)\} \\ &\quad + \frac{\Theta_{km}}{n^{\frac{1}{2}(k-2)}}\left\{\sum_{i=1}^{m}\beta_{ki}^{3(k-2)/k}(|t_i|^k\right. \\ &\quad \left. + |t_i|^{k+1} + \cdots + |t_i|^{3(k-2)})\right\}e^{-\Delta/4m^{m-1}\sum_{i=1}^{m}t_i^2}\end{aligned} \tag{24}$$

where $\psi(it_1, \cdots, it_m)$ is a polynomial each of whose terms has the form

$$\frac{1}{n^{\nu/2}}a_{\nu_1\cdots\nu_m}(it_1)^{\nu_1}\cdots(it_m)^{\nu_m},$$

with $1 \le \nu \le k-3$, $3 \le \nu_1 + \cdots + \nu_m \le 3(k-3)$, and $a_{\nu_1\cdots\nu_m}$ depending only on k and the moments $\epsilon(\zeta_1^{\mu_1}\cdots\zeta_m^{\mu_m})$, $3 \le \mu_1 + \cdots + \mu_m \le k-1$. If $k = 3$, then $\psi = 0$.

PROOF. If $|t_i| \le m^{-2+(1/k)}\beta_{ki}^{-3/k}\Delta\sqrt{n}$, then $|t_i| \le m^{-2+(1/k)}\beta_{ki}^{-1/k}\sqrt{n}$ since $\Delta \le 1$ and $\beta_{ki} \ge 1$. It follows from Lemma 2 and the fact $U_2 = \Sigma\rho_{ij}t_it_j$ that

$$\begin{aligned}\left\{p\left(\frac{t_1}{\sqrt{n}}, \cdots, \frac{t_m}{\sqrt{n}}\right)\right\}^n &= \varphi(t_1, \cdots, t_m)e^s \\ &= \varphi(t_1, \cdots, t_m)\left\{1 + \sum_{j=1}^{k-3}\frac{s^j}{j!} + \frac{\vartheta|s|^{k-2}e^{|s|}}{(k-2)!}\right\}\end{aligned} \tag{25}$$

where

$$s = \frac{i^3}{\sqrt{n}}\sum_{r=0}^{k-3}\frac{i^r U_{r+3}}{(r+3)!\,n^{r/2}} + \frac{\Theta_k V_k}{n^{\frac{1}{2}(k-2)}}. \tag{26}$$

Regarding s formally as a polynomial in $n^{-\frac{1}{2}}$ let us expand each $(j!)^{-1}s^j$ $(1 \leq j \leq k-3)$ to get a polynomial s_j of degree $k-3$ in $n^{-\frac{1}{2}}$ and a remainder r_j. For the formal polynomial s we have the majorating relation

$$s \ll \frac{A_k}{\sqrt{n}} \sum_{r=0}^{k-3} \frac{V_{r+3}}{r!\, n^{r/2}} \ll \frac{A_k}{\sqrt{n}} \sum_{r=0}^{k-3} \frac{V_k^{(r+3)/k}}{r!\, n^{r/2}} \ll \frac{A_k V_k^{3/k}}{\sqrt{n}} e^{V_k^{1/k} n^{-\frac{1}{2}}}, \tag{27}$$

whence

$$\frac{1}{j!} s^j \ll A_k \frac{V_k^{3j/k}}{n^{j/2}} e^{jV_k^{1/k} n^{-\frac{1}{2}}},$$

which gives

$$|r_j| \leq \frac{A_k V_k^{3j/k}}{n^{j/2}} \sum_{\nu=k-2-j}^{\infty} \frac{j^\nu V_k^{\nu/k}}{\nu!\, n^{\nu/2}} \leq \frac{A_k V_k^{(k-2+2j)/k}}{n^{\frac{1}{2}(k-2)}} e^{j(V_k^{1/k}/\sqrt{n})}.$$

Since $V_k^{1/k} n^{-\frac{1}{2}} \leq 1$ as shown in the proof of Lemma 2, we have

$$|r_j| \leq \frac{A_k V_k^{(k-2+2j)/k}}{n^{\frac{1}{2}(k-2)}} \leq \frac{A_{km}(\sum_i \beta_{ki} |t_i|^k)^{(k-2+2j)/k}}{n^{\frac{1}{2}(k-2)}}$$

$$\leq \frac{A_{km}(\sum_i \beta_{ki}^{1/k} |t_i|)^{k-2+2j}}{n^{\frac{1}{2}(k-2)}} \leq \frac{A_{km} \sum \beta_{ki}^{(k-2+2j)/k} |t_i|^{k-2+2j}}{n^{\frac{1}{2}(k-2)}}.$$

Since $\beta_{ki} \geq 1$ we have $\beta_{ki}^{(k-2+2j)/k} \leq \beta_{ki}^{3(k-2)/k}$. Hence

$$|r_j| \leq \frac{A_{km} \sum_i \beta_{ki}^{3(k-2)/k} |t_i|^{k-2+2j}}{n^{\frac{1}{2}(k-2)}}. \tag{28}$$

Similarly

$$\frac{|s|^{k-2}}{(k-2)!} \leq \frac{A_{km} \sum_i \beta_{ki}^{3(k-2)/k} |t_i|^{3(k-2)}}{n^{\frac{1}{2}(k-2)}}. \tag{29}$$

From (25), (28), (29) we get

$$\left\{p\left(\frac{t_1}{\sqrt{n}}, \cdots, \frac{t_m}{\sqrt{n}}\right)\right\}^n = \varphi(t_1, \cdots, t_m)\left\{1 + \sum_{j=1}^{k-3} s_j + \sum_{j=1}^{k-3} r_j + \frac{\vartheta |s|^{k-2}}{(k-2)!} e^{|s|}\right\}$$

$$= \varphi(t_1, \cdots, t_m) \{1 + \psi(it_1, \cdots, it_m)\}$$

$$+ \frac{\Theta_{km}}{n^{\frac{1}{2}(k-2)}} \{\Sigma \beta_{ki}^{3(k-2)/k}(|t_i|^k + |t_i|^{k+1} + \cdots + |t_i|^{3(k-2)})\} \varphi(t_1, \cdots, t_m) e^{|s|}$$

where $\psi(it_1, \cdots, it_m)$ stands for Σs_j. The assertion about $\psi(it_1, \cdots, it_m)$ announced in the lemma can now be seen without difficulty. It remains to show that with suitable B_{km} in the lemma, we have

$$\varphi(t_1, \cdots, t_m) e^{|s|} \leq e^{-\Delta/4m^{m-1} \sum_{i=1}^m t_i^2}$$

i.e.

$$(30)\qquad -\frac{1}{2}\sum_{i,j=1}^{m}\rho_{ij}t_it_j + |s| \le -\frac{\Delta}{4m^{m-1}}\sum_{i=1}^{m}t_i^2 .$$

From (27) we have

$$(31)\qquad \begin{aligned} |s| &\le \frac{A_k}{\sqrt{n}}V_k^{3/k} \le \frac{A_{km}}{\sqrt{n}}\left(\sum_i \beta_{ki}|t_i|^k\right)^{3/k} \\ &\le \frac{A_{km}}{\sqrt{n}}\left(\sum_i \beta_{ki}^{1/k}|t_i|\right)^3 \le \frac{A_{km}}{\sqrt{n}}\sum_i \beta_{ki}^{3/k}|t_i|^3. \end{aligned}$$

If we choose $B_{km} \lessapprox (4m^{m-1}A_{km})^{-1}$ (and $B_{km} \le m^{-2+(1/k)}$ in order that the earlier results may not be affected), the A_{km} here coinciding with the last written A_{km} in (31), we have, for $|t_i| \le B_{km}\beta_{ki}^{-3/k}\Delta\sqrt{n}$,

$$(32)\qquad |s| \le \frac{\Delta}{4m^{m-i}}\sum_{i=1}^{m}t_i^2 .$$

On the other hand, if $\lambda_1, \lambda_2, \cdots, \lambda_m$ are the latent roots of $\|\rho_{ij}\|$ then each $\lambda_i \le m$ since their sum is m. Letting λ_1 be the smallest one we have

$$(33)\qquad \frac{1}{2}\sum_{i,j}\rho_{ij}t_it_j \ge \frac{1}{2}\lambda_1\sum t_i^2 = \frac{\lambda_1\lambda_2\cdots\lambda_m}{2\lambda_2\cdots\lambda_m}\sum t_i^2 \ge \frac{\Delta}{2m^{m-1}}\sum t_i^2 .$$

(32) and (33) imply (30). Hence the lemma is proved.

Let us write down the particular cases $m = 1$ and $m = 2$ of (24):

$$(34)\qquad \begin{aligned} \left\{p\left(\frac{t}{\sqrt{n}}\right)\right\}^n &= e^{-\frac{1}{2}t^2}(1 + \psi(it)) \\ &+ \frac{\Theta_k}{n^{\frac{1}{2}(k-2)}}\beta_k^{3(k-2)/k}\{|t|^k + |t|^{k+1} + \cdots + |t|^{3(k-2)}\}e^{-t^2/4}, \left(|t| \le \frac{A_k\sqrt{n}}{\beta_k^{3/k}}\right) \end{aligned}$$

$$(35)\qquad \begin{aligned} \left\{p\left(\frac{t_1}{\sqrt{n}}, \frac{t_2}{\sqrt{n}}\right)\right\}^n &= e^{-\frac{1}{2}(t_1^2+t_2^2+2\rho t_1t_2)}\{1 + \psi(it_1, it_2)\} \\ &+ \frac{\Theta_k}{n^{\frac{1}{2}(k-2)}}\left\{\sum_{i=1}^{2}\beta_{ki}^{3(k-2)/k}(|t_i|^k + |t_i|^{k+1} + \cdots + |t_i|^{3(k-2)})\right\}e^{-(1-\rho^2)(t_1^2+t_2^2)/8} \\ &\left(|t_i| \le \frac{A_k(1-\rho^2)\sqrt{n}}{\beta_{ki}^{3/k}}, \quad \rho = \epsilon(\zeta_1\zeta_2)\right). \end{aligned}$$

More specially let us rewrite (34) and (35) with $k = 3$:

$$(36)\qquad \left\{p\left(\frac{t}{\sqrt{n}}\right)\right\}^n = e^{-\frac{1}{2}t^2} + \frac{\Theta}{\sqrt{n}}\beta_3|t|^3e^{-\frac{1}{2}t^2}, \qquad \left(|t| \le \frac{A\sqrt{n}}{\beta_3}\right);$$

$$(37)\qquad \begin{aligned} \left\{p\left(\frac{t_1}{\sqrt{n}}, \frac{t_2}{\sqrt{n}}\right)\right\}^n &= e^{-\frac{1}{2}(t_1^2+t_2^2+2\rho t_1t_2)} \\ &+ \frac{\Theta}{\sqrt{n}}(\beta_{31}|t_1|^3 + \beta_{32}|t_2|^3)e^{-(1-\rho^2)(t_1^2+t_2^2)/8}, \qquad \left(|t_i| \le \frac{A(1-\rho^2)\sqrt{n}}{\beta_{3i}}\right). \end{aligned}$$

In this paper only these last four formulae are needed; they are used in the proofs of Theorems 2, 1, 3, 4 respectively. Cases of $m > 2$ of (24) will be needed for the works on other functions alluded to in the introduction.

1.2. In the following group of lemmas, which culminate in Lemma 7, one finds a generalization of the Riemann-Lebesgue theorem, viz. Lemma 6.

LEMMA 4. *Let $f(x)$ be a polynomial of degree $m > 0$, with real coefficients:*

$$f(x) = \sum_{i=0}^{m} a_i x^{m-i} \qquad (a_0 \neq 0) \tag{38}$$

Then

$$\left| \int_0^1 e^{if(x)}\, dx \right| \leq \frac{A_m}{|a_0|^{1/m}}. \tag{38}$$

PROOF: It is sufficient to prove the inequality for $\int_0^1 \cos f(x)\, dx$. Divide the interval into A_m sub-intervals in each of whose interior none of the derivatives $f^{(i)}(x)$ $(i = 1, \cdots, m)$ vanishes. It is sufficient to consider one of these sub-intervals, say (a, b). Consequently each of the polynomials $f^{(i)}(x)$ are monotonic in (a, b). Let

$$I = \int_a^b \cos f(x)\, dx. \tag{39}$$

Suppose first that $f'(x)$ is positive and increasing for $a < x \leq b$. Then

$$|I| \leq \epsilon + \left| \int_{a+\epsilon}^{b} \frac{f'(x) \cos f(x)\, dx}{f'(x)} \right|$$

$$= \epsilon + \frac{1}{f'(a+\epsilon)} \left| \int_{a+\epsilon}^{b_1} f'(x) \cos f(x)\, dx \right|, \quad (a + \epsilon \leq b_1 \leq b),$$

by the second mean-value theorem. Hence

$$|I| \leq \epsilon + \frac{2}{f'(a + \epsilon)}. \tag{40}$$

Now $0 < f'(a + \frac{1}{2}\epsilon) = f'(a + \epsilon) - \epsilon f''(a + \theta\epsilon)/2$, $\frac{1}{2} \leq \theta \leq 1$. Hence $f'(a + \epsilon) > \frac{1}{2}\epsilon f''(a + \theta\epsilon)$. Since $f''(x)$ is monotonic, we have either $f'(a + \epsilon) > \frac{1}{2}\epsilon f''(a + \epsilon)$ or $f'(a + \epsilon) > \frac{1}{2}\epsilon f''(a + \frac{1}{2}\epsilon)$. In other words, there exists a constant C_2, independent of a or ϵ, such that $\frac{1}{2} \leq C_2 \leq 1$ and $f'(a + \epsilon) > \frac{1}{2}\epsilon f(a + C_2\epsilon)$.

If $f'''(x) \geq 0$, we have, as before $f''(a + C_2\epsilon) > \frac{1}{2}C_2\epsilon f'''(a + C_3\epsilon)$, where C_3 is independent of a or ϵ and $\frac{1}{4} \leq C_3 \leq 1$. If $f'''(x) < 0$, then, since $0 < f''(a + 2C_2\epsilon) = f''(a + C_2\epsilon) + C_2\epsilon f'''(a + \theta_1 C_2\epsilon)$, $\frac{1}{2} \leq \theta_1 \leq 1$, we have $f''(a + C_2\epsilon) > -C_2\epsilon f'''(a + 2\theta_1 C_2\epsilon)$. As $f'''(x)$ is monotonic, either $f''(a + C_2\epsilon) > -C_2\epsilon f'''(a + C_2\epsilon)$ or $f''(a + C_2\epsilon) > -C_2\epsilon f'''(a + 2C_2\epsilon)$. In all cases we obtain $f''(a + C_2\epsilon) > B_3\epsilon\, |f'''(a + C_3\epsilon)|$, where B_3 and C_3 are independent of a or ϵ, and $\frac{1}{4} \leq C_3 \leq 2$. Hence $f'(a + \epsilon) > \frac{1}{2}B_3\epsilon^2\, |f'''(a + C_3\epsilon)|$. Arguing with $\pm f'''(a + C_3\epsilon)$ as we did with $f''(a + C_2\epsilon)$, and so on until we come to $f^{(m)}$, we obtain $f'(a + \epsilon) > B_m \epsilon^{m-1} | f^{(m)}(a + C_m \epsilon) | = B_m \epsilon^{m-1} | a_0 |$. Substituting in (40) and putting $\epsilon = | a_0 |^{-1/m}$ we obtain $| I | \leq A_m | a_0 |^{-1/m}$. The proof presupposes that $C_m \epsilon < b - a$. If the reverse inequality is true, then $| I | \leq b - a < C_m | a_0 |^{-1/m}$. Hence the lemma is true for $f'(x)$ positive and increasing in (a, b).

If $f'(x)$ is positive and decreasing in (a, b), then $I = \int_0^{b-a} \cos(-f(b - y))\, dy$, $-f(b - y)$ being a polynomial with the leading coefficient $\pm a_0$ and the first derivative $f'(b - y)$, which is positive and increasing. This case reduces therefore to the preceding one. Finally, if $f'(x)$ is negative, we have only to notice that $I = \int_a^b \cos(-f(x))\, dx$. Hence the lemma is proved.

LEMMA 5. *Let $f(x)$ be the polynomial (38a), and let $a_r \neq 0$ for some r, $0 \leq r < m$. Then*

$$\left| \int_0^1 e^{if(x)}\, dx \right| \leq \frac{A_m}{| a_r |^{A_m}}. \tag{41}$$

PROOF: We may assume that $| a_r | \geq 1$, (41) being trivial if $| a_r | < 1$. If $r = 0$ this reduces to Lemma 4. Suppose that the lemma is true for $a_0, a_1, \cdots, a_{r-1}$. Let $f_1(x) = a_0 x^m + \cdots + a_{r-1} x^{m-r+1}$, $f_2(x) = f(x) - f_1(x)$ and divide (0, 1) into A_m sub-intervals in each of which $f_1(x)$ is monotonic. It is sufficient to consider one of these sub-intervals, say, (a, b). We have

$$I = \int_a^b \cos\{f_1(x) + f_2(x)\}\, dx$$

$$= \int_a^b \cos f_1(x) \cos f_2(x)\, dx - \int_a^b \sin f_1(x) \sin f_2(x)\, dx.$$

We have only to consider the integral of cosines, say J. Divide (a, b) into subintervals in each of whose interior $\cos f_1(x)$ is monotonic and does not vanish. The number of such intervals does not exceed $(\frac{1}{2}\pi)^{-1} | f_1(b) - f_1(a) | \leq (\frac{1}{2}\pi)^{-1}(| f_1(b) | + | f_1(a) |) < 2(| a_0 | + \cdots + | a_{r-1} |)$. Then, by the second mean-value theorem,

$$| J | \leq 2(| a_0 | + \cdots + a_{r-1} |) \left| \int_a^{b_1} \cos f_2(x)\, dx \right| \qquad (a \leq b_1 \leq b).$$

Hence, applying Lemma 4 to $f_2(x)$, we get

$$| I | \leq \frac{A_m(| a_0 | + \cdots + | a_{r-1} |)}{| a_r |^{1/(m-r)}} \leq \frac{A_m(| a_0 | + \cdots + | a_{r-1} |)}{| a_r |^{1/m}}. \tag{42}$$

On the hypothesis of induction we have $| I | \leq A_m | a_i |^{-B_m}$ $(i = 0, \cdots, r - 1)$. If $| a_i | \geq | a_r |^{1/2m}$ for some $i < r$, then $| I | \leq A_m | a_r |^{-B_m/2m}$; if $| a_i | < | a_r |^{1/2m}$, then by (42), $| I | \leq A_m | a_r |^{-1/2m}$. The proof is therefore complete.

LEMMA 6. *Let $f(x)$ be the polynomial (38a) and $g(x)$ be summable over $(-\infty, \infty)$. Then for every r we have*

$$\lim_{|a_r| \to \infty} \int_{-\infty}^{\infty} e^{if(x)} g(x)\, dx = 0, \quad \text{uniformly in } a_i (i \neq r). \tag{43}$$

PROOF: By Lemma 5 We have

$$\lim_{|a_r| \to \infty} \int_0^1 e^{if(x)}\, dx = 0, \quad \text{uniformly in } a_i(i \neq r).$$

Hence

$$\lim_{|a_r| \to \infty} \int_a^b e^{if(x)}\, dx = 0, \quad \text{uniformly in } a_i(i \neq r) \tag{44}$$

for if $a \neq 0$ and $b \neq 0$, then (a, b) is the sum or the difference of two intervals of the form $(0, c)$ or $(c, 0)$, and for the latter intervals the transformation $x = \pm cy$ reduces the interval of integration to $(0, 1)$.

Let G be any open set of finite measure. Then G is the sum of a sequence $\{I_\nu\}$ of non-overlapping intervals. Since $\Sigma mI_\nu = mG < \infty$, we have

$$\sum_{\nu \geq n} mI_\nu < \epsilon, \qquad n \geq N.$$

Hence

$$\left| \int_G e^{if(x)}\, dx \right| < \epsilon + \sum_{\nu=1}^{N} \left| \int_{I_\nu} e^{if(x)}\, dx \right|$$

which, together with (44), implies

$$\lim_{|a_r| \to \infty} \int_G e^{if(x)}\, dx = 0 \quad \text{uniformly in } a_i(i \neq r). \tag{45}$$

Let S be any set of finite measure. Then there is an open set G such that $G \supset S$ and $m(G - S) < \epsilon$. Hence

$$\left| \int_S e^{if(x)}\, dx \right| < \epsilon + \left| \int_G e^{if(x)}\, dx \right|.$$

Hence, by (45),

$$\lim_{|a_r| \to \infty} \int_S e^{if(x)}\, dx = 0 \quad \text{uniformly in } a_i(i \neq r). \tag{46}$$

Now let $h(x)$ be any positive "simple" summable function, i.e. $h(x) = a_\nu > 0$ for $x \in S$ $(\nu = 1, 2, \cdots, n)$ and $h(x) = 0$ otherwise. Since $h(x)$ is summable, each S_ν must be of finite measure. Hence

$$\left| \int_{-\infty}^{\infty} e^{if(x)} h(x)\, dx \right| \leq \sum_{\nu=1}^{n} a_\nu \left| \int_{S_\nu} e^{if(x)}\, dx \right|$$

which, together with (46), implies

$$\lim_{|a_r| \to \infty} \int_{-\infty}^{\infty} e^{if(x)} h(x)\, dx = 0 \quad \text{uniformly in } a_i(i \neq r).$$

Finally, let $g(x)$ be any summable function ≥ 0. Then by a well-known theorem[6] we have $g(x) = \lim h_n(x)$, where $\{h_n(x)\}$ is an ascending sequence of positive summable simple functions. Hence

$$\left|\int_{-\infty}^{\infty} e^{if(x)} g(x)\, dx\right| \leq \left|\int_{-\infty}^{\infty} e^{if(x)} h_n(x)\, dx\right| + \int_{-\infty}^{\infty} (g(x) - h_n(x))\, dx.$$

By monotonic convergence the last integral tends to 0 as $n \to \infty$. Hence

$$\left|\int_{-\infty}^{\infty} e^{if(x)} g(x)\, dx\right| \leq \epsilon + \left|\int_{-\infty}^{\infty} e^{if(x)} h_n(x)\, dx\right|,$$

which implies (43). If $g(x)$ is any summable function, we have only to consider the customary expression of $g(x)$ as the difference of two non-negative functions. This completes the proof.

LEMMA 7. *Let $P(x)$ be a non-singular distribution function of a random variable X, and let*

$$p(t_1, t_2, \cdots, t_m) = \int_{-\infty}^{\infty} e^{i\sum_{r=1}^{m} t_r x^r}\, dP. \tag{47}$$

Then for every r and every positive constant c we have

$$\underset{|t_r| \geq c}{\text{l.u.b.}} |p(t_1, \cdots, t_m)| < 1. \tag{48}$$

PROOF: We have $P(x) = a_1P_1(x) + a_2P_2(x)$, where $P_1(x)$ is absolutely continuous, P_2 is singular, $a_1 > 0$, $a_1 + a_2 = 1$. Hence

$$|p(t_1, t_2, \cdots, t_m)| \leq a_1\left|\int_{-\infty}^{\infty} e^{i\sum_{r=1}^{m} t_r x^r} P_1'(x)\, dx\right| + a_2 .$$

By Lemma 6 we may find $C > 0$ such that

$$|p(t_1, t_2, \cdots, t_m)| \leq \tfrac{1}{2}a_1 + a_2 < 1, \quad \text{if any} \quad |t_i| > C.$$

Suppose that

$$\underset{|t_r| \geq c}{\text{l.u.b.}}\, p(t_1, \cdots, t_m) = 1,$$

then $c < C$ and we must have

$$\underset{c \leq |t_r| \leq C, |t_i| \leq C (i \neq r)}{\text{l.u.b.}} |p(t_1, \cdots, t_m)| = 1. \tag{49}$$

Since $p(t_1, \cdots, t_m)$ is a continuous function, it must attain its least upper bound in any bounded closed set. It follows that there is a point $(t_1^0, \cdots, t_m^0)$ such that[7] $t_r^0 \neq 0$ ($|t_r^0| \geq c$) and $p(t_1^0, \cdots, t_m^0) = 1$. But this implies that the distribution of $\Sigma t_i^0 X^i$ is discrete, i.e. that the distribution of X itself is discrete,

[6] H. Kestelman: *Modern Theories of Integration* (1937), p. 108.
[7] Cf. (C), p. 26.

which contradicts the non-singularity of $P(x)$. Hence (49) is false and (48) is true.

1.3. In his cited work Berry[8] shows that if $F(x)$ is any distribution function and if $\Phi(x)$ is the function (6), then there is a constant a such that

(50)
$$\left| \int_{\infty}^{\infty} \frac{1 - \cos Tx}{x^2} \{F(x + a) - \Phi(x + a)\}\, dx \right| \geq \sqrt{\frac{2}{\pi}}\, T\delta \left\{ 3 \int_0^{P\delta} \frac{1 - \cos x}{x^2}\, dx - \pi \right\}$$

where $\delta = \sqrt{\frac{\pi}{2}}$ l.u.b. $| F(x) - \Phi(x) |$. This is easily extended to the following lemma, which needs no further proof.

LEMMA 8. *Let $F(x)$ be a distribution function and $F_1(x)$ be a function having the following properties:* (i) *$F_1(x)$ is bounded for all x,* (ii) *$F_1(x) \to 1$ as $x \to \infty$, $F_1(x) \to 0$ as $x \to -\infty$,* (iii) *$F_1(x)$ has a bounded derivative, $| F_1'(x) | \leq M$. Let*

$$\delta = \frac{1}{2M} \text{ l.u.b.} | F(x) - F_1(x) | .$$

Then there exists a constant a such that

(51)
$$\left| \int_{\infty}^{\infty} \frac{1 - \cos Tx}{x^2} \{F(x + a) - F_1(x + a)\}\, dx \right| \geq 2MT\delta \left\{ 3 \int_0^{T\delta} \frac{1 - \cos x}{x^2}\, dx - \pi \right\}.$$

1.4. In section 3 we define, for given ϵ, k, λ and z, a function

(52) $$G(x, y) = e^{-\epsilon y^{2k}} \text{ if } z < x \leq z + \lambda y^2, \quad G(x, y) = 0 \text{ otherwise.}$$

The introduction of $G(x, y)$ and the appraisal of its Fourier transform constitute the essence of our method of solving the problem of the asymptotic expansion of the distribution function $G(x)$. The solution of the same problem about other functions of (1) alluded to in section 3 is based on the introduction of functions playing the role of $G(x, y)$. We now prove the following lemma:

LEMMA 9. *Let $G(x, y)$ be defined by* (52) *and let*

(53) $$g(t_1, t_2) = \int_{\infty}^{\infty} \int_{\infty}^{\infty} e^{-it_1 x - it_2 y} G(x, y)\, dx\, dy.$$

Then

(i) $$| g(t_1, t_2) | \leq \frac{\lambda A_k}{\epsilon^{3/2k}}$$

(ii) $$| g(t_1, t_2) | \leq \frac{A}{| t_2 |^3} \left(\lambda + \frac{\lambda^2 | t_1 |}{\epsilon^{1/3}} + \frac{\lambda^3 | t_1 |^2}{\epsilon^{2/3}} \right) \quad \text{if } k = 3,$$

(iii) $$| g(t_1, t_2) | \leq \frac{A_k}{| t_2 |^2} \left(\frac{\lambda}{\epsilon^{1/2k}} + \frac{\lambda^2 | t_1 |}{\epsilon^{3/2k}} \right).$$

---

[8] (B), p. 128.

PROOF:

(i) $$|g(t_1, t_2)| \leq \int_{R_2} G(x, y)\, dx\, dy = \lambda \int_{-\infty}^{\infty} y^2 e^{-\epsilon y^{2k}}\, dy = \frac{A_k \lambda}{\epsilon^{3/2k}}$$

(ii) Putting $k = 3$ we have

$$g(t_1, t_2) = \frac{e^{-it_1x}}{it_1} \int_{-\infty}^{\infty} e^{-\epsilon y^6 - it_2 y}(1 - e^{-it_1\lambda y^2})\, dy,$$

$$|g(t_1, t_2)| \leq \frac{1}{|t_1||t_2|^3} \left| \int_{-\infty}^{\infty} u(y)v'''(y)\, dy \right|,$$

where $u(y) = e^{-\epsilon y^6}(1 - e^{-it_1\lambda y^2})$, $v(y) = e^{-it_2 y}$ On integrating by parts we obtain

(54) $$|g(t_1, t_2)| \leq \frac{1}{|t_1||t_2|^3} \left| \int_{-\infty}^{\infty} v(y)u'''(y)\, dy \right| \leq \frac{1}{|t_1||t_2|^3} \int_{-\infty}^{\infty} |u'''(y)|\, dy.$$

Elementary calculation establishes that

$$\frac{|u'''(y)|}{|t_1|} \leq e^{-\epsilon y^6}(216\lambda\epsilon^3 |y|^{17} + 756\lambda\epsilon^2 |y|^{11} + 336\lambda\epsilon |y|^5 + 8\lambda^3 |t_1|^2 |y|^3 + 12\lambda^2 |t_1||y|).$$

Substituting in (54) and making the transformation $y = \epsilon^{-1/6}x$ we get the result.

(iii) We have

$$|g(t_1, t_2)| \leq \frac{1}{|t_1|} \left| \int_{-\infty}^{\infty} e^{-\epsilon y^{2k} - it_2 y}(1 - e^{-it_1\lambda y^2})\, dy \right|.$$

Integrating by parts twice we obtain

$$|g(t_1, t_2)| \leq \frac{1}{|t_1||t_2|^2} \int_{-\infty}^{\infty} \left| \frac{d^2}{dy^2} \{e^{-\epsilon y^{2k}}(1 - e^{-it_1\lambda y^2})\} \right| dy.$$

By elementary calculations we get

$$|g(t_1, t_2)| \leq \frac{1}{|t_2|^2} \int_{-\infty}^{\infty} (4k^2\lambda\epsilon y^{4k} + 2k(k+3)\lambda\epsilon y^{2k} + 4\lambda^2 |t_1| y^2 + 2\lambda)e^{-\epsilon y^{2k}}\, dy$$

which, on the transformation $y = \epsilon^{-1/2k}x$, gives the result.

1.5. We prove a few additional lemmas used in the proof of Theorems 3 and 4.

LEMMA[9] 10. *Let $u(x_1, \cdots, x_m) \geq 0$ be summable in the m-dimensional space and let*

(55) $$v(t_1, \cdots, t_m) = \int_{-\infty}^{\infty} \cdots \int_{-\infty}^{\infty} e^{-it_1x_1 - \cdots - it_mx_m} u(x_1, \cdots, x_m)\, dx_1 \cdots dx_m.$$

[9] Although the author believes that this lemma is almost classical, a proof is given owing to lack of reference.

If $v(t_1, \cdots, t_m)$ *is summable in the m-dimensional space, then*

$$(56)\qquad u(x_1, \cdots, x_m) = \frac{1}{(2\pi)^m} \int_{-\infty}^{\infty} \cdots \int_{-\infty}^{\infty} e^{it_1x_1+\cdots+it_mx_m} v(t_1, \cdots t_m)\, dt_1 \cdots dt_m.$$

PROOF: Except for a constant factor the function $u(x_1, \cdots, x_m)$ may be regarded as a probability density function. Hence by the well-known inversion formula of (55),

$$(57)\qquad \begin{aligned} &\underset{a_i \le x_i \le b_i \ (i=1,\cdots,m)}{\int \cdots \int} u(x_1, \cdots, x_m)\, dx_1 \cdots dx_m \\ &= \frac{1}{(2\pi)^m} \int_{-\infty}^{\infty} \cdots \int_{-\infty}^{\infty} \left(\prod_{j=1}^{m} \frac{e^{it_jb_j} - e^{it_ja_j}}{it_j} \right) v(t_1, \cdots, t_m)\, dt_1 \cdots dt_m . \end{aligned}$$

Now $u(x_1, \cdots, x_m)$ is almost everywhere the symmetric derivative of the interval function in the left-hand side of (57):

$$u(x_1, \cdots, x_m) = \lim_{\epsilon \to 0} \frac{1}{(2\epsilon)^m} \underset{x_i - \epsilon \le y_i \le x_i + \epsilon \ (i=1,2,\cdots,m)}{\int \cdots \int} u(y_1, \cdots, y_m)\, dy_1 \cdots dy_m .$$

Hence

$$(58)\qquad \begin{aligned} u(x_1, \cdots, x_m) &= \frac{1}{(2\pi)^m} \lim_{\epsilon \to 0} \frac{1}{(2\epsilon)^m} \int_{-\infty}^{\infty} \cdots \int_{-\infty}^{\infty} \\ &\cdot \left(\prod_{j=1}^{m} \frac{e^{it_j\epsilon} - e^{-it_j\epsilon}}{it_j} \right) e^{it_1x_1+\cdots+it_mx_m} v(t_1, \cdots, t_m)\, dt_1 \cdots dt_m . \end{aligned}$$

Owing to dominated convergence the order of the limit sign and the integration sign in (58) may be inverted: Hence (56) is true.

LEMMA 11. *We have*

$$(59)\qquad \int_{-\infty}^{\infty} e^{-itu} \frac{1 - \cos Tu}{u^2}\, du = \begin{cases} \pi(T - |t|) & \text{if } |t| \le T, \\ 0 & \text{if } |t| > T. \end{cases}$$

PROOF: The Fourier transform of the function in the right-hand side of (59) is

$$\pi \int_{-T}^{T} e^{itu}(T - |t|)\, dt = \frac{2\pi}{u^2}(1 - \cos Tu).$$

Hence (59) follows from (56).

LEMMA 12.

$$(60)\qquad |\epsilon(\xi_1 + \cdots + \xi_n)^k| \le A_k n^{k/2} \beta_k$$

PROOF. As (60) is true for $k = 1$, let us assume, for induction, that it is true for 1, 2, $\cdots$, k. Then, by symmetry,

$$\epsilon(\xi_1 + \cdots + \xi_n)^{k+1} = n\epsilon\{\xi_1(\xi_1 + \cdots + \xi_n)^k\} = n \sum_{r=0}^{k} \binom{k}{r} \epsilon(\xi_1^{r+1} U^{k-r})$$

where $U = \xi_2 + \cdots + \xi_k$. Since $\epsilon(\xi_1) = 0$, we have

$$\epsilon(\xi_1 + \cdots + \xi_n)^{k+1} = n \sum_{r=1}^{k} \binom{k}{r} \epsilon(\xi_1^{r+1} U^{k-r}).$$

On the hypotheses of induction we have $|\epsilon(U^{k-r})| \leq A_k(n-1)^{\frac{1}{2}(k-r)}\beta_{k-r} < A_k n^{\frac{1}{2}(k-1)}\beta_{k-r}$. Hence

$$|\epsilon(\xi_1 + \cdots + \xi_n)^{k+1}| \leq k! A_k n^{\frac{1}{2}(k+1)} \Sigma \beta_{r+1}\beta_{k-r} \leq A_{k+1} n^{\frac{1}{2}(k+1)} \beta_{k+1}.$$

Therefore the induction is complete.

3. Elementary Proof of Theorem 1. 2.1 We have defined

$$F(x) = Pr\{\sqrt{n}\xi \leq x\}, \quad \Phi(x) = \frac{1}{\sqrt{2\pi}} \int_{-\infty}^{x} e^{-\frac{1}{2}y^2}\, dy \tag{61}$$

with the characteristic functions

$$f(t) = \left\{p\left(\frac{t}{\sqrt{n}}\right)\right\}^n, \qquad \varphi(t) = e^{-\frac{1}{2}t^2} \tag{62}$$

Following Berry[10] we use the equation

$$\int_{-\infty}^{\infty} \{F(x) - \Phi(x)\} e^{itx}\, dx = \frac{f(t) - \varphi(t)}{-it}. \tag{63}$$

Let $\psi(it)$ be the polynomial in (34), and let us define $\Psi(x)$ as the function obtained from $\psi(it)$ through the replacement of each power $(it)^\nu$ by $(-1)^\nu \Phi^{(\nu)}(x)$. Integration by parts shows $(-1)^{\nu-1} \int_{-\infty}^{\infty} e^{itx}\Phi^{(\nu)}(x)\, dx = (it)^{\nu-1}\varphi(t)$, whence

$$\int_{-\infty}^{\infty} \Psi(x) e^{itx}\, dx = \frac{\psi(it)\varphi(t)}{-it}. \tag{64}$$

From (63) and (64) we obtain

$$\int_{-\infty}^{\infty} \{F(x) - \Phi(x) - \Psi(x)\} e^{itx}\, dx = \frac{f(t) - \varphi(t)\{1 + \psi(it)\}}{-it}. \tag{65}$$

The function $\Psi(x)$ defined here is precisely the $\Psi(x)$ appearing in (5) under Theorem 1. Our task is to prove that

$$|F(x) - \Phi(x) - \Psi(x)| \leq \frac{Q_k}{n^{(k-2)/2}}. \tag{66}$$

Following Berry[11] we replace x by $x + a$ in (65), getting

$$\begin{aligned} &\int_{-\infty}^{\infty} \{F(x+a) - \Phi(x+a) - \Psi(x+a)\} e^{itx}\, dx \\ &\qquad = \frac{e^{-ita}[f(t) - \varphi(t)\{1 + \psi(it)\}]}{-it} \end{aligned} \tag{67}$$

[10] (B), p. 127, Equation (23).
[11] (B), p. 127.

multiply both sides of (67) by $T - |t|$ and integrate with respect to t in $(-T, T)$:

$$2\int_{-\infty}^{\infty} \frac{1-\cos Tx}{x^2}\{F(x+a) - \Phi(x+a) - \Psi(x+a)\}\,dx$$
$$= \int_{-T}^{T} \frac{(T-|t|)e^{-ita}[f(t) - \varphi(t)\{1+\psi(it)\}]}{-it}\,dt$$

the reversion of order of integration involved is obviously justifiable. Hence

$$\left|\int_{-\infty}^{\infty} \frac{1-\cos Tx}{x^2}\{F(x+a) - \Phi(x+a) - \Psi(x+a)\}\,dx\right| \tag{68}$$
$$\leq T\int_0^T \frac{|f(t) - \varphi(t)\{1+\psi(it)\}|}{t}\,dt.$$

2.2. When in particular $k = 3$, (68) becomes

$$\left|\int_{-\infty}^{\infty} \frac{1-\cos Tx}{x^2}\{F(x+a) - \Phi(x+a)\}\,dx\right| \leq T\int_0^T \frac{|f(t) - \varphi(t)|}{t}\,dt. \tag{69}$$

If we choose a to be the a in (50), the left-hand side of (69) is not less than

$$\sqrt{\frac{2}{\pi}}\,T\delta\left\{3\int_0^{T\delta}\frac{1-\cos x}{x^2}\,dx - \pi\right\}, \qquad \delta = \sqrt{\frac{\pi}{2}}\,\text{l.u.b.}\,|F(x) - \Phi(x)|.$$

On the other hand, taking $T = \dfrac{A\sqrt{n}}{\beta_3}$ as in (36) the right-hand side of (69) is not greater than

$$A\int_0^{\infty} t^2 e^{-\frac{1}{4}t^2}\,dt = A.$$

Hence

$$T\delta\left\{3\int_0^{T\delta}\frac{1-\cos x}{x^2}\,dx - \pi\right\} \leq A. \tag{70}$$

Now the left-hand side of (70), as a function of $T\delta$, is positive and increasing for sufficiently large $T\delta$, and becomes infinite as $T\delta \to \infty$. Hence (70) implies that $T\delta \leq A$, i.e.

$$\text{l.u.b.}\,|F(x) - \Phi(x)| \leq \frac{A}{T} = \frac{A\beta_3}{\sqrt{n}},$$

giving Theorem 2.

2.3. Coming back to the general case, we see that the function $\Phi(x) + \Psi(x)$ has a bounded derivative: $|\Phi'(x) + \Psi'(x)| \leq Q_k$, and also has all the properties of the function $F_1(x)$ in Lemma 8. On choosing a in (69) to be the a in (51) we obtain

$$(71)\qquad Q_k T\delta\left\{3\int_0^{T\delta}\frac{1-\cos x}{x^2}\,dx-\pi\right\}\leq T\int_0^T\frac{|f(t)-\varphi(t)\{1+\psi(it)\}|}{t}\,dt,$$

where

$$\delta = Q_k \text{ l.u.b. } |F(x)-\Phi(x)-\Psi(x)|.$$

Let us take $T=(A_k\beta_k^{-3/k}\sqrt{n})^{k-2}$ with A_k in accordance with (34). Then

$$(72)\qquad \begin{aligned} T\int_0^T&\frac{|f(t)-\varphi(t)\{1+\psi(it)\}|}{t}\,dt \\ &= Q_k n^{\frac12(k-2)}\int_0^{T^{1/(k-2)}} + Q_k n^{\frac12(k-2)}\int_{Q_k\sqrt{n}}^T = J_1+J_2 \quad\text{say.}\end{aligned}$$

By (34) we have

$$(73)\qquad J_1\leq Q_k\int_0^\infty(t^{k-1}+\cdots+t^{3k-7})e^{-\frac14 t^2}\,dt=Q_k.$$

Also,

$$(74)\qquad J_2\leq Q_k n^{\frac12(k-2)}\int_{Q_k\sqrt{n}}^T\frac{|p(t/\sqrt{n})|^n}{t}\,dt+Q_k n^{\frac12(k-2)}\int_{Q_k\sqrt{n}}^T\frac{\varphi(t)\,|1+\psi(it)|}{t}\,dt.$$

The second term in the right-hand side of (74) is evidently $\leq Q_k$. The first term does not exceed

$$(75)\qquad Q_k n^{\frac12(k-3)}\,T\ \underset{t\geq Q_k}{\text{l.u.b.}}\ |p(t)|^n.$$

At this step we make use of the non-singularity of $P(x)$ and apply Lemma 7 for $m = 1$. We have

$$\underset{t\geq Q_k}{\text{l.u.b.}}\ |p(t)| = e^{-Q_k}.$$

Hence (75) does not exceed $Q_k n^{\frac12(2k-5)}e^{-Q_k n}\leq Q_k$. We have therefore

$$(76)\qquad T\delta\left\{3\int_0^{T\delta}\frac{1-\cos x}{x^2}\,dx-\pi\right\}\leq Q_k,\qquad T=Q_k n^{\frac12(k-2)}.$$

Arguing with (76) as we did with (70) we conclude that

$$\text{l.u.b.}\,|F(x)-\Phi(x)-\Psi(x)|\leq\frac{Q_k}{T}=\frac{Q_k}{n^{\frac12(k-2)}}.$$

(72) is valid for $T\geq 1$. If $T<1$, we have only to suppress the term J_2. Hence Theorem 1 is proved.

4. Proof of Theorem 3 and Theorem 4. **3.1.** In connection with the random variables (1), we assume that $\beta_{2k}<\infty$ for some integer $k\geq 3$ and define

$$(77)\qquad \eta=\frac1n\sum_{r=1}^n(\xi_r-\bar\xi)^2,\qquad G(z)=Pr\left\{\frac{\sqrt{n}(\eta-1)}{\sqrt{\alpha_4-1}}\leq z\right\}.$$

Now,

$$\eta = \frac{1}{n} \sum \xi_r^2 - \bar{\xi}^2 = 1 + \sqrt{\frac{\alpha_4 - 1}{n}}\, X - \frac{Y^2}{n}$$

where

(78) $$X = \frac{1}{\sqrt{n}} \sum \frac{(\xi_r^2 - 1)}{\sqrt{\alpha_4 - 1}}, \qquad Y = \sqrt{n}\, \bar{\xi}.$$

Hence

(79) $$G(z) = Pr\{X - \lambda Y^2 \leq z\}$$

with

(80) $$\lambda = \frac{1}{\sqrt{n(\alpha_4 - 1)}}.$$

Let W be the probability function of the distribution of the random point (X, Y) and $f(t_1, t_2)$ be the characteristic function:

(81) $$W(S) = Pr\{(X, Y)\epsilon S\} \quad \text{for every Borel set } S \text{ in } R_2,$$

(82) $$f(t_1, t_2) = \epsilon(e^{it_1X+it_2Y}) = \left\{p\left(\frac{t_1}{\sqrt{n}}, \frac{t_2}{\sqrt{n}}\right)\right\}^n$$

(83) $$p(t_1, t_2) = \int_{-\infty}^{\infty} e^{it_1(x^2-1)/(\sqrt{\alpha_4-1})+it_2x}\, dP.$$

Let $G_1(z)$ be the distribution function of X. Then

(84) $$G(z) - G_1(z) = \iint\limits_{z<x\leq z+\lambda y^2} dW = K(z), \quad \text{say.}$$

Let

(85) $$K_\epsilon(z) = \iint\limits_{z<x\leq z+\lambda y^2} e^{-\epsilon y^{2k}}\, dW.$$

If we define (for fixed z) the function $G(x, y)$ by

(86) $$G(x, y) = e^{-\epsilon y^{2k}} \quad \text{if} \quad z < x \leq z + \lambda y^2, \qquad G(x, y) = 0 \quad \text{otherwise,}$$

then

(87) $$K_\epsilon(z) = \int_{-\infty}^{\infty}\int_{-\infty}^{\infty} G(x, y)\, dW.$$

Letting

(88) $$\int_{-\infty}^{\infty}\int_{-\infty}^{\infty} e^{-it_1x-it_2y}\, G(x, y)\, dx\, dy = g(t_1, t_2),$$

we replace x by $x - u$ in the integral and get

$$(89)\qquad \int_{-\infty}^{\infty}\int_{-\infty}^{\infty} e^{-it_1x-it_2y}G(x-u, y)\,dx\,dy = e^{-it_1u}g(t_1, t_2).$$

Multiplying both sides by $\frac{1-\cos Tu}{u^2}$ and integrating with respect to u we obtain, with the help of (59), Lemma 11,

$$(90)\qquad \begin{aligned}\int_{-\infty}^{\infty}\int_{-\infty}^{\infty} e^{-it_1x-it_2y}\,dx\,dy\int_{-\infty}^{\infty}\frac{1-\cos Tu}{u^2}G(x-u, y)\,du \\ = \begin{cases}\pi(T-|t_1|)g(t_1, t_2) & \text{if } |t_1| \leq T,\\ 0 & \text{if } |t_1| > T;\end{cases}\end{aligned}$$

the reversion of order of integration in the left-hand side is obviously justifiable. By Lemma 9 the right-hand side of (90) is summable in the whole plane of (t_1, t_2). Hence, by Lemma 10,

$$(91)\qquad \begin{aligned}\int_{-\infty}^{\infty}\frac{1-\cos Tu}{u^2}G(x-u, y)\,du \\ = \frac{1}{4\pi}\iint_{|t_1|\leq T}(T-|t_1|)g(t_1, t_2)e^{it_1x+it_2y}\,dt_1\,dt_2\,.\end{aligned}$$

If we integrate both sides with respect to the probability function W, we obtain, on reversing the order of integration,

$$(92)\qquad \begin{aligned}\int_{-\infty}^{\infty}\frac{1-\cos Tu}{u^2}\,du\iint_{R_2}G(x-u, y)\,dW \\ = \frac{1}{4\pi}\iint_{|t_1|\leq T}(T-|t_1|)g(t_1, t_2)f(t_1, t_2)\,dt_1\,dt_2\,.\end{aligned}$$

By (86) and (87),

$$(93)\qquad \int_{-\infty}^{\infty}\int_{-\infty}^{\infty} G(x-u, y)\,dW = K_\epsilon(u+z).$$

Hence

$$(94)\qquad \int_{-\infty}^{\infty}\frac{1-\cos Tu}{u^2}K_\epsilon(u+z)\,du = \frac{1}{4\pi}\iint_{|t_1|\leq T}(T-|t_1|)g(t_1, t_2)f(t_1, t_2)\,dt_1dt_2\,.$$

We now take the functions

$$(95)\qquad \varphi(t_1, t_2) = c^{-\frac{1}{2}(t_1^2+t_2^2+2\rho t_1t_2)}$$

and $\psi(it_1, it_2)$ as in (35), where

(96) $$\rho = \int_{-\infty}^{\infty} \frac{(x^2 - 1)x}{\sqrt{\alpha_4 - 1}}\, dP = \frac{\alpha_3}{\sqrt{\alpha_4 - 1}}.$$

Since the condition $\alpha_4 - 1 - \alpha_3^2 \neq 0$ is assumed in Theorem 3 and implied in Theorem 4, we have $|\rho| < 1$. Let

(97) $$w(x, y) = \frac{1}{2\pi\sqrt{1 - \rho^2}}\, e^{-(1/2(1-\rho^2))(x^2+y^2-2\rho xy)}$$

and let $\gamma(x, y)$ be the function obtained from $\psi(it_1, it_2)$ through the replacement of each power $(it_1)^{\nu_1}\ (it_2)^{\nu_2}$ by $(-1)^{\nu_1+\nu_2} W_{\nu_1\nu_2}(x, y) = (-1)^{\nu_1+\nu_2} \dfrac{\partial^{\nu_1+\nu_2} w(x, y)}{\partial x^{\nu_1} \partial y^{\nu_2}}$.

Since

(98) $$w(x, y) = \frac{1}{(2\pi)^2} \int_{-\infty}^{\infty} \int_{-\infty}^{\infty} e^{-it_1x - it_2y} \varphi(t_1, t_2)\, dt_1 dt_2,$$

we have

(99) $$w_{\nu_1\nu_2}(x, y) = \frac{(-1)^{\nu_1+\nu_2}}{(2\pi)^2} \int_{-\infty}^{\infty} \int_{-\infty}^{\infty} (it_1)^{\nu_1}(it_2)^{\nu_2} e^{-it_1x - it_2y} \varphi(t_1, t_2)\, dt_1\, dt_2,$$

whence, by Fourier inversion,

(100) $$(it_1)^{\nu_1}(it_2)^{\nu_2}\varphi(t_1, t_2) = \int_{-\infty}^{\infty} \int_{-\infty}^{\infty} e^{it_1x + it_2y} w_{\nu_1\nu_2}(x, y)\, dxdy.$$

From the definition of $\gamma(x, y)$ it follows therefore

(101) $$\int_{-\infty}^{\infty} \int_{-\infty}^{\infty} e^{it_1x + it_2y} \{w(x, y) + \gamma(x, y)\}\, dx\, dy = \varphi(t_1, t_2)\{1 + \psi(it_1, it_2)\}.$$

A comparison of (101) with $\int_{-\infty}^{\infty} \int_{-\infty}^{\infty} e^{it_1x+it_2y}\, dW = f(t_1, t_2)$ shows that (94) will remain true if $K_\epsilon(u)$ be replaced by

(102) $$\iint\limits_{u < x \leq u + \lambda y^2} e^{-\epsilon y^{2k}}(w(x, y) + \gamma(x, y))\, dx\, dy = L_\epsilon(u), \text{ say,}$$

and $f(t_1, t_2)$ be replaced by $\varphi(t_1, t_2)\{1 + \psi(it_1, it_2)\}$. Hence

(103) $$\begin{aligned} &\int_{-\infty}^{\infty} \frac{1 - \cos Tu}{u^2} \{K_\epsilon(u + z) - L_\epsilon(u + z)\}\, du \\ &= \frac{1}{4\pi} \iint\limits_{|t_1| \leq T} (T - |t_1|) g(t_1, t_2) \{f(t_1, t_2) \\ &\qquad - \varphi(t_1, t_2)[1 + \psi(it_1, it_2)]\}\, dt_1 dt_2. \end{aligned}$$

Let also

$$(104)\qquad H(z) = \iint_{x-\lambda y^2 \leq z} \{w(x, y) + \gamma(x, y)\}\, dx\, dy,$$

$$H_1(z) = \iint_{x \leq z} \{w(x, y) + \gamma(x, y)\}\, dx\, dy,$$

$$(105)\qquad L(z) = H(z) - H_1(z) = \iint_{z < x \leq z+\lambda y^2} \{w(x, y) + \gamma(x, y)\}\, dx\, dy.$$

3.2. We now consider the particular case $k = 3$ and prove Theorem 3. For $k = 3$ we have $\psi \equiv \gamma \equiv 0$ and so

$$(106)\qquad \begin{aligned} H(z) &= \iint_{x-\lambda y^2 \leq z} w(x, y)\, dx\, dy, \\ H_1(z) &= \iint_{x \leq z} w(x, y)\, dx\, dy = \Phi(z), \end{aligned}$$

$$(107)\qquad \begin{aligned} L(z) &= H(z) - H_1(z), \\ L_\epsilon(z) &= \iint_{z < x \leq z+\lambda y^2} e^{-\epsilon y^6} w(x, y)\, dx\, dy, \end{aligned}$$

$$(108)\qquad \begin{aligned} &\int_{-\infty}^{\infty} \frac{1 - \cos Tu}{u^2} \{K_\epsilon(u + x) - L_\epsilon(u + x)\}\, du \\ &= \frac{1}{4\pi} \iint_{|t_1| \leq T} (T - |t_1|) g(t_1, t_2) \{f(t_1, t_2) - \varphi(t_1, t_2)\} dt_1\, dt_2 . \end{aligned}$$

Now

$$\begin{aligned} K_\epsilon(u) - L_\epsilon(u) = \{G(u) - \Phi(u)\} - \{H(u) - \Phi(u)\} - \{G_1(u) - \Phi(u)\} \\ - \{K(u) - K_\epsilon(u)\} + \{L(u) - L_\epsilon(u)\}, \end{aligned}$$

$$\begin{aligned} 0 \leq H(u) - \Phi(u) &= \frac{1}{2\pi\sqrt{1 - \rho^2}} \int_{-\infty}^{\infty} e^{-\frac{1}{2}y^2}\, dy \int_u^{u+\lambda y^2} e^{-(1/2(1-\rho^2))(x-\rho y)^2}\, dx \\ &\leq \frac{\lambda}{2\pi\sqrt{1 - \rho^2}} \int_{-\infty}^{\infty} y^2 e^{-\frac{1}{2}y^2}\, dy = \frac{\lambda}{\sqrt{2\pi(1 - \rho^2)}}, \end{aligned}$$

$$|G_1(u) - \Phi(u)| \leq \frac{A}{\sqrt{n}} \int_{-\infty}^{\infty} \left| \frac{x^2 - 1}{\sqrt{\alpha_4 - 1}} \right|^3 dP \leq \frac{A\alpha_6}{(\alpha_4 - 1)^{3/2}\sqrt{n}} \text{ by Theorem 2,}$$

$$0 \leq K(u) - K_\epsilon(u) \leq \epsilon\epsilon(Y^6) \leq A\alpha_6\epsilon \text{ by Lemma 12,}$$

$$0 \leq L(u) - L_\epsilon(u) \leq A\epsilon.$$

Hence

$$\int_{-\infty}^{\infty} \frac{1 - \cos Tu}{u^2} \{G(u + \lambda) - \Phi(u + \lambda)\}\, du$$

(109)
$$= \theta T \left\{ \alpha_6 \epsilon + \frac{\alpha_6}{(\alpha_4 - 1)^{3/2} \sqrt{n}} + \frac{1}{\sqrt{n}\sqrt{(\alpha_4 - 1)(1 - \rho^2)}} \right.$$
$$+ \theta T \iint_{|t_1| \le T} |g(t_1, t_2)| \cdot |f(t_1, t_2) - \varphi(t_1, t_2)|\, dt_1 dt_2 .$$

It is easy to verify that

$$\frac{\alpha_6}{(\alpha_4 - 1)^{3/2}} + \frac{1}{\sqrt{(\alpha_4 - 1)(1 - \rho^2)}} \le \left(\frac{\alpha_6}{\alpha_4 - 1 - \alpha_3^2} \right)^{3/2} .$$

For the left-hand side of (109) we refer to (50) and take x to be the number a therein. Hence

(110)
$$T\delta \left\{ 3 \int_0^{T\alpha} \frac{1 - \cos u}{u^2}\, du - \pi \right\} \le AT \left\{ \alpha_6 \epsilon + \frac{1}{\sqrt{n}} \left(\frac{\alpha_6}{\alpha_4 - 1 - \alpha_3^2} \right)^{3/2} \right\}$$
$$+ AT \iint_{|t_1| \le T, |t_2| \le T} |g(t_1, t_2)| \cdot |f(t_1, t_2) - \varphi(t_1, t_2)|\, dt_1 dt_2$$
$$+ AT \iint_{|t_1| \le T, |t_2| > T} |g(t_1, t_2)|\, dt_1 dt_2 .$$

By Lemma 9 (ii) we have

(111)
$$T \iint_{|t_1| \le T, |t_2| > T} |g(t_1, t_2)|\, dt_1\, dt_2$$
$$\le AT \iint_{|t_1| \le T, |t_2| > T} \frac{1}{|t_2|^3} \left(\lambda + \frac{\lambda^2 |t_1|}{\epsilon^{\frac{1}{2}}} + \frac{\lambda^3 |t_1|^2}{\epsilon^{\frac{3}{2}}} \right) dt_1\, dt_2 .$$
$$\le A \left(\lambda + \frac{\lambda^2 T}{\epsilon^{\frac{1}{2}}} + \frac{\lambda^3 T^2}{\epsilon^{\frac{3}{2}}} \right).$$

Hence

(112)
$$T\delta \left\{ 3 \int_0^{T\delta} \frac{1 - \cos u}{u^2}\, du - \pi \right\}$$
$$\le A \left\{ \alpha_6 T \epsilon + \left(\frac{\alpha_6}{\alpha_4 - 1 - \alpha_3^2} \right)^{3/2} \frac{T}{\sqrt{n}} + \lambda + \frac{\lambda^2 T}{\epsilon^{\frac{1}{2}}} + \frac{\lambda^3 T^2}{\epsilon^{\frac{3}{2}}} \right\}$$
$$+ AT \iint_{|t_1| \le T, |t_2| \le T} |g(t_1, t_2)|\, |f - \varphi|\, dt_1\, dt_2 .$$

By Lemma 9 (i) with $k = 3$ we have

$$(113)\quad T \iint\limits_{|t_1| \le T, |t_2| \le T} |g| \cdot |f - \varphi| \, dt_1 \, dt_2 \le \frac{AT\lambda}{\epsilon^{1/2}} \iint\limits_{|t_1| \le T, |t_2| \le T} |f - \varphi| \, dt_1 \, dt_2 .$$

By (37) under Lemma 3,

$$(114)\quad |f - \varphi| \le \frac{A}{\sqrt{n}} (\beta_{31} | t_1 |^3 + \beta_{32} | t_2 |^3) e^{-\frac{1}{4}(1-\rho^2)(t_1^2+t_2^2)} \quad \text{for } | t_i | \le \frac{A(1 - \rho^2)\sqrt{n}}{\beta_{3i}}$$

with

$$(115)\quad \begin{aligned} \beta_{31} &= \int_{-\infty}^{\infty} \left| \frac{x^2 - 1}{\sqrt{\alpha_4 - 1}} \right|^3 dP \le \frac{4}{(\alpha_4 - 1)^{3/2}} \int_{-\infty}^{\infty} (x^6 + 1) \, dP \\ &\le \frac{8\alpha_6}{(\alpha_4 - 1)^{3/2}}, \qquad \beta_{32} = \int_{-\infty}^{\infty} | x |^3 \, dP = \beta_3 . \end{aligned}$$

We now take

$$(116)\quad T = \frac{A}{8} \left(\frac{\alpha_4 - 1 - \alpha_3^2}{\alpha_6} \right)^{3/2} \sqrt{n},$$

the A coinciding with that in (114). Then

$$(117)\quad \begin{aligned} \frac{A(1 - \rho^2)\sqrt{n}}{\beta_{31}} &\ge \frac{A(1 - \rho^2)(\alpha_4 - 1)^{3/2}\sqrt{n}}{8\alpha_6} \\ &= \frac{A(\alpha_4 - 1 - \alpha_3^2)\sqrt{\alpha_4 - 1}\,\sqrt{n}}{8\alpha_6} \ge \frac{A(\alpha_4 - 1 - \alpha_3^2)^{3/2}\sqrt{n}}{8\alpha_6^{3/2}} = T \end{aligned}$$

$$(118)\quad \begin{aligned} \frac{A(1 - \rho^2)\sqrt{n}}{\beta_{32}} &= \frac{A(\alpha_4 - 1 - \alpha_3^2)\sqrt{n}}{(\alpha_4 - 1)\beta_3} \\ &\ge \frac{A(\alpha_4 - 1 - \alpha_3^2)^{3/2}\sqrt{n}}{\alpha_4^{3/2}\beta_3} \ge \frac{A(\alpha_4 - 1 - \alpha_3^2)^{3/2}\sqrt{n}}{\alpha_6^{3/2}} > T. \end{aligned}$$

Hence (114) is true for $| t_1 | \le T$ and $| t_2 | \le T$. Using this fact on (113) we obtain

$$(119)\quad \begin{aligned} T &\iint\limits_{|t_1| \le T, |t_2| \le T} |g||f - \varphi| \, dt_1 \, dt_2 \\ &\le \frac{AT\lambda}{\epsilon^{1/2}} \frac{1}{\sqrt{n}} \int_{-\infty}^{\infty} \int_{-\infty}^{\infty} \left\{ \frac{\alpha_6}{(\alpha_4 - 1)^{3/2}} | t_1 |^3 + \beta_3 | t_2 |^3 \right\} e^{-\frac{1}{4}(1-\rho^2)(t_1^2+t_2^2)} \, dt_1 \, dt_2 \\ &\le \frac{AT\lambda}{\epsilon^{1/2}\sqrt{n}} \left(\frac{\alpha_6}{(\alpha_4 - 1)^{3/2}} + \beta_3 \right) \frac{1}{(1 - \rho^2)^{5/2}} \\ &= \frac{AT\lambda}{\sqrt{n\epsilon}} (\alpha_6(\alpha_4 - 1) + \beta_3(\alpha_4 - 1)^{5/2}) \frac{1}{(\alpha_4 - 1 - \alpha_3^2)^{5/2}} \\ &= \frac{AT}{n\sqrt{\epsilon}} (\alpha_6 \sqrt{\alpha_4 - 1} + \beta_3(\alpha_4 - 1)^2) \frac{1}{(\alpha_4 - 1 - \alpha_3^2)^{5/2}} \\ &\le \frac{AT\alpha_6^{11/6}}{n\sqrt{\epsilon}(\alpha_4 - 1 - \alpha_3^2)^{5/2}} . \end{aligned}$$

Substituting in (112), setting $\epsilon = (\alpha_6 T)^{-1}$ and using (116) we obtain after some easy reduction

$$\begin{aligned} &T\delta\left\{3\int_0^{T\delta}\frac{1-\cos u}{u^2}\,du-\pi\right\} \\ (120)\qquad &\leq A\left[1+\frac{1}{\sqrt{n(\alpha_4-1)}}+\left(\frac{\alpha_6}{n(\alpha_4-1-\alpha_3^2)}\right)^{\frac{1}{2}}+\left(\frac{\alpha_6}{n(\alpha_4-1-\alpha_3^2)}\right)^{\frac{1}{4}}\right]. \end{aligned}$$

If $n \geq (\alpha_4 - 1 - \alpha_3^2)^{-1}\alpha_6$, then the right-hand side of (120) is $\leq A$, and so, arguing with (120), as we did with (70), we obtain

$$\text{(121)}\qquad \text{l.u.b.}\,|G(u)-\Phi(u)| \leq \frac{A}{T} = \frac{A}{\sqrt{n}}\left(\frac{\alpha_6}{\alpha_4-1-\alpha_3^2}\right)^{\frac{1}{2}}.$$

For $n < (\alpha_4 - 1 - \alpha_3^2)^{-1}\alpha_6$, however, the right-hand side of (121) $\geq A(\alpha_4 - 1 - \alpha_3^2)^{-1}\alpha_6 \geq A$ and (121) becomes a triviality. Hence Theorem 3 is proved.

3.3. To prove Theorem 4, we start again with the identity (103). We have

$$\begin{aligned} &K_\epsilon(u) - L_\epsilon(u) = \{G(u) - H(u)\} - \{G_1(u) - H_1(u)\} \\ (122)\qquad &\qquad - \{K(u) - K_\epsilon(u)\} + \{L(u) - L_\epsilon(u)\}, \end{aligned}$$

$$\text{(123)}\qquad 0 \leq K(u) - K_\epsilon(u) \leq \epsilon\epsilon(Y^{2k}) \leq Q_k\epsilon \quad \text{by Lemma 12,}$$

$$\text{(124)}\qquad 0 \leq L(u) - L_\epsilon(u) \leq \epsilon\int_{-\infty}^{\infty}\int_{-\infty}^{\infty} y^{2k}(w(x, y) + |\gamma(x, y)|)\,dx\,dy \leq Q_k\,\epsilon.$$

Let us show that

$$\text{(125)}\qquad |G_1(u) - H_1(u)| \leq Q_k/n^{\frac{1}{2}(k-1)}.$$

The function $X = \frac{1}{\sqrt{n}}\sum_{i=1}^{n}\left(\frac{\xi_i^2 - 1}{\sqrt{\alpha_4 - 1}}\right)$ has the same structure as $\sqrt{n}\,\bar{\xi}$ (with $(\alpha_4 - 1)^{-\frac{1}{2}}(\xi_i^2 - 1)$ playing the role of ξ_i); hence, by Theorem 1, there exists an asymptotic expansion of the distribution function $G_1(u)$. We shall see that the terms of this asymptotic expansion are precisely $H_1(u)$, whence (125) follows from Theorem 1.

It is obvious that for the polynomial $\psi(it_1, it_2)$ in (35) $\psi(it, 0)$ coincides with the polynomial $\psi(it)$ in (34). Hence the terms of the asymptotic expansion of $G_1(u)$ are the inversion of $e^{-\frac{1}{2}t^2}\{1 + \psi(it, 0)\}$ viz.

$$\text{(126)}\qquad \Phi(u) + \frac{1}{2\pi}\int_{-\infty}^{u}dx\int_{-\infty}^{\infty}e^{-itx-\frac{1}{2}t^2}\psi(it, 0)\,dt.$$

On the other hand, by (104),

$$\text{(127)}\qquad H_1(u) = \Phi(u) + \int_{-\infty}^{u}dx\int_{-\infty}^{\infty}\gamma(x, y)\,dy,$$

and by (101) with $t_2 = 0$,

$$\text{(128)}\qquad \int_{-\infty}^{\infty}e^{itx}\,dx\int_{-\infty}^{\infty}\gamma(x, y)\,dy = e^{-\frac{1}{2}t^2}\psi(it, 0).$$

Inversion of (118) gives

$$\int_{-\infty}^{\infty} \gamma(x, y)\, dy = \frac{1}{2\pi} \int_{-\infty}^{\infty} e^{-itx - \frac{1}{2}t^2} \psi(it, 0)\, dt \tag{129}$$

which establishes the equality of $H_1(u)$ and (126).

Using (122), (123), (124), (125) on (103) we get

$$\begin{aligned} \int_{-\infty}^{\infty} \frac{1 - \cos Tu}{u^2} \{G(u + z) - H(u + z)\}\, du &= \Lambda_k T \left(\epsilon + \frac{1}{n^{\frac{1}{2}(k-2)}}\right) \\ &+ \theta T \int\int |g(t_1, t_2)| \cdot |f(t_1, t_2) - \varphi(t_1, t_2)[1 + \psi(it_1, it_2)]|\, dt_1\, dt_2 . \end{aligned} \tag{130}$$

If we expand

$$H(u) = \iint_{-\lambda_{\nu}^2 \leq u} \{w(x, y) + \gamma(x, y)\}\, dx\, dy \tag{131}$$

in powers of $n^{-\frac{1}{2}}$ up to and including the term $n^{-\frac{1}{2}(k-3)}$, the remainder is obviously $\Lambda_k n^{-\frac{1}{2}(k-2)}$. Hence

$$H(u) = \Phi(u) + \chi(u) + \Lambda_k / n^{\frac{1}{2}(k-2)}, \tag{132}$$

where $\Phi(u) + \chi(u)$ is the group of terms of the Taylor expansion of (131) in powers of $n^{-\frac{1}{2}}$ up to and including the term $n^{-\frac{1}{2}(k-3)}$. From (130) and (132) we get

$$\begin{aligned} \left| \int_{-\infty}^{\infty} \frac{1 - \cos Tu}{u^2} \{G(u + z) - \Phi(u + z) - \chi(u + z)\}\, du \right| \\ \leq Q_k T \left(\epsilon + \frac{1}{n^{\frac{1}{2}(k-2)}}\right) + AI, \end{aligned} \tag{133}$$

where

$$I = T \iint_{|t_1| \leq T} |g(t_1, t_2)| \cdot |f(t_1, t_2) - \varphi(t_1, t_2)\{1 + \psi(it_1, it_2)\}|\, dt_1\, dt_2 . \tag{134}$$

We are going to prove that the function $\chi(u)$ here defined satisfies all the requirements of the function $\chi(u)$ in Theorem 4. The structure of $\chi(u)$ announced in Theorem 4 is easily verifiable. It remains to prove the inequalities (15) and (16) satisfied by

$$| G(u) - \Phi(u) - \chi(u) | .$$

It is obvious that the function $\Phi(u) + \chi(u)$ has all the properties of the function $F_1(u)$ in Lemma 8, having a bounded derivative $| \Phi'(u) + \chi'(u) | \leq Q_k$. Hence, on taking z in (133) to be the number a in (51), the left-hand side of (133) does not exceed

$$Q_k T\delta \left(3 \int_0^{T\delta} \frac{1 - \cos u}{u^2}\, du - \pi\right), \qquad \delta = Q_k \, \text{l.u.b.} | G(u) - \Phi(u) - \chi(u) |.$$

Hence

$$T\delta\left(3\int_0^{T\delta}\frac{1-\cos u}{u^2}\,du-\pi\right)\le Q_k T\left(\epsilon+\frac{1}{n^{\frac{1}{2}(k-2)}}\right)+Q_k I. \tag{135}$$

In order to appraise I we recall (35) under Lemma 3 (replacing therein each β_{ki} by the larger number $\beta_{k1}\beta_{k2}$, and merging the latter into Q_k)

$$|f(t_1, t_2)-\varphi(t_1, t_2)\{1+\psi(it_1, it_2)\}|\le\frac{Q_k}{n^{\frac{1}{2}(k-2)}}\{\Sigma(|t_i|^k+\cdots+|t_i|^{3(k-2)})\}e^{-(1-\rho^2)(t_1^2+t_2^2)/8} \tag{136}$$

for

$$|t_i|\le Q_k\sqrt{n}. \tag{137}$$

Put $T=(Q_k\sqrt{n})^l$, with Q_k here coinciding with that in (137) and then (136) is valid for $|t_1|\le T^{1/l}$ and $|t_2|\le T^{1/l}$. Write

$$I=T\iint_{|t_1|\le T^{1/l},|t_2|\le T^{1/l}}+T\iint_{|t_1|\le T,|t_2|>T^{1/l}}+T\iint_{\substack{T^{1/l}<|t_1|\le T\\|t_2|\le T^{1/l}}}=I_1+I_2+I_3.$$

By Lemma 9 (i),

$$I_1\le\frac{Q_k T}{n^{\frac{1}{2}}\epsilon^{3/2k}}\iint|f-\varphi(1+\psi)|\,dt_1\,dt_2, \tag{138}$$

whence, by (136)

$$I_1\le\frac{Q_k T}{n^{\frac{1}{2}(k-1)}\epsilon^{3/2k}}\int_{-\infty}^{\infty}\int_{-\infty}^{\infty}\left(\sum_{i=1}^{2}(|t_i|^k+\cdots+|t_i|^{3(k-2)})\right)\cdot e^{-(1-\rho^2)(t_1^2+t_2^2)/8}\,dt_1\,dt_2\le\frac{Q_k T}{n^{\frac{1}{2}(k-1)}\epsilon^{3/2k}}. \tag{139}$$

By Lemma 9 (iii) we have

$$I_2\le Q_k T\iint_{|t_1|\le T,|t_2|>T^{1/l}}\frac{1}{|t_2|^2}\left(\frac{1}{\sqrt{n}\epsilon^{1/2k}}+\frac{|t_1|}{n\epsilon^{3/2k}}\right)\{|f(t_1, t_2)|+\varphi(t_1, t_2)|1+\psi(it_1, it_2)|\}\,dt_1\,dt_2.$$

Obviously,

$$\underset{t_2>T^{1/k-2}}{\text{l.u.b.}}\ \varphi(t_1, t_2)|1+\psi(it_1, it_2)|=e^{-nQ_k}. \tag{140}$$

On the assumption of non-singularity of $P(x)$ we have, by Lemma 7,

$$\underset{|t_2|>T^{1/k-2}}{\text{l.u.b.}}|f(t_1, t_2)|=\underset{|t_2|>Q_k\sqrt{n}}{\text{l.u.b.}}\left|p\left(\frac{t_1}{\sqrt{n}},\frac{t_2}{\sqrt{n}}\right)\right|^n=\underset{|t_2|\ge Q_k}{\text{l.u.b.}}\left|p\left(\frac{t_1}{\sqrt{n}}, t_2\right)\right|^n=e^{-nQ_k}. \tag{141}$$

Hence

$$I_2 \leq Q_k T e^{-nQ_k} \iint_{|t_1| \leq T, |t_2| > T^{1/l}} \frac{1}{|t_2|^2} \left(\frac{1}{\sqrt{n}\epsilon^{1/2k}} + \frac{|t_1|}{n\epsilon^{3/2k}} \right) dt_1 \, dt_2 \tag{142}$$

$$= Q_k \left(\frac{n^{l-1}}{\epsilon^{1/2k}} + \frac{n^{(3/2)(l-1)}}{\epsilon^{3/2k}} \right) e^{-nQ_k}.$$

For I_3 we have $|t_1| > T^{1/l} = Q_k\sqrt{n}$, and so Lemma 7 is applicable to I_3 in the same manner as to I_2. Using Lemma 9 (i) on the factor $|g(t_1, t_2)|$ we get

$$I_3 \leq \frac{Q_k n^l e^{-nQ_k}}{\epsilon^{3/2k}}. \tag{143}$$

Combining (135), (138), (139), (142), (143) we obtain

$$T\delta \left(3 \int_0^{T\delta} \frac{1 - \cos u}{u^2} du - \pi \right) \leq Q_k \left(n^{l/2}\epsilon + \frac{n^{l/2}}{n^{\frac{1}{2}(k-2)}} + \frac{n^{l/2}}{n^{\frac{1}{2}(k-1)}\epsilon^{3/2k}} \right) \tag{144}$$

$$+ Q_k \left(\frac{n^{l-1}}{\epsilon^{1/2k}} + \frac{n^{3/2(l-1)}}{\epsilon^{3/2k}} + \frac{n^l}{\epsilon^{3/2k}} \right) e^{-nQ_k}.$$

Putting $\epsilon = \dfrac{1}{n^{k(k-1)/(2k+3)}}$ we get, as the last term in (144) is $\leq Q_k$,

$$T\delta \left(3 \int_0^{T\delta} \frac{1 - \cos u}{u^2} du - \pi \right) \leq Q_k + Q_k n^{l/2} \left(\frac{1}{n^{k(k-1)/(2k+3)}} + \frac{1}{n^{\frac{1}{2}(k-2)}} \right).$$

If $4 \leq k \leq 6$, we take $l = k - 2$ and get

$$T\delta \left(3 \int_0^{T\delta} \frac{1 - \cos u}{u^2} du - \pi \right) \leq Q_k + Q_k \left(\frac{1}{n^{(6-k)/(2(2k+3))}} + 1 \right) \leq Q_k.$$

Hence, by the argument following (70),

$$\text{l.u.b.} |G(u) - \Phi(u) - \chi(u)| \leq \frac{Q_k}{T} = \frac{Q_k}{n^{\frac{1}{2}(k-2)}},$$

giving (15). If $k \geq 7$, we take $l = \dfrac{2k(k-1)}{2k+3}$ and get

$$T\delta \left(3 \int_0^{T\delta} \frac{1 - \cos u}{u^2} du - \pi \right) \leq Q_k + Q_k \left(1 + \frac{1}{n^{(k-6)/(2(k+3))}} \right) \leq Q_k.$$

Hence

$$\text{l.u.b.} |G(u) - \Phi(u) - \chi(u)| \leq \frac{Q_k}{T} = \frac{Q_k}{n^{k(k-1)/2(k+3)}},$$

giving (16). Therefore Theorem 4 is proved.

5. When $\alpha_4 - 1 - \alpha_3^2 = 0$. If $\alpha_4 - 1 - \alpha_3^2 = 0$, then there is unit probability that ξ_i assumes exactly two values:

$$Pr\{\xi_i = a\} = p, \qquad Pr\{\xi_i = b\} = q, \qquad p + q = 1.$$

Let $\zeta_i = 1$ with probability p and $\zeta_i = 0$ with probability q. Then $\xi_i = b + (a-b)\zeta_i$, $\eta = (a-b)^2 \frac{1}{n} \Sigma(\zeta_i - \bar{\zeta})^2$. Hence it is sufficient to consider the variable $\frac{1}{n} \sum (\zeta_i - \bar{\zeta})^2 = \eta$. Letting $\Sigma\zeta_i = r = np + \sqrt{npq}\, X$ we have $\eta_1 = r - \frac{r^2}{n} = npq + (q-p)\sqrt{npq}\, X - pqX^2$. We now consider two distinct cases:

Case (*i*). $p \neq q$. Here

$$F(z) = Pr\left\{\frac{\eta_1 - n/\partial q}{|p-q|\sqrt{npq}} \leq z\right\}$$

$$= Pr\{(X + c\sqrt{n})^2 \geq c^2 n - 2|c|\sqrt{n}\, z\}, \quad c = \frac{p-q}{2\sqrt{pq}}.$$

Thus $F(z) = 1$ if $z \geq \frac{1}{2}|c|\sqrt{n}$. If $z < \frac{1}{2}|c|\sqrt{n}$, then

$$F(z) = Pr\{X \leq -cn - (c^2 n - 2|c|\sqrt{n}z)^{\frac{1}{2}}\}$$

$$+ Pr\{X \geq -c\sqrt{n} + (c^2 n - 2|c|\sqrt{n}z)^{\frac{1}{2}}\} = F_1(z) + F_2(z).$$

To the random variable X Theorem 2 can be applied. Suppose that $c < 0$; then, by Tchebycheff's inequality,

$$F_2(z) \leq Pr\{X \geq -cn\} \leq \frac{1}{c^2 n} \leq \frac{1}{(p-q)^2 n}.$$

By Theorem 2,

$$F_1(z) = Pr\{X \leq -cn - (c^2 n - 2|c|\sqrt{n}\, z)^{\frac{1}{2}}\}$$

$$= \Phi(z) + \frac{\Theta z^2}{\sqrt{n}|p-q|} + \frac{\Theta(p^2+q^2)}{\sqrt{npq}}.$$

Hence

$$(145) \qquad |F(z) - \Phi(z)| \leq A\left\{\frac{p^2+q^2}{\sqrt{npq}} + \frac{z^2}{\sqrt{n}|p-q|} + \frac{1}{n(p-q)^2}\right\}.$$

The same inequality holds also for $c > 0$.

Case (*ii*). $p = q = 1/2$. Here $\eta_1 = \frac{1}{4}(n - X^2)$; hence

$$(146) \qquad Pr\left\{\eta_1 \geq \frac{n-z}{4}\right\} = Pr\{X^2 \leq z\} = \frac{1}{\sqrt{2\pi}}\int_0^z x^{-\frac{1}{2}} e^{-x/2}\, dx + \frac{\Theta}{\sqrt{n}}.$$

There is no asymptotic expansion for the distribution function of η_1. (See (C), p. 83.)

Comments by K. L. Chung

The application of Hsu's method to the asymptotic expansion (with error estimate) of Student's statistic was carried out by me in "The approximate distribution of Student's statistic", *Ann. Math. Stat.* **17** (1946), 447–465. A reader informed me many years ago that he found errors in the explicit formula exhibited there. I am sorry for that but such a formula can be computed by the heuristic method of Cornish and Fisher mentioned in Hsu's paper. Some other applications were carried out by students in Hsu's class. Unfortunately, I was unable to locate the publications.

Reprinted from
Ann. Math. Statist.
16, 204–210 (1945).

ON THE APPROXIMATE DISTRIBUTION OF RATIOS

BY P. L. HSU
National University of Peking

The purpose of this paper is to apply Cramer's theorem of asymptotic expansion[1] and Berry's theorem[2] to study the approximate distribution of ratios of the following two types:

$$\text{(I)} \qquad Z = \frac{1}{n}(Y_1 + \cdots + Y_n) \Big/ \frac{1}{m}(\bar{X}_1 + \cdots + \bar{X}_m) = \bar{Y}/\bar{X},$$

$$\text{(II)} \qquad Z = Y \Big/ \frac{1}{m}(X_1 + \cdots + X_m) = Y/\bar{X}.$$

In (I) the X_i, Y_j are independent, the Y_j are equi-distributed,[3] and the X_i are equi-distributed and positive. In (II) $X_1, \cdots, X_n$, Y are independent and positive, and the X_i are equi-distributed.

1. The ratio (I). Assume that (I1) the absolute kth moment of X_i and that of Y_j are finite and positive, where k is a fixed integer ≥ 3,
(I2) the distribution of X_i and that of Y_j are non-singular.

Let

$$\xi = \epsilon(X_i), \qquad \eta = \epsilon(Y_j), \qquad \sigma^2 = \epsilon(X_i^2) - \xi^2, \qquad \tau^2 = \epsilon(Y_j^2) - \eta^2$$

and

$$U = \frac{\sqrt{m}}{\sigma}(\bar{X} - \xi), \qquad V = \frac{\sqrt{n}}{\tau}(\bar{Y} - \eta).$$

Let $F(x)$, $G(x)$ and $H(x)$ be respectively the distribution functions of Z, U and V. Let

$$b = \left(\frac{\sigma^2 x^2}{m} + \frac{\tau^2}{n}\right)^{\frac{1}{2}}, \qquad u = \frac{\xi n - \eta}{b}.$$

Then the relation $Z \leq x$ is equivalent to

$$-\frac{x\sigma U}{b\sqrt{m}} + \frac{\tau V}{b\sqrt{n}} \leq u.$$

---

[1] H. CRAMÉR. *Random Variables and Probability Distributions* (1937), Chap. 7.

[2] A. C. BERRY. "The accuracy of the Gaussian approximation to the sum of independent variates", *Trans. Amer. Math. Soc.*, Vol. 49 (1941), pp. 122–136.

[3] The Y_j are said to be equi-distributed if all Y_j have the same distribution function.

For simplicity we shall assume $x > 0$; the results are, however, general. Then the distribution functions of $-\frac{x\sigma U}{b\sqrt{m}}$ and $\frac{\tau V}{b\sqrt{n}}$ are

$$Pr\left\{-\frac{x\sigma U}{b\sqrt{m}} < y\right\} = 1 - G\left(-\frac{b\sqrt{m}y}{\sigma x}\right), \qquad Pr\left\{\frac{\tau V}{b\sqrt{n}} \leq y\right\} = H\left(\frac{b\sqrt{n}y}{\tau}\right).$$

Hence, by the theorem of convolution,

$$F(x) = \int_{\infty}^{\infty}\left\{1 - G\left(-\frac{b\sqrt{m}(u-y)}{\sigma x}\right)\right\} dH\left(\frac{b\sqrt{n}y}{\tau}\right). \tag{1}$$

Here we recall the theorems of Cramér and Berry: Under the conditions (I1) and (I2)

$$G(x) = \Phi(x) + \sum_{\nu=1}^{k-3}\frac{P_\nu(x)}{m^{\nu/2}} + \frac{D_k}{m^{\frac{1}{2}(k-2)}}, \tag{2}$$

where

$$\Phi(x) = \frac{1}{\sqrt{2\pi}}\int_{-\infty}^{x} e^{-\frac{1}{2}y^2}\, dy, \qquad P_\nu(x) = \sum_{j=1}^{\nu} c_{j\nu}\,\Phi^{(\nu+2j)}(x),$$

and $| D_k |$ is less than a positive number which depends only on k and the distribution of X_i. If $k = 3$, condition (I2) may be removed.[4]

Analogously,

$$H(x) = \Phi(x) + \sum_{\nu=1}^{k-3}\frac{Q_\nu(x)}{n^{\nu/2}} + \frac{D_k'}{m^{\frac{1}{2}(k-2)}}, \tag{3}$$

where

$$Q_\nu(x) = \sum_{j=1}^{\nu} d_{j\nu}\,\Phi^{(\nu+2j)}(x).$$

In the sequel we shall use the letter Δ_k to denote an unspecified quantity such that $| \Delta_k |$ is less than a positive number which depends only on k, the distribution of X_i and the distribution of Y_j.

Using (2) we have

$$1 - G(-x) = \Phi(x) + \sum_{\nu=1}^{k-3}\frac{(-1)^\nu P_\nu(x)}{m^{\nu/2}} + \frac{D_k}{m^{\frac{1}{2}(k-2)}} \tag{4}$$

and this making this substitution in (1) we get

$$F(x) = \int_{\infty}^{\infty}\Phi\left(\frac{b\sqrt{m}(u-y)}{\sigma x}\right) dH\left(\frac{b\sqrt{n}y}{\tau}\right)$$
$$+ \sum_{\nu=1}^{k-3}\frac{(-1)^\nu}{m^{\nu/2}}\int_{\infty}^{\infty} P_\nu\left(\frac{b\sqrt{m}(u-y)}{\sigma x}\right) dH\left(\frac{b\sqrt{n}y}{\tau}\right) + \frac{\Delta_k}{m^{\frac{1}{2}(k-2)}},$$

[4] This last assertion constitutes Berry's theorem.

and so by partial integration,

$$F(x) = \int_{-\infty}^{\infty} H\left(\frac{b\sqrt{n}(u-y)}{\tau}\right) d\Phi\left(\frac{b\sqrt{m}y}{\sigma x}\right)$$

$$+ \sum_{\nu=1}^{k-3} \frac{(-1)^\nu}{m^{\nu/2}} \int_{-\infty}^{\infty} H\left(\frac{b\sqrt{n}(u-y)}{\tau}\right) dP_\nu\left(\frac{b\sqrt{m}y}{\sigma x}\right) + \frac{\Delta_k}{m^{\frac{1}{2}(k-2)}}.$$

Making the transformation $y = \sigma x v/b\sqrt{m}$ and writing

$$\alpha = \frac{b\sqrt{n}}{\tau}, \qquad \beta = \frac{\sigma\sqrt{n}\,x}{\tau\sqrt{m}} \tag{5}$$

we get

$$F(x) = \int_{-\infty}^{\infty} H(du - \beta v)\Phi'(v)\, dv + \sum_{\nu=1}^{k-3} \frac{(-1)^\nu}{m^{\nu/2}} \int_{-\infty}^{\infty} H(\alpha u - \beta y)P_\nu'(v)\, dv + \frac{\Delta_k}{m^{\frac{1}{2}(k-2)}}$$

$$= I_0 + \sum_{\nu=1}^{k-3} \frac{(-1)^\nu}{m^{\nu/2}} I_\nu + \frac{\Delta_k}{m^{\frac{1}{2}(k-2)}}.$$

For I_0 we use (3) and obtain

$$I_0 = \int_{-\infty}^{\infty} \Phi(\alpha u - \beta v)\Phi'(v)\, dv + \sum_{\nu=1}^{k-3} \frac{1}{n^{\nu/2}} \int_{-\infty}^{\infty} Q_\nu(\alpha u - \beta v)\Phi'(v)\, dv + \frac{\Delta_k}{n^{\frac{1}{2}(k-2)}}.$$

For I_ν we use (3) with k replaced by $k - \nu$. Thus

$$I_\nu = \int_{-\infty}^{\infty} \Phi(\alpha u - \beta v)P_\nu'(v)\, dv + \sum_{\mu=1}^{k-3-\nu} \frac{1}{n^{\mu/2}} \int_{-\infty}^{\infty} Q_\mu(\alpha u - \beta v)P_\nu'(v)\, dv + \frac{\Delta_k}{n^{\frac{1}{2}(k-2-\nu)}}.$$

Combining these results we get

$$F(x) = \int_{-\infty}^{\infty} \Phi(\alpha u - \beta v)\Phi'(v)\, dv + \sum_{\nu=1}^{k-3} \frac{1}{n^{\nu/2}} \int_{-\infty}^{\infty} Q_\nu(\alpha u - \beta v)\Phi'(v)\, dv \tag{6}$$

$$+ \sum_{\nu=1}^{k-3} \frac{(-1)^\nu}{m^{\nu/2}} \int_{-\infty}^{\infty} \Phi(\alpha u - \beta v)P_\nu'(v)\, dv$$

$$+ \sum_{\nu=1}^{k-3} \sum_{\mu=1}^{k-3-\nu} \frac{(-1)^\nu}{m^{\nu/2} n^{\mu/2}} \int_{-\infty}^{\infty} Q_\mu(\alpha u - \beta u)P_\nu'(v)\, dv + R_k,$$

where

$$R_k = \frac{\Delta_k}{m^{\frac{1}{2}(k-2)}} + \frac{\Delta_k}{n^{\frac{1}{2}(k-2)}} + \sum_{\nu=1}^{k-3} \frac{\Delta_k}{m^{\nu/2} n^{\frac{1}{2}(k-2-\nu)}} = \Delta_k\left(\frac{1}{\sqrt{m}} + \frac{1}{\sqrt{n}}\right)^{k-2}.$$

Now by (5), $\alpha > 0$ and $\alpha^2 - \beta^2 = 1$. For such values of α and β, however, it follows easily from the theorem of convolution that

$$\int_{-\infty}^{\infty} \Phi(\alpha u - \beta v)\Phi'(v)\, dv = \Phi'(u).$$

As differentiation under the integration sign is justified by the boundedness of the derivatives of Φ we have

$$\alpha^p \int_{-\infty}^{\infty} \Phi^{(p)}(\alpha u - \beta v)\Phi'(v)\, dv = \Phi^{(p)}(u).$$

Repeated partial integration then gives

$$\int_{-\infty}^{\infty} \Phi^{(p)}(\alpha u - \beta v)\Phi^{(q)}(v)\, dv = \beta^{q-1} \int_{-\infty}^{\infty} \Phi^{(p+q-1)}(\alpha u - \beta v)\Phi'(v)\, dv$$

$$= \frac{\beta^{q-1}}{\alpha^{p+q-1}} \Phi^{(p+q-1)}(u).$$

Hence

$$\int_{-\infty}^{\infty} Q_\nu(\alpha u - \beta v)\Phi'(v)\, dv = \sum_{j=1} d_{j\nu} \int_{-\infty}^{\infty} \Phi^{(\nu+2j)}(\alpha u - \beta v)\Phi'(v)\, dv$$

$$= \sum_{j=1}^{\nu} \frac{d_{j\nu}}{\alpha^{\nu+2j}} \Phi^{(\nu+2j)}(u),$$

$$\int_{-\infty}^{\infty} \Phi(\alpha u - \beta v)P_\nu'(v)\, dv = \sum_{j=1}^{\nu} c_{j\nu} \int_{-\infty}^{\infty} \Phi(\alpha u - \beta v)\Phi^{(\nu+2j+1)}(v)\, dv$$

$$= \sum_{j=1}^{\nu} \frac{\beta^{\nu+2j} c_{j\nu}}{\alpha^{\nu+2j}} \Phi^{(\nu+2j)}(u),$$

$$\int_{-\infty}^{\infty} Q_\mu(\alpha u - \beta v)P_\nu'(v)\, dv = \sum_{i=1}^{\mu}\sum_{j=1}^{\nu} d_{i\mu} c_{j\nu} \int \Phi^{(\mu+2j)}(\alpha u - \beta v)\Phi^{(\nu+2j+1)}(v)\, dv$$

$$= \sum_{i=1}^{\mu}\sum_{j=1}^{\nu} d_{i\mu} c_{j\nu} \frac{\beta^{\nu+2j}}{\alpha^{\mu+\nu+2i+2j}} \Phi^{(\mu+\nu+2i+2j)}(u).$$

Making all these substitutions in (6) we obtain the final result

$$F(x) = \Phi(u) + \sum_{\nu=1}^{k-3} \frac{(-1)^\nu}{m^{\nu/2}} \sum_{j=1}^{\nu} \frac{\beta^{\nu+2j} c_{j\nu}}{\alpha^{\nu+2j}} \Phi^{(\nu+2j)}(u) + \sum_{\nu=1}^{k-3} \frac{1}{n^{\nu/2}} \sum_{j=1}^{\nu} \frac{d_{j\nu}}{\alpha^{\nu+2j}} \Phi^{(\nu+2j)}(u)$$

$$+ \sum_{\nu=1}^{k-3} \sum_{\mu=1}^{k-3-\nu} \frac{(-1)^\nu}{m^{\nu/2} n^{\mu/2}} \sum_{i=1}^{\mu}\sum_{j=1}^{\nu} d_{i\mu} c_{j\nu} \frac{\beta^{\nu+2j}}{\alpha^{\mu+\nu+2i+2j}} \phi^{(\mu+\nu+2i+2j)}(u)$$

$$+ \Delta_k \left(\frac{1}{\sqrt{m}} + \frac{1}{\sqrt{n}}\right)^{k-2}$$

If $k = 3$, the result remains true without the condition (I2).

2. The ratio (II). Here we make the following assumptions:
(II1) The kth moment of X_i is finite and positive, where k is a fixed integer $\geq k$, $\epsilon(X_i) = 1$,[5] $\epsilon(X_i^2) - 1 = \sigma^2$.
(II2) The distribution of X_i is non-singular.

[5] As the case $\epsilon(X_i) = 0$ is excluded, there is no loss of generality in this assumption.

Let $U = \sqrt{m}(\bar{X} - 1)/\sigma$, and $F(x)$, $G(x)$ and $H(x)$ be respectively the distribution functions of Z, U and Y. Then

$$F(x) = Pr\left\{Y - \frac{\sigma x U}{\sqrt{m}} \leq x\right\}.$$

Because of the positiveness of X_i and Y we may always assume $x > 0$. Then, by the theorem of convolution,

$$F(x) = \int_{-\infty}^{\infty}\left\{1 - G\left(-\frac{\sqrt{m}(x-y)}{\sigma x}\right)\right\} dH(y).$$

Using (4) we have

$$F(x) = \int_{-\infty}^{\infty}\left\{\Phi\left(\frac{\sqrt{m}(x-y)}{\sigma x}\right) + \sum_{\nu=1}^{k-3}\frac{(-1)^\nu}{m^{\nu/2}} P_\nu\left(\frac{\sqrt{m}(x-y)}{\sigma x}\right)\right\} dH(y) + \frac{A_k}{m^{\frac{1}{2}(k/2)}},$$

where, as throughout the rest of this paper, A_k represents an unspecified quantity such that $|A_k|$ is less than a positive number depending only on k, the distribution of X_i and the distribution of Y. By partial integration we get

$$(7)\quad F(x) = \int_{-\infty}^{\infty} H(x-y)\, d\left\{\Phi\left(\frac{\sqrt{m}y}{\sigma x}\right) + \sum_{\nu=1}^{k-3}\frac{(-1)^\nu P_\nu\left(\frac{\sqrt{m}y}{\sigma x}\right)}{m^{\nu/2}}\right\} + \frac{A_k}{m^{\frac{1}{2}(k-2)}}$$

$$= \int_{-\infty}^{\infty} H\left(x - \frac{\sigma x z}{\sqrt{m}}\right)\left(\Phi'(z) + \sum_{\nu=1}^{k-3}\frac{(-1)^\nu P_\nu'(z)}{m^{\frac{1}{2}(k-2)}}\right) dz + \frac{A_k}{m^{\frac{1}{2}(k-2)}}.$$

An interesting special case is the following: Suppose that (II3) $H^{(k-2)}(x)$ exists and is continuous for all $x \geq 0$; (II4) the functions

$$\xi_\nu(x) = x^\nu H^{(\nu)}(x) \qquad (\nu = 1, \cdots, k-3)$$

are bounded, i.e.

$$\xi_\nu(x) = A_k;$$

(II3) there is a positive constant $c < 1$ such that

$$x^{k-2} H^{(k-2)}(y) = A_k$$

for all $x \geq 0$ and $(1 - c)x \leq y \leq (1 + c)x$. Under these conditions we have

$$H\left(x - \frac{\sigma x z}{\sqrt{m}}\right) = \sum_{\nu=0}^{k=3}\frac{(-1)^\nu \sigma^\nu x^\nu z^\nu H^{(\nu)}(x)}{\nu!\, m^{\nu/2}}$$

$$+ \frac{(-1)^{k-2}\sigma^{k-2}x^{k-2}z^{k-2}}{(k-2)!\, m^{\frac{1}{2}(k-2)}} H^{(k-2)}\left(x + \frac{\delta \sigma x z}{\sqrt{m}}\right) \quad (|\delta| \leq 1),$$

and so, for $|z| \leq \dfrac{c\sqrt{m}}{\sigma}$ we have

$$(8)\qquad H\left(x - \frac{\sigma x z}{\sqrt{m}}\right) = \sum_{\nu=0}^{k-3}\frac{(-1)^\nu \sigma^\nu \xi_\nu z^\nu}{\nu!\, m^{\nu/2}} + \frac{A_k z^{k-2}}{m^{\frac{1}{2}(k-2)}}.$$

Separate now the integral in (7) into two parts:

$$I_1 = \int_{|z| \leq c\sqrt{m}/\sigma}, \qquad I_2 = \int_{|z| > c\sqrt{m}/\sigma}$$

Now

$$|I_2| \leq \int_{|z| > c\sqrt{m}/\sigma} \left| \Phi'(z) + \sum_{\nu=1}^{k-3} \frac{(-1)^\nu P_\nu'(z)}{m^{\nu/2}} \right| dz.$$

Evidently this last integral is exponially small and so is $A_k/m^{\frac{1}{2}(k-2)}$. By (8),

$$I_1 = \int_{|z| \leq c\sqrt{m}/\sigma} \left(\sum_{\nu=0}^{k-3} \frac{(-1)^\nu \sigma^\nu \xi_\nu z^\nu}{\nu! m^{\nu/2}} \right) \left(\Phi'(z) + \sum_{\nu=1}^{k-3} \frac{(-1)^\nu P_\nu'(z)}{m^{\nu/2}} \right) dz + \frac{A_k}{m^{\frac{1}{2}(k-2)}}$$

$$= \int_{-\infty}^{\infty} \left(\sum_{\nu=0}^{k-3} \frac{(-1)^\nu \sigma^\nu \xi_\nu z^\nu}{\nu! m^{\nu/2}} \right) \left(\Phi'(z) + \sum_{\nu=1}^{k-3} \frac{(-1)^\nu P_\nu'(z)}{m^{\nu/2}} \right) dz + \frac{A_k}{m^{\frac{1}{2}(k-2)}}.$$

Combining these results we obtain

$$F(x) = \int_{-\infty}^{\infty} \left(\sum_{\nu=0}^{k-3} \frac{(-1)^\nu \sigma^\nu \xi_\nu z^\nu}{\nu! m^{\nu/2}} \right) \left(\Phi'(z) + \sum_{\nu=1}^{k-3} \frac{(-1)^\nu}{m^{\nu/2}} \sum_{j=1}^{\nu} c_{j\nu} \Phi^{(\nu+2j+1)}(z) \right) dz + \frac{A_k}{m^{\frac{1}{2}(k-2)}}$$

$$= \sum_{\nu=0}^{k-3} \frac{d_\nu \xi_\nu}{m^{\nu/2}} I_{\nu 1} + \sum_{\nu=0}^{k-3} \sum_{\mu=1}^{k-3} \sum_{j=1}^{\mu} \frac{d_{j\mu\nu} \xi_\nu}{m^{\frac{1}{2}(\mu+\nu)}} \Phi_{\nu,\mu+2j+1} + \frac{A_k}{m^{\frac{1}{2}(k-2)}}$$

$$= \sum_1 + \sum_2 + \frac{A_k}{m^{\frac{1}{2}(k-2)}},$$

where

$$I_{\alpha\beta} = \int_{-\infty}^{\infty} z^\alpha \Phi^{(\beta)}(z)\, dz.$$

Now the following facts can easily be established by means of partial integration:

(9) $$I_{\alpha\beta} = 0 \quad \text{when} \quad \alpha - \beta \text{ is even},$$

(10) $$I_{\alpha\beta} = 0 \quad \text{when} \quad \beta - \alpha > 1.$$

By (9), the non-vanishing terms in $\sum_1$ are the even terms and the non-vanishing terms in $\sum_2$ are those for which $\mu + \nu$ is even. Hence

$$\sum_1 = \sum_{\nu=0}^{[\frac{1}{2}(k-3)]} \frac{e_\nu \xi_{2\nu}}{m^\nu},$$

$$\sum_2 = \sum_{\nu=0}^{[\frac{1}{2}(k-3)]} \sum_{\mu=1}^{[\frac{1}{2}(k-3)]} \sum_{j=1}^{2\mu} \frac{e_{j\mu\nu} \xi_{2\nu}}{m^{\mu+\nu}} I_{2\nu,2\mu+2j+1} + \sum_{\nu=0}^{[\frac{1}{2}(k-4)]} \sum_{\mu=0}^{[\frac{1}{2}(k-4)]} \sum_{j=1}^{2\mu+1} \frac{e_{j\mu\nu}' \xi_{2\nu+1}}{m^{\mu+\nu+1}} I_{2\nu+1,2\mu+2j+2}.$$

Using (10) to reduce $\sum_2$ further we get

$$\begin{aligned}\sum_2 &= \sum_{\nu=2}^{[\frac12(k-3)]}\sum_{\mu=1}^{\nu-1}\sum_{j=1}^{2\mu}\frac{c_{j\mu\nu}\,\xi_{2\nu}}{m^{\mu+\nu}}\, I_{2\nu,\,2\mu+2j+1} + \sum_{\nu=1}^{[\frac12(k-4)]}\sum_{\mu=0}^{\nu-1}\sum_{j=1}^{2\mu+1}\frac{e'_{j\mu\nu}\,\xi_{2\nu+1}}{m^{\mu+\nu+1}}\, I_{2\nu+1,\,2\mu+2j+2}\\
&= \sum_{\nu=0}^{[\frac12(k-7)]}\sum_{\mu=0}^{\nu}\frac{g_{\mu\nu}\,\xi_{2\nu+4}}{m^{\mu+\nu+3}} + \sum_{\nu=0}^{[\frac12(k-6)]}\sum_{\mu=0}^{\nu}\frac{g'_{\mu\nu}\,\xi_{2\nu+3}}{m^{\mu+\nu+2}}\\
&= \sum_{\alpha=0}^{[\frac12(k-9)]}\frac{1}{m^{\alpha+3}}\sum_{\beta=[\frac12(\alpha+1)]}^{\alpha} h_{\alpha\beta}\,\xi_{2\beta+4} + \sum_{\alpha=0}^{[\frac12(k-6)]}\frac{1}{m^{\alpha+2}}\sum_{\beta=[\frac12(\alpha+1)]}^{\alpha} h'_{\alpha\beta}\,\xi_{2\beta+3} + \frac{A_k}{m^{\frac12(k-2)}}\\
&= \sum_{i=3}^{[\frac12(k-3)]}\frac{1}{m^{i}}\sum_{j=[\frac12(i-2)]}^{i-3} l_{ij}\,\xi_{2j+4} + \sum_{i=2}^{[\frac12(k-3)]}\frac{1}{m^{i}}\sum_{j=[\frac12(i-1)]}^{i-2} l'_{ij}\,\xi_{2j+3} + \frac{A_k}{m^{\frac12(k-2)}}.\end{aligned}$$

Hence

$$\begin{aligned}\sum_1 + \sum_2 &= \xi_0 + \frac{e_1\xi_2}{m} + \frac{e_2\xi_4 + l'_{20}\xi_3}{m^2} + \sum_{\nu=3}^{[\frac12(k-3)]}\frac{1}{m^{\nu}}\\
&\qquad\left(e_\nu\xi_{2\nu} + \sum_{\mu=[\frac12(\nu-2)]}^{\nu-3} l_{\mu\nu}\,\xi_{2\nu+4} + \sum_{\mu=[\frac12(\nu-2)]}^{\nu-2} l'_{\mu\nu}\,\xi_{2\nu+3}\right) + \frac{A_k}{m^{\frac12(k-2)}}\\
&= \xi_0 + \sum_{\nu=1}^{k-3}\frac{1}{m^{\nu}}\sum_{j=\nu+1}^{2\nu} p_{j\nu}\,\xi_j + \frac{A_k}{m^{\frac12(k-2)}}.\end{aligned}$$

Hence

$$F(x) = \xi_0 + \sum_{\nu=1}^{k-3}\frac{1}{m^{\nu}}\sum_{j=\nu+1}^{2\nu} p_{j\nu}\,\xi_j + \frac{p_k}{m^{\frac12(k-2)}}.$$

Our final conclusion is: Under the conditions (II1)–(II5) formula (11) is true; if $k = 3$, (11) remains true without the condition (II2).

Reprinted from
Ann. Math. Statist.
16 (1945), 278–286.

ON THE POWER FUNCTIONS OF THE E^2-TEST AND THE T^2-TEST

The General Linear Hypothesis

Every linear hypothesis about a p-variate normal population or several such populations having common variances and covariances is reducible to the canonical form [4]: the sample distribution, when nothing whatever has been discarded from the whole sample, being

$$(2\pi)^{-(1/2)p(m+n)}|\alpha_{ij}|^{(1/2)(m+n)}\exp\left\{-\frac{1}{2}\sum_{i,j=1}^{p}\alpha_{ij}\sum_{r=1}^{m}(y_{ir}-\eta_{ir})(y_{jr}-\eta_{jr})\right.$$
$$\left.-\frac{1}{2}\sum_{i,j=1}^{p}\alpha_{ij}\sum_{s=1}^{n}z_{is}z_{js}\right\}\Pi\, dy\, dz \qquad (n \geqslant p), \tag{1}$$

where the η_{ir} and the α_{ij} are unknown, and the hypothesis to be tested is

$$H:\eta_{ir}=0 \qquad (i=1,\ldots,p;\ r=1,\ldots,n_1,\ n_1\leqslant m).$$

It is clear that the y_{ir} $(i=1,\ldots,p;\ r=n_1+1,\ldots,m)$ can have no use. Also, the only useful quantities supplied by the set z_{is} are the statistics

$$b_{ij}=\sum_{s=1}^{n}z_{is}z_{js},$$

because the remaining quantities may be regarded as a set of angles which are independent of y_{ir} and the b_{ij} and which has a known distribution free from any unknown parameter in (1) [2]. After discarding the irrelevant y's and the angles there results the reduced sample distribution

$$K|\alpha_{ij}|^{(1/2)(n_1+n)}|b_{ij}|^{(1/2)(n-p-1)}\exp\left\{-\frac{1}{2}\sum_{i,j=1}^{p}\alpha_{ij}\cdot\sum_{r=1}^{n_1}(y_{ir}-\eta_{ir})(y_{jr}-\eta_{jr})\right.$$
$$\left.-\frac{1}{2}\sum_{i,j=1}^{p}\alpha_{ij}b_{ij}\right\}\Pi\, dy\, db.$$

Hereafter the indices i, j, and r shall have the following ranges:

$$i,j=1,\ldots,p, \qquad r=1,\ldots,n_1,$$

and the convention that repetition of an index indicates summation will be adopted.

Writing

$$a_{ij} = y_{ir}y_{jr}\,, \qquad c_{ij} = a_{ij} + b_{ij}\,,$$

we obtain the distribution of the y_{ir} and the c_{ij}:

$$K|\alpha_{ij}|^{(1/2)(n_1+n)}|c_{ij} - a_{ij}|^{(1/2)(n-p-1)}\exp(-\tfrac{1}{2}\alpha_{ij}c_{ij} + \alpha_{ij}y_{ir}\eta_{jr} - \tfrac{1}{2}\alpha_{ij}\eta_{ir}\eta_{jr})\Pi\, dy\, dc. \quad (2)$$

In the remaining two sections of this paper we deal exclusively with the special cases $p = 1$ and $n_1 = 1$. According as $p = 1$ or $n_1 = 1$ we shall drop the indices i and j or the index r.

The Case $p = 1$

When $p = 1$, (2) reduces to

$$K\alpha^{(1/2)(n_1+n)}(c - y_ry_r)^{(1/2)n-1}\exp(-\tfrac{1}{2}\alpha c + \alpha y_r\eta_r - \tfrac{1}{2}\alpha\eta_r\eta_r)\, dc\, \Pi\, dy.$$

Putting $y_r = c^{1/2}x_r$ we obtain

$$K\alpha^{(1/2)(n_1+n)}c^{(1/2)(n_1+n)-1}(1 - x_rx_r)^{(1/2)n-1}\exp\left(-\tfrac{1}{2}\alpha c + \alpha c^{1/2}x_r\eta_r - \tfrac{1}{2}\alpha\eta_r\eta_r\right) dc\, \Pi\, dx. \quad (3)$$

The hypothesis H is now

$$H' : \eta_r = 0 \qquad (r = 1, \ldots, n_1).$$

If w is any critical region for the rejection of H', denote by $w(c)$ the cross section of w for every fixed c. Then the power function of w is

$$\begin{aligned}\beta_w(\eta, \alpha) &= \beta_w(\eta_1, \ldots, \eta_{n_1}, \alpha)\\ &= K\alpha^{(1/2)(n_1+n)}e^{-(1/2)\alpha\eta_r\eta_r}\int_0^{\infty} c^{(1/2)(n_1+n)-1}e^{-(1/2)\alpha c}\, dc\\ &\quad\times\int_{w(c)}(1 - x_rx_r)^{(1/2)n-1}e^{\alpha c(1/2)x_r\eta_r}\Pi\, dx. \end{aligned} \quad (4)$$

It is known [3] that, in order to have

$$\beta_w(0, \alpha) = \epsilon \quad (5)$$

for all α, it is necessary and sufficient that

$$\int_{w(c)}(1 - x_rx_r)^{(1/2)n-1}\Pi\, dx = A\epsilon, \quad (6)$$

where A is a constant.

The E^2-test is the test based on the critical region

$$w_0 : x_rx_r = c^{-1/2}y_ry_r = E^2 \geqslant \text{const.}$$

The author has proved [3] that of all the critical regions which satisfy (5) and whose power function is a function of $\alpha\eta_r\eta_r$ alone, the region w_0 is the uniformly most powerful one. This result is generalized by Wald [7], who proved that, of all the regions satisfying (5), the surface integral

$$\gamma_w(\alpha, \lambda) = \int_{\eta_r\eta_r=\lambda}\beta_w(\eta, \alpha)\, dA$$

is maximum when w is w_0. The author gives here another proof of Wald's theorem which is easier as it dispenses with the somewhat intricate Lemma 1 of Wald. From (4) we have

$$\gamma_w(\alpha,\lambda) = K\alpha^{(1/2)(n_1+n)}\int_0^\infty c^{(1/2)(n_1+n)-1}e^{-(1/2)\alpha c}\,dc$$

$$\int_{w(c)}(1-x_rx_r)^{(1/2)n-1}\Pi\,dx\int_{\eta_r\eta_r=\lambda}\exp(-\tfrac{1}{2}\alpha\eta_r\eta_r+\alpha c^{1/2}x_r\eta_r)\,dA.$$

By means of a rotation in the space of $(\eta_1,\ldots,\eta_{n_1})$ we can obtain

$$\int_{\eta_r\eta_r=\lambda}\exp\left[-\tfrac{1}{2}\alpha\eta_r\eta_r+\alpha c^{1/2}x_r\eta_r\right]dA = \int_{\zeta_r\zeta_r=\lambda}\exp\left[-\tfrac{1}{2}\alpha\zeta_r\zeta_r+\alpha c^{1/2}(x_rx_r)^{1/2}\zeta_1\right]dA$$
$$= \sum_{k=0}^\infty a_k\alpha^{2k}(cx_rx_r)^k,$$

where a_k depends only on α, k and λ. Hence

$$\gamma_w(\alpha,\lambda) = \sum_{k=0}^\infty b_k\int_0^\infty c^{(1/2)(n_1+n)-1}e^{-(1/2)\alpha c}\,dc\int_{w(c)}(x_rx_r)^k(1-x_rx_r)^{(1/2)n-1}\Pi\,dx, \tag{7}$$

where b_k depends only on k, α and λ. Since $w(c)$ satisfies (6), it follows from a lemma of Neyman and Pearson [5] that

$$\int_{w(c)}(x_rx_r)^k(1-x_rx_r)^{(1/2)n-1}\Pi\,dx$$

is maximum, for all c and k, when $w(c)$ is the region $x_rx_r \geqslant$ const., i.e., when w is itself the region $x_rx_r \geqslant$ const. This proves Wald's theorem.

Still another optimum property of the E^2-test may be established on using the volume integral instead of the surface integral. This is stated in the following theorem.

Theorem 1. *Let S be any linear set and let*

$$\varphi_w(\alpha,S) = \int_{\eta_r\eta_r\in S}\beta_w(\eta,\alpha)\Pi\,d\eta.$$

Of all the regions satisfying (5), *the region w_0 has the maximum $\varphi_\omega(\alpha,S)$.*

For, by the same computation which leads to (7), we easily obtain

$$\varphi_w(\alpha,S) = \sum_{k=0}^\infty c_k\int_0^\infty c^{(1/2)(n_1+n)-1}e^{-(1/2)\alpha c}\,dc\int_{w(c)}(x_rx_r)^k(1-x_rx_r)^{(1/2)n-1}\Pi\,dx,$$

where c_k depends only on k, α and S. Hence the result follows.

This theorem also contains my previous result as a consequence. For, writing

$$\beta_w(\eta_i\alpha) = f(\alpha\eta_r\eta_r),\qquad \beta_{w_0}(\eta,\alpha) = f_0(\alpha\eta_r\eta_r),$$

we have

$$0 \leqslant \int_{\eta_r\eta_r \in S} (f_0(\alpha\eta_r\eta_r) - f(\alpha\eta_r\eta_r))\Pi\, d\eta = \frac{\pi^{(1/2)n_1}}{\Gamma(\frac{1}{2}\eta_1)} \int_S t^{(1/2)n_1 - 1}(f_0(\alpha t) - f(\alpha t))\, dt.$$

Since S is arbitrary, we must have $f(\alpha t) \leqslant f_0(\alpha t)$.

The Case $n_1 = 1$

When $n_1 = 1$, (2) and H become, respectively,

$$K|\alpha_{ij}|^{(1/2)(n+1)}|c_{ij} - y_i y_j|^{(1/2)(n-p-1)}\exp(-\tfrac{1}{2}\alpha_{ij}c_{ij} + \alpha_{ij}y_i\eta_j - \tfrac{1}{2}\alpha_{ij}\eta_i\eta_j)\Pi\, dy\, dc, \tag{8}$$

$$H'' : \eta_i = 0 \qquad (i = 1, \dots, p).$$

There is a unique real matrix

$$\mathbf{T} = \begin{bmatrix} t_{11} & & & \\ t_{12} & t_{22} & & \\ \cdots & \cdots & \cdots & \cdots \\ t_{1p} & t_{2p} & \cdots & t_{pp} \end{bmatrix} \qquad (t_{ii} > 0;\ \text{zeros above the principal diagonal})$$

such that $[c_{ij}] = \mathbf{TT}'$ [2]. Introducing the new variables $x_1, \dots, x_p$ by means of the transformation

$$[y_1, \dots, y_p] = [x_1, \dots, x_p]\mathbf{T}' \tag{9}$$

with the Jacobian $|\mathbf{T}| = |c_{ij}|^{1/2}$ we obtain the distribution

$$\begin{aligned} f(x,c)\Pi\, dx\, dc = K|\alpha_{ij}|^{(1/2)(n+1)}|c_{ij}|^{(1/2)(n-p)}(1 - x_i x_i)^{(1/2)(n-p-1)} \\ \cdot \exp(-\tfrac{1}{2}\alpha_{ij}c_{ij} + \alpha_{ij}t_{ki}x_k\eta_j - \tfrac{1}{2}\alpha_{ij}\eta_i\eta_j)\Pi\, dx\, dc \end{aligned} \tag{10}$$

$$(k = 1, \dots, p; \qquad t_{ki} = 0 \quad \text{when} \quad k > i).$$

If w is any region, we write

$$\beta_w(\eta, \alpha) = \beta_w(\eta_1, \dots, \eta_p, \alpha_{11}, \alpha_{12}, \dots, \alpha_{pp}) = \int_w f(x,c)\Pi\, dx\, dc,$$

so that $\beta_w(\eta, \alpha)$ is the power function if w serves as a critical region for rejecting H''. We have, symbolically,

$$w = D \times w(c),$$

where D is the set of points (c_{ij}) for which $[c_{ij}]$ is positive definite and $w(c)$ is the cross section of w for fixed c_{ij}. Then

$$\begin{aligned} \beta_w(\eta, \alpha) = K|\alpha_{ij}|^{(1/2)(n+1)}e^{-(1/2)\alpha_{ij}\eta_i\eta_j}\int_D |c_{ij}|^{(1/2)(n-p)}e^{-(1/2)\alpha_{ij}c_{ij}}\Pi\, dc \\ \cdot \int_{w(c)} (1 - x_i x_i)^{(1/2)(n-p-1)}e^{\alpha_{ij}t_{ki}x_r\eta_j}\Pi\, dx. \end{aligned}$$

It is known [6] that, in order to have

$$\beta_w(0, \alpha) = \epsilon \tag{11}$$

for all α_{ij}, it is necessary and sufficient that

$$\int_{w(c)} (1 - x_i x_i)^{(1/2)(n-p-1)} \Pi\, dx = B\epsilon, \tag{12}$$

where

$$B = \int_{x_i x_i \leqslant 1} (1 - x_i x_i)^{(1/2)(n-p-1)} \Pi\, dx.$$

The T^2-test is the test based on the critical region

$$w_0 : x_i x_i = c^{ij} y_i y_j = T^2/(1 + T^2) \geqslant \text{const.}, \quad \text{or} \quad T^2 \geqslant \text{const.},$$

where c^{ij} is the general element of $[c_{ij}]^{-1}$ and T^2 is, except for a constant factor, Hotelling's generalization of "Student's" ratio.

In order to establish an optimum property of T^2 analogous to that of E^2 given in Theorem 1, we define, for any linear set S and any region R in the sample space,

$$\psi_R(S) = \int_{\alpha_{ij} \eta_i \eta_j \in S} \beta_R(\eta, \alpha) \Pi\, d\eta\, d\alpha.$$

$\Psi_R(S)$ does not necessarily have a finite value, and it is this fact which renders the following theorem less satisfactory than Theorem 1.

Theorem 2. *Let ρ_p be the smallest latent root of $[c_{ij}]$ and let E be any subset of D in which ρ_p is at least equal to a fixed positive constant. Of all the critical regions w which satisfy* (11), *the region w_0 has the maximum $\psi_{wE}(S)$.*

In order to prove this theorem we need the following two lemmas.

Lemma 1. *If c is a positive constant, the integral*

$$I = \int_{\rho_p \geqslant c} |c_{ij}|^{-(p+1/2)} \Pi\, dc$$

has a finite value.

Proof. Let $\rho_1, \ldots, \rho_p$ be the latent roots of $[c_{ij}]$ in the descending order of magnitude. From a known theorem [1] we get

$$I = C \int_{c \leqslant \rho_p \leqslant \cdots \leqslant \rho_1 < \infty} (\rho_1 \ldots \rho_p)^{-(p+1/2)} \prod_{i<j} (\rho_i - \rho_j) \Pi\, d\rho$$

$$\leqslant C \int_c^\infty \cdots \int_c^\infty \left(\prod_{i=1}^p \rho_i^{-(i+1/2)} \right) d\rho_1 \ldots d\rho_p .$$

Hence I is finite.

Lemma 2.

$$\psi_{wE}(S) = \sum_{k=0}^\infty g_k \int_E |c_{ij}|^{-(p+1/2)} \Pi\, dc \int_{w(c)} (1 - x_i x_i)^{(1/2)(n-p-1)} (x_i x_i)^k \Pi\, dx \tag{13}$$

and $\psi_{wE}(S)$ is finite, where g_k depends only on k and S.

Proof. Let Δ be the set of points (α_{ij}) for which $[\alpha_{ij}]$ is positive definite. By (8), we have

$$\psi_{wE}(S) = K\int_{wE}|c_{ij}-y_iy_j|^{(1/2)(n-p-1)}\Pi\, dy\, dc\int_{\Delta}|\alpha_{ij}|^{(1/2)(n+1)}e^{-(1/2)c_{ij}\alpha_{ij}}J\,\Pi\, d\alpha,$$

where

$$J=\int_{\alpha_{ij}\eta_i\eta_j\in S}\exp(-\tfrac{1}{2}\alpha_{ij}\eta_i\eta_j+\alpha_{ij}y_i\eta_j)\Pi\, d\eta.$$

There is a real non-singular matrix $\mathbf{G}=[g_{ij}]$ such that $[\alpha_{ij}]=\mathbf{GG}'$. Using the transformation

$$[\eta_1,\ldots,\eta_p]\mathbf{G}=[\zeta_1,\ldots,\zeta_p],$$

whose Jacobian is $|\mathbf{G}|^{-1}=|\alpha_{ij}|^{-1/2}$, we have

$$J=|\alpha_{ij}|^{-1/2}\int_{\zeta_i\zeta_i\in S}\exp(-\tfrac{1}{2}\zeta_i\zeta_i+g_{ji}\zeta_iy_j)\Pi\, d\zeta.$$

This is reducible by means of a rotation to

$$\begin{aligned}J&=|\alpha_{ij}|^{-1/2}\int_{\tau_i\tau_i\in S}\exp\left[-\tfrac{1}{2}\tau_i\tau_i+(\alpha_{ij}y_iy_j)^{1/2}\tau_1\right]\Pi\, d\tau\\&=|\alpha_{ij}|^{-1/2}\sum_{k=0}^{\infty}d_k(\alpha_{ij}y_iy_j)^k,\end{aligned}\tag{14}$$

where

$$\begin{aligned}d_k&=\frac{1}{(2k)!}\int_{\tau_i\tau_i\in S}\tau_1^{2k}e^{-(1/2)\tau_i\tau_i}\Pi\, d\tau\\&\leqslant\frac{1}{(2k)!}\int_{-\infty}^{\infty}\cdots\int_{-\infty}^{\infty}\tau_1^{2k}e^{-(1/2)\tau_i\tau_i}d\tau_1\ldots d\tau_p=\frac{(2\pi)^{p/2}}{2^kk!}\end{aligned}$$

and d_k depends only on k and S. Hence

$$\int_{\Delta}|\alpha_{ij}|^{(1/2)(n+1)}e^{-(1/2)c_{ij}\alpha_{ij}}J\,\Pi\, d\alpha=\sum_{k=0}^{\infty}d_kI_k,$$

where

$$I_k=\int_{\Delta}|\alpha_{ij}|^{(1/2)n}(\alpha_{ij}y_iy_j)^ke^{-(1/2)\alpha_{ij}c_{ij}}\Pi\, d\alpha.\tag{15}$$

Now

$$I_k=\frac{d^k}{dt^k}f(t)\bigg|_{t=0},$$

where

$$\begin{aligned}f(t)\int_{\Delta}|\alpha_{ij}|^{(1/2)n}e^{-(1/2)(c_{ij}-2ty_iy_j)\alpha_{ij}}\Pi\, d\alpha&=K_1|c_{ij}-2ty_iy_j|^{-(1/2)(n+p+1)}\\&=K_1|c_{ij}|^{-(1/2)(n+p+1)}\left(1-2tc^{ij}y_iy_j\right)^{-(1/2)(n+p+1)}\end{aligned}$$

Hence

$$I_k = e_k|c_{ij}|^{-(1/2)(n+p+1)}\left(c^{ij}y_iy_j\right)^k, \tag{16}$$

where

$$e_k = \frac{K_1 2^k \Gamma((n+p+1)/2+k)}{k!\Gamma((n+p+1)/2)}.$$

Hence

$$\psi_{wE}(S) = K\sum_{k=0}^{\infty} d_k e_k \int_{wE} |c_{ij}|^{-(1/2)(n+p+1)}|c_{ij}-y_iy_j|^{(1/2)(n-p-1)}\left(c^{ij}y_iy_j\right)^k \Pi\, dy\, dc$$

$$= \sum_{k=0}^{\infty} g_k \int_E |c_{ij}|^{-(p+1/2)}\Pi\, dc \int_{w(c)} (1-x_ix_i)^{(1/2)(n-p-1)}(x_ix_i)^k \Pi\, dx,$$

where $g_k = K_1 d_k e_k$ depends only on k and S.

Now

$$\int_{w(c)} (1-x_ix_i)^{(1/2)(n-p-1)}(x_ix_i)^k \Pi\, dx \leqslant \int_{x_ix_i \leqslant 1} \Pi\, dx,$$

$$\int_E |c_{ij}|^{-(p+1/2)}\Pi\, dc \leqslant \int_{\rho_p \geqslant c > 0} |c_{ij}|^{-(1/2)(p+1/2)}\Pi\, dc$$

is finite by Lemma 1. Hence

$$\psi_{wE}(S) \leqslant \text{const.}\sum_{k=0}^{\infty} d_k e_k = \text{const.}\sum_{k=0}^{\infty}\frac{\Gamma((n+p+1)/2+k)}{(k!)^2}$$

and so $\varphi_{wE}(S)$ is finite. This proves Lemma 2.

Proof of Theorem 2. Since $\psi_{wE}(S)$ is expressible as (13) and is always finite, it follows from (12) and the Neyman–Pearson lemma that $\psi_{wE}(S)$ is maximum when w is w_0. This proves Theorem 2.

Simaika [6] proved that of all the critical regions w which satisfy the conditions

(a) $\beta_w(0,\alpha) = \epsilon$ for all α_{ij},
(b) $\beta_w(\eta,\alpha) = f(\alpha_{ij}\eta_i\eta_j)$,

w_0 is the uniformly most powerful one. Strangely enough, this result cannot be deduced as a consequence from our Theorem 2.

The difficulty in dealing with the integral $\psi_w(S)$ is that it is not always finite. In order to have a finite integral let us consider the following.

$$\Gamma_w(\theta,S) = \int_{\alpha_{ij}\eta_i\eta_j \in S} e^{-(1/2)\theta_{ij}\alpha_{ij}}\beta_w(\eta,\alpha)\Pi\, d\eta\, d\alpha,$$

where $[\theta_{ij}]$ is a positive definite matrix. As an immediate consequence of Simaika's theorem, we have

$$\Gamma_w(\theta,S) \leqslant \Gamma_{w_0}(\theta,S) \tag{17}$$

for any region w satisfying (a) and (b). Now the question arises whether (17) remains true if the condition (b) on w is removed. The following theorem answers this question in the negative.

Theorem 3. *Let* $[\theta_{ij}]$ *be a positive definite matrix* $[\rho_{ij}] = [c_{ij} + \theta_{ij}]^{-1}$ *and* $\lambda_1, \ldots, \lambda_p$ *be the roots of the equation* $|c_{ij} - \lambda\theta_{ij}| = 0$. *There is a function* $g = g(\lambda_1, \ldots, \lambda_p)$ *such that the region*

$$w_1 : \rho_{ij}y_iy_j \geqslant g(\lambda_1, \ldots, \lambda_p)$$

satisfies (a) *and has the maximum* $\Gamma_w(\theta, S)$.

Proof. From (10) and (14) we obtain

$$\Gamma_w(\theta, S) = K\sum_{k=0}^{\infty} d_k \int_w |c_{ij} - y_iy_j|^{(1/2)(n-p-1)} \Pi\, dy\, dc$$
$$\cdot \int_\Delta |\alpha_{ij}|^{(1/2)n} (\alpha_{ij}y_iy_j)^k e^{-(1/2)(c_{ij}+\theta_{ij})\alpha_{ij}} \Pi\, d\alpha.$$

Comparing the inner integral with (15) and using (16) we get

$$\Gamma_w(\theta, S) = \sum_{k=0}^{\infty} g_k \int_w |c_{ij} + \theta_{ij}|^{-(1/2)(n+p+1)} |c_{ij} - y_iy_j|^{(1/2)(n-p-1)} (\rho_{ij}y_iy_j)^k \Pi\, dy\, dc$$
$$= \sum_{k=0}^{\infty} g_k \int_D |c_{ij} + \theta_{ij}|^{-(1/2)(n+p+1)} |c_{ij}|^{(1/2)(n-p)} \Pi\, dc$$
$$\cdot \int_{w(c)} (1 - x_ix_i)^{(1/2)(n-p-1)} (\gamma_{ij}x_ix_j)^k \Pi\, dx, \tag{18}$$

where $\gamma_{ij}x_ix_j$ is the result of applying the transformation (9) on $\rho_{ij}y_iy_j$. We shall show that, for every fixed set of c_{ij}, a unique number $g = g(\lambda_1, \ldots, \lambda_p)$ exists such that the region $\rho_{ij}y_iy_j = \gamma_{ij}x_ix_j \geqslant g$ satisfies (12), i.e.,

$$\int_{\gamma_{ij}x_ix_j \geqslant g} (1 - x_ix_j)^{(1/2)(n-p-1)} \Pi\, dx = B\epsilon. \tag{19}$$

Since $[\gamma_{ij}] = T'[c_{ij} + \theta_{ij}]^{-1}T$, the latent roots of $[\gamma_{ij}]$ are $\lambda_i/(1 + \lambda_i)$ $(i = 1, \ldots, p)$. Hence by a rotation the equation (19) is reduced to

$$\int_{(\lambda_i/1+\lambda_i)\xi_i\xi_i \geqslant g} (1 - \xi_i\xi_i)^{(1/2)(n-p-1)} \Pi\, d\xi = B\epsilon. \tag{20}$$

As g increases from 0 onwards, the left member of (20) decreases steadily from B to 0. Hence there is a unique $g = g(\lambda_1, \ldots, \lambda_p)$ which satisfies (20).

For this $g(\lambda_1, \ldots, \lambda_p)$ the region w_1 satisfies (a). Hence, applying the Neyman–Pearson lemma on (18) we obtain the result.

From Theorem 3 we learn that there actually exist other exact tests for H'' which have some optimum property not possessed by T^2, viz., the tests based on the critical regions w_1 corresponding to various values of the θ_{ij}. However, the great difficulty in numerical computation prohibits their application and the T^2-test stands out as the only test which is both simple and good.

References

1. P. L. Hsu, On the distribution of roots of certain determinantal equations. *Annals of Eugenics* **9** (1939), pp. 250–258.
2. P. L. Hsu, An algebraic derivation of the distribution of rectangular coordinates. *Proc. Edin. Math. Soc.* **6** (1940), pp. 185–189.
3. P. L. Hsu, Analysis of variance from the power function standpoint. *Biometrika* **32** (1941), pp. 62–69.
4. P. L. Hsu, Canonical reduction of the general regression problem. *Annals of Eugenics* **11** (1941), pp. 42–46.
5. J. Neyman and E. S. Pearson, Contribution to the theory of testing statistical hypotheses. *Stat. Res. Mem.* **16** (1936).
6. J. B. Simaika, On an optimum property of two important statistical tests. *Biometrika* **32** (1941), pp. 70–80.
7. A. Wald, On the power function of the analysis of variance tests. *Annals of Math. Stat.* **33** (1942), pp. 434–439.

Reprinted from
Quart. J. Math. Oxford Ser.
17, 162–165 (1946).

ON A FACTORIZATION OF PSEUDO-ORTHOGONAL MATRICES

By P. L. HSU (*Peiping*)

[Received 7 July 1945]

A PSEUDO-ORTHOGONAL matrix is defined to be a matrix† A which satisfies

$$A\{I_p\dot{+}(-I_q)\}A' = I_p\dot{+}(-I_q), \tag{1}$$

where the accent denotes the transpose, $\dot{+}$ the direct sum, and I_p denotes the p-rowed unit matrix. For definiteness we assume that $p \leqslant q$. A factorization of A into a product of $\frac{1}{2}(p+q)(p+q-1)-p$ plane rotations and p pseudo-plane rotations has recently been worked out by H. C. Lee.‡ In this note I give a more explicit factorization which is unique in general.

LEMMA 1. *Let U be any $p\times q$ matrix of rank r and let the latent roots of UU' be $\alpha_1,\ldots,\alpha_p$, where $\alpha_i > 0$ $(i \leqslant r)$, $\alpha_i = 0$ $(i > r)$. Then there is a $p\times p$ orthogonal matrix Ω_1 and a $q\times q$ orthogonal matrix Ω_2 such that*

$$U = \Omega_1'[\alpha_1^{\frac{1}{2}}\dot{+}\ldots\dot{+}\alpha_p^{\frac{1}{2}}, 0]\Omega_2.$$

Proof. Choose Ω_1 so that

$$\Omega_1 UU'\Omega_1' = \alpha_1\dot{+}\alpha_2\dot{+}\ldots\dot{+}\alpha_r\dot{+}0. \tag{2}$$

Let S be the matrix formed by the first r rows, and T that formed by the last $p-r$ rows, of Ω_1. Then (2) gives

$$SUU'S' = \alpha_1\dot{+}\ldots\dot{+}\alpha_r, \qquad TUU'T' = 0,$$

so that

$$TU = 0.$$

Hence the $r\times q$ matrix $G = (\alpha_1^{-\frac{1}{2}}\dot{+}\ldots\dot{+}\alpha_r^{-\frac{1}{2}})SU$ has mutually orthogonal rows, i.e. $GG' = I_r$. Taking Ω_2 to be an orthogonal matrix whose first r rows are those of G we have

$$U = (I-T'T)U = S'SU = S'(\alpha_1^{\frac{1}{2}}\dot{+}\ldots\dot{+}\alpha_r^{\frac{1}{2}})G$$

$$= \Omega_1'\begin{bmatrix}\alpha_1^{\frac{1}{2}}\dot{+}\ldots\dot{+}\alpha_r^{\frac{1}{2}} & 0\\ 0 & 0\end{bmatrix}\Omega_2 = \Omega_1'[\alpha_1^{\frac{1}{2}}\dot{+}\ldots\dot{+}\alpha_p^{\frac{1}{2}}, 0]\Omega_2.$$

† All the quantities in this paper are real.

‡ *Quart. J. of Math.* (Oxford), 15 (1944), 7–10. Owing to my inaccessibility to literature Lee's paper is the only one citable in this connexion. For the same reason lemmas, even when embodying known results, are given with proof.

LEMMA 2. *If B is any $m\times m$ matrix, then there is an $m\times m$ orthogonal matrix Ω such that $B\Omega$ is of the triangular type, viz. a matrix whose super-diagonal elements are zero and whose diagonal elements are non-negative.*

The proof may be found in Lee's paper.† The elements $(1,2)$, $(1,3),\ldots,(1,m)$, $(2,3)$, etc., are annihilated successively by means of appropriate orthogonal matrices representing plane rotations.

Having established the two lemmas, consider a pseudo-orthogonal matrix

$$A = \begin{bmatrix} B & C \\ D & E \end{bmatrix} \tag{3}$$

satisfying (1), where B is $p\times p$ and E is $q\times q$. Using (1) we have

(i) $BB' = I_p+CC'$, (ii) $EE' = I_q+DD'$, (iii) $BD' = CE'$.

By (i) and (ii) B and E are non-singular. Hence, by (iii),

$$C = BU, \qquad D = EU', \tag{4}$$

where U is some $p\times q$ matrix. Substituting in (i) and (ii) we get‡

$$(B'B)^{-1} = I_p-UU', \qquad (E'E)^{-1} = I_q-U'U. \tag{5}$$

The latent roots of UU' being necessarily smaller than 1 in virtue of (5), let them be $\lambda_i/(1+\lambda_i)$ $(i = 1,\ldots, p)$. Following Lemma 1 we can write

$$U = \Omega_1'\left[\left(\frac{\lambda_1}{1+\lambda_1}\right)^{\frac12}\dotplus\ldots\dotplus\left(\frac{\lambda_p}{1+\lambda_p}\right)^{\frac12}, 0\right]\Omega_2, \tag{6}$$

whence, by (5),

$$B'B = \Omega_1'\{(1+\lambda_1)\dotplus\ldots\dotplus(1+\lambda_p)\}\Omega_1,$$
$$E'E = \Omega_2'\{(1+\lambda_1)\dotplus\ldots\dotplus(1+\lambda_p)\dotplus I_{q-p}\}\Omega_2.$$

Hence

$$B = \Gamma_1\{(1+\lambda_1)^{\frac12}\dotplus\ldots\dotplus(1+\lambda_p)^{\frac12}\}\Omega_1,$$
$$E = \Gamma_2\{(1+\lambda_1)^{\frac12}\dotplus\ldots\dotplus(1+\lambda_p)^{\frac12}\dotplus I_{q-p}\}\Omega_2, \tag{7}$$

where Γ_1 is a $p\times p$ orthogonal matrix and Γ_2 is a $q\times q$ orthogonal matrix.

Taking (3), (7), (4), and (6) together we get

$$A = (\Gamma_1\dotplus\Gamma_2)(\Lambda\dotplus I_{q-p})(\Omega_1\dotplus\Omega_2), \tag{8}$$

where

$$\Lambda = \begin{bmatrix} (1+\lambda_1)^{\frac12}\dotplus\ldots\dotplus(1+\lambda_p)^{\frac12} & \lambda_1^{\frac12}\dotplus\ldots\dotplus\lambda_p^{\frac12} \\ \lambda_1^{\frac12}\dotplus\ldots\dotplus\lambda_p^{\frac12} & (1+\lambda_1)^{\frac12}\dotplus\ldots\dotplus(1+\lambda_p)^{\frac12} \end{bmatrix}.$$

† Ibid., p. 8.

‡ Routine steps, such as pre- and post-multiplication by appropriate matrices, are left to the reader.

Since each of Γ_1 and Ω_1 depends on $\frac{1}{2}p(p-1)$ independent parameters, each of Γ_2 and Ω_2 depends on $\frac{1}{2}q(q-1)$ independent parameters, and Λ depends on p independent parameters, apparently the right-hand side of (8) depends on $p(p-1)+q(q-1)+p$ independent parameters, whilst A depends on $\frac{1}{2}(p+q)(p+q-1)$ independent parameters by its definition, the discrepancy being $\frac{1}{2}(q-p)(q-p-1)$. The factorization (8) cannot be unique.

If Δ is any $(q-p)$-rowed orthogonal matrix, $I_{2p}\dot{+}\Delta'$ is commutative with the middle factor on the right of (8). Hence

$$\begin{aligned} A &= (\Gamma_1\dot{+}\Gamma_2)(I_{2p}\dot{+}\Delta)(\Lambda\dot{+}I_{q-p})(I_{2p}\dot{+}\Delta')(\Omega_1\dot{+}\Omega_2) \\ &= \{\Gamma_1\dot{+}\Gamma_2(I_p\dot{+}\Delta)\}(\Lambda\dot{+}I_{q-p})\{\Omega_1\dot{+}(I_p\dot{+}\Delta')\Omega_2\}. \end{aligned} \tag{9}$$

The last factor in (9) is of the same nature as that in (8), so it can still be written as $\Omega_1\dot{+}\Omega_2$, with a changed Ω_2.

Writing $$\Gamma_2 = \begin{bmatrix} \Pi_1 & \Pi_2 \\ \Pi_3 & \Pi_4 \end{bmatrix},$$

where Π_1 is $p\times p$ and Π_2 is $(q-p)\times(q-p)$, we have

$$\Gamma_2(I_p\dot{+}\Delta) = \begin{bmatrix} \Pi_1 & \Pi_2\Delta \\ \Pi_3 & \Pi_4\Delta \end{bmatrix}.$$

We select Δ, in accordance with Lemma 2, so that $\Pi_4\Delta$ is of the triangular type. In this way the number of parameters in Γ_2 is reduced by $\frac{1}{2}(q-p)(q-p-1)$, which accounts for the discrepancy mentioned above. We summarize all the foregoing results in the following theorem.

THEOREM 1. *The following factorization exists for every pseudo-orthogonal matrix that leaves $I_p\dot{+}(-I_q)$ $(p \leqslant q)$ invariant:*

$$A = (\Gamma_1\dot{+}\Gamma_2)(\Lambda\dot{+}I_{q-p})(\Omega_1\dot{+}\Omega_2), \qquad \Gamma_2 = \begin{bmatrix} \Pi_1 & \Pi_2 \\ \Pi_3 & \Pi_4 \end{bmatrix}, \tag{10}$$

where Γ_1 and Ω_1 are $p\times p$ orthogonal matrices, Γ_2 and Ω_2 are $q\times q$ orthogonal matrices, and Π_4 is of the triangular type.

We now examine the uniqueness of the factorization (10).

THEOREM 2. *The factorization* (10) *is unique under the following conditions:*

(i) $\lambda_1,\dots,\lambda_p$ *are all distinct and arranged in descending order*;

(ii) $\lambda_i > 0$ $(i = 1,\dots, p)$;

(iii) Π_4 *is non-singular.*

Proof. The middle factor of every factorization (10) is the same, because $1+\lambda_1,\ldots, 1+\lambda_p$ are the latent roots of BB', where B is the sub-matrix in (3), arranged in descending order. Let

$$(\Gamma_1^*\dot{+}\Gamma_2^*)(\Lambda\dot{+}I_{q-p})(\Omega_1^*\dot{+}\Omega_2^*), \quad \text{where} \quad \Gamma_2^* = \begin{bmatrix} \Pi_1^* & \Pi_2^* \\ \Pi_3^* & \Pi_4^* \end{bmatrix}, \tag{11}$$

be a second factorization of A. Then

$$(\Gamma_1^*\dot{+}\Gamma_2^*)'(\Gamma_1\dot{+}\Gamma_2)(\Lambda\dot{+}I_{q-p}) = (\Lambda\dot{+}I_{q-p})(\Omega_1^*\dot{+}\Omega_2^*)(\Omega_1\dot{+}\Omega_2)'. \tag{12}$$

Denoting by

$$\begin{bmatrix} G & 0 & 0 \\ 0 & H_1 & H_2 \\ 0 & H_3 & H_4 \end{bmatrix} \quad \text{and} \quad \begin{bmatrix} L & 0 & 0 \\ 0 & M_1 & M_2 \\ 0 & M_3 & M_4 \end{bmatrix}$$

respectively the pre-multiplier and the post-multiplier of $\Lambda\dot{+}I_{q-p}$ in (12) we obtain, on equating corresponding blocks,

$$\begin{aligned} GD_1 &= D_1 L, & GD_2 &= D_2 M_1, & 0 &= D_2 M_2, \\ H_1 D_2 &= D_2 L, & H_1 D_1 &= D_1 M_1, & H_2 &= D_1 M_2, \\ H_3 D_2 &= 0, & H_3 D_1 &= M_3, & H_4 &= M_4, \end{aligned}$$

where $D_1 = (1+\lambda_1)^{\frac{1}{2}}\dot{+}\ldots\dot{+}(1+\lambda_p)^{\frac{1}{2}}$ and $D_2 = \lambda_1^{\frac{1}{2}}\dot{+}\ldots\dot{+}\lambda_p^{\frac{1}{2}}$. By condition (ii), D_2 is non-singular, hence $H_2 = H_3 = M_2 = M_3 = 0$. From the first equation we get $GD_1^2 G' = D_1 LL'D_1 = D_1^2$, whence $GD_1^2 = D_1^2 G$, giving $G = I_p$ since the λ_i are all distinct. Then $L = H_1 = M_1 = I_p$. Hence

$$\Gamma_1\dot{+}\Gamma_2 = (\Gamma_1^*\dot{+}\Gamma_2^*)(I_{2p}\dot{+}H_4).$$

In particular $\Pi_4 = \Pi_4^* H_4$. Since Π_4 is non-singular, so is Π_4^* and we have $H_4 = (\Pi_4^*)^{-1}\Pi_4$, a matrix of the triangular type. Hence $H_4 = I_{q-p}$ and so the three factors in (11) are identical with the corresponding factors in (10). This completes the proof.

As a final remark we add that Lee's factorization may be deduced from (10). For the matrix A is factorized into the factors

$$\Gamma_1\dot{+}I_q, \qquad I_p\dot{+}\begin{bmatrix} \Pi_1 & \Pi_2 \\ \Pi_3 & \Pi_4 \end{bmatrix}, \qquad \Lambda\dot{+}I_{q-p}, \qquad \Omega_1\dot{+}I_q, \qquad I_p\dot{+}\Omega_2.$$

Clearly the first and the last matrices are each factorizable into $\frac{1}{2}p(p-1)$ plane rotations, the second into $\frac{1}{2}q(q-1)-\frac{1}{2}(q-p)(q-p-1)$ plane rotations, the fourth into $\frac{1}{2}q(q-1)$ plane rotations, and, finally, Λ is factorizable into p pseudo-plane rotations.

Reprinted from
C. R. Acad. Sci. Paris
467–469 (1946).

CALCUL DES PROBABILITÉS. — *Sur un théorème de probabilités dénombrables.*

Note [1] de MM. **P. L. Hsu** et **K. L. Chung**.

Soit M un module, c'est-à-dire un ensemble de nombres [2] tels que $a \in M$ et $(a+b) \in M$ soient équivalentes à $a \in M$ et $b \in M$.

En cherchant à étendre la validité d'une proposition d'Émile Borel sur le *retour à l'équilibre*, nous avons éprouvé l'utilité du théorème suivant :

Théorème. — *Soit* $\{X_\nu\}$ *une suite de variables aléatoires indépendantes ayant la même fonction de répartition. Alors en posant* $S_n = \sum_{\nu=1}^{n} X_\nu$, *la probabilité*

$$P(S_n \in M \text{ pour une infinité de valeurs de } n)$$

est égale à 0 *ou* 1 *suivant que la série*

$$\sum_{n=1}^{\infty} P(S_n \in M)$$

est convergente ou divergente.

Démonstration. — Le cas de convergence est une conséquence immédiate du théorème classique de Borel-Cantelli.

Considérons donc le cas de divergence. Désignons par E_n l'événement $S_n \in M$ et en général par E' le complément de E. Nous avons, d'une manière générale,

$$E_n = (E_1 + E'_1 E_2 + \ldots + E'_1 E'_2 \ldots E'_{n-1} E_n) E_n,$$

où les produits et les sommes veulent dire des conjonctions et des disjonctions, respectivement. Si nous écrivons $P(E_n) = p_n$, $P(E'_1 E'_2 \ldots E'_{k-1} E_k E_n) = p_{1'2'\ldots(k-1)'kn}$, nous avons

$$(1) \qquad p_n = p_{1n} + p_{1'2n} + \ldots + p_{1'2'\ldots(n-2)'(n-1)n} + p_{1'2'\ldots(n-1)'n}.$$

Puisque la conjonction de E_k et E_n équivaut à la conjonction de E_k et de l'événement $X_{k+1} + \ldots + X_n \in M$, par définition même de M, et puisque le dernier événement est indépendant des événements $E_1, \ldots, E_k$, nous avons

$$(2) \qquad p_{1'2'\ldots(k-1)'kn} = p_{1'2'\ldots(k-1)'k} P(X_{k+1} + \ldots + X_n \in M).$$

[1] Séance du 16 septembre 1946.

[2] Nous pouvons même considérer un ensemble abstrait et une opération abstraite définie pour les variables aléatoires.

Mais les variables aléatoires ont la même fonction de répartition, donc

$$P(X_{k+1}+\ldots+X_n\in M)=P(S_{n-k}\in M)=p_{n-k}.$$

Si nous écrivons de plus $p_{1'2'\ldots(k-1)'k}=\pi_k$, la formule (2) ci-dessus s'écrit

$$p_{1'2'\ldots(k-1)'kn}=\pi_k p_{n-k}.$$

En substituant dans (1), nous obtenons la relation suivante

$$(3)\qquad p_n=\sum_{k=1}^{n}\pi_k p_{n-k},\qquad p_0=1.$$

Il s'ensuit que

$$(4)\qquad \begin{cases} \displaystyle\sum_{j=0}^{n}p_j\sum_{j=1}^{u}\pi_k\geq\sum_{m=1}^{n}\Big(\sum_{j+k=m}p_j\pi_k\Big)=\sum_{m=1}^{n}p_m,\\ \displaystyle\sum_{\kappa=1}^{n}\pi_k\geq\sum_{j=1}^{n}p_j:\sum_{i=0}^{n}p_i=\frac{1}{1+\Big(\sum_{j=1}^{n}p_j\Big)^{-1}}. \end{cases}$$

Puisque dans notre hypothèse $\sum_{j=1}^{n}p_j\to\infty$ quand $n\to\infty$, le passage à la limite nous apprend que

$$(5)\qquad \sum_{k=1}^{\infty}\pi_k\geq 1.$$

D'autre part, il est évident que

$$(6)\qquad \sum_{k=1}^{\infty}\pi_k=P(S_n\in M\text{ pour au moins une valeur de }n)\leq 1.$$

De (5) et (6) nous conclurons donc que

$$\sum_{k=1}^{\infty}\pi_k=1,$$

c'est-à-dire que

$$P\Big(\sum_{n=1}^{\infty}E_n\Big)=1.$$

D'une manière semblable, nous aurons

$$P\Big(\sum_{n=N}^{\infty}E_n\Big)=1,$$

pour n'importe quel entier $N>0$. Il s'ensuit que

$$P\Big(\prod_{N=1}^{\infty}\sum_{n=N}^{\infty}E_n\Big)=\lim_{N\to\infty}P\Big(\sum_{n=N}^{\infty}E_n\Big)=1.$$

Le théorème se trouve démontré.

En prenant pour le module M celui qui consiste en le seul élément o et en ne considérant que le cas de divergence, nous obtenons :

Corollaire 1. — *Soit* $\{x_\nu\}$ *une suite de variables aléatoires indépendantes ayant la même fonction de répartition. Alors si* $\sum_{n=1}^{\infty} P(S_n = 0) = \infty$, *il est presque certain que* S_n *s'annule une infinité de fois.*

En raisonnant d'une manière semblable à celle qui nous a conduit aux formules (3) et (4), nous obtenons aussi le corollaire suivant :

Corollaire 2. — *Dans les conditions du corollaire* 1, *si pour un nombre quelconque c nous avons de plus*

$$\frac{P(S_n = c)}{P(S_n = 0)} \to 1 \qquad \text{avec } n \to \infty,$$

alors il est presque certain que $S_n = c$ *une infinité de fois.*

L'application des résultats ci-dessus aux jeux de hasard dits *bernoulliens avec une probabilité* constante commensurable est immédiate.

(Extrait des *Comptes rendus des séances de l'Académie des Sciences*, t. 223, p. 467-469, séance du 23 septembre 1946.)

Dépôt légal d'éditeur. — 1946. — N° d'ordre 64.
Dépôt légal d'imprimeur. — 1946. — N° d'ordre 144.

GAUTHIER-VILLARS IMPRIMEUR-LIBRAIRE DES COMPTES RENDUS DES SÉANCES DE L'ACADÉMIE DES SCIENCES.
124556-46 Paris. — Quai des Grands-Augustins, 55.

Comments by K. L. Chung

The origin of this note may be interesting as an "epoch" story. In 1943 or 1944 I read Borel's "Valeur pratique et philosophie des probabilités" (Gauthier-Villars, Paris, 1939; the last volume of Borel's grand collection of treatises on probability) with fascination. In a section entitled "L'Illusion de retour à l'équilibre," he treats what is now known as the recurrence of a symmetric Bernoullian random walk to the origin, in order to refute a proposed winning strategy. Let $\{X_n\}$ be independent random variables with $P(X_n = 1) = P(X_n = -1) = 1/2$ for all n, $S_n = \sum_{k=1}^{n} X_k$, and $\pi_n = P\,(S_k \neq 0$ for $1 \leqslant k < n;\ S_n = 0)$, then π_n is known explicitly and $\sum_{n=1}^{\infty} \pi_n = 1$ follows (actually this is shown in the first volume of Borel's collection). I attempted the generalization to the case where for two positive integers a and b, $a < b$, we have $P(X_n = b - a) = a/b$, $P(X_n = -a) = (b - a)/b$. For $a = 1$, I obtained by a graphic counting (almost forgotten) the explicit expression for π_n but was able to prove $\sum_{n=1}^{\infty} \pi_n = 1$ only when $b = 3$. Hsu proved this for all b by an elaborate calculation using contour integration (his proof is kept in my file). But soon afterward the general result in this Note was found. Equation (3) is the so-called "renewal equation", and must have been known before our time. The derivation of $\sum_{n=1}^{\infty} \pi_n = 1$ was first done by using generating functions, but Hsu discarded this as spurious. The proof was shown to a noted probabilist in the U.S., who claimed that the result was obvious in the Bernoullian case, but was unable to produce a special proof for it. Looking back over a span of nearly forty years, I had to chuckle over this: how can there be a simpler proof than the one given in the Note? And it is perfectly general. Let me add that if the X_n's are independent and identically distributed with zero mean, and if all integers are "possible values" for the sums, then the additional condition in Corollary 2 of the Note holds true. This is proved in Chung and Erdös, "Probability limit theorems assuming only the first moment", *Mem. Amer. Math. Soc.*, No. 6 (1951). Thus the recurrence of such a random walk follows. While a more general result is known (see Chung and Fuchs, "On the distribution of values of sums of random variables", *loc. cit.*), the connection is amusing.

Reprinted from
Ann. Math. Statist.
17, 350–354 (1946).

ON THE ASYMPTOTIC DISTRIBUTIONS OF CERTAIN STATISTICS USED IN TESTING THE INDEPENDENCE BETWEEN SUCCESSIVE OBSERVATIONS FROM A NORMAL POPULATION

BY P. L. HSU

Columbia University

1. The statistics to be considered here have the general expression

$$T = \frac{Q}{S}, \qquad Q = \sum_{i=1}^{N} a_{ij}(x_i - \bar{x})(x_j - \bar{x}), \qquad S = \sum_{i=1}^{N} (x_i - \bar{x})^2,$$

where $(x_1, \cdots, x_N)$ is a sample from a normal population whose mean and variance can evidently be assumed to be 0 and 1 respectively.[1] The purpose of this note is to study the asymptotic distribution of T assuming that the x_i are independent. The whole work may be regarded as a straightforward application of Cramér's theory of asymptotic expansion (see [1], pp. 69–88).

If $A = [a_{ij}]$ and γ is the row vector $N^{-\frac{1}{2}}[1, 1, \cdots, 1, 1]$ the quadratic form Q has the matrix $(I - \gamma'\gamma)A(I - \gamma'\gamma)$. The latent roots of this matrix, which are also the latent roots of $A(I - \gamma'\gamma)^2 = A(I - \gamma'\gamma)$, will be denoted by $0, \lambda_1, \cdots, \lambda_n$, with $n = N - 1$. Then Q and S can be simultaneously diagonalized (by a rotation of the N-dimensional space), so that

$$Q = \sum_{r=1}^{N} \lambda_r y_r^2, \qquad S = \sum_{r=1}^{n} y_r^2,$$

where the y_r are again independently and normally distributed with zero mean and unit variance.

We shall make the following assumptions

(a) $|\lambda_r| \leq 1$ for all r.

(b) There is a positive number c independent of n such that

$$\sum_{r=1}^{n} (\lambda_r - \bar{\lambda})^2 > cn, \quad \text{where } \bar{\lambda} = \frac{1}{n}\sum_{r=1}^{n} \lambda_r.$$

Write

$$z = \frac{\sqrt{2\sum_{r=1}^{n} (\lambda_r - \bar{\lambda})^2 x}}{\sqrt{n^2 - 2nx^2}}, \qquad s_m(x) = \sum_{r=1}^{n} (\lambda_r - \bar{\lambda} - z)^m,$$

$$X_r = (\lambda_r - \bar{\lambda} - z)(y_r^2 - 1), \qquad G(x) = Pr\{T \leq \bar{\lambda} + z\}.$$

[1] The exact and the approximate distribution of such statistics were a recent subject of study by a number of statisticians. See W. J. Dixon, "Further contributions to the problem of serial correlation," *Annals of Math. Stat.*, Vol. 15 (1944), pp. 119–144. Further references are listed in Dixon's paper.

Then it can easily be verified that

$$G(x) = Pr\left\{\frac{\sum_{i=1}^{r} X_r}{\sqrt{2s_2(x)}} \le x\right\}.$$

This expression of $G(x)$ shows that the application of Cramér's expansion is at hand, since $E(X_r) = 0$ and $2s_2(x)$ is the variance of ΣX_r. Let ρ_{kn} and T_{kn} stand for the same quantities as defined in Cramér's work (see [1], pp. 70–71). Since moments of all order of X_r exist, we may use $2k + 2$ in place of k. We have

$$\rho_{2k+2,n} = \frac{\frac{1}{n}\, m_k\, s_{2k+2}(x)}{\left(\frac{2}{n}\, s_2(x)\right)^{k+1}}, \qquad T_{2k+2,n} = \frac{\sqrt{n}}{4\rho_{2k+2,n}^{3/2k+2}},$$

where $m_k = E(y^2 - 1)^{2k+2}$ and y is a normal variate with mean 0 and variance 1.

By virtue of assumption (a) $|T| \le 1$. Therefore we may confine ourselves to the range of values for which $|\bar{\lambda} + z| \le 1$. Then $|\lambda_r - \bar{\lambda} - z| \le 2$. Also, by assumption (b), $s_2(x) \ge \Sigma(\lambda_r - \bar{\lambda})^2 > cn$. Hence $\rho_{2k+2,n}$, and in consequence $\sqrt{n}T_{2k+2,n}^{-1}$, are less than some constant independent of n and x. The remainder of Cramer s expansion, if it is justifiable, will therefore be less than Mn^{-k}, where M is independent of n and x. The justification consists in verifying that the following condition is satisfied: if $f_r(t)$ is the characteristic function of X_r and A is any positive number, then

$$\text{l.u.b.} \prod_{r=1}^{n} |f_r(t)| \quad \text{for} \quad |t| > \frac{T_{2k+2,n}}{\sqrt{2s_2(x)}}$$

is less than $M_1 T_{2k+2,n}^{-A}$, where M_1 is independent of n and x (see [1], p. 85). Since $T_{2k+2,n} \le \frac{1}{4}\sqrt{n}^{\,2}$ and $s_2(x) > c\sqrt{n}$, it is sufficient to show that, if a and A are any positive numbers and if

$$U = \text{l.u.b.} \prod_{r=1}^{n} |f_r(t)| \quad \text{for} \quad |t| > a,$$

then $U \le M_2 n^{-A}$, where M_2 is independent of n and x. Now

$$|f_r(t)| = \{1 + 4t^2(\lambda_r - \bar{\lambda} - z)^2\}^{-\frac{1}{4}}$$

whence

$$U = \prod_{r=1}^{n} \{1 + 4a^2(\lambda_r - \bar{\lambda} - z)^2\}^{-\frac{1}{4}}.$$

Let μ be the number of λ_r for which $(\lambda_r - \bar{\lambda} - z)^2 < \frac{1}{2}c$. Then $cn < s_2(x) \le \frac{1}{2}c(n - \mu) + 4\mu$; hence $cn < (8 - c)\mu$ and

$$U \le (1 + 2a^2c)^{-\frac{1}{4}\mu} < (1 + 2a^2c)^{-(cn/4(8-c))}$$

This shows that the desired condition on U is satisfied, and that therefore Cramér's procedure can be adopted.

---

[3] This follows from the fact that $P_{2k+2,n} > 1$. Cf. Cramér, [1], p. 70.

Wherever Cramér's asymptotic expansion is valid, the terms in the expansion are most conveniently obtained with the help of Cornish and Fisher's symbolic expression (see [2]):

$$e^{-(1/3!)\gamma_3(d^3/dx^3)+(1/4!)\gamma_4(d^4/dx^4)-\cdots}\Phi(x),$$

where

$$\Phi(x) = \frac{1}{\sqrt{2\pi}}\int_{-\infty}^{x} e^{-\frac{1}{2}y^2}\,dy$$

and γ_j is the jth semi-invariant of the random variable whose distribution is under asymptotic expansion. In the present case we have

$$\frac{\gamma_j}{j!} = \frac{\beta_j(x)}{n^{\frac{1}{2}(j-2)}},$$

where

$$\beta_j(x) = \frac{2^{\frac{1}{2}(j-2)}}{j}\,\frac{\frac{1}{n}\,s_j(x)}{\left(\frac{1}{n}\,s_2(x)\right)^{\frac{1}{2}j}}.$$

Hence we may express our result as follows:

$$(1)\qquad G(x) = \exp\left[\sum_{j=3}^{2k+1}\frac{(-1)^j\beta_j(x)}{n^{\frac{1}{2}(j-2)}}\left(\frac{d}{dx}\right)^j\right]\Phi(x) + R_k(x),$$

where $|R_k(x)| \leq Mn^{-k}$, and M is independent of n and x. The symbolic exponential in (1) is to be expanded as far as and including the term in $n^{-\frac{1}{2}(2k-1)}$.

2. Let us apply the result (1) to the following three statistics: $T_\alpha = Q_\alpha/S$, $(\alpha = 1, 2, 3)$, where

$$Q_1 = \sum_{i=1}^{N}(x_i - \bar{x})(x_{i+1} - \bar{x}) \quad \text{with} \quad x_{N+1} = x_1,$$

$$Q_2 = \tfrac{1}{2}(x_1 - \bar{x})^2 + \tfrac{1}{2}(x_N - \bar{x})^2 + \sum_{i=1}^{N-1}(x_i - \bar{x})(x_{i+1} - \bar{x}),$$

$$Q_3 = \sum_{i=1}^{N-1}(x_i - \bar{x})(x_{i+1} - \bar{x}).$$

T_2 is simply related with $T^* = Q^*/S$, where

$$Q^* = \sum_{i=1}^{N-1}(x_i - x_{i+1})^2;$$

for we have $Q_2 = S - \frac{1}{2}Q^*$, whence $T_2 = 1 - \frac{1}{2}T^*$. We shall write $\lambda_r^{(\alpha)}$ for the λ's corresponding to Q_α, and

$$b_{m\alpha} = \sum_{r=1}^{n}(\lambda_r^{(\alpha)})^m, \qquad (\alpha = 1, 2, 3).$$

(i) For Q_1 we have $\lambda_r^{(1)} = \cos \frac{2\pi r}{N}$ (see [3]). Since

$$\cos^m \theta = \frac{1}{2^m}(e^{i\theta} + e^{-i\theta})^m = \frac{1}{2^m}\sum_{j=0}^{m}\binom{m}{j} e^{i(2j-m)\theta},$$

we have

$$b_{m1} = \frac{1}{2^m}\sum_{j=0}^{m}\binom{m}{j}\sum_{r=1}^{n}\xi^r, \quad \text{where} \quad \xi = e^{2\pi(2j-m)i/N}.$$

If $m < n$, then

$$\sum_{r=1}^{n}\xi^r = -1 \quad \text{if} \quad j \neq \tfrac{1}{2}m, \qquad = n \quad \text{if} \quad j = \tfrac{1}{2}m.$$

Hence, for $m < n$, $b_{m1} = -1$ if m is odd, $b_{m1} = \frac{N}{2^m}\binom{m}{\frac{1}{2}m} - 1$ if m is even.

In particular

$$\bar{\lambda}^{(1)} = -\frac{1}{n}, \qquad \sum_{r=1}^{n}(\lambda_r^{(1)} - \bar{\lambda}^{(1)})^2 = \frac{n^2 - n - 2}{2n} > 0.4n \quad \text{if} \quad n \geq 7.$$

Hence assumptions (a) and (b) are true (for $n \geq 7$). The $s_j(x)$ are conveniently computed with the help of b_{m1}. The $\beta_j(x)$ are then computed to yield the terms in (1).

(ii) The λ's corresponding to Q^* are $4 \sin^2 \frac{r\pi}{2N}$ (see [4]). Hence

$$\lambda_r^{(2)} = \cos \frac{r\pi}{N}.$$

By a computation similar to that in (i) we easily obtain $b_{m2} = \frac{N}{2^m}\binom{m}{\frac{1}{2}m} - 1$ for even m and $b_{m2} = 0$ for odd m, provided $m < 2n$. In particular, $\bar{\lambda}^{(2)} = 0$, $\Sigma(\lambda_r^{(2)} - \bar{\lambda}^{(2)})^2 = \frac{n-1}{2} \geq \cdot 4n$ for $n \geq 5$. Hence assumptions (a) and (b) are true (for $n \geq 5$).

(iii) In the case of Q_3 the matrix A is

$$A = \begin{Vmatrix} 0 & \frac{1}{2} & & & & 0 \\ \frac{1}{2} & 0 & \frac{1}{2} & & & \\ & \frac{1}{2} & \cdot & & & \\ & & \cdot & \cdot & & \\ & & & \cdot & & \\ & & & & 0 & \frac{1}{2} \\ 0 & & & & \frac{1}{2} & 0 \end{Vmatrix}$$

whose latent roots are $\cos \pi t/(N + 1)$, $(t = 1, \cdots, N)$ (see [5]), all less than or equal to unity in absolute value. It follows that the same is true for the $\lambda_r^{(3)}$.

Hence assumption (a) is true. Unlike the two previous cases, there is no simple expression for b_{m3}. With the help of the formula

$$b_{m3} = \operatorname{tr} \{A(I - \gamma'\gamma)\}^m$$

we may compute b_{m3} for small values of m. Thus

$$b_{13} = -\frac{n}{n+1}$$

$$b_{23} = \frac{n}{2} - \frac{2n-1}{n+1} + \frac{n^2}{(n+1)^2}$$

$$b_{33} = -\frac{3(n-1)}{n+1} + \frac{3n(2n-1)}{2(n+1)^2} - \frac{n^3}{(n+1)^3}$$

$$b_{43} = \frac{3n-2}{8} - \frac{8n-11}{2(n+1)} + \frac{4n(n-1)}{(n+1)^2} + \frac{(2n-1)^2}{2(n+1)^2} - \frac{2n^2(2n-1)}{(n+1)^3} + \frac{n^4}{(n+1)^4}$$

$$b_{53} = -\frac{5(4n-7)}{4(n+1)} + \frac{5n(8n-11)}{8(n+1)^2} + \frac{5(2n-1)(n-1)}{2(n+1)^2} - \frac{5n^2(n-1)}{(n+1)^3}$$
$$- \frac{5n(2n-1)^2}{4(n+1)^3} + \frac{5n^3(2n-1)}{2(n+1)^5} - \frac{n^5}{(n+1)^5}$$

$$\overline{\lambda^{(3)}} = -\frac{1}{n+1}, \sum_{r=1}^{n} (\lambda_r^{(3)} - \overline{\lambda^{(3)}})^2 = \frac{n}{2} - \frac{2n-1}{n+1} + \frac{n^2 - n}{(n+1)^2} > 0.4n \text{ for } n \geq 10.$$

Hence assumption (b) is true (for $n \geq 10$). Using these values of b_{m3} we may compute $\beta_3(x)$, $\beta_4(x)$ and $\beta_5(x)$. By (1) we have

$$G(x) = \Phi(x) - \frac{1}{n^{\frac{1}{2}}} \beta_3(x)\Phi^{(3)}(x) + \frac{1}{n} (\beta_4(x)\Phi^{(4)}(x) + \tfrac{1}{2}\beta_3^2(x)\Phi^{(6)}(x))$$
$$- \frac{1}{n^{\frac{3}{2}}} (\beta_5(x)\Phi^{(5)}(x) - \beta_3(x)\beta_4(x)\Phi^{(7)}(x) + \tfrac{1}{6}\beta_3^3(x)\Phi^{(9)}(x)) + R(x),$$

where $| R(x) | \leq Mn^{-2}$ and M is independent of n and x.

REFERENCES

[1] H. Cramér, *Random Variables and Probability Distributions*, Cambridge Tract No. 36, 1937.

[2] E. A. Cornish and R. A. Fisher, *Revue de l'Institut International de Statistique*, 1937.

[3] R. L. Anderson, "Distribution of the serial correlation coefficient", *Annals of Math. Stat.*, Vol. 13 (1942), pp. 1–13.

[4] T. Koopmans, "Serial correlation and quadratic forms in normal variables", *Annals of Math. Stat.* Vol. 13 (1942), pp. 14–33.

[5] J. von Neumann, "A further remark concerning the distribution of the ratio of the mean square successive difference to the variance", *Annals of Math. Stat.*, Vol. 13 (1942), pp. 86–88.

PROCEEDINGS
OF THE
NATIONAL ACADEMY OF SCIENCES

Volume 33 February 15, 1947 Number 2

COMPLETE CONVERGENCE AND THE LAW OF LARGE NUMBERS

By P. L. Hsu and Herbert Robbins

Department of Mathematical Statistics, University of North Carolina

Communicated January 7, 1947

1. We begin by listing some standard definitions in the theory of probability. A *probability space* is a set Ω of elements ω together with a σ-field m of subsets of Ω on which is defined a completely additive measure P such that $P(\Omega) = 1$. A real-valued P-measurable function $X = X(\omega)$ is a *random variable*, and the function $F(x) = P\{X \leq x\}$, where $\{\ \}$ denotes the set of all ω such that the relation within the braces holds, is the *distribution function* of X. The sets of a sequence $A_1, A_2, \ldots$ are *independent* if for every finite set $i_1, \ldots, i_n$ of distinct integers, $P(\prod_{r=1}^{n} A_{i_r}) = \prod_{r=1}^{n} P(A_{i_r})$, and the random variables of a sequence $X_1, X_2, \ldots$ are independent if, for every sequence $x_1, x_2, \ldots$ of real numbers, the sets $\{X_1 \leq x_1\}, \{X_2 \leq x_2\}, \ldots$ are independent.

For purposes of comparison we list the following modes in which a sequence

$$X_1, X_2, \ldots \tag{1}$$

of random variables defined on Ω may converge to 0.

(*i*) The sequence (1) converges to 0 *in probability* if for every $\epsilon > 0$,

$$\lim_{n\to\infty} P\{|X_n| > \epsilon\} = 0.$$

(*ii*) The sequence (1) converges to 0 *with probability* 1 if for every $\epsilon > 0$,

$$\lim_{n\to\infty} P\{\{|X_n| > \epsilon\} + \{|X_{n+1}| > \epsilon\} + \ldots\} = 0.$$

It is easily seen that this is equivalent to the usual condition, $P\{\lim_{n\to\infty} X_n = 0\} = 1$, and that (*ii*) implies (*i*) but not conversely.

2. We shall be concerned with a third mode of convergence, which, for want of a better name, we call *complete.*

(*iii*) The sequence (1) converges to 0 *completely* if for every $\epsilon > 0$,

$$\lim_{n \to \infty} [P\{|X_n| > \epsilon\} + P\{|X_{n+1}| > \epsilon\} + \ldots] = 0.$$

Clearly, (*iii*) implies (*ii*). The example: Ω = unit interval $0 < \omega < 1$, P = Lebesgue measure, $X_n = 1$ for $0 < \omega < \frac{1}{n}$ and 0 otherwise, shows that (*ii*) does not imply (*iii*).

Let us call two sequences of random variables $X_1, X_2, \ldots$ and $Y_1, Y_2, \ldots$, defined, respectively, on probability spaces Ω and Ω_1, *F-equivalent,* if for every n the distribution function of Y_n is identical with that of X_n. Definitions (*i*) and (*iii*) are invariant under F-equivalence, while (*ii*) is not. However, *a sequence* $X_1, X_2, \ldots$ *of random variables converges to* 0 *completely if and only if every F-equivalent sequence converges to* 0 *with probability* 1. The necessity is obvious; to prove sufficiency consider a sequence $Y_1, Y_2, \ldots$ of independent random variables F-equivalent to the given sequence. If the sequence $Y_1, Y_2, \ldots$ converges to 0 with probability 1 then for any $\epsilon > 0$, $P\{\limsup_{n \to \infty}\{|Y_n| > \epsilon\}\} = 0$. Since the sets $\{|Y| > \epsilon\}$ are independent, it follows from a theorem of Borel-Cantelli[1] that $\sum_{n=1}^{\infty} P\{|Y_n| > \epsilon\} = \sum_{n=1}^{\infty} P\{|X_n| > \epsilon\} < \infty$.

It follows from this proof that if $X_1, X_2, \ldots$ is a sequence of *independent* random variables, then definitions (*ii*) and (*iii*) are equivalent.

3. Let the random variables X_n in (1) be independent with the same distribution function $F(x) = P\{X_n \leq x\}$ and such that the expectation $E(X_n) = \int_{-\infty}^{\infty} x dF(x) = 0$. The *strong law of large numbers* for identically distributed random variables states that the sequence of random variables $Y_1, Y_2, \ldots$, where for each n

$$Y_n = (X_1 + \ldots + X_n)/n \tag{2}$$

converges to 0 with probability 1. We shall show in Theorems 1 and 2 that under the same hypotheses the sequence (2) need not converge to 0 completely, but that it will do so under the further hypothesis that $\int_{-\infty}^{\infty} x^2 d F(x) < \infty$.

4. THEOREM 1. *Let (1) be a sequence of independent random variables with the same distribution function* $F(x)$ *and such that*

$$\int_{-\infty}^{\infty} x dF(x) = 0, \quad \sigma^2 = \int_{-\infty}^{\infty} x^2 dF(x) < \infty. \tag{3}$$

Then the sequence (2) converges to 0 completely; i.e., the series

$$\sum_{n=1}^{\infty} P\{|Y_n| > \epsilon\} \tag{4}$$

converges for every $\epsilon > 0$.

Proof. We shall prove the theorem for $\epsilon = 2$. This is no restriction since we can always consider $\frac{2}{\epsilon}X_n$ instead of X_n. Moreover, we may assume that $\sigma^2 > 0$.

Let $f(t) = \int_{-\infty}^{\infty} e^{itx}dF(x)$ be the characteristic function of the distribution $F(x)$. From (3) it follows that constants α, α', α'' exist such that

$$|1 - f(t)| \leq \alpha t^2, \quad |f'(t)| \leq \alpha' t, \quad |f''(t)| \leq \alpha''. \tag{5}$$

Choose and fix a positive δ so small that for $|t| \leq 4\delta$ the following conditions (6) and (7) are satisfied:

$$|\sin {}^1/_2 t| \geq Bt, \ |(1 - f(t))^2 - 4(1 - f(t)) \sin^2 {}^1/_2 t + 4 \sin^2 {}^1/_2 t| \geq Ct^2, \tag{6}$$

where B and C are constants,

$$|f(\ )| \neq 1 \qquad \text{except at } t = 0. \tag{7}$$

Let Z be a random variable distributed with the density $3(2\pi)^{-1}x^{-4} \sin^4 x$ and hence the characteristic function

$$\varphi(t) = \begin{cases} 1 - {}^3/_8 t^2 + {}^3/_{32}|t|^3, & |t| \leq 2, \\ {}^1/_{32}(4 - |t|)^2, & 2 \leq |t| \leq 4, \\ 0 & 4 \leq |t|. \end{cases} \tag{8}$$

We regard Z as independent of Y_n and use addition in this sense. Since

$$P\{|Y_n| > 2\} \leq P\left\{\left| Y_n + \frac{Z}{n\delta}\right| > 1\right\} + P\left\{\left|\frac{Z}{n\delta}\right| > 1\right\} =$$

$$P\left\{\left|\frac{Z}{n\delta}\right| \leq 1\right\} - P\left\{\left| Y_n + \frac{Z}{n\delta}\right| \leq 1\right\} + 2P\left\{\left|\frac{Z}{n\delta}\right| > 1\right\},$$

and since

$$\sum_{n=1}^{\infty} P\left\{\left|\frac{Z}{n\delta}\right| > 1\right\} \leq \frac{3}{\pi} \sum_{n=1}^{\infty} \int_{n\delta}^{\infty} \frac{dx}{x^4} = \frac{1}{\pi\delta^3} \sum_{n=1}^{\infty} \frac{1}{n^3} < \infty,$$

it is sufficient to prove that

$$\sum_{n=1}^{N-1} \left[P\left\{\left|\frac{Z}{n\delta}\right| \leq 1\right\} - P\left\{\left| Y_n + \frac{Z}{n\delta}\right| \leq 1\right\}\right] = 0(1), \tag{9}$$

where $0(1)$ always denotes a quantity bounded with respect to N.

The characteristic function $f^n\left(\frac{t}{n}\right) \varphi\left(\frac{t}{n\delta}\right)$ of $Y_n + \frac{Z}{n\delta}$ vanishes for $|t| > 4n\delta$; hence by a well-known inversion formula,[2]

$$P\left\{\left| Y_n + \frac{Z}{n\delta}\right| \leq 1\right\} = \frac{1}{\pi}\int_{-4n\delta}^{4n\delta} f^n\left(\frac{t}{n}\right)\varphi\left(\frac{t}{n\delta}\right)\frac{\sin t}{t}dt =$$
$$\frac{1}{\pi}\int_{-4\delta}^{4\delta} f^n(t)\,\varphi\left(\frac{t}{\delta}\right)\frac{\sin nt}{t}dt.$$

Also,

$$P\left\{\left|\frac{Z}{n\delta}\right| \leq 1\right\} = \frac{1}{\pi}\int_{-4\delta}^{4\delta}\varphi\left(\frac{t}{\delta}\right)\frac{\sin nt}{t}dt.$$

Hence, by subtraction the left side of (9) is equal to

$$\frac{1}{\pi}\int_{-4\delta}^{4\delta}\frac{1}{t}\varphi\left(\frac{t}{\delta}\right)\sum_{n=1}^{N-1}(1 - f^n(t))\sin nt\,dt = \frac{1}{\pi}A_N, \text{ say.} \tag{10}$$

From now on we write f for $f(t)$. Direct computation gives the result

$$\sum_{n=1}^{N-1}(1 - f^n)\sin nt = \frac{(1-f)^2 \sin {}^1/_2Nt \sin {}^1/_2(N-1)t}{q(t)\sin {}^1/_2t}$$
$$+ \frac{(1-f)\sin t}{q(t)} - \frac{4(1-f)\sin {}^1/_2t \sin {}^1/_2Nt \sin {}^1/_2(N-1)t}{q(t)}$$
$$+ \frac{(1-f)f^N \sin Nt}{q(t)} - \frac{2(1-f^{N+1})\sin {}^1/_2t \cos (N - {}^1/_2)t}{q(t)}, \tag{11}$$

where

$$q(t) = (f - e^{it})(f - e^{-it}) = (1-f)^2 - 4(1-f)\sin^2 {}^1/_2t + 4\sin^2 {}^1/_2t. \tag{12}$$

By (6) we have $|q(t)| \geq Ct^2$, $|q(t)\sin {}^1/_2t| \geq C'|t|^3$, where C and C' are constants. Hence when (11) is substituted into (10) and the first inequality of (5) is used, we see that the first three terms merely contribute $0(1)$. Consequently,

$$A_N = \int_{-4\delta}^{4\delta}\varphi\left(\frac{t}{\delta}\right)\frac{(1-f)f^N \sin Nt - 2(1-f^{N+1})\sin {}^1/_2t\cos(N - {}^1/_2)t}{tq(t)}dt + 0(1). \tag{13}$$

For $\delta \leq |t| \leq 4\delta$ we have, by (7), $|f| \neq 1$, hence $q(t) = |f - e^{it}|^2 \geq (1 - |f|)^2 \geq a > 0$. Therefore the part of the integral in (13) extended over the range $\delta \leq |t| \leq 4\delta$ is $0(1)$, so that (13) holds with 4δ replaced by δ in the two limits of integration. For $|t| \leq \delta$, however, $\varphi\left(\frac{t}{\delta}\right) = 1 - \frac{3t^2}{8\delta^2} + \frac{3|t|^3}{32\delta^3}$, and the terms with t^2 and $|t|^3$ are easily seen to contribute $0(1)$. Hence,

$$A_N = \int_{-\delta}^{\delta} \frac{(1-f)f^N \sin Nt - 2(1-f^{N+1}) \sin {}^1/_2 t \cos (N - {}^1/_2)t}{tq(t)} dt$$
$$+ 0(1). \tag{14}$$

Since

$$\left| \frac{1}{tq(t)} - \frac{1}{t^3} \right| = \left| \frac{t^2 - q(t)}{t^3 q(t)} \right| \leq k \frac{t^4}{|t|^5} = \frac{k}{|t|},$$

where k is a constant, the replacement of $tq(t)$ by t^3 in (14) will make a difference of only 0(1), so that

$$A_N = \int_{-\delta}^{\delta} \frac{(1-f)f^N \sin Nt}{t^3} dt - \int_{-\delta}^{\delta} \frac{2(1-f^{N+1}) \sin {}^1/_2 t \cos (N - {}^1/_2)t}{t^3} dt$$
$$+ 0(1). \tag{15}$$

Let the two integrals in (15) be denoted by I_N and J_N, respectively. We have

$$I_N = \frac{2}{N} \int_{-\delta}^{\delta} \frac{(1-f)f^N}{t^3} d(\sin^2 {}^1/_2 Nt) = 0(1) + \frac{2}{N} \int_{-\delta}^{\delta} \left\{ \frac{3(1-f)f^N}{t^4} + \frac{f^N f'}{t^3} \right.$$
$$\left. - \frac{Nf^{N-1}(1-f)f'}{t^3} \right\} \sin^2 {}^1/_2 Nt dt.$$

Using the first two inequalities of (5) we obtain the result

$$|I_N| \leq 0(1) + (6\alpha + 2\alpha') \int_{-\infty}^{\infty} \frac{\sin^2 {}^1/_2 Nt}{Nt^2} dt + 2 \int_{-\infty}^{\infty} \frac{|1-f|\,|f'|}{|t^3|} dt = 0(1),$$

since the integral involving N is independent of N.

To deal with J_N we observe first that in J_N, $\sin {}^1/_2 t$ may be replaced by ${}^1/_2 t$ and f^{N+1} by f^N, the difference thus made being 0(1). Hence

$$J_N = \int_{-\delta}^{\delta} \frac{(1-f^N) \cos (N - {}^1/_2)t}{t^2} dt + 0(1) =$$
$$\int_{-\infty}^{\infty} \frac{(1-f^N) \cos (N - {}^1/_2)t}{t^2} dt + 0(1).$$

We may replace $\cos (N - {}^1/_2)t$ by $\cos Nt$, since

$$\frac{2}{\pi} \int_{-\infty}^{\infty} \frac{1-f^N}{t^2} (\cos (N - {}^1/_2)t - \cos Nt) dt = \frac{1}{\pi} \int_{-\infty}^{\infty} (1-f^N) \frac{\sin {}^1/_4 t}{{}^1/_4 t}$$
$$\frac{\sin (N - {}^1/_4)t}{t} dt = P\{|U| \leq N - {}^1/_4\} - P\{|U + NY_N| \leq N - {}^1/_4\}$$
$$= 0(1),$$

where U is a random variable independent of Y_N and whose characteristic function is $4/t \sin {}^1/_4t$. Hence

$$J_N = 0(1) + \frac{1}{N}\int_{-\infty}^{\infty}\frac{1-f^N}{t^2}\,d\sin Nt = 0(1) + \frac{2}{N^2}\int_{-\infty}^{\infty}\left\{\frac{2(1-f^N)}{t^3} + \frac{Nf^{N-1}f'}{t^2}\right\}d\sin^2 {}^1/_2t$$

$$= 0(1) + \frac{2}{N^2}\int_{-\infty}^{\infty}\left\{\frac{6(1-f^N)}{t^4} + \frac{4Nf^{N-1}f'}{t^3} - \frac{N(N-1)f^{N-2}f'^2}{t^2} - \frac{Nf^{N-1}f''}{t^2}\right\}\sin^2 {}^1/_2Ntdt.$$

Using all the inequalities (5) we have

$$|J_N| \leq 0(1) + (12\alpha + 8\alpha' + 2\alpha'')\int_{-\infty}^{\infty}\frac{\sin^2 {}^1/_2Nt}{Nt^2}dt + 2\int_{-\infty}^{\infty}\frac{|f'|^2}{t^2}dt = 0(1)$$

The proof is now complete.

5. By following the essential steps of the proof of Theorem 1 we obtain the following theorem, the proof of which is omitted from the present communication.

THEOREM 2. *If instead of conditions* (3) *we have*

$$\int_{-\infty}^{\infty} xdF(x) = 0, \int_{-\infty}^{\infty} |x|^a dF(x) < \infty, \int_{-\infty}^{\infty} x^2dF(x) = \infty \qquad (16)$$

where a is some constant such that $\frac{1}{2}(1 + 5^{1/2}) \leq a < 2$, then the series (4) diverges for every $\epsilon > 0$. (Example: Let X_n be distributed with the density $|x|^{-3}$ for $|x| \geq 1$ and 0 elsewhere.)

Since the finiteness of the second integral in (16) would seem rather to favor than to oppose the convergence of (4), it may be conjectured that given the first condition of (3), the finiteness of σ^2 is not only sufficient but also necessary for the convergence of (4). We have not been able to prove this.

6. The following generalization[8] of the strong law of large numbers is an immediate consequence of Theorem 1 and the remarks in section 2.

THEOREM 3 *Let* $X_r^{(n)} (n = 1, 2, \ldots, r = 1, \ldots, n)$ *be an array of random variables with the same distribution function* $F(x)$ *and such that* (1) $\int_{-\infty}^{\infty} xdF(x) = 0$, $\int_{-\infty}^{\infty} x^2dF(x) < \infty$, *and* (2) *for each* n *the random variables* $X_1^{(n)}, \ldots X_n^{(n)}$ *are independent. Then the sequence of random variables* $Y_1, Y_2, \ldots,$ *where for each* n, $Y_n = (X_1^{(n)} + \ldots + X_n^{(n)})/n$, *converges to 0 with* *probability 1. (Note that we do not assume any relation of dependence or independence between $X_n^{(r)}$ and $X_m^{(s)}$ for $r \neq s$.)*

[1] See M. Fréchet, *Recherches theoriques modernes*, Vol. 1, Paris, 1937, p. 27.

[2] See H. Cramér, *Mathematical methods of statistics*, Princeton, 1946, p. 93.

[3] Compare F. P. Cantelli, Considerazioni sulla legge uniforme dei grandi numeri ecc., *Giornale dell 'Istituto Italiano degli Attuari*, IV (1933), pp. 331–332; also H. Cramér, Su un theorema relativo alla leggi uniforme dei grandi numeri, *Ibid.*, V (1934), pp. 1–13.

Editors note: The proof of Theorem 2 is given in handwritten notes by Hsu. It is reproduced below with minor editing.

Proof of Theorem 2. Let $\frac{1}{2}(1+\sqrt{5}) \leqslant a < 2$. (The number $\frac{1}{2}(1+\sqrt{5})$ is the positive root of $a^2 - a - 1 = 0$.) Let

$$\int x\,dF = 0, \qquad \int |x|^a\,dF < \infty, \qquad \int x^2\,dF = \infty.$$

Let $X_1, X_2, \ldots$ be a sequence of independent random variables having the same distribution $F(x)$. Let $Y_n = (1/n)(X_1 + \cdots + X_n)$, and suppose that $\sum_{n=1}^{\infty} P\{|Y_n| > \epsilon\}$ were convergent. We shall obtain a contradiction.

Consider $Wn = X_n - Z_n$, in the sense of convolution, where Z_n has the same distribution F. Then

$$E(W_n) = 0, \qquad E|W_n|^a \leqslant 2^{a-1}(E|X_n|^a + E|Z_n|^a) < \infty,$$

$$E|W_n|^2 = \lim_{A\to\infty} \int_{-A}^{A}\int_{-A}^{A} (x-z)^2\,dF(x)\,dF(z)$$

$$= 2 \lim_{A\to\infty} \left\{ \int_{-A}^{A} x^2\,dF \cdot \int_{-A}^{A} dF - \left(\int_{-A}^{A} x\,dF \right)^2 \right\} = \infty.$$

Hence the conditions on moments remain unchanged for the W_n. Moreover, we have

$$\sum_{n=1}^{\infty} P\left\{ \frac{1}{n}|W_1 + \cdots + W_n| > \epsilon \right\} \leqslant \sum_{n=1}^{\infty} P\left\{ \frac{1}{n}|X_1 + \cdots + X_n| > \frac{\epsilon}{2} \right\}$$

$$+ \sum_{n=1}^{\infty} P\left\{ \frac{1}{n}|Z_1 + \cdots + Z_n| > \frac{\epsilon}{2} \right\}$$

$$= 2\sum_{n=1}^{\infty} P\left\{ |Y_n| > \frac{\epsilon}{2} \right\} < \infty.$$

Hence the situation in the first paragraph remains the same if we replace X_n by $X_n - Z_n$. This amounts to the additional assumption that $F(x)$ *is a symmetric distribution.*

Given a symmetric F, we have for its characteristic function

$$f(t) = \int \cos tx\, dF, \qquad 1 - f(t) = 2\int \sin^2 \frac{tx}{2}\, dF, \qquad f'(t) = -\int x \sin tx\, dF,$$

whence

$$|1 - f(t)| = 1 - f(t) \leqslant 2\int \left|\sin \frac{tx}{2}\right|^a dF \leqslant \frac{|t|^a}{2^{a-1}} \int |x|^a\, dF \leqslant \alpha |t|^a,$$

$$|f'(t)| \leqslant \int |x|\, |\sin tx|^{a-1}\, dF \leqslant |t|^{a-1} \int |x|^a\, dF \leqslant \beta |t|^{a-1}.$$

These inequalities,

$$|1 - f(t)| \leqslant \alpha |t|^a, \qquad |f'(t)| \leqslant \beta |t|^{a-1}, \tag{5*}$$

play the parts of inequalities (5).

It is then easy to retrace all the steps (ohne weiteres) until we arrive at (15), or, what is the same thing,

$$A_N = \int_{-\infty}^{\infty} \frac{(1-f)f^N \sin Nt}{t^3}\, dt - \int_{-\infty}^{\infty} \frac{(1-f^N)\cos Nt}{t^2}\, dt + O(1). \tag{15*}$$

Let the two integrals in (15*) be I_N and J_N. Then

$$\begin{aligned}
I_N &= \frac{2}{N} \int \frac{(1-f)f^N}{t^3}\, d\left(\sin^2 \frac{Nt}{2}\right) \\
&= -\frac{2}{N} \int \sin^2 \frac{Nt}{2} \left\{ \frac{-3(1-f)f^N}{t^4} - \frac{f^N f'}{t^3} + \frac{N(1-f)f^{N-1}f'}{t^3} \right\} dt \\
&= \frac{6}{N} \int \frac{(1-f)f^N \sin^2(Nt/2)}{t^4}\, dt + \frac{2}{N} \int \frac{f^N f' \sin^2(Nt/2)}{t^3}\, dt + O(1) \\
&= \frac{6}{N} \int \frac{(1-f)\sin^2(Nt/2)}{t^4}\, dt + \frac{2}{N} \int \frac{f' \sin^2(Nt/2)}{t^3}\, dt \\
&\quad - \frac{6}{N} \int \frac{(1-f)(1-f^N)\sin^2(Nt/2)}{t^4}\, dt \\
&\quad - \frac{2}{N} \int \frac{(1-f^N)f' \sin^2(Nt/2)}{t^3}\, dt + O(1) \\
&= \frac{6}{N} \int \frac{(1-f)\sin^2(Nt/2)}{t^4}\, dt + \frac{2}{N} \int \frac{f' \sin^2(Nt/2)}{t^3}\, dt + 6\theta \int \frac{(1-f)^2}{t^4}\, dt \\
&\quad + 2\theta \int \frac{(1-f)|f'|}{|t|^3}\, dt + O(1) \\
&= \frac{6}{N} \int \frac{(1-f)\sin^2(Nt/2)}{t^4}\, dt + \frac{2}{N} \int \frac{f' \sin^2(Nt/2)}{t^3}\, dt + O(1),
\end{aligned}$$

where θ denotes any quantity such that $|\theta| \leqslant 1$. We obtain these θ terms because $|1 - f^N| = 1 - f^N = (1-f)(1 + f + \cdots + f^{N-1}) \leqslant N(1-f)$.

Hence

$$I_N = \frac{6}{N}\int \frac{(1-f)\sin^2(Nt/2)}{t^4}\,dt - \frac{2}{N}\int \frac{\sin(Nt/2)}{t^3}\,d(1-f) + O(1)$$

$$= \int \frac{(1-f)\sin Nt}{t^3}\,dt + O(1).$$

Again

$$J_N = \frac{1}{N}\int \frac{1-f^N}{t^2}\,d(\sin Nt) = -\frac{1}{N}\int\left(-\frac{Nf^{n-1}f'}{t^2} - \frac{2(1-f)^N}{t^3}\right)\sin Nt\,dt$$

$$= \int \frac{f^{N-1}f'}{t^2}\sin Nt\,dt + \frac{2}{N}\int \frac{1-f^N}{t^3}\sin Nt\,dt$$

$$= \int \frac{f^{N-1}f'}{t^2}\sin Nt\,dt + \frac{4}{N^2}\int \frac{1-f}{t^3}\,d\left(\sin^2\frac{Nt}{2}\right)$$

$$= O(1) + \int \frac{f^Nf'}{t^2}\sin Nt\,dt - \frac{4}{N^2}\int\left(\sin^2\frac{Nt}{2}\right)\left(-\frac{4(1-f^N)}{t^4} - \frac{Nf^{N-1}f'}{t^3}\right)dt$$

$$= \frac{16}{N^2}\int \frac{1-f^N}{t^4}\sin^2\frac{Nt}{2}\,dt + \frac{4}{N}\int \frac{f^{N-1}f'\sin^2(Nt/2)}{t^3}\,dt$$

$$+ \int \frac{f^Nf'}{t^2}\sin Nt\,dt + O(1)$$

$$= \frac{16}{N}\int \frac{1-f}{t^4}\sin^2\frac{Nt}{2}\,dt - 16\int \frac{\sin^2(Nt/2)}{t^4}\left(\frac{1-f}{N} - \frac{1-f^N}{N^2}\right)dt$$

$$+ \frac{4}{N}\int \frac{f'\sin^2(Nt/2)}{t^3}\,dt - \frac{4}{N}\int \frac{(1-f^{N-1})f'\sin^2(Nt/2)}{t^3}\,dt$$

$$+ \int \frac{f^Nf'\sin Nt}{t^2} + O(1)$$

$$= \frac{16}{N}\int \frac{1-f}{t^4}\sin^2\frac{Nt}{2}\,dt + \frac{4}{N}\int \frac{f'\sin^2(Nt/2)}{t^3}\,dt + \int \frac{f^Nf'\sin Nt}{t^2}\,dt$$

$$+ 16\theta\int \frac{(1-f)^2}{2t^4}\,dt + 4\theta\int \frac{(1-f)|f'|}{|t|^3}\,dt + O(1)$$

$$= \frac{16}{N} \int \frac{1-f}{t^4} \sin^2 \frac{Nt}{2}\, dt + \frac{4}{N} \int \frac{f' \sin^2(Nt/2)}{t^3}\, dt$$

$$+ \int \frac{f^N f' \sin Nt}{t^2}\, dt + O(1).$$

The θ terms come from the inequalities

$$\frac{1}{N}|1 - f^{N-1}| \leqslant 1 - f,$$

$$\left| \frac{1-f}{N} - \frac{1-f^N}{N^2} \right| = \frac{1}{N^2}(1-f)|N - (1 + f + \cdots + f^{N-1})|$$

$$\leqslant \frac{1}{N^2}(1-f)\{(1-f) + (1-f^2) + \cdots + (1-f^{N-1})\}$$

$$= \frac{1}{N^2}(1-f)^2\{1 + (1+f) + (1+f+f^2) + \cdots$$

$$+ (1 + f + \cdots + f^{N-2})\}$$

$$\leqslant \frac{1}{N^2}(1-f)^2(1 + 2 + 3 + \cdots + N - 1) \leqslant \tfrac{1}{2}(1-f)^2.$$

Hence

$$J_N = \frac{16}{N} \int \frac{1-f}{t^4} \sin^2 \frac{Nt}{2}\, dt - \frac{4}{N} \int \frac{\sin^2(Nt/2)}{t^3}\, d(1-f)$$

$$+ \int \frac{f^N f' \sin Nt}{t^2}\, dt + O(1)$$

$$= 2 \int \frac{1-f}{t^3} \sin Nt\, dt + \int \frac{f^N f' \sin Nt}{t^2}\, dt + O(1). \tag{18}$$

Substituting (17) and (18) into (15*), we have

$$A_N = -\int \frac{1-f}{t^3} \sin Nt\, dt - \int \frac{f^N f' \sin Nt}{t^2}\, dt + O(1)$$

$$= \pi B_N + C_N + O(1), \tag{19}$$

where

$$B_N = -\frac{1}{\pi} \int \frac{1-f}{t^3} \sin Nt\, dt - \frac{1}{\pi} \int \frac{f' \sin Nt}{t^2}\, dt,$$

$$C_N = \int \frac{(1-f^N) f' \sin Nt}{t^2}\, dt.$$

Now,

$$B_N = -\frac{1}{\pi}\int \frac{1-f}{t^3}\sin Nt\,dt + \frac{1}{\pi}\int \frac{\sin Nt}{t^2}\,d(1-f)$$

$$= -\frac{1}{\pi}\int \frac{1-f}{t^3}\sin Nt\,dt - \frac{1}{\pi}\int (1-f)\left(-\frac{2}{t^3}\sin Nt + \frac{N\cos Nt}{t^2}\right)dt$$

$$= \frac{1}{\pi}\int \frac{1-f}{t^3}\sin Nt\,dt - \frac{N}{\pi}\int \frac{(1-f)\cos Nt}{t^2}\,dt$$

$$= \frac{2}{\pi}\int_0^\infty \frac{1-f}{t^3}\sin Nt\,dt - \frac{2N}{\pi}\int_0^\infty \frac{(1-f)\cos Nt}{t^2}\,dt$$

$$= 2\int_{-\infty}^{\infty} dF(x)\left\{\frac{2}{\pi}\int_0^\infty \frac{\sin^2(tx/2)\sin Nt}{t^3}\,dt - \frac{2N}{\pi}\int_0^\infty \frac{\sin^2(tx/2)\cos Nt}{t^2}\,dt\right\}$$

But

$$\frac{2}{\pi}\int_0^\infty \frac{\sin^2(tx/2)\sin Nt}{t^3}\,dt = \begin{cases} \frac{N}{4}(2|x| - N), & |x| > N, \\ \frac{x^2}{4}, & |x| \leqslant N, \end{cases}$$

$$\frac{2}{\pi}\int_0^\infty \frac{\sin^2(tx/2)\cos Nt}{t^2}\,dt = \begin{cases} \frac{1}{2}(|x| - N), & |x| > N, \\ 0, & |x| \leqslant N. \end{cases}$$

Hence, after some reduction, we obtain

$$B_N = \frac{1}{2}\int_{|x|\leqslant N} x^2\,dF(x) + \frac{N^2}{2}\int_{|x|>N} dF(x) \to \infty \qquad \text{as} \quad N\to\infty.$$

It remains to prove that $C_n = O(1)$. For this purpose we write

$$f'(t) = -\int_{|x|\leqslant N} x\sin tx\,dF - \int_{|x|>N} x\sin tx\,dF = g_N(t) + h_N(t)$$

and

$$C_N = \int \frac{(1-f^N)g_N\sin Nt}{t^2}\,dt + \int \frac{(1-f^N)h_N\sin Nt}{t^2}\,dt$$

$$= D_N + E_N.$$

We have

$$|f'| \leqslant \beta|t|^{a-1}, \qquad |g_N| \leqslant \beta|t|^{a-1}, \qquad \text{as before,} \tag{20}$$

$$|g_N'| = \left|\int_{|x|\leqslant N} x^2 \cos tx\, dF\right| \leqslant \int_{|x|\leqslant N} x^2\, dF \leqslant N^{2-a}\int |x|^a\, dF \leqslant \gamma N^{2-a}, \tag{21}$$

$$|h_N| \leqslant \int_{|x|>N} |x|\, dF \leqslant N^{1-a}\int_{|x|>N} |x|^a\, dF \leqslant \delta N^{1-a}, \tag{22}$$

$$|1 - f^N| \leqslant N(1-f) \leqslant \alpha N t^a, \qquad \text{as before.} \tag{23}$$

Hence

$$D_N = \frac{2}{N}\int \frac{(1-f^N)g_N}{t^2}\, d\left(\sin^2 \frac{Nt}{2}\right)$$

$$= -\frac{2}{N}\int \sin^2\frac{Nt}{2}\left\{-\frac{3}{t^3}(1-f^N)g_N - \frac{Nf^{N-1}f'g_N}{t^2} + \frac{(1-f^N)g_N'}{t^2}\right\} dt$$

$$= O(1) - \frac{2}{N}\int \frac{(1-f^N)g_N'}{t^2}\sin^2\frac{Nt}{2}\, dt$$

$$= O(1) + \frac{2\theta}{N}\gamma N^{2-a}\int \frac{1-f^N}{t^2}\, dt, \qquad \text{by (21).}$$

Also,

$$E_N = \theta\delta_N^{1-a}\int \frac{(1-f^N)}{t^2}\, dt, \qquad \text{by (22).}$$

Thus it is sufficient to prove that

$$G_N = N^{1-a}\int \frac{1-f^N}{t^2}\, dt = O(1).$$

Now

$$G_N = N^{1-a}\left\{\int_{|t|\leqslant A} \frac{1-f^N}{t^2}\, dt + \int_{|t|>A} \frac{1-f^N}{t^2}\, dt\right\}$$

$$\leqslant N^{1-a}\left\{\int_{|t|\leqslant a} \alpha N t^{a-2}\, dt + 2\int_{|t|>A} \frac{dt}{t^2}\right\}, \qquad \text{by (23),}$$

$$= 2N^{1-a}\left\{\frac{\alpha}{a-1} N A^{a-1} + \frac{2}{A}\right\}.$$

Putting $A = N^{-1/a}$, we get

$$G_N \leqslant 2\left(\frac{\alpha}{a-1} + 2\right)N^{1-a+1/a} = O(1), \qquad \text{since} \quad a \geqslant \frac{1+\sqrt{5}}{2}.$$

Reprinted from
B. V. Gnedenko and A. N. Kolmogorov:
Limit Theorems of Sums of Independent Random Variables,
Addison-Wesley, 1968.

A General Weak Limit Theorem for Independent Distributions

by P. L. Hsu

1. Introduction

For every integer n, let there be n independent random variables* (R.V.): X_{nj} ($j = 1, 2, \ldots, n$). The distribution function (D.F.) and the characteristic function (C.F.) of X_{nj} will be denoted by $F_{nj}(x)$ and $f_{nj}(t)$ respectively, so that†

$$F_{nj}(x) = P(X_{nj} \leq x), \tag{1.1}$$

$$f_{nj}(t) = Ee^{itX_{nj}}. \tag{1.1'}$$

The D.F. $F_n(x)$ and the C.F. $f_n(t)$ of the sum $X_{n1} + X_{n2} + \cdots + X_{nn}$ are then, respectively,

$$F_n(x) = F_{n1}(x) * F_{n2}(x) * \cdots * F_{nn}(x), \tag{1.2}$$

$$f_{nj}(t) = f_{n1}(t) f_{n2}(t) \cdots f_{nn}(t), \tag{1.2'}$$

where the star denotes convolution. The weak limit problem in its full generality requires to find conditions necessary and sufficient under which

$$\lim_{n\to\infty} F_n(x) = F(x) \text{ at every continuity point of } F(x), \tag{1.3}$$

where $F(x)$ is some given D.F. This is equivalent to the following:**

$$\lim_{n\to\infty} f_n(t) = f(t), \tag{1.3'}$$

where $f(t)$ is the C.F. of $F(x)$. In a highly elaborate paper Feller [1]‡ solved the important case where $F(x)$ is the Gaussian D.F., under the following assumption:

$$\lim_{n\to\infty} \max_{1\leq j\leq n} \int_{|y|>x} dF_{nj}(y) = 0 \quad \text{for every } x > 0, \tag{0}$$

---

*Random variables are used here for convenience rather than necessity. The weak law in the theory of probability being purely function-theoretic, the concept of random variables may be dispensed with.

† $P(\ )$ denotes the probability of the relation inside ( ); E denotes the mathematical expectation.

** This equivalence was first established by Glivenko [2].

‡ Numbers in square brackets indicate the references cited at end of this appendix.

and in the nonessentially restricted form $F_{nj}(x) = V_j(a_n x)$. Gnedenko [3] and Marcinkiewicz [7] investigated the general problem, and in the latter [7], with the aid of Khintchine's results [4], proved that under the assumption (0) the limit law $F(x)$ is necessarily infinitely divisible (Paul Lévy [6], Khintchine [4]). Marcinkiewicz [7] also dealt with the case where the limit $f(t)$ in (1.3′) is an integral function of a complex variable, and [9] obtained explicit conditions necessary and sufficient for (1.3) for the cases where $F(x)$ is the Gaussian law, the Poisson law, and the D.F. of the "sure number" zero (the last case being the case of the weak law of large numbers).

In this paper we shall also take (0) as our assumption and shall solve the problem for the general case where $F(x)$ is an infinitely divisible law, specified by the fact that its C.F. $f(t)$ is the following one:*

$$f(t) = \exp\left\{mit + \int_{-\infty}^{\infty} \left(e^{itx} - 1 - \frac{itx}{1+x^2}\right) \frac{1+x^2}{x^2}\, dG(x)\right\}, \tag{1.4}$$

where $G(x)$ is a nondecreasing function with $G(\infty) - G(-\infty) < \infty$.

2. Statement of the Theorem

THEOREM. *Let $F_n(x)$ be defined in* (1.2), *$f(t)$ be the function* (1.4), *and let $F(x)$ be the D.F. having $f(t)$ as the C.F. Let condition* (0) *hold. Then in order to have*

$$\lim_{n\to\infty} F_n(x) = F(x), \tag{2.1}$$

at all continuity points of $F(x)$, it is necessary and sufficient that the following relations hold at every $x > 0$ such that $\pm x$ are continuity points of $G(y)$:

$$\lim_{n\to\infty} \sum_{j=1}^{n} \int_{|y|>x} dF_{nj}(y) = \int_{|y|>x} \frac{1+y^2}{y^2}\, dG(y), \tag{I}$$

$$\lim_{n\to\infty} \sum_{j=1}^{n} \left\{ \int_{|y|<x} y^2\, dF_{nj}(y) - \left(\int_{|y|<x} y\, dF_{nj}(y) \right)^2 \right\} = \int_{|y|<x} (1+y^2)\, dG(y), \tag{II}$$

$$\lim_{n\to\infty} \sum_{j=1}^{n} \int_{|y|<x} y\, dF_{nj}(y) = m + \int_{|y|<x} y\, dG(y) - \int_{|y|>x} \frac{1}{y}\, dG(y)\,. \tag{III}$$

The proof will be given in the next two sections. In order to have a condition on $f_{nj}(t)$ which is equivalent to (0), we prove the following lemma in spite of the fact that the proof may be found in Feller's paper [1].

* This form is due to Lévy and Khintchine [4].

LEMMA 2.1. *Condition* (0) *is equivalent to the following condition:*

$$\lim_{n\to\infty} \max_{1\le j\le n} |1 - f_{nj}(t)| = 0 \tag{α}$$

uniformly in every finite interval.

Proof. We have

$$|1 - f_{nj}(t)| \le \int_{-\infty}^{\infty} (1 - \cos tx)\, dF_{nj}(x) \le \int_{|x|\le\epsilon} |tx|\, dF_{nj} + 2 \int_{|x|>\epsilon} dF_{nj}$$
$$\le \epsilon|t| + 2 \int_{|x|>\epsilon} dF_{nj}.$$

Hence

$$\max_{1\le j\le n} |1 - f_{nj}(t)| \le \epsilon|t| + \max_{1\le j\le n} \int_{|x|>\epsilon} dF_{nj}$$

for every $\epsilon > 0$. If (0) is true, then it is easy to deduce (α) from the last inequality.

Suppose next that (α) is true. Then, since

$$|1 - f_{nj}(t)| \ge \int_{-\infty}^{\infty} (1 - \cos ty)\, dF_{nj}(y) \ge \int_{|y|>x} (1 - \cos ty)\, dF_{nj}(y),$$

we can find $N = N(x;\epsilon)$ so that

$$\int_{|y|>x} (1 - \cos ty)\, dF_{nj}(y) \le \tfrac{1}{2}\epsilon, \quad \text{for} \quad |t| \le 2x^{-1}, \quad n \ge N(x;\epsilon), \quad j \le n.$$

Integrating both sides with respect to t on the interval $(0, 2x^{-1})$, we get

$$\frac{\epsilon}{x} \ge \int_{|y|>x} \left(\frac{2}{x} - \frac{1}{y} \sin \frac{2y}{x}\right) dF_{nj}(y) \ge \int_{|y|>x} \left(\frac{2}{x} - \frac{1}{|y|}\right) dF_{nj}(y)$$
$$\ge \frac{1}{x} \int_{|y|>x} dF_{nj}(x),$$

whence

$$\int_{|y|>x} dF_{nj}(y) \le \epsilon \quad \text{for} \quad n \ge N(x;\epsilon), \quad j \le n.$$

Hence (0) is true. Q.E.D.

By virtue of (α), $\log f_{nj}(t)$ is defined in every finite interval of t, for all sufficiently large n and all $j \leq n$, if we fix $\log f_{nj}(0) = 0$. We then have the following simple lemma:

LEMMA 2.2. *Equations* (0) *and* (1.3) *are equivalent to* (α) *and* (β), *where* (β) *is the following condition:*

$$\lim_{n\to\infty} \sum_{j=1}^{n} \log f_{nj}(t) = mit + \int_{-\infty}^{\infty} \left(e^{itx} - 1 - \frac{itx}{1+x^2}\right) \frac{1+x^2}{x^2}\, dG(x). \quad (\beta)$$

This follows directly from (1.3′) *on taking* (1.4) *for* $f(t)$ *and taking logarithms.*

3. Proof of the Sufficiency Part of the Theorem

We shall call any positive number a a continuity number if all the F_{nj} and G are continuous at both a and $-a$. Throughout the following pages we use θ to denote any, not necessarily the same, quantity such that $|\theta| \leq 1$. We write

$$C_{nj}(x) = \int_{|y|<x} y\, dF_{nj}(y), \qquad C_n(x) = \max_{1\leq j\leq n} |C_{nj}(x)|,$$

$$A_n(x) = \sum_{j=1}^{n} \int_{|y|>x} dF_{nj}(y),$$

$$B_n(x) = \sum_{j=1}^{n} \int_{|y|<x} (y - C_{nj}(x))^2\, dF_{nj}(y),$$

$$C_n(x) = \sum_{j=1}^{n} \int_{|y|<x} y\, dF_{nj}(y),$$

$$A(x) = \int_{|y|>x} \frac{1+y^2}{y^2}\, dG(y),$$

$$B(x) = \int_{|y|<x} (1+y^2)\, dG(y),$$

$$C(x) = m + \int_{|y|<x} y\, dG(y) - \int_{|y|>x} \frac{1}{y}\, dG(y).$$

In this section our hypotheses are (0) [or its equivalent (α)], (I), (II), (III), and it is required to prove (β). We proceed with a series of Lemmas.

LEMMA 3.1. *We have*

$$\lim_{n\to\infty} C_n(x) = 0 \quad \text{for every } x > 0. \tag{3.1}$$

Proof.

$$|C_{nj}(x)| \leq \int_{|y|\leq\epsilon} |y|\, dF_{nj}(y) + \int_{\epsilon<|y|<x} |y|\, dF_{nj}(y)$$
$$\leq \epsilon + x \max_{1\leq j\leq n} \int_{|y|>\epsilon} dF_{nj}(y).$$

Using (0), we get

$$\overline{\lim_{n}}\, C_n(x) \leq \epsilon,$$

whence the result. Q.E.D.

LEMMA 3.2. *Equations* (I), (II), (III) *imply the following:*

$$\lim_{n\to\infty} A_n(x) = A(x), \tag{I$'$}$$
$$\lim_{n\to\infty} B_n(x) = B(x), \tag{II$'$}$$
$$\lim_{n\to\infty} C_n(x) = C(x), \tag{III$'$}$$

for all $x > 0$ such that both x and $-x$ are continuity points of $G(y)$.

Proof. The only thing new is (II$'$). To prove it we have

$$B_n(x) = \sum_{j=1}^{n} \left\{ \int_{|y|<x} y^2\, dF_{nj}(y) - 2C_{nj}^2(x) + C_{nj}^2(x) \int_{|y|<x} dF_{nj}(y) \right\}$$
$$= \sum_{j=1}^{n} \left\{ \int_{|y|<x} y^2\, dF_{nj}(y) - C_{nj}^2(x) \right\} - \sum_{j=1}^{n} C_{nj}^2(x) \int_{|y|\geq x} dF_{nj}(y)$$
$$= \sum_{j=1}^{n} \left\{ \int_{|y|<x} y^2\, dF_{nj}(y) - C_{nj}^2(x) \right\} + \theta C_n^2(x) A_n(x).$$

Hence the result follows from (3.1), (I), and (II). Q.E.D.

LEMMA 3.3. *Let ϵ be any positive number, and let*

$$f_{nj}^*(t;\epsilon) = f_{nj}(t) e^{-itC_{nj}(\epsilon)}$$

Then

$$\lim_{n\to\infty} \max_{1\leq j\leq n} |1 - f^*_{nj}(t;\epsilon)| = 0 \tag{i}$$

uniformly in every finite interval. If ϵ is a continuity number, then

$$\sum_{j=1}^{n} |1 - f^*_{nj}(t;\epsilon)| \tag{ii}$$

is bounded in n for every fixed t.

Proof. For (i),

$$|1 - f^*_{nj}(t;\epsilon)| \leq |1 - e^{-itC_{nj}(\epsilon)}| + |1 - f_{nj}(t)| \leq C_n(\epsilon)|t| + \max_j |1 - f_{nj}(t)|.$$

Hence the result follows from (α) and (3.1). For (ii),

$$\begin{aligned}
\sum_j |1 - f^*_{nj}(t;\epsilon)| \\
&\leq \sum_j \left| \int_{|x|<\epsilon} (1 - e^{itx - itC_{nj}(\epsilon)})\, dF_{nj} + \int_{|x|\geq\epsilon} \right| \\
&= \theta \sum_j \left| \int_{|x|<\epsilon} \{-it(x - C_{nj}(\epsilon)) + \tfrac{1}{2}\theta t^2 (x - C_{nj}(\epsilon))^2\}\, dF_{nj} \right| \\
&\quad + 2\theta A_n(\epsilon) \\
&= \theta \sum_j \left| tC_{nj}(\epsilon) \int_{|x|\geq\epsilon} dF_{nj} \right| + \tfrac{1}{2}\theta t^2 B_n(\epsilon) + 2\theta A_n(\epsilon) \\
&= \theta |t| C_n(\epsilon) A_n(\epsilon) + \tfrac{1}{2}\theta t^2 B_n(\epsilon) + 2\theta A_n(\epsilon),
\end{aligned}$$

whence the result follows from (3.1), (I′), and (II′).

LEMMA 3.4. *Let ϵ be any continuity number. Then*

$$\lim_{n\to\infty} \sum_{j=1}^{n} \int_{|x|>\epsilon} e^{itx}\, dF_{nj}(x) = \int_{|x|>\epsilon} e^{itx} \frac{1+x^2}{x^2}\, dG(x). \tag{3.2}$$

Proof. For the real part:

$$\sum_{j=1}^{n} \int_{|x|>\epsilon} \cos tx\, dF_{nj}(x) = -\int_{\epsilon}^{\infty} \cos tx\, dA_n(x),$$

$A_n(x)$ is a nonincreasing function of bounded total variation $A_n(\epsilon)$ which is also bounded in n, by (I′). Hence

$$\lim_{n\to\infty} \sum_{j=1}^{n} \int_{|x|>\epsilon} \cos tx \, dF_{nj}(x) = -\int_{\epsilon}^{\infty} \cos tx \, dA(x)$$

$$= \int_{|x|>\epsilon} \cos tx \, \frac{1+x^2}{x^2} \, dG(x).$$

For the imaginary part: let l be any continuity number. Then

$$\sum_{j=1}^{n} \int_{|x|>\epsilon} \sin tx \, dF_{nj}(x) = \sum_{j=1}^{n} \int_{\epsilon<|x|<l} \sin tx \, dF_{nj}(x) + \theta A_n(l)$$

$$= \int_{\epsilon}^{l} \frac{\sin tx}{x} \, dC_n(x) + \theta A_n(l)$$

$$= \left[\frac{\sin tx}{x} C_n(x)\right]_{\epsilon}^{l} - \int_{\epsilon}^{l} C_n(x) d \, \frac{\sin tx}{x} + \theta A_n(l).$$

For $\epsilon < x < l$, however,

$$C_n(x) = C_n(\epsilon) + \theta \sum_{j} \int_{\epsilon\le|y|\le l} |y| \, dF_{nj}(y) = C_n(\epsilon) + \theta l A_n(\epsilon).$$

Thus $|C_n(x)|$ is bounded in both x and n, because $C_n(\epsilon)$ and $A_n(\epsilon)$ are convergent sequences. Hence using (III′) and denoting by $\overline{\underline{\lim}}$ either the upper or the lower limit,

$$\overline{\underline{\lim}}_{n} \sum_{j} \int_{|x|>\epsilon} \sin tx \, dF_{nj}(x) = \left[C(x) \, \frac{\sin tx}{x}\right]_{\epsilon}^{l} - \int_{\epsilon}^{l} C(x) \, d \, \frac{\sin tx}{x} + \theta A(l)$$

$$= \int_{\epsilon}^{l} \frac{\sin tx}{x} \, dC(x) + \theta A(l)$$

$$= \int_{\epsilon<|x|<l} \sin tx \, \frac{1+x^2}{x^2} \, dG(x) + \theta \, A(l).$$

Since $A(l) \to 0$ as $l \to \infty$, we get

$$\lim_{n} \sum_{j} \int_{|x|>\epsilon} \sin tx \, dF_{nj}(x) = \int_{|x|>\epsilon} \sin tx \, \frac{1+x^2}{x^2} \, dG(x). \qquad \text{Q.E.D.}$$

Proof of the sufficiency of the conditions. Let ϵ be any continuity number. We have

$$\sum_j \{1 - f^*_{nj}(t;\epsilon)\} = \sum_j \int_{|x|<\epsilon} (1 - e^{itx - itC_{nj}(\epsilon)})\, dF_{nj}(x) + \sum_j \int_{|x|\geq\epsilon}$$

$$= \sum_j \int_{|x|<\epsilon} \{-it(x - C_{nj}(\epsilon)) + \tfrac{1}{2}t^2(x - C_{nj}(\epsilon))^2$$

$$+ \tfrac{1}{6}\theta t^3(x - C_{nj}(\epsilon))^3\}\, dF_{nj}$$

$$+ \sum_j \int_{|x|>\epsilon} (1 - e^{itx})\, dF_{nj}$$

$$+ \sum_j \int_{|x|>\epsilon} e^{itx}(1 - e^{-itC_{nj}(\epsilon)})\, dF_{nj}(x)$$

$$= \theta t C_n(\epsilon) A_n(\epsilon) + \tfrac{1}{2}t^2 B_n(\epsilon) + \tfrac{1}{6}\theta t^3(\epsilon + C_n(\epsilon))B_n(\epsilon)$$

$$+ \sum_j \int_{|x|>\epsilon} (1 - e^{itx})\, dF_{nj} + \theta t C_n(\epsilon) A_n(\epsilon)$$

$$= \tfrac{1}{2}t^2 B_n(\epsilon) + A_n(\epsilon) - \sum_j \int_{|x|>\epsilon} e^{itx}\, dF_{nj}$$

$$+ 2\theta t C_n(\epsilon) A_n(\epsilon) + \tfrac{1}{6}\theta t^3(\epsilon + C_n(\epsilon))B_n(\epsilon).$$

Hence, using (II′), (I′), (3.2), and (3.1), we get

$$\overline{\lim_n} \sum_j \{1 - f^*_{nj}(t;\epsilon)\}$$

$$= \tfrac{1}{2}t^2 B(\epsilon) + \int_{|x|>\epsilon} (1 - e^{itx}) \frac{1 + x^2}{x^2}\, dG(x) + \tfrac{1}{6}\theta\epsilon t^3 B(\epsilon). \tag{3.3}$$

But, for n sufficiently large so that $|1 - f^*_{nj}(t;\epsilon)| < \frac{1}{2}$ for all $j \leq n$ (cf. Lemma 3.2(i)),

$$-\sum_j \log f^*_{nj}(t;\epsilon) = \sum_j \{1 - f^*_{nj}(t;\epsilon)\} + \tfrac{1}{2}\theta \sum_j \frac{|1 - f^*_{nj}(t;\epsilon)|^2}{1 - |1 - f^*_{nj}(t;\epsilon)|}$$

$$= \sum_j \{1 - f^*_{nj}(t;\epsilon)\}$$

$$+ \theta \max_j |1 - f^*_{nj}(t;\epsilon)| \sum_j |1 - f^*_{nj}(t;\epsilon)|,$$

and so

$$-\sum_j \log f_{nj}(t) = -itC_n(\epsilon) + \sum_j \{1 - f^*_{nj}(t;\epsilon)\}$$
$$+ \theta \max_j |1 - f^*_{nj}(t;\epsilon)| \sum_j |1 - f^*_{nj}(t;\epsilon)|.$$

Using (III′), (3.3), and Lemma 3.3, we get

$$-\overline{\lim_n} \sum_j \log f_{nj}(t) = -itC(\epsilon) + \tfrac{1}{2}t^2B(\epsilon) + \int_{|x|>\epsilon} (1 - e^{itx}) \frac{1+x^2}{x^2}\, dG(x)$$
$$+ \tfrac{1}{6}\theta t^3 B(\epsilon)$$
$$= -mit + \tfrac{1}{2}t^2B(\epsilon) + \int_{|x|>\epsilon} \left(1 - e^{itx} + \frac{itx}{1+x^2}\right)$$
$$\times \frac{1+x^2}{x^2}\, dG(x) - it \int_{|x|<\epsilon} y\, dG(y) + \tfrac{1}{6}\theta\epsilon t^3 B(\epsilon).$$

As ϵ tends to 0, the term involving θ tends to 0. Hence

$$\lim_{n\to\infty} \sum_j \log f_{nj}(t) = mit + \lim_{\epsilon\to 0} \left\{ \int_{|x|\geq\epsilon} \left(e^{itx} - 1 - \frac{itx}{1+x^2}\right) \right.$$
$$\left. \times \frac{1+x^2}{x^2}\, dG(x) - \tfrac{1}{2}t^2 B(\epsilon) \right\}$$
$$= mit + \int_{-\infty}^{\infty} \left(e^{itx} - 1 - \frac{itx}{1+x^2}\right) \frac{1+x^2}{x^2}\, dG(x)$$
$$-\lim_{\epsilon\to 0} \int_{|x|<\epsilon} \left(e^{itx} - 1 - \frac{itx}{1+x^2} + \frac{t^2x^2}{2}\right) \frac{1+x^2}{x^2}\, dG(x)$$
$$= mit + \int_{-\infty}^{\infty} \left(e^{itx} - 1 - \frac{itx}{1+x^2}\right) \frac{1+x^2}{x^2}\, dG(x),$$

which gives (β). Hence the sufficiency part of the theorem is proved.

4. Proof of the Necessity Part of the Theorem

We first prove four lemmas which together with their proofs, may be found in Feller's paper [1]. We give their proofs here partly because in Feller's paper he does not state these lemmas separately, which makes it hard to refer to, and partly because Feller restricts (nonessentially) himself to the form $F_{nj}(x) = V_j(A_n x)$. In Lemmas 4.1 through 4.4, we consider a system of D.F.'s $F'_{nj}(x)$ together with their C.F.'s $f'_{nj}(t) = u'_{nj}(t) + iv'_{nj}(t)$ $(j = 1, 2, \ldots, n;\ n = 1, 2, 3, \ldots)$ satisfying the following three conditions:

$$\lim_{n\to\infty} \max_{1\le j\le n} |1 - f'_{nj}(t)| = 0 \quad \text{uniformly in every finite interval} \tag{α'}$$

$$\lim_{n\to\infty} \prod_{j=1}^{n} |f'_{nj}(t)| = \exp\left\{\int_{-\alpha}^{\alpha} (\cos tx - 1)\, \frac{1+x^2}{x^2}\, dG(x)\right\} \tag{β'}$$

$$0 \text{ is a median of every } F'_{nj}, \text{ that is } F'_{nj}(-0) \le \tfrac{1}{2},\ 1 - F'_{nj}(0) \le \tfrac{1}{2}. \tag{M}$$

Throughout Lemmas 4.1 to 4.4, it is understood that (α'), (β'), and (M) are satisfied.

Lemma 4.1. *We have*

$$v'_{nj}(t) \le \tfrac{4}{3}(1 - u'_{nj}(t)) \quad \text{for} \quad |t| \le T, \quad n \ge N(T), \quad j \le n.$$

Proof. We need only consider the case $t > 0$. Let J' be the set of values x for which $\sin tx \ge 0$ and J'' be the set of values x for which $\sin tx \le 0$. Then

$$\begin{aligned} v'^2_{nj}(t) &= \Big(\int_{J'} \sin tx\, dF'_{nj} + \int_{J''}\Big)^2 \\ &\le \max\left\{\Big(\int_{J'} \sin tx\, dF'_{nj}\Big)^2, \Big(\int_{J''} \sin tx\, dF'_{nj}\Big)^2\right\}. \end{aligned} \tag{4.2}$$

By Schwarz' inequality we have, if J denotes either J' or J'',

$$\begin{aligned} \Big(\int_{J} \sin tx\, dF'_{nj}\Big)^2 &\le \int_J dF'_{nj} \int_{-\infty}^{\infty} \sin^2 tx\, dF'_{nj} \\ &= \int_J dF'_{nj} \int_{-\infty}^{\infty} (1 - \cos tx)(1 + \cos tx)\, dF'_{nj} \\ &\le 2\int_J dF'_{nj} \int_{-\infty}^{\infty} (1 - \cos tx)\, dF'_{nj} \\ &= 2(1 - u'_{nj}(t)) \int_J dF'_{nj}. \end{aligned} \tag{4.3}$$

Since $|t| \le T$, J'' contains the interval $[-\pi/T, 0]$. Hence

$$\begin{aligned} \int_{J'} dF'_{nj} &\le \int_{|x|>\pi/T} dF'_{nj} + \int_{0<x\le\pi/T} dF'_{nj} \\ &\le 1 - F'_{nj}(0) + \int_{|x|>\pi/T} dF'_{nj} \le \tfrac{1}{2} + \int_{|x|>\pi/T} dF'_{nj}. \end{aligned}$$

Similarly, because J' contains the interval $[0, \pi/T]$, we have

$$\int_{J''} dF'_{nj} \leq F'_{nj}(-0) + \int_{|x|>\pi/T} dF'_{nj} \leq \tfrac{1}{2} + \int_{|x|>\pi/T} dF'_{nj}.$$

Since $\{F'_{nj}\}$ satisfies (0) because of (α'), we can find $N(T)$ so that

$$\int_{|x|>\pi/T} dF'_{nj} < \tfrac{1}{6}$$

for $n \geq N(T)$, $j \leq n$. Hence for J equal to either J' or J'', we have

$$\int_J dF'_{nj} \leq \tfrac{2}{3} \quad \text{for} \quad n \geq N(T), \quad j \leq n.$$

Substituting into (4.3) and using (4.2), we get (4.1). Q.E.D.

LEMMA 4.2. *We have*

$$\sum_{j=1}^{n} (1 - u'_{nj}(t)) \leq M(T) \quad \text{for} \quad |t| \geq T. \tag{4.4}$$

Proof. By Lemma 4.1 we have

$$-\log (u'^2_{nj} + v'^2_{nj}) \geq 1 - u'^2_{nj} - v'^2_{nj} \geq (1 - u'_{nj})(1 + u'_{nj} - \tfrac{4}{3})$$

for

$$|t| \leq T, \quad n \geq N_1(T), \quad j \leq n.$$

By (α) [which is equivalent to (0)], we have $1 - u'_{nj} < \frac{1}{3}$, whence $u'_{nj} > \frac{2}{3}$ for $|t| \leq T$, $n = N_2(T)$, $j \leq n$. Then

$$-\log (u'^2_{nj} + v'^2_{nj}) \geq \tfrac{1}{3}(1 - u'_{nj}),$$

whence

$$\frac{1}{3} \sum_{j=1}^{n} (1 - u'_{nj}) \leq -2 \sum_{j=1}^{n} \log |f'_{nj}| \quad \text{for} \quad |t| \leq T, \quad n \geq N(T). \tag{4.5}$$

By (β) the right side of (4.5) tends to a limit which is a function bounded in t for $|t| \leq T$. Q.E.D.

LEMMA 4.3. *We have for all* n,

$$\sum_{j=1}^{n} \int_{|x|>\eta} dF'_{nj}(x) \leq M_1(\eta), \qquad \sum_{j=1}^{n} \int_{|x|<\eta} x^2\, dF'_{nj}(x) < M_2(\eta). \tag{4.6}$$

Proof. If η is given, we have, by Lemma 4.2,

$$\sum_{j=1}^{n} \int_{-\infty}^{\infty} (1 - \cos tx)\, dF'_{nj}(x) < K(\eta) \quad \text{for} \quad |t| \leq 2\eta^{-1},$$

whence

$$\sum_{j=1}^{n} \int_{|x|>\eta} (1 - \cos tx)\, dF'_{nj}(x) < K(\eta) \quad \text{for} \quad |t| \leq 2\eta^{-1}.$$

Integrating with respect to t on the interval $(0, 2\eta^{-1})$ we get, as in the proof of Lemma 2.1,

$$\frac{1}{\eta} \sum_{j=1}^{n} \int_{|x|>\eta} dF'_{nj}(x) < \frac{2}{\eta} K(\eta),$$

which is the first assertion in (4.6). To prove the second assertion let η be given and let $c = c(\eta)$ be so determined that $1 - \cos cx > \frac{1}{3}c^2x^2$ for all $|x| < \eta$. Then, by (4.4),

$$M(c) \geq \sum_{j=1}^{n} \int_{-\infty}^{\infty} (1 - \cos cx)\, dF'_{nj}(x) \geq \sum_{j=1}^{n} \int_{|x|<\eta} (1 - \cos cx)\, dF'_{nj}(x)$$

$$\geq \tfrac{1}{3}c^2 \sum_{j=1}^{n} \int_{|x|<\eta} x^2\, dF'_{nj}(x). \qquad \text{Q.E.D.}$$

LEMMA 4.4. *We have*

$$\lim_{n\to\infty} \sum_{j=1}^{n} \left\{ v'^2_{nj}(t) - t^2 \left(\int_{|y|<x} y\, dF'_{nj}(y) \right)^2 \right\} = 0 \quad \text{for every } x > 0. \qquad (4.7)$$

Proof.

$$\sum_{j=1}^{n} \left\{ v'^2_{nj}(t) - t^2 \left(\int_{|y|<x} y\, dF'_{nj}(y) \right)^2 \right\}$$

$$= \sum_{j=1}^{n} \left\{ \left(\int_{-\infty}^{\infty} \sin ty\, dF'_{nj}(y) \right)^2 - \left(\int_{|y|<x} \sin ty\, dF'_{nj}(y) \right)^2 \right\}$$

$$+ \sum_{j=1}^{n} \left\{ \left(\int_{|y|<x} \sin ty\, dF'_{nj}(y) \right)^2 - \left(\int_{|y|<x} ty\, dF'_{nj}(y) \right)^2 \right\}$$

$$= \Sigma'_n + \Sigma''_n;$$

$$\left|\sum_n'\right| = \left|\sum_{j=1}^{n}\left(\int_{-\infty}^{\infty} \sin ty\, dF'_{nj}(y) + \int_{|y|<x} \sin ty\, dF'_{nj}(y)\right)\int_{|y|\geq x} \sin ty\, dF'_{nj}(y)\right|$$

$$\leq 2\sum_{j=1}^{n}\int_{-\infty}^{\infty} |\sin ty|\, dF'_{nj}(y)\int_{|y|\geq x} dF'_{nj}(y)$$

$$\leq 2^{3/2}\sum_{j=1}^{n}(1-u'_{nj}(t))^{1/2}\int_{|y|\geq x} dF'_{nj}(y)$$

$$\leq 2^{3/2}\max_j (1-u'_{nj}(t))^{1/2}\sum_j\int_{|y|\geq x} dF'_{nj}(y).$$

Using (α') and Lemma (4.3), we get $\sum_n' \to 0$. Also,

$$\left|\sum_n''\right| = \left|\sum_j \int_{|y|<x} \sin ty\, dF'_{nj}\int_{|y|<x} (\sin ty - ty)\, dF'_{nj}\right|$$

$$\leq \sum_j \int_{|y|<x} 2|ty|\, dF'_{nj}\int_{|y|<x} \tfrac{1}{2}t^2y^2\, dF'_{nj}$$

$$\leq |t|^3\sum_j\left(\int_{|y|\leq\epsilon} |y|\, dF'_{nj} + \int_{\epsilon<|y|<x} |y|\, dF'_{nj}\right)\int_{|y|<x} y^2\, dF'_{nj}$$

$$\leq |t|^3\sum_j\left(\epsilon + x\int_{|y|\geq\epsilon} dF'_{nj}\right)\int_{|y|<x} y^2\, dF'_{nj}$$

$$\leq |t|^3\left(\epsilon + x\max_j\int_{|y|>\epsilon} dF'_{nj}\right)\sum_j\int_{|y|<x} y^2\, dF'_{nj}.$$

Using (0) and Lemma (4.3), we get

$$\overline{\lim}\left|\sum_n''\right| = |t|^3 M_2(x)\epsilon,$$

whence $\sum_n'' \to 0$. Q.E.D.

Now we return to our original $\{F_{nj}\}$ and $\{f_{nj}\}$. Our hypotheses are now (α) [or its equivalent (0)] and (β). Let μ_{nj} be a median of F_{nj}, that is, $F(\mu_{nj}-0) \leq \frac{1}{2}$ and $1 - F(\mu_{nj}) \leq \frac{1}{2}$. Let $X'_{nj} = X_{nj} - \mu_{nj}$; then $F'_{nj}(x) = P(X'_{nj} \leq x) = F_{nj}(x+\mu_{nj})$, $f'_{nj}(t) = f_{nj}(t)e^{-it\mu_{nj}}$. By definition $\{F'_{nj}\}$ satisfies (M). Also (β) implies that $\{f'_{nj}\}$ satisfies (β'). We shall prove that $\{f'_{nj}\}$ satisfies (α'). This follows from the following lemma.

LEMMA 4.5.

$$\lim_{n\to\infty} \max_{1\le j\le n} |\mu_{nj}| = 0. \tag{4.8}$$

Proof. By (0) we have, for any $\epsilon > 0$,

$$\int_{|x|>\epsilon} dF_{nj} < \tfrac{1}{2} \quad \text{for} \quad n \ge N(\epsilon), \quad j \le n.$$

This implies that $|\mu_{nj}| \le \epsilon$ for $n \ge N(\epsilon), j \le n$. Q.E.D.

That $\{f'_{nj}\}$ satisfies (α') then follows, as in the proof of Lemma 3.3(i). Since all the conditions (α'), (β'), and (M) are satisfied, we shall make use of the Lemmas 4.2, 4.3, and 4.4. We write

$$\mu_n = \mu_{n1} + \cdots + \mu_{nn}, \quad n = 1, 2, 3, \ldots \tag{4.9}$$

LEMMA 4.6. *We have*

$$\lim_{n\to\infty} \sum_{j=1}^{n} \left\{1 - \mu'_{nj}(t) - \tfrac{1}{2}t^2 \left(\int_{|y|<x} y\, dF'_{nj}(y) \right)^2 \right\}$$
$$= \int_{-\infty}^{\infty} (1 - \cos ty) \frac{1 + y^2}{y^2} dG(y) \quad \text{for every } x, \tag{4.10}$$

$$\lim_{n\to\infty} \left(\sum_{j=1}^{n} v'_{nj}(t) + \mu_n t \right) = mt + \int_{-\infty}^{\infty} \sin ty\, dG(y) + \int_{-\infty}^{\infty} \frac{\sin ty - ty}{y^2} dG(y). \tag{4.11}$$

Proof. As is easy to verify, the right sides of (4.10) and (4.11) are, respectively, the real part (with the sign changed) and the imaginery part of the right side of (β). Call them $U(t)$ and $V(t)$. Then, by (β),

$$-\lim_{n\to\infty} \sum_{j=1}^{n} \log f_{nj}(t) = -\lim_{n\to\infty} \left\{ \sum_{j=1}^{n} \log f'_{nj}(t) + i\mu_n t \right\} = U(t) - iV(t), \tag{4.12}$$

t being fixed, we can have N so that

$$|1 - f'_{nj}(t)| < \tfrac{2}{3}, \quad \text{for} \quad n \ge N, \quad j \le n,$$

on account of (α'). Since

$$-\log f'_{nj}(t) = -\log\,(1 - (1 - f'_{nj}(t)))$$
$$= 1 - f'_{nj}(t) + \tfrac{1}{2}(1 - f'_{nj}(t))^2 + \frac{\theta}{3} \frac{|1 - f'_{nj}(t)|^3}{1 - |1 - f'_{nj}(t)|},$$

we have

$$\begin{aligned}-\log f'_{nj} &= 1 - u'_{nj} - iv'_{nj} + \tfrac{1}{2}(1 - u'_{nj} - iv'_{nj})^2 \\ &\quad + \theta|1 - f'_{nj}|\{(1 - u'_{nj})^2 + v'^2_{nj}\} \\ &= 1 - u'_{nj} - \tfrac{1}{2}v'^2_{nj} - iv'_{nj} + \tfrac{1}{2}(1 - u'_{nj})^2 - iv'_{nj}(1 - u'_{nj}) \\ &\quad + 3\theta|1 - f'_{nj}|(1 - u'_{nj}) \\ &= 1 - u'_{nj} - \tfrac{1}{2}v'^2_{nj} - iv'_{nj} + \tfrac{9}{2}\theta|1 - f'_{nj}|(1 - u'_{nj}).\end{aligned}$$

Hence

$$\begin{aligned}-\sum_j \log f'_{nj} &= \sum_j (1 - u'_{nj} - \tfrac{1}{2}v'^2_{nj}) - i\sum_j v'_{nj} \\ &\quad + \tfrac{9}{2}\theta \max_j |1 - f'_{nj}| \sum_j (1 - u'_{nj}).\end{aligned} \tag{4.13}$$

By (α') and Lemma 4.2, the term involving θ tends to zero. Combining (4.13) and (4.12), we get

$$\begin{aligned}\lim_{n\to\infty} \sum_j (1 - u'_{nj} - \tfrac{1}{2}v'^2_{nj}) &= U(t), \\ \lim_{n\to\infty} \left(\sum_j v'_{nj} + \mu_n t\right) &= V(t).\end{aligned} \tag{4.14}$$

Thus (4.11) is proved. Using (4.7) on the first equation in (4.14), we get (4.10). Q.E.D.

LEMMA 4.7. *We have*

$$\left|\sum_{j=1}^{n}\left\{1 - u'_{nj}(t) - \tfrac{1}{2}t^2\left(\int_{|y|<x} y\, dF'_{nj}(y)\right)^2\right\}\right| \le a(x)t^2 + b(x), \quad \text{for every } x > 0, \tag{4.15}$$

$$\left|\sum_{j=1}^{n} v'_{nj}(t) + \mu_n t\right| \le at^2 + b|t| + c, \tag{4.16}$$

where $a(x)$, $b(x)$, a, b, c are constants, the first two depending on x.

Proof.

$$\begin{aligned}\sum_j (1 - u'_{nj}) &= \sum_j \int_{-\infty}^{\infty} (1 - \cos ty)\, dF'_{nj}(y) \\ &\le \sum_j \int_{|y|<1} \tfrac{1}{2}t^2y^2\, dF'_{nj} + \sum_j \int_{|y|>1} 2\, dF'_{nj}.\end{aligned} \tag{4.17}$$

Equation (4.15) now follows from (4.17) and (4.6). Also,

$$\sum_j v'_{nj} + \mu_n t = \sum_j \int\limits_{|y|<1} (\sin ty - ty)\, dF'_{nj} + t \sum_j \int\limits_{|y|<1} y\, dF'_{nj}$$
$$+ \sum_j \int\limits_{|y|>1} \sin ty\, dF'_{nj} + \mu_n t, \tag{4.18}$$

whence, putting $t = 1$,

$$\left| \sum_j \int\limits_{|y|<1} y\, dF'_{nj} + \mu_n \right| \le \left| \sum_j v'_{nj}(1) + \mu_n \right| + \frac{1}{2} \sum_j \int\limits_{|y|<1} y^2\, dF'_{nj}$$
$$+ \sum_j \int\limits_{|y|>1} dF'_{nj}. \tag{4.19}$$

Since $|\sum_j v'_{nj}(1) + \mu_n|$ tends to a limit by (4.11), the right side of (4.19) is bounded. Hence

$$\left| \sum_j \int\limits_{|y|<1} y\, dF'_{nj} + \mu_n \right| < b, \text{ a constant.}$$

Then (4.18) gives

$$\left| \sum_j v'_{nj}(t) + \mu_n t \right| \le \frac{t^2}{2} \sum_j \int\limits_{|y|<1} y^2\, dF'_{nj} + b|t| + \sum_j \int\limits_{|y|>1} dF'_{nj},$$

which implies (4.16), by Lemma 4.6. Q.E.D.

Throughout the following pages we shall adopt the following notation: $g(t) = (\sin t)/t$. Notice that $g(t)$ is the C.F. of the uniform distribution on the interval $(-1, 1)$. We shall use U to denote a R.V. independent of all the X_{nj} and whose distribution is the convolution of three uniform distributions on the interval $(-1, 1)$. It follows that $g^3(\epsilon t)$ is the C.F. of ϵU, $g^3(\epsilon t) f'_{nj}(t)$ is the C.F. of $\epsilon U + X'_{nj}$, and $|U| \le 3$.

Lemma 4.8. *Let $\varphi(z)$ be summable on $(-\infty, \infty)$, and let*

$$\psi(t) = \frac{1}{2\pi} \int\limits_{-\infty}^{\infty} e^{-itz} \varphi(z)\, dz. \tag{4.20}$$

Let ϵ be any constant. Then

$$E\varphi(\epsilon U) - E\varphi(\epsilon U + X'_{nj}) = \int\limits_{-\infty}^{\infty} \psi(t) g^3(\epsilon t) \big(1 - u'_{nj}(t)\big)\, dt \tag{4.21}$$

if φ is an even function

$$E\varphi(\epsilon U + X'_{nj}) = i \int_{-\infty}^{\infty} \psi(t) g^3(\epsilon t) v'_{nj}(t)\, dt \tag{4.22}$$

if φ is an odd function.

Proof. For any summable $\varphi(z)$, we have by (4.20),

$$\int_{-\infty}^{\infty} \psi(t) g^3(\epsilon t) f'_{nj}(t)\, dt = \int_{-\infty}^{\infty} \varphi(z)\, dz \frac{1}{2\pi} \int_{-\infty}^{\infty} e^{-itz} g^3(\epsilon t) f'_{nj}(t)\, dt.$$

But the inner integral represents the density of the distribution of $\epsilon U + X'_{nj}$. Hence

$$\int_{-\infty}^{\infty} \psi(t) g^3(\epsilon t) f'_{nj}(t)\, dt = E\varphi(\epsilon U + X'_{nj}).$$

Similarly

$$\int_{-\infty}^{\infty} \psi(t) g^3(\epsilon t)\, dt = E\varphi(\epsilon U).$$

Hence

$$E\varphi(\epsilon U) - E\varphi(\epsilon U + X'_{nj}) = \int_{-\infty}^{\infty} \psi(t) g^3(\epsilon t) \big(1 - f'_{nj}(t)\big)\, dt.$$

From this follow (4.21) and (4.22) immediately because $\psi(t)$ is even or odd according as $\varphi(z)$ is even or odd. Q.E.D.

LEMMA 4.9. *Let $\varphi(z)$ and $\psi(t)$ be as in Lemma 4.8. Let*

$$\lambda_n(x) = \frac{1}{2} \sum_{j=1}^{n} \left(\int_{|y|<x} y\, dF'_{nj}(y) \right)^2, \qquad k(\epsilon) = i \int_{-\infty}^{\infty} \psi(t) g^3(\epsilon t) t\, dt,$$

$$l(\epsilon) = \int_{-\infty}^{\infty} \psi(t) g^3(\epsilon t) t^2\, dt.$$

Let $\psi(t) = 0(t^{-1})$ for large $|t|$. Then

$$\lim_{n\to\infty} \left[\sum_{j=1}^{n} \{E\varphi(\epsilon U) - E\varphi(\epsilon U + X'_{nj})\} - \lambda_n(x) l(\epsilon) \right]$$

$$= \int_{-\infty}^{\infty} \psi(t) g^3(\epsilon t)\, dt \int_{-\infty}^{\infty} (1 - \cos ty) \frac{1 + y^2}{y^2}\, dG(y) \tag{4.23}$$

if φ is even;

$$\lim_{n\to\infty}\left\{\sum_{j=1}^{n} E\varphi(\epsilon U + X'_{nj}) + \mu_n k(\epsilon)\right\}$$

$$= i\int_{-\infty}^{\infty} \psi(t)g^3(\epsilon t)\left\{mt + \int_{-\infty}^{\infty} \sin ty\, dG(y) + \int_{-\infty}^{\infty} \frac{\sin ty - ty}{y^2}\, dG(y)\right\} dt \tag{4.24}$$

if φ is odd.

Proof. From (4.21) and (4.22), we get immediately

$$\sum_{j=1}^{n} \{E\varphi(\epsilon U) - E\varphi(\epsilon U + X'_{nj})\} - \lambda_n(x)l(\epsilon)$$

$$= \int_{-\infty}^{\infty} \psi(t)g^3(\epsilon t)\left\{\sum_{j=1}^{n} (1 - u'_{nj}(t)) - \lambda_n(x)t^2\right\} dt \tag{4.25}$$

if φ is even;

$$\sum_{j=1}^{n} E\varphi(\epsilon U + X'_{nj}) + \mu_n k(\epsilon) = i\int_{-\infty}^{\infty} \psi(t)g^3(\epsilon t)\left\{\sum_{j=1}^{n} v'_{nj}(t) + \mu_n t\right\} dt \tag{4.26}$$

if φ is odd. By (4.10) and (4.11) the integrands in (4.25) and (4.26) tend to limits as $n \to \infty$. If $\psi(t) = 0(t^{-1})$, then

$$\psi(t)g^3(\epsilon t) = 0(t^{-4}).$$

Hence, by (4.15) and (4.16), these integrands are dominated by summable functions. It follows that the integrals tend to limits equal to the integrals of the respective limiting integrands. Q.E.D.

LEMMA 4.10. *For $x > 3\epsilon$ we have*

$$\frac{x}{\pi}\int_{-\infty}^{\infty} g(xt)g^3(\epsilon t)\, dt = 1, \tag{4.27}$$

$$\frac{x}{\pi}\int_{-\infty}^{\infty} g(xt)g^3(\epsilon t)t^2\, dt = 0, \tag{4.28}$$

$$\frac{x^2}{\pi}\int_{-\infty}^{\infty} g'(xt)g^3(\epsilon t)t\, dt = 1, \tag{4.29}$$

$$\frac{x^3}{\pi}\int_{-\infty}^{\infty} g''(xt)g^3(\epsilon t)t^2\, dt = 2. \tag{4.30}$$

Proof. Equation (4.27) is true because its left side is equal to $P(|\epsilon U| \leq x)$ while $|\epsilon U| \leq 3\epsilon$. Differentiating (4.27) twice with respect to x we get (4.28). Write (4.27) in the following form:

$$\frac{1}{\pi}\int_{-\infty}^{\infty} g(xt)g^3(\epsilon t)\,dt = \frac{1}{x}\cdot$$

Differentiation with respect to x once and twice yield (4.29) and (4.30), respectively. Q.E.D.

We are now in the position to prove the necessity of the conditions (I), (II), and (III). Consider the following particular functions $\varphi(z)$:

$$\varphi_1(z;x) = \begin{cases} 1 & \text{for} \quad |z| \leq x, \\ 0 & \text{for} \quad |z| > x, \end{cases}$$

$$\varphi_2(z;x) = \begin{cases} z^2 & \text{for} \quad |z| < x, \\ 0 & \text{for} \quad |z| \geq x, \end{cases}$$

$$\varphi_3(z;x) = \begin{cases} z & \text{for} \quad |z| < x, \\ 0 & \text{for} \quad |z| \geq x. \end{cases}$$

Correspondingly,

$$\psi_1(t;x) = \frac{x}{\pi}\,g(xt), \qquad \psi_2(t;x) = -\frac{x^3}{\pi}\,g''(xt),$$

$$\psi_3(t;x) = \frac{ix^2}{\pi}\,g'(xt),$$

and, by (4.28) and (4.30),

$$l_1(\epsilon) = 0, \qquad l_2(\epsilon) = -2, \qquad k_3(\epsilon) = 1 \quad \text{for} \quad 3\epsilon < x.$$

We take $\varphi(z)$ to be $\varphi_1(z;x)$ and $\varphi_2(z;x)$ successively in (4.23), and take $\varphi(z)$ to be $\varphi_3(z;x)$ in (4.24). Then we get the following three equations, valid for $x > 3\epsilon$:

$$\lim_n \sum_j P(|\epsilon U + X'_{nj}| > x)$$

$$= \int_{-\infty}^{\infty} \psi_1(t;x)g^3(\epsilon t)\,dt \int_{-\infty}^{\infty} (1 - \cos ty)\,\frac{1+y^2}{y^2}\,dG(y), \tag{4.31}$$

$$\lim_n \left[\sum_j \{E\varphi_2(\epsilon U;x) - E\varphi_2(\epsilon U + X'_{nj};x)\} + 2\lambda_n(x) \right]$$
$$= \int_{-\infty}^{\infty} \psi_2(t;x) g^3(\epsilon t)\, dt \int_{-\infty}^{\infty} (1 - \cos ty) \frac{1 + y^2}{y^2}\, dG(y), \qquad (4.32)$$

$$\lim_n \left\{ \sum_j E\varphi_3(\epsilon U + X'_{nj}) + \mu_n \right\}$$
$$= m + i \int_{-\infty}^{\infty} \psi_3(t;x) g^3(\epsilon t)\, dt \left\{ \int_{-\infty}^{\infty} \sin ty\, dG(y) + \int_{-\infty}^{\infty} \frac{\sin ty - ty}{y^2}\, dG(y) \right\}. \qquad (4.33)$$

There is no restriction in assuming $G(-\infty) = 0$. For convenience we assume that $G(\infty) > 0$, but it is easily seen that the results hold equally true for $G(\infty) = 0$. Let

$$K(\eta) = \int_{-\infty}^{\infty} \frac{1 + u^2}{\eta + u^2}\, dG(u), \qquad H(y;\eta) = \frac{1}{K(\eta)} \int_{-\infty}^{y} \frac{1 + u^2}{\eta + u^2}\, dG(u),$$
$$\eta > 0.$$
$$K_1(\eta) = \int_{-\infty}^{\infty} \frac{1}{\eta + u^2}\, dG(u), \qquad H_1(y;\eta) = \frac{1}{K_1(\eta)} \int_{-\infty}^{y} \frac{dG(u)}{\eta + u^2},$$

Let Z, V, and W be the R.V.'s independent of U and whose D.F.'s are, respectively, $H(y;\eta)$, $H_1(y;\eta)$, and $G(y)/G(\infty)$. Then we may write (4.31), (4.32), (4.33) as follows:

$$\lim_n \sum_j P(|\epsilon U + X'_{nj}| > x)$$
$$= \lim_{\eta \to 0} K(\eta) \int_{-\infty}^{\infty} \psi_1(t;x) g^3(\epsilon t)\, dt \int_{-\infty}^{\infty} (1 - \cos ty)\, dH(y;\eta)$$
$$= \lim_{\eta \to 0} K(\eta) \{E\varphi_1(\epsilon U;x) - E\varphi_1(\epsilon U + Z;x)\}$$
$$= \lim_{\eta \to 0} K(\eta) P(|\epsilon U + Z| > x), \qquad (4.34)$$

$$\lim_n \left[\sum_j \{E\varphi_2(\epsilon U;x) - E\varphi_2(\epsilon U + X'_{nj};x)\} + 2\lambda_n(x) \right]$$
$$= \lim_{\eta \to 0} K(\eta) \int_{-\infty}^{\infty} \psi_2(t;x) g^3(\epsilon t)\, dt \int_{-\infty}^{\infty} (1 - \cos ty)\, dH(y;\eta)$$
$$= \lim_{\eta \to 0} K(\eta) \{E\varphi_2(\epsilon U;x) - E\varphi_2(\epsilon U + Z;x)\}, \qquad (4.35)$$

$$\lim_n \left(\sum_j E\varphi_3(\epsilon U + X'_{nj};x) + \mu_n \right)$$

$$= m + iG(\infty) \int_{-\infty}^{\infty} \psi_3(t;x) g^3(\epsilon t)\, dt \int_{-\infty}^{\infty} \sin ty\, d\,\frac{G(y)}{G(\infty)} + i \lim_{\eta\to 0} K_1(\eta)$$

$$\times \left\{ \int_{-\infty}^{\infty} \psi_3(t;x) g^3(\epsilon t)\, dt \int_{-\infty}^{\infty} \sin ty\, dH_1(y;\eta) \right.$$

$$\left. - \int_{-\infty}^{\infty} \psi_3(t;x) g^3(\epsilon t) t\, dt \int_{-\infty}^{\infty} y\, dH_1(y;\eta) \right\}$$

$$= m + G(\infty) E\varphi_3(\epsilon U + X'_{nj};x) + \lim_{\eta\to 0} K_1(\eta)$$

$$\times \left\{ E\varphi_3(\epsilon U + V;x) - \int_{-\infty}^{\infty} y\, dH_1(y;\eta) \right\}, \tag{4.36}$$

using (4.21), (4.22) (with X'_{nj} replaced by Z, V, or W as the case arises). For any R.V. X, we have

$$P(|\epsilon U + X| > x + 3\epsilon) \le P(|X| > x) \le P(|\epsilon U + X| > x - 3\epsilon).$$

Using this on (4.34) we get, for $6\epsilon < x$,

$$\begin{aligned}
\overline{\lim_n} \sum_j P(|X'_{nj}| > x) &\le \lim_n \sum_j P(|\epsilon U + X_{nj}| > x - 3\epsilon) \\
&= \lim_{\eta\to 0} K(\eta) P(|\epsilon U + Z| > x - 3\epsilon) \\
&\le \lim_{\eta\to 0} K(\eta) P(|Z| > x - 6\epsilon) \\
&= \lim_{\eta\to 0} \int_{|y|>x-6\epsilon} \frac{1+y^2}{\eta + y^2}\, dG(y) \\
&= \int_{|y|>x-6\epsilon} \frac{1+y^2}{y^2}\, dG(y),
\end{aligned}$$

$$\begin{aligned}
\underline{\lim_n} \sum_j P(|X'_{nj}| > x) &\ge \lim_n \sum_j P(|\epsilon U + X'_{nj}| > x + 3\epsilon) \\
&= \lim_{\eta\to 0} K(\eta) P(|\epsilon U + Z| > x + 3\epsilon) \\
&\ge \lim_{\eta\to 0} K(\eta) P(|Z| > x + 6\epsilon) \qquad \textit{(cont.)}
\end{aligned}$$

$$= \lim_{\eta \to 0} \int_{|y| > x + 6\epsilon} \frac{1 + y^2}{\eta + y^2} \, dG(y)$$

$$= \int_{|y| > x + 6\epsilon} \frac{1 + y^2}{y^2} \, dG(y).$$

If both x and $-x$ are continuity points of $G(y)$, we conclude from the above, on letting $\epsilon \to 0$, that

$$\lim_{n \to \infty} \sum_{j=1}^{n} \int_{|y| > x} dF'_{nj}(y) = \int_{|y| > x} \frac{1 + y^2}{y^2} \, dG(y). \tag{4.37}$$

For any R.V. X with D.F. $Q(y)$ and independent of U, it is true that for $3\epsilon < x$, if $V(u)$ denotes the D.F. of U,

$$\begin{aligned} E\varphi_2(\epsilon U;x) &- E\varphi_2(\epsilon U + X'_{nj};x) \\ &= \int_{|\epsilon u| < x} \epsilon^2 u^2 \, dV(u) - \int_{|\epsilon u + y| < x} (\epsilon u + y)^2 \, dQ(y) \, dV(u) \\ &= \epsilon^2 EU^2 - \int_{|y| < x} (\epsilon u + y)^2 \, dQ(y) \, dV(u) \\ &\quad + \theta \int_{x - 3\epsilon \leq |y| \leq x + 3\epsilon} (\epsilon u + y)^2 \, dQ(y) \, dV(u) \\ &= \epsilon^2 EU^2 \int_{|y| \geq x} dQ(y) - \int_{|y| < x} y^2 \, dQ(y) \\ &\quad + \theta(x + 6\epsilon)^2 \int_{x - 3\epsilon \leq |y| \leq x + 3\epsilon} dQ(y), \end{aligned} \tag{4.38}$$

and that

$$\begin{aligned} &E\varphi_3(\epsilon U + X_{nj};x) \\ &= \int_{|\epsilon u + y| < x} (\epsilon u + y) \, dQ(y) \, dV(u) \\ &= \int_{|y| < x} (\epsilon u + y) \, dQ(y) \, dV(u) + \theta \int_{x - 3\epsilon \leq |y| \leq x + 3\epsilon} |\epsilon u + y| \, dQ(y) \, dV(u) \\ &= \int_{|y| < x} y \, dQ(y) + \theta(x + 6\epsilon) \int_{x - 3\epsilon \leq |y| \leq x + 3\epsilon} dQ(y). \end{aligned}$$

Applying the above to (4.35) and (4.36), we have

$$\lim_{n\to\infty}\left[\sum_{j=1}^{n}\left\{\epsilon^2 EU^2\int_{|y|\geq x} dF'_{nj}(y)-\int_{|y|<x} y^2\,dF'_{nj}(y)\right.\right.$$
$$\left.\left.+\,\theta(x+6\epsilon)^2\int_{x-3\epsilon\leq|y|\leq x+3\epsilon} dF'_{nj}(y)\right\}+2\lambda_n(x)\right]$$
$$=\lim_{\eta\to 0}K(\eta)\left\{\epsilon^2 EU^2\int_{|y|\geq x} dH(y;\eta)-\int_{|y|<x} y^2\,dH(y;\eta)\right.$$
$$\left.+\,\theta(x+6\epsilon)^2\int_{x-3\epsilon\leq|y|\leq x+3\epsilon} dH(y;\eta)\right\},\tag{4.39}$$

$$\lim_{n}\left[\sum_{j}\left\{\int_{|y|<x} y\,dF'_{nj}(y)+\theta(x+6\epsilon)\int_{x-3\epsilon\leq|y|\leq x+3\epsilon} dF'_{nj}(y)\right\}+\mu_n\right]$$
$$=m+G(\infty)\left\{\int_{|y|<x} yd\,\frac{G(y)}{G(\infty)}+\theta(x+6\epsilon)\int_{x-3\epsilon\leq|y|\leq x+3\epsilon} d\,\frac{G(y)}{G(\infty)}\right\}$$
$$+\lim_{\eta\to 0}K_1(\eta)\left\{\int_{|y|<x} y\,dH_1(y;\eta)+\theta(x+6\epsilon)\int_{x-3\epsilon\leq|y|\leq x+3\epsilon} dH_1(y;\eta)\right.$$
$$\left.-\int_{-\infty}^{\infty} y\,dH_1(y;\eta)\right\}.\tag{4.40}$$

If $\pm x$, $\pm(x+3\epsilon)$, $\pm(x-3\epsilon)$ are continuity points of $G(x)$, then, on using (4.38), (4.39) and (4.40) are reduced to the following:

$$-\overline{\lim_{n}}\left(\sum_{j}\int_{|y|<x} y^2\,dF'_{nj}(y)-2\lambda_n(x)\right)+\epsilon^2 EU^2\int_{|y|>x}\frac{1+y^2}{y^2}\,dG(y)$$
$$+\,\theta(x+6\epsilon)^2\int_{x-3\epsilon\leq|y|\leq x+3\epsilon}\frac{1+y^2}{y^2}\,dG(y)$$
$$=\lim_{\eta\to 0}\left(\epsilon^2 EU^2\int_{|y|>x}\frac{1+y^2}{\eta+y^2}\,dG(y)-\int_{|y|<x}\frac{y^2}{\eta+y^2}\,(1+y^2)\,dG(y)\right.$$
$$\left.+\,\theta(x+6\epsilon)^2\int_{x-3\epsilon\leq|y|\leq x+3\epsilon}\frac{1+y^2}{\eta+y^2}\,dG(y)\right)$$

(cont.)

$$= -\int_{|y|<x} (1+y^2)\,dG(y) + \epsilon^2 EU^2 \int_{|y|>x} \frac{1+y^2}{y^2}\,dG(y)$$
$$+ \theta(x+6\epsilon)^2 \int_{x-3\epsilon\le|y|\le x+3\epsilon} \frac{1+y^2}{y^2}\,dG(y),$$

$$\overline{\lim_n}\left\{\sum_j \int_{|y|<x} y\,dF'_{nj}(y) + \theta(x+6\epsilon)\int_{x-3\epsilon\le|y|\le x+3\epsilon} \frac{1+y^2}{y^2}\,dG(y) + \mu_n\right\}$$
$$= m + \int_{|y|<x} y\,dG(y) + \theta(x+6\epsilon)\int_{x-3\epsilon\le|y|\le x+3\epsilon} dG(y)$$
$$+ \lim_{\eta\to 0}\left(-\int_{|y|>x} \frac{y}{\eta+y^2}\,dG(y) + \theta(x+6\epsilon)\int_{x-3\epsilon\le|y|\le x+3\epsilon} \frac{dG(y)}{\eta+y^2}\right)$$
$$= m + \int_{|y|<x} y\,dG(y) - \int_{|y|>x} \frac{1}{y}\,dG(y)$$
$$+ \theta(x+6\epsilon)\int_{x-3\epsilon\le|y|\le x+3\epsilon} \frac{1+y^2}{y^2}\,dG(y).$$

Hence

$$\overline{\lim_n}\left(\sum_j \int_{|y|<x} y^2\,dF'_{nj}(y) - 2\lambda_n(x)\right)$$
$$= \int_{|y|<x} (1+y^2)\,dG(y) + 2\theta(x+6\epsilon)^2 \int_{x-3\epsilon\le|y|\le x+3\epsilon} \frac{1+y^2}{y^2}\,dG(y),$$
$$\overline{\lim_n}\left(\sum_j \int_{|y|<x} y\,dF'_{nj}(y) + \mu_n\right) = m + \int_{|y|<x} y\,dG(y) - \int_{|y|>x} \frac{1}{y}\,dG(y)$$
$$+ 2\theta(x+6\epsilon)\int_{x-3\epsilon\le|y|\le x+3\epsilon} \frac{1+y^2}{y^2}\,dG(y).$$

Since the terms involving θ tend to zero with ϵ, we obtain the following:

$$\lim_n \sum_j \left\{\int_{|y|<x} y^2\,dF'_{nj}(y) - \left(\int_{|y|<x} y\,dF'_{nj}(y)\right)^2\right\} = \int_{|y|<x} (1+y^2)\,dG(y), \tag{4.41}$$

$$\lim_n \left(\sum_j \int_{|y|<x} y\,dF'_{nj}(y) + \mu_n\right) = m + \int_{|y|<x} y\,dG(y) - \int_{|y|>x} \frac{1}{y}\,dG(y). \tag{4.42}$$

We shall now deduce (I), (II), and (III), respectively, from (4.38), (4.41), and (4.42); ϵ being chosen, Lemma 4.5 asserts that there is $N(\epsilon)$ so that $\max_j |\mu_{nj}| < \epsilon$ for $n \geq N(\epsilon)$. Then

$$\sum_j \int_{|y|>x} dF'_{nj}(y) = \sum_j \int_{|y+\mu_{nj}|>x} dF'_{nj}(y)$$
$$= \sum_j \int_{|y|>x} dF'_{nj}(y) + \theta \sum_j \int_{x-\epsilon\leq|y|\leq x+\epsilon} dF'_{nj}(y).$$

If $\pm x$, $\pm(x \pm \epsilon)$ are all continuity points of $G(y)$, then by (4.38) we have

$$\overline{\lim_n}\, A_n(x) = A(x) + \theta\big(A(x+\epsilon) - A(x-\epsilon)\big),$$

whence, letting $\epsilon \to 0$, $\lim_n A_n(x) = A(x)$, which proves (I). Again

$$\sum_j \left\{ \int_{|y|<x} y^2\, dF'_{nj}(y) - \left(\int_{|y|<x} y\, dF'_{nj}(y) \right)^2 \right\}$$
$$= \sum_j \left\{ \int_{|y+\mu_{nj}|<x} (y+\mu_{nj})^2\, dF'_{nj}(y) - \left(\int_{|y+\mu_{nj}|<x} (y+\mu_{nj})\, dF'_{nj}(y) \right)^2 \right\}$$
$$= \sum_j \left\{ \int_{|y|<x} y^2\, dF'_{nj}(y) - \left(\int_{|y|<x} y\, dF'_{nj}(y) \right)^2 \right\} + \sum\nolimits_n^{(1)} + \sum\nolimits_n^{(2)}, \tag{4.43}$$

where

$$\sum\nolimits_n^{(1)} = \sum_j \left\{ \int_{|y+\mu_{nj}|<x} (y+\mu_{nj})^2\, dF'_{nj} - \left(\int_{|y+\mu_{nj}|<x} (y+\mu_{nj})\, dF'_{nj} \right)^2 \right\}$$
$$- \sum_j \left\{ \int_{|y|<x} (y+\mu_{nj})^2\, dF'_{nj} - \left(\int_{|y|<x} (y+\mu_{nj})\, dF'_{nj} \right)^2 \right\}$$
$$= \theta \sum_j \int_{x-\epsilon\leq|y|\leq x+\epsilon} (y+\mu_{nj})^2\, dF'_{nj}$$
$$+ \theta \sum_j \left\{ \int_{|y+\mu_{nj}|<x} |y+\mu_{nj}|\, dF'_{nj} + \int_{|y|<x} |y+\mu_{nj}|\, dF'_{nj} \right\}$$
$$\times \int_{x-\epsilon\leq|y|\leq x+} |y+\mu_{nj}|\, dF'_{nj}$$
$$= \theta(x+2\epsilon)^2 \sum_j \int_{x-\epsilon\leq|y|\leq x+\epsilon} dF'_{nj}$$
$$+ \theta(x+\epsilon)(x+2\epsilon) \sum_j \int_{x-\epsilon\leq|y|\leq x+\epsilon} dF'_{nj}\,,$$

$$\sum_n^{(2)} = \sum_j \left\{ \int\limits_{|y|<x} (y + \mu_{nj})^2 \, dF'_{nj} - \left(\int\limits_{|y|<x} (y + \mu_{nj}) \, dF'_{nj} \right)^2 - \int\limits_{|y|<x} y^2 \, dF'_{nj} + \left(\int\limits_{|y|<x} y \, dF'_{nj} \right)^2 \right\}$$

$$= \sum_j \left\{ 2\mu_{nj} \int\limits_{|y|<x} y \, dF'_{nj} + \mu_{nj}^2 \int\limits_{|y|<x} dF'_{nj} - \left(\int\limits_{|y|<x} (2y + \mu_{nj}) \, dF'_{nj} \right) \mu_{nj} \int\limits_{|y|<x} dF'_{nj} \right\}$$

$$= \sum_j \mu_{nj} \int\limits_{|y|<x} (2y + \mu_{nj}) \, dF'_{nj} \int\limits_{|y|\geq x} dF'_{nj}$$

$$= \theta\epsilon(2x + \epsilon) \sum_j \int\limits_{|y|\geq x} dF'_{nj}.$$

Using (4.38) we get

$$\overline{\lim_n} \left| \sum_n^{(1)} \right| \leq (x + 2\epsilon)(3x + 4\epsilon)\big(A(x + \epsilon) - A(x - \epsilon)\big).$$

$$\overline{\lim_n} \left| \sum_n^{(2)} \right| \leq \epsilon(2x + \epsilon) A(x).$$

Hence, letting $\epsilon \to 0$,

$$\lim_n \sum_n^{(1)} = \lim_n \sum_n^{(2)} = 0. \tag{4.44}$$

Equations (4.41), (4.43), and (4.44) give

$$\lim_{n\to\infty} \sum_j \left\{ \int\limits_{|y|<x} y^2 \, dF_{nj}(y) - \left(\int\limits_{|y|<x} y \, dF_{nj} \right)^2 \right\} = B(x),$$

which is (II). Finally,

$$\sum_j \int\limits_{|y|<x} y \, dF_{nj}(y) = \sum_j \int\limits_{|y+\mu_{nj}|<x} (y + \mu_{nj}) \, dF'_{nj}$$

$$= \sum_j \int\limits_{|y|<x} (y + \mu_{nj}) \, dF'_{nj} + \theta \sum_j \int\limits_{x-\epsilon\leq|y|\leq x+\epsilon} |y + \mu_{nj}| \, dF'_{nj}$$

(cont.)

$$= \sum_j \int_{|y|<x} y\, dF'_{nj} + \mu_n - \sum_j \mu_{nj} \int_{|y|\geq x} dF'_{nj}$$

$$+ \theta(x + 2\epsilon) \sum_j \int_{x-\epsilon\leq |y| \leq x+\epsilon} dF'_{nj}$$

$$= \sum \int_{|y|<x} dF'_{nj} + \mu_n + \theta\epsilon \sum_j \int_{|y|\geq x} dF'_{nj}$$

$$+ \theta(x + 2\epsilon) \sum_j \int_{x-\epsilon\leq |y| \leq x+\epsilon} dF'_{nj}.$$

Hence, by (4.42)

$$\overline{\underline{\lim}}\, C_n(x) = C(x) + \theta\epsilon A(x) + \theta(x + 2\epsilon)\big(A(x + \epsilon) - A(x - \epsilon)\big).$$

Letting $\epsilon \to 0$, we get

$$\lim_n C_n(x) = C(x)$$

This proves (III). Q.E.D.

The proof of the necessity part of the theorem is now complete.

5. Some Special Cases

We shall consider a few interesting cases in which $A(x)$, $B(x)$, and $C(x)$ can be explicitly computed.

A) $G(x) = 0$ for $x < 0$ and 1 for $x > 0$, $m = 0$. Then

$$A(x) = 0, \qquad B(x) = 1, \qquad C(x) = 0.$$

This is the central limit theorem, for which $f(t) = e^{-t^2/2}$.

B) $G(x) = 0$, $m = 0$. Then

$$A(x) = 0, \qquad B(x) = 0, \qquad C(x) = 0.$$

This is the weak law of large numbers, for which $f(t) = 1$.

C) $G(x) = 0$ for $x < 1$ and $\frac{1}{2}a$ for $x > 1$, $m = \frac{1}{2}a$, where $a > 0$. Then

$$A(x) = \begin{cases} a & \text{for} \quad 0 < x < 1, \\ 0 & \text{for} \quad x > 1, \end{cases} \qquad B(x) = C(x) = \begin{cases} 0 & \text{for} \quad 0 < x < 1, \\ a & \text{for} \quad x > 1. \end{cases}$$

This is the Poisson case, where $f(t) = e^{a(e^{it}-1)}$.

D) $G(x) = \lambda\int_0^x (ye^{-y}\,dy)/(1+y^2)$ for $x > 0$ and $G(x) = 0$ for $x < 0$,

$$m = \lambda\int_0^\infty (e^{-x}\,dx)/(1+x^2),$$

where $\lambda > 0$. Then

$$A(x) = \lambda\int_x^\infty \frac{e^{-y}}{y}\,dy, \qquad B(x) = \lambda - \lambda(1+x)e^{-x}, \qquad C(x) = \lambda(1-e^{-x}).$$

In this case,

$$f(t) = (1-it)^{-\lambda}, \qquad F(x) = \frac{1}{\Gamma(\lambda)}\int_0^x y^{\lambda-1}e^{-y}\,dy, \quad x > 0.$$

REFERENCES

[1] W. Feller, "Über den zentralen Grenzwertsatz der Wahrscheinlichkeitsrechnung," Math. Z. **40,** 521–559 (1935).

[2] V. Glivenko, "Sul teorema limite della teoria delle funjioni caratteristische," Giornale d. Istituto Italiano d. Attuari **7,** 160–167 (1936).

[3] B. Gnedenko, "Über die Konvergenz der Verteilungsgesetze von Summen voneinander unabhängiger Summanden," Comptes Rendus U.R.S.S. **18,** 231–234 (1938).

[4] A. Khintchine, "Zur Theorie der unbeschränkt teilbaren Verteilungsgesetze," Recueil Math. **2(44)**, 79–119 (1937).

[5] P. Lévy, Calcul des probabilités, Gauthier-Villars, Paris, 1925.

[6] P. Lévy, Théorie de l'addition des variables aléatoires, Gauthier-Villars, Paris, 1937.

[7] J. Marcinkiewicz, "Quelques théorèmes de la théorie des probabilités," Travaux de la Société des Sciences et des Lettres de Wilno, Classe des Sciences Mathématique et Naturelles, **13,** 1–13 (1939).

[8] J. Marcinkiewicz, "Sur les fonctions indépendantes. I. Fund. Math. **30,** 202–214 (1938).

[9] J. Marcinkiewicz, "Sur les fonctions indépendantes. II. Fund. Math. **30,** 349–364 (1938).

Reprinted from
Proc. Berkeley Symp. Math. Statist. Probability, 359–402,
University of California Press, Berkeley and Los Angles, 1949.

THE LIMITING DISTRIBUTION OF FUNCTIONS OF SAMPLE MEANS AND APPLICATION TO TESTING HYPOTHESES

P. L. HSU

NATIONAL UNIVERSITY OF PEKING

Introduction

In 1935 J. L. Doob published a paper **[2]**[1] in which he derived the limiting distribution of a function of four sample means from one homogeneous sample. This work is susceptible to an easy generalization and supplies a powerful weapon with which to find the limiting distribution of a vast number of statistics. But since publication its importance seems to have been overlooked. A generalization of Doob's theorem to any number of sample means was given by the author **[7]**.

In the first part of this paper two theorems are proved which embody a further generalization of Doob's result to the case of several samples of different sizes, and numerous examples are given to illustrate their wide applicability.

These examples are confined to the limiting distributions of given statistics, but in the second part a much more important constructive application is made. Two hypotheses of a general character, concerning one sample and several samples respectively, are formulated, and a systematic method of constructing a test function for each hypothesis included in the two general ones is given. The construction is done in such a manner that, as a consequence of the results obtained in the first part, (i) the test function has for its limiting distribution the χ^2 distribution with a known degree of freedom when the hypothesis tested is true, and (ii) the power of the test tends in general to unity as its limit. Special hypotheses and their large sample tests are treated as examples in the second part of the paper.

I

The limiting distribution of functions of sample means

1. *The mathematical model of* k *samples.*—Let there be given k random vectors of m components each,

$$\mathbf{u}_a = [U_{1a}, U_{2a}, \cdots, U_{ma}], \qquad a = 1, \cdots, k, \tag{1}$$

possessing finite second moments. Let

$$E(U_{ia}) = \mu_{ia}, \qquad E(U_{ia}U_{ja}) - \mu_{ia}\mu_{ja} = \eta_{ija}. \tag{2}$$

[1] Boldface numbers in brackets refer to references at the end of the paper (p. 402).

By a sample of size N_a of $\mathbf{u}_a$ is meant a system of N_a mutually independent random vectors,

$$\mathbf{u}_{ar} = [U_{1ar}, U_{2ar}, \cdots, U_{mar}], \qquad r = 1, \cdots, N_a, \tag{3}$$

each of which is distributed the same as U_a. Thus the k vectors (1) give rise to k samples, namely, the vectors (3) wherein a takes the values $1, \cdots, k$. The total number of such vectors is

$$N = N_1 + N_2 + \cdots + N_k.$$

We shall also assume that two vectors belonging to two different samples are always independent. Then about the distribution of the N vectors $\mathbf{u}_{ar}$ we know the following facts: (i) $\mathbf{u}_{ar}$ and $\mathbf{u}_{\beta s}$ are independent if either $a \neq \beta$ or $r \neq s$; (ii) for every fixed a the vectors $\mathbf{u}_{ar}$ ($r = 1, \cdots, N_a$) are equi-distributed; (iii) each $\mathbf{u}_{ar}$ has finite moments of the first two orders given by (2).

2. *Sample mean and normalized sample mean.*—If U is any random variable and if $U_1, \cdots, U_n$ are a sample of size n, we shall term the quantities $\bar{U} = \frac{1}{n}(U_1 + \cdots + U_n)$ and $n^{\frac{1}{2}}\{\bar{U} - E(U)\}$ the sample mean and the normalized sample mean of U respectively. Thus the samples (3) give rise to the sample means

$$\bar{U}_{ia} = \frac{1}{N_a} \sum_{r=1}^{N_a} U_{iar}$$

and the normalized sample means

$$Z_{ia} = N^{\frac{1}{2}}_a (\bar{U}_{ia} - \mu_{ia}).$$

Hence

$$\bar{U}_{ia} = \mu_{ia} + N_a^{-\frac{1}{2}} Z_{ia}. \tag{4}$$

We recall here the well-known central limit theorem:[2]

As $N_a \to \infty$, *the distribution law of the vector* $[Z_{1a}, \cdots, Z_{ma}]$ *tends to the* m-*dimensional normal law with zero means and the dispersion matrix* $[\eta_{ija}]$.

3. *The statistic* T.—Consider a function of mk real variables,

$$f(x_{11}, \cdots, x_{m1}; \cdots; x_{1k}, \cdots, x_{mk}), \tag{5}$$

defined in the whole mk-dimensional space and possessing continuous derivatives of every kind of order two or three, as the case may be, in the neighborhood

$$| x_{ia} - \mu_{ia} | \leqq \delta, \qquad i = 1, \cdots, m; a = 1, \cdots, k. \tag{6}$$

[2] Cf. Cramér [1], Chap. 10, theorem 20-a.

Write

$$f_{ia} = \frac{\partial f}{\partial x_{ia}}, \qquad f_{ija\beta} = \frac{\partial^2 f}{\partial x_{ia}\partial x_{j\beta}}, \qquad f_{ijha\beta\gamma} = \frac{\partial^3 f}{\partial x_{ia}\partial x_{j\beta}\partial x_{h\gamma}},$$

$$a = f(\mu_{11}, \cdots, \mu_{m1}; \cdots; \mu_{1k}, \cdots, \mu_{mk}),$$

$$b_{ia} = f_{ia}(\mu_{11}, \cdots, \mu_{m1}; \cdots; \mu_{1k}, \cdots, \mu_{mk}),$$

$$c_{ija\beta} = f_{ija\beta}(\mu_{11}, \cdots, \mu_{m1}; \cdots; \mu_{1k}, \cdots, \mu_{mk}).$$

If in f each argument x_{ia} is replaced by $\bar{U}_{ia}$, the result is a statistic,

$$T = f(\bar{U}_{11}, \cdots, \bar{U}_{m1}; \cdots; \bar{U}_{1k}, \cdots, \bar{U}_{mk}). \tag{7}$$

By (4) we have

$$T = f(\mu_{11} + N_1^{-\frac{1}{2}} Z_{11}, \cdots; \cdots; \cdots, \mu_{mk} + N_k^{-\frac{1}{2}} Z_{mk}). \tag{8}$$

The main purpose of the first part of this paper is to derive the limiting distribution of T when the sample sizes become infinite simultaneously. It is necessary to impose a restriction on the manner in which these sizes grow. We put

$$N_a = Ng_a, \qquad a = 1, \cdots, k; \qquad g_1 + \cdots + g_k = 1, \tag{9}$$

regard the g_a as fixed, and allow N to grow indefinitely. The method is based on the Taylor expansion of (8) in the neighborhood of

$$| N_a^{-\frac{1}{2}} Z_{ia} | \leqq \delta, \qquad i = 1, \cdots, m; \; a = 1, \cdots, k. \tag{10}$$

If all the second derivatives exist and are continuous in (6), then in (10) we have

$$T = a + N^{-\frac{1}{2}} R + N^{-1} \sum_{i,j,a,\beta} \varphi_{ija\beta} Z_{ia} Z_{j\beta}, \tag{11}$$

where

$$R = \sum_{i,a} g_a^{-\frac{1}{2}} b_{ia} Z_{ia},$$

$$\varphi_{ija\beta} = \tfrac{1}{2}(g_a g_\beta)^{-\frac{1}{2}} f_{ija\beta}(\mu_{11} + \theta N_1^{-\frac{1}{2}} Z_{11}, \cdots; \cdots; \cdots, \mu_{mk} + \theta N_k^{-\frac{1}{2}} Z_{mk}), \qquad |\theta| \leqq 1.$$

Again, if all the third derivatives exist and are continuous in (6), then in (10) we have

$$T = a + N^{-\frac{1}{2}} R + N^{-1} S + N^{-\frac{3}{2}} \sum_{i,j,h,a,\beta,\gamma} \varphi_{ijha\beta\gamma} Z_{ia} Z_{j\beta} Z_{h\gamma}, \tag{12}$$

where

$$S = \tfrac{1}{2} \sum_{i,j,\alpha,\beta} (g_\alpha g_\beta)^{-\frac{1}{2}} c_{ij\alpha\beta} Z_{i\alpha} Z_{j\beta},$$

$$\varphi_{ijh\alpha\beta\gamma} = \tfrac{1}{6} (g_\alpha g_\beta g_\gamma)^{-\frac{1}{2}} f_{ijh\alpha\beta\gamma} (\mu_{11} + \theta N_1^{-\frac{1}{2}} Z_{11}, \cdots ; \cdots ; \cdots , \mu_{mk} + \theta N_k^{-\frac{1}{2}} Z_{mk}), \qquad |\theta| \leqq 1.$$

4. *The limiting distributions of* R *and* S.—In view of the central limit theorem (sec. 2) and the independence of the vectors $[Z_{1\alpha}, \cdots, Z_{m\alpha}]$ for different values of α, we obtain immediately the following lemma:

Lemma 1. *As* $N \to \infty$, *the distribution law of* R *tends to the limit*

$$\frac{1}{\sqrt{2\pi}} \int_{-\infty}^{\frac{x}{\sigma}} e^{-\frac{1}{2}y^2} dy, \tag{13}$$

where

$$\sigma^2 = \sum_{i,j,\alpha} g_\alpha^{-1} b_{i\alpha} b_{j\alpha} \eta_{ij\alpha}, \tag{14}$$

provided $\sigma^2 \neq 0$. *If* $\sigma^2 = 0$, *then* R $= 0$ *with unit probability.*

On the same ground we conclude:

Lemma 2. *As* $N \to \infty$, *the distribution law of* S *tends to a limit which is the distribution law of the quadratic form*

$$\sum_{i,j,\alpha,\beta} (g_\alpha g_\beta)^{-\frac{1}{2}} c_{ij\alpha\beta} W_{i\alpha} W_{j\beta}, \tag{15}$$

where the $W_{i\alpha}$ *are normal variates having zero means and the same second moments as the variables* $U_{i\alpha} - \mu_{i\alpha}$.

It turns out that the limiting distribution of S is the distribution of a certain quadratic form in normal variates. In most of the actual cases that we encounter this form is semi-definite. Hence we shall complete the solution of the limiting distribution of S by a lemma, given in the next section, about the distribution of semi-definite quadratic forms in normal variates.

5. *Distribution of semi-definite quadratic forms in normal variates.*—Suppose that a semi-definite form Q in normal variates with zero means is reduced in any manner to a sum of squares,[3]

$$Q = W_1^2 + W_2^2 + \cdots + W_q^2, \tag{16}$$

where the W's are themselves normal variates with zero means. Let

$$E(W_i W_j) = \omega_{ij}. \tag{17}$$

Let the dispersion matrix $[\omega_{ij}]$ be of rank $\rho > 0$ and let its non-vanishing latent roots, which are necessarily positive, be $\lambda_1, \cdots, \lambda_\rho$. Then it is always possible to apply an orthogonal transformation on $W_1, \cdots, W_q$ to get a new

[3] If Q is negative, we have only to give the right-hand side of (16) a minus sign.

set of normal variates with zero means, $W'_1, \cdots, W'_q$, such that $Q = W'^2_1 + \cdots + W'^2_q$ and

$$(18) \qquad E(W'_iW'_j) = 0, (i \neq j); \qquad E(W'^2_i) = \lambda_i, (i = 1, \cdots, \rho); \quad E(W'^2_i) = 0, (i = \rho + 1, \cdots, q).$$

The last equation of (18) implies that all the W'_i $(i>\rho)$ vanish with unit probability. Hence Q is essentially equal to $W'^2_1 + \cdots + W'^2_\rho$ and so its distribution law is

$$(19) \quad (2\pi)^{-\frac{1}{2}\rho}(\lambda_1 \cdots \lambda_\rho)^{-\frac{1}{2}} \int_{y_1^2+\cdots+y_\rho^2 \leq z} \exp\left(-\frac{y_1^2}{2\lambda_1} - \cdots - \frac{y_\rho^2}{2\lambda_\rho}\right) dy_1 \cdots dy_\rho.$$

If, further, the relations

$$(20) \qquad \sum_{h=1}^{q} \omega_{ih}\,\omega_{jh} = \omega_{ij}, \qquad i,j = 1, \cdots, q,$$

are satisfied by the ω_{ij}, then $[\omega_{ij}]^2 = [\omega_{ij}]$ and so all the λ_i are unity. Then (19) reduces to the familiar χ^2 distribution with ρ degrees of freedom,

$$\left\{2^{\frac{1}{2}\rho}\,\Gamma(\tfrac{1}{2}\rho)\right\}^{-1} \int_0^z y^{\frac{1}{2}\rho-1} e^{-\frac{1}{2}y} dy.$$

In this case it is also easy to find ρ. In fact, $\rho = \Sigma\lambda_i = \Sigma\omega_{ii}$.

We have therefore established the following lemma:

Lemma 3. *The distribution law of* Q *is* (19) *in general. If, in particular, the relations* (20) *are satisfied, then the distribution is the* χ^2 *distribution with* ρ *degrees of freedom, where* $\rho = \omega_{11} + \cdots + \omega_{qq}$.

6. *Limiting distribution of* T.—We shall use $\bar{E}$ to denote the negation of an event E, $(E_1; E_2)$ the conjunction of two events E_1 and E_2, and $P(E)$ the probability of E.

Theorem 1. *If the function* f *in* (5) *possesses continuous second derivatives of every kind in the neighborhood* (6), *then the limiting distribution of* $N^{\frac{1}{2}}(T-a)$ *is the same as the limiting distribution of* R. *Consequently this limit is the normal law* (13), *provided the quantity* σ^2 *in* (14) *does not vanish.*

Proof. We have seen that, when $Z_{i\alpha}$ satisfy the inequalities (10), T may be expressed as in (11), namely,

$$(21) \qquad T = a + N^{-\frac{1}{2}} R + N^{-1} R_1,$$

where $N^{-1}R_1$ denotes the last term in (11). Let us denote by E the event that all the inequalities (10) are true, and by $F(x)$ the distribution law of $N^{\frac{1}{2}}(T-a)$. Then

$$F(x) = P\{N^{\frac{1}{2}}(T-a) \leq x; E\} + P\{N^{\frac{1}{2}}(T-a) \leq x; \bar{E}\}.$$

But

$$P(E_1;\bar{E}) \leqq P(\bar{E}) \leqq \sum_{i,a} P(Z_{ia}^2 \geqq N_a\delta^2) \leqq \frac{1}{N\delta^2} \sum_{i,a} g_a^{-1}\eta_{iia} = o(1) \tag{22}$$

for every event E_1; hence

$$F(x) = P\{N^{\frac{1}{2}}(T-a) \leqq x; E\} + o(1). \tag{23}$$

Using (21), we have

$$N^{\frac{1}{2}}(T-a) = R + N^{-\frac{1}{2}}R_1 \text{ in conjunction with E.}$$

Besides, in the neighborhood (10) the functions $\varphi_{ija\beta}$ are continuous and therefore bounded. Let A be a common upper bound of the absolute values of all these functions. Then we have

$$|R_1| \leqq A\left(\sum_{i,a}\left|Z_{ia}\right|\right)^2 \text{ in conjunction with } E.$$

Hence

$$P\left\{R + N^{-\frac{1}{2}}A\left(\sum_{i,a}\left|Z_{ia}\right|\right)^2 \leqq x; E\right\} \leqq P\left\{N^{\frac{1}{2}}(T-a) \leqq x; E\right\} \leqq P\left\{R - N^{-\frac{1}{2}}A\left(\sum_{i,a}\left|Z_{ia}\right|\right)^2 \leqq X; E\right\}.$$

Using (22), we get

$$P\left\{R + N^{-\frac{1}{2}}A\left(\sum_{i,a}\left|Z_{ia}\right|\right)^2 \leqq x\right\} + o(1) \leqq F(x) \leqq P\left\{R - N^{-\frac{1}{2}}A\left(\sum_{i,a}\left|Z_{ia}\right|\right)^2 \leqq x\right\} + o(1). \tag{24}$$

Now it has been shown by Doob [2] that if X has a limiting distribution and if Y tends to zero in probability, then $X + Y$ has the same distribution as X. This theorem may be applied to the two extreme terms in (24), because evidently $N^{-\frac{1}{2}}(\Sigma|Z_{ia}|)^2$ tends to zero in probability. Hence both these terms are equal to $P(R \leqq x) + o(1)$ and consequently

$$F(x) = P(R \leqq x) + o(1), \quad \text{q.e.d.}$$

Theorem 2. *If the function* f *in* (5) *possesses continuous third derivatives of every kind in the neighborhood* (6), *and if quantity* σ^2 *in* (14) *vanishes, then the limiting distribution of* N(T–a) *is the same as the limiting distribution of* S. *Consequently this limit is the distribution law of the quadratic form* (15).

We shall merely sketch the proof, which is similar to that of theorem 1. Denoting the distribution law of $N(T-a)$ by $F_1(x)$ we have, as analogy of (23),

$$F_1(x) = P\{N(T - a) \leqq x; E\} + o(1).$$

In conjunction with E, T may be expressed as in (12), whereby the second term may be dropped, since now $\sigma^2 = 0$ and so R is essentially zero. Hence

$$N(T - a) = S + N^{-\frac{1}{2}}S_1,$$

where $N^{-\frac{1}{2}} S_1$ denotes the last term in (12). As before, we have

$$\left|S_1\right| \leqq B\left(\sum_{i,a}\left|Z_{ia}\right|\right)^3$$

in conjunction with E, where B is some constant. Then we obtain the analogy of (24),

$$P\left\{S + N^{-\frac{1}{2}}B\left(\sum_{i,a}\left|Z_{ia}\right|\right)^3 \leqq x\right\} + o(1) \leqq F_1(x)$$
$$\leqq P\left\{S - N^{-\frac{1}{2}}B\left(\sum_{i,a}\left|Z_{ia}\right|\right)^3 \leqq x\right\} + o(1),$$

which leads as before to the result

$$F_1(x) = P(S \leqq x) + o(1), \quad \text{q.e.d.}$$

With the help of lemma 3 the limiting distribution of T is completely solved, provided the quadratic form S is semi-definite.

Let us summarize the results contained in theorems 1 and 2: In order to obtain the limiting distribution of T, which is a function of the sample means $\bar{U}_{ia}$, make the substitution (4) and compute the Taylor expansion in powers of $N^{-\frac{1}{2}}$ to three terms,

$$a + N^{-\frac{1}{2}}R + N^{-1}S. \tag{25}$$

If the quantity σ^2 in (14) does not vanish, the limiting distribution of $N^{\frac{1}{2}}(T-a)$ is the normal distribution with mean zero and variance σ^2. If $\sigma^2 = 0$, then $N(T-a)$ has the same limiting distribution as that of S, and this latter is the distribution of a certain quadratic form in normal variates. If the form in question is semi-definite, the explicit formula of its distribution law is given in lemma 3.

In what follows, when we are dealing with cases of a single sample ($k = 1$), we shall drop the index a from all the letters.

7. *Probabilities of events.*—Consider a set of events, $E_1, \cdots, E_m$, forming a complete disjunction and having the probabilities $p_1, \cdots, p_m$. Let X_i be

the random variable such that $X_i = 1$ or zero according as E_i happens or does not happen. Then we have a random vector $[X_1, \cdots, X_m]$ with

$$(26) \qquad E(X_i) = p_i, \qquad E(X_i^2) = p_i, \qquad E(X_i X_j) = 0, \qquad i \neq j.$$

A sample of size N corresponds to N trials of experiment, and the sample means $\overline{X}_1, \cdots, \overline{X}_m$ are the relative frequencies $n_1/N, \cdots, n_m/N$, where n_i denotes the number of happenings of E_i in N trials. The quantity σ^2 in (14) has a simple expression. We have, by (26),

$$(27) \qquad \sigma^2 = \sum_i b_i^2 p_i(1-p_i) - \sum_{i \neq j} b_i b_j p_i p_j = \sum_i p_i b_i^2 - \left(\sum_i p_i b_i\right)^2$$
$$= \sum_i p_i (b_i - \sum_i p_i b_i)^2.$$

8. *Example 1: The* χ^2 *statistic.*—This classical statistic is defined as

$$T_1 = \sum_{i \neq 1}^{m} \frac{(n_i - p_i^0 N)^2}{p_i^0 N}$$

and is used to test the hypothesis that $p = p_i^0$, $(i = 1, \cdots, m)$. As explained in section 7, we have $n_i = N\overline{X}_i$. Hence

$$\frac{T_1}{N} = \sum_{i=1}^{m} \frac{(\overline{X}_i - p_i^0)}{p_i^0}.$$

The expansion (25) of T_1/N is

$$a + N^{-\frac{1}{2}} \sum_{i=1}^{m} b_i Z_i + N^{-1} \sum_{i=1}^{m} \frac{Z_i^2}{p_i^0},$$

where

$$a = \sum_{i=1}^{m} \frac{(p_i - p_i^0)^2}{p_i^0}, \qquad b_i = \frac{p_i - p_i^0}{p_i^0}.$$

If the hypothesis is false, $p_i \neq p_i^0$ for some i. Then, by (27), the quantity σ^2 in (14) takes the value

$$\sigma_1^2 = \sum_{i=1}^{m} p_i \left(b_i - \sum_{i=1}^{m} p_i b_i\right)^2 \neq 0.$$

For, if $\sigma_1^2 = 0$, b_i would be independent of i and so $p_i = \lambda p_i^0$ for all i. Since $\Sigma p_i = 1 = \Sigma p_i^0$, we would have $p_i = p_i^0$ for all i, contrary to our assumption. Hence the limiting distribution of $N^{\frac{1}{2}}(N^{-1}T_1 - a)$ is the normal distribution with mean zero and variance σ_1^2.

If the hypothesis is true, then a and all the b_i vanish. Hence the limiting distribution of T_1 is the same as that of

$$\sum_{i=1}^{m} \left(\frac{Z_i}{\sqrt{p_i^0}}\right)^2.$$

This limit is the distribution law of $W_1^2 + \cdots + W_m^2$, where $[W_1, \cdots, W_m]$ is a normal vector having zero means and the same dispersion matrix as the vector

$$\left[\frac{X_1 - p_1^0}{\sqrt{p_1^0}}, \cdots, \frac{X_m - p_m^0}{\sqrt{p_m^0}}\right].$$

Hence

$$\omega_{ij} = E(W_iW_j) = 1 - p_i^0, \ (i = j), \ = -\sqrt{p_i^0 p_j^0}, \ (i \neq j).$$

It is easy to verify that the relations (20) are satisfied, and that $\Sigma\omega_{ii} = m - 1$. Hence the limiting distribution of T_1 is the χ^2 distribution with m–1 degrees of freedom.

9. *Example 2: The mean square contingency.*—Let $E_1, \cdots, E_s$ and $E'_1, \cdots, E'_t$ be two sets of events, each forming a complete disjunction. Then the st events $E_{ij} = (E_i; E'_j)$ form a complete disjunction. Let

$$P(E_{ij}) = p_{ij}, \qquad P(E_i) = \sum_j p_{ij} = p_i, \qquad P(E'_j) = \sum_i p_{ij} = p'_j.$$

Let n_{ij} be the number of occurrences of E_{ij} in N trials, and let

$$n_i = \sum_j n_{ij}, \qquad n'_j = \sum_i n_{ij}.$$

The mean square contingency is defined as

$$T_2 = N \sum_{i=1}^{s} \sum_{j=1}^{t} \frac{\left(n_{ij} - \dfrac{n_i n'_j}{N}\right)^2}{n_i n'_j}.$$

It is used to test the hypothesis of complete independence of the two sets of events, that is, that $p_{ij} = p_i p'_j$ for all i and j.

We define st random variables X_{ij} such that $X_{ij} = 1$ or zero according as E_{ij} happens or does not happen. A sample of size N gives the sample means

$$\overline{X}_{ij} = \frac{n_{ij}}{N}.$$

Let also

$$\overline{X}_i = \sum_j \overline{X}_{ij} = \frac{n_i}{N}, \qquad \overline{X}'_j = \sum_i \overline{X}_{ij} = \frac{n'_j}{N}.$$

Then

$$\frac{T_2}{N} = \sum_{i,j} \frac{(\overline{X}_{ij} - \overline{X}_i\overline{X}'_j)^2}{\overline{X}_i\overline{X}'_j}.$$

Upon substituting $p_{ij} + N^{-\frac{1}{2}}Z_{ij}$ for $\overline{X}_{ij}$ we obtain the three-term expansion of T_2/N,

$$a + N^{-\frac{1}{2}}R + N^{-1}S,$$

where

$$a = \sum_{i,j} \frac{(p_{ij} - p_i p'_j)^2}{p_i p'_j},$$

$$R = \sum_{i,j} \frac{p_{ij} - p_i p'_j}{p_i^2 p'^2_j} \left\{ 2p_i p'_i Z_{ij} - (p_{ij} + p_i p'_i)(p_i Z'_j + p'_j Z_i) \right\},$$

$$S = \sum_{i,j} \left\{ \frac{(Z_{ij} - p_i Z'_j - p'_j Z_i)^2}{p_i p'_j} + (p_{ij} - p_i p'_j) Q_{ij} \right\},$$

$$Z_i = \sum_j Z_{ij}, \qquad Z'_j = \sum_i Z_{ij},$$

and the Q_{ij} are certain quadratic forms in the Z_{ij}.

We have

$$R = \sum_{i,j} b_{ij} Z_{ij},$$

where

$$b_{ij} = \frac{2p_{ij}}{p_i p_j} - \frac{1}{p_i^2} \sum_{\nu=1}^{t} \frac{p^2_{i\nu}}{p'_\nu} - \frac{1}{p'^2_j} \sum_{\mu=1}^{s} \frac{p^2_{\mu j}}{p_\mu}.$$

According to (27) the quantity σ^2 in (14) has the value

$$\sigma_2^2 = \sum_{i,j} p_{ij} \left(b_{ij} - \sum_{i,j} p_{ij} b_{ij} \right)^2.$$

But

$$\sum_{i,j} p_{ij} b_{ij} = 2 \sum_{i,j} \frac{p^2_{ij}}{p_i p'_j} - \sum_{i,\nu} \frac{p^2_{i\nu}}{p_i p'_\nu} - \sum_{\mu,j} \frac{p^2_{\mu j}}{p_\mu p'_j} = 0.$$

Hence

$$\sigma_2^2 = \sum_{i,j} p_{ij} b^2_{ij}.$$

If the hypothesis is false, $p_{ij} \neq p_i p'_j$ for some (i, j). Then $\sigma_2^2 \neq 0$. For σ_2^2 can vanish only when all the $b_{ij} = 0$, and this implies that

$$0 = \sum_{i,j} p_i p'_j b_{ij} = 2 - \sum_{i,\nu} \frac{p^2_{i\nu}}{p_i p'_\nu} - \sum_{\mu,j} \frac{p^2_{\mu j}}{p_\mu p'_j} = 2 - 2 \sum_{i,j} \frac{p^2_{ij}}{p_i p'_j}$$
$$= -2 \sum_{i,j} \frac{(p_{ij} - p_i p'_j)^2}{p_i p'_j},$$

that is, $p_{ij} = p_i p'_j$ for all (i,j). Therefore in this case the limiting distribution of $N^{\frac{1}{2}}(N^{-1}T - a)$ is the normal distribution about zero with variance σ_2^2.

If the hypothesis is true, $p_{ij} = p_i p'_j$ for all (i, j). Then a and all the b_{ij} vanish. Hence the limiting distribution of T_2 is the distribution of

$$\sum_{i,j} W_{ij}^2,$$

where the W_{ij} are normal variates with zero means and the same second moments as the system

$$\frac{X_{ij}-p_{ij}-p_i(X'_j-p'_j)-p'_j(X_i-p_i)}{\sqrt{p_i p'_j}} \qquad X_i = \sum_j X_{ij}, \qquad X'_j = \sum_i X_{ij}\,.$$

Direct computation gives the values of these moments.:

$$\omega_{ij\mu\nu} = E(W_{ij}W_{\mu\nu}) = \begin{cases} (1-p_i)(1-p'_j) & \text{if } i=\mu, j=\nu, \\ -(1-p_i)\sqrt{p'_j p'_\nu} & \text{if } i=\mu, j\neq\nu, \\ -(1-p'_j)\sqrt{p_i p_\mu} & \text{if } i\neq\mu, j=\nu, \\ \sqrt{p_i p_\mu p'_j p'_\nu} & \text{if } i\neq\mu, j\neq\nu. \end{cases}$$

It may easily be verified that relations (20) are satisfied:

$$\sum_{g,h} \omega_{ijgh}\omega_{\mu\nu gh} = \omega_{ij\mu\nu}, \qquad i,\mu = 1,\cdots,s; \qquad j,\nu = 1,\cdots,t,$$

and that $\Sigma\omega_{ijij} = (s-1)(t-1)$. Hence the limiting distribution of T_2 is the χ^2 distribution with $(s-1)(t-1)$ degrees of freedom.

10. *Example 3*: *"Student's" t-statistic.*—Let X be a random variable having

$$E(X) = \xi, \qquad E\{(X-\xi)^2\} = 1, \quad E\{(X-\xi)^3\} = \alpha_3, \quad E\{(X-\xi)^4\} = \alpha_4 < \infty.$$

Let $X_1, \cdots, X_N$ be a sample. Then "Student's" is defined, except a factor depending on N, as

$$T_3 = \frac{\bar{X}}{V^{\frac{1}{2}}},$$

where $\bar{X}$ and V are the mean and the variance of the sample.

Consider the random variables

$$U_1 = X, \qquad U_2 = X^2.$$

They have the means

$$E(U_1) = \xi, \qquad E(U_2) = 1 + \xi^2,$$

and the sample means

$$\bar{U}_1 = \bar{X}, \qquad \bar{U}_2 = \frac{1}{N}\sum_r X'^2_r = V + \bar{U}_1^2.$$

Hence

$$T_3 = \bar{U}_1(\bar{U}_2 - \bar{U}_1^2)^{-\frac{1}{2}}.$$

The two terms of the Taylor expansion are

$$\xi + N^{-\frac{1}{2}}\{(1+\xi^2)Z_1 - \tfrac{1}{2}\xi Z_2\}.$$

The quantity σ^2 in (14) is the expectation of the square of

$$(1+\xi^2)(X-\xi) - \tfrac{1}{2}\xi(X^2-1-\xi^2) \tag{28}$$

and has the value

$$\sigma_3^2 = \frac{1}{4}(a_4-1)\xi^2 - a_3\xi + 1.$$

If (28) does not vanish with unit probability, then $\sigma_3^2 \neq 0$ and the limiting distribution of $N^{\frac{1}{2}}(T_3 - \xi)$ is the normal distribution about zero with the variance σ_3^2.

If (28) is essentially zero, then $\sigma_3^2 = 0$ and therefore $\xi \neq 0$. The random variable X can take precisely two values, namely,

$$a = \frac{1+\xi^2+\sqrt{1+\xi^2}}{\xi} \quad \text{and} \quad b = \frac{1+\xi^2-\sqrt{1+\xi^2}}{\xi}.$$

Let

$$P(X=a) = p, \qquad P(X=b) = 1-p.$$

Then we must have

$$\xi = pa + (1-p)b = \frac{1+\xi^2+(2p-1)\sqrt{1+\xi^2}}{\xi},$$

whence

$$p = \frac{\sqrt{1+\xi^2}-1}{2\sqrt{1+\xi^2}}.$$

Among the N numbers $X_1, \cdots, X_N$, let n have the value a and $N-n$ have the value b.
Then

$$T_3 = \frac{b + (a-b)\dfrac{n}{N}}{|a-b|\left(\dfrac{n}{N} - \dfrac{n^2}{N^2}\right)^{\frac{1}{2}}}.$$

As explained in section 7, n/N is the sample mean of a random variable which takes the value one with probability p and zero with probability $1 - p$. On substituting $p + N^{-\frac{1}{2}}Z$ for n/N, we obtain the expansion

$$\xi + \frac{2(1+\xi^2)^2}{N\xi^3} Z^2 .$$

Hence the limiting distribution of $N(T_3 - \xi)$ is the same as that of $2\xi^{-3}(1+\xi^2)^2Z^2$. But limiting distribution of Z is the normal distribution with mean zero and the variance $p(1-p) = \frac{1}{4}\xi^2(1+\xi^2)^{-1}$. Hence the limiting distribution of $2N\xi(1+\xi^2)^{-1}(T_3 - \xi)$ is the χ^2 distribution with one degree of freedom.

11. *Example 4: The ratio of moments.*—Let X be a random variable having

$E(X) = 0, \qquad E(X^2) = 1, \qquad E(X^i) = a_i, \qquad a_{2m} < \infty$ from some integer

$m \geq 3$, and $X_1, \cdots, X_N$ be a sample. Consider the statistic

$$T_4 = \frac{S_m}{S_2^{\frac{m}{2}}},$$

where

$$S_i = \frac{1}{N}\sum_{r=1}^{N}(X_r - \bar{X})^i.$$

When $m=3$ and $m=4$, T_4 becomes the familiar b_1 and b_2 of K. Pearson.

The random variables

$$U_i = X^i, \qquad i = 1, \cdots, m,$$

have the means a_i and the sample means

$$\bar{U}_i = \frac{1}{N}\sum_{r=1}^{N} X_r^i.$$

We have

$$T_4 = \frac{\bar{U}_m - m\,\bar{U}_1\bar{U}_{m-1} + \cdots}{(\bar{U}_2 - \bar{U}_1^2)^{\frac{1}{2}m}} .$$

Making the substitution

$$\bar{U}_1 = N^{-\frac{1}{2}}Z_1, \qquad \bar{U}_2 = 1 + N^{-\frac{1}{2}}Z_2, \qquad \bar{U}_i = a_i + N^{-\frac{1}{2}}Z_i, \qquad i = 3, \cdots, m,$$

and computing the two-term expansion we obtain

$$a_m + N^{-\frac{1}{2}}(Z_m - \tfrac{1}{2}m a_m Z_2 - m a_{m-1}Z_1).$$

The quantity σ^2 in (14) is the expectation of the square of

$$X^m - a_m - \tfrac{1}{2}m a_m(X^2 - 1) - m a_{m-1}X, \tag{29}$$

and has the value

$$\sigma_4^2 = a_{2m} - m a_m a_{m+2} - 2m a_{m-1} a_{m+1} - \frac{1}{4}(m-2)^2 a_m^2 + \frac{1}{4} m^2 a_4 a_m^2 + m^2 a_3 a_{m-1} a_m + m^2 a^2_{m-1}\,.$$

Hence, if (29) is not essentially zero, the limiting distribution of $N^{\frac{1}{2}}(T_4 - a_m)$ is the normal distribution about zero with the variance σ_4^2.

12. *Functions of variances and covariances; a simplification.*—We are going to study a pair of statistics, denoted by T_5 and T_6, which are formed of one homogeneous multivariate sample and are functions of the variances and covariances.

Let

$$[X_1, \cdots, X_p] \tag{30}$$

be a random vector having finite fourth moments and not satisfying any linear or quadratic relation with unit probability. Let

$$E(X_i) = \xi_i, \qquad E(X_i X_j) - \xi_i \xi_j = \sigma_{ij}\,.$$

Let

$$[X_{1r}, \cdots, X_{pr}],\ r = 1, \cdots, N,$$

be a sample of size N and let

$$\overline{X}_i = \frac{1}{N}\sum_{r=1}^{N} X_{ir}, \qquad v_{ij} = \frac{1}{N}\sum_{r=1}^{N} X_{ir}X_{jr} - \overline{X}_i\overline{X}_j\,.$$

Let T be any statistic which is a function of the v_{ij} only:

$$T = F(v_{11}, v_{12}, \cdots, v_{p-1,p}, v_{pp}). \tag{31}$$

The $\frac{1}{2}p(p+1)$ random variables

$$U_i = X_i - \xi_i, \qquad U_{ij} = (X_i - \xi_i)(X_j - \xi_j), \qquad i \leqq j,$$

have the means

$$E(U_i) = 0, \qquad E(U_{ij}) = \sigma_{ij},$$

and the sample means

$$\bar{U} = \overline{X}_i - \xi_i, \qquad \bar{U}_{ij} = v_{ij} + \bar{U}_i\bar{U}_j.$$

Hence

$$T = F(\bar{U}_{11} - \bar{U}_1^2, \bar{U}_{12} - \bar{U}_1\bar{U}_2, \cdots, \bar{U}_{pp} - \bar{U}_p^2).$$

On substituting $N^{-\frac{1}{2}}Z_i$ for $\bar{U}_i$ and $\sigma_{ij}+N^{-\frac{1}{2}}Z_{ij}$ for $\bar{U}_{ij}$ we obtain the three-term expansion of T,

$$(32)\quad A + N^{-\frac{1}{2}} \sum_{i \leq j} B_{ij}Z_{ij} + N^{-1}\Big(\tfrac{1}{2} \sum_{\substack{i \leq j \\ \mu \leq \nu}} C_{ij\mu\nu}Z_{ij}Z_{\mu\nu} - \sum_{i \leq j} B_{ij}Z_iZ_j \Big),$$

where

$$A = F(\sigma_{11}, \sigma_{12}, \cdots, \sigma_{pp}),$$

$$B_{ij} = \frac{\partial}{\partial \sigma_{ij}} F(\sigma_{11}, \sigma_{12}, \cdots, \sigma_{pp}),$$

$$C_{ij\mu\nu} = \frac{\partial^2}{\partial \sigma_{ij} \partial \sigma_{\mu\nu}} F(\sigma_{11}, \sigma_{12}, \cdots, \sigma_{pp}).$$

If $B_{ij} \neq 0$ for some (i, j) then the term

$$\sum_{i \leq j} B_{ij}Z_{ij},$$

being the normalized sample mean of

$$\sum_{i \leq j} B_{ij} \Big\{ (X_i - \xi_i)(X_j - \xi_j) - \sigma_{ij} \Big\},$$

cannot vanish with unit probability. Therefore it is sufficient to have the two-term expansion,

$$(33)\qquad A + N^{-\frac{1}{2}} \sum_{i \leq j} B_{ij}Z_{ij}.$$

If $B_{ij} = 0$ for all (i, j), then (32) becomes

$$(34)\qquad A + \frac{1}{2N} \sum_{\substack{i \leq j \\ \mu \leq \nu}} C_{ij\mu\nu}Z_{ij}Z_{\mu\nu}.$$

But (33) and (34) are precisely the expansions that we shall obtain if we make the direct substitution $v_{ij} = \sigma_{ij} + N^{-\frac{1}{2}}Z_{ij}$ in F. Hence we have the following rule of simplification:

In order to obtain the limiting distribution of (31), *make the substitution* $v_{ij} = \sigma_{ij} + N^{-\frac{1}{2}}Z_{ij}$ *and then follow the steps described at the end of section* 6.

This rule of simplification can be extended immediately to the case of k samples.

13. *Example 5: The hypothesis of independence and Wilks's test function.*— Consider again the random vector (30). The hypothesis of independence is the following: $X_1, \cdots, X_p$ are classified into κ mutually independent sets consisting of $s_1, \cdots, s_\kappa$ members respectively:

$$(35)\quad [X_1, \cdots, X_{s_1}], [X_{s_1+1}, \cdots, X_{s_1+s_2}], \cdots, [X_{s_1+\cdots+s_{\kappa-1}+1}, \cdots, X_p]$$

This hypothesis was first studied by Wilks [**10**], who applied the principle of likelihood on the assumption of normality of (30) and obtained the test function which we now define.

Let the matrices

$$V = [v_{ij}], \qquad M = [\sigma_{ij}]$$

be so partitioned that

$$V = \begin{bmatrix} V_{11}\, V_{12} \cdots V_{1\kappa} \\ V_{21}\, V_{22} \cdots V_{2\kappa} \\ \text{-----------} \\ V_{\kappa 1}\, V_{\kappa 2} \cdots V_{\kappa\kappa} \end{bmatrix}, \qquad M = \begin{bmatrix} M_{11}\, M_{12} \cdots M_{1\kappa} \\ M_{21}\, M_{22} \cdots M_{2\kappa} \\ \text{-----------} \\ M_{\kappa 1}\, M_{\kappa 2} \cdots M_{\kappa\kappa} \end{bmatrix}$$

where $V_{\mu\nu}$ and $M_{\mu\nu}$ have s_μ rows and s_ν columns, $(\mu, \nu = 1, \cdots, \kappa)$. Let also

$$V_1 = \begin{bmatrix} V_{11} & & O \\ & \ddots & \\ O & & V_{\kappa\kappa} \end{bmatrix}, \qquad M_1 = \begin{bmatrix} M_{11} & & O \\ & \ddots & \\ O & & M_{\kappa\kappa} \end{bmatrix}.$$

Then Wilks's test function is

$$T_5 = \frac{|V|}{|V_1|}.$$

Let us now study the limiting distribution of T_5. Suppose first that not only is the hypothesis of independence false but actually some of the covariances σ_{ij} lying in the matrices $M_{\mu\nu}$, $(\mu \neq \nu)$, are not zero. Following the rule of simplification in section 12 we make the substitution $v_{ij} = \sigma_{ij} + N^{-\frac{1}{2}} Z_{ij}$ in T_5 and obtain the two-term expansion,

$$a + aN^{-\frac{1}{2}} \sum_{i,j=1}^{p} (\alpha_{ij} - \beta_{ij}) Z_{ij},$$

where

$$a = \frac{|M|}{|M_1|},$$

α_{ij} is the element (i, j) of the matrix M^{-1} and β_{ij} that of M_1^{-1}. By our assumption $M \neq M_1$, hence $\alpha_{ij} - \beta_{ij}$ cannot vanish for all (i,j). The quantity σ^2 in (14) is the expectation of the square of

$$(36) \qquad a \sum_{i,j} (\alpha_{ij} - \beta_{ij})(U_{ij} - \sigma_{ij}) = a \sum_{i,j} (\alpha_{ij} - \beta_{ij}) U_{ij}$$

and has the value

$$\sigma_5^2 = a^2 \sum_{i,j,\mu,\nu} (\alpha_{ij} - \beta_{ij})(\alpha_{\mu\nu} - \beta_{\mu\nu}) \sigma_{ij\mu\nu},$$

where

$$\sigma_{ij\mu\nu} = E\{(X_i - \xi_i)(X_j - \xi_j)(X_\mu - \xi_\mu)(X_\nu - \xi_\nu)\}\,. \tag{37}$$

Since (36) cannot be essentially zero, the limiting distribution of $N^{\frac{1}{2}}(T_5 - a)$ is the normal distribution about zero with the variance $\sigma_5^2 \neq 0$.

Suppose next that the hypothesis is true, so that the sets (30) are independent. Then we can assume without loss of generality that $\sigma_{ii} = 1$ and $\sigma_{ij} = 0$ for $i \neq j$, $(i, j = 1, \cdots, p)$. For, if these are not true, we can subject each set in (30) to a linear transformation so that for the new variables the variances are one and the covariances are zero. The new sets of variables are still independent whereas T_5 is invariant under such a transformation. Hence our problem reduces to finding the limiting distribution of T_5 under the assumption that

$$M = I \tag{38}$$

and that the sets (30) are independent.

Remembering the rule of simplification (sec. 12) we make the substitution $v_{ii} = 1 + N^{-\frac{1}{2}}Z_{ii}$, $v_{ij} = N^{-\frac{1}{2}}Z_{ij}$, $(i \neq j)$, in T_5 and obtain

$$T_5 = \frac{|I + N^{-\frac{1}{2}}Z\,|}{|I + N^{-\frac{1}{2}}Z_1|}\,, \tag{39}$$

where Z and Z_1 are the matrices obtained on replacing each v_{ij} by Z_{ij} in V and V_1. The three-term expansion of (39) is the same as that of

$$\frac{1 + N^{-\frac{1}{2}}b + N^{-1}c}{1 + N^{-\frac{1}{2}}b_1 + N^{-1}c_1}\,,$$

where b is the sum of the diagonal elements of Z, c is the sum of the two-rowed principal minors of Z, b_1 and c_1 are the same functions of the elements of Z_1. Since evidently $b = b_1$, we have the expansion

$$1 - \frac{1}{N}(c_1 - c)\,.$$

Obviously

$$c_1 - c = \sum_{i<j}{}' Z_{ij}^2\,, \tag{40}$$

where Σ' denotes summation extended to those (i, j) for which the position of Z_{ij} in Z is the position of a zero in Z_1. Hence the limiting distribution of $N(1 - T_5)$ is the same as that of (40), that is, the distribution of

$$\sum_{i<j}{}' W_{ij}^2, \tag{41}$$

where the W_{ij} are normal varieties having zero means and the same second moments as the set U_{ij}. Under the assumption of independence and (38) we have, for all the W_{ij} in (41),

$$E(W_{ij}^2) = \sigma_{iijj} = \sigma_{ii}\sigma_{jj} = 1,$$

$$E(W_{ij}W_{\mu\nu}) = \sigma_{ij\mu\nu} = \sigma_{i\mu}\sigma_{j\nu} = 0, \qquad (i, j) \neq (\mu, \nu).$$

Hence the W_{ij} in (41) are independent unit normal variates. Thus the limiting distribution of $N(1 - T_5)$ is the χ^2 distribution whose degree of freedom is the number of terms in the sum (41), namely, $\Sigma s_i s_j$ $(i \leqq j; i, j = 1, \cdot \cdot \cdot, \kappa)$.

14. *Example 6: The hypothesis of independence and homoscedasticity and the likelihood ratio test.*—Consider the following hypothesis: The random variables X_i in (30) are independent and $\sigma_{11} = \sigma_{22} = \cdot \cdot \cdot = \sigma_{pp}$. If we regard the distribution of (30) as normal and apply the principle of likelihood we easily obtain the test function

$$T_6 = \frac{|V|^{\frac{1}{p}}}{\frac{1}{p}(v_{11} + \cdot \cdot \cdot + v_{pp})},$$

which is the ratio of the geometric mean of the latent roots of V to their arithmetic mean.

Suppose first that not only is the hypothesis false but the relations

$$(42) \qquad \sigma_{11} = \cdot \cdot \cdot = \sigma_{pp}, \qquad \sigma_{ij} = 0, \qquad i \neq j,$$

are not all true. Making the substitution $v_{ij} = \sigma_{ij} + N^{-\frac{1}{2}}Z_{ij}$ we get the two-term expansion

$$a + aN^{-\frac{1}{2}} \sum_{i,j=1}^{p} b_{ij}Z_{ij}, \qquad Z_{ji} = Z_{ij},$$

where

$$a = \frac{|M|^{\frac{1}{p}}}{\frac{1}{p}(\sigma_{11} + \cdot \cdot \cdot + \sigma_{pp})},$$

$$b_{ii} = \frac{1}{p}a_{ii} - \frac{1}{\sigma_{11} + \cdot \cdot \cdot + \sigma_{pp}}, \qquad b_{ij} = \frac{1}{p}a_{ij}, \qquad i \neq j,$$

and the a_{ij} are the elements of M^{-1}. Since some of relations (42) are not true, the b_{ij} cannot all vanish. The quantity σ^2 in (14) is the expectation of the square of

$$(43) \qquad a \sum_{i,j} b_{ij}(U_{ij} - \sigma_{ij}) = a \sum_{i,j} b_{ij}U_{ij}$$

and has the value

$$\sigma_6^2 = a^2 \sum_{i,j,\mu,\nu} b_{ij}b_{\mu\nu}\sigma_{ij\mu\nu},$$

where $\sigma_{ij\mu\nu}$ is defined in (37). Since (43) cannot be essentially zero, the limiting distribution of $N^{\frac{1}{2}}(T_6 - a)$ is the normal distribution about zero with the variance $\sigma_6^2 \neq 0$.

Suppose next that the hypothesis is true, so that $X_1, \cdots, X_p$ are independent and have a common variance η. Making the substitution $v_{ii} = \eta + N^{-\frac{1}{2}}Z_{ii}$, $v_{ij} = N^{-\frac{1}{2}}Z_{ij}, (i \neq j)$, in T_6 we get the three-term expansion

$$1 - \frac{1}{pN\eta^2}\left\{\frac{1}{2}\sum_i Z_{ii}^2 - \frac{1}{2p}\left(\sum_i Z_{ii}\right)^2 + \sum_{i<j} Z_{ij}^2\right\}.$$

Hence $pN(1 - T_6)$ has the same limiting distribution as that of

$$\frac{1}{2\eta^2}\sum_i\left(Z_{ii} - \frac{1}{p}\sum_i Z_{ii}\right)^2 + \frac{1}{\eta^2}\sum_{i\wedge j} Z_{ij}^2 = \sum_{i\leqq j} Y_{ij}^2,$$

where

$$Y_{ii} = \frac{1}{\sqrt{2}\eta}\left(Z_{ii} - \frac{1}{p}\sum_i Z_{ii}\right), \qquad Y_{ij} = \frac{1}{\eta}Z_{ij}, \qquad i<j.$$

The Y's are the normalized sample means of the following system of variates:

$$\frac{1}{\sqrt{2}\eta}\left(U_{ii} - \frac{1}{p}\sum_i U_{ii}\right), \quad \frac{1}{\eta}U_{ij}, \qquad i<j. \tag{44}$$

The limiting distribution in question is the distribution of

$$\sum_{i\leqq j} W_{ij}^2, \tag{45}$$

where the W_{ij} are normal variates with zero means and the same second moments as (44). Under the assumption that the hypothesis is true, we have

$$E(U_{ii}^2) = \mu_{4i} - \eta^2, \qquad E(U_{ij}^2) = \eta^2, \qquad i<j,$$
$$E(U_{ij}U_{\mu\nu}) = 0, \qquad (i,j) \neq (\mu,\nu),$$

where

$$\mu_{4i} = E\{(X_i - \xi_i)^4\}.$$

Then it is easy to compute the second moments of the W_{ij}:

$$E(W_{ii}^2) = \frac{1}{2p^2}\sum_i \frac{\mu_{4i}}{\eta^2} + \frac{p-2}{2p}\frac{\mu_{4i}}{\eta^2} - \frac{p-1}{2p},$$

$$E(W_{ii}W_{jj}) = \frac{1}{2p^2}\sum_i \frac{\mu_{4i}}{\eta^2} - \frac{1}{2p}\left(\frac{\mu_{4i} + \mu_{4i}}{\eta^2}\right) + \frac{1}{2p}, \qquad i \neq j,$$

$$E(W_{ii}W_{\mu\nu}) = 0, \qquad \mu < \nu,$$

$$E(W_{ij}^2) = 1, \qquad i < j,$$

$$E(W_{ij}W_{\mu\nu}) = 0, \qquad i<j,\ \mu < \nu;\ (i,j) \neq (\mu,\nu).$$

It is thus seen that the distribution of (45) is the composition of two independent parts: the part contributed by ΣW_{ij}^2, $(i < j)$, and that contributed by ΣW_{ii}^2. The former is χ^2 distribution with $\frac{1}{2}p(p-1)$ degrees of freedom; for the latter we apply lemma 3. Let $\omega_{ij} = E(W_{ii}W_{jj})$ and $\lambda_1, \cdots, \lambda_{p-1}$ be the non-vanishing latent roots of (ω_{ij}), which is of rank $p-1$. Then the distribution of (45), that is, the limiting distribution of $pN(1-T_6)$ is

$$\frac{1}{(2\pi)^{\frac{1}{2}(p-1)}2^{\frac{1}{4}p(p-1)}\Gamma\left(\frac{1}{4}p(p-1)\right)\left(\lambda_1\cdots\lambda_{p-1}\right)^{\frac{1}{2}}}\times$$

$$\int_{y_1^2+\ldots+y_{p-1}^2+z\leqq x} z^{\frac{1}{4}p(p-1)-1}\exp\left(-\frac{y_1^2}{2\lambda_1}-\cdots-\frac{y_{p-1}^2}{2\lambda_{p-1}}-\frac{z}{2}\right)dy_1\cdots dy_{p-1}dz.$$

A sufficient condition for this distribution to be the χ^2 distribution with $\frac{1}{2}(p+2)(p-1)$ degrees of freedom is that $\mu_{4i} = 3\eta^2$ for all i, for then $\lambda_1 = \cdots = \lambda_{p-1} = 1$.

It may be noticed that, although the limiting distribution of $N(1-T_5)$ is always the χ^2 distribution when the hypothesis of independence is true, regardless of the distribution of (30), the limiting distribution of $pN(1-T_6)$, even when the hypothesis tested is true, still depends on the fourth moments of (30), and becomes the χ^2 distribution under the condition that

$$E\{(X_i-\xi_i)^4\} = 3\,[E\{(X_i-\xi_i)^2\}]^2 \text{ for all } i.$$

15. *Problems of* k *samples and the statistics* L *and* L_1.—Let

$$X_1, \cdots, X_k$$

be k random variables having

$$E(X_a) = \xi_a, \qquad E\{(X_a-\xi_a)^2\} = \eta_a \neq 0, \qquad \eta_a^{-\frac{3}{2}}E\{(X_a-\xi_a)^3\} = a_a,$$
$$\eta_a^{-2}E\{(X_a-\xi_a)^4\} = b_a < \infty.$$

Let

$$X_{a1}, \cdots, X_{aN_a}, \qquad a = 1, \cdots, k,$$

be k samples of sizes $N_1, \cdots, N_k$. Consider the following two hypotheses:

$$H:\ \xi_1 = \cdots = \xi_k \text{ and } \eta_1 = \cdots = \eta_k,$$

$$H':\ \eta_1 = \cdots = \eta_k.$$

With the help of their method of likelihood ratio applied to normal distributions, Neyman and Pearson [9] obtain the following test functions for H and H' respectively:

$$L = \left(\prod_{a=1}^{k} Y_a^{g_a}\right)\left\{\sum_{a=1}^{k} g_a Y_a + \sum_{a=1}^{k} g_a(\bar{X}_a - \bar{X})^2\right\}^{-1},$$

$$L_1 = \left(\prod_{a=1}^{k} Y_a^{g_a}\right)\left(\sum_{a=1}^{k} g_a Y_a\right)^{-1},$$

where the g_a are defined in (9),

$$\bar{X}_a = \frac{1}{N_a}\sum_{r=1}^{N_a} X_{ar}, \qquad \bar{X} = \sum_{a=1}^{k} g_a \bar{X}_a, \qquad Y_a = \frac{1}{N}\sum_{r=1}^{N_a} (X_{ar} - \bar{X}_a)^2 .$$

We shall call L and L_1 respectively T_7 and T_8 and find their limiting distributions when the sample sizes N_a become infinite in the manner specified in section 3, from arbitrary parent distributions.

16. *Example 7: The L-statistic.*—The random variables

$$U_{1a} = X_a - \xi_a, \qquad U_{2a} = (X_a - \xi_a)^2,$$

have the means

$$E(U_{1a}) = 0, \qquad E(U_{2a}) = \eta_a,$$

and sample means

$$\bar{U}_{1a} = \bar{X}_a - \xi_a, \qquad \bar{U}_{2a} = Y_a + \bar{U}_{1a}^2.$$

Hence

$$(46) \quad T_7 = L = \left\{\prod_{a=1}^{k} (\bar{U}_{2a} - \bar{U}_{1a}^2)^{g_a}\right\}\left\{\bar{U}_2 - \bar{U}_1^2 - 2\sum_{a=1}^{k} g_a(\xi_a - \bar{\xi})\,\bar{U}_{1a} + \sigma_\xi^2\right\}^{-1},$$

where

$$\bar{U}_i = \sum_a g_a U_{ia}, \qquad i = 1, 2,$$

$$\bar{\xi} = \sum_a g_a \xi_a, \qquad \sigma_\xi^2 = \sum_a g_a(\xi_a - \bar{\xi})^2.$$

If the hypothesis H is false, we make the substitution

$$\bar{U}_{1a} = N_a^{-\frac{1}{2}} Z_{1a}, \qquad \bar{U}_{2a} = \eta_a + N_a^{-\frac{1}{2}} Z_{2a} ,$$

in T_7 and obtain the two-term expansion

$$a + aN^{-\frac{1}{2}} \sum_a (A_a Z_{1a} + B_a Z_{2a}),$$

where

$$a = \left(\prod_a \eta_a^{g_a}\right)\left(\sum_a g_a\eta_a + \sigma_\xi^{12}\right)^{-1},$$

$$A_a = \frac{2g_a^{\frac{1}{2}}(\xi_a - \bar{\xi})}{\sum_a g_a\eta_a + \sigma_\xi^2}, \qquad B_a = g_a^{\frac{1}{2}}\left(\frac{1}{\eta_a} - \frac{1}{\sum_a g_a\eta_a + \sigma\xi^2}\right).$$

The A_a and B_a cannot all vanish, for otherwise H would be true. Let

$$V_a = A_a(X_a - \xi_a) + B_a\{(X_a - \xi_a)^2 - \eta_a\}.$$

The quantity σ^2 in (14) is equal to $a^2\Sigma E(V_a^2)$, $(a = 1, \cdots, k)$, and has the value

$$\sigma_7^2 = a^2\sum_a\{A_a^2\eta_a + 2A_aB_a a_a\eta_a^{\frac{1}{2}} + B_a\eta_a^2(b_a - 1)\}.$$

Suppose that one at least of the V_a is not essentially zero. Then the limiting distribution of $N^{\frac{1}{2}}(T_7 - a)$ is the normal distribution about zero with the variance $\sigma_7^2 \neq 0$.

If the hypothesis H is true, so that

$$\xi_a = \xi, \qquad \eta_a = \eta, \qquad a = 1, \cdots, k,$$

then (46) becomes

$$T_7 = \left\{\prod_{a=1}^{k}(\bar{U}_{2a} - \bar{U}_{1a}^2)g_a\right\}(\bar{U}_{2a} - \bar{U}_1^2)^{-1}.$$

Making the substitution $\bar{U}_{1a} = N_a^{-\frac{1}{2}}Z_{1a}$ and $\bar{U}_{2a} = \eta + N^{-\frac{1}{2}}Z_{2a}$ we obtain the three-term expansion

$$1 - \frac{1}{N}\left\{\frac{1}{\eta}\prod_a Z_{1a}^2 - \frac{1}{\eta}\left(\sum_a g_a^{\frac{1}{2}}Z_{1a}\right)^2 + \frac{1}{2\eta^2}\sum_a Z_{2a}^2 - \frac{1}{2\eta^2}\left(\sum_a g_a^{\frac{1}{2}}Z_{2a}\right)^2\right\}.$$

Hence the limiting distribution of $N(1 - T_7)$ is the distribution of

$$Q = \sum_a W_{1a}^2 - \left(\sum_a g_a^{\frac{1}{2}}W_{1a}\right)^2 + \sum_a W_{2a}^2 - \left(\sum_a g_a^{\frac{1}{2}}W_{2a}\right)^2,$$

where W_{ia}, W_{2a} are normal variates with zero means and the following second moments:

$$E(W_{ia}W_{j\beta}) = 0, \qquad a \neq \beta; i, j = 1, 2,$$

$$E(W_{1a}^2) = \frac{1}{\eta}E(U_{1a}^2) = 1, \qquad E(W_{2a}^2) = \frac{1}{2\eta^2}E\{(U_{2a}^2 - \eta)^2\} = \frac{1}{2}(b_a - 1).$$

$$E(W_{1a}W_{2a}) = \frac{1}{\sqrt{2}\eta^{\frac{3}{2}}}E(U_{1a}^3) = \frac{1}{\sqrt{2}}a_a.$$

Let us treat in detail the particular case where the a_α and the b_α are independent of α:

$$\frac{1}{\sqrt{2}} a_\alpha = A, \qquad \frac{1}{2}(b_\alpha - 1) = B, \qquad \alpha = 1, \cdots, k.$$

It is possible to perform an orthogonal transformation to each of the vectors $[W_{i1}, \cdots, W_{ik}]$:

$$[W_{i1}, \cdots, W_{ik}] \to [W'_{i1}, \cdots, W'_{ik}] \qquad i = 1, 2),$$

so that

$$E(W'_{i\alpha}W'_{j\beta}) = 0, \qquad \alpha \neq \beta; i, j = 1, 2,$$

$$E(W'^2_{1\alpha}) = 1, \quad E(W'_{1\alpha}W'_{2\alpha}) = A, \quad E(W'^2_{2\alpha}) = B, \quad Q = \sum_{\alpha=1}^{k-1}(W'^2_{1\alpha} + W'^2_{2\alpha}).$$

We now apply lemma 3. The dispersion matrix being

$$\begin{bmatrix} I & AI \\ AI & BI \end{bmatrix},$$

its λ-equation is easily reduced to

$$\{\lambda^2 - (1 + B)\lambda + B - A^2\}^{k-1} = 0. \tag{47}$$

If each X_α can take essentially two values, then $B = A^2$ and so the only non-vanishing root of (47) is $1 + B$ of multiplicity $k - 1$. In this case $Q/(1 + B)$ has the χ^2 distribution with $k - 1$ degrees of freedom. In the contrary case, $B > A^2$ and the equation (47) has the roots

$$\gamma_1 = \tfrac{1}{2}[1 + B + \{(1 - B)^2 + 4A^2\}^{\frac{1}{2}}], \qquad \gamma_2 = \tfrac{1}{2}[1 + B - \{(1 - B)^2 + 4A^2\}^{\frac{1}{2}}],$$

both of multiplicity $k - 1$. The distribution law of Q is then

$$(4\gamma_1\gamma_2)^{-\frac{1}{2}(k-1)}\left\{\Gamma\left(\frac{k-1}{2}\right)\right\}^{-k}\int_R (y_1y_2)^{\frac{1}{2}(k-3)}\exp\left(-\frac{y_1}{2\gamma_1} - \frac{y_2}{2\gamma_2}\right)dy_1dy_2, \tag{48}$$

where R is the region $0 \leqq y_1, 0 \leqq y_2, y_1 + y_2 \leqq x$. The necessary and sufficient condition for $\gamma_1 = \gamma_2$ is that $B = 1$ and $A = 0$, that is, $b_\alpha = 3$ and $a_\alpha = 0$ for all α. If this condition is satisfied, then $\gamma_1 = \gamma_2 = 1$ and (48) becomes the χ^2 distribution with $2k - 2$ degrees of freedom.

17. *Example 8: The* L_1*-statistic.*—We have

$$T_8 = L_1 = \left(\prod_{\alpha=1}^{h} Y_\alpha^{g_\alpha}\right)\left(\sum_{\alpha=1}^{k} g_\alpha Y_\alpha\right)^{-1}.$$

Following the rule of simplification in section 12 we make the substitution $Y_\alpha = \eta_\alpha + N_\alpha^{-\frac{1}{2}}Z_\alpha$ and expand the result.

If the hypothesis H' is false, the two-term expansion is

$$a + aN^{-\frac{1}{2}} \sum_{\alpha=1}^{k} A_\alpha Z_\alpha ,$$

where

$$a = \left(\prod_\alpha \eta_\alpha^{g_\alpha}\right) \left(\sum_\alpha g_\alpha \eta_\alpha\right)^{-1},$$

$$A_\alpha = g_\alpha^{\frac{1}{2}} \left(\frac{1}{\eta_\alpha} - \frac{1}{\sum_\alpha g_\alpha \eta_\alpha}\right).$$

The A_α cannot all vanish, for otherwise H' would be true. Let

$$V_\alpha = A_\alpha \{(X_\alpha - \xi_\alpha)^2 - \eta_\alpha\} .$$

The quantity σ^2 in (14) has the value

$$\sigma_8^2 = a^2 \sum_\alpha E(V_\alpha^2) = a^2 \sum_\alpha A_\alpha^2 \eta_\alpha^2 (b_\alpha - 1) .$$

If the V_α are not all essentially zero, the limiting distribution of $N^{\frac{1}{2}}(T_8 - a)$ is the normal distribution about zero with the variance $\sigma_8^2 \neq 0$.

If the hypothesis H' is true, then $\eta_\alpha = \eta$, $(\alpha = 1, \cdots, k)$. Making the substitution $Y_\alpha = \eta + N_\alpha^{-\frac{1}{2}} Z_\alpha$ in T_8 we obtain the three-term expansion

$$1 - \frac{1}{2N\eta^2}\left\{\sum_\alpha Z_\alpha^2 - \left(\sum_\alpha g_\alpha^{\frac{1}{2}} Z_\alpha\right)^2\right\}.$$

Hence the limiting distribution of $N(1 - T_8)$ is the distribution of $\Sigma W_\alpha^2 - (\Sigma g_\alpha^{\frac{1}{2}} W_\alpha)^2$, $(\alpha = 1, \cdots, k)$, where the W_α are independent normal variates with zero means and $E(W_\alpha^2) = (2\eta^2)^{-1} E[\{(X_\alpha - \xi_\alpha)^2 - \eta\}^2] = \frac{1}{2}(b_\alpha - 1)$. In particular, if each $b_\alpha = 3$, the limiting distribution of $N(1 - T_8)$ is the χ^2 distribution with $k - 1$ degrees of freedom.

II

Application to testing hypotheses

18. Lemma 4. *Let* $\mathrm{w} = [\mathrm{W}_1, \cdots, \mathrm{W}_l]$ *be a normally distributed vector such that each* $\mathrm{E}(\mathrm{W}_\lambda) = 0$ *and the dispersion matrix* Φ *is non-singular. Let* C *be any real matrix of order* $\mathrm{h} \times \mathrm{l}$, $(\mathrm{h} < \mathrm{l})$, *and rank* h. *Then the quadratic form*

$$X = -\begin{vmatrix} \Phi & C' & w' \\ C & O & o' \\ w & o & 0 \end{vmatrix} : \begin{vmatrix} \Phi & C' \\ C & O \end{vmatrix} \qquad (49)$$

has the χ^2 *distribution with* $\mathrm{l} - \mathrm{h}$ *degrees of freedom.*

This lemma becomes familiar when $h = 0$, for then C does not appear and X is the quadratic form $w\, \Phi^{-1}\, w'$.

PROOF. By (49) we have

$$X = [w, o]\begin{bmatrix}\Phi & C' \\ C & O\end{bmatrix}^{-1}\begin{bmatrix}w' \\ o'\end{bmatrix} = w\Phi^{-1}w' - w\Phi^{-1}C'(C\Phi^{-1}C')^{-1}C\Phi^{-1}w' .$$

Since Φ is positive definite, there is a real non-singular G such that $G\Phi^{-1}G' = I$. Hence

$$X = yy' - yB'(BB')^{-1}By' , \tag{50}$$

where $y = wG^{-1}$ is again a normally distributed vector and where $B = C\Phi^{-1}G'$ has the same order and rank as C.

The components of y are independent unit normal variates, because the dispersion matrix is $E(y'y) = E(G'^{-1}w'wG^{-1}) = G'^{-1}\Phi G^{-1} = I$. The matrix of the quadratic form (50), $I - B'(BB')^{-1}B = A$ say, has the property that $A^2 = A$. Hence the latent roots of A are either zero or unity. This shows that by an orthogonal transformation X can be reduced to a sum of squares. Hence the χ^2 distribution is established. The number of such squares is $trA = l - trB'(BB')^{-1}B = l - tr(BB')^{-1}BB'$, which gives the degree of freedom, q.e.d.

19. *The case of one sample: the hypothesis* **H**.—Let

$$[U_1, \cdots, U_m]$$

be a random vector, possessing finite second moments and a non-singular dispersion matrix. Let

$$E(U_i) = \mu_i , \qquad E(U_iU_j) - \mu_i\mu_j = \eta_{ij} .$$

Let

$$[U_{1r}, \cdots, U_{mr}] , \qquad r = 1, \cdots, N ,$$

be a sample of size N, and $\bar{U}_i$, Z_i be respectively the sample means and the normalized sample means.

We call the hypothesis **H** the following hypothesis,

$$\mathbf{H}:\ f_\lambda(\mu_1, \cdots, \mu_m) = \sum_{q=1}^{h} c_{q\lambda}\rho_q , \qquad \lambda = 1, \cdots, l, \tag{51}$$

which asserts that l given functions of m populational constants are expressible as linear combinations of h unspecified parameters ρ_q with known coefficients $c_{q\lambda}$. The three integers m, l, and h shall satisfy the relation

$$0 \leqq h < l \leqq m .$$

If $h = 0$, the right-hand side of (51) means zero.

Concerning f_λ and $c_{q\lambda}$, we make the following assumptions:

(i) Each $f_\lambda(x_1, \cdots, x_m)$ is defined in the whole m-dimensional space and possesses continuous third derivatives of every kind in the neighborhood of $(\mu_1, \cdots, \mu_m)$.

(ii) The matrix

$$F = \begin{bmatrix} f_1^{(1)} & \cdots & f_1^{(m)} \\ - & - & - \\ f_l^{(1)} & \cdots & f_l^{(m)} \end{bmatrix}$$

where

$$f_\lambda^{(i)} = \frac{\partial}{\partial \mu_i} f_\lambda(\mu_1, \cdots, \mu_m)$$

is of rank l for all $\mu_1, \cdots, \mu_m$ satisfying (51).

(iii) The matrix

$$C = \begin{bmatrix} c_{11} & \cdots & c_{1l} \\ - & - & - \\ c_{h1} & \cdots & c_{hl} \end{bmatrix}$$

is of rank h.

20. *The unstudentized statistic* D.—Writing

$$Y_\lambda = f_\lambda(\bar{U}_1, \cdots, \bar{U}_m), \quad y = [Y_1, \cdots, Y_l],$$

$$\varphi_{\lambda\nu} = \sum_{i,j=1}^{m} \eta_{ij} f_\lambda^{(i)} f_\nu^{(j)}, \qquad \Phi = [\varphi_{\lambda\nu}] = F[\eta_{ij}]F',$$

we define the statistic D as

$$D = -\begin{vmatrix} \Phi & C' & y' \\ C & O & 0' \\ y & 0 & 0 \end{vmatrix} : \begin{vmatrix} \Phi & C' \\ C & O \end{vmatrix}. \tag{52}$$

We shall show that, when the hypothesis **H** is true, the limiting distribution of ND is the χ^2 distribution with $l - h$ degrees of freedom. For this purpose we follow the procedure described at the end of section 6. Expanding

$$Y_\lambda = f_\lambda(\mu_1 + N^{-\frac{1}{2}}Z_1, \cdots, \mu_m + N^{-\frac{1}{2}}Z_m)$$

to two terms and using (51) we obtain

$$\sum_{q=1}^{h} c_{q\lambda}\rho_q + N^{-\frac{1}{2}}R_\lambda, \tag{53}$$

where

$$R_\lambda = \sum_{i=1}^{m} f_\lambda^{(i)} Z_i.$$

When (53) is substituted for Y_λ in (52), the terms $\theta_\lambda = \Sigma_q c_{q\lambda}\rho_q$ may be canceled, because $[\theta_1, \cdots, \theta_l]$ is a linear combination of the rows of C. Then we get

$$(54) \qquad -\frac{1}{N}\begin{vmatrix} \Phi & C' & r' \\ C & O & 0' \\ r & 0 & 0 \end{vmatrix} : \begin{vmatrix} \Phi & C' \\ C & O \end{vmatrix},$$

where

$$r = [R_1, \cdots, R_l]\,.$$

The expansion (54) represents the three-term expansion (25) of D. Hence the limiting distribution of ND is the distribution of

$$-\begin{vmatrix} \Phi & C' & w' \\ C & O & 0' \\ w & 0 & 0 \end{vmatrix} : \begin{vmatrix} \Phi & C' \\ C & O \end{vmatrix},$$

where w is a normally distributed vector having zero means and the same dispersion matrix as the system

$$\sum_{i=1}^{m} f_\lambda^{(i)}(U_i - \mu_i)\,, \qquad \lambda = 1, \cdots, l\,.$$

This dispersion matrix is Φ, which is non-singular under our assumptions. The result now follows from lemma 4.

21. *Studentization of* D.—In order to construct a test function for the hypothesis **H**, we still have to studentize D, that is, to replace the unknown populational constants in Φ by quantities computable from the sample. For this purpose we may replace the set $\varphi_{\lambda\nu}$ by any functions $\psi_{\lambda\nu}$ of sample means (so that the procedure described at the end of section 6 may be applied), provided that when each argument of $\psi_{\lambda\nu}$ is replaced by its expectation the result is $\varphi_{\lambda\nu}$. The studentized statistic

$$ND_1 = -N\begin{vmatrix} \Psi & C' & y' \\ C & O & 0' \\ y & 0 & 0 \end{vmatrix} : \begin{vmatrix} \Psi & C' \\ C & O \end{vmatrix}, \qquad \Psi = (\psi_{ij})\,,$$

thus obtained has the same limiting distribution as ND when the hypothesis **H** is true, for evidently the expansion (54) is not affected through the replacement of $\varphi_{\lambda\nu}$ by $\psi_{\lambda\nu}$. In their generality the functions $\psi_{\lambda\nu}$ cannot be specified by any fixed rule. In concrete cases, as manifested by the examples given below, most natural functions playing the roles of the $\psi_{\lambda\nu}$ suggest themselves.

A practical difficulty in significance tests is that there are many conceivable composite hypotheses for which one does not know how to construct a test criterion even to satisfy the single requirement of exactness. The hypothesis **H** has many special cases of this kind. When the sample is large, ND_1 may be employed as a test function as its distribution in the limit is known and is independent of any nuisance parameters. The actual test consists in computing ND_1 and referring to the χ^2 distribution with $l - h$ degrees of freedom, large values of ND_1 being significant. A further justification of the test is that its power tends generally to unity as its limit, as will be shown in the next section.

22. *Power of the test.*—If the hypothesis **H** is false, then in the expansion of D_1 the constant term is

$$a = -\begin{vmatrix} \Phi & C' & a' \\ C & O & 0' \\ a & 0 & 0 \end{vmatrix} : \begin{vmatrix} \Phi & C' \\ C & O \end{vmatrix} > 0 ,$$

where

$$a = [a_1, \cdots, a_l] , \qquad a_\lambda = f_\lambda(\mu_1, \cdots, \mu_m) ,$$

and the term with $N^{-\frac{1}{2}}$ is not in general essentially zero. Hence $N^{\frac{1}{2}}(D_1 - a)$ tends to be normally distributed about zero with a dispersion $\sigma^2 > 0$. If the test criterion based on D_1 is to reject **H** when $ND_1 \geqq c$, then the power of the test is asymptotically equal to

$$\frac{1}{\sqrt{2\pi}\,\sigma}\int_{N^{\frac{1}{2}}(N^{-1}c-a)}^{\infty} e^{-\frac{1}{2}y^2}dy ,$$

which tends to one as $N\rightarrow\infty$.

In the following six examples we consider a random vector

$$[X_1, \cdots, X_p] \tag{55}$$

as in section 12. The letters ξ_i, σ_{ij}, $\overline{X}_i$, v_{ij}, U_i, U_{ij} have the same meaning as in that section. Besides, we write

$$\sigma_{ijkl} = E(U_iU_jU_kU_l), \qquad v_{ijkl} = \frac{1}{N}\sum_{r=1}^{N}(X_{ir}-\overline{X}_i)(X_{jr}-\overline{X}_j)(X_{kr}-\overline{X}_k)(X_{lr}-\overline{X}_l).$$

The relations

$$\sigma_{ijkl} = \sigma_{ij}\sigma_{kl} + \sigma_{ik}\sigma_{jl} + \sigma_{il}\sigma_{jk} , \qquad i, j, k, l = 1, \cdots, p , \tag{56}$$

which hold true if the distribution of (55) is normal, will be called the normal moment relations. In the first two examples the conditions that the fourth moments are finite and that the X_i do not satisfy any quadratic relation with unit probability may be removed.

23. *Example 1: To test the hypothesis* H_1.[4]

$$H_1:\ \xi_i = 0 , \qquad i = 1, \cdots, p .$$

[4] The hypothesis $\xi_i = \xi_i^\circ$, $(i = 1, \ldots, p)$, may be reduced to this by using $X_i - \xi_i^\circ$ instead of X_i.

We have

$$D = -\begin{vmatrix} M & \bar{x}' \\ \bar{x} & 0 \end{vmatrix} : |M|,$$

where $M = [\sigma_{ij}]$ and $\bar{x} = [\overline{X}_1, \cdots \overline{X}_p]$. In order to studentize D it is natural to employ v_{ij} in the place of σ_{ij}. Then

$$ND_1 = -N\begin{vmatrix} V & \bar{x}' \\ \bar{x} & 0 \end{vmatrix} : |V|, \qquad V = [v_{ij}]. \tag{57}$$

The expression (57) is Hotelling's T^2-statistic (see [3] and [6]) except for a factor depending on N. Its limiting distribution when H_1 is true is the χ^2 distribution with p degrees of freedom, valid for arbitrary parent distribution.

24. *Example 2: To test the hypothesis* H₂.

$$H_2\text{:}\quad \xi_i = \xi, \qquad i = 1, \cdots, p.$$

Here

$$D = -\begin{vmatrix} M & j' & \bar{x}' \\ j & 0 & 0 \\ \bar{x} & 0 & 0 \end{vmatrix} : \begin{vmatrix} M & j' \\ j & 0 \end{vmatrix}, \qquad j = [1, 1, \cdots, 1],$$

$$ND_1 = -N\begin{vmatrix} V & j' & \bar{x}' \\ j & 0 & 0 \\ \bar{x} & 0 & 0 \end{vmatrix} : \begin{vmatrix} V & j' \\ j & 0 \end{vmatrix}. \tag{58}$$

The statistic (58) has been studied elsewhere (see [6]). Its limiting distribution when H_2 is true is the χ^2 distribution with $p - 1$ degrees of freedom, valid for arbitrary parent distribution.

25. *Example 3: To test the hypothesis* H₃.

$$H_3\text{:}\quad \sigma_{11} = \cdots = \sigma_{pp}.$$

Here

$$D = -\begin{vmatrix} \gamma_{11} & \cdots & \gamma_{1p} & 1 & v_{11} \\ \hline \gamma_{p1} & \cdots & \gamma_{pp} & 1 & v_{pp} \\ 1 & \cdots & 1 & 0 & 0 \\ v_{11} & \cdots & v_{pp} & 0 & 0 \end{vmatrix} : \begin{vmatrix} \gamma_{11} & \cdots & \gamma_{1p} & 1 \\ \hline \gamma_{p1} & \cdots & \gamma_{pp} & 1 \\ 1 & \cdots & 1 & 0 \end{vmatrix},$$

where

$$\gamma_{ij} = \sigma_{iijj} - \sigma_{ii}\sigma_{jj}.$$

If no further knowledge is assumed about the parent distribution, we may use $v_{iijj} - v_{ii}v_{jj}$ in place of γ_{ij} for studentization. If the normal moment relations are assumed, then

$$\gamma_{ij} = 2\sigma_{ij}^2.$$

We may studentize D by using v_{ij} for σ_{ij}, $(i \neq j)$, and $v = \frac{1}{p}(v_{11} + \cdots + v_{pp})$ for each σ_{ii}. Then

$$
(59)\qquad ND_1 = -N \left|\begin{array}{cccccc} 2\bar{v}^2 & 2v_{12}^2 & \cdots & 2v_{1p}^2 & 1 & v_{11} \\ 2v_{21}^2 & 2\bar{v}^2 & \cdots & 2v_{2p} & 1 & v_{22} \\ - & - & - & - & - & - \\ 2v_{p1}^2 & 2v_{p2}^2 & \cdots & 2\bar{v}^2 & 1 & v_{pp} \\ 1 & 1 & \cdots & 1 & 0 & 0 \\ v_{11} & v_{22} & \cdots & v_{pp} & 0 & 0 \end{array}\right| : \left|\begin{array}{ccccc} 2\bar{v}^2 & 2v_{12}^2 & \cdots & 2v_{1p}^2 & 1 \\ 2v_{21}^2 & 2\bar{v}^2 & \cdots & 2v_{2p}^2 & 1 \\ - & - & - & - & - \\ 2v_{p1}^2 & 2v_{p2}^2 & \cdots & 2\bar{v}^2 & 1 \\ 1 & 1 & \cdots & 1 & 0 \end{array}\right| .
$$

The limiting distribution of (59) when H_3 is true is the χ^2 distribution with $p - 1$ degrees of freedom. If $p = 2$, (59) reduces to

$$
\frac{N(v_{11} - v_{22})^2}{(v_{11} + v_{22})^2 - 4v_{12}^2},
$$

which is the test function obtained by C. T. Hsu [**5**].

26. *Example 4: To test the hypothesis,* H$_4$, *that* X$_1, \cdots,$ X$_p$ *are independent and homoscedastic.*

As a consequence of H_4 we have

$$
(60)\qquad \sigma_{ii} = \eta, \qquad \sigma_{ij} = 0, \qquad i \neq j; \qquad i, j = 1, \cdots, p.
$$

Then

$$
(61)\qquad D = -\left|\begin{array}{ccccccccccc} \gamma_{11} & & & & & & & 0 & 1 & v_{11} \\ & \ddots & & & & & & & \vdots & \vdots \\ & & \gamma_{pp} & & & & & & 1 & v_{pp} \\ & & & \sigma_{11}\sigma_{22} & & & & & 0 & v_{12} \\ & & & & \sigma_{11}\sigma_{33} & & & & 0 & v_{13} \\ & & & & & \ddots & & & \vdots & \vdots \\ 0 & & & & & & \sigma_{p-1,p-1}\sigma_{pp} & & 0 & v_{p-1,p} \\ 1 & \cdots & 1 & 0 & 0 & \cdots & 0 & & 0 & 0 \\ v_{11} & \cdots & v_{pp} & v_{12} & v_{13} & \cdots & v_{p-1,p} & & 0 & 0 \end{array}\right| : \left|\begin{array}{ccccccccc} \gamma_{11} & & & & & & & 0 & 1 \\ & \ddots & & & & & & & \vdots \\ & & \gamma_{pp} & & & & & & 1 \\ & & & \sigma_{11}\sigma_{22} & & & & & 0 \\ & & & & \sigma_{11}\sigma_{33} & & & & 0 \\ & & & & & \ddots & & & \vdots \\ 0 & & & & & & \sigma_{p-1,p-1}\sigma_{pp} & & 0 \\ 1 & \cdots & 1 & 0 & 0 & \cdots & 0 & & 0 \end{array}\right|
$$

because under the hypothesis H_4 the dispersion matrix of the vector

$$
[U_{11} - \sigma_{11}, \cdots, U_{pp} - \sigma_{pp}, U_{12}, U_{13}, \cdots, U_{p-1,p}]
$$

is the diagonal matrix figuring in (61).

We may studentize D by using v_{iiii} and v_{ii} in place of σ_{iiii} and σ_{ii}, assuming nothing further about the parent distribution. If the normal moment relations are assumed, then, taking into account (60), we have

$$\sigma_{ii} = \eta , \qquad \gamma_{ii} = 2\eta^2 .$$

Hence

$$D = \frac{1}{\eta^2}\left(\frac{1}{2}\sum_{i=1}^{p} (v_{ii} - \bar{v})^2 + \sum_{i<j} v_{ij}^2\right).$$

In order to studentize D we may replace η by $\bar{v}$. Then

$$\text{(62)} \qquad ND_1 = \frac{N}{\bar{v}^2}\left(\frac{1}{2}\sum_{i} (v_{ii} - \bar{v})^2 + \sum_{i<j} v_{ij}^2\right) = \frac{N}{2}\left(\frac{1}{\bar{v}^2}\sum_{i,j=1}^{p} v_{ij}^2 - p\right).$$

When H_4 is true, the limiting distribution of (62) is the χ^2 distribution with $\frac{1}{2}(p+2)(p-1)$ degrees of freedom.

27. *Example 5: Given that* p $= 4$, *to test the hypothesis that the three tetrad differences are zero.*

This is equivalent to

$$H_5\colon \quad \sigma_{12}\sigma_{34} = \sigma_{13}\sigma_{24} = \sigma_{14}\sigma_{23} = \theta .$$

We have

$$\text{(63)} \qquad D = -\begin{vmatrix} \varphi_{11} & \varphi_{12} & \varphi_{13} & 1 & Y_1 \\ \varphi_{21} & \varphi_{22} & \varphi_{23} & 1 & Y_2 \\ \varphi_{31} & \varphi_{32} & \varphi_{33} & 1 & Y_3 \\ 1 & 1 & 1 & 0 & 0 \\ Y_1 & Y_2 & Y_3 & 0 & 0 \end{vmatrix} : \begin{vmatrix} \varphi_{11} & \varphi_{12} & \varphi_{13} & 1 \\ \varphi_{21} & \varphi_{22} & \varphi_{23} & 1 \\ \varphi_{31} & \varphi_{32} & \varphi_{33} & 1 \\ 1 & 1 & 1 & 0 \end{vmatrix},$$

where

$$Y_1 = v_{12}v_{34}, \qquad Y_2 = v_{13}v_{24} , \qquad Y_3 = v_{14}v_{23} .$$

In the expansions of the Y_i the coefficients of $N^{-\frac{1}{2}}$ are

$$\sigma_{34}Z_{12} + \sigma_{12}Z_{34} , \qquad \sigma_{24}Z_{13} + \sigma_{13}Z_{24} , \qquad \sigma_{23}Z_{14} + \sigma_{14}Z_{23} ,$$

where Z_{ij} is the normalized sample mean of $U_{ij} - \sigma_{ij}$. Hence $[\varphi_{ij}]$ is the dispersion matrix of the three variables

$$\sigma_{34}U_{12} + \sigma_{12}U_{34} - 2\sigma_{12}\sigma_{34} , \qquad \sigma_{24}U_{13} + \sigma_{13}U_{24} - 2\sigma_{13}\sigma_{24} ,$$

$$\sigma_{23}U_{14} + \sigma_{14}U_{23} - 2\sigma_{14}\sigma_{23} .$$

$$\begin{aligned}
\varphi_{11} &= \sigma_{34}^2\sigma_{1122} + \sigma_{12}^2\sigma_{3344} + 2\sigma_{12}\sigma_{34}\sigma_{1234} - 4\sigma_{12}^2\sigma_{34}^2\,,\\
\varphi_{22} &= \sigma_{24}^2\sigma_{1133} + \sigma_{13}^2\sigma_{2244} + 2\sigma_{13}\sigma_{24}\sigma_{1324} - 4\sigma_{13}^2\sigma_{24}^2\,,\\
\varphi_{33} &= \sigma_{23}^2\sigma_{1144} + \sigma_{14}^2\sigma_{2233} + 2\sigma_{14}\sigma_{23}\sigma_{1423} - 4\sigma_{14}^2\sigma_{23}^2\,,\\
\varphi_{12} &= \sigma_{12}\sigma_{13}\sigma_{2344} + \sigma_{12}\sigma_{24}\sigma_{1433} + \sigma_{13}\sigma_{34}\sigma_{1422} + \sigma_{24}\sigma_{34}\sigma_{2311} - 4\sigma_{12}\sigma_{13}\sigma_{24}\sigma_{34}\,,\\
\varphi_{13} &= \sigma_{12}\sigma_{14}\sigma_{2433} + \sigma_{12}\sigma_{23}\sigma_{1344} + \sigma_{23}\sigma_{34}\sigma_{2411} + \sigma_{14}\sigma_{34}\sigma_{1322} - 4\sigma_{12}\sigma_{14}\sigma_{23}\sigma_{34}\,,\\
\varphi_{23} &= \sigma_{13}\sigma_{14}\sigma_{3422} + \sigma_{23}\sigma_{24}\sigma_{3411} + \sigma_{13}\sigma_{23}\sigma_{1244} + \sigma_{14}\sigma_{24}\sigma_{1233} - 4\sigma_{13}\sigma_{14}\sigma_{23}\sigma_{24}\,.
\end{aligned}$$

With no further knowledge on the parent distribution we can only studentize D by means of the fourth moments. If the normal moment relations are assumed, then

$$\begin{aligned}
\varphi_{11} &= \sigma_{11}\sigma_{22}\sigma_{34}^2 + \sigma_{33}\sigma_{44}\sigma_{12}^2 + 2a\sigma_{12}\sigma_{34}\,,\\
\varphi_{22} &= \sigma_{11}\sigma_{33}\sigma_{24}^2 + \sigma_{22}\sigma_{44}\sigma_{13}^2 + 2a\sigma_{13}\sigma_{24}\,,\\
\varphi_{33} &= \sigma_{11}\sigma_{44}\sigma_{23}^2 + \sigma_{22}\sigma_{33}\sigma_{14}^2 + 2a\sigma_{14}\sigma_{33}\,,\\
\varphi_{12} &= b + 4\sigma_{12}\sigma_{13}\sigma_{24}\sigma_{34}\,,\\
\varphi_{13} &= b + 4\sigma_{12}\sigma_{14}\sigma_{23}\sigma_{34}\,,\\
\varphi_{23} &= b + 4\sigma_{13}\sigma_{14}\sigma_{23}\sigma_{24}\,.
\end{aligned}$$

Where

$$a = \sigma_{12}\sigma_{34}+\sigma_{13}\sigma_{24}+\sigma_{14}\sigma_{23}\,, \quad b = \sigma_{11}\sigma_{23}\sigma_{24}\sigma_{34}+\sigma_{22}\sigma_{13}\sigma_{14}\sigma_{34}+\sigma_{33}\sigma_{12}\sigma_{14}\sigma_{24}+\sigma_{44}\sigma_{12}\sigma_{13}\sigma_{23}.$$

If H_5 is true, then

$$\begin{aligned}
\varphi_{11} &= \sigma_{11}\sigma_{22}\sigma_{34}^2 + \sigma_{33}\sigma_{44}\sigma_{12}^2 + 6\theta^2\,,\\
\varphi_{22} &= \sigma_{11}\sigma_{33}\sigma_{24}^2 + \sigma_{22}\sigma_{44}\sigma_{13}^2 + 6\theta^2\,,\\
\varphi_{33} &= \sigma_{11}\sigma_{44}\sigma_{23}^2 + \sigma_{22}\sigma_{33}\sigma_{14}^2 + 6\theta^2\,,\\
\varphi_{12} &= \varphi_{13} = \varphi_{23} = b + 4\theta^2\,.
\end{aligned}$$

Substituting in (63) we get by an easy computation

$$D = \frac{c_1(Y_2 - Y_3)^2 + c_2(Y_3 - Y_1)^2 + c_3(Y_1 - Y_2)^2}{c_2c_3 + c_3c_1 + c_1c_2},$$

where

$$\begin{aligned}
c_1 &= \sigma_{11}\sigma_{22}\sigma_{34}^2 + \sigma_{33}\sigma_{44}\sigma_{12}^2 - b + 2\theta^2\,,\\
c_2 &= \sigma_{11}\sigma_{33}\sigma_{24}^2 + \sigma_{22}\sigma_{44}\sigma_{13}^2 - b + 2\theta^2\,,\\
c_3 &= \sigma_{11}\sigma_{44}\sigma_{23}^2 + \sigma_{22}\sigma_{33}\sigma_{14}^2 - b + 2\theta^2\,.
\end{aligned}$$

If now we replace σ_{ij} by v_{ij} and θ by $\frac{1}{3}(Y_1 + Y_2 + Y_3)$ for studentization, we obtain after an easy reduction

$$(64)\quad ND_1 = N\frac{d_1(r_{13}r_{24}-r_{14}r_{23})^2 + d_2(r_{14}r_{23}-r_{12}r_{34})^2 + d_3(r_{12}r_{34}-r_{13}r_{24})^2}{d_2d_3 + d_3d_1 + d_1d_2},$$

where the r_{ij} are the correlation coefficients of the sample and

$$\begin{aligned}
d_1 &= r_{12}^2 + r_{34}^2 + g\,,\\
d_2 &= r_{13}^2 + r_{24}^2 + g\,,\\
d_3 &= r_{14}^2 + r_{23}^2 + g\,.
\end{aligned}$$

$$g = \tfrac{2}{9}\,(r_{12}r_{34}+r_{13}r_{24}+r_{14}r_{23})^2 - (r_{12}r_{13}r_{23}+r_{12}r_{14}r_{24}+r_{13}r_{14}r_{34}+r_{23}r_{24}r_{34})\,.$$

When H_5 is true, the limiting distribution of (64) is the χ^2 distribution with two degrees of freedom.

The following example is taken from a paper by D. N. Lawley: a sample of size N yields the correlational matrix

$$\begin{bmatrix} 1 & \cdot 4 & \cdot 4 & \cdot 2 \\ \cdot 4 & 1 & \cdot 7 & \cdot 3 \\ \cdot 4 & \cdot 7 & 1 & \cdot 3 \\ \cdot 2 & \cdot 3 & \cdot 3 & 1 \end{bmatrix}.$$

The expression (64) has the value $0.001085N$. If the hypothesis H_5 is true and N is large, the probability that (64) may exceed this is approximately $e^{-0.000543N}$, which will be significantly small only when N is several thousand. In his paper Lawley proposed another test criterion whose limiting distribution under the hypothesis H_5 is also the χ^2 distribution with two degrees of freedom and whose value for this example is $0.00113N$.

28. *Example 6: To test the hypothesis,* H_6, *that the first* s *and the last* t, (s+t=p), *of the variables* X_i *in* (55) *are independent.*

Under this hypothesis we have

$$(65) \qquad \sigma_{ij} = 0\,, \qquad i = 1, \cdots, s; \qquad j = s+1, \cdots, s+t.$$

Let the dispersion matrices of $[X_1, \cdots, X_s]$ and $[X_{s+1}, \cdots, X_{s+t}]$ be respectively M_1 and M_2, and let the matrix $V = [v_{ij}]$ be partitioned:

$$V = \begin{bmatrix} V_{11} & V_{12} \\ V_{21} & V_{22} \end{bmatrix},$$

where V_{11} has s rows and columns, V_{22} has t rows and columns.

On the basis of (65) we construct

$$D = -\begin{vmatrix} \Phi_{11} & \Phi_{12} & \cdots & \Phi_{1s} & v'_1 \\ \Phi_{21} & \Phi_{22} & \cdots & \Phi_{2s} & v'_2 \\ \text{---} & \text{---} & \text{---} & \text{---} & \text{---} \\ \Phi_{s1} & \Phi_{s2} & \cdots & \Phi_{ss} & v'_s \\ v_1 & v_2 & \cdots & v_s & 0 \end{vmatrix} : \begin{vmatrix} \Phi_{11} & \cdots & \Phi_{1s} \\ \text{---} & \text{---} & \text{---} \\ \Phi_{s1} & \cdots & \Phi_{ss} \end{vmatrix},$$

where

$$v_i = [v_{i,s+1}, \cdots, v_{i,s+t}]\,, \qquad i = 1, \cdots, s\,,$$

and Φ_{ij} is the covariance matrix of the vectors $[U_{i,s+1} - \sigma_{i,s+1}\,, \cdots, U_{i,s+t} - \sigma_{i,s+t}]$ and $[U_{j,s+1} - \sigma_{j,s+1}, \cdots, U_{j,s+t} - \sigma_{j,s+t}]$. If H_6 is true, then

$$\Phi_{ij} = \sigma_{ij} M_2\,, \qquad i, j = 1, \cdots, s\,.$$

Hence

$$\begin{bmatrix} \Phi_{11} & \cdots & \Phi_{1s} \\ \hline \Phi_{s1} & \cdots & \Phi_{ss} \end{bmatrix}^{-1} = \begin{bmatrix} a_{11}M_2^{-1} & \cdots & a_{1s}M_2^{-1} \\ \hline a_{s1}M_2^{-1} & \cdots & a_{ss}M_2^{-1} \end{bmatrix},$$

where the a_{ij} are the elements of M_1^{-1}. Then

$$D = [v_1, \cdots, v_s] \begin{bmatrix} \Phi_{11} & \cdots & \Phi_{1s} \\ \hline \Phi_{s1} & \cdots & \Phi_{ss} \end{bmatrix}^{-1} \begin{bmatrix} v'_1 \\ \cdot \\ \cdot \\ \cdot \\ v'_s \end{bmatrix}$$

$$= \sum_{i,j=1}^{s} a_{ij} v_i M_2^{-1} v'_j = tr\, M_1^{-1} V_{12} M_2^{-1} V'_{12} ;$$

in other words, D is equal to the sum of the roots of the equation

$$| V_{12} M_2^{-1} V'_{12} - \lambda M_1 | = 0 .$$

In order to studentize D we have merely to substitute v_{ij} for σ_{ij}. Then

$$D_1 = tr V_{11}^{-1} V_{12} V_{22}^{-1} V'_{12} ,$$

or the sum of the roots of the equation

$$| V_{12} V_{22}^{-1} V'_{12} - \lambda V_{11} | = 0 .$$

Hence D_1 is the sum of the canonical correlation coefficients of Hotelling [**4**]. The limiting distribution of ND_1 when H_6 is true is the χ^2 distribution with st degrees of freedom.

Let us consider the particular case of two sets of events. Let $E_1, \cdots, E_{s+1}$ and $E'_1, \cdots, E'_{t+1}$ be two sets of events each forming a complete disjunction, and let the letters p_{ij}, p_i, p'_j, n_{ij}, n_i, n'_j represent the same quantities as in section 9. If X_i, $(i = 1 \cdots, s)$, is one or zero according as E_i happens or does not happen, and if X_{s+j} $(j = 1, \cdots, t)$, is one or zero according as E'_j happens or does not happen, then H_6 becomes the following hypothesis:

$$p_{ij} = p_i p'_j , \qquad i = 1, \cdots, s+1; j = 1, \cdots, t+1 .$$

Set

$$d_{ij} = n_{ij} - \frac{n_i n'_j}{N}, \qquad i = 1, \cdots, s+1; j = 1, \cdots, t+1 ,$$

$$D = \begin{bmatrix} d_{11} \cdots d_{1t} \\ \text{— — —} \\ d_{s1} \cdots d_{st} \end{bmatrix}, \qquad \Delta_1 = \frac{1}{N}\begin{bmatrix} n_1 & & 0 \\ & \ddots & \\ 0 & & n_s \end{bmatrix}, \qquad \Delta_2 = \frac{1}{N}\begin{bmatrix} n'_1 & & 0 \\ & \ddots & \\ 0 & & n'_t \end{bmatrix},$$

$$a = \frac{1}{N}[n_1, \cdots, n_s], \qquad b = \frac{1}{N}[n'_1, \cdots, n'_t].$$

Then clearly

$$V_{11} = \Delta_1 - a'a, \qquad V_{22} = \Delta_2 - b'b, \qquad V_{12} = \frac{1}{N}D.$$

Hence

$$V_{11}^{-1} = \Delta_1^{-1} + \frac{N}{n_{s+1}}j'_s j_s, \qquad V_{22}^{-1} = \Delta_2^{-1} + \frac{N}{n_{t+1}'}j'_t j_t,$$

where

$$j_q = [1, 1, \cdots, 1], \qquad (q \text{ elements}).$$

Hence

$$ND_1 = N tr V_{11}^{-1}V_{12}V_{22}^{-1}V'_{12} = \frac{1}{N} tr\left(\Delta_1^{-1} + \frac{N}{n_{s+1}}j'_s j_s\right) D \left(\Delta_2^{-1} + \frac{N}{n'_{t+1}}j'_t j_t\right) D'$$

$$= \frac{1}{N}\left(tr\Delta_1^{-1}D\Delta_2^{-1}D' + \frac{N}{n_{s+1}}j_s D\Delta_2^{-1}D'j'_s + \frac{N}{n'_{t+1}}j_t D'\Delta_1^{-1}Dj'_t \right.$$

$$\left. + \frac{N^2}{n_{s+1}n'_{t+1}}[j_s D j'_t]^2\right)$$

$$= N\left(\sum_{i=1}^{s}\sum_{j=1}^{t}\frac{d_{ij}^2}{n_i n'_j} + \frac{1}{n_{s+1}}\sum_{j=1}^{t}\frac{d^2_{s+1,j}}{n'_j} + \frac{1}{n'_{t+1}}\sum_{i=1}^{s}\frac{d^2_{i,s+1}}{n_i} + \frac{d^2_{s+1,t+1}}{n_{s+1}n'_{t+1}}\right)$$

$$= N\sum_{i=1}^{s+1}\sum_{j=1}^{t+1}\frac{d_{ij}^2}{n_i n'_j} = \textit{mean square contingency}.$$

29. *Extension of example* 6 *to several sets of variables.*—If the hypothesis is that

$$(X_1, \cdots, X_{s_1}), (X_{s_1+1}, \cdots, X_{s_1+s_2}), \cdots, (X_{s_1+\cdots+s_{\kappa-1}+1}, \cdots, X_{s_1+\cdots+s_\kappa})$$

are mutually independent vectors, then our method of construction gives the test function $D_1 = \Sigma D_1(ij)$, $(i, j = 1, \cdots, \kappa; i < j)$, where $D_1(ij)$ is the D_1 in example 6 for the ith and jth vectors. The details of the construction are omitted. The limiting distribution of ND_1 when the hypothesis is true is the χ^2 distribution with $\Sigma s_i s_j$, $(i, j = 1, \cdots, \kappa; i < j)$, degrees of freedom.

30. *The case of* k *samples, the hypothesis* **H′**, *and the test function* $N\Delta_1$.—In this section we consider again the k random vectors (1); the notation used here is the same as in section 1. Let $f_\lambda(x_1, \cdots, x_m)$, $(\lambda = 1, \cdots, l; l < m)$, be l functions defined in the whole m-dimensional space and possessing continuous third derivatives of every kind in each of the neighborhoods of the points $(\mu_{1a}, \cdots, \mu_{ma})$, $(a = 1, \cdots, k)$. It is assumed that the matrices

$$F_a = \begin{bmatrix} f_{1a}{}^{(1)} & \cdots & f_{1a}{}^{(m)} \\ \text{---} & \text{---} & \text{---} \\ f_{la}{}^{(1)} & \cdots & f_{la}{}^{(m)} \end{bmatrix}, \qquad a = 1, \cdots, k,$$

where

$$f_{\lambda a}{}^{(i)} = \frac{\partial}{\partial \mu_{ia}} f_\lambda(\mu_{1a}, \cdots, \mu_{ma}),$$

are of rank l. Let

$$\theta_{\lambda a} = f_\lambda(\mu_{1a}, \cdots, \mu_{ma}), \qquad \lambda = 1, \cdots, l; a = 1, \cdots, k.$$

We call the hypothesis **H′** the following hypothesis:

$$\mathbf{H}': \ \theta_{\lambda 1} = \theta_{\lambda 2} = \cdots = \theta_{\lambda k}, \qquad \lambda = 1, \cdots, l.$$

Let

$$Y_{\lambda a} = f_\lambda(\bar{U}_{1a}, \cdots, \bar{U}_{ma}), \qquad y_a = [Y_{1a}, \cdots, Y_{la}],$$

$$\Delta = -\begin{vmatrix} \frac{1}{g_1}\Phi_1 & & O & I & y'_1 \\ & \ddots & & \vdots & \vdots \\ O & & \frac{1}{g_k}\Phi_k & I & y'_k \\ I & \cdots & I & O & 0' \\ y_1 & \cdots & y_k & 0 & 0 \end{vmatrix} : \begin{vmatrix} \frac{1}{g_1}\Phi_1 & & O & I \\ & \ddots & & \vdots \\ O & & \frac{1}{g_k}\Phi_k & I \\ I & \cdots & I & O \end{vmatrix},$$

where the g_a are defined in (9) and $\Phi_a = F_a[\eta_{ij}]F'_a$. Φ_a is the dispersion matrix of $[R_{1a}, \cdots, R_{la}]$, where $R_{\lambda a}$ is the coefficient of $N^{-\frac{1}{2}}$ in the expansion of $Y_{\lambda a}$ and is non-singular under our assumptions. If the hypothesis **H′** is true, the limiting distribution of $N\Delta$ (as the sample sizes become infinite in the manner specified in section 3) is the χ^2 distribution with $l(k - 1)$ degrees of freedom. This proposition is a consequence of theorem 2 and lemma 4. Its proof is similar to that for the limiting distribution of ND as set forth in section 20, and is omitted.

Now, if we write

$$y = [y_1, \cdots, y_k],$$

$$\Phi = \begin{vmatrix} \frac{1}{g_1}\Phi_1 & & O & I \\ & \cdot & & \cdot \\ & & \cdot & \cdot \\ & & & \cdot \\ O & & \frac{1}{g_k}\Phi_k & I \\ I & \cdots & I & O \end{vmatrix},$$

then

$$\Delta = [y, 0]\,\Phi^{-1}\,[y, 0]^{-1}$$

$$= y\left(\begin{bmatrix} g_1\Phi_1^{-1} & & O \\ & \cdot & \\ & \cdot & \\ & \cdot & \\ O & & g_k\Phi_k^{-1} \end{bmatrix} - \begin{bmatrix} g_1\Phi_1^{-1} \\ \cdot \\ \cdot \\ \cdot \\ g_k\Phi_k^{-1} \end{bmatrix} (g_1\Phi_1^{-1} + \cdots + g_k\Phi_k^{-1})^{-1}[g_1\Phi_1^{-1}, \cdots, g_k\Phi_k^{-1}]\right) y'.$$

Hence

$$N\Delta = \sum_a N_a y_a \Phi_a^{-1} y'_a - \left(\sum_a N_a y_a \Phi_a^{-1}\right)\left(\sum_a N_a \Phi_a^{-1}\right)^{-1}\left(\sum_a N_a \Phi_a^{-1} y'_a\right). \tag{66}$$

If $k = 2$, we write

$$y_1 + y_2 = s, \qquad y_1 - y_2 = d,$$

so that

$$y_1 = \tfrac{1}{2}(s + d), \quad y_2 = \tfrac{1}{2}(s - d),$$

and substitute in (66). Direct computation shows that the result is independent of s and is equal to

$$\tfrac{1}{4}d\{N_1\Phi_1^{-1} + N_2\Phi_2^{-1} - (N_1\Phi_1^{-1} - N_2\Phi_2^{-1})(N_1\Phi_1^{-1} + N_2\Phi_2^{-1})^{-1}(N_1\Phi_1^{-1} - N_2\Phi_2^{-1})\}d'.$$

But

$$N_1\Phi_1^{-1} + N_2\Phi_2^{-1} - (N_1\Phi_1^{-1} - N_2\Phi_2^{-1})(N_1\Phi_1^{-1} + N_2\Phi_2^{-1})^{-1}(N_1\Phi_1^{-1} - N_2\Phi_2^{-1})$$

$$= N_1\Phi_1^{-1} + N_2\Phi_2^{-1} - (\overline{N_1\Phi_1^{-1} + N_2\Phi_2^{-1}} - 2N_2\Phi_2^{-1})(N_1\Phi_1^{-1} + N_2\Phi_2^{-1})^{-1}(2N_1\Phi_1^{-1} - \overline{N_1\Phi_1^{-1} + N_2\Phi_2^{-1}}$$

$$= 4N_1N_2\Phi_2^{-1}(N_1\Phi_1^{-1} + N_2\Phi_2^{-1})^{-1}\Phi_1^{-1} = 4\left(\frac{1}{N_1}\Phi_1 + \frac{1}{N_2}\Phi_2\right)^{-1}.$$

Hence (66) reduces to

$$N\Delta = (y_1 - y_2)\left(\frac{\Phi_1}{N_1} + \frac{\Phi_2}{N_2}\right)^{-1}(y'_1 - y'_2)\,, \tag{67}$$

a result which is to be expected.

If $\Phi_a = \Phi$ for all a, then (66) and (67) reduce to

$$N\Delta = \sum_a N_a y_a \Phi^{-1} y'_a - \frac{1}{N}\left(\sum_a N_a y_a\right)\Phi^{-1}\left(\sum_a N_a y'_a\right), \tag{68}$$

$$N\Delta = \left(\frac{1}{N_1} + \frac{1}{N_2}\right)(y_1 - y_2)\Phi^{-1}(y'_1 - y'_2)\,, \quad \text{if} \quad k = 2\,. \tag{69}$$

When the unknown populational constants involved in the Φ_a in (68) are replaced by appropriate functions of sample means, we have a studentized statistic whose limiting distribution is the same as that of $N\Delta$ when the hypothesis $\mathbf{H}'$ is true. This statistic we shall call $N\Delta_1$ and propose to use as a test function for $\mathbf{H}'$ when the samples are large. The actual test consists in computing $N\Delta_1$, referring to the χ^2 distribution with $l(k - 1)$ degrees of freedom, and rejecting $\mathbf{H}'$ if $N\Delta_1$ is significantly large. The power of the test tends in general to unity as its limit, a fact which may be deduced in the same manner as done in section 22.

In the four examples which follow we consider k random vectors

$$[X_{1a}, \cdots, X_{pa}]\,, \qquad a = 1, \cdots, k\,, \tag{70}$$

each having the properties of (30) described in section 12. In example 1′ only the finiteness of the second moments and the non-singularity of the dispersion matrices need be assumed. The meanings of the symbols ξ_{ia}, σ_{ija}, σ_{ijkla}, $\overline{X}_{ia}$, v_{ija}, v_{ijkla} are self-evident.

31. *Example* 1′: *To test the hypothesis* H_1':

$$H'_1: \quad \xi_{i1} = \xi_{i2} = \cdots = \xi_{ik}\,, \qquad i = 1, \cdots, p\,.$$

Here

$$\Phi_a = M_a = [\sigma_{ija}]\,.$$

Using (66) and (67) we have

$$N\Delta = \sum_a N_a \bar{x}_a M_a^{-1} x_a^{-1} - \left(\sum_a N_a \bar{x}_a M_a^{-1}\right)\left(\sum_a N_a M_a^{-1}\right)^{-1}\left(\sum_a N_a M_a^{-1} x_a^{-1}\right),$$

where

$$\bar{x}_a = [\overline{X}_{1a}, \cdots, \overline{X}_{pa}]\,,$$

and

$$N\Delta = (\bar{x}_1 - \bar{x}_2)\left(\frac{1}{N_1}M_1 + \frac{1}{N_2}M_2\right)^{-1}(\bar{x}'_1 - \bar{x}'_2)\,, \quad \text{if } k = 2\,.$$

If no further knowledge is assumed about the parent distributions, we may studentize $N\Delta$ by employing $V_a = [v_{ija}]$ for M_a. Then

$$(71)\begin{cases} N\Delta_1 = \sum_a N_a\bar{x}_a V_a^{-1}\bar{x}'_a - \left(\sum_a N_a\bar{x}_a V_a^{-1}\right)\left(\sum_a N_a V_a^{-1}\right)^{-1}\left(\sum_a N_a V_a^{-1}\bar{x}'_a\right), \\ N\Delta_1 = (\bar{x}_1 - \bar{x}_2)\left(\frac{1}{N_1}V_1 + \frac{1}{N_2}V_2\right)^{-1}(\bar{x}'_1 - \bar{x}'_2), \quad \text{if } k = 2. \end{cases}$$

The limiting distribution of (71) when H'_1 is true is the χ^2 distribution with $p(k-1)$ degrees of freedom.

If $\sigma_{ija} = \sigma_{ij}$ for all i, j, and a, we write $M = [\sigma_{ij}]$ and get

$$N\Delta = \sum_a N_a\bar{x}_a M^{-1}\bar{x}'_a - N\bar{x}M^{-1}\bar{x}',$$

where $\bar{x} = \frac{1}{N}\sum_a N_a\bar{x}_a$ is the row vector whose components are the grand means $\overline{X}_1, \cdots, \overline{X}_p$. Hence, writing $[a_{ij}] = M^{-1}$, we have

$$(72)\ N\Delta = \sum_{a=1}^{k} N_a \sum_{i,j=1}^{p} a_{ij}\overline{X}_{ia}\overline{X}_{ja} - N\sum_{i,j=1}^{p} a_{ij}\overline{X}_i\overline{X}_j = \sum_{i,j} a_{ij}\left(\sum_a N_a\overline{X}_{ia}\overline{X}_{ja} - N\overline{X}_i\overline{X}_j\right)$$

$$= \sum_{i,j} a_{ij}\sum_a N_a(\overline{X}_{ia} - \overline{X}_i)(\overline{X}_{ja} - \overline{X}_j).$$

In order to studentize (72) we use

$$v_{ij} = \frac{1}{N}\sum_{a=1}^{k}\sum_{r=1}^{N_a}(X_{iar} - \overline{X}_i)(X_{jar} - \overline{X}_j)$$

in place of σ_{ij}. Setting $[a_{ij}] = [v_{ij}]^{-1}$ we have

$$(73)\qquad N\Delta_1 = \sum_{i,j} a_{ij}\sum_a N_a(\overline{X}_{ia} - \overline{X}_i)(\overline{X}_{ja} - \overline{X}_j).$$

Consider now the particular case of k sets of events. Let

$$(74)\qquad E_{1a}, E_{2a}, \cdots, E_{ma}, \qquad a = 1, \cdots, k,$$

be k sets of events, each forming a complete disjunction. Let $P(E_{ia}) = p_{ia}$. If X_{ia}, $(i = 1, \cdots, m-1;\ a = 1, \cdots, k)$, is one or zero according as E_{ia} happens or does not happen, then H'_1 becomes the hypothesis

$$(75)\qquad p_{i1} = p_{i2} = \cdots = p_{ik} = p_i, \qquad i = 1, \cdots, m.$$

Since

$$\sigma_{iia} = p_{ia}(1 - p_{ia}) , \qquad \sigma_{ija} = -p_{ia}p_{ja} , \qquad i \neq j ,$$

we have, when (75) is true,

$$\sigma_{iia} = p_i(1 - p_i) , \qquad \sigma_{ija} = -p_i p_j , \qquad i \neq j .$$

Hence (73) can be used. Now if N_a trials of experiment are made on the ath set (74), if the number of happenings of E_{ia} is n_{ia}, and if $n_i = \sum_a n_{ia}$, then

$$\overline{X}_{ia} = \frac{n_{ia}}{N_a} , \qquad \overline{X}_i = \frac{n_i}{N} ,$$

$$v_{ii} = \frac{n_i}{N}\left(1 - \frac{n_i}{N}\right), \qquad v_{ij} = -\frac{n_i n_j}{N^2} , \qquad i \neq j; \qquad i, j = 1, \cdots, m - 1 .$$

Hence

$$a_{ii} = \frac{N}{n_i} + \frac{N}{n_m} , \qquad a_{ij} = \frac{N}{n_m} , \qquad i \neq j ,$$

$$N\Delta_1 = \sum_{i=1}^{m} \sum_{a=1}^{k} \frac{N\left(n_{ia} - \dfrac{n_i N_a}{N}\right)^2}{n_i N_a} . \tag{76}$$

The limiting distribution of (76) when (75) is true is the χ^2 distribution with $(m - 1)(k - 1)$ degrees of freedom.

32. *Example* 2′: *To test the hypothesis* H'_2:

$$H'_2: \quad \sigma_{ija} = \sigma_{ij} , \qquad i, j = 1, \cdots, p; \; a = 1, \cdots, k .$$

Here

$$y_a = [v_{11a}, v_{12a}, \cdots, v_{22a}, v_{23a}, \cdots, v_{p-1,pa}, v_{ppa}] ,$$

and therefore Φ_a is dispersion matrix of the system

$$U_{11a} - \sigma_{11a}, U_{12a} - \sigma_{12a}, \cdots, U_{p-1,pa} - \sigma_{p-1,pa}, U_{ppa} - \sigma_{ppa} ,$$

where

$$U_{ija} = (X_{ia} - \xi_{ia})(X_{ja} - \xi_{ja}) . \tag{77}$$

The elements of Φ_a are

$$E\{(U_{ija} - \sigma_{ija})(U_{kla} - \sigma_{kla})\} = \sigma_{ijkla} - \sigma_{ija}\sigma_{kla} . \tag{78}$$

If nothing is assumed about the parent distributions, we may employ v_{ijkla} for studentization. If the normal moment relations (56) are assumed for each vector (70), then (78) reduces to $\sigma_{ika}\sigma_{jla} + \sigma_{ila}\sigma_{jka}$, which under the hypothesis H'_2 is equal to $\sigma_{ik}\sigma_{jl} + \sigma_{il}\sigma_{jk}$. Hence $\Phi_a = \Phi$ and the formula (68) can be used. Replacing each σ_{ij} by $\bar{v}_{ij} = \frac{1}{N}\sum_a N_a v_{ija}$ in Φ we get the studentized statistic $N\Delta_1$ whose limiting distribution when H'_2 is true is the χ^2 distribution with $\frac{1}{2}p(p+1)(k-1)$ degrees of freedom.

33. *Example* 3′: *To test the hypothesis* H′₃:

$$H'_3:\quad \rho_{ija} = \rho_{ij}\,, \qquad i,j = 1,\cdots,p\colon a = 1,\cdots,k\,,$$

where ρ_{ija} *is the correlation coefficient of* X_{ia} *and* X_{ja}.

Here

$$y_a = [r_{12a}, r_{13a}, \cdots, r_{23a}, \cdots, r_{p-1,pa}]\,,$$

where the r_{ija} are the sample correlation coefficients.

Now

$$r_{ija} = \frac{\bar{U}_{ija} - \bar{U}_{ia}\bar{U}_{ja}}{(\bar{U}_{iia} - \bar{U}_{ia}^2)^{\frac{1}{2}}\,(\bar{U}_{jja} - \bar{U}_{ja}^2)^{\frac{1}{2}}}\,,$$

where $U_{ia} = X_{ia} - \xi_{ia}$ and U_{ija} is defined in (77). Setting $\bar{U}_{ia} = N_a^{-\frac{1}{2}}Z_{ia}$, $\bar{U}_{ija} = \rho_{ija}(\sigma_{iia}\sigma_{jja})^{\frac{1}{2}} + N_a^{-\frac{1}{2}}Z_{ija}$ in r_{ija} and expanding in powers of $N_a^{-\frac{1}{2}}$ we obtain the following coefficients of $N_a^{-\frac{1}{2}}$:

$$\frac{Z_{ija}}{\sqrt{\sigma_{iia}\sigma_{jja}}} - \tfrac{1}{2}\rho_{ija}\left(\frac{Z_{iia}}{\sigma_{iia}} + \frac{Z_{jja}}{\sigma_{jja}}\right),$$

which is the normalized sample mean of

$$T_{ija} = \frac{U_{ija}}{\sqrt{\sigma_{iia}\sigma_{jja}}} - \tfrac{1}{2}\rho_{ija}\left(\frac{U_{iia}}{\sigma_{iia}} + \frac{U_{jja}}{\sigma_{jja}}\right).$$

Hence Φ is the dispersion matrix of the system

$$T_{12a}, T_{13a}, \cdots, T_{23a}, \cdots, T_{p-1,pa}\,.$$

The elements of Φ_a are

$$(79)\quad E(T_{ija}T_{kla}) = \tau_{ijkla} - \tfrac{1}{2}\rho_{ija}(\tau_{iikla} + \tau_{jjkla}) - \tfrac{1}{2}\rho_{kla}(\tau_{kkija} + \tau_{llija}) + \tfrac{1}{4}\rho_{ija}\rho_{kla}(\tau_{iikka} + \tau_{iilla} + \tau_{jjkka} + \tau_{jjkka})\,,$$

where

$$\tau_{ijkla} = \frac{\sigma_{ijkla}}{\sqrt{\sigma_{iia}\sigma_{jja}\sigma_{kka}\sigma_{lla}}}\,.$$

With no further knowledge on the parent distributions we can only studentize by means of v_{ijkla}. If the normal moment relations (56) are assumed for each vector (70), then

$$\tau_{ijkl} = \rho_{ija}\rho_{kla} + \rho_{ika}\rho_{jla} + \rho_{ila}\rho_{jka} .$$

If the hypothesis H'_3 is also taken into account, (79) becomes

$$\rho_{ik}\rho_{jl} + \rho_{il}\rho_{jk} - (\rho_{ij}\rho_{ik}\rho_{il} + \rho_{ij}\rho_{jk}\rho_{jl} + \rho_{ik}\rho_{jk}\rho_{kl} + \rho_{il}\rho_{jl}\rho_{kl}) + \tfrac{1}{2}\rho_{ij}\rho_{kl}(\rho_{ik}^2 + \rho_{il}^2 + \rho_{jk}^2 + \rho_{jl}^2) .$$

Hence in this case $\Phi_a = \Phi$ and the formula (68) can be used. The studentization consists in replacing ρ_{ij} by one of the following functions:

$$\frac{1}{k}\sum_a r_{ija} , \qquad \frac{1}{N}\sum N_a r_{ija} , \qquad \sum_a N_a v_{ija} / \sum_a N_a (v_{iia} v_{jja})^{\frac{1}{2}} .$$

Then we get a test function $N\Delta_1$ whose limiting distribution when H'_3 is true is the χ^2 distribution with $\frac{1}{2}p(p-1)(k-1)$ degrees of freedom.

34. *Example* 4′: *Given that*

$$(80) \qquad \sigma_{iia} = \sigma_{ii} , \qquad i = 1, \cdots, p; a = 1, \cdots, k ,$$

to test the hypothesis H'_4:

$$H'_4: \quad \sigma_{ija} = \sigma_{ij} , \qquad i \neq j; i, j = 1, \cdots, p; a = 1, \cdots, k .$$

Here

$$y_a = [v_{12a}, v_{13a}, \cdots, v_{23a}, \cdots, v_{p-1,pa}] .$$

Hence Φ_a is the dispersion matrix of the system

$$U_{12a} - \sigma_{12a}, U_{13a} - \sigma_{13a}, \cdots, U_{23a} - \sigma_{23a}, \cdots, U_{p-1,pa} - \sigma_{p-1,pa} .$$

Φ_a is a certain arrangement of the elements

$$(81) \qquad \sigma_{ijkla} - \sigma_{ija}\sigma_{kla} .$$

Without any further knowledge about the parent distribution we have to employ v_{ijkla} for studentization. If the normal moment relations (56) are assumed for each vector (70), then (81) becomes $\sigma_{ika}\sigma_{jla} + \sigma_{ila}\sigma_{jka} = \sigma_{ik}\sigma_{jl} + \sigma_{il}\sigma_{jk}$ under the assumption (80) and the hypothesis H'_4. Hence formula (68) can be used. Replacing each σ_{ij} by $\bar{v}_{ij} = \frac{1}{N}\sum_a N_a v_{ija}$, we get a test function $N\Delta_1$ whose limiting distribution when H'_4 is true is the χ^2 distribution with $\frac{1}{2}p(p-1)(k-1)$ degrees of freedom.

A final remark

One of our assumptions on the statistic T in (7) is that the function f is defined in the whole mk-dimensional space. But we give examples in which the functions playing the role of f have less extensive domains of definition. This difficulty may be overcome by observing that we can extend the definition of the functions in question by assigning any constant value, for example zero, as the value of the functions outside their natural domains of definition. The same consideration applies to the functions in sections 19 and 20.

Reprinted from
J. Chinese Math. Soc. (New Series)
1 (1951), 257–280.

ABSOLUTE MOMENTS AND CHARACTERISTIC FUNCTION*†

1. Introduction and Definitions

In this paper we study the relation between the absolute moments and the characteristic function (c.f.) of a distribution law. A distribution law (which for brevity will be called "law" throughout this paper) is a function $F(x)$ defined and non-decreasing on the whole line $-\infty < x < \infty$ and satisfying the condition

$$F(-\infty) = 0, \qquad F(\infty) = 1. \tag{1}$$

The c.f. of a law $F(x)$ is the function

$$f(t) = \int_{-\infty}^{\infty} e^{itx}\, dF(x), \tag{2}$$

and the absolute moment of order β ($\beta \gtreqless 0$) of a law $F(x)$ is the quantity

$$M_\beta(F) = \int_{-\infty}^{\infty} |x|^\beta\, dF(x).^1 \tag{3}$$

We shall also consider the quantity

$$L_\beta(F) = \int_{|x|>e} |x|^\beta \log|x|\, dF(x),^2 \tag{4}$$

but only for non-negative integral values of β.

The evident facts that $f(-t) = \overline{f(t)}$, $|f(t)| \leqq 1 = f(0)$, that $\Re\{f(t)\}$ and $\mathfrak{g}\{f(t)\}$ are respectively even and odd functions, will be used frequently in this paper without any explicit mention. For a fuller knowledge of laws and c.f.'s we refer to the books by P. Levy [4] and H. Cramér [1].

In the following investigation we shall establish various identity relations between $M_\beta(F)$ and $f(t)$, and various necessary and sufficient conditions, in terms of the c.f., for the finiteness of the absolute moment of a given positive order [including one for

*Received January 19, 1951.

† *Editor's note*: A fairly long summary in Chinese is not reprinted here.

[1] $M_0(F)$ is defined as $F(\infty) \cdot F(-\infty) = 1$. If $\beta < 0$ and $F(x)$ has a discontinuity at $x = 0$, then $M_\beta(F)$ is defined to be ∞. If $\beta < 0$ and $F(x)$ is continuous at $x = 0$, then in the definition (3) of $M_\beta(F)$ the point $x = 0$ is omitted from the domain of integration.

[2] If $\beta > 0$, we may as well consider $\int_{-\infty}^{\infty} |x|^\beta \log|x|\, dF(x)$ so far as the finiteness of the latter is concerned. The precaution $|x| > e$ is made only to avoid the anomaly at $x = 0$, when $\beta = 0$.

the finiteness of $L_{2n}(F)$]. Some results about moments of negative order are also touched upon.

If $M_k(F) < \infty$, where k is a non-negative integer, then, as is immediately seen from the permissibility of differentiation under the integration sign in (2), $f(t)$ is k-time differentiable, and

$$f(t) = 1 + tf'(0) + \frac{1}{2}t^2f''(0) + \cdots + \frac{1}{k!}t^kf^{(k)}(0) + o(t^k) \qquad (t\to 0), \tag{5}$$

with

$$f^{(\nu)}(0) = i^\nu \int_{-\infty}^{\infty} x^\nu \, dF(x) \qquad (\nu = 1,2, \ldots, k). \tag{6}$$

Fortet [2] has shown that if k is even and if $f(t)$ admits an expansion of the form

$$f(t) = 1 + a_1 it + \cdots + a_k(it)^k + o(t^k) \qquad (t\to 0), \tag{7}$$

then $M_k(F) < \infty$; if k is odd and if (7) is replaced by the stronger condition

$$f(t) = 1 + a_i it + \cdots + a_k(it)^k + O(|t|^{k+\epsilon}) \qquad (\epsilon > 0), \tag{8}$$

then $M_k(F) < \infty$. Although several of our theorems include Fortet's theorem as a special case, their proofs are much simpler.

At the end of this paper we shall establish a relation between the Stieltjes transform and the c.f.

2. Absolute Moments of Positive Fractional Order

We shall briefly indicate the derivation of the following known formula:

$$x^n e^{-(1/4)x^2} = \frac{(-i)^n}{\sqrt{\pi}} \int_{-\infty}^{\infty} e^{itz - t^2} H_n(t)\, dt, \tag{9}$$

where $H_n(t)$ is the Hermite polynomial:

$$H_n(t) = (-1)^n e^{t^2} \frac{d^n}{dt^n} e^{-t^2}. \tag{10}$$

Starting from the identity

$$\int_{-\infty}^{\infty} e^{-itx} e^{-(1/4)x^2}\, dx = 2\sqrt{\pi}\, e^{-t^2}, \tag{11}$$

which may be verified by expanding e^{-itx} and integrating term by term, we differentiate it n times and obtain

$$(-i)^n \int_{-\infty}^{\infty} x^n e^{-itx-(1/4)x^2}\, dx = (-1)^n 2\sqrt{\pi}\, H_n(t) e^{-t^2}. \tag{12}$$

Formula (9) is then derived from (12) by the Fourier reciprocity formula.

In (9), replace x by xy and n by $2n$:

$$x^{2n} e^{-(1/4)y^2x^2} = \frac{(-1)^n y^{-2n}}{\sqrt{\pi}} \int_{-\infty}^{\infty} e^{itxy - t^2} H_{2n}(t)\, dt. \tag{13}$$

Let $F(x)$ be a law; integrate both sides of (13) with respect to $F(x)$:

$$\int_{-\infty}^{\infty} x^{2n} e^{-(1/4)y^2x^2}\, dF(x) = \frac{(-1)^n y^{-2n}}{\sqrt{\pi}} \int_{-\infty}^{\infty} dF(x) \int_{-\infty}^{\infty} e^{ityx-t^2} H_{2n}(t)\, dt$$

$$= \frac{(-1)^n y^{-2n}}{\sqrt{\pi}} \int_{-\infty}^{\infty} e^{-t^2} H_{2n}(t)\, dt \int_{-\infty}^{\infty} e^{ityx}\, dF(x)$$

$$= \frac{(-1)^n y^{-2n}}{\sqrt{\pi}} \int_{-\infty}^{\infty} e^{-t^2} H_{2n}(t) f(yt)\, dt, \tag{14}$$

the change of order of integration involved in the above computation being easily justifiable.

Let β be any positive number; multiply both sides of (14) by $y^{\beta-1}$ and integrate with respect to y over $0 < y < \infty$: If either $2n - \beta > 0$ or $F(x)$ is continuous at $x = 0$, then,

$$\int_0^{\infty} y^{\beta-1}\, dy \int_{-\infty}^{\infty} x^{2n} e^{-(1/4)y^2x^2}\, dF(x) = \int_0^{\infty} y^{\beta-1}\, dy \int_{|x|>0} x^{2n} e^{-(1/4)y^2x^2}\, dF(x)$$

$$= \int_{|x|>0} x^{2n}\, dF(x) \int_0^{\infty} y^{\beta-1} e^{-(1/4)y^2x^2} dy = \int_{|x|>0} |x|^{2n-\beta} dF(x) \int_0^{\infty} u^{\beta-1} e^{-(1/4)u^2}\, du$$

$$= 2^{\beta-1}\Gamma\left(\frac{\beta}{2}\right) M_{2n-\beta}(F),$$

which is to be equated to the result of the same operation applied to the right side of (14). Therefore, *if* $\beta > 0$, n *is a non-negative integer, and if either* $2n - \beta > 0$ *or* $F(x)$ *is continuous at* $x = 0$, we have

$$M_{2n-\beta}(F) = \frac{(-1)^n 2^{1-\beta}}{\sqrt{\pi}\,\Gamma(\beta/2)} \int_0^{\infty} y^{-2n+\beta-1}\, dy \int_{-\infty}^{\infty} e^{-t^2} H_{2n}(t) f(yt)\, dt. \tag{15}$$

By means of (15) the absolute moment of any order is expressed in terms of the c.f.

Let $k + \gamma$ ($k \geqq 0$ and integral, $0 < \gamma < 1$) be any positive fraction. Setting $n = [k/2] + 1$ and $\beta = 2[k/2] + 2 - k - \gamma$ in (15) we get

$$M_{k+\gamma}(F) = \frac{(-1)^{[k/2]+1} 2^{k-2[k/2]+\gamma-1}}{\sqrt{\pi}\,\Gamma([k/2] - k/2 + 1 - \gamma/2)}$$

$$\times \int_0^{\infty} y^{-k-\gamma-1}\, dy \int_{-\infty}^{\infty} e^{-t^2} H_{2[k/2]+2}(t) f(yt)\, dt. \tag{16}$$

Both formula (16) and the following lemma are preparatory for the proof of Theorem 2.1.

If a c.f. $f(t)$ is k-time differentiable, we shall always denote by $P_k(t)$ the following polynomial.

$$P_k(t) = \sum_{\nu=0}^{k} \frac{1}{\nu!} t^{\nu} f^{(\nu)}(0). \tag{17}$$

Lemma. *Let $F(x)$ be a law, $f(t)$ be its c.f., k be a non-negative integer, and $0 < \gamma < 1$. If $M_{k+\gamma}(F) < \infty$, then*

$$f(t) = P_k(t) + O\big(|t|^{k+\gamma}\big). \tag{18}$$

Proof. The finiteness of $M_{k+\gamma}(F)$ implies that of $M_k(F)$ and hence the k-time differentiability of $f(t)$. We have

$$f^{(k)}(t) = \int_{-\infty}^{\infty} (ix)^k e^{itx}\, dF(x),$$

$$f^{(k)}(t) - f^{(k)}(0) = \int_{-\infty}^{\infty} (ix)^k (e^{itx} - 1)\, dF(x),$$

$$|f^{(k)}(t) - f^{(k)}(0)| \leqq \int_{-\infty}^{\infty} |x|^k |e^{itx} - 1|\, dF(x) = 2\int_{-\infty}^{\infty} |x|^k |\sin \tfrac{1}{2} tx|\, dF(x)$$

$$\leqq 2\int_{-\infty}^{\infty} |x|^k |\sin \tfrac{1}{2} tx|^{\gamma}\, dF(x) \leqq 2^{1-\gamma}|t|^{\gamma}\int_{-\infty}^{\infty} |x|^{k+\gamma}\, dF(x) \leqq C|t|^{\gamma}.$$ [3]

The above inequality proves the lemma in case $k = 0$. For $k > 0$ we have

$$|f(t) - P_k(t)| = \left| \frac{1}{(k-1)!} \int_0^t (t-u)^{k-1} \{ f^{(k)}(u) - f^{(k)}(0) \}\, du \right|$$

$$\leqq C\int_0^{|t|} (|t| - u)^{k-1} |f^{(k)}(u) - f^{(k)}(0)|\, du$$

$$\leqq C\int_0^{|t|} (|t| - u)^{k-1} u^{\gamma}\, du = C|t|^{k+\gamma}.$$

Hence the lemma is proved.

Theorem 2.1. *Let $F(x)$, $f(t)$, k, γ have the same meaning as in the preceding lemma. In order that $M_{k+\gamma}(F) < \infty$, it is necessary and sufficient that, in a neighborhood $|t| < \delta$, $f(t)$ should admit an expansion of the following form.*

$$f(t) = Q_k(t) + O\big(|t|^{k+\gamma}\psi(t)\big), \tag{19}$$

where $Q_k(t)$ is a polynomial of degree k, and $\psi(t)$ is a function defined, non-negative and bounded for $0 < |t| < \delta$ and satisfying the condition[4]

$$\int_{-\delta}^{\delta} |t|^{-1}\psi(t)\, dt < \infty. \tag{20}$$

Further, if (19) *and* (20) *are fulfilled, then $Q_k(t) = P_k(t)$ and*

$$M_{k+\gamma}(F) = (-1)^{[k/2]+1} 2\pi^{-1}\Gamma(k+\gamma+1)\sin\left(\frac{k}{2} - \left[\frac{k}{2}\right] + \frac{\gamma}{2}\right)\pi$$

$$\times \int_0^{\infty} \frac{\Re\{f(t) - P_k(t)\}}{t^{k+\gamma+1}}\, dt. \tag{21}$$

[3] From now on we shall let the letter C denote any positive constant, not necessarily the same at each occurrence.

[4] Examples of $\psi(t)$ are $(\log(1/|t|))^{1+\epsilon}$, $\log(1/|t|)(\log\log(1/|t|))^{1+\epsilon}$, etc., with $\epsilon > 0$.

Proof. The condition is necessary. We need only show that, assuming $M_{k+\gamma}(F) < \infty$, we have

$$\int_{-\infty}^{\infty} |t|^{-k-\gamma-1} |f(t) - P_k(t)|\, dt < \infty, \tag{22}$$

for then the functions

$$Q_k(t) = P_k(t), \qquad \psi(t) = |t|^{-k-\gamma} |f(t) - P_k(t)|$$

will satisfy (19) and (20) and, besides, $\psi(t)$ is bounded, by the lemma just proved. To establish (22), set

$$E_k(t) = \sum_{\nu=0}^{k} \frac{1}{\nu!} (it)^\nu.\text{[5]} \tag{23}$$

Then

$$\begin{aligned}\int_{-\infty}^{\infty} \frac{|f(t) - P_k(t)|}{|t|^{k+\gamma+1}}\, dt &= \int_{-\infty}^{\infty} \frac{dt}{|t|^{k+\gamma+1}} \left| \int_{-\infty}^{\infty} \{e^{itx} - E_k(tx)\}\, dF(x) \right| \\ &\leqq \int_{-\infty}^{\infty} \frac{dt}{|t|^{k+\gamma+1}} \int_{-\infty}^{\infty} |e^{itx} - E_k(tx)|\, dF(x) \\ &= \int_{-\infty}^{\infty} dF(x) \int_{-\infty}^{\infty} \frac{|e^{itx} - E_k(tx)|}{|t|^{k+\gamma+1}}\, dt \\ &= \int_{-\infty}^{\infty} |x|^{k+\gamma}\, dF(x) \int_{-\infty}^{\infty} \frac{|e^{iu} - E_k(u)|}{|u|^{k+\gamma+1}}\, du < \infty,\end{aligned}$$

which proves (22).

The condition is sufficient. The equations (19) and (20) together imply that

$$\int_{-\delta}^{\delta} |t|^{-k-\gamma-1} |f(t) - Q_k(t)|\, dt < \infty.$$

Then,

$$\int_{-\infty}^{\infty} |t|^{-k-\gamma-1} |f(t) - Q_k(t)|\, dt < \infty, \tag{24}$$

because the part of the integral extended over $|t| > \delta$ is evidently finite.

By virtue of the fact that

$$\int_{-\infty}^{\infty} e^{-t^2} H_m(t) t^l\, dt = 0 \qquad \text{for} \quad 0 \leqq l < m \tag{25}$$

(which results from the orthogonality property of the Hermite polynomials), we may write (16) as

$$M_{k+\gamma}(F) = C \int_0^{\infty} y^{-k-\gamma-1}\, dy \int_{-\infty}^{\infty} e^{-t^2} H_{2[k/2]+2}(t) \{f(yt) - Q_k(yt)\}\, dt. \tag{26}$$

---

[5]We shall stick to this notation throughout the rest of this paper.

Therefore,

$$M_{k+\gamma}(F) \leqq C\int_{\theta}^{\infty} y^{-k-\gamma-1}\,dy \int_{-\infty}^{\infty} e^{-t^2}|H_{2[k/2]+2}(t)|\,|f(yt)-Q_k(yt)|\,dt$$

$$= C\int_{-\infty}^{\infty} e^{-t^2}|H_{2[k/2]+2}(t)|\,dt \int_{0}^{\infty} y^{-k-\gamma-1}|f(yt)-Q_k(yt)|\,dy$$

$$= C\int_{-\infty}^{\infty} e^{-t^2}|H_{2[k/2]+2}(t)|\,|t|^{k+\gamma}\,dt \int_{0}^{\infty} u^{-k-\gamma-1}|f(u)-Q_k(u)|\,du.$$

By (24), $M_{k+\gamma}(F) < \infty$. Thus the sufficiency of the condition is proved. Then $f(t)$ is k-time differentiable and so we must have $Q_k(t) = P_k(t)$. Replacing $Q_k(yt)$ by $P_k(yt)$ in (26) and reversing the order of integration we obtain

$$M_{k+\gamma}(F) = C\int_{-\infty}^{\infty} e^{-t^2}H_{2[k/2]+2}(t)\,dt \int_{0}^{\infty} y^{-k-\gamma-1}\{f(yt)-P_k(yt)\}\,dy$$

$$= C\int_{0}^{\infty} e^{-t^2}H_{2[k/2]+2}(t)\,dt \int_{0}^{\infty} y^{-k-\gamma-1}\Re\{f(yt)-P_k(yt)\}\,dy$$

$$= C\int_{0}^{\infty} e^{-t^2}H_{2[k/2]+2}(t)t^{k+\gamma}\,dt \int_{0}^{\infty} u^{-k-\gamma-1}\Re\{f(u)-P_k(u)\}\,du$$

$$= A\int_{0}^{\infty} u^{-k-\gamma-1}\Re\{f(u)-P_k(u)\}\,du, \tag{27}$$

where A is some constant independent of $F(x)$. We choose the particular pair

$$F(x) = \frac{1}{2}\int_{-\infty}^{x} e^{-|y|}\,dy, \qquad f(t) = \frac{1}{2}\int_{-\infty}^{\infty} e^{itx-|x|}\,dx = \frac{1}{1+t^2}.$$

For this pair we have

$$M_{k+\gamma}(F) = \frac{1}{2}\int_{-\infty}^{\infty} |x|^{k+\gamma}e^{-|x|}\,dx = \Gamma(k+\gamma+1),$$

$$f(t) - P_k(t) = (-1)^{[k/2]+1}(1+t^2)^{-1}t^{2[k/2]+2}$$

Substituting in (27) we obtain

$$\Gamma(k+\gamma+1) = A\int_{0}^{\infty}(-1)^{[k/2]+1}t^{2[k/2]-k+1-\gamma}(1+t^2)^{-1}\,dt$$

$$= \frac{(-1)^{[k/2]+1}\pi A}{2\sin(k/2-[k/2]+\gamma/2)\pi}. \tag{28}$$

(27) and (28) together lead to (21). The proof of Theorem 2.1 is thus complete.

We note the following special case of Theorem 2.1:

Theorem 2.2. *If in a neighborhood* $|t| < \delta$ *we have*

$$f(t) = Q_k(t) + O(|t|^{k+\alpha}), \qquad (\alpha > 0), \tag{29}$$

where $Q_k(t)$ *is a polynomial of degree* k, *then* $M_{k+\gamma}(F) < \infty$ *for every* $0 < \gamma < \alpha$.

This theorem, which follows from Theorem 2.1 on putting $\psi(t) = |t|^{\alpha-\gamma}$, contains Fortet's result, referred to in §1, that (29) implies the finiteness of $M_k(F)$.

3. Moments of Non-negative Even Order

If $\phi(t)$ is any function defined for $-\infty < t < \infty$, we write

$$T_n(\phi) = \sum_{\nu=0}^{n} (-1)^\nu \binom{n}{\nu} \phi\left(\left(\frac{n}{2} - \nu\right)t\right) \tag{30}$$

The quantity $\phi^{[n]}(0) = \lim_{t\to 0} t^{-n} T_n(\phi)$, if it exists, is known as the generalized nth derivative of $\phi(t)$ at $t = 0$. In this connection two familiar facts are that $\phi^{[n]}(0) = \phi^{(n)}(0)$ provided the latter exists, and that $T_n(\phi) = 0$ for an even or an odd function $\phi(t)$, according as n is odd or even.

When we apply T_n to e^{itx}, regarding x as constant, we get

$$T_n(e^{itx}) = \sum_{\nu=0}^{n} (-1)^\nu \binom{n}{\nu} e^{(n/2-\nu)itx} = \left(e^{(1/2)itx} - e^{-(1/2)itx}\right)^n = (2i)^n \sin^n \tfrac{1}{2}tx. \tag{31}$$

We also have

$$T_n(t^l) = 0 \qquad \text{for} \quad 0 \leqq l < n. \tag{32}$$

To see this, observe that the right side of (31) is a power series beginning with the term t^n. If we apply T_n term by term to the exponential series e^{itx}, the resulting series must also begin with the term t^n, and this implies (32).

Theorem 3.1.[6] *In order that $M_{2n}(F)$ be finite, it is necessary and sufficient that we should have*

$$T_{2n}(f) = O(t^{2n}), \tag{33}$$

where $f = f(t)$ is the c.f. of the law $F(x)$.

Proof. The necessity of the condition follows from the existence of $f^{[2n]}(0) = f^{(2n)}(0)$. To prove the sufficiency, we have, by (31),

$$t^{-2n} T_{2n}(f) = (-1)^n \int_{-\infty}^{\infty} \left(\frac{\sin \frac{1}{2}tx}{\frac{1}{2}t}\right)^{2n} dF(x). \tag{34}$$

If now $t^{-2n}|T_{2n}(f)| \leqq K$, then

$$K \geqq \int_{-\infty}^{\infty} \left(\frac{\sin \frac{1}{2}tx}{\frac{1}{2}t}\right)^{2n} dF(x) \geqq \int_{|x|<\pi|t|^{-1}} \left(\frac{\sin \frac{1}{2}tx}{\frac{1}{2}tx}\right)^{2n} x^{2n}\, dF(x). \tag{35}$$

But

$$\left|\frac{\sin\theta}{\theta}\right| > \frac{2}{\pi} \qquad \text{for} \quad |\theta| < \frac{\pi}{2}.$$

[6]Compare with Levy [4], p. 174.

Applying this to (35) we get

$$K \geqq \left(\frac{2}{\pi}\right)^{2n} \int_{|x| < \pi |t|^{-1}} x^{2n}\, dF(x). \tag{36}$$

Making $|t| \to 0$ in (36) we conclude that $M_{2n}(F)$ is finite.

The following two theorems are simple corollaries to Theorem 3.1.

Theorem 3.2. *A necessary and sufficient condition for the finiteness of* $M_{2n}(F)$ *is that*

$$f(t) = Q_{2n-1}(t) + O(t^{2n}), \tag{37}$$

where $Q_{2n-1}(t)$ *is a polynomial of degree* $2n - 1$.

For we have, by (32), $T_{2n}(Q_{2n-1}) = 0$, whence $T_{2n}(f) = O(t^{2n})$.

Theorem 3.3. *A necessary and sufficient condition for the finiteness of* $M_{2n}(F)$ *is that*

$$f(t) = Q_{2n}(t) + o(t^{2n}) \qquad (t \to 0), \tag{38}$$

where $Q_{2n}(t)$ *is a polynomial of degree* $2n$.

This is Fortet's criterion for the finiteness of an even-ordered moment, as referred to in §1.

Theorem 3.4. *In order that* $M_{2n}(F) < \infty$ $(n > 0)$, *it is necessary that, in a neighborhood* $|t| < \delta$,

$$\mathfrak{g}\{f(t)\} = \mathfrak{g}\{P_{2n}(t)\} + O\left(t^{2n}\psi(t)\right), \tag{39}$$

where $\psi(t)$ *satisfies the same conditions as in Theorem* 2.1.

Proof. We need only verify that

$$\int_{-\infty}^{\infty} |t|^{-2n-1} |\mathfrak{g}\{f(t) - P_{2n}(t)\}|\, dt = \int_{-\infty}^{\infty} |t|^{-2n-1} |\mathfrak{g}\{f(t) - P_{2n-1}(t)\}|\, dt < \infty.$$

Now,

$$\begin{aligned}
&\int_{-\infty}^{\infty} |t|^{-2n-1} |\mathfrak{g}\{f(t) - P_{2n-1}(t)\}|\, dt \\
&\quad = \int_{-\infty}^{\infty} |t|^{-2n-1}\, dt \left| \int_{-\infty}^{\infty} \mathfrak{g}\left\{e^{itx} - E_{2n-1}(tx)\right\} dF(x) \right| \\
&\quad \leqq \int_{-\infty}^{\infty} |t|^{-2n-1}\, dt \int_{-\infty}^{\infty} \left|\mathfrak{g}\left\{e^{itx} - E_{2n-1}(tx)\right\}\right| dF(x) \\
&\quad = \int_{-\infty}^{\infty} dF(x) \int_{-\infty}^{\infty} |t|^{-2n-1} \left|\mathfrak{g}\left\{e^{itx} - E_{2n-1}(tx)\right\}\right| dt \\
&\quad = \int_{-\infty}^{\infty} x^{2n}\, dF(x) \int_{-\infty}^{\infty} |u|^{-2n-1} \left|\mathfrak{g}\left\{e^{iu} - E_{2n-1}(u)\right\}\right| du < \infty,
\end{aligned}$$

which gives the proof.

Theorem 2.1, which gives a criterion for the finiteness of a fractional-ordered absolute moment $M_{k+\gamma}(F)$, is no longer true for integral-ordered absolute moments ($\gamma = 0$). The counterpart of Theorem 2.1 for the case $k = 2n$, $\gamma = 0$ is the following.

Theorem 3.5. *In order that $L_{2n}(F)$*[7] *be finite, it is necessary and sufficient that, in a neighborhood $|t| < \delta$, we should have*

$$f(t) = Q_{2n}(t) + O\big(t^{2n}\psi(t)\big), \tag{40}$$

where $Q_{2n}(t)$ is a polynomial of degree $2n$ and $\psi(t)$ satisfies the same conditions as in Theorem 2.1.

Proof. The condition is necessary. Since the finiteness of $L_{2n}(F)$ implies that of $M_{2n}(F)$, we may set $Q_{2n}(t) = P_{2n}(t)$. We need only verify that

$$\int_{-1}^{1} |t|^{-2n-1} |f(t) - P_{2n}(t)|\, dt = 2\int_0^1 t^{-2n-1} |f(t) - P_{2n}(t)|\, dt < \infty. \tag{41}$$

We have

$$\begin{aligned}
\int_0^1 t^{-2n-1} |f(t) - P_{2n}(t)|\, dt &= \int_0^1 t^{-2n-1}\, dt \left| \int_{-\infty}^{\infty} \{e^{itx} - E_{2n}(tx)\}\, dF(x) \right| \\
&\leqq \int_0^1 t^{-2n-1}\, dt \int_{-\infty}^{\infty} |e^{itx} - E_{2n}(tx)|\, dF(x) \\
&= \int_{-\infty}^{\infty} dF(x) \int_0^1 t^{-2n-1} |e^{itx} - E_{2n}(tx)|\, dt \\
&= \int_{-\infty}^{\infty} x^{2n}\, dF(x) \int_0^{|x|} u^{-2n-1} |e^{iu} - E_{2n}(u)|\, du \\
&= C + \int_{|x|>e} x^{2n}\, dF(x) \int_e^{|x|} u^{-2n-1} |e^{iu} - E_{2n}(u)|\, du \\
&\leqq C + \int_{|x|>e} x^{2n}\, dF(x) \int_e^{|x|} C u^{-1}\, du \\
&= C + C \int_{|x|>e} x^{2n} \log|x|\, dF(x) < \infty,
\end{aligned}$$

which constitutes proof.

The condition is sufficient. By virtue of Theorem 3.2, the condition (40) implies the finiteness of $M_{2n}(F)$. Therefore, $Q_{2n}(t) = P_{2n}(t)$ and we have

$$\int_0^1 t^{-2n-1} |f(t) - P_{2n}(t)|\, dt < \infty.$$

whence, *a fortiori*

$$\int_0^1 t^{-2n-1} |\Re\{f(t) - P_{2n}(t)\}|\, dt < \infty. \tag{41a}$$

[7] For definition, see equation (4).

Since $\Re\{e^{it} - E_{2n}(t)\}$ is a function which does not change sign, we have by (41a),

$$\infty > \int_0^1 t^{-2n-1}\,dt \left| \int_{-\infty}^{\infty} \Re\{e^{itx} - E_{2n}(tx)\}\,dF(x) \right|$$

$$= \int_0^1 t^{-2n-1}\,dt \int_{-\infty}^{\infty} |\Re\{e^{itx} - E_{2n}(tx)\}|\,dF(x)$$

$$= \int_{-\infty}^{\infty} dF(x) \int_0^1 t^{-2n-1} |\Re\{e^{itx} - E_{2n}(tx)\}|\,dt$$

$$= \int_{-\infty}^{\infty} x^{2n}\,dF(x) \int_0^{|x|} u^{-2n-1} |\Re\{e^{iu} - E_{2n}(u)\}|\,du$$

$$\geqq \int_{|x|>c} x^{2n}\,dF(x) \int_e^{|x|} u^{-2n-1} |\Re\{e^{iu} - E_{2n}(u)\}|\,du$$

$$\geqq \int_{|x|>c} x^{2n}\,dF(x) \left\{ -C + C\int_e^{|x|} u^{-1}\,du \right\} = -C + C\int_{|x|>e} x^{2n} \log|x|\,dF(x),$$

which implies the finiteness of $L_{2n}(F)$.

4. Absolute Moments of Positive Odd Order

We begin by giving an example showing that the $(2n+1)$-time differentiability of the c.f. at $t = 0$ does not imply the finiteness of $M_{2n+1}(F)$.

Let

$$p(x) = \frac{1}{x^{2n+2}\log|x|} \quad (|x| \geqq e), \quad p(x) = 0 \quad (|x| < e),$$

$$F(x) = c\int_{-\infty}^{x} p(y)\,dy,$$

where c is a constant so determined as to make $F(\infty) = 1$. Then

$$f(t) = 2c\int_e^{\infty} \frac{\cos tx}{x^{2n+2}\log x}\,dx, \qquad f^{(2n)}(t) = \pm 2c\int_e^{\infty} \frac{\cos tx}{x^2\log x}\,dx,$$

$$\left| \frac{f^{(2n)}(t) - f^{(2n)}(0)}{t} \right| = \frac{2c}{|t|}\int_e^{\infty} \frac{1-\cos tx}{x^2\log x}\,dx$$

$$\leqq \frac{4c}{|t|}\int_e^{|t|^{-1}} \frac{\sin^2(tx/2)}{x^2\log x}\,dx + \frac{4c}{|t|}\int_{|t|^{-1}}^{\infty} \frac{dx}{x^2\log x}$$

$$\leqq c|t|\int_e^{|t|^{-1}} \frac{dx}{\log x} - \frac{4c}{\log|t|} \leqq c|t|\int_e^{|t|^{-1/2}} \frac{dx}{\log x} - \frac{2c}{\log|t|} - \frac{4c}{\log|t|}$$

$$\leqq c|t|^{1/2} - \frac{6c}{\log|t|}\,.$$

Therefore $f^{(2n+1)}(0) = 0$; but plainly $M_{2n+1}(F) = \infty$.

Theorem 4.1. *In order that $M_{2n+1}(F)$ be finite, it is necessary and sufficient that, in a neighborhood $|t| < \delta$, we should have*

$$\Re\{f(t)\} = Q_n(t^2) + O\{|t|^{2n+1}\psi(t)\}, \tag{42}$$

where $Q_n(t)$ is a polynomial of degree n and $\psi(t)$ satisfies the same conditions as in Theorem 2.1.

If (42) is satisfied, then $Q_n(t^2) = \Re\{P_{2n}(t)\}$ and $M_{2n+1}(F)$ is given by (21) with $k = 2n+1$, $\gamma = 0$:

$$M_{2n+1}(F) = (-1)^{n+1}2\pi^{-1}(2n+1)!\int_0^{\infty} t^{-2n-2}\Re\{f(t) - P_2n(t)\}\,dt. \tag{43}$$

Proof. The condition is necessary. If $M_{2n+1}(F) < \infty$, then we set $Q_n(t^2) = \Re\{P_{2n}(t)\}$ and verify that

$$\begin{aligned}&\int_{-\infty}^{\infty} t^{-2n-2}|\Re\{f(t) - P_{2n}(t)\}|\,dt\\ &\leqq \int_{-\infty}^{\infty} dF(x)\int_{-\infty}^{\infty} t^{-2n-2}|\Re\{e^{itx} - E_{2n}(tx)\}|\,dt\\ &= M_{2n+1}(F)\int_{-\infty}^{\infty} u^{-2n-2}|\Re\{e^{iu} - E_{2n}(u)\}|\,du < \infty.\end{aligned}$$

The condition is sufficient. Since $\Re\{f(t)\}$ is the c.f. of the law $G(x) = \frac{1}{2}\{F(x) + 1 - F(-x)\}$, it follows from (42) and Theorem 3.2 that $M_{2n}(G) = M_{2n}(F) < \infty$. Therefore $Q_n(t^2) = \Re\{P_{2n}(t)\}$, and our condition asserts that

$$\int_{-\infty}^{\infty} t^{-2n-2}|\Re\{f(t) - P_{2n}(t)\}|\,dt = \int_{|t|<\delta} + \int_{|t|>\delta} < \infty.$$

Since $\Re\{e^{it} - E_{2n}(t)\}$ is a function which keeps a constant sign, routine computation leads to

$$\int_{-\infty}^{\infty} \frac{|\Re\{f(t) - P_{2n}(t)\}|}{t^{2n+2}}\,dt = M_{2n+1}(F)\int_{-\infty}^{\infty} \frac{|\Re\{e^{iu} - E_{2n}(u)\}|}{u^{2n+2}}\,du,$$

wherefrom the finiteness of $M_{2n+1}(F)$ follows. This proves the sufficiency of the condition (42).

A similar computation gives

$$\int_{-\infty}^{\infty} t^{-2n-2}\Re\{f(t) - P_{2n}(t)\}\,dt = M_{2n+1}(F)\int_{-\infty}^{\infty} u^{-2n-2}|\Re\{e^{iu} - E_{2n}(u)\}|\,du,$$

whence

$$M_{2n+1}(F) = A_1\int_{-\infty}^{\infty} t^{-2n-2}\Re\{f(t) - P_{2n}(t)\}\,dt,$$

where A_1 is independent of $F(x)$. The device we used to determine the constant A in (27) can be used to determine the constant A_1, and it leads to (43). This completes the proof.

The analogy to Theorem 2.1 for the case $k = 2n+1$ and $\gamma = 0$ is the following.

Theorem 4.2. *In order that* $L_{2n+1}(F)$[8] $< \infty$, *it is necessary that, in a neighborhood* $|t| < \delta$, *we should have*

$$f(t) = P_{2n+1}(t) + O\left\{|t|^{2n+1}\psi(t)\right\}, \tag{44}$$

where $\psi(t)$ *satisfies the same conditions as in Theorem* 2.1.

Conversely, if in a neighborhood $|t| < \delta$ *we have*

$$f(t) = Q_{2n+1}(t) + O\left\{|t|^{2n+1}\psi(t)\right\}, \tag{45}$$

where $Q_{2n+1}(t)$ *is a polynomial of degree* $2n + 1$ *and* $\psi(t)$ *satisfies the same conditions as in Theorem* 2.1 *and if further* $F(x) = 0$ *for* $x < 0$, *then* $L_{2n+1}(F) < \infty$.

Proof. To prove the first part of the theorem, we need only verify the relation

$$\int_0^1 \frac{|f(t) - P_{2n+1}(t)|}{t^{2n+2}}\, dt < \infty$$

under the assumption that $L_{2n+1}(F) < \infty$ [which implies the finiteness of $M_{2n+1}(F)$ and hence the $(2n+1)$-time differentiability of $f(t)$].

We have

$$\begin{aligned}
\int_0^1 t^{-2n-2}|f(t) - P_{2n+1}(t)|\, dt &\leqq \int_{-\infty}^{\infty} |x|^{2n+1}\, dF(x) \int_0^{|x|} u^{-2n-2}|e^{iu} - E_{2n+1}(u)|\, du \\
&\leqslant C + \int_{|x|>e} |x|^{2n+1}\, dF(x) \int_e^{|x|} u^{-2n-2}|e^{iu} - E_{2n+1}(u)|\, du \\
&\leqq C + \int_{|x|>e} |x|^{2n+1}\, dF(x) \int_e^{|x|} Cu^{-1}\, du \\
&= C + CL_{2n+1}(F) < \infty,
\end{aligned}$$

which gives the proof.

To prove the second part, observe that, by virtue of Theorem 4.1, the condition (45) implies the finiteness of $M_{2n+1}(F)$. Therefore, $Q_{2n+1}(t) = P_{2n+1}(t)$ and our condition asserts that

$$\int_0^1 t^{-2n-2}|f(t) - P_{2n+1}(t)|\, dt < \infty,$$

whence, taking into account the condition that $F(x) = 0$ for $x < 0$,

$$\begin{aligned}
\int_0^1 t^{-2n-2}\, dt &\left|\int_0^{\infty} \mathfrak{g}\left\{e^{itx} - E_{2n+1}(tx)\right\} dF(x)\right| \\
&= \int_0^1 t^{-2n-2}|\mathfrak{g}\{f(t) - P_{2n+1}(t)\}|\, dt < \infty.
\end{aligned} \tag{46}$$

But, when $x \geqq 0$ and $t \geqq 0$, the function $\mathfrak{g}\{e^{itx} - E_{2n+1}(tx)\}$ keeps a constant

[8] For definition, see equation (4).

sign. Therefore, by (46),

$$\begin{aligned}
\infty &> \int_0^1 t^{-2n-2}\,dt \left| \int_0^\infty \mathfrak{g}\{e^{itx} - E_{2n+1}(tx)\}\,dF(x) \right| \\
&= \int_0^1 t^{-2n-2}\,dt \int_0^\infty |\mathfrak{g}\{e^{itx} - E_{2n+1}(tx)\}|\,dF(x) \\
&= \int_0^\infty x^{2n+1}\,dF(x) \int_0^x u^{-2n-2} |\mathfrak{g}\{e^{iu} - E_{2n+1}(u)\}|\,du \\
&\geqq \int_e^\infty x^{2n+1}\,dF(x) \int_e^x u^{-2n-2} |\mathfrak{g}\{e^{iu} - E_{2n+1}(u)\}|\,du \\
&\geqq \int_e^\infty x^{2n+1}\,dF(x) \left\{ \int_e^x C u^{-1}\,du - C \right\} = C L_{2n+1}(F) - C,
\end{aligned}$$

which implies the finiteness of $L_{2n+1}(F)$.

5. Absolute Moments of Negative Order

If $F(x)$ is continuous at $x = 0$, then, putting $n = 0$ in (15), we obtain

$$\begin{aligned}
M_{-\beta}(F) &= \frac{2^{1-\beta}}{\sqrt{\pi}\,\Gamma(\beta/2)} \int_0^\infty y^{\beta-1}\,dy \int_{-\infty}^\infty e^{-t^2} f(yt)\,dt \\
&= \frac{2^{2-\beta}}{\sqrt{\pi}\,\Gamma(\beta/2)} \int_0^\infty y^{\beta-1}\,dy \int_0^\infty e^{-t^2} \Re\{f(yt)\}\,dt \qquad (\beta > 0), \qquad (47)
\end{aligned}$$

a formula expressing a general absolute moment of negative order in terms of the c.f. We shall not investigate further along this line, except to note the following readily proved but somewhat curious theorem.

Theorem 5.1. *If $f(t)$ has a non-negative real part, then $M_{-1}(F) = \infty$.*

For if $\Re\{f(t)\} \geqq 0$, then we have

$$\begin{aligned}
M_{-1}(F) &= \frac{2}{\pi} \int_0^\infty e^{-t^2}\,dt \int_0^\infty \Re\{f(yt)\}\,dy \\
&= \frac{2}{\pi} \int_0^\infty t^{-1} e^{-t^2}\,dt \int_0^\infty \Re\{f(t)\}\,dt = \infty.
\end{aligned}$$

6. The Stieltjes Transform

If $F(x)$ is a law, the function

$$L(t) = \int_{-\infty}^\infty \frac{dF(x)}{x + it} \qquad (-\infty < t < \infty) \qquad (48)$$

is known as the Stieltjes transform of $F(x)$. We have

$$\frac{i}{t}L\left(\frac{1}{t}\right)=\int_{-\infty}^{\infty}\frac{dF(x)}{1-itx}=\int_{-\infty}^{\infty}dF(x)\int_0^{\infty}e^{-(1-itx)s}\,ds$$
$$=\int_0^{\infty}e^{-s}\,ds\int_{-\infty}^{\infty}e^{itsx}\,dF(x)=\int_0^{\infty}e^{-s}\,ds\int_{-\infty}^{\infty}e^{itx}\,dF\left(\frac{x}{s}\right)$$
$$=\int_{-\infty}^{\infty}e^{itx}\,dF_1(x), \tag{49}$$

where

$$F_1(x)=\int_0^{\infty}e^{-s}F\left(\frac{x}{s}\right)ds \tag{50}$$

is again a law. Thus $(i/t)L(1/t)$ (which is defined to be 1 at $t=0$) is the c.f. of $F_1(x)$. Besides, for $\beta>0$ we have

$$M_\beta(F_1)=\int_0^{\infty}e^{-s}\,ds\int_{-\infty}^{\infty}|x|^\beta\,dF\left(\frac{x}{s}\right)$$
$$=\int_0^{\infty}s^\beta e^{-s}\,ds\int_{-\infty}^{\infty}|x|^\beta\,dF(x)=\Gamma(\beta+1)M_\beta(F). \tag{51}$$

It follows that $M_\beta(F)$ is finite if and only if $M_\beta(F_1)$ is so. Accordingly, corresponding to every theorem [except the two theorems about $L_k(F)$] in §§2, 3, 4 there is a theorem in which $L(t)$ plays the part of $f(t)$. For example, the theorem corresponding to Theorem 2.1 reads:

In order that $M_{k+\gamma}(F)<\infty$ ($k\geqq 0$ *and integral*, $0<\gamma<1$), *it is necessary and sufficient that for* $|t|>K$ *we should have*

$$L(t)=\frac{1}{it}+\frac{c_1}{(it)^2}+\cdots+\frac{c_k}{(it)^{k+1}}+O\left\{|t|^{-k-\gamma-1}\chi(t)\right\}, \tag{52}$$

where $\chi(t)$ *is defined, non-negative, and bounded for* $|t|>K$ *and satisfies the condition*

$$\int_{|t|>K}\frac{\chi(t)}{|t|}\,dt<\infty. \tag{53}$$

If (52) *and* (53) *are satisfied, then*

$$M_{k+\gamma}(F)=(-1)^{[k/2]}2\pi^{-1}\sin\left(\frac{k}{2}-\left[\frac{k}{2}\right]+\frac{\gamma}{2}\right)\pi$$
$$\times\int_0^{\infty}t^{-k-\gamma}\mathfrak{g}\left\{L(t)-\frac{1}{it}-\frac{c_1}{(it)^2}-\cdots-\frac{c_k}{(it)^{k+1}}\right\}dt. \tag{54}$$

The two theorems corresponding to Theorem 3.4 and Theorem 2.2 (taking the special case $\gamma=0$) constitute Hamburger's lemma on the Stieltjes transform.[9]

[9]Cf. Hamburger [3], p. 267, or Fortet, loc. cit, p. 118.

References

1. H. Cramér, *Random Variables and Probability Distributions*. Cambridge Univ. Press, 1937.
2. R. Fortet, Calcul des Moments d'une Fonction de Répartition à Partir de la Characté-ristique. *Bull des Sci. Math.* **68** (2) (1944), 117–131.
3. H. Hamburger, Über eine Erweiterung des Stieltjesschen Momentenproblems. *Math. Ann.* **81** (1920), 255–319.
4. P. Levy, *Calcul des Probabilités*. Paris, 1925.

Reprinted from
Science Record **4** (1951), 197–200.

A LEMMA ON THE COEFFICIENT OF REDUCTION OF A SUM OF INDEPENDENT VARIATES*

The purpose of this note is to prove Lemma A, which within our knowledge does not seem to have been a published result.

Lemma A. *Let $\{X_n\}$ be a sequence of independent random variables and $\{a_n\}$ be a sequence of positive numbers such that the distribution law of*

$$\frac{1}{a_n}\sum_{j=l}^{n} X_j$$

tends to a limit law which is not the law of any sure number. Then the following alternatives are the only ones possible: either (i) *a_n tends to a finite positive limit, or* (ii) *there is a sequence $\{a_n'\}$ such that $a_n' \geqslant a_n$, $a_n'/a_n \to 1$, $a_n' \uparrow \infty$.*

Let $f_n(t)$ be the characteristic function of X_n. Then the hypothesis in Lemma A is equivalent to the following:

$$\lim_{n\to\infty}\prod_{j=1}^{n} f_j\left(\frac{t}{a_n}\right) = g(t), \tag{1}$$

where $g(t)$ is a characteristic function, $|g(t)| \not\equiv 1$. Thus it is required to deduce the conclusion in Lemma A, given (1).

We need the following lemma.

Lemma 1. *In order that the conclusion in Lemma A hold for a sequence of positive numbers $\{a_n\}$, it is necessary and sufficient that for every pair of ascending sequences of positive integers $\{m_k\}$ and $\{n_k\}$ with $m_k \leqslant n_k$ $(k = 1, 2, 3, \cdots)$ we have*

$$\limsup_{k\to\infty}\frac{a_{m_k}}{a_{n_k}} \leqslant 1. \tag{2}$$

Proof of Lemma 1. The necessity of the condition being obvious, we prove its sufficiency only. The condition implies immediately the existence of $\lim a_n$. This limit cannot be zero, as then there would be a subsequence $\{a_{m_k}\}$ such that

*Received April 13, 1951.

Note added in proof: The part of the proof of Lemma 1 given in the text for the case $a_n \to \infty$ is totally unnecessary. In fact, it suffices to take $a_n' = \max\{a_1, \cdots, a_n\}$.

$a_{m_k}/a_{m_{k+1}} \to \infty$. The case of a finite positive limit corresponds to (i). It remains to prove that $a_n \to \infty$ leads to the conclusion (ii). Since in this case a_n is unbounded, we may define an ascending sequence of integers $\{m_k\}$ as follows: $m_1 = 1$, m_k is the smallest index for which $a_{m_k} > a_{m_{k-1}}$. Let further n_k be an index such that

$$m_k \leqslant n_k < m_{k+1}, \qquad a_{n_k} = \min_{m_k \leqslant j < m_{k+1}} a_j .$$

Then from (2) and the obvious fact that $a_{m_k} \geqslant a_{n_k}$, we conclude that

$$\lim_{k\to\infty} \frac{a_{m_k}}{a_{n_k}} = 1. \tag{3}$$

We now define the numbers a'_n as follows:

$$a'_n = a_{m_k} \qquad \text{for} \quad m_k \leqslant n < m_{k+1} \qquad (k = 1, 2, 3, \cdots).$$

Evidently $a'_n \geqslant a_n$, $a'_n\uparrow$. Moreover, using the fact

$$\frac{a'_n}{a_n} = \frac{a_{m_k}}{a_n} \leqslant \frac{a_{m_k}}{a_{n_k}} \qquad \text{for} \quad m_k \leqslant n < m_{k+1},$$

together with (3), we see easily that $a'_n/a_n \leqslant 1 + \varepsilon$, with any given positive ε for all sufficiently large n. This proves $a'_n/a_n \to 1$ and thereby Lemma 1.

Coming now to the proof of Lemma A, it suffices to show that the condition in Lemma 1 holds for the a_n appearing in (1). Suppose to the contrary that this were not true. Then there exist two subsequences $\{a_{p_k}\}$ and $\{a_{q_k}\}$ such that, on putting $c_k = a_{p_k}/a_{q_k}$,

$$p_k < q_k \quad (k = 1, 2, 3, \cdots), \quad \text{either} \quad \lim_{k\to\infty} c_k = c > 1 \quad \text{or} \quad \lim_{k\to\infty} c_k = \infty.$$

Set

$$\phi_k(t) = \prod_{j=1}^{p_k} f_j\left(\frac{t}{a_{p_k}}\right), \qquad \psi_k(t) = \prod_{j=p_k+1}^{q_k} f_j\left(\frac{t}{q_{q_k}}\right).$$

By (1), we have

$$\lim_{k\to\infty} |\phi_k(c_k t)|^2 |\psi_k(t)|^2 = |g(t)|^2, \tag{4}$$

$$\lim_{k\to\infty} |\phi_k(t)|^2 = |g(t)|^2. \tag{5}$$

First, suppose $c_k \to c > 1$. Then by (5) and the fact that the convergence of any sequence of characteristic functions to a characteristic function is uniform on every finite interval, we have

$$\lim_{k\to\infty} |\phi_k(c_k t)|^2 = |g(ct)|^2. \tag{6}$$

Equations (6) and (4) give $|g(ct)| \geqslant |g(t)|$. Replacing t by t/c, we get $|g(t)| \geqslant |g(t/c)|$, whence, successively, $|g(t)| \geqslant g(c^{-n}t)|$. Allowing here $n \to \infty$ we have $|g(t)| \geqslant 1$, whence $|g(t)| = 1$, which contradicts our hypothesis.

Next, suppose $c_k \to \infty$. Consider random variables Z, U_k, V_k, the last two being independent, corresponding to the characteristic functions $|g(t)|^2, |\phi_k(t)|^2, |\psi_k(t)|^2$.

Then (4) and (5) assert that

$$\lim_{k\to\infty} \Pr\{|c_k U_k + V_k| > x\} = \Pr\{|Z| > x\}, \tag{7}$$

$$\lim_{k\to\infty} \Pr\{|U_k| > x\} = \Pr\{|Z| > x\}, \tag{8}$$

for all $x > 0$ such that $\Pr\{|Z| = x\} = 0$. Since the distribution of V_k is symmetric, we have

$$\Pr\{c_k U_k + V_k > x\} \geqslant \Pr\{c_k U_k > x\}\Pr\{V_k \geqslant 0\} \geqslant \tfrac{1}{2}\Pr\{c_k U_k > x\},$$

$$\Pr\{c_k U_k + V_k < -x\} \geqslant \Pr\{c_k U_k < -x\}\Pr\{V_k \leqslant 0\} \geqslant \tfrac{1}{2}\Pr\{c_k U_k < -x\},$$

whence by adding

$$\Pr\{|c_k U_k + V_k| > x\} \geqslant \tfrac{1}{2}\Pr\left\{|U_k| > \frac{x}{c_k}\right\}. \tag{9}$$

Let x and y be any positive numbers such that $\Pr\{|Z| = x\} = \Pr\{|Z| = y\} = 0$. From (9) we get

$$\Pr\{|c_k U_k + V_k| > x\} \geqslant \tfrac{1}{2}\Pr\{|U_k| > y\} \tag{10}$$

for all sufficiently large n.

Combining (7), (8), and (10) we obtain

$$\Pr\{|Z| > x\} \geqslant \tfrac{1}{2}\Pr\{|Z| > y\}.$$

Since we can have x arbitrarily large and y arbitrarily small, the last inequality on passing to limits gives $0 \geqslant \Pr\{|Z| > 0\}$. This implies $\Pr\{|Z| = 0\} = 1$ or, what is the same thing, $|g(t)| = 1$. Thus our hypothesis is again contradicted, and this contradiction completes the proof.

Reprinted from
Proc. Edinburgh Math. Soc.
(2), 37–44 (1953).

On Symmetric, Orthogonal, and Skew-Symmetric Matrices

By P. L. Hsu

(*Received 9th July* 1948. *Read 5th November* 1948.)

1. Introduction and Notation. In this paper all the scalars are real and all matrices are, if not stated to be otherwise, p-rowed square matrices. The diagonal and superdiagonal elements of a symmetric matrix, and the superdiagonal elements of a skew-symmetric matrix, will be called the distinct elements of the respective matrices. Σ will denote both the set of all symmetric matrices and the $\frac{1}{2}p\,(p+1)$-dimensional space whose coordinates are the distinct elements arranged in some specific order. K will denote both the set of all skew-symmetric matrices and the $\frac{1}{2}p\,(p-1)$-dimensional space whose coordinates are the distinct elements arranged in some specific order. Any sub-set of Σ (K) will mean both the sub-set of symmetric (skew-symmetric) matrices and the set of points of Σ (K). Any point function defined in Σ (K) will be written as a function of a symmetric (skew-symmetric) matrix. D_a will denote the diagonal matrix whose diagonal elements are $a_1, a_2, \ldots, a_p$. The characteristic roots of a symmetric matrix will be called its roots.

The orthogonal matrices Γ with $|\,\Gamma + I\,| \neq 0$ and the skew-symmetric matrices X are in (1, 1)-correspondence, on account of the following pair of equivalent equations:

$$\Gamma = 2\,(I + X)^{-1} - I, \tag{1}$$

$$X = 2\,(I + \Gamma)^{-1} - I. \tag{1'}$$

That the skew-symmetry of X implies the orthogonality of Γ, and vice versa,[1] is the direct result of the following computation:

$$\begin{aligned}[2\,(I + X)^{-1} - I]\;[2\,(I + X)^{-1} - I]' \\ &= (I + X)^{-1}\;(I - X)\;\;(I + X)\,(I - X)^{-1} \\ &= (I + X)^{-1}\,(I + X)\,(I - X)\,(I - X)^{-1} = I,\end{aligned}$$

$$\begin{aligned}&[2\,(I + \Gamma)^{-1} - I] + [2\,(I + \Gamma)^{-1} - I]' \\ &= 2\,(I + \Gamma)^{-1} + 2\,(I + \Gamma)^{-1}\,\Gamma - 2I = 2\,(I + \Gamma)^{-1}\,(I + \Gamma) - 2I = 0.\end{aligned}$$

If $S \in \Sigma$, and if its roots are $\theta_1, \ldots, \theta_p$, it is well known that

$$S = \Delta D_\theta \Delta' \tag{2}$$

where Δ is some orthogonal matrix.

---

[1] This result, together with a lengthy derivation, is given in Kowalewski, *Einführung in die Determinantentheorie* (Leipzig, 1909), pp. 171-175.

LEMMA 1. *If A is any $p \times p$ matrix, there is a matrix D_ϵ, with $\epsilon_i = +1$ or -1 $(i = 1, \ldots, p)$, such that* $| A + D_\epsilon | \neq 0$.

Proof. The lemma is evidently true for $p = 1$. Assume it is true for $p - 1$. Write

$$A = \begin{pmatrix} A_1 & b' \\ c & d \end{pmatrix},$$

where A_1 is $(p-1) \times (p-1)$ and b, c are rows. By assumption there is a D_η, with $\eta_i = +1$ or -1 $(i = 1, \ldots, p-1)$, such that $| A_1 + D_\eta | \neq 0$. Since

$$(3) \qquad \begin{vmatrix} A_1 + D_\eta & b' \\ c & d+1 \end{vmatrix} - \begin{vmatrix} A_1 + D_\eta & b' \\ c & d-1 \end{vmatrix} = 2 \, | A_1 + D_\eta | \neq 0,$$

the two determinants on the left side of (3) cannot both vanish. Hence the lemma is proved.

LEMMA 2. *If $S \in \Sigma$, there is an orthogonal Γ with* $| \Gamma + I | \neq 0$ *such that $S = \Gamma D_\theta \Gamma'$.*

Proof. For the Δ in (2) there is, by Lemma 1, a D_ϵ with $| \Delta + D_\epsilon | \neq 0$. Hence $| \Delta D_\epsilon + I | \neq 0$. Moreover, ΔD_ϵ also satisfies (2). Hence the matrix $\Gamma = \Delta D_\epsilon$ answers all the requirements.

Combining Lemma 2 with (1) we obtain

THEOREM 1. *If $S \in \Sigma$, there is an $X \in K$ such that*

$$(4) \qquad S = [2\,(I + X)^{-1} - I]\, D_\theta \,[2\,(I + X)^{-1} - I]'.$$

In § 2 we investigate the uniqueness of the expression (4) for a given S. (4) expresses each distinct element of S as a function of $\frac{1}{2}p\,(p+1)$ arguments, viz., the distinct elements of X and the θ's. In § 3 we evaluate the functional determinant of these functions or, to use another term, the Jacobian of the transformation (4). In § 4 we apply the results to prove a formula for the integral of some function of S.

2. The symmetric matrices S which have multiple roots satisfy the equation $f(S) = 0$, where $f(S)$ is the discriminant of the characteristic equation of S. Let F be the surface $f(S) = 0$. If $S \in \Sigma - F$,

we shall so name the roots $\theta_1, \ldots, \theta_p$ that $\theta_1 > \theta_2 > \ldots > \theta_p$. Thus each θ_i is a well-defined function of S. The equation

$$S = \Gamma D_\theta \Gamma', \tag{5}$$

where $\Gamma = (\gamma_{ij})$ is orthogonal, then uniquely determines every column of Γ except for a sign. In particular,

$$|\gamma_{1i}| = g_i(S) \qquad (i=2,\ldots,p)$$

are well-defined functions of S. Let F_i be the surface $g_i(S) = 0$. If $S \in \Sigma - F - F_2 - \ldots - F_p$, we take $\gamma_{1i} = -g_i(S)$ $(i = 2, \ldots, p)$. Thus all the columns except the first one of Γ are uniquely determined, since the signs of the top elements of these columns are determined. Hence we have

LEMMA 3. *If $S \in \Sigma - F - F_2 - \ldots - F_p$, equation* (5) *has exactly two solutions for Γ: Γ_1 and Γ_2, whose elements* (1, 2), (1, 3), ..., (1, p) *are all negative. Γ_1 and Γ_2 differ only in the respect that the first column of one is the negative of the first column of the other.*

By virtue of Lemma 3 all the elements γ_{ij} $(i, j = 2, \ldots, p)$ of Γ are well-defined functions of S. Writing $\Gamma^{(1)} = (\gamma_{ij})_{(i,\, j=2,\ldots,p)}$ we have $|\Gamma^{(1)} + I| = h(S)$, a well-defined function of S. Let F' be the surface $h(S) = 0$. Let $E = \Sigma - F - F_2 - \ldots - F_p - F'$.

LEMMA 4. *If $S \in E$, one of the determinants $|\Gamma_1 + I|$ and $|\Gamma_2 + I|$ in Lemma* 3 *is zero while the other is not zero.*

Proof. By Lemma 3 and the defiinition of E we have

$$|\Gamma_1 + I| + |\Gamma_2 + I| = 2\,|\Gamma^{(1)} + I| \neq 0.$$

Hence one of the determinants is not zero. Suppose $|\Gamma_1 + I| \neq 0$. Then, by (1), $\Gamma_1 = 2(I + X)^{-1} - I$, where X is skew-symmetric. Letting H be the diagonal matrix $[-1, 1, \ldots, 1]$ we have $\Gamma_2 = \Gamma_1 H$. Hence

$$\begin{aligned} |\Gamma_2 + I| = |\Gamma_1 H + I| &= |2(I + X)^{-1} H + I - H| \\ &= |X + I|^{-1}\,|2H + (I + X)(I - H)|. \end{aligned}$$

It is easily seen that the first row of the matrix $2H + (I + X)(I - H)$ consists of zeros. Hence $|\Gamma_2 + I| = 0$. This completes the proof.

If we take Γ to be Γ_1 or Γ_2 according as $|\Gamma_1 + I| \neq 0$ or $|\Gamma_2 + I| \neq 0$ we obtain

LEMMA 5. *If $S \in E$, there is a unique Γ which satisfies* (5) *and the following conditions:*

$$\text{(i)} \quad \gamma_{1i} < 0 \qquad (i = 2, \ldots, p)$$

$$\text{(ii)} \quad |\Gamma + I| \neq 0.$$

By combining Lemma 5 with (1) and noticing that in (1) the non-diagonal elements of Γ and $(I + X)^{-1}$ must have the same sign, we obtain

THEOREM 2. *If $S \in E$, the equation (4) has a unique solution in X such that the elements $(1, 2), (1, 3), \ldots, (1, p)$ of $(I + X)^{-1}$ are all negative.*

Let M be the sub-set of K defined by the condition that the elements $(1, 2), (1, 3), \ldots, (1, p)$ of $(I + X)^{-1}$ are all negative, let Θ be the sub-set of the space of $(\theta_1, \ldots, \theta_p)$ defined by the condition $\theta_1 > \theta_2 > \ldots > \theta_p$, and let $E^* = M \times \Theta$, a sub-set of the $\frac{1}{2}p(p+1)$-dimensional space. Theorem 2 asserts that the equation (4) effects a (1, 1)-mapping of E^* on E. Notice also that by the definition of E, it differs from Σ by a set of measure zero.

3. In order to facilitate the computation of the functional determinant[1] for (4) we shall adopt the following notation. If $x = (x_1, \ldots, x_m)$ and $y = (y_1, \ldots, y_m)$ and if

$$x_i = \sum_{j=1}^{m} a_{ij} y_j \qquad (i = 1, \ldots, m),$$

then we define the symbol $D(x; y)$ to be the discriminant $|a_{ij}|$.

LEMMA 6.

$$\text{(i)} \quad D(x; y) = \frac{1}{D(y; x)}$$

$$\text{(ii)} \quad D(x; y) = D(x; z)\, D(z; y).$$

Proof. (i) is an immediate consequence of definition; (ii) is a special case of the multiplicative law of functional determinants.

The extension of (ii) to more than two factors is obvious.

In the equation

$$X = AYA', \tag{6}$$

X is symmetric or skew-symmetric according as Y is symmetric or skew-symmetric. In either case (6) expresses each distinct element of X as a linear function of the distinct elements of Y, with coefficients depending on A. The following lemma gives the discriminant $D(X; Y)$ of this set of linear functions.

[1] Functional determinants are considered as determined up to a sign. In all the computations in this section signs of functional determinants are neglected.

LEMMA 7. *For* (6) *we have*

$$D(X;\,Y) = \begin{cases} |A|^{p+1} & \text{if } Y \text{ is symmetric} \\ |A|^{p-1} & \text{if } Y \text{ is skew-symmetric.} \end{cases} \tag{7}$$

Proof. It is sufficient to prove (7) for a non-singular A. Now every non-singular matrix is a product of a finite number of matrices of either of the following two types, (a) a diagonal matrix whose diagonal elements are 1 with the exception of one, which is, say, a; (b) a matrix whose diagonal elements are 1 and whose non-diagonal elements are 0 with the exception of one. Further, if we denote $D(X;\,Y)$ for (6) by $P(A)$, a moment of reflection will show that $P(AB) = P(A)\,P(B)$. Hence it is sufficient to prove (7) for A belonging to either of the types (a) and (b), i.e. $D(X;\,Y) = \begin{cases} a^{p+1} \\ a^{p-1} \end{cases}$ if A is of type (a) and $D(X;\,Y) = 1$ for both cases if A is of type (b). The proof of these assertions is easy and is left to the reader.

If $x_1, \ldots, x_m$ are functions of $y_1, \ldots, y_m$, then by definition the functional determinant is equal to $D(dx;\,dy)$, where dx denotes the system $dx_1, \ldots, dx_m$. For brevity we write dA for the matrix whose elements are the differentials of the elements of A.

In order to differentiate (1) we use the formula

$$dA^{-1} = -A^{-1}(dA)A^{-1}.$$

Then

$$d\Gamma = -2(I+X)^{-1}dX(I+X)^{-1} = -\tfrac{1}{2}(I+\Gamma)\,dX\,(I+\Gamma). \tag{8}$$

Let J be the functional determinant for (4), i.e. for

$$S = \Gamma D_\theta \Gamma' \tag{9}$$

where Γ is given by (1). Differentiating (9) and using (8) we have

$$\begin{aligned} dS = \Gamma D_\theta d\Gamma' + (d\Gamma)\,D_\theta\Gamma' + \Gamma D_\theta \Gamma' &= \tfrac{1}{2}\Gamma D_\theta(I+\Gamma')\,dX\,(I+\Gamma') \\ &\quad - \tfrac{1}{2}(I+\Gamma)\,dX\,(I+\Gamma)\,D_\theta\Gamma' + \Gamma D_{d\theta}\Gamma'. \end{aligned} \tag{10}$$

Also,

$$J = D(dS;\,dX,\,d\theta). \tag{11}$$

From (10) we get

$$\Gamma' dS\,\Gamma = \tfrac{1}{2}D_\theta(I+\Gamma')\,dX\,(I+\Gamma) - \tfrac{1}{2}(I+\Gamma')\,dX\,(I+\Gamma)\,D_\theta + D_{d\theta}. \tag{12}$$

If we write ξ_i for $d\theta_i$, and define U and Y by

$$U = \Gamma' dS\,\Gamma \tag{13}$$

$$Y = (I+\Gamma')\,dX\,(I+\Gamma), \tag{14}$$

then (12) gives

$$U = \tfrac{1}{2}D_\theta Y - \tfrac{1}{2}YD_\theta + D_\xi,$$

i.e.,

(15) $$u_{ii} = \xi_i \qquad (i = 1, \ldots, p), \\ u_{ij} = \tfrac{1}{2}(\theta_i - \theta_j)\, y_{ij} \qquad (i < j).$$

(13), (14) and (15) give respectively

$$D(dS;\, U) = \frac{1}{D(U;\, dS)} = 1, \text{ by Lemmas 6 and 7};$$

$$D(Y;\, dX) = 2^{p(p-1)}\, |\, I + X\, |^{-(p-1)}, \text{ by Lemma 6 and (1)};$$

$$D(U;\, Y,\, \xi) = 2^{-\frac{1}{2}p(p-1)} \prod_{i<j} (\theta_i - \theta_j).$$

Hence, by (11) and Lemma 7,

$$J = D(dS;\, U)\, D(U;\, Y,\, \xi)\, D(Y,\, \xi;\, dX,\, \xi) \\ = D(dS;\, U)\, D(U;\, Y,\, \xi)\, D(Y;\, dX),$$

and so, finally,

(16) $$J = 2^{\frac{1}{2}p(p-1)}\, |\, I + X\, |^{-(p-1)} \prod_{i<j} (\theta_i - \theta_j).$$

4. Let us first evaluate the following integrals:

(17) $$A_p = \int_K |\, I + X\, |^{-(p-1)}\, dm$$

(18) $$B_p = \int_M |\, I + X\, |^{-(p-1)}\, dm$$

where X is a p-rowed skew-symmetric matrix, dm denotes the volume element, K is the $\tfrac{1}{2}p\,(p-1)$-dimensional space, and M is defined in the last paragraph of § 2.

We write

$$X = \begin{pmatrix} 0 & x \\ -x' & Y \end{pmatrix}, \qquad x = (x_1, \ldots, x_{p-1})$$

so that

$$|\, I + X\, | = \begin{vmatrix} 1 & X \\ -x' & I+Y \end{vmatrix} = |\, I + Y\, |\, (1 + x\,(I + Y)^{-1}\, x').$$

Also, the elements (1, 2), (1, 3), ..., (1, p) of $(I+X)^{-1}$ are the elements of the row $-(1 + x\,(I + Y)^{-1}\, x')^{-1}\, x\,(I + Y)^{-1}$. Hence M is the set of points such that every element of the row $x\,(I + Y)^{-1}$ is positive. Making the transformation

$$x' = (I + Y)\, u', \qquad u = (u_1, \ldots, u_{p-1})$$

whose Jacobian is $|\, I + Y\, |$, we obtain

$x(I+Y)^{-1}x' = u(I+Y)u' = uu' = u_1^2 + \ldots + u_{p-1}^2.$

Hence

$$(19)\quad A_p = \int_{K1} |I+Y|^{-(p-2)}\,dm \int_{-\infty}^{\infty} \ldots \int_{-\infty}^{\infty} (1 + u_1^2 + \ldots + u_{p-1}^2)^{-(p-1)}\,du_1 \ldots du_{p-1},$$

$$(20)\quad B_p = \int_{K_1} |I+Y|^{-(p-2)}\,dm \int_{0}^{\infty} \ldots \int_{0}^{\infty} (1 + u_1^2 + \ldots + u_{p-1}^2)^{-(p-1)}\,du_1 \ldots du_{p-1},$$

where K_1 is the $\frac{1}{2}(p-1)(p-2)$-dimensional space. It follows from (19) and (20) that

$$(21)\quad B_p = 2^{-(p-1)} A_p,$$

$$(22)\quad A_p = A_{p-1} \int_{-\infty}^{\infty} \ldots \int_{-\infty}^{\infty} (1 + u_1^2 + \ldots + u_{p-1}^2)^{-(p-1)} du_1 \ldots du_{p-1} = \frac{\pi^{p/2}}{2^{p-2}\Gamma(p/2)} A_{p-1},$$

and the easy computation $A_2 = \pi$ leads to the result

$$(23)\quad A_p = \pi \prod_{r=3}^{p} \frac{\pi^{r/2}}{2^{r-2}\Gamma(r/2)} = \frac{\pi^{p(p+1)/4}}{2^{(p-1)(p-2)/2} \prod_{r=1}^{p} \Gamma(r/2)},$$

whence also

$$(24)\quad B_p = \frac{\pi^{p(p+1)/4}}{2^{p(p-1)/2} \prod_{r=1}^{p} \Gamma(r/2)}.$$

We can now prove

THEOREM 3. *Let $f(S)$ be a function of the distinct elements of the p-rowed symmetric matrix S, such that $f(S)$ is a function only of the roots $\theta_1, \ldots, \theta_p$ of S, $\theta_1 \geqq \theta_2 \geqq \ldots \geqq \theta_p$:*

$$f(S) = g(\theta_1, \ldots, \theta_p).$$

Let Σ be the $\frac{1}{2}p(p+1)$-dimensional space. Then

$$(25)\quad \int_{\Sigma} f(S)\,dm = \frac{\pi^{p(p+1)/4}}{\prod_{r=1}^{p} \Gamma(r/2)} \int_{\Theta} g(\theta_1, \ldots, \theta_p) \prod_{i<j} (\theta_i - \theta_j)\, d\theta_1 \ldots d\theta_p,$$

where Θ is the domain $\theta_1 > \theta_2 > \ldots > \theta_p$.

Proof. By the remark at the end of §2 we have

$$\int_{\Sigma} f(S)\,dm = \int_{E} f(S)\,dm.$$

By Theorem 2 we may use the transformation (4), with the Jacobian (16). Hence

$$\int_{\Sigma} f(S)\, dm = 2^{\frac{1}{2}p(p-1)} \int_{E^*} |\, I + X\,|^{-(p-1)} g(\theta_1, \ldots, \theta_p) \prod_{i<j} (\theta_i - \theta_j)\, dm$$

where $E^* = M \times \Theta$. Hence

$$\int_{\Sigma} f(S)\, dm = 2^{\frac{1}{2}p(p-1)} B_p \int_{\Theta} g(\theta_1, \ldots, \theta_p) \prod_{i<} (\theta_i - \theta_j)\, d\theta_1 \ldots d\theta_p.$$

Using (24) we get (25).

Theorem 3 implies the following theorem on the probability distribution of roots, which is an important subject in statistics.

Theorem 4. *Let the distinct elements of the p-rowed symmetric matrix S be random variables whose joint distribution in* Σ *has a probability density* $f(S)$, *such that* $f(S)$ *is a function only of the roots* $\theta_1 \geqq \ldots \geqq \theta_p$ *of S:*

$$f(S) = g(\theta_1, \ldots, \theta_p).$$

Then the joint distribution of roots in Θ *has the probability density function*

$$\frac{\pi^{p(p+1)/4}}{\prod_{r=1}^{p} \Gamma(r/2)} g(\theta_1, \ldots, \theta_p) \prod_{i<j} (\theta_i - \theta_j).$$

Proof. If A is any sub-set of Θ, let B be the sub-set of Σ such that $(\theta_1, \ldots, \theta_p) \in A$. Then

$$\text{(26)} \qquad \Pr\left\{ \theta_1, \ldots, \theta_p) \in A \right\} = \int_{\Sigma} \phi(S) f(S)\, dm,$$

where $\phi(S) = 1$ or 0 according as $S \in B$ or $S \bar{\in} B$. Now $\phi(S) = \psi(\theta_1, \ldots, \theta_p)$, where $\psi = 1$ or 0 according as $(\theta_1, \ldots, \theta_p) \in A$ or not. Hence $\phi(S) f(S) = \psi(\theta_1, \ldots, \theta_p) g(\theta_1, \ldots, \theta_p)$. Applying Theorem 3 to (26) we get

$$\Pr\{(\theta_1, \ldots, \theta_p) \in A\}$$

$$= \frac{\pi^{p(p+1)/4}}{\prod_{r=1}^{p} \Gamma(r/2)} \int_{\Theta} \psi(\theta_1, \ldots, \theta_p) g(\theta_1, \ldots, \theta_p) \prod_{i<j} (\theta_i - \theta_j)\, d\theta_1 \ldots d\theta_p$$

$$= \frac{\pi^{p(p+1)/4}}{\prod_{r=1}^{p} \Gamma(r/2)} \int_{A} g(\theta_1, \ldots, \theta_p) \prod_{i<j} (\theta_i - \theta_j)\, d\theta_1 \ldots d\theta_p,$$

which proves the theorem.

University of North Carolina,
U.S.A.

Reprinted from
Acta Math. Sinica
4 (1954), 21–32.

On Characteristic Functions which Coincide in a Neighborhood of Zero*†

§1. Introduction

We all know two characteristic functions can be equal in a neighborhood of zero without being identically equal.[1] We say in this paper that a characteristic function (c.f.) belongs to the class $(\overline{U})$ if it can be equal to another c.f. in a neighborhood of zero without being equal to it identically. A c.f. is said to belong to (U) if it does not belong to $(\overline{U})$.

Marcinkiewicz proved that if a distribution function $F(x)$ satisfies

$$\int_{-\infty}^{0} e^{-rx}\,dF(x) < \infty \qquad \left(\text{or} \quad \int_{0}^{\infty} e^{rx}\,dF(x) < \infty\right), \tag{1}$$

in which r is a positive constant, then its c.f. belongs to (U).[2] Esseen pointed out if all the moments of a distribution function are finite and its corresponding moment problem is determined, then its c.f. belongs to (U).[3] It follows that in the finite moment case the condition "its c.f. belongs to $(\overline{U})$" is stronger than the condition "its moment problem is not determined".

In §§2, 3, 4 of this paper, we shall give 3 classes of c.f.'s belonging to $(\overline{U})$. The first class is the simplest; however, it includes all the past examples. The second class shows that a subclass of the stable distributions has corresponding c.f.'s belonging to $(\overline{U})$. The third class is the most interesting case. Its distribution functions all have density functions and the tails of the density functions can be as small as

$$O\left[\exp\left(-\frac{|x|}{\psi(|x|)}\right)\right],$$

where $\psi(x)$ is any one of the following functions.

$$(\ln x)^{\lambda},\ \ln x(\ln\ln x)^{\lambda},\ \ldots \qquad (\lambda > 1). \tag{2}$$

---

*Received April 24, 1953.

† Translated by Tze-chien Sun, Wayne State University.

[1] Gnedenko, Khinchine and Esseen provided examples of this type. See references [2] and [3], p. 190, and [1], p. 22.

[2] See [4], Theorem 3. It contains this result implicitly.

[3] See [1], p. 21.

From this result, we can see that it is very hard to further improve Marcinkiewicz's result above.

In §5, we shall clarify several points about the convergence of a sequence of c.f.'s in a neighborhood of zero.

§2

Theorem 1. *Suppose $g(t)$ is an even function defined at all t which is concave, nonnegative, and nonincreasing in $0 < t < \infty$, and $g(0) = 1$. Then $g(t)$ is a c.f.*

Proof. We all know that if we add one more condition, $g(\infty) = 0$, to the condition stated in this theorem, then $g(t)$ is a c.f.[4] Now, suppose $g(t)$ satisfies all the conditions stated in this theorem; then, unless $g(t) \equiv 0$, the function

$$g_1(t) = \frac{g(t) - g(\infty)}{1 - g(\infty)}$$

also satisfies the extra condition $g_1(\infty) = 0$. Hence it is a c.f. Therefore $g(t) = g(\infty) + (1 - g(\infty))g_1(t)$ is also a c.f.

Theorem 2. *All the c.f.'s satisfying the conditions in Theorem 1 belong to $(\overline{U})$.*

Proof. Suppose $g(t)$ is a c.f. and suppose $a > 0$ is a point of decrease of $g(t)$. Then the function

$$g^*(t) = \begin{cases} g(t), & |t| < a, \\ g(a), & |t| \geqslant a, \end{cases}$$

is another c.f., according to Theorem 1. $g^*(t)$ is not identically equal to $g(t)$ (because $g^*(t) > g(t)$, $|t| > a$) and is equal to $g(t)$ in the interval $|t| < a$. Hence $g(t)$ belongs to $(\overline{U})$.

The most interesting example in this class is the c.f. $\exp(-|t|^\alpha)$, $0 < \alpha \leqslant 1$; its corresponding distribution functions are the symmetric stable distributions. In the next section, we shall show that there also exist some asymmetric stable distributions belonging to the class $(\overline{U})$.

§3

The c.f. of a stable distribution of order $\alpha \neq 1$ can be expressed as

$$\exp\{-(c_0 - ic_1 \operatorname{sgn} t)|t|^\alpha\}, \quad 0 < \alpha \leqslant 2, \quad \alpha \neq 1, \quad c_0 > 0, \quad |c_1| \leqslant c_0 \left|\tan \frac{\pi\alpha}{2}\right|.$$

Without loss of generality, we can assume $c_0 = 1$. Thus, we shall only consider c.f.'s

[4]See [5], p. 170.

of the form

$$\phi(t) = \exp\{-(1 - ic\,\mathrm{sgn}\,t)|t|^{\alpha}\}, \qquad 0 < \alpha \leqslant 2, \qquad \alpha \neq 1, \qquad |c| \leqslant \left|\tan\frac{\pi\alpha}{2}\right|. \quad (3)$$

Theorem 3. *For each* $0 < \alpha < 1$, *there exists a positive constant* b, *depending only on* α, *such that for each* c, $|c| \leqslant b$, *the c.f.* $\phi(t)$ *in* (3) *belongs to* $(\overline{U})$.

This theorem follows immediately from the next theorem.

Theorem 4. *For each* $0 < \alpha < 1$, *there exists a positive constant* b, *depending only on* α, *such that for each* c, $|c| \leqslant b$, *the following function* $\phi_1(t)$ *is a c.f.*

$$\phi_1(t) = \begin{cases} \phi(t) & |t| < 1 \\ \exp\{ic\,\mathrm{sgn}\,t(2 - |t|) - 1\}, & 1 \leqslant t < 2, \\ e^{-1}, & |t| \geqslant 2. \end{cases}$$

Now, we are going to prove Theorem 4. Through the whole proof α is a fixed value between 0 and 1 while the condition $|c| \leqslant b$ is used only at the end of the proof to make $\phi_1(t)$ a c.f.

We write

$$\phi_1(t) = \exp\{u(t) - 1 + icv(t)\} \quad (4)$$

and use it to define the functions $u(t)$ and $v(t)$. Let

$$g(x) = \frac{1}{2\pi}\int_{-\infty}^{\infty} e^{-itx}u(t)\,dt,$$
$$h(x) = \frac{i}{2\pi}\int_{-\infty}^{\infty} e^{-itx}v(t)\,dt. \quad (5)$$

So

$$g(x) = -\frac{1}{\pi}\int_0^1 t^{\alpha}\cos tx\,dt + \frac{1}{\pi}\frac{\sin x}{x},$$

$$h(x) = \frac{1}{\pi}\int_0^1 t^{\alpha}\sin tx\,dt + \frac{1}{\pi}\int_1^2 (2 - t)\sin tx\,dt.$$

Using the method of integration by parts twice, we have

$$g(x) = \frac{\alpha(1 - \cos x)}{\pi x^2} + \frac{\alpha(1 - \alpha)}{\pi x^2}\int_0^1 t^{\alpha - 2}(1 - \cos tx)\,dt,$$

$$h(x) = \frac{(1 + \alpha)\sin x - \sin 2x}{\pi x^2} + \frac{\alpha(1 - \alpha)}{\pi x^2}\int_0^1 t^{\alpha - 2}\sin tx\,dt.$$

By the transformation $t = |x|^{-1}u$,

$$g(x) = \frac{\alpha(1 - \cos x)}{\pi x^2} + \frac{\alpha(1 - \alpha)}{\pi |x|^{1+\alpha}}\int_0^{|x|} u^{\alpha - 2}(1 - \cos u)\,du, \quad (6)$$

$$h(x) = \frac{(1 + \alpha)\sin x - \sin 2x}{\pi x^2} + \frac{\alpha(1 - \alpha)}{\pi |x|^{1+\alpha}}\int_0^{|x|} u^{\alpha - 2}\sin u\,du. \quad (7)$$

Here we can see that $g(x)$ and $h(x)$ are integrable. Hence, by (5) and the continuity of $u(t)$ and $v(t)$, we have

$$u(t) = \int_{-\infty}^{\infty} e^{itx} g(x)\,dx,$$

$$iv(t) = \int_{-\infty}^{\infty} e^{itx} h(x)\,dx.$$

In particular, we have, by letting $t = 0$,

$$\int_{-\infty}^{\infty} g(x)\,dx = u(0) = 1,$$

$$\int_{-\infty}^{\infty} h(x)\,dx = iv(0) = 0.$$

Therefore, we obtain

$$u(t) - 1 + icv(t) = \int_{-\infty}^{\infty} (e^{itx} - 1)(g(x) + ch(x))\,dx. \tag{8}$$

It is easy to see $g(x)$ has a positive lower bound in each bounded interval. Clearly it has a positive lower bound in a neighborhood of zero, say $|x| < a$, and for $a \leqslant x \leqslant A$ we have, by (6),

$$g(x) \geqslant \frac{\alpha(1-\alpha)}{\pi A^{\alpha+1}} \int_0^{|x|} u^{\alpha-2}(1 - \cos u)\,du.$$

From (6) and (7), we can have, by direct computation,

$$\lim_{|x|\to\infty} \frac{|h(x)|}{g(x)} = \frac{\int_0^\infty u^{\alpha-2} \sin u\,du}{\int_0^\infty u^{\alpha-2}(1-\cos u)\,du}.$$

This relation, together with the boundedness of $h(x)$ and the fact that $g(x)$ has a positive lower bound in each bounded interval, implies that $h(x)/g(x)$ is also bounded, i.e.,

$$b|h(x)| \leqslant g(x),$$

where b is a positive constant, depending only on α. Thus, if $|c| \leqslant b$, then, for each x,

$$f(x) = g(x) + ch(x) \geqslant 0. \tag{9}$$

From (4) and (8), we obtain

$$\phi_1(t) = \exp\left\{ \int_{-\infty}^{\infty} (e^{itx} - 1) f(x)\,dx \right\}. \tag{10}$$

We all know, under the condition (9), (10) is a c.f. Hence, the theorem is proved.

§4

Theorem 5. *Suppose $q(x)$ is an integrable function of x, is hermitian (i.e., $q(-x) = \overline{q(x)}$) and is not equivalent to zero. Suppose the Fourier transform of $q(x)$*

$$Q(t) = \int_{-\infty}^{\infty} d^{itx} q(x)\,dx$$

is equal to zero in an interval of t. Then the c.f. of the density function[5] $|q(x)|$ *belongs to* $(\overline{U})$.

Proof. Choose a point t_0 from the interval in which $Q(t)$ vanishes such that the function

$$r(x) = \mathrm{Re}\{e^{it_0x}q(x)\}$$

is not equivalent to zero. This choice is clearly possible. According to our assumption, we have

$$Q(t+t_0) = \int_{-\infty}^{\infty} e^{itx}e^{it_0x}q(x)\,dx = 0,$$

which holds in an interval containing $t=0$, e.g., $|t|<\delta$. Replacing t by $-t$, the above equality still holds. Adding these two equalities, we have

$$\int_{-\infty}^{\infty} \cos tx \cdot e^{it_0x}q(x)\,dx = 0, \qquad |t|<\delta.$$

Hence

$$\int_{-\infty}^{\infty} \cos tx \cdot r(x)\,dx = 0, \qquad |t|<\delta.$$

But $r(x)$ is an even function (because $q(x)$ is hermitian), so

$$\int_{-\infty}^{\infty} e^{itx}r(x)\,dx = 0, \qquad |t|<\delta. \tag{11}$$

Since $|r(x)| \leqslant |q(x)|$, the function $|q(x)| + r(x)$ is nonnegative and integrable, and is, therefore, a density function.

By (11), its c.f. is equal to the c.f. of $|q(x)|$ in $|t|<\delta$. But these two c.f.'s are not identically equal because $r(x)$ is not equivalent to zero. The theorem is thus proved.

As an example, the function $(1-ix)^{-1}$, $\lambda>1$, satisfies all the assumptions in Theorem 5, because its Fourier transform is equal to zero for all $t<0$. Hence the c.f. of the density function $(1+x^2)^{-\lambda/2}$, $\lambda>1$, belongs to $(\overline{U})$.

Corollary 1. *Suppose $q(x+iy)$ is an entire function of the exponential type and $q(x)$ is integrable and hermitian. Then the c.f. of $|q(x)|$ belongs to $(\overline{U})$.*

Proof. We all know the Fourier transform of $q(x)$ is equal to zero outside a finite interval. Hence $q(x)$ satisfies all the assumptions given in Theorem 5.

An example which satisfies Corollary 1 is the density function $|x|^{-n}|\sin x|^n$, $n = 2, 3, 4, \ldots$.

[5]In this section, we omit the normalization constants for all the density functions.

Corollary 2. *Suppose $M(t)$ is a real-valued function of bounded variation for t in $(0, \infty)$, and suppose the function*

$$p(x) = \exp\left\{ \int_0^\infty \frac{\cos tx - 1}{t} (1 + t)\, dM(t) \right\}$$

is integrable. Then the c.f. of the density function $p(x)$ belongs to $(\overline{U})$.

Proof. Since $p(x)$ is integrable, the function

$$q(x) = \exp\left\{ \int_0^\infty \frac{e^{itx} - 1}{t} (1 + t)\, dM(t) \right\} \tag{12}$$

is integrable and hermitian and $|q(x)| = p(x)$. According to the theory of Fourier–Stieltjes transform, a function given by (12) is the Fourier–Stieltjes transform of a function $G(t)$ such that $G(t)$ is of bounded variation for all t and has zero variation for $t < 0$. Because $q(x)$ is integrable, we have

$$G'(t) = \frac{1}{2\pi} \int_0^\infty e^{-itx} q(x)\, dx.$$

But when $t < 0$, $G'(t) = 0$; hence

$$\int_{-\infty}^\infty e^{itx} q(x)\, dx = 0, \qquad t > 0.$$

So $q(x)$ satisfies all the assumptions in Theorem 5 and the corollary is proved.

Corollary 3. *Suppose $m(t)$ is a real-valued integrable function on $(0, \infty)$ and the function*

$$p(x) = \exp\left\{ \int_0^\infty \frac{\cos tx - 1}{t} (1 + t) m(t)\, dt \right\} \tag{13}$$

is integrable. Then the c.f. of the density function $p(x)$ belongs to $(\overline{U})$.

This is a special case of Corollary 2 when $M(t)$ is absolutely continuous.

In order to explain how to construct examples using Corollary 3, we shall prove another corollary.

Corollary 4. *Suppose $\theta(t)$ is a real-valued measurable function on $(0, \infty)$, satisfying*

$$\int_0^e t|\theta(t)|\, dt + \int_e^\infty |\theta(t)| \ln t\, dt < \infty.$$

Let

$$f(x) = \int_0^{|x|} du \int_u^\infty \theta(v)\, dv.$$

Suppose $e^{-f(x)}$ is integrable. Then the c.f. of $e^{-f(x)}$ belongs to $(\overline{U})$.

Proof. We define

$$m(t) = \frac{2}{\pi} \frac{1}{t(1 + t)} \int_0^\infty (1 - \cos ut) \theta(u)\, du, \qquad t > 0,$$

and we shall prove the integrability of $m(t)$ through the following estimations.

$$\begin{aligned}\frac{\pi}{2}\int_0^\infty |m(t)|\,dt &\leqslant \int_0^1 \frac{dt}{t}\int_0^\infty (1-\cos ut)|\theta(u)|\,du + \int_1^\infty \frac{dt}{t^2}\int_0^\infty (1-\cos ut)|\theta(u)|\,du \\ &= \int_0^\infty |\theta(u)|\,du \int_0^u \frac{1-\cos t}{t}\,dt + \int_0^\infty u|\theta(u)|\,du\int_u^\infty \frac{1-\cos t}{t^2}\,dt \\ &\leqslant \int_0^e |\theta(u)|\,du\int_0^u \frac{t}{2}\,dt + \int_e^\infty |\theta(u)|\,du\left\{\int_0^e \frac{1-\cos t}{t}\,dt + 2\int_e^u \frac{dt}{t}\right\} \\ &\quad + \int_0^e u|\theta(u)|\,du\int_0^\infty \frac{1-\cos t}{t^2}\,dt + 2\int_e^\infty u|\theta(u)|\,du\int_u^\infty \frac{dt}{t^2} \\ &\leqslant \text{const.}\left(\int_0^e u|\theta(u)|\,du + \int_e^\infty |\theta(u)|\ln u\,du + \int_e^\infty |\theta(u)|\,du\right) < \infty.\end{aligned}$$

We only have to compute out the integral in (13) for the $m(t)$ defined here and apply Corollary 3:

$$\begin{aligned}\int_0^\infty \frac{\cos tx - 1}{t}(1+t)m(t)\,dt &= -\frac{2}{\pi}\int_0^\infty \frac{1-\cos tx}{t^2}\,dt\int_0^\infty (1-\cos ut)\theta(u)\,du \\ &= -\frac{2}{\pi}\int_0^\infty \theta(u)\,du\int_0^\infty \frac{(1-\cos tx)(1-\cos tu)}{t^2}\,du \\ &= -\int_0^\infty \theta(u)\min\{u,|x|\}\,du \\ &= -\int_0^{|x|} u\theta(u)\,dx - |x|\int_{|x|}^\infty \theta(u)\,du \\ &= -\int_0^{|x|} du\int_u^\infty \theta(v)\,dv = -f(x).\end{aligned}$$

The corollary is proved.

The following are examples of Corollary 4 corresponding to $\theta(t) = \alpha(1-\alpha)t^{\alpha-2}$, $0<\alpha<1$. We have the density function $\exp(-|x|^\alpha)$. If

$$\theta(t) = \begin{cases} 0, & 0<t<e, \\ \dfrac{2}{t(\ln t)^3} - \dfrac{6}{t(\ln t)^4}, & t \geqslant e, \end{cases}$$

then its corresponding density function is $\exp\{-|x|\min[1,(\ln|x|)^{-2}]\}$. It can also be shown that if we choose $\theta(t)$ properly, we can obtain the density function

$$\exp\left\{-|x|\min\left[1,\frac{1}{\psi(|x|)}\right]\right\},$$

where $\psi(x)$ is any one of functions given in (2). We shall not carry out the details of the computation here.

§5

Now we shall consider the sequence $F_n(x)$ of distribution functions and their c.f.'s $f_n(t)$. We say the sequence $\{F_n(x)\}$ is *compact* (or *normal*) if each of its convergent subsequences converges to a distribution function.[6] We shall prove the following two theorems.

Theorem 6. *Suppose $\{F_n(x)\}$ is a sequence of distribution functions and $\{f_n(t)\}$ is the corresponding sequence of c.f.'s. Suppose $\{f_n(t)\}$ converges for $|t| < \delta$. Also suppose the limit function $h(t)$ is continuous at $t = 0$. Then*

(i) *$\{F_n(x)\}$ is compact;*
(ii) *in the interval $|t| < \delta$, $h(t)$ is equal to the limit function of every convergent subsequence (this limit function must be a c.f.);*
(iii) *the convergence of $\{f_n(t)\}$ to $h(t)$ is uniform in the interval $|t| < \delta$.*

We need the following lemma to prove Theorem 6.

Lemma. *Suppose the sequence of distribution functions $\{G_n(x)\}$ converges to the limit $G(x)$. Then*

$$\int_{-\infty}^{\infty} \frac{e^{itx}-1}{ix}\, dG_n(x) \to \int_{-\infty}^{\infty} \frac{e^{itx}-1}{ix}\, dG(x). \tag{14}$$

This lemma is known. However, the author cannot give a good reference for it.[7] Here we just want to point out that the proof of (14) is similar to the well-known proof of

$$\int_{-\infty}^{\infty} e^{itx}\, dG_n \to \int_{-\infty}^{\infty} e^{itx}\, dG(x) \tag{15}$$

under the condition "$G(x)$ is itself a distribution function". The only difference is that to prove (15), we need the estimation

$$\left| \int_{|x|>A} e^{itx}\, dG_n(x) \right| \leqslant \int_{|x|>A} dG_n(x) \to \int_{|x|>A} dG(x) \tag{16}$$

to show that the left side of (16) is uniformly small with respect to large values of A; and to prove (14), we need the estimation

$$\left| \int_{|x|>A} \frac{e^{itx}-1}{ix}\, dG_n(x) \right| \leqslant \frac{2}{A} \int_{|x|>A} dG_n(x) \leqslant \frac{2}{A} \tag{17}$$

to decide that the left side of (17) is uniformly small.

[6] In the definition of convergence of distribution functions, we adopt the usual convention about the discontinuities of the limit functions, i.e., at the discontinuities of the limiting distribution function, the sequence does not have to converge.

[7] *Translator's note.* Since $(e^{itx} - 1)/ix$ is a bounded continuous function of x, the proof of this lemma is a standard technique nowadays, which can be found in any probability theory textbook; see, e.g., K. L. Chung, *A Course in Probability*, Academic, 1974. If the paper were written today, the following outlines would not be necessary.

Now we shall prove Theorem 6. Suppose $\{F_{n_k}\}$ is a convergent subsequence of $\{F_n\}$ and $F_{n_k} \to G$. Then, by the above lemma,

$$\int_{-\infty}^{\infty} \frac{e^{itx}-1}{ix}\, dF_{n_k}(x) \to \int_{-\infty}^{\infty} \frac{e^{itx}-1}{ix}\, dG(x). \tag{18}$$

The left side of (18) is

$$\int_0^t f_{n_k}(u)\, du.$$

Its limit, in the interval $0 < t < \delta$, is $\int_0^t h(u)\, du$ by the bounded convergence theorem. Therefore,

$$\int_{-\infty}^{\infty} \frac{e^{itx}-1}{ix}\, dG(x) = \int_0^t h(u)\, du, \qquad 0 < t < \delta.$$

Let $t \to 0$ and use the continuity of $h(t)$ at $t = 0$. We have

$$\int_{-\infty}^{\infty} dG(x) = h(0) = 1.$$

Thus, $G(x)$ is a distribution function and this proves (i).

Also, $f_{n_k}(t)$ converges to the c.f. of $G(x)$ and this proves (ii). Suppose the convergence of $f_n(t)$ to $h(t)$ is not uniform. We can find a subsequence of $\{f_n(t)\}$ which converges not uniformly to a c.f. in the interval $|t| < \delta$. We all know this is impossible and this proves (iii).

Theorem 7. *For each c.f. $f(t)$, the following two statements are equivalent.*

1°. *$f(t)$ belongs to (U);*
2°. *every sequence of c.f.'s which converges to $f(t)$ in a neighborhood of zero must converge to $f(t)$ on the whole t-axis.*

Proof. Suppose $f(t)$ belongs to $(\overline{U})$. Then there exists a c.f. $f_1(t)$ satisfying the conditions $f_1(t) = f(t)$ for $|t| < \delta$, and $f_1(t) \not\equiv f(t)$. The sequence $\{f, f_1, f, f_1, \ldots\}$ converges to f in the interval $|t| < \delta$, but does not converge to f at all t. This proves that part 2° implies 1°.

Suppose there exists a sequence $\{f_n(t)\}$ of c.f.'s which converges to $f(t)$ in the interval $|t| < \delta$ and does not converge to $f(t)$ for all t. By Theorem 6, its corresponding sequence of distribution functions is compact. Therefore, $\{f_n(t)\}$ contains a subsequence which converges to a c.f. $f_1(t)$ different from $f(t)$. But $f_1(t) = f(t)$ for $|t| < \delta$, so $f(t)$ belongs to $(\overline{U})$. This proves the other part (1° implies 2°).

The above two theorems can all be generalized to multi-dimensional cases.

Zygmund ([6]) proved the following theorem.

Suppose each element of the sequence $\{F_n\}$ of distribution functions has zero variations in a fixed half of the real line and suppose its corresponding sequence $\{f_n\}$ of c.f.'s converges to $h(t)$ in the interval $|t| < \delta$, and $h(t)$ is continuous at $t = 0$. Then $\{f_n\}$ converges to a c.f.

Based on our Theorem 6 and Theorem 7, this theorem is just a special case of the result of Marcinkiewicz stated in §1. In fact, by Theorem 6, we can use a c.f. $f(t)$ to replace $h(t)$, where $f(t)$ is the limit function of convergence subsequence of $\{f_n(t)\}$. Clearly, the corresponding distribution function of $f(t)$ has zero variation in the same half-line and hence it satisfies condition (1). Thus, the result of Marcinkiewicz says that $f(t)$ belongs to (U). Then our Theorem 7 implies that $f_n(t)$ converges to $f(t)$ at all t.

References

1. C. Esseen, *Acta Math.* **77** (1945), 1–125.
2. B. Gnedenko, *Bull. Moscow Univ. A*, **1** (1937).
3. P. Levy, *Additions des Variables Aléatoires*. Paris, 1937.
4. J. Marcinkiewicz, *Fund. Math.* **31** (1938), 86–102.
5. E. C. Titchmarsh, *Introduction to the Theory of Fourier Integrals*. Oxford, 1937.
6. A. Zygmund, *Second Berkeley Symposium* (1951), 365–372.

Reprinted from
Acta Math. Sinica **5** (1955), 333–346.

On a Kind of Transformations of Matrices*

I. Introduction and Notation

In this paper, scalar quantities are complex numbers. The transformations studied here are those of the form

$$A \to B = PA\overline{P}^{-1}, \tag{1}$$

carrying a square matrix A into a square matrix B, where P is an arbitrary nonsingular matrix and $\overline{P}$ is formed of the conjugates of the elements of P. Obviously, all such transformations constitute a group. Any such transformation will be called a transformation ($\mathfrak{A}$). If two square matrices A and B can be transformed into each other by a transformation ($\mathfrak{A}$), then we say A and B are similar ($\mathfrak{A}$).

A natural extension of transformation ($\mathfrak{A}$) is a transformation of the following form:

$$A_1 \to B_1 = PA_1Q, \qquad A_2 \to B_2 = PA_2\overline{Q}, \tag{2}$$

mapping a matric pair A_1, A_2 into a matric pair B_1, B_2, where P, Q are two arbitrary nonsingular matrices. Here A_1, A_2 have the same size but are not necessarily square. Such a transformation is called a transformation ($\mathfrak{B}$). If matric pairs A_1, A_2 and B_1, B_2 can be transformed into each other by a transformation ($\mathfrak{B}$), then we say that the two matric pairs are equivalent ($\mathfrak{B}$).

We have obtained a complete solution to fundamental problems of transformations ($\mathfrak{A}$) and transformations ($\mathfrak{B}$), including canonical forms and criteria for determining similarity or equivalence. The present paper is confined to the discussion of transformations ($\mathfrak{A}$). Results for transformations ($\mathfrak{B}$) will be published elsewhere.

These two transformations have the following three applications:

Application I. General solutions of the following matrix equation:

$$\begin{aligned} XA &= \pm A\overline{X}, \\ XA &= \pm \overline{A}\overline{X}, \\ XA &= \pm A'\overline{X}, \\ XA &= \pm \overline{A}'\overline{X}. \end{aligned}$$

*Translated by Ching-shui Cheng, University of California at Berkeley.

Application II. The classification of matric pairs H, W (H is a Hermitian matrix and W is symmetric or skew-symmetric) under the transformation

$$H \to PH\overline{P}', \qquad W \to PWP',$$

where P is nonsingular, but H and W are not assumed to be nonsingular.

Application III. The problem of decomposing an arbitrary square matrix as the product of a Hermitian matrix H and a symmetric or skew-symmetric matrix W. (For symmetric W, the decomposition is always possible; but if W is skew-symmetric, then some conditions need to be satisfied).

Now we shall define some notation.

Matrices are denoted by capital (occasionally lowercase) Roman letters; real numbers are denoted by lowercase Roman letters; complex numbers are denoted by lowercase Greek letters; zero matrices (of any size) are denoted by 0; the identity matrix of order m is denoted by I_m; and I is the general notation for an identity matrix (of any order). We will use J_{mn} and M_{mn} to denote the following two special matrices:

$$J_{mn} = \begin{bmatrix} 0 & I_n & & & 0 \\ & 0 & I_n & & \\ & & \ddots & \ddots & \\ & & & \ddots & I_n \\ & & & & \ddots \\ 0 & & & & 0 \end{bmatrix} \qquad (m \times m \text{ blocks}), \tag{3}$$

$$M_{mn} = I + J_{mn}. \tag{4}$$

Note that when $m = 1$, J_{mn} is the $n \times n$ zero matrix. We also write

$$J_m = J_{m1}, \qquad M_m = M_{m1}. \tag{5}$$

The transpose of a matrix A is denoted by A'. Both $\dotplus$ and $\sum'$ denote a direct sum. When a matrix is partitioned into blocks, all of the irrelevant submatrices (blocks) are denoted by *. Finally, throughout this paper, "the elementary divisors of $A - \lambda I$" is abbreviated as "the elementary divisors of A".

II. Transformations ($\mathfrak{A}$)

1. Definition and Lemma

Definition. Transformation (1) is called a transformation ($\mathfrak{A}$). If two square matrices A and B can be obtained from each other through a transformation ($\mathfrak{A}$), then we say that A and B are similar ($\mathfrak{A}$).

If A and B are similar ($\mathfrak{A}$), then $A\bar{A}$ and $B\bar{B}$ are similar. This is because that if (1) holds, then

$$B\bar{B} = PA\bar{P}^{-1}\bar{P}\bar{A}P^{-1} = PA\bar{A}P^{-1}.$$

Later on (Theorem 1), we shall see that if A is nonsingular, then the similarity of $A\bar{A}$ and $B\bar{B}$ implies the similarity ($\mathfrak{A}$) of A and B. But in general, this is not true. For example, let

$$A = \begin{bmatrix} i & 1 \\ -1 & i \end{bmatrix}, \qquad B = \begin{bmatrix} 0 & 0 \\ 0 & 0 \end{bmatrix}.$$

Then $A\bar{A} = B\bar{B} = 0$, but since A and B have different ranks, they are not similar ($\mathfrak{A}$). Now we shall prove a lemma, which can reduce the general case to the combination of two special cases.

Lemma 1. *Any square matrix A is similar ($\mathfrak{A}$) to a direct sum*

$$B_1 \dotplus B_2,$$

where B_1 is nonsingular and $B_2\bar{B}_2$ is nilpotent.

Proof. Use a similar transformation to change $A\bar{A}$ to

$$PA\bar{A}P^{-1} = C = C_1 \dotplus C_2,$$

where C_1 is nonsingular and C_2 is nilpotent. Let $B = PA\bar{P}^{-1}$. Then $B\bar{B} = C$. Therefore $CB = B\bar{B}B = B\bar{C}$. Partition B into blocks:

$$B = \begin{bmatrix} B_1 & F \\ G & B_2 \end{bmatrix}.$$

Then

$$\begin{bmatrix} C_1 & 0 \\ 0 & C_2 \end{bmatrix}\begin{bmatrix} B_1 & F \\ G & B_2 \end{bmatrix} = \begin{bmatrix} B_1 & F \\ G & B_2 \end{bmatrix}\begin{bmatrix} \bar{C}_1 & 0 \\ 0 & \bar{C}_2 \end{bmatrix}.$$

Thus

$$C_1F = F\bar{C}_2, \qquad C_2G = G\bar{C}_1.$$

Since C_1 and $\bar{C}_2$ have no common characteristic roots, we must have $F = 0$, $G = 0$, then $B = B_1 \dotplus B_2$. Substituting this into $B\bar{B} = C$, we get $B_1\bar{B}_1 = C_1$, $B_2\bar{B}_2 = C_2$. Thus B_1 is nonsingular and $B_2\bar{B}_2$ is nilpotent. This proves the lemma. □

By Lemma 1, we only need to consider two special cases: (i) A is nonsingular, (ii) $A\bar{A}$ is nilpotent. Combining these two cases, one can solve the general case. Now let us proceed in this way.

2. The First Special Case

Theorem 1. *Two nonsingular matrices A and B are similar ($\mathfrak{A}$) if and only if $A\bar{A}$ and $B\bar{B}$ are similar.*

Proof. The necessity of the condition has been proved earlier. Now we shall prove the sufficiency. Suppose

$$B\bar{B} = PA\bar{A}P^{-1}. \tag{6}$$

Since B is nonsingular, there is a real number c such that

$$\det(e^{2ci}B + PA\bar{P}^{-1}) \neq 0.$$

Let $Q = (e^{ci}B + e^{-ci}PA\bar{P}^{-1})\bar{P}\bar{A} = e^{ci}B\bar{P}\bar{A} + e^{-ci}PA\bar{A}$. Then Q is nonsingular. It follows from (6) that

$$Q = e^{ci}B\bar{P}\bar{A} + e^{-ci}B\bar{B}P.$$

Therefore

$$QA = e^{ci}B\bar{P}\bar{A}A + e^{-ci}B\bar{B}PA = B(e^{ci}\bar{P}\bar{A}A + e^{-ci}\bar{B}PA) = B\bar{Q},$$

and hence A and B are similar ($\mathfrak{A}$). The theorem is proved. □

Theorem 1 tells us that in order to construct the canonical form B of a nonsingular matrix A under transformations ($\mathfrak{A}$), we only need to construct a B such that the elementary divisors of $B\bar{B}$ and $A\bar{A}$ coincide. Then $B\bar{B}$ and $A\bar{A}$ are similar, and hence (by Theorem 1) B and A are similar ($\mathfrak{A}$). But the elementary divisors of $A\bar{A}$ are not arbitrary; they must satisfy certain conditions. The next theorem is concerned with the establishment of these conditions.

Theorem 2. *A nonsingular matrix G can be expressed in the form $A\bar{A}$ if and only if the elementary divisors of G satisfy the following conditions*:

(i) *Any elementary divisor of the form $(\lambda + c)^k$, where $c > 0$, must appear an even number of times.*

(ii) *The complex elementary divisors must appear in conjugate pairs as follows*:

$$(\lambda - \gamma)^l, (\lambda - \bar{\gamma})^l.$$

Proof. We first prove the necessity. If $G = A\bar{A}$, then $\bar{G} = \bar{A}A = \bar{A}G\bar{A}^{-1}$, i.e., G and $\bar{G}$ have the same elementary divisors. This shows the necessity of condition (ii). Now we prove the necessity of condition (i). Suppose an elementary divisor $(\lambda + c)^{m_1}$ of G, where $c > 0$, appears n_1 times. We want to show that n_1 is even. Without loss of generality, assume

$$G = G_1 \dotplus G_2, \tag{7}$$

where $-c$ is not a characteristic root of G_2, and

$$G_1 = -c\sum_{i=1}^{r}{}' M_{m_i n_i} \qquad (i \neq j \Rightarrow m_i \neq m_j). \tag{8}$$

By (7) and the assumption that $A\bar{A} = G$, imitating the proof of Lemma 1, one can easily show that, corresponding to the decomposition of G into $G_1 \dotplus G_2$, A is also a direct sum: $A = A_1 \dotplus A_2$. Thus

$$G_1 = A_1\bar{A}_1. \tag{9}$$

Since G_1 is a real matrix, $G_1 = \bar{A}_1A_1$. Therefore $A_1G_1 = A_1\bar{A}_1A_1 = G_1A_1$, i.e., A_1 and G_1 are commutative.

Now we partition A_1 as follows:

$$A_1 = \begin{bmatrix} U_{11} & U_{12} & \cdots & U_{1r} \\ U_{21} & U_{22} & \cdots & U_{2r} \\ \cdots & & & \\ U_{r1} & U_{r2} & \cdots & U_{rr} \end{bmatrix}, \tag{10}$$

where U_{ij} is $m_in_i \times m_jn_j$. By (8), (9), and (10), we have

$$-cM_{m_1n_1} = U_{11}\bar{U}_{11} + U_{12}\bar{U}_{21} + \cdots + U_{1r}\bar{U}_{r1}. \tag{11}$$

But we have just shown that A_1 and G_1 are commutative. A matrix which is commutative with (8) must have the following structures:

$$U_{ii} = \begin{bmatrix} U_{ii}^{(1)} & U_{ii}^{(2)} & \cdots\cdots & U_{ii}^{(m_i)} \\ & U_{ii}^{(1)} & U_{ii}^{(2)} \ddots & \\ & & \ddots\ddots & U_{ii}^{(2)} \\ 0 & & & U_{ii}^{(1)} \end{bmatrix}, \tag{12}$$

$$U_{ij} = \begin{bmatrix} 0 & U_{ij}^{(1)} & U_{ij}^{(2)} & \cdots\cdots & U_{ij}^{(m_i)} \\ 0 & & U_{ij}^{(1)} & U_{ij}^{(2)} \ddots & \\ \vdots & & & \ddots\ddots & U_{ij}^{(2)} \\ 0 & 0 & & & U_{ij}^{(1)} \end{bmatrix}, \quad \text{if} \quad m_i < m_j, \tag{13}$$

$$U_{ij} = \begin{bmatrix} U_{ij}^{(1)} & U_{ij}^{(2)} & \cdots\cdots & U_{ij}^{(m_j)} \\ & U_{ij}^{(1)} & U_{ij}^{(2)} \ddots & \\ & & \ddots\ddots & U_{ij}^{(2)} \\ 0 & & & U_{ij}^{(1)} \\ 0 & 0 & \cdots & 0 \end{bmatrix}, \quad \text{if} \quad m_i > m_j, \tag{14}$$

where each $U_{ij}^{(k)}$ is $n_i \times n_j$; in (13), there are $(m_j - m_i)n_j$ zero columns to the left, and in (14), there are $(m_i - m_j)n_i$ zero rows at the bottom.

If we substitute the U_{ij}'s into (11), then it is not hard to see that the $n_1 \times n_1$ submatrices on the upper-left corners of the two sides are $-cI_{n_1}$ and $U_{11}^{(1)}\bar{U}_{11}^{(1)}$, respectively. Equating the determinants of these two square matrices, we conclude that $(-c)^{n_1} > 0$. Therefore n_1 is even and the necessity of (i) is proved.

Now we shall prove the sufficiency. Under conditions (i), (ii), we can write down all the elementary divisors of G:

$$\left.\begin{array}{lll} (\lambda - a_r^2)^{h_r} & (r = 1, 2, \ldots), & \\ (\lambda + b_s^2)^{k_s}, & (\lambda + b_s^2)^{k_s}, & (s = 1, 2, \ldots), \\ (\lambda - \alpha_t^2)^{l_t}, & (\lambda - \bar{\alpha}_t^2)^{l_t}, & (t = 1, 2, \ldots). \end{array}\right\} \tag{15}$$

where $a_r > 0$, $b_s > 0$, and α_t is a complex number with positive real and imaginary parts. Construct the matrix

$$A^* = \sum_r{}' a_r M_{h_r} \dotplus \sum_s{}' b_s \begin{bmatrix} 0 & M_{k_s} \\ -M_{k_s} & 0 \end{bmatrix} \dotplus \sum_t{}' \begin{bmatrix} 0 & \alpha_t M_{l_t} \\ \bar{\alpha}_t M_{l_t} & 0 \end{bmatrix}. \tag{16}$$

Obviously, all the elementary divisors of $A^*\bar{A}^*$ are the same as those of G. This shows the sufficiency of conditions (i) and (ii) and the theorem is proved. □

Based on what we have concluded after the proof of Theorem 1 we may take A^* as the canonical form of A under transformations ($\mathfrak{A}$). Thus we obtain the following.

Theorem 3. *Any nonsingular matrix A can be transformed into canonical form* (16) *through a transformation* ($\mathfrak{A}$), *where $a_r > 0$, $b_s > 0$, and α_t is a complex number with positive real and imaginary parts. Furthermore, A uniquely determines the canonical form* (16) *due to the fact that the expressions in* (15) *are exactly the elementary divisors of $A\bar{A}$.*

Conditions (i), (ii) in Theorem 2 are also the necessary and sufficient conditions for matrix G to be similar to the square of some real matrix R. The necessity is obvious, since $R^2 = R\bar{R}$. To show sufficiency, note that each complex submatrix

$$\begin{bmatrix} 0 & \alpha M_t \\ \bar{\alpha} M_t & 0 \end{bmatrix}$$

in (16) can be transformed into a real matrix as follows:

$$\begin{bmatrix} C & I_2 & & 0 \\ & C & I_2 & \\ & & \ddots & I_2 \\ 0 & & & C \end{bmatrix} \qquad (l \times l \text{ blocks}), \tag{17}$$

where

$$C = \begin{bmatrix} 0 & 1 \\ -\alpha\bar{\alpha} & \alpha + \bar{\alpha} \end{bmatrix}.$$

The elementary divisors of the square of (17) are $(\lambda - \alpha^2)^l$, $(\lambda - \bar{\alpha}^2)^l$. Thus, we have found a real matrix whose square has the same elementary divisors as G.

Summarizing the above observations, we conclude:

(1) A nonsingular matrix can be expressed in the form $A\bar{A}$ if and only if it is similar to the square of a real matrix.
(2) Any nonsingular matrix can be transformed into a real matrix through a transformation $(\mathfrak{A})$.

In Section 4, we will eliminate the condition of nonsingularity from the above propositions (1) and (2).

3. The Second Special Case

Now we shall investigate the second special case, i.e., $A\bar{A}$ is nilpotent.

Theorem 4. *If $A\bar{A}$ is nilpotent, then A is similar $(\mathfrak{A})$ to a direct sum as follows*:

$$\sum_i{}' J_{m_i}. \tag{18}$$

Proof. We shall use mathematical induction. Notice that the present theorem is true if the order of A is 1. Assume that the order of A is $n > 1$, and that the theorem is true for all square matrices of order less than n. Since A is singular, there is a nonsingular matrix Q such that the first column of AQ is zero. Then the first column of $\bar{Q}^{-1}AQ$ is also zero. Therefore A is similar $(\mathfrak{A})$ to

$$B = \begin{bmatrix} 0 & * \\ 0 & B_1 \end{bmatrix},$$

where $*$ represents a row, and B_1 is a square matrix of order $n-1$. Since

$$B\bar{B} = \begin{bmatrix} 0 & * \\ 0 & B_1\bar{B}_1 \end{bmatrix},$$

and $B\bar{B}$ is nilpotent, $B_1\bar{B}_1$ is also nilpotent. Therefore, by the induction hypothesis, there is a nonsingular matrix P such that the matrix $PB_1\bar{P}^{-1} = C$ is of the form (18). Then we have the following transformation $(\mathfrak{A})$:

$$\begin{bmatrix} 1 & 0 \\ 0 & P \end{bmatrix} B \begin{bmatrix} 1 & 0 \\ 0 & \bar{P}^{-1} \end{bmatrix} = \begin{bmatrix} 0 & * \\ 0 & C \end{bmatrix}, \tag{19}$$

where $*$ still represents a row. Therefore A is similar $(\mathfrak{A})$ to the right side of (19), and the submatrix C of the latter has form (18). Now we consider two cases separately:

(1) C has only one component. Then the right side of (19) is

$$E = \begin{bmatrix} 0 & y \\ 0 & J_{n-1} \end{bmatrix}. \tag{20}$$

We may apply a transformation $(\mathfrak{A})$ to E so that all the elements of y except the first one become zero, and the other elements of E are unchanged. In fact, if

$$y=[\eta_1,\eta_2,\cdots,\eta_{n-1}],$$

let

$$z=[\eta_2,\cdots,\eta_{n-1},0],$$

then

$$y-zJ_{n-1}=[\eta_1,0,\cdots,0].$$

Therefore the transformation

$$\begin{bmatrix}1 & -z\\ 0 & I\end{bmatrix}E\begin{bmatrix}1 & \bar{z}\\ 0 & I\end{bmatrix}$$

does the job. Assume that this has been done, then in (20), all the elements of y except the first one are zero. If the first element is also zero, then $E=J_1\dotplus J_{n-1}$. If the first element is $\eta\neq 0$, then it can be changed to 1 through the transformation ($\mathfrak{A}$) $DE\overline{D}^{-1}$, where

$$D=\operatorname{diag}\{1,\bar{\eta},\eta,\bar{\eta},\cdots\}.$$

Now, when $y=[1,0,\ldots,0]$, E is J_n. So case (1) is proved.

(2) C has at least two components. Then we can write the right side of (19) as

$$F=\begin{bmatrix}0 & y_1 & y_2 & *\\ 0 & J_{m_1} & 0 & *\\ 0 & 0 & J_{m_2} & *\\ 0 & 0 & 0 & *\end{bmatrix}\qquad (m_1\leqslant m_2).$$

Using the method in the previous case, we can transform all the elements of rows y_1 and y_2, except the first element, into zeros. Assume that this has been done. If one of the two first elements is zero, then F is a direct sum in which one component is J_{m_1} or J_{m_2}. The order of the other component is less than n so by the induction hypothesis, it can be transformed into (18) through a transformation ($\mathfrak{A}$). Therefore we only have to consider the case where both of the two first elements of y_1 and y_2 are not zero. Similar to case (1), we may transform these two elements into 1. Then in matrix F, we have

$$y_1=[1,\overbrace{0,\cdots,0}^{m_1-1}],\qquad y_2=[1,\overbrace{0,\cdots,0}^{m_2-1}].$$

Let

$$N=\begin{bmatrix}I_{m_1}\\ 0\end{bmatrix},$$

where 0 represents m_2-m_1 zero rows. It is easy to verify that

$$y_1=y_2N,\qquad NJ_{m_1}=J_{m_2}N.$$

Thus we have the following equation of transformation ($\mathfrak{A}$):

$$\begin{bmatrix} 1 & 0 & 0 & 0 \\ 0 & I & 0 & 0 \\ 0 & N & I & 0 \\ 0 & 0 & 0 & I \end{bmatrix} F \begin{bmatrix} 1 & 0 & 0 & 0 \\ 0 & I & 0 & 0 \\ 0 & -N & I & 0 \\ 0 & 0 & 0 & I \end{bmatrix} = \begin{bmatrix} 0 & 0 & y_2 & * \\ 0 & J_{m_1} & 0 & 0 \\ 0 & 0 & J_{m_2} & 0 \\ 0 & 0 & 0 & * \end{bmatrix}. \tag{21}$$

The right side of (21) is a direct sum in which one component is J_{m_1}, and by the induction hypothesis, the other component can be transformed into (18).

Thus the induction is done and the theorem is proved. □

The canonical form (18) is still uniquely determined by A. This will be proved in the next section when we discuss the general case. Now we state the following two corollaries of Theorem 4.

Corollary 1. *A square matrix is nilpotent and can be expressed in the form $A\bar{A}$ if and only if it is similar to the square of a nilpotent real matrix.*

Proof. The sufficiency is self-evident. The necessity is a direct consequence of Theorem 4 since matrix (18) is real. □

The condition in Corollary 1 can also be explicitly expressed in terms of the elementary divisors. The set of elementary divisors of the square of a nilpotent matrix consists of the elementary divisors of several matrices of the form J_m^2. The elementary divisors of J_m^2 are easy to calculate. They are

For J_1^1: λ,
For J_{2k}^2: $\lambda^k, \lambda^k \quad (k > 0)$,
For J_{2k+1}^2: $\lambda^k, \lambda^{k+1} \quad (k > 0)$.

Therefore the set of elementary divisors of the square of a nilpotent matrix can be characterized by the following condition which we call Condition (C):

Condition (C). These elementary divisors can be grouped into several pairs. Each pair is (λ^k, λ^k) or $(\lambda^k, \lambda^{k+1})$. Besides these pairs, there may or may not be an extra elementary divisor λ.

Thus, we have the following corollary.

Corollary 2. *A nilpotent matrix can be expressed in the form $A\bar{A}$ if and only if its elementary divisors satisfy Condition (C).*

4. General Case

By Lemma 1, a series of results regarding the general case can be obtained by combining the two special cases in the two previous sections through direct sums.

First of all, we shall solve the problem of when an arbitrary square matrix can be expressed in the form $A\bar{A}$. Combining conclusion (1) at the end of Section 2 and Corollary 1 of Theorem 4 in Section 3, we obtain

Theorem 5. *A square matrix can be expressed in the form $A\bar{A}$ if and only if it is similar to the square of a real matrix.*

If we want to express the condition in Theorem 5 in terms of elementary divisors, then we can combine Theorem 2 and the Corollary of Theorem 4 to obtain

Theorem 6. *A square matrix can be expressed in the form $A\bar{A}$ if and only if its elementary divisors satisfy the following condition: the elementary divisors corresponding to negative characteristic roots satisfy condition (i) of Theorem 2, the elementary divisors corresponding to complex characteristic roots satisfy condition (ii) of Theorem 2, and those corresponding to a zero characteristic root satisfy condition (C).*

Combining conclusion (2) at the end of Section 2 and Theorem 4 in Section 3, we obtain

Theorem 7. *Any square matrix can be transformed into a real matrix through a transformation* ($\mathfrak{A}$).

It is interesting to note that when the order of the square matrix is 1, Theorem 7 says that any complex number can be transformed into a real number through a rotation.

Now we shall start to discuss the fundamental problem of transformation ($\mathfrak{A}$), i.e., canonical forms and the necessary and sufficient conditions for similarity ($\mathfrak{A}$). Theorem 3 and Theorem 4 together provide the canonical form, i.e., the direct sum of matrix (16) and matrix (18). In other words, any square matrix A is similar ($\mathfrak{A}$) to a direct sum as follows:

$$\sum_r{}' a_r M_{h_r} \dotplus \sum_s{}' b_s \begin{bmatrix} 0 & M_{k_s} \\ -M_{k_s} & 0 \end{bmatrix} \dotplus \sum_t{}' \begin{bmatrix} 0 & \alpha_t M_{l_t} \\ \bar{\alpha}_t M_{l_t} & 0 \end{bmatrix} \dotplus \sum_i{}' J_{m_i}. \tag{22}$$

But we still have to show the uniqueness of the canonical form (22). Each of the nonsingular components of (22) has been shown to be uniquely determined by A (see Theorem 3), but the uniqueness of its singular components has not been solved. In order to fill this gap, we proceed as follows.

If matrix A and matrix B are similar ($\mathfrak{A}$):

$$PA\bar{P}^{-1} = B,$$

then matrices

$$\begin{bmatrix} 0 & A \\ \bar{A} & 0 \end{bmatrix} \quad \text{and} \quad \begin{bmatrix} 0 & B \\ \bar{B} & 0 \end{bmatrix}$$

are similar, since

$$\begin{bmatrix} P & 0 \\ 0 & \bar{P} \end{bmatrix} \begin{bmatrix} 0 & A \\ \bar{A} & 0 \end{bmatrix} \begin{bmatrix} P^{-1} & 0 \\ 0 & \bar{P}^{-1} \end{bmatrix} = \begin{bmatrix} 0 & B \\ \bar{B} & 0 \end{bmatrix}. \tag{23}$$

Let B be the canonical form (22) of A, then the right side of (23) is similar (by rearranging rows and columns) to the following direct sum:

$$\sum_r{}' a_r \begin{bmatrix} 0 & M_{h_r} \\ M_{h_r} & 0 \end{bmatrix} \dotplus \sum_s{}' b_s \begin{bmatrix} 0 & 0 & 0 & M_{k_s} \\ 0 & 0 & -M_{k_s} & 0 \\ 0 & M_{k_s} & 0 & 0 \\ -M_{k_s} & 0 & 0 & 0 \end{bmatrix}$$

$$\dotplus \sum_t{}' \begin{bmatrix} 0 & 0 & 0 & \bar{\alpha}_t M_{l_t} \\ 0 & 0 & \alpha_t M_{l_t} & 0 \\ 0 & \alpha_t M_{l_t} & 0 & 0 \\ \bar{\alpha}_t M_{l_t} & 0 & 0 & 0 \end{bmatrix} \dotplus \sum_i{}' \begin{bmatrix} 0 & J_{m_i} \\ J_{m_i} & 0 \end{bmatrix}. \tag{24}$$

Now let us calculate the elementary divisors of (24). The elementary divisors contributed by the nonsingular components of (24) obviously are the following expressions:

$$\left.\begin{array}{ll} (\lambda + a_r)^{h_r}, (\lambda + a_r)^{h_r} & (r = 1, 2, \cdots) \\ (\lambda + b_s i)^{k_s}, (\lambda + b_s i)^{k_s}, (\lambda - b_s i)^{k_s}, (\lambda - b_s i)^{k_s} & (s = 1, 2, \cdots) \\ (\lambda + \alpha_t)^{l_t}, (\lambda + \bar{\alpha}_t)^{l_t}, (\lambda - \alpha_t)^{l_t}, (\lambda - \bar{\alpha}_t)^{l_t} & (t = 1, 2, \cdots) \end{array}\right\}. \tag{25}$$

As to the singular components of (24), each of them has the following form:

$$\begin{bmatrix} 0 & J_m \\ J_m & 0 \end{bmatrix}. \tag{26}$$

Since

$$\frac{1}{2}\begin{bmatrix} I & I \\ -I & I \end{bmatrix}\begin{bmatrix} 0 & J_m \\ J_m & 0 \end{bmatrix}\begin{bmatrix} I & -I \\ I & I \end{bmatrix} = \begin{bmatrix} J_m & 0 \\ 0 & -J_m \end{bmatrix},$$

the elementary divisors of (26) are λ^m, λ^m. Thus the elementary divisors of matrix (24) consist of (25) and

$$\lambda^{m_i}, \lambda^{m_i} \qquad (i = 1, 2, \ldots). \tag{27}$$

These should be the same as the elementary divisors of

$$\begin{bmatrix} 0 & A \\ \bar{A} & 0 \end{bmatrix}.$$

Thus canonical form (24) is uniquely determined by matrix A.

Now let us state the above results in two theorems.

Theorem 8. *Any square matrix A can be transformed into the canonical form (22) through a transformation ($\mathfrak{A}$), where $a_r > 0$, $b_s > 0$, and α_t is a complex number with positive real and imaginary parts. The canonical form (22) is uniquely determined by A due to the fact that expressions (25) and (27) constitute the elementary divisors of* *matrix*

$$\begin{bmatrix} 0 & A \\ \bar{A} & 0 \end{bmatrix}. \tag{28}$$

Theorem 9. *Two matrices A and B are similar* ($\mathfrak{A}$) *if and only if*

$$\begin{bmatrix} 0 & A \\ \bar{A} & 0 \end{bmatrix} \quad \text{and} \quad \begin{bmatrix} 0 & B \\ \bar{B} & 0 \end{bmatrix}$$

have the same elementary divisors.

Accidentally, we have proved the following theorem.

Theorem 10. *A square matrix is similar to a matrix of the form*

$$\begin{bmatrix} 0 & A \\ \bar{A} & 0 \end{bmatrix}$$

if and only if its elementary divisors satisfy the following conditions:

(i) *The elementary divisors corresponding to real characteristic roots (including zero) can be grouped into the following pairs:*

$$(\lambda + a)^h, (\lambda - a)^h.$$

(ii) *The elementary divisors corresponding to complex characteristic roots (including pure imaginary roots) can be grouped into sets of four of the following form:*

$$(\lambda + \alpha)^k, (\lambda + \bar{\alpha})^k, (\lambda - \alpha)^k, (\lambda - \bar{\alpha})^k.$$

If we use matrix

$$\begin{bmatrix} 0 & A \\ -\bar{A} & 0 \end{bmatrix}$$

instead of matrix (28), then we can obtain results similar to Theorems 8, 9, and 10. These results are stated below. The proofs are omitted.

Theorem 8′. *The canonical form* (24) *of a square matrix A under transformations* ($\mathfrak{A}$) *is uniquely determined by A due to the fact that the expressions*

$$\begin{array}{ll} (\lambda + a_r i)^{h_r}, (\lambda - a_r i)^{h_r} & (r = 1, 2, \cdots), \\ (\lambda + b_s)^{k_s}, (\lambda + b_s)^{k_s}, (\lambda - b_s)^{k_s}, (\lambda - b_s)^{k_s} & (s = 1, 2, \cdots), \\ (\lambda + i\alpha_t)^{l_t}, (\lambda + i\bar{\alpha}_t)^{l_t}, (\lambda - i\alpha_t)^{l_t}, (\lambda - i\bar{\alpha}_t)^{l_t} & (t = 1, 2, \cdots), \\ \lambda^{m_i}, \lambda^{m_i} & (i = 1, 2, \cdots) \end{array}$$

constitute the elementary divisors of matrix

$$\begin{bmatrix} 0 & A \\ -\bar{A} & 0 \end{bmatrix}.$$

Theorem 9′. *Two square matrices A and B are similar ($\mathfrak{A}$) if and only if*

$$\begin{bmatrix} 0 & A \\ -\bar{A} & 0 \end{bmatrix} \quad \text{and} \quad \begin{bmatrix} 0 & B \\ -\bar{B} & 0 \end{bmatrix}$$

have the same elementary divisors.

Theorem 10′. *A square matrix is similar to a matrix of the form*

$$\begin{bmatrix} 0 & A \\ -\bar{A} & 0 \end{bmatrix}$$

if and only if its elementary divisors satisfy the following conditions:

(i) *The elementary divisors corresponding to purely imaginary characteristic roots (including zero) can be grouped into the following pairs:*

$$(\lambda + ai)^h, (\lambda - ai)^h.$$

(ii) *The elementary divisors corresponding to non-purely imaginary characteristic roots (including real roots) can be grouped into sets of four of the following form:*

$$(\lambda + \alpha)^k, (\lambda + \bar{\alpha})^k, (\lambda - \alpha)^k, (\lambda - \bar{\alpha})^k.$$

III. Decomposing an Arbitrary Square Matrix into the Product of a Hermitian Matrix and a Symmetric Matrix

We shall prove the following theorem as an application of Theorem 7.

Theorem 11. *Any square matrix A can be expressed as HS (or SH), where H is Hermitian and S is symmetric. Furthermore, we can prespecify which one of H and S is nonsingular.*

Proof. By Theorem 7, we can write $A = PR\bar{P}^{-1}$, where R is a real matrix. We know that the real matrix R can be expressed as $R = S_1S_2$, where S_1 and S_2 are real symmetric matrices and one of which can be preassigned to be nonsingular. Let $H = PS_1\bar{P}'$, $S = \bar{P}'^{-1}S_2\bar{P}^{-1}$. Then $A = HS$. Corresponding to nonsingular S_1 and S_2, we have nonsingular H and S. The same method can be used to prove the decomposition into SH. □

Reprinted from
Acta Sci. Natur. Univ. Pekinensis **1** (1955), 1–16.

On a Kind of Transformations of Matrix Pairs*

1. Definitions and Notation

In this paper, the scalar quantities are complex numbers. We will study transformations of the form

$$A_1 \to B_1 = PA_1Q, \qquad A_2 \to B_2 = PA_2\overline{Q}, \tag{1}$$

where A_1 and A_2 are two matrices of the same size, and P, Q are two arbitrary nonsingular matrices of appropriate sizes. Hereafter, two matrices of the same size are called a matric pair. Transformation (1) maps the matric pair (A_1, A_2) into (B_1, B_2). Such a transformation is called a transformation $(\mathfrak{B})$. Obviously, all such transformations constitute a group. If two matric pairs (A_1, A_2) and (B_1, B_2) can be obtained from each other through a transformation $(\mathfrak{B})$, then we say that the two matric pairs are equivalent $(\mathfrak{B})$. If in the second expression of (1), $\overline{Q}$ is replaced with Q (i.e., the classical transformation of matric pairs), then we say that the two matric pencils $\lambda A_1 + \mu A_2$ and $\lambda B_1 + \mu B_2$ are equivalent. The equivalence $(\mathfrak{B})$ of two matric pairs is denoted by $\sim$.

We shall use the same notations as in Hsu (1955). Moreover, we shall use K_m and L_m to denote the following two special matrices:

$$K_m = \begin{bmatrix} 1 & 0 & & & 0 & 0 \\ & 1 & 0 & & & 0 \\ & & \ddots & \ddots & & \vdots \\ & & & & 0 & \\ 0 & & & & 1 & 0 \end{bmatrix}, \quad L_m = \begin{bmatrix} 0 & 1 & & & & 0 \\ 0 & 0 & 1 & & & \\ \vdots & & 0 & \ddots & & \\ & & & & \ddots & \\ 0 & 0 & & & 0 & 1 \end{bmatrix} \quad (m \times \overline{m+1}) \tag{2}$$

For simplicity, one more notation is adopted. For an arbitrary matrix A, we will use

$$A \dotplus 0$$

to denote a matrix which is A itself, or is obtained by adding several zero columns to the right of A (i.e., $[A \quad 0]$), or is obtained by adding several zero rows to the

*Received September 6, 1955. (Translated by Ching-shui Cheng, University of California at Berkeley.)

bottom of A (i.e., $[{A \atop 0}]$), or obtained by adding several zero columns and zero rows to the right and bottom of A, respectively (i.e., $A \dotplus 0$). We will use $A \ddot{+} 0_{gh}$ to represent that g zero rows and h zero columns have been added (it is possible that $g=0$, or $h=0$, or both are equal to zero).

2. A Transformation Law for Matric Pairs

The purpose of this section is to transform a given matric pair into the form of (3) below.

Theorem 1. *Any matric pair* (A_1, A_2) *is equivalent* ($\mathfrak{B}$) *to a matric pair of the following form*:

$$\left.\begin{aligned} &I \dotplus B_1 \dotplus \sum_\alpha{}' K_{m_\alpha} \dotplus \sum_\beta{}' K'_{n_\beta} \ddot{+} 0_{gh}, \\ &B_2 \dotplus I \dotplus \sum_\alpha{}' L_{m_\alpha} \dotplus \sum_\beta{}' L'_{n_\beta} \ddot{+} 0_{gh}, \end{aligned}\right\}, \tag{3}$$

where B_1 (B_2, *respectively*) *is a square matrix of the same order as the component* I *of the second* (*first, respectively*) *matrix, and* $B_2\overline{B}_2$ *is nilpotent.*

Theorem 1 can be proved by a series of lemmas. For convenience, we say that the matric pair (A_1, A_2) has an annihilating row (or column) with respect to $\mathfrak{B}$ if it is possible to simultaneously transform a certain row (or column) of A_1 and the corresponding row (or column) of A_2 into zero through a transformation ($\mathfrak{B}$).

Lemma 1. *Any matric pair* (A_1, A_2) *is equivalent* ($\mathfrak{B}$) *to the following matric pair*:

$$C_1 \ddot{+} 0, C_2 \ddot{+} 0,$$

where the matric pair (C_1, C_2) *has neither annihilating row nor annihilating column with respect to* $\mathfrak{B}$.

Proof. We shall first show that the matric pair (A_1, A_2) is equivalent ($\mathfrak{B}$) to the matric pair

$$[D_1, 0], [D_2, 0],$$

where (D_1, D_2) has no annihilating column with respect to $\mathfrak{B}$. For this purpose, take a nonsingular matrix X such that

$$A_1 X = [E_1, 0],$$

where all the columns of E_1 are linearly independent; also, write

$$A_2 \overline{X} = [E_2, E_3].$$

Then

$$A_1, A_2 \sim [E_1, 0], [E_2, E_3].$$

Take a nonsingular matrix Y such that $E_3 Y = [F, 0]$ and all the columns of F are linearly independent. Then we have

$$[E_1, 0]\begin{bmatrix} I & 0 \\ 0 & \bar{Y} \end{bmatrix} = [E_1, 0] = [E_1, 0, 0],$$

$$[E_2, E_3]\begin{bmatrix} I & 0 \\ 0 & Y \end{bmatrix} = [E_2, E_3 Y] = [E_2, F, 0].$$

Therefore

$$A_1, A_2 \sim [D_1, 0], [D_2, 0],$$

where

$$D_1 = [E_1, 0], \qquad D_2 = [E_2, F].$$

We claim that the matric pair (D_1, D_2) has no annihilating column with respect to $\mathfrak{B}$. If this is not the case, then there exist two nonsingular matrices P and Q such that the last column of $PD_1 Q$ and $PD_2 \bar{Q}$ are zero. Then the last column of $D_1 Q$ and $D_2 \bar{Q}$ are also zero. So if q is the last column of Q, then $D_1 q = 0$, and $D_2 \bar{q} = 0$. Let $q = [{x \atop y}]$. Then

$$E_1 x = 0, \qquad E_2 \bar{x} + F\bar{y} = 0.$$

Since all the columns of E_1 are linearly independent, we have $x = 0$. Substituting this into the second equation, we get $F\bar{y} = 0$. Then $y = 0$ since all the columns of F are linearly independent. Thus $q = 0$ which contradicts the nonsingularity of Q. Therefore (D_1, D_2) has no annihilating column with respect to $\mathfrak{B}$.

By the same argument, the matric pair (D_1, D_2) is equivalent ($\mathfrak{B}$) to

$$\begin{bmatrix} C_1 \\ 0 \end{bmatrix}, \begin{bmatrix} C_2 \\ 0 \end{bmatrix},$$

where (C_1, C_2) has no annihilating row with respect to $\mathfrak{B}$. Therefore

$$A_1, A_2 \sim C_1 \dotplus 0, C_2 \dotplus 0,$$

where (C_1, C_2) has neither annihilating rows nor annihilating columns with respect to $\mathfrak{B}$. The Lemma is proved.

Lemma 2. *If the matric pair* (A_1, A_2) *has no annihilating column with respect to* $\mathfrak{B}$, *and all the rows of* A_2 *are linearly independent, then* (A_1, A_2) *is equivalent* ($\mathfrak{B}$) *to the following matric pair*:

$$D \dotplus \sum_{\alpha}{}' K_{m_\alpha}, I \dotplus \sum_{\alpha}{}' L_{m_\alpha}. \tag{4}$$

Proof. We shall first consider the case where both A_1 and A_2 have only one row. If they also have one column only, then A_2 is a nonzero number and can be transformed into 1. Then (4) is achieved. If A_1 and A_2 both have two columns, we may assume

$$A_1 = [a, b], \qquad A_2 = [0, 1].$$

Then $a \neq 0$, otherwise there would be an annihilating column. Let

$$Q = \begin{bmatrix} \frac{1}{a} & -\frac{b}{a} \\ 0 & 1 \end{bmatrix}.$$

Then

$$A_1 Q = [1,0], \qquad A_2 \overline{Q} = [0,1].$$

Thus $(A_1, A_2) \sim (K_1, L_1)$; again (4) is achieved. Finally, it is not possible that both A_1 and A_2 have more than two columns, otherwise the matric pair (A_1, A_2) would clearly have an annihilating column with respect to $\mathfrak{B}$. Hence the Lemma is true when both A_1 and A_2 have one row.

Now we shall proceed by induction. Assume A_1 and A_2 have $n > 1$ rows, and the lemma holds when the number of rows is less than n.

If A_1 and A_2 are square matrices, then A_2 is nonsingular. Thus we have $(A_1, A_2) \sim (A_2^{-1} A_1, I)$ and (4) is achieved. Therefore we may assume that A_1 and A_2 are not square matrices. The first column of A_2 can be transformed into zero through an appropriate column transformation ($\mathfrak{B}$)*:

$$A_1, A_2 \sim [g, G_1], [0, G_2].$$

Since the given matric pair has no annihilating column with respect to $\mathfrak{B}$, column g is not zero and can be transformed into a column with the first element equal to one and all the other elements equal to zero through an appropriate row transformation ($\mathfrak{B}$). Thus

$$A_1, A_2 \sim \begin{bmatrix} 1 & x_1 \\ 0 & H_1 \end{bmatrix}, \begin{bmatrix} 0 & x_2 \\ 0 & H_2 \end{bmatrix},$$

where x_1 and x_2 are two rows. Furthermore, since

$$\begin{bmatrix} 1 & x_1 \\ 0 & H_1 \end{bmatrix} \begin{bmatrix} 1 & -x_1 \\ 0 & I \end{bmatrix} = \begin{bmatrix} 1 & 0 \\ 0 & H_1 \end{bmatrix}, \begin{bmatrix} 0 & x_2 \\ 0 & H_2 \end{bmatrix} \begin{bmatrix} 1 & -\bar{x}_1 \\ 0 & I \end{bmatrix} = \begin{bmatrix} 0 & x_2 \\ 0 & H_2 \end{bmatrix},$$

we have

$$A_1, A_2 \sim \begin{bmatrix} 1 & 0 \\ 0 & H_1 \end{bmatrix}, \begin{bmatrix} 0 & x_2 \\ 0 & H_2 \end{bmatrix}. \tag{5}$$

Suppose the matric pair (H_1, H_2) has an annihilating column with respect to $\mathfrak{B}$. Then we can use a transformation ($\mathfrak{B}$) to annihilate the first column of H_1 and H_2 and obtain

$$A_1, A_2 \sim \begin{bmatrix} 1 & 0 & 0 \\ 0 & 0 & T_1 \end{bmatrix}, \begin{bmatrix} 0 & \xi & y \\ 0 & 0 & T_2 \end{bmatrix}. \tag{6}$$

The number ξ cannot be zero, otherwise the matric pair (A_1, A_2) would have an

*Applying a "column transformation ($\mathfrak{B}$)" means to multiply Q and $\overline{Q}$ (Q is nonsingular) to the right of two matrices, respectively. If we multiply the same nonsingular matrix to the left of two matrices, then we have a row transformation ($\mathfrak{B}$).

annihilating column with respect to $\mathfrak{B}$. Multiplying the matrix

$$\begin{bmatrix} 1 & 0 & 0 \\ 0 & \frac{1}{\bar{\xi}} & -\frac{1}{\bar{\xi}}\bar{y} \\ 0 & 0 & I \end{bmatrix}$$

and its conjugate matrix to the right of the two matrices on the right side of (6), respectively, we obtain

$$A_1, A_2 \sim \begin{bmatrix} 1 & 0 & 0 \\ 0 & 0 & T_1 \end{bmatrix}, \begin{bmatrix} 0 & 1 & 0 \\ 0 & 0 & T_2 \end{bmatrix},$$

i.e.,

$$A_1, A_2 \sim K_1 \dotplus T_1, L_1 \dotplus T_2. \tag{7}$$

Now the matric pair (T_1, T_2) with $n-1$ rows obviously satisfies the condition of the theorem (i.e., this matric pair has no annihilating columns with respect to $\mathfrak{B}$ and all the rows of T_2 are linearly independent). By the induction hypothesis, the matric pair (T_1, T_2) is equivalent ($\mathfrak{B}$) to a matric pair of the form (4), and so is (A_1, A_2).

So the case that remains to be considered is when the matric pair (H_1, H_2) in (5) has no annihilating column with respect to $\mathfrak{B}$. In this case, the matric pair (H_1, H_2) with $n-1$ rows satisfies the condition of the theorem, and by the induction hypothesis, this matric pair is equivalent ($\mathfrak{B}$) to a matric pair of the form (4). Thus we may assume that the matric pair (H_1, H_2) in (5) itself is of the form (4):

$$\left.\begin{aligned} H_1 &= D \dotplus \sum_{\alpha}{}' K_{m_\alpha} \\ H_2 &= I \dotplus \sum_{\alpha}{}' L_{m_\alpha} \end{aligned}\right\}. \tag{8}$$

Now we shall separate the discussion into three cases.

Case (i). The component pair (D, I) indeed appears in (8). Then we have

$$H_1 = D \dotplus *, \qquad H_2 = I \dotplus *.$$

Therefore we may write (5) as

$$A_1, A_2 \sim \begin{bmatrix} 1 & 0 & 0 \\ 0 & D & 0 \\ 0 & 0 & * \end{bmatrix}, \begin{bmatrix} 0 & y & * \\ 0 & I & 0 \\ 0 & 0 & * \end{bmatrix}.$$

Since

$$\begin{bmatrix} 1 & -y & 0 \\ 0 & I & 0 \\ 0 & 0 & 1 \end{bmatrix} \begin{bmatrix} 1 & 0 & 0 \\ 0 & D & 0 \\ 0 & 0 & * \end{bmatrix} \begin{bmatrix} 1 & yD & 0 \\ 0 & I & 0 \\ 0 & 0 & 1 \end{bmatrix} = \begin{bmatrix} 1 & 0 & 0 \\ 0 & D & 0 \\ 0 & 0 & * \end{bmatrix},$$

and

$$\begin{bmatrix} 1 & -y & 0 \\ 0 & I & 0 \\ 0 & 0 & 1 \end{bmatrix} \begin{bmatrix} 0 & y & * \\ 0 & I & 0 \\ 0 & 0 & * \end{bmatrix} \begin{bmatrix} 1 & \bar{y}\bar{D} & 0 \\ 0 & I & 0 \\ 0 & 0 & 1 \end{bmatrix} = \begin{bmatrix} 0 & 0 & * \\ 0 & I & 0 \\ 0 & 0 & * \end{bmatrix},$$

we have

$$A_1, A_2 \sim D \dotplus *, I \dotplus * \tag{9}$$

Thus one component pair (D, I) has been isolated. Since the number of rows in the other component pair in (9) is less than n, it can be transformed into the form of (4) through a transformation $(\mathfrak{B})$. Accordingly, (A_1, A_2) is equivalent $(\mathfrak{B})$ to a matric pair of the form in (4).

Case (ii). The matric pair (8) has at least two component pairs (K_{m_1}, L_{m_1}) and (K_{m_2}, L_{m_2}) $(m_1 \leqslant m_2)$. In this case, we have

$$H_1 = K_{m_1} \dotplus K_{m_2} \dotplus *, \qquad H_2 = L_{m_1} \dotplus L_{m_2} \dotplus *.$$

Therefore we may write (5) as

$$A_1, A_2 \sim \begin{bmatrix} 1 & 0 & 0 & 0 \\ 0 & K_{m_1} & 0 & 0 \\ 0 & 0 & K_{m_2} & 0 \\ 0 & 0 & 0 & * \end{bmatrix}, \begin{bmatrix} 0 & y & z & * \\ 0 & L_{m_1} & 0 & 0 \\ 0 & 0 & L_{m_2} & 0 \\ 0 & 0 & 0 & * \end{bmatrix} \tag{10}$$

Denote the matric pair on the right side of (10) as (E_1, E_2). It is easy to verify that if η and ζ are the first elements of row y and row z, respectively, then

$$y - yL'_{m_1}L_{m_1} = [\eta, 0, \ldots, 0], \qquad z - zL'_{m_2}L_{m_2} = [\zeta, 0, \ldots, 0].$$

Thus, if we let

$$P = \begin{bmatrix} 1 & -yL'_{m_1} & -zL'_{m_2} & 0 \\ 0 & I & 0 & 0 \\ 0 & 0 & I & 0 \\ 0 & 0 & 0 & I \end{bmatrix}, \qquad Q = \begin{bmatrix} 1 & yL'_{m_1}K_{m_1} & zL'_{m_2}K_{m_2} & 0 \\ 0 & I & 0 & 0 \\ 0 & 0 & I & 0 \\ 0 & 0 & 0 & I \end{bmatrix},$$

then

$$PE_1Q = E_1, \qquad PE_2Q = \begin{bmatrix} 0 & \eta a_{m_1} & \zeta a_{m_2} & * \\ 0 & L_{m_1} & 0 & 0 \\ 0 & 0 & L_{m_2} & 0 \\ 0 & 0 & 0 & * \end{bmatrix},$$

where

$$a_m = [1, \overbrace{0, \ldots, 0}^{m}]. \tag{11}$$

Therefore

$$A_1, A_2 \sim \begin{bmatrix} 1 & 0 & 0 & 0 \\ 0 & K_{m_1} & 0 & 0 \\ 0 & 0 & K_{m_2} & 0 \\ 0 & 0 & 0 & * \end{bmatrix}, \begin{bmatrix} 0 & \eta a_{m_1} & \zeta a_{m_2} & * \\ 0 & L_{m_1} & 0 & 0 \\ 0 & 0 & L_{m_2} & 0 \\ 0 & 0 & 0 & * \end{bmatrix}.$$

If η or ζ is zero, then a component pair (K_{m_1}, L_{m_1}) or (K_{m_2}, L_{m_2}) is isolated. By the induction hypotheses, the other component pair can be transformed into the form of (4) through a transformation $(\mathfrak{B})$ and the proof is completed. So we only have to consider the case where $\eta \neq 0$ and $\zeta \neq 0$. We may transform η and ζ into 1 by the following method. Construct the following diagonal matrix

$$U_1 = \mathrm{diag}(\overbrace{\bar{\eta}, \eta, \bar{\eta}, \eta, \ldots}^{m_1}), \qquad U_2 = \mathrm{diag}(\overbrace{\bar{\zeta}, \zeta, \bar{\zeta}, \zeta, \ldots}^{m_2}),$$

$$V_1 = \mathrm{diag}(\overbrace{\bar{\eta}^{-1}, \eta^{-1}, \bar{\eta}^{-1}, \eta^{-1}, \ldots}^{m_1+1}), \qquad V_2 = \mathrm{diag}(\overbrace{\bar{\zeta}^{-1}, \zeta^{-1}, \bar{\zeta}^{-1}, \zeta^{-1}, \ldots}^{m_2+1}).$$

It is easy to verify that

$$\begin{bmatrix} 1 & & & \\ & U_1 & & \\ & & U_2 & \\ & & & I \end{bmatrix} \begin{bmatrix} 1 & 0 & 0 & 0 \\ 0 & K_{m_1} & 0 & 0 \\ 0 & 0 & K_{m_2} & 0 \\ 0 & 0 & 0 & * \end{bmatrix} \begin{bmatrix} 1 & & & \\ & V_1 & & \\ & & V_2 & \\ & & & I \end{bmatrix} = \begin{bmatrix} 1 & 0 & 0 & 0 \\ 0 & K_{m_1} & 0 & 0 \\ 0 & 0 & K_{m_2} & 0 \\ 0 & 0 & 0 & * \end{bmatrix},$$

$$\begin{bmatrix} 1 & 0 & 0 & 0 \\ 0 & U_1 & 0 & 0 \\ 0 & 0 & U_2 & 0 \\ 0 & 0 & 0 & I \end{bmatrix} \begin{bmatrix} 0 & \eta a_{m_1} & \zeta a_{m_2} & * \\ 0 & L_{m_1} & 0 & 0 \\ 0 & 0 & L_{m_2} & 0 \\ 0 & 0 & 0 & * \end{bmatrix} \begin{bmatrix} 1 & 0 & 0 & 0 \\ 0 & \bar{V}_1 & 0 & 0 \\ 0 & 0 & \bar{V}_2 & 0 \\ 0 & 0 & 0 & I \end{bmatrix} = \begin{bmatrix} 0 & a_{m_1} & a_{m_2} & * \\ 0 & L_{m_1} & 0 & 0 \\ 0 & 0 & L_{m_2} & 0 \\ 0 & 0 & 0 & * \end{bmatrix}.$$

Therefore

$$A_1, A_2 \sim \begin{bmatrix} 1 & 0 & 0 & 0 \\ 0 & K_{m_1} & 0 & 0 \\ 0 & 0 & K_{m_2} & 0 \\ 0 & 0 & 0 & * \end{bmatrix}, \begin{bmatrix} 0 & a_{m_1} & a_{m_2} & * \\ 0 & L_{m_1} & 0 & 0 \\ 0 & 0 & L_{m_2} & 0 \\ 0 & 0 & 0 & * \end{bmatrix}. \tag{12}$$

Let

$$M = [I_{m_1}, \overbrace{0}^{m_2-m_1}], \qquad N = [I_{m_1+1}, \overbrace{0}^{m_2-m_1}].$$

Then

$$a_{m_2} = a_{m_1}N, \qquad MK_{m_2} = K_{m_1}N, \qquad ML_{m_2} = L_{m_1}N.$$

So we have the following transformation $(\mathfrak{B})$ on the matric pair on the right side of (12):

$$\begin{bmatrix} 1 & 0 & 0 & 0 \\ 0 & I & M & 0 \\ 0 & 0 & I & 0 \\ 0 & 0 & 0 & I \end{bmatrix}\begin{bmatrix} 1 & 0 & 0 & 0 \\ 0 & K_{m_1} & 0 & 0 \\ 0 & 0 & K_{m2} & 0 \\ 0 & 0 & 0 & * \end{bmatrix}\begin{bmatrix} 1 & 0 & 0 & 0 \\ 0 & I & -N & 0 \\ 0 & 0 & I & 0 \\ 0 & 0 & 0 & I \end{bmatrix} = \begin{bmatrix} 1 & 0 & 0 & 0 \\ 0 & K_{m_1} & 0 & 0 \\ 0 & 0 & K_{m_2} & 0 \\ 0 & 0 & 0 & * \end{bmatrix},$$

$$\begin{bmatrix} 1 & 0 & 0 & 0 \\ 0 & I & M & 0 \\ 0 & 0 & I & 0 \\ 0 & 0 & 0 & I \end{bmatrix}\begin{bmatrix} 0 & a_{m_1} & a_{m_2} & * \\ 0 & L_{m_1} & 0 & 0 \\ 0 & 0 & L_{m_2} & 0 \\ 0 & 0 & 0 & * \end{bmatrix}\begin{bmatrix} 1 & 0 & 0 & 0 \\ 0 & I & -N & 0 \\ 0 & 0 & I & 0 \\ 0 & 0 & 0 & I \end{bmatrix} = \begin{bmatrix} 0 & a_{m_1} & 0 & * \\ 0 & L_{m_1} & 0 & 0 \\ 0 & 0 & L_{m_2} & 0 \\ 0 & 0 & 0 & * \end{bmatrix}.$$

Then again a component pair (K_{m_2}, L_{m_2}) is isolated. It is enough to transform the other matric pair into the form of (4), which is possible by the induction hypothesis. Then at the same time (A_1, A_2) is transformed into the form of (4). This completes the proof of case (ii).

Case (iii). The last case is when the matric pair (H_1, H_2) in (8) is (K_{n-1}, L_{n-1}). Then relation (5) becomes

$$A_1, A_2 \sim \begin{bmatrix} 1 & 0 \\ 0 & K_{n-1} \end{bmatrix}, \begin{bmatrix} 0 & x_2 \\ 0 & L_{n-1} \end{bmatrix}. \tag{13}$$

Let ξ be the first element of row x_2. Then

$$\begin{bmatrix} 1 & -x_2 L'_{n-1} \\ 0 & I \end{bmatrix}\begin{bmatrix} 1 & 0 \\ 0 & K_{n-1} \end{bmatrix}\begin{bmatrix} 1 & x_2 L'_{n-1} K_{n-1} \\ 0 & I \end{bmatrix} = \begin{bmatrix} 1 & 0 \\ 0 & K_{n-1} \end{bmatrix},$$

$$\begin{bmatrix} 1 & -x_2 L'_{n-1} \\ 0 & I \end{bmatrix}\begin{bmatrix} 0 & x_2 \\ 0 & L_{n-1} \end{bmatrix}\begin{bmatrix} 1 & x_2 L'_{n-1} K_{n-1} \\ 0 & I \end{bmatrix} = \begin{bmatrix} 0 & \xi a_{n-1} \\ 0 & L_{n-1} \end{bmatrix},$$

where a_{n-1} was defined as in (11). Therefore

$$A_1, A_2 \sim \begin{bmatrix} 1 & 0 \\ 0 & K_{n-1} \end{bmatrix}, \begin{bmatrix} 0 & \xi a_{n-1} \\ 0 & L_{n-1} \end{bmatrix}. \tag{14}$$

Since the rows of A_2 are linearly independent, $\xi \neq 0$. Imitating the method used in case (ii), we can transform ξ into 1. Thus

$$A_1, A_2 \sim \begin{bmatrix} 1 & 0 \\ 0 & K_{n-1} \end{bmatrix}, \begin{bmatrix} 0 & a_{n-1} \\ 0 & L_{n-1} \end{bmatrix},$$

i.e.,

$$A_1, A_2 \sim K_n, L_n,$$

which are of the form in (4).

This completes the proof of the Lemma.

Corollary. *If the matric pair* (A_1, A_2) *has no annihilating row with respect to* $\mathfrak{B}$, *and the columns of* A_2 *are linearly independent, then* (A_1, A_2) *is equivalent* ($\mathfrak{B}$) *to the* *following matric pair*:

$$D \dotplus \sum_{\alpha}{}' K'_{m_\alpha}, I \dotplus \sum_{\alpha}{}' L'_{m_\alpha}.$$

Proof. We only need to apply Lemma 2 to the matric pair $(A'_1, \bar{A}'_2)$.

Lemma 3. *Suppose A, B, C, D are four matrices, one of A, B is nonsingular. Then the two equations*

$$XA + K_m Y = C, \qquad XB + L_m \bar{Y} = D \tag{15}$$

have solutions for the unknown matrices X and Y.

Proof. We shall only prove the case where A is nonsingular; the proof for nonsingular B is similar. Let

$$C = \begin{bmatrix} c_1 \\ \vdots \\ c_m \end{bmatrix}, \qquad D = \begin{bmatrix} d_1 \\ \vdots \\ d_m \end{bmatrix}, \qquad X = \begin{bmatrix} x_1 \\ \vdots \\ x_m \end{bmatrix}, \qquad Y = \begin{bmatrix} y_1 \\ \vdots \\ y_{m+1} \end{bmatrix},$$

where c_i, d_i, x_i, y_i are row vectors. Equations (15) in fact are the following system of equations:

$$x_i A + y_i = c_i, \qquad x_i B + \bar{y}_{i+1} = d_i \qquad (i = 1, 2, \ldots, m). \tag{16}$$

Choose an arbitrary row as y_1. Then the system of equations (16) has the following solution.

$$\begin{aligned} x_1 &= (c_1 - y_1)A^{-1}, & y_2 &= \bar{d}_1 - \bar{x}_1 \bar{B}, \\ x_2 &= (c_2 - y_2)A^{-1}, & y_3 &= \bar{d}_2 - \bar{x}_2 \bar{B}, \\ &\vdots & &\vdots \\ x_m &= (c_m - y_m)A^{-1}, & y_{m+1} &= \bar{d}_m - \bar{x}_m \bar{B}. \end{aligned}$$

Lemma 4. *If the matrix pair (A_1, A_2) has neither annihilating row nor annihilating column with respect to $\mathfrak{B}$, then it is equivalent $(\mathfrak{B})$ to the following matric pair.*

$$\left.\begin{aligned} I \dotplus D_1 \dotplus \sum_{\alpha}{}' K_{m_\alpha} \dotplus \sum_{\beta}{}' K'_{n_\beta} \\ D_2 \dotplus I \dotplus \sum_{\alpha}{}' L_{m_\alpha} \dotplus \sum_{\beta} L'_{n_\beta} \end{aligned}\right\}, \tag{17}$$

where D_1 (respectively, D_2) is a square matrix of the same order as the component I of the second (respectively, first) matrix.

Proof. We shall prove by induction on the number of rows of A_1 and A_2. First of all, consider the case where the number of rows is 1.

Suppose both A_1 and A_2 have only one row. If $A_2 \neq 0$, we have shown in the proof of Lemma 2 that either $A_1, A_2 \sim a, 1$ or $A_1, A_2 \sim K_1, L_1$, both of which satisfy (17). If $A_2 = 0$, then A_1 can be transformed into $[1, 0, \ldots, 0]$. Since (A_1, A_2) has no annihilating column with respect to $\mathfrak{B}$, both A_1 and A_2 only have one column. In other words, $A_1, A_2 \sim 1, 0$. Again (17) is satisfied.

Now assume that both A_1 and A_2 have $n > 1$ rows and that the lemma is true if the number of rows is less than n.

If the columns of A_2 are linearly independent, then the Lemma follows from the corollary of Lemma 2. So we may assume that the columns of A_2 are linearly dependent. Then similar to the derivation of (5), we have

$$A_1, A_2 \sim \begin{bmatrix} 1 & 0 \\ 0 & H_2 \end{bmatrix}, \begin{bmatrix} 0 & x \\ 0 & H_2 \end{bmatrix} \tag{18}$$

Now, (A_1, A_2) has no annihilating row with respect to $\mathfrak{B}$, so (H_1, H_2) also has no annihilating row with respect to $\mathfrak{B}$. If (H_1, H_2) has an annihilating column with respect to $\mathfrak{B}$, then similar to the derivation of (7), we have

$$A_1, A_2 \sim K_1 \dotplus T_1, L_1 \dotplus T_2 .$$

By the induction hypothesis, the matric pair (T_1, T_2) can be transformed into form (17) through a transformation ($\mathfrak{B}$), and so can (A_1, A_2).

Thus the case that remains to be shown is when the matric pair (H_1, H_2) in (18) has neither annihilating row nor annihilating column with respect to $\mathfrak{B}$. In this case, by the induction hypothesis, the matric pair (H_1, H_2) can be transformed into form (17) through a transformation ($\mathfrak{B}$). Therefore we may assume that the matric pair (H_1, H_2) in (18) itself is of the form (17):

$$\left.\begin{aligned} H_1 &= I \dotplus D_1 \dotplus \sum_\alpha{}' K_{m_\alpha} \dotplus \sum_\beta{}' K'_{n_\beta}, \\ H_2 &= D_2 \dotplus I \dotplus \sum_\alpha{}' L_{m_\alpha} \dotplus \sum_\beta{}' L'_{n_\beta} . \end{aligned}\right\} . \tag{19}$$

Now we shall separate the discussion into several cases.

Case (i). The component pair (D_1, I) indeed appears in (19). By exactly the same argument as in the proof of case (i) of Lemma 2, we obtain

$$A_1, A_2 \sim D_1 \dotplus *, I \dotplus * .$$

Since the number of rows of the component pair represented by $*$ is less than n, it can be transformed into form (17) through a transformation ($\mathfrak{B}$), and so can the matric pair (A_1, A_2).

Case (ii). (19) has a component pair (K'_m, L'_m). In this case, we have

$$H_1 = K'_m \dotplus *, \qquad H_2 = L'_m \dotplus * .$$

Then (18) can be written as

$$A_1, A_2 \sim \begin{bmatrix} 1 & 0 & 0 \\ 0 & K'_m & 0 \\ 0 & 0 & * \end{bmatrix}, \begin{bmatrix} 0 & y & * \\ 0 & L'_m & 0 \\ 0 & 0 & * \end{bmatrix}.$$

Let $z = [0, y]$. Then $y = zL'_m$. Therefore we have the following equation of transformation ($\mathfrak{B}$):

$$\begin{bmatrix} 1 & -z & 0 \\ 0 & I & 0 \\ 0 & 0 & I \end{bmatrix} \begin{bmatrix} 1 & 0 & 0 \\ 0 & K'_m & 0 \\ 0 & 0 & * \end{bmatrix} \begin{bmatrix} 1 & zK'_m & 0 \\ 0 & I & 0 \\ 0 & 0 & I \end{bmatrix} = \begin{bmatrix} 1 & 0 & 0 \\ 0 & K'_m & 0 \\ 0 & 0 & * \end{bmatrix},$$

$$\begin{bmatrix} 1 & -z & 0 \\ 0 & I & 0 \\ 0 & 0 & I \end{bmatrix} \begin{bmatrix} 0 & y & * \\ 0 & L'_m & 0 \\ 0 & 0 & * \end{bmatrix} \begin{bmatrix} 1 & \bar{z}K'_m & 0 \\ 0 & I & 0 \\ 0 & 0 & I \end{bmatrix} = \begin{bmatrix} 0 & 0 & * \\ 0 & L'_m & 0 \\ 0 & 0 & * \end{bmatrix}.$$

Thus a component pair (K'_m, L'_m) is isolated:

$$A_1, A_2 \sim K'_m \dotplus *, L'_m \dotplus * .$$

As before, the proof is completed by induction.

Case (iii) (19) has at least two component pairs $(K_{m_1}, L_{m_1}), (K_{m_2}, L_{m_2})$. In this case, by exactly the same argument as in the proof of case (ii) of Lemma 2, we have

$$A_1, A_2 \sim K_m \dotplus *, L_m \dotplus *,$$

where $m = m_1$ or m_2. Again induction can proceed as before.

Case (iv). The matric pair (H_1, H_2) is the same as (K_{n-1}, L_{n-1}). In this case, (18) becomes

$$A_1, A_2 \sim \begin{bmatrix} 1 & 0 \\ 0 & K_{n-1} \end{bmatrix}, \begin{bmatrix} 0 & x \\ 0 & L_{n-1} \end{bmatrix}.$$

By exactly the same method used in the derivation of (14) from (13), now we have

$$A_1, A_2 \sim \begin{bmatrix} 1 & 0 \\ 0 & K_{n-1} \end{bmatrix}, \begin{bmatrix} 0 & \xi a_{n-1} \\ 0 & L_{n-1} \end{bmatrix}, \tag{20}$$

where a_{n-1} was defined in (11). If $\xi = 0$, then the right side of (20) has form (17); if $\xi \neq 0$, then as before, we can transform it into 1 and get $(A_1, A_2) \sim (K_n, L_n)$ which still satisfies (17).

Case (v). The matric pair (H_1, H_2) is the same as (I, D_2). Then (18) becomes

$$A_1, A_2 \sim I, *,$$

and (17) is achieved.

Case (vi). The last case is

$$H_1 = I \dotplus K_m, \qquad H_2 = D_2 \dotplus L_m .$$

Then relation (18) can be written as

$$A_1, A_2 \sim \begin{bmatrix} 1 & 0 & 0 \\ 0 & K_m & 0 \\ 0 & 0 & I \end{bmatrix}, \begin{bmatrix} 0 & y & * \\ 0 & L_m & 0 \\ 0 & 0 & * \end{bmatrix}. \tag{21}$$

Exactly as what we have done several times before, we can transform all the elements of y except the first one into zero. Thus we may assume that $y = \eta a_m$ (a_m was defined in (11)). If $\eta = 0$, then (21) becomes

$$A_1, A_2 \sim K_m \dotplus I, L_m \dotplus *,$$

which is of form (17). If $\eta \neq 0$, then as before, we can transform it into 1. Then

$$A_1, A_2 \sim \begin{bmatrix} 1 & 0 & 0 \\ 0 & K_m & 0 \\ 0 & 0 & I \end{bmatrix}, \begin{bmatrix} 0 & a_m & * \\ 0 & L_m & 0 \\ 0 & 0 & * \end{bmatrix}. \tag{22}$$

The matric pair on the right side of (22) can be written as

$$\begin{bmatrix} K_{m+1} & 0 \\ 0 & I \end{bmatrix}, \begin{bmatrix} L_{m+1} & U \\ 0 & V \end{bmatrix}. \tag{23}$$

By Lemma 3, there are matrices X and Y such that

$$X + K_{m+1} Y = 0, \qquad XV + L_{m+1} \bar{Y} = U.$$

Using such X and Y, we apply the following transformation $(\mathfrak{B})$ to matric pair (23):

$$\begin{bmatrix} 0 & I \\ I & -X \end{bmatrix} \begin{bmatrix} K_{m+1} & 0 \\ 0 & I \end{bmatrix} \begin{bmatrix} -Y & I \\ I & 0 \end{bmatrix} = I \dotplus K_{m+1},$$

$$\begin{bmatrix} 0 & I \\ I & -X \end{bmatrix} \begin{bmatrix} L_{m+1} & U \\ 0 & V \end{bmatrix} \begin{bmatrix} -\bar{Y} & I \\ I & 0 \end{bmatrix} = V \dotplus L_{m+1}.$$

Therefore we have

$$A_1, A_2 \sim I \dotplus K_{m+1}, V \dotplus L_{m+1},$$

which is of form (17).

This completes the proof of Lemma 4.

Proof of Theorem 1. Combining Lemma 1 and Lemma 4, we see immediately that any matric pair (A_1, A_2) can be transformed through a transformation $(\mathfrak{B})$ into the following matric pair:

$$\left.\begin{aligned} &I \dotplus D_1 \dotplus \sum_{\alpha}{}' K_{m_\alpha} \dotplus \sum_{\beta}{}' K'_{n_\beta} \dotplus 0, \\ &D_2 \dotplus I \dotplus \sum_{\alpha}{}' L_{m_\alpha} \dotplus \sum_{\beta}{}' L'_{n_\beta} \dotplus 0. \end{aligned}\right\}. \tag{24}$$

Consider the first two component pairs, i.e.,

$$I \dotplus D_1, D_2 \dotplus I. \tag{25}$$

By Lemma 1 of Hsu (1955), there is a nonsingular matrix P such that $PD_2\bar{P}^{-1} = E \dotplus B_2$, where E is nonsingular and $B_2\bar{B}_2$ is nilpotent. Therefore we may apply the following transformation $(\mathfrak{B})$ to matric pair (25):

$$(P \dotplus I)(I \dotplus D_1)(P^{-1} \dotplus I) = I \dotplus D_1 = I \dotplus I \dotplus D_1,$$

$$(P \dotplus I)(D_2 \dotplus I)(\bar{P}^{-1} \dotplus I) = PD_2\bar{P}^{-1} \dotplus I = E \dotplus B_2 \dotplus I.$$

Rearranging rows and columns, we may transform the two matrices on the extreme right of the above into

$$I \dotplus (I \dotplus D_1), B_2 \dotplus (E \dotplus I).$$

Multiplying $I \dotplus (E^{-1} \dotplus I)$ to the left of the two matrices above, we obtain the matric pair

$$I \dotplus B_1, B_2 \dotplus I. \tag{26}$$

Thus, we may transform the component pair (25) of (24) into (26). Since $B_2\bar{B}_2$ is nilpotent, the theorem is proved.

3. Canonical forms of matric pairs under transformation ($\mathfrak{B}$)

From Theorem 1, we can easily establish a canonical form for any matric pair under transformation ($\mathfrak{B}$). It is enough to consider matric pair (3). Since

$$I, B_2 \sim XIX^{-1} = I, XB_2\bar{X}^{-1},$$

$$B_1, I \sim YB_1\bar{Y}^{-1}, YIY^{-1} = I,$$

we may transform the matrices B_1, B_2 in (3) into their canonical forms under transformations ($\mathfrak{A}$) (see Theorem 8 of Hsu (1955)). Doing this and recalling the nilpotence of $B_2\bar{B}_2$, we have

$$B_1 = \sum_r{}' a_r M_{hr} \dotplus \sum_s{}' b_s \begin{bmatrix} 0 & M_{k_s} \\ -M_{k_s} & 0 \end{bmatrix} \dotplus \sum_t{}' \begin{bmatrix} 0 & \alpha_t M_{1_t} \\ \bar{\alpha}_t M_{1_t} & 0 \end{bmatrix} \dotplus \sum_i{}' J_{p_i}, \tag{27}$$

$$B_2 = \sum_j{}' J_{q_j}. \tag{28}$$

Now we use matric pair (3), where B_1 and B_2 have the forms given in (27) and (28), as the canonical form of a matric pair (A_1, A_2) under transformations ($\mathfrak{B}$). We still have to show the uniqueness of the canonical form.

If two matric pairs (A_1, A_2) and (A_1^*, A_2^*) are equivalent ($\mathfrak{B}$),

$$PA_1Q = A_1^*, \qquad PA_2\bar{Q} = A_2^*,$$

then the two pencils of matrices

$$\begin{bmatrix} \lambda A_1 & \mu A_2 \\ \mu \bar{A}_2 & \lambda \bar{A}_1 \end{bmatrix} \quad \text{and} \quad \begin{bmatrix} \lambda A_1^* & \mu A_2^* \\ \mu \bar{A}_2^* & \lambda \bar{A}_1^* \end{bmatrix}$$

are equivalent since

$$\begin{bmatrix} P & 0 \\ 0 & \bar{P} \end{bmatrix} \begin{bmatrix} \lambda A_1 & \mu A_2 \\ \mu \bar{A}_2 & \lambda \bar{A}_1 \end{bmatrix} \begin{bmatrix} Q & 0 \\ 0 & \bar{Q} \end{bmatrix} = \begin{bmatrix} \lambda A_1^* & \mu A_2^* \\ \mu \bar{A}_2^* & \lambda \bar{A}_1^* \end{bmatrix}. \tag{29}$$

Now let A_1^* and A_2^* be the two matrices in (3), where B_1 and B_2 are matrices (27) and (28), respectively. Then, by rearranging rows and columns, we see that the matric pencil on the right side of (29) is equivalent to the following matric pencil.

$$\begin{bmatrix} \lambda I & \mu B_2 \\ \mu B_2 & \lambda I \end{bmatrix} \dotplus \begin{bmatrix} \mu I & \lambda B_1 \\ \lambda \bar{B}_1 & \mu I \end{bmatrix} \dotplus M(\lambda, \mu) \ddot{+} 0_{2g,2h}, \tag{30}$$

where

$$M(\lambda, \mu) = \sum_{\alpha}{}' \begin{bmatrix} \lambda K_{m\alpha} & \mu L_{m_\alpha} \\ \mu L_{m_\alpha} & \lambda K_{m_\alpha} \end{bmatrix} \dot{+} \sum_{\beta}{}' \begin{bmatrix} \lambda K'_{n_\beta} & \mu L'_{n_\beta} \\ \mu L'_{n_\beta} & \lambda K'_{n_\beta} \end{bmatrix}.$$

Now we will calculate the elementary divisors and minimum numbers of matric pencil (30). The elementary divisors of the first two component pencils of (30) have been calculated in Hsu (1955). The elementary divisors of the second component pencil (where B_1 is matrix (27)) are

$$\left.\begin{array}{ll} (\lambda a_r + \mu)^{h_r}(\lambda a_r - \mu)^{h_r} & (r = 1, 2, \ldots) \\ (\lambda b_s i + \mu)^{k_s}, (\lambda b_s i + \mu)^{k_s}, (\lambda b_s i - \mu)^{k_s}, (\lambda b_s i - \mu)^{k_s} & (s = 1, 2, \ldots) \\ (\lambda \alpha_t + \mu)^{l_t}, (\lambda \bar{\alpha}_t + \mu)^{l_t}, (\lambda \alpha_t - \mu)^{l_t}, (\lambda \bar{\alpha}_t - \mu)^{l_t} & (t = 1, 2, \ldots) \\ \mu^{p_i}, \mu^{p_i} & (i = 1, 2, \ldots) \end{array}\right\}. \tag{31}$$

The elementary divisors of the first component pencil (where B_2 is matrix (28)) are

$$\lambda^{q_j}, \lambda^{q_j} \qquad (j = 1, 2, \ldots). \tag{32}$$

Now let us examine the component pencil $M(\lambda, \mu)$ of (30). It is the direct sum of several matric pencils of the following form.

$$\begin{bmatrix} \lambda K_m & \mu L_m \\ \mu L_m & \lambda K_m \end{bmatrix}, \begin{bmatrix} \lambda K'_n & \mu L'_n \\ \mu L'_n & \lambda K'_n \end{bmatrix}. \tag{33}$$

It is enough to consider the first matric pencil in (33), since the second one is simply the transpose of the first. Let D_m be an $m \times m$ diagonal matrix with diagonal elements $1, -1, 1, -1, \ldots$. Then it is easy to verify the following relationship:

$$\frac{1}{2}\begin{bmatrix} I_m & I_m \\ D_m & -D_m \end{bmatrix}\begin{bmatrix} \lambda K_m & \mu L_m \\ \mu L_m & \lambda K_m \end{bmatrix}\begin{bmatrix} I_{m+1} & D_{m+1} \\ I_{m+1} & -D_{m+1} \end{bmatrix} = \begin{bmatrix} \lambda K_m + \mu L_m & 0 \\ 0 & \mu K_m + \mu L_m \end{bmatrix}. \tag{34}$$

This relation tells us that the component pencil $M(\lambda, \mu)$ of matric pencil (30) does not contribute any elementary divisor. Therefore expressions (31) and (32) together constitute the set of elementary divisors of matric pencil (30). Furthermore, from (34), we know that the minimum numbers of the matric pencil $M(\lambda, \mu)$ are $m_1, m_1, m_2, m_2, \ldots; n_1, n_1, n_2, n_2, \ldots$

We shall summarize the above results in two theorems.

Theorem 2. *Any matric pair* (A_1, A_2) *can be transformed by a transformation* $(\mathfrak{B})$ *into canonical form* (3), *where* B_1 *and* B_2 *are given by* (27) *and* (28), *respectively* (*and in* (27), $a_r > 0$, $b_s > 0$, *and* α_t *is a complex number with positive real and imaginary parts*). *Furthermore, this canonical form is uniquely determined by* (A_1, A_2) *due to the fact that expressions* (31) *and* (32) *together constitute the set of elementary divisors of the matric pencil*

$$\begin{bmatrix} \lambda A_1 & \mu A_2 \\ \mu \bar{A}_2 & \lambda \bar{A}_1 \end{bmatrix} \tag{35}$$

and that the set of integers

$$\overbrace{0, \ldots, 0}^{2g}, m_1, m_1, m_2, m_2, \ldots; \overbrace{0, \ldots, 0}^{2h}, n_1, n_1, n_2, n_2, \ldots \tag{36}$$

coincides with the set of minimum numbers of matric pencil (35).

Theorem 3. *Two matric pairs* (A_1, A_2) *and* (A_1^*, A_2^*) *are equivalent* ($\mathfrak{B}$) *if and only if the following two matric pencils are equivalent*:

$$\begin{bmatrix} \lambda A_1 & \mu A_2 \\ \mu \bar{A}_2 & \lambda \bar{A}_1 \end{bmatrix}, \begin{bmatrix} \lambda A_1^* & \mu A_2^* \\ \mu \bar{A}_2^* & \lambda \bar{A}_1^* \end{bmatrix},$$

i.e., they have the same elementary divisors and minimum numbers.

Accidentally, we have proved the following theorem.

Theorem 4. *A matric pencil is equivalent to a matric pencil of the form*

$$\begin{bmatrix} \lambda A_1 & \mu A_2 \\ \mu \bar{A}_2 & \lambda \bar{A}_1 \end{bmatrix}$$

if and only if its elementary divisors and minimum numbers satisfy the following conditions:

(i) *The elementary divisors of the form* $(\lambda a + \mu b)^h$ (*a and b are real numbers, one of which can be zero*) *can be grouped into pairs as the following*:

$$(\lambda a + \mu b)^h, (\lambda a - \mu b)^h.$$

(ii) *The elementary divisors of the form* $(\lambda \alpha + \mu)^k$ (α *is a nonzero complex number*) *can be grouped into sets of four of the following form*:

$$(\lambda \alpha + \mu)^k, (\lambda \bar{\alpha} + \mu)^k, (\lambda \alpha - \mu)^k, (\lambda \bar{\alpha} - \mu)^k.$$

(iii) *Each minimum number appears an even number of times.*

Remark 1. If we use the matric pencil

$$\begin{bmatrix} \lambda A_1 & \mu A_2 \\ -\mu \bar{A}_2 & \lambda \bar{A}_1 \end{bmatrix} \tag{37}$$

instead of (35), then we can obtain results similar to Theorems 2, 3, and 4. In particular, the equivalence of the two matric pencils

$$\begin{bmatrix} \lambda A_1 & \mu A_2 \\ -\mu \bar{A}_2 & \lambda \bar{A}_1 \end{bmatrix} \quad \text{and} \quad \begin{bmatrix} \lambda A_1^* & \mu A_2^* \\ -\mu \bar{A}_2^* & \lambda \bar{A}_1^* \end{bmatrix}$$

is also a necessary and sufficient condition for the equivalence ($\mathfrak{B}$) of (A_1, A_2) and (A_1^*, A_2^*). Furthermore, the set of integers (36) generated by the integers g, h, m_α, n_β in canonical form (3) is the same as the set of minimum numbers of matric pencil (37).

Remark 2. We have shown that any square matrix can be transformed into a real matrix by a transformation ($\mathfrak{A}$) (see Theorem 7 of Hsu (1955)). Therefore in matric pair (3), B_1 and B_2 can be chosen to be real matrices. Thus we conclude that *any matric pair can be transformed into a real matric pair through a transformation* ($\mathfrak{B}$).

4. Nonsingular Matric Pair with Respect to $\mathfrak{B}$

Based on the discussion in Section 3, now we introduce the concept of nonsingular matric pairs with respect to $\mathfrak{B}$. A matric pair (A_1, A_2) is said to be nonsingular with respect to $\mathfrak{B}$ if the matric pencil

$$\begin{bmatrix} \lambda A_1 & \mu A_2 \\ \mu \bar{A}_2 & \lambda \bar{A}_1 \end{bmatrix} \tag{35}$$

is nonsingular, i.e., it is square and has a determinant not always equal to zero.

The nonsingularity of a matric pair with respect to $\mathfrak{B}$ clearly is an invariant property under transformations ($\mathfrak{B}$). Now we transform the matric pair (A_1, A_2) in (35) into the equivalent ($\mathfrak{B}$) matric pair (3). Then we can see that (35) is a nonsingular matric pencil if and only if in (3), all the components except the first two do not appear. Thus we have

Theorem 5. *A matric pair is a nonsingular with respect to $\mathfrak{B}$ if and only if it is equivalent ($\mathfrak{B}$) to the following matric pair:*

$$I \dotplus B_1, B_2 \dotplus I,$$

where B_1 is a square matrix of the same order as the component I of the second matrix, B_2 is a square matrix of the same order as the component I of the first matrix, and $B_2\bar{B}_2$ is nilpotent.

The same method can be used to prove

Theorem 6. *If the matric pencil*

$$\begin{bmatrix} \lambda A_1 & \mu A_2 \\ -\mu \bar{A}_2 & \lambda \bar{A}_1 \end{bmatrix}$$

is nonsingular, then the matric pair (A_1, A_2) is nonsingular with respect to $\mathfrak{B}$, and vice versa.

From Theorem 1 and Theorem 5, we immediately obtain

Theorem 7. *Any matric pair is equivalent ($\mathfrak{B}$) to the following matric pair:*

$$C_1 \dotplus \sum_{\alpha}{}' K_{m_\alpha} \dotplus \sum_{\beta}{}' K'_{n_\beta} \ddot{+}\, 0,$$

$$C_2 \dotplus \sum_{\alpha}{}' L_{m_\alpha} \dotplus \sum_{\beta}{}' L'_{n_\beta} \ddot{+}\, 0,$$

where the matric pair (C_1, C_2) is nonsingular with respect to $\mathfrak{B}$.

5. Conclusion

We now point out that a transformation ($\mathfrak{B}$) in fact is a special transformation of quaternion matrices (i.e., matrices with quaternions as entries). As a matter of fact, if as usual we use $1, i, j, k(=ij)$ to denote the quaternion units, then any quaternion matrix can be written as $A_1 + A_2 j$, where A_1, A_2 are two complex matrices of the same size. Since $jQ = \overline{Q} j$ for any complex matrix Q, transformation (1) can be rewritten as

$$A_1 + A_2 j \to B_1 + B_2 j = P(A_1 + A_2 j)\, Q.$$

Thus, a transformation ($\mathfrak{B}$) is a transformation from quaternion matrices to quaternion matrices, wherein the (nonsingular) matrices P and Q are confined to ordinary complex matrices.

Reference

1. P. L. Hsu (1955). On a kind of transformations of matrices. *Acta Math. Sinica* **5**, 333–346.

Simultaneous Transformation of a Hermitian Matrix and a Symmetric or Skew-Symmetric Matrix*

Abstract Let A_1, A_2 be two matrices of the same size over the complex field, with $A_1' = \overline{A}_1$ and $A_2' = sA_2$ ($s = \pm 1$). Such a pair of matrices is called a ϕ_s-pair. We speak of a ϕ_+-pair or a ϕ_--pair according to whether $s = 1$ or $s = -1$. By a transformation $\mathcal{C}$ of a ϕ_s-pair (A_1, A_2) we mean a transformation which takes the ϕ_s-pair (A_1, A_2) into the ϕ_s-pair $(PA_1\overline{P}', PA_2P')$, where P is some nonsingular matrix. If (A_1, A_2) is a ϕ_s-pair, the hermitian pencil of matrices

$$\begin{pmatrix} \lambda A_1 & \mu A_2 \\ s\mu\overline{A}_2 & \lambda\overline{A}_1 \end{pmatrix}$$

is called the associated pencil of the pair (A_1, A_2).

In this paper we constructed canonical forms for ϕ_+-pairs and ϕ_--pairs under transformations $\mathcal{C}$. In both cases, the canonical form of a ϕ_s-pair (A_1, A_2) depends only on certain parameters which are completely and explicitly determined by the conjunctive invariants of the associated pencil. As immediate consequences, the canonical form of a ϕ_s-pair depends only on the pair itself and two ϕ_s-pairs can be taken into each other by a transformation $\mathcal{C}$ if and only if their associated pencils are conjunctive.

Although the above-mentioned results are worked out in the last two sections (Section 5 and 6), we have to carry out the computations through three lengthy sections (Section 3, 4 and 5) in order to get the desired canonical forms. These three sections, which are responsible for the bulkiness of the paper, contain computational details and may be read as an appendix. Nevertheless, let us mention a few points of interest that are provided by the material in these three sections. (i) the separation of the singular component (identical in form with the classical singular component of Kronecker), as accomplished in Section 2, has methodological interest; (ii) Section 3 contains, as a matter of fact, a solution of the equation $AX = \pm\overline{X}\overline{A}'$ for the unknown matrix X, with an arbitrarily given matrix A; (iii) Lemma 4.7 (in Section 4) actually provides a diagonalization process for quaternion matrices of some special kind (but we suspect that this lemma is known).

1. Terminology and Notation

In this paper we work on the complex number field. Two matrices of the same size will be called a matrix pair. The object of study in this paper is a matrix pair (A_1, A_2) consisting of a hermitian matrix A_1 and a symmetric or skew-symmetric matrix A_2: $A_1' = \overline{A}_1$, $A_2' = sA_2$ ($s = \pm 1$). Such a pair is called a ϕ_s-pair; we speak of a ϕ_+-pair or a ϕ_--pair according as $s = 1$ or $s = -1$.

---

*Translated by Hsu Pei, Stanford University and Academia Sinica.

The transformation

$$A_1 \to PA_1\bar{P}', \qquad A_2 \to PA_2P' \qquad (P \text{ non-singular}) \tag{1}$$

preserves the type of the pair. By a transformation $\mathcal{C}$ we mean a transformation like (1). If two pairs can be taken into each other by a transformation $\mathcal{C}$, then we say that they are $\mathcal{C}$ congruent; the relation of $\mathcal{C}$ congruence will be denoted by $\approx$. When performing the transformation (1), we say: transform (A_1, A_2) by P.

In this paper, two fundamental problems about transformations $\mathcal{C}$ are solved; namely, canonical forms and criteria for $\mathcal{C}$ congruence. The work of this paper depends on the results obtained in two previous papers [1], [2] by the present author.

We introduce some new notations in addition to those used in [1], [2], which will also be adopted here. For the sake of clarity and convenience we write down both the old and new notation here.

$\dot{+}$ and $\sum'$ denote direct sum. A' is the transpose of A. If $A = A_1 + iA_2$ (A_1, A_2 are matrices with real entries), then $\bar{A}$ denotes $A_1 - iA_2$.

0 stands for a zero matrix of any size, I a unit matrix of any order. I_m denotes the unit matrix of order m.

Symbols J_{mn}, J_m, M_{mn}, M_m, N_{mn}, N_n, K_m, L_m, F, and F_m are used specially for the following matrices.

$$J_{mn} = \begin{bmatrix} 0 & I_n & & & 0 \\ & 0 & I_n & & \\ & & \ddots & \ddots & \\ & & & \ddots & I_n \\ 0 & & & & 0 \end{bmatrix} \quad (m \times m \text{ blocks}), \qquad J_m = J_{m1},$$

$$M_{mn} = I + J_{mn}, \qquad M_m = M_{m1} = I + J_m,$$

$$N_{mn} = \begin{bmatrix} 0 & & & I_n \\ & & I_n & \\ & \cdot^{\cdot^{\cdot}} & & \\ I_n & & & 0 \end{bmatrix} \quad (m \times m \text{ blocks}), \qquad N_m = N_{m1},$$

$$K_m = \begin{bmatrix} 1 & 0 & & & 0 & 0 \\ & 1 & 0 & & \vdots & 0 \\ & & \ddots & \ddots & & \vdots \\ & & & \ddots & 0 & \\ 0 & & & & 1 & 0 \end{bmatrix}, \qquad L_m = \begin{bmatrix} 0 & 1 & & & 0 \\ 0 & 0 & 1 & & \\ & & \ddots & \ddots & \\ 0 & 0 & & 0 & 1 \end{bmatrix} \quad (m \times \overline{m+1}),$$

$$F_m = \begin{bmatrix} 0 & I_m \\ -I_m & 0 \end{bmatrix}, \qquad F = \begin{bmatrix} 0 & I \\ -I & 0 \end{bmatrix} \quad (\text{two identical } I\text{'s of any order}).$$

We will also introduce some temporary notation in the course of discussion, which are not recorded here.

Now a few words about the content of the five sections of this paper. Sections 2, 3 and 4 are the preparatory sections for determining canonical forms of ϕ_s-pairs under transformations $\mathcal{C}$ (so-called $\mathcal{C}$ canonical forms), which are worked out in Sections 5 and 6. In Sections 5 and 6, after determining the canonical forms, we relate the parameters of these canonical forms to the classical invariants of the hermitian matrix pencils associated with the matrix pairs. This enables us to prove the uniqueness of $\mathcal{C}$ canonical forms and find out the necessary and sufficient conditions for $\mathcal{C}$ congruence. The computations carried out in Sections 2, 3 and 4 are quite long and involved, although the principles leading to these computations are very clear. If the reader does not care about the details of the computations he may skip these three sections and start directly from Section 5.

2. Preliminary Simplification

In this section we make some preliminary simplification by using $\mathcal{C}$ transformations and reduce any ϕ_s-pair to a direct sum of matrices with certain structures. The first nine lemmas are of a preparatory nature. Lemma 2.10 sums up the results of this section. In this section, K, L, J always denote the matrices

$$K = \sum_\alpha{}' K_{r_\alpha}, \qquad L = \sum_\alpha{}' L_{r_\alpha}, \qquad J = \sum_\alpha{}' J_{r_\alpha}.$$

The following relation is easy to check.

$$LK' = J.$$

Lemma 2.1. *If $L'X = 0$, then $X = 0$.*

Lemma 2.2. *If $KX = 0$, and $LX = 0$, then $X = 0$.*

We omit the easy proofs of these two lemmas.

Lemma 2.3. *If $X = Q\bar{X}J_r$ (Q is arbitrary), then $X = 0$. If $X = J_r\bar{X}Q$ (Q is arbitrary), then $X = 0$.*

Proof. The first assertion is proved as follows (the second is proved similarly). Let $X_1, X_2, \ldots$ be the columns of X. From $X = Q\bar{X}J_r$ we have

$$X_1 = 0, \qquad X_{i+1} = Q\bar{X}_i \qquad (i = 1, 2, \ldots).$$

Hence $X_i = 0$, $(i = 1, 2, \ldots)$.

Lemma 2.4. *If $X = QXJ$ (Q arbitrary), then $X = 0$. If $X = J\bar{X}Q$ (Q arbitrary), then $X = 0$.*

Proof. Let $X = [X_1, X_2, \ldots]$. $X = Q\bar{X}J$ implies $X_\alpha = Q\bar{X}_\alpha J_{r_\alpha}$. The same proof applies to the second assertion.

Lemma 2.5. *If* $XK = Q\overline{X}L$ *(* Q *arbitrary), then* $X = 0$. *If* $L'X = K'\overline{X}Q$ *(* Q *arbitrary), then* $X = 0$.

Proof. Multiplying from the right both sides of the first equation by K', we get $X = Q\overline{X}LK = Q\overline{X}J$. This and Lemma 2.4 imply $X = 0$. Thus the first assertion is proved. To prove the second assertion, multiply from the left both sides of the second equation by L.

Lemma 2.6. *The system of equations*

$$\begin{aligned} YK_n' + K_m\overline{Z}' &= U, \\ sYL_n' + L_mZ' &= V \qquad (s = \pm 1), \end{aligned}$$

(U, V *are arbitrarily given matrices) is solvable for* Y *and* Z.

Proof. This can be proved, say, by induction on m.

Lemma 2.7. *Suppose* (U, V) *is an arbitrarily given* ϕ_s*-pair of order* m. *Then the system of equations*

$$\left.\begin{aligned} XK_m' + K_m\overline{X}' &= U \\ sXL_m' + L_mX' &= V \end{aligned}\right\} \tag{2}$$

is solvable for X.

Proof. Write

$$U = \begin{bmatrix} \alpha_{11} & \cdots & \alpha_{1m} \\ \cdots & \cdots & \cdots \\ \alpha_{m1} & \cdots & \alpha_{mm} \end{bmatrix}, \qquad V = \begin{bmatrix} \beta_{11} & \cdots & \beta_{1m} \\ \cdots & \cdots & \cdots \\ \beta_{m1} & \cdots & \beta_{mm} \end{bmatrix},$$

$$X = \begin{bmatrix} \xi_{11} & \cdots & \xi_{1,m+1} \\ \cdots & \cdots & \cdots \\ \xi_{m1} & \cdots & \xi_{m,m+1} \end{bmatrix}.$$

The system (2) becomes

$$\xi_{ij} + \overline{\xi}_{ji} = \alpha_{ij}, \qquad s\xi_{i,j+1} + \xi_{j,i+1} = \beta_{ij} \qquad (i, j = 1, 2, \ldots, m). \tag{3}$$

It is enough to consider the equations in (3) for which $i \leqslant j$. In case $j = i$,

$$\xi_{ii} + \overline{\xi}_{ii} = \alpha_{ii}, \qquad (1 + s)\xi_{i,i+1} = \beta_{ii}.$$

These equations are clearly solvable (for example, $\xi_{ii} = \frac{1}{2}\alpha_{ii}$, and $\xi_{i,i+1} = \frac{1}{2}\beta_{ii}$ if $s = 1$ or arbitrarily if $s = -1$). Thus ξ_{ii} and $\xi_{i,i+1}$ can be regarded as known. Next, let $j = i + 1$ in (3),

$$\xi_{i,i+1} + \overline{\xi}_{i+1,i} = \alpha_{i,i+1}, \qquad s\xi_{i,i+2} + \xi_{i+1,i+1} = \beta_{i,i+1}.$$

This system has the solution

$$\xi_{i+1,i} = \bar{\alpha}_{i,i+1} - \bar{\xi}_{i,i+1}, \qquad \xi_{i,i+2} = s\beta_{i,i+1} - s\xi_{i+1,i+1}.$$

Thus we have determined $\xi_{i+1,i}$ and $\xi_{i,i+2}$. Letting $j = i + 2$ in (3), we determine $\xi_{i+2,i}$, $\xi_{i,i+3}$ and so on.

Lemma 2.8. *If (U, V) is an arbitrarily given ϕ_s-pair, then the system*

$$\left.\begin{aligned} XK' + K\bar{X}' &= U, \\ sXL' + LX' &= V \end{aligned}\right\} \tag{4}$$

is solvable for X.

Proof. Let

$$U = \begin{bmatrix} U_{11} & U_{12} & \cdots \\ U_{21} & U_{22} & \cdots \\ \cdots & \cdots & \cdots \\ \cdots & \cdots & \cdots \end{bmatrix}, \qquad V = \begin{bmatrix} V_{11} & V_{12} & \cdots \\ V_{21} & V_{22} & \cdots \\ \cdots & \cdots & \cdots \\ \cdots & \cdots & \cdots \end{bmatrix},$$

where the minors $U_{\alpha\beta}$ and $V_{\alpha\beta}$ are of the size $r_\alpha \times r_\beta$. According to Lemma 2.7 and Lemma 2.6, we can find matrices $X_\alpha, Y_{\alpha\beta}, Z_{\alpha\beta}$ ($\alpha < \beta = 1, 2, \ldots$) such that

$$X_\alpha K'_{r_\alpha} + K_{r_\alpha}\bar{X}'_\alpha = U_{\alpha\alpha}, \qquad sX_\alpha L'_{r_\alpha} + L_{r_\alpha}X'_\alpha = V_{\alpha\alpha},$$

$$Y_{\alpha\beta}K'_{r_\beta} + K_r\bar{Z}'_{\alpha\beta} = U_{\alpha\beta}, \qquad sY_{\alpha\beta}L'_{r_\beta} + L_{r_\alpha}Z'_{\alpha\beta} = V_{\alpha\beta}.$$

Thus the system (4) is easily checked to have the following solution

$$X = \begin{bmatrix} X_1 & Y_{12} & Y_{13} & \cdots \\ Z_{12} & X_2 & Y_{23} & \cdots \\ Z_{13} & Z_{23} & X_3 & \cdots \\ \cdots & \cdots & \cdots & \cdots \\ \cdots & \cdots & \cdots & \cdots \end{bmatrix}.$$

Lemma 2.9. *Every ϕ_s-pair (A_1, A_2) is $\mathfrak{C}$ congruent to a pair of the form*

$$B_1C \dotplus \bar{D} + \begin{bmatrix} 0 & K \\ K' & 0 \end{bmatrix} \dotplus 0, \qquad C_1 + B_2D \dotplus \begin{bmatrix} 0 & L \\ sL' & 0 \end{bmatrix} \dotplus 0, \tag{5}$$

where C, D are non-singular matrices,

$$K = \sum_\alpha{}' K_{r_\alpha}, \qquad L = \sum_\alpha{}' L_{r_\alpha}, \tag{6}$$

$$B_1 = \sum_r{}' a_r M_{hr} \dotplus \sum_s{}' b_s \begin{bmatrix} 0 & M_{ks} \\ -M_{ks} & 0 \end{bmatrix} \dotplus \sum_t{}' \begin{bmatrix} 0 & \alpha + M_{1t} \\ \bar{\alpha}_t M_{1t} & 0 \end{bmatrix} \dotplus \sum_i J_{p_i}, \tag{7}$$

$$B_2 = \sum_j{}' J_{q_j}. \tag{8}$$

$a_r > 0$, $b_r > 0$ in (7), α_t are complex numbers with positive real and imaginary parts.

Proof. We proved the following theorem (see [2], p. 14, Theorem 2): for any pair (A_1, A_2), there are non-singular matrices P and Q, such that

$$PA_1Q = B_1 \dotplus I \dotplus \sum_\alpha K_{m_\alpha} \dotplus \sum_\beta{}' K_{n_\beta} \dotplus 0_{g_h}, \tag{9}$$

$$PA_2\bar{Q} = I \dotplus B_2 + \sum_\alpha{}' I_{m_\alpha} \dotplus \sum_\beta{}' L'_{n_\beta} \dotplus 0_{g_h}, \tag{10}$$

where B_1 and the summand I of the right-hand side of (10) have the same size; B_2 and the summand I of the right-hand side of (9) have the same size. B_1 and B_2 have the form (7) and (8), respectively. We proved also that the set of integers

$$\overbrace{0\cdots 0}^{2g}, m_1, m_1, m_2, m_2, \ldots,$$

$$\overbrace{0\cdots 0}^{2h}, n_1, n_1, n_2, n_2, \ldots$$

are exactly the minimal indices of the pencil

$$\begin{bmatrix} \lambda A_1 & \mu A_2 \\ s\mu\bar{A}_2 & \lambda\bar{A}_1 \end{bmatrix} \qquad (s = \pm 1) \tag{11}$$

In our present case (i.e., $A_1' = \bar{A}_1$, $A_2' = sA_2$), the pencil (11) is hermitian, hence $g = h$, $m_\alpha = n_\alpha$ $(\alpha = 1, 2, \ldots)$. Taking these facts into account, we can write (9) and (10) as

$$\left.\begin{aligned} PA_1Q &= B_1 \dotplus I \dotplus K \dotplus K' \dotplus 0, \\ PA_2\bar{Q} &= I \dotplus B_2 \dotplus L \dotplus L' \dotplus 0, \end{aligned}\right\} \tag{12}$$

where K and L are given by (6). Put $R = \bar{Q}^{-1}P$. Multiply from the right both sides of the first equation of (12) by $\bar{R}$, and multiply from the right both sides of the second equation of (12) by R,

$$\left.\begin{aligned} PA_1\bar{P}' &= (B_1 \dotplus I \dotplus K \dotplus K' \dotplus 0)\bar{R} \\ PA_2P' &= (I \dotplus B_2 + L \dotplus L' + 0)R \end{aligned}\right\}. \tag{13}$$

In other words, we obtain a new ϕ_s-pair, namely, two matrices on the right-hand side of (13), which is $\mathcal{C}$ congruent to the ϕ_s-pair (A_1, A_2).

Now divide R into blocks (corresponding to the division in (12)):

$$R = \begin{bmatrix} R_{11} & R_{12} & R_{13} & R_{14} & R_{15} \\ R_{21} & R_{22} & R_{23} & R_{24} & R_{25} \\ R_{31} & R_{32} & R_{33} & R_{34} & R_{35} \\ R_{41} & R_{42} & R_{43} & R_{44} & R_{45} \\ R_{51} & R_{52} & R_{53} & R_{54} & R_{55} \end{bmatrix}. \tag{14}$$

The pair (13) becomes

$$\begin{bmatrix} B_1\overline{R}_{11} & B_1\overline{R}_{12} & B_1\overline{R}_{13} & B_1\overline{R}_{14} & B_1\overline{R}_{15} \\ \overline{R}_{21} & \overline{R}_{22} & \overline{R}_{23} & \overline{R}_{24} & \overline{R}_{25} \\ K\overline{R}_{31} & K\overline{R}_{32} & K\overline{R}_{33} & K\overline{R}_{34} & K\overline{R}_{35} \\ K'\overline{R}_{41} & K'\overline{R}_{42} & K'\overline{R}_{43} & K'\overline{R}_{44} & K'\overline{R}_{45} \\ 0 & 0 & 0 & 0 & 0 \end{bmatrix},$$

$$\begin{bmatrix} R_{11} & R_{12} & R_{13} & R_{14} & R_{15} \\ B_2R_{21} & B_2R_{22} & B_2R_{23} & B_2R_{24} & B_2R_{25} \\ LR_{31} & LR_{32} & LR_{33} & LR_{34} & LR_{35} \\ L'R_{41} & L'R_{42} & L'R_{43} & L'R_{44} & L'R_{45} \\ 0 & 0 & 0 & 0 & 0 \end{bmatrix}. \tag{15}$$

Remember that (15) is a ϕ_s-pair, we have the following relations:

$$R_{15} = 0, \qquad R_{25} = 0, \tag{16}$$

$$K\overline{R}_{35} = LR_{35} = 0, \qquad K'\overline{R}_{45} = L'R_{45} = 0, \tag{17}$$

$$\overline{R}'_{41}K = B_1R_{44}, \qquad R'_{41}L = sR_{14}, \tag{18}$$

$$\overline{R}'_{42}K = R_{24}, \qquad R'_{42}L = sB_2R_{24}, \tag{19}$$

$$R'_{44}K = K'\overline{R}_{44}, \qquad R'_{44}L = sL'R_{44}, \tag{20}$$

$$\overline{R}_{21} = R'_{12}B'_1, \qquad B_2R_{21} = sR'_{12}. \tag{21}$$

According to Lemmas 2.1 and 2.2, (17) implies

$$R_{35} = 0, \qquad R_{45} = 0. \tag{22}$$

Equations (18) and (19) yield the relations

$$\overline{R}'_{41}K = sB_1R'_{41}L,$$

$$L'R_{42} = sK'\overline{R}_{42}B'_2.$$

By this and Lemma 2.5,

$$R_{41} = 0, \qquad R_{42} = 0, \tag{23}$$

consequently

$$R_{14} = 0, \qquad R_{24} = 0. \tag{24}$$

Multiplying from the left both sides of the second equality in (20) by L, we have $R_{44} = sLR'_{14}L$. Substituting this into the first equation in (20), we get $R'_{44}K = sK'LR'_{44}L$. By Lemma 2.5, this implies

$$R_{44} = 0. \tag{25}$$

Finally, two equations (21) yield $R'_{12} = sB_2\overline{R}'\overline{B}'$. By Lemma 2.4 (notice that J has the form (8)), we have

$$R_{12} = 0, \qquad \text{consequently} \quad R_{21} = 0. \tag{26}$$

Substituting (16), (22), (23), (24), (25), (26) into (14) and remember that R is non-singular, we find that R_{11}, R_{22} and R_{34} are non-singluar.

By virtue of (16), (22), (23), (24), (25), (26), the pair (15) can be written as

$$\begin{bmatrix} B_1\bar{R}_{11} & 0 & B_1\bar{R}_{13} & 0 \\ 0 & \bar{R}_{22} & \bar{R}_{23} & 0 \\ K\bar{R}_{31} & K\bar{R}_{32} & K\bar{R}_{33} & K\bar{R}_{34} \\ 0 & 0 & K'\bar{R}_{43} & 0 \end{bmatrix} \dot{+} 0, \qquad \begin{bmatrix} R_{11} & 0 & R_{13} & 0 \\ 0 & B_2R_{22} & B_2R_{23} & 0 \\ LR_{31} & LR_{32} & LR_{33} & LR_{34} \\ 0 & 0 & L'R_{43} & 0 \end{bmatrix} \dot{+} 0. \tag{27}$$

Since (27) is a ϕ_s-pair, it can be written as

$$\begin{bmatrix} B_1\bar{R}_{11} & 0 & B_1\bar{R}_{13} & 0 \\ 0 & \bar{R}_{22} & \bar{R}_{23} & 0 \\ R'_{13}B'_1 & R'_{23} & * & K\bar{R}_{34} \\ 0 & 0 & R'_{34}K' & 0 \end{bmatrix} \dot{+} 0, \qquad \begin{bmatrix} R_{11} & 0 & R_{13} & 0 \\ 0 & B_2R_{22} & B_2R_{23} & 0 \\ sR'_{13} & sR'_{23}B'_2 & * & LR_{34} \\ 0 & 0 & sR'_{34}L' & 0 \end{bmatrix} \dot{+} 0. \tag{28}$$

Let C_1 and C_2 denote the two matrices in (28) after deleting the zero summands. Using the fact that R_{11}, R_{22}, R_{34} are non-singular, we put

$$P = \begin{bmatrix} I & 0 & 0 & 0 \\ 0 & I & 0 & 0 \\ -R'_{13}R'^{-1}_{11} & -R'_{23}R'^{-1}_{22} & 0 & 0 \\ 0 & 0 & 0 & R'^{-1}_{34} \end{bmatrix}.$$

It can be verified by computation that

$$PC_1\bar{P}' = B_1\bar{R}_{11} \dot{+} \bar{R}_{22} \dot{+} \begin{bmatrix} U & K \\ -K' & 0 \end{bmatrix},$$

$$PC_2P' = R_{11} \dot{+} B_2R_{22} \dot{+} \begin{bmatrix} V & L \\ sL' & 0 \end{bmatrix}$$

(where (U, V) is a certain ϕ_s-pair). Denote the non-singular matrices $\bar{R}_{11}$, R_{22} by C and D, we obtain

$$A_1, A_2 \approx B_1C \dot{+} \bar{D} \dot{+} \begin{bmatrix} U & K \\ K' & 0 \end{bmatrix} \dot{+} 0, \qquad \bar{C} \dot{+} B_2D \dot{+} \begin{bmatrix} V & L \\ sL' & 0 \end{bmatrix} \dot{+} 0. \tag{29}$$

If X is a matrix satisfying the system (4), then

$$\begin{bmatrix} I & -X \\ 0 & I \end{bmatrix}\begin{bmatrix} U & K \\ K' & 0 \end{bmatrix}\begin{bmatrix} I & 0 \\ -\bar{X}' & I \end{bmatrix} = \begin{bmatrix} 0 & K \\ K' & 0 \end{bmatrix},$$
$$\begin{bmatrix} I & -X \\ 0 & I \end{bmatrix}\begin{bmatrix} V & L \\ sL' & 0 \end{bmatrix}\begin{bmatrix} I & 0 \\ -X' & I \end{bmatrix} = \begin{bmatrix} 0 & L \\ sL' & 0 \end{bmatrix}. \tag{30}$$

The assertion of the lemma follows from (29) and (30). Q.E.D.

Lemma 2.10. *Every ϕ_s-pair (A_1, A_2) is $\mathcal{C}$ congruent to a pair of the following form.*

$$\sum_\nu{}' E_{1\nu} C_\nu \dotplus \bar{D} \dotplus \begin{bmatrix} 0 & K \\ K' & 0 \end{bmatrix} \dotplus 0,$$

$$\sum_\nu{}' \bar{C}_\nu \dotplus E_2 D \dotplus \begin{bmatrix} 0 & L \\ sL' & 0 \end{bmatrix} \dotplus 0, \tag{31}$$

where C_ν, D are nonsingular. Each $E_{1\nu}$ has one of the following four forms.[1]

$$a \sum_i{}' M_{m_i n_i} \qquad (a < 0;\ m_1 > m_2 > \cdots), \tag{32}$$

$$b \sum_i{}' \begin{bmatrix} 0 & M_{m_i n_i} \\ -M_{m_i n_i} & 0 \end{bmatrix} \qquad (b > 0;\ m_1 > m_2 > \cdots), \tag{33}$$

$$\sum_i{}' \begin{bmatrix} 0 & \alpha M_{m_i n_i} \\ \bar{\alpha} M_{m_i n_i} & 0 \end{bmatrix} \qquad (\textit{real and imaginary parts of } \alpha \textit{ are positive}, m_i > m_2 > \cdots), \tag{34}$$

$$\sum_i{}' J_{m_i n_i} \qquad (m_1 > m_2 > \cdots). \tag{35}$$

E_2 has the form

$$E_2 = \sum_i{}' J_{m_i n_i} \qquad (m_1 > m_2 > \cdots). \tag{36}$$

K and L have the form (6).

Proof. We begin with the pair (5) in Lemma 2.9. In (5), we are allowed to make a symmetric but otherwise arbitrary rearrangement on the rows and columns of B_2 in the summand $(\bar{D}, B_2 D)$, if only we make a corresponding one on D. Now we rearrange (8) into (36), the matrix after rearrangement is still called D. We thus obtain a summand $(\bar{D}, E_2 D)$ in (31). In the same way, we are allowed to rearrange symmetrically the rows and columns of B_1 in the summand $(B_1 C, \bar{C})$ of (5). It is readily seen from the definition (7) of B_1 that its first direct sum can be rearranged in such a way that after rearrangement each summand has the form (32), and each different summand has a different coefficient a. Similarly, the second [third] direct sum in (7) can be rearranged in such a way that each summand has the form (33) [(34)] after the rearrangement and a different summand has a different coefficient $b[\alpha]$. Finally, the fourth direct sum in (7) can be rearranged into (35). From now on, we assume all these rearrangements have been done. Then B_1 is a direct sum, each summand of which has one of forms (32), (33), (34), or (35) and the summands (32), (33), and (34) have different coefficients a, b, α. Denote these summands by $E_{1\nu}$,

[1](32), (33), (34), (35), and (36) only indicate the forms the matrices take. The indices in these expressions do not necessarily represent the same integers in different cases and the number of terms in the summations are not necessarily the same.

i.e.,

$$B_1 = \sum_\nu{}' E_{1\nu}.$$

Since $(B_1\overline{C}, C)$ is a ϕ_s-pair,

$$B_1 C = \overline{C}'\overline{B}_1', \qquad \overline{C} = s\overline{C}',$$

we have

$$B_1 C = s\overline{C}\overline{B}_1', \tag{37}$$

hence

$$\overline{B}_1 B_1 C = s\overline{B}_1$$

or, equivalently,

$$\Big(\sum_\nu{}' E_{1\nu}' E_{1\nu}\Big)C = C\Big(\sum_\nu{}' \overline{E}_{1\nu} E_{1\nu}\Big)'. \tag{38}$$

However, it is obvious that for distinct ν, $\nu_1 \neq \nu_2$, $\overline{E}_{1\nu}$, $E_{1\nu}$, and $(\overline{E}_{1\nu_2} E_{1\nu_2})'$ have no common eigenvalues. Hence after a well-known computation we are able to claim from (38) that C itself splits into a direct sum

$$C = \sum_\nu{}' C_\nu,$$

where C_ν and $E_{1\underline{\nu}}$ have the same order. Up to this stage, we have proved that the summand $(B_1 C, \overline{C})$ in the pair (5) can be written in the following form.

$$\sum_\nu{}' E_{1\nu} C_\nu, \qquad \sum_\nu{}' C_\nu$$

Thus the Lemma is proved.

3. Further Simplification

In order to obtain canonical forms of ϕ_s-pairs we must further simplify the pair (31) in Lemma 2.10. This is the object of the present section. The last result in this section is Lemma 3.3; other lemmas are preparatory. For the pair in (31) it is enough to consider its summand

$$\sum_\nu{}' E_{1\nu} C_\nu \dotplus \overline{D}, \qquad \sum_\nu{}' C_\nu' + E_2 D.$$

Lemma 2.10 tells us that this summand is a direct sum, each summand of which has one of the following five forms.[2]

(I) ϕ_s-pair $(BC, \overline{C})$; B is a square matrix (32) and C is nonsingular.

[2]From here on, the same letters might not represent the same matrices. The context will prevent any confusion.

(II) ϕ_s-pair $(BC, \overline{C})$; B is a square matrix (33), C is nonsingular.
(III) ϕ_s-pair $(BC, \overline{C})$; B is a square matrix (34), C is nonsingular.
(IV) ϕ_s-pair $(BC, \overline{C})$; B is a square matrix (35), C is nonsingular.
(V) ϕ_s-pair $(\overline{C}, BC)$; B is a square matrix (36), C is nonsingular.

For pairs (I) to (IV), we have

$$BC = s\overline{C}\overline{B}' \tag{39}$$

(see (37)); for (V), (39) holds because $\overline{C} = C'$ and $BC = sC'B'$. Now we are going to find a general formula for C's in (I) to (V).

(I) When B is the square matrix (32), the coefficient a can be canceled by using (39). Let $C = C_1 + iC_2$, where C_1, C_2 are real. We have

$$\Big(\sum_i{}' M_{m_in_i}\Big)C_1 = sC_1\sum_i{}' M'_{m_in_i},$$

$$\Big(\sum_i{}' M_{m_in_i}\Big)C_2 = -sC_2\sum_i{}' M_{m_in_i}.$$

If $s = +1$, the second equation gives $C_2 = 0$, since the two matrices obtained by multiplying both sides by C have no common eigenvalues. Subtracting C_1 from both sides of the first equation, we have:

$$\Big(\sum_i{}' J_{m_in_i}\Big)C_1 = C_1\sum_i J'_{m_in_i}; \tag{40}$$

similarly, if $s = -1$, then $C_1 = 0$, and (40) holds with C_1 replaced by C_2. The solution of (40) is well-known, we have therefore the following conclusion: The matrix C in the ϕ_s-pair (I) has the form:

$$C = \begin{bmatrix} C_{11} & C_{12} & \cdots \\ C_{21} & C_{22} & \cdots \\ \cdots & \cdots & \cdots \\ \cdots & \cdots & \cdots \\ \cdots & \cdots & \cdots \end{bmatrix} \quad \text{when} \quad s = +1, \tag{41}$$

$$C = \begin{bmatrix} C_{11} & C_{12} & \cdots \\ C_{21} & C_{22} & \cdots \\ \cdots & \cdots & \cdots \\ \cdots & \cdots & \cdots \\ \cdots & \cdots & \cdots \end{bmatrix} \quad \text{when} \quad s = -1. \tag{42}$$

Here the minors C_{ij} are real of the size $m_in_i \times m_in_i$, and have the following special forms.

$$C_{ii} = \begin{bmatrix} X_{m_i} & X_{m_i-1} & \cdots & & X_2 & X_1 \\ X_{m_i-1} & \cdots & \cdots & X_2 & X_1 & \\ \vdots & & \cdot^{\cdot^{\cdot}} & \cdot^{\cdot^{\cdot}} & & \\ \vdots & \cdot^{\cdot^{\cdot}} & \cdot^{\cdot^{\cdot}} & & & \\ X_2 & X_1 & & & & \\ X_1 & & & & & 0 \end{bmatrix}, \tag{43}$$

$$C_{ij} = \begin{bmatrix} X_{m_j} & X_{m_j-1} & \cdots & X_2 & X_1 \\ \vdots & & \cdot^{\cdot^{\cdot}} & \cdot^{\cdot^{\cdot}} & \\ \vdots & \cdot^{\cdot^{\cdot}} & \cdot^{\cdot^{\cdot}} & & \\ X_2 & X_1 & & & \\ X_1 & & & & \\ 0 & 0 & \cdots & 0 & 0 \end{bmatrix} \quad (i < j), \tag{44}$$

$$C_{ij} = \begin{bmatrix} X_{m_i} & X_{m_i-1} & \cdots & X_2 & X_1 & 0 \\ X_{m_i-1} & & X_2 & X_1 & & 0 \\ & \cdot^{\cdot^{\cdot}} & \cdot^{\cdot^{\cdot}} & & & \vdots \\ \cdot^{\cdot^{\cdot}} & \cdot^{\cdot^{\cdot}} & & & & \vdots \\ X_2 & X_1 & & & & 0 \\ X_1 & & & & 0 & 0 \end{bmatrix} \quad (i > j). \tag{45}$$

In (43), the size of each minor X_α is $n_i \times n_i$; in (44) and (45), the size of each X_α is $n_i \times n_j$. In (44), the last zero row at the bottom represents $(m_i - m_j)n_i$ rows of zeros. In (45), the last zero column represents $(m_j - m_i)n_j$ columns of zeros.

In order to find the general form of C in (II) and (III), we first need a lemma.

Lemma 3.1. *If M is a direct sum of square matrices of the form M_{mn}, then $M^2X = XM'^2$ implies $MX = XM'$.*

Proof. Under the given condition, we have

$$M(MX - XM') = -(MX - XM')M'.$$

It follows that $MX - XM' = 0$, since M and $-M'$ has no common eigenvalues.

(II) Consider Eq. (39), in which B is the square matrix (33). After canceling the common factor b, we get

$$EC = s\overline{C}E', \tag{46}$$

where

$$E = \sum_i \begin{bmatrix} 0 & M_{m_in_i} \\ -M_{m_in_i} & 0 \end{bmatrix} \qquad (m_1 > m_2 > \cdots).$$

Take a transforming matrix P such that

$$PEP' = \begin{bmatrix} 0 & M \\ -M & 0 \end{bmatrix}, \qquad \text{where} \quad M = \sum_i M_{m_in_i}. \tag{47}$$

Let

$$X = PCP' = \begin{bmatrix} Y & Z \\ Z_1 & Y_1 \end{bmatrix}. \tag{48}$$

Equation (46) becomes

$$\begin{bmatrix} 0 & M \\ -M & 0 \end{bmatrix}\begin{bmatrix} Y & Z \\ Z_1 & Y_1 \end{bmatrix} = s\begin{bmatrix} \overline{Y} & \overline{Z} \\ \overline{Z}_1 & \overline{Y}_1 \end{bmatrix}\begin{bmatrix} 0 & -M' \\ M' & 0 \end{bmatrix}$$

or, equivalently,

$$MY_1 = -s\overline{Y}M', \qquad -M\overline{Y} = s\overline{Y}_1M', \tag{49}$$

$$MZ_1 = s\overline{Z}M', \qquad MZ = s\overline{Z}_1M'. \tag{50}$$

From two equations in (49) we get

$$M^2Y = -sM\overline{Y}_1M' = YM'^2.$$

Hence, by Lemma 3.1,

$$MY = YM'. \tag{51}$$

Substituting into the second equation of (49),

$$Y_1 = -s\overline{Y}. \tag{52}$$

From (50) we get, in the same way,

$$MZ = ZM', \tag{53}$$

$$Z_1 = s\overline{Z}. \tag{54}$$

Substituting (52) and (54) in (48), we have

$$X = \begin{bmatrix} Y & Z \\ s\overline{Z} & -s\overline{Y} \end{bmatrix}. \tag{55}$$

Moreover, Y and Z satisfy (51) and (53). Subtracting Y and Z from (51) and (53), we have

$$\Big(\sum_i{}' J_{m_in_i}\Big)Y = Y\sum_i{}' J'_{m_in_i},$$

$$\Big(\sum_i{}' J_{m_in_i}\Big)Z = Z\Big(\sum_i{}' J_{m_in_i}\Big),$$

respectively. From the well-known solutions of these equations, we find that

$$Y = \begin{bmatrix} Y_{11} & Y_{12} & \cdots \\ Y_{21} & Y_{22} & \cdots \\ \cdots & \cdots & \cdots \\ \cdots & \cdots & \cdots \\ \cdots & \cdots & \cdots \end{bmatrix}, \qquad Z = \begin{bmatrix} Z_{11} & Z_{12} & \cdots \\ Z_{21} & Z_{22} & \cdots \\ \cdots & \cdots & \cdots \\ \cdots & \cdots & \cdots \\ \cdots & \cdots & \cdots \end{bmatrix}, \tag{56}$$

where the minors Y_{ij} and Z_{ij} have the forms indicated in (43), (44) and (45).

From (47) and (48) we see clearly that C changes to X at the same time when we rearrange E. Reversing this rearrangement we arrange X into C. This fact enables us to obtain the general form of C by using the formula (55) for X:

$$C = \begin{bmatrix} Y_{11} & Z_{11} & Y_{12} & Z_{12} & \cdots \\ s\overline{Z}_{11} & -s\overline{Y}_{11} & s\overline{Z}_{12} & -s\overline{Y}_{12} & \cdots \\ Y_{21} & Z_{21} & Y_{22} & Z_{22} & \cdots \\ s\overline{Z}_{21} & -s\overline{Y}_{21} & s\overline{Y}_{22} & -s\overline{Z}_{22} & \cdots \\ \cdots & \cdots & \cdots & \cdots & \cdots \\ \cdots & \cdots & \cdots & \cdots & \cdots \\ \cdots & \cdots & \cdots & \cdots & \cdots \end{bmatrix}. \tag{57}$$

Here Y_{ij}, Z_{ij} have the forms (43), (44), and (45).

(III) Now we consider equation (39) with B equal to the square matrix (34). As before, rearrange B into

$$\begin{pmatrix} 0 & \alpha M \\ \alpha M & 0 \end{pmatrix}, \qquad M = \sum_i M_{m_i n_i} \qquad (m_1 > m_2 > \cdots).$$

Rearrange C correspondingly into the square matrix

$$X = \begin{bmatrix} Y & Z \\ Z_1 & Y_1 \end{bmatrix}. \tag{58}$$

Equation (39) becomes

$$\begin{bmatrix} 0 & \alpha M \\ \overline{\alpha} M & 0 \end{bmatrix} \begin{bmatrix} Y & Z \\ Z_1 & Y_1 \end{bmatrix} = s \begin{bmatrix} \overline{Y} & \overline{Z} \\ \overline{Z}_1 & \overline{Y}_1 \end{bmatrix} \begin{bmatrix} 0 & \alpha M' \\ \overline{\alpha} M' & 0 \end{bmatrix}$$

or, equivalently,

$$MY_1 = s\overline{Y}M', \qquad MY = s\overline{Y}_1 M', \tag{59}$$

$$\alpha M Z_1 = s\overline{\alpha}\overline{Z}M', \qquad \overline{\alpha} M Z = s\alpha \overline{Z}_1 M'. \tag{60}$$

The two equations in (60) yield

$$\overline{\alpha}^2 M^2 Z = \overline{\alpha} M\left(s\alpha \overline{Z}_1 M\right) = s\alpha\left(\overline{\alpha} M \overline{Z}_1\right) M' = \alpha^2 Z M'^2.$$

Hence $Z = 0$, since $\overline{\alpha}^2 M^2$ and $\alpha^2 M'^2$ have no common eigenvalues (notice that if the real and imaginary parts of α are positive, $\alpha = \pm\overline{\alpha}$ is impossible). The two equations in (59) give

$$M^2Y = sM\overline{Y}_1M' = YM'^2,$$

hence

$$MY = YM' \tag{61}$$

by Lemma 3.1. Substituting into the second equation of (59), we have $Y_1 = s\overline{Y}$; then substituting into (58), we get

$$X = \begin{pmatrix} Y & 0 \\ 0 & s\overline{Y} \end{pmatrix}. \tag{62}$$

Here Y satisfies (61). Subtract Y from this equation, then

$$\left(\sum_i{}' J_{m_in_i}\right)Y = Y\sum_i{}' J'_{m_in_i}.$$

This equation determines the general form of Y, hence also of X. The matrix X will change back to C if we reverse the rearrangement at the beginning of this proof. Thus the general form of C is obtained from that of X, namely (62):

$$C = \begin{bmatrix} Y_{11} & 0 & Y_{12} & 0 & \cdots \\ 0 & s\overline{Y}_{11} & 0 & s\overline{Y}_{12} & \cdots \\ Y_{21} & 0 & Y_{22} & 0 & \cdots \\ 0 & s\overline{Y}_{21} & 0 & s\overline{Y}_{22} & \cdots \\ \cdots & \cdots & \cdots & \cdots & \cdots \\ \cdots & \cdots & \cdots & \cdots & \cdots \\ \cdots & \cdots & \cdots & \cdots & \cdots \end{bmatrix}, \tag{63}$$

where the minors Y_{ij} have the form (43), (44), or (45).

(IV) and (V). Finally, we consider (39), in which B is the matrix (35). Let $C = C_1 + iC_2$, C_1, C_2 real. From (39),

$$\left(\sum_i{}' J_{m_in_i}\right)C_1 = sC_1\sum_i{}' J'_{m_in_i},$$

$$\left(\sum_i{}' J_{m_in_i}\right)C_2 = -sC_2\sum_i{}' J'_{m_in_i}.$$

By using the well-known solutions of these two equations, we obtain the general form of C as follows.

$$C = \begin{bmatrix} C_{11} & C_{12} & \cdots \\ C_{21} & C_{22} & \cdots \\ \cdots & \cdots & \cdots \\ \cdots & \cdots & \cdots \end{bmatrix},$$

where the structures of the minors C_{ij} are similar to (43), (44), and (45), the only difference being that the matrices on each diagonal are not the same matrix X_α, but $X_\alpha, s\overline{X}_\alpha, X_\alpha, s\overline{X}_\alpha, \ldots$ alternately.

We have found the general forms of ϕ_s-pairs (I)–(V).

Now let B_1 denote the first summand of B in (I)–(V), no matter which one it is. In other words, B_1 denotes the following matrices in each case (I)–(V), respectively:

$$aM_{m_1n_1}, \quad b\begin{pmatrix} 0 & M_{m_in_i} \\ -M_{m_in_i} & 0 \end{pmatrix}, \quad \begin{pmatrix} 0 & \alpha M_{m_1n_1} \\ \bar{\alpha} M_{m_1n_1} & 0 \end{pmatrix},$$

$$J_{m_1n_1}, \quad J_{m_1n_1}.$$

Let C_1 denote the matrix on the upper left corner of C in (I)–(V) which have the same size as B_1 (in (I), for example, the size of C is $2m_1n_1 \times 2m_1n_1$). We claim that $C1$ is nonsingular. In fact, in Case (I), C_1 is just C_{11} in (41) or iC_{11} in (42). Substituting the formulas (43), (44) and (45) for the blocks in C_{ij} in (41) and (42), we obtain the minors $X_1, 0, 0, \ldots$ on the m_1th row. The nonsingularity of X_1, hence of C_{11}, follows from that of C. The proofs of Case (IV) and (V) are similar. For Case (III), C_1 is the minor $Y_{11} + s\bar{Y}_{11}$ in (63). Since C is nonsingular,

$$\begin{bmatrix} Y_{11} & Y_{12} & \cdots \\ Y_{21} & Y_{22} & \cdots \\ \cdots & \cdots & \cdots \\ \cdots & \cdots & \cdots \end{bmatrix}$$

is nonsingular. But this matrix is not different from (41) as far as the structure is concerned. Hence Y_{11} is nonsingular; so is C_1. Finally, in Case (II), C_1 is just the minor

$$\begin{pmatrix} Y_{11} & Z_{11} \\ s\bar{Z}_{11} & -s\bar{Y}_{11} \end{pmatrix}$$

in (57). Putting the formulas (43), (44), and (45) for the blocks of Y_{ij} and Z_{ij} in (57) and looking at the minors on the m_1th row and $2m_1$th row, we find that the nonsingularity of C implies that of C. Hence the C_1 in the five matrices are nonsingular.

Now we proceed to prove:

Lemma 3.2. *ϕ_s-pair (I) is $\mathcal{C}$ congruent to a pair of the form*

$$a\sum_i{}' M_{m_in_i}C_i, \qquad \sum_i{}' \bar{C}_i \qquad (C_i \text{ nonsingular}). \tag{65}$$

ϕ_s-pair (II) is $\mathcal{C}$ congruent to a pair of the form

$$\sum_i{}' \begin{pmatrix} 0 & M_{m_in_i} \\ -M_{m_in_i} & 0 \end{pmatrix} C_i, \qquad \sum_i{}' \bar{C}_i \qquad (C_i \text{ nonsingular}). \tag{66}$$

ϕ_s-pair (III) is $\mathcal{C}$ congruent to a pair of the form

$$\sum_i{}' \begin{bmatrix} 0 & \alpha M_{m_in_i} \\ \alpha \bar{M}_{m_in_i} & 0 \end{bmatrix} C_i, \qquad \sum_i{}' \bar{C}_i \qquad (C_i \text{ nonsingular}). \tag{67}$$

ϕ_s-pair (IV) is $\mathcal{C}$ congruent to a pair of the form

$$\sum_i{}' J_{m_in_i}C_i, \qquad \sum_i{}' \bar{C}_i \qquad (C_i \text{ nonsingular}). \tag{68}$$

ϕ_s-pair (V) is $\mathcal{C}$ congruent to a pair of the form

$$\sum_i{}' \bar{C}_i, \qquad \sum_i{}' J_{m_i n_i} C_i \qquad (C_i \text{ nonsingular}). \tag{69}$$

Proof. We keep the notations B_1 and C_1 just introduced and write

$$B = B_1 \dotplus B_2. \tag{70}$$

Here, in the five cases, B_2 represents the direct sum of the summands of B except for the first one. For example, $B_2 = aM_{m_2 n_2} \dotplus \ldots$ in Case (I).

Since $C' = sC$, we can write

$$C = \begin{pmatrix} C_1 & U \\ sU' & V \end{pmatrix}. \tag{71}$$

Substituting (70) and (71) in (39), we have

$$B_1 C_1 = s\bar{C}_1 \bar{B}_1', \qquad B_2 U' = s\bar{U}' \bar{B}_1'. \tag{72}$$

C_1 being nonsingular, we define a matrix

$$Q = \begin{pmatrix} I & 0 \\ -s\bar{U}'\bar{C}_1^{-1} & I \end{pmatrix}.$$

Using (72), it is easy to check that

$$QB = B\bar{Q}. \tag{73}$$

Now transform ϕ_s pair (I)–(V) by Q:

$$BC_1\bar{C} \approx QBC\bar{Q}', \; Q\bar{C}Q'.$$

By (73), we have

$$BC_1\bar{C} \approx B\bar{Q}C\bar{Q}', \; Q\bar{C}Q'. \tag{74}$$

Direct computation shows that

$$Q\bar{C}Q' = \bar{C}_1 \dotplus \bar{D},$$

where, by virtue of the nonsingularity of C, D is nonsingular. Substituting into (74), we have

$$BC_1\bar{C} \approx B_1C_1 \dotplus B_2 D, \; \bar{C}_1 \dotplus \bar{D}.$$

At this step a summand $(B_1C_1, \bar{C}_1)$ is split out. The remaining summand $(B_2D, \bar{D})$ has the same form as the original pair $(BC, \bar{C})$ from which we started. The same method can be used repeatedly until the whole process is completed. The assertion in the Lemma is proved for (I)–(IV). A similar proof applies to (V) and will not be repeated here.

Combining Lemma 2.10 and Lemma 3.2, we obtain

Lemma 3.3. *Every ϕ_s-pair (A_1, A_2) is $\mathcal{C}$ congruent to a pair of the form*

$$\sum_\nu{}' A_{1\nu} \dotplus \begin{pmatrix} 0 & K \\ K' & 0 \end{pmatrix} \dotplus 0, \qquad \sum_\nu{}' A_{2\nu} \dotplus \begin{pmatrix} 0 & L \\ sL' & 0 \end{pmatrix} \dotplus 0,$$

where K and L are defined by (6), *and each summand* $(A_{1\nu}, A_{2\nu})$ *takes one of the following five forms.*

(i) ϕ_s*-pair* $(aM_{mn}C, \overline{C})$ $(a > 0$, *C nonsingular*).

(ii) ϕ_s*-pair*

$$b\begin{pmatrix} 0 & M_{mn} \\ -M_{mn} & 0 \end{pmatrix} C, \overline{C} \qquad (b > 0, C \text{ nonsingular})$$

(iii) ϕ_s*-pair*

$$\begin{pmatrix} 0 & \alpha M_{mn} \\ \alpha M_{mn} & 0 \end{pmatrix} C, \overline{C} \qquad \begin{matrix} (\textit{both real and imaginary parts of } \alpha \\ \textit{are positive, C nonsingular}) \end{matrix}$$

(iv) ϕ_s*-pair* $(J_{mn}C, \overline{C})$ (*C nonsingular*)

(v) ϕ_s*-pair* $(\overline{C}, J_{mn}C)$ (*C nonsingular*)

4. Canonical Forms of the Five ϕ_s-Pairs of the Special Type

In the present section, we are going to produce a canonical form for each pair (I)–(V) in Lemma 3.3, and hence solve the problem of canonical forms for general ϕ_s-pairs. First of all, we need a few new notations and lemmas.

The two matrices

$$\begin{bmatrix} X_1 & X_2 & \cdots & X_{m-1} & X_m \\ & X_1 & \ddots & & X_{m-1} \\ & & \ddots & \ddots & \vdots \\ & & & \ddots & X_2 \\ 0 & & & & X_1 \end{bmatrix},$$

$$\begin{bmatrix} X_1 & & & & 0 \\ X_2 & X_1 & & & \\ \ddots & X_2 & \ddots & & \\ X_{m-1} & \ddots & \ddots & \ddots & \\ X_m & Y_{m-1} & \ddots & X_2 & X_1 \end{bmatrix}$$

(X_i's are square matrices of order n)

will be denoted by

$$\psi(X_1, X_2, \ldots, X_m), \qquad \psi'(X_1, X_2, \ldots, X_m),$$

respectively. It is easily seen that

$$(\psi(X_1,X_2,\ldots,X_m))' = \psi'(X_1',X_2',\ldots,X_m'),$$
$$N_{mn}\psi'(X_1,X_2,\ldots,X_m) = \psi(X_1,X_2,\ldots,X_m)N_{mn},$$
$$M_{mn}\psi(X_1,X_2,\ldots,X_m) = \psi(X_1,X_2,\ldots,X_m)N_{mn}.$$

These three equalities will be used implicitly.

Lemma 4.1. *We have*

$$\psi(X_1,\ldots,X_m)\psi(Y_1,\ldots,Y_m) = \psi(Z_1,\ldots,Z_m),$$

where Z_i is determined by comparing the coefficients in

$$\sum_{i=1}^{m} Z_i\lambda^{i-1} = \left(\sum_{i=1}^{m} X_i\lambda^{i-1}\right)\left(\sum_{i=1}^{m} Y_i\lambda^{i-1}\right) \qquad (\operatorname{mod}\lambda^m).$$

Proof. Easy by direct verification, or using direct product.

Lemma 4.2.[3] *Assume that $X_1,\ldots,X_m$ are all real symmetric or all skew-symmetric and X_1 is nonsingular. There are real matrices $P_1,\ldots,P_m$ (P_1 nonsingular) such that*

$$\psi(P_1,\ldots,P_m)\psi(X_1,\ldots,X_m)\psi(P_1',\ldots,P_m') = \psi(X,0,\ldots,0),$$

where

$$X = \operatorname{diag}(\pm 1,\ldots,\pm 1) \qquad \text{if } X_i\text{'s are symmetric,}$$
$$X = F \qquad \text{if } X_i\text{'s are skew-symmetric.}$$

Proof. Using the product rule in Lemma 4.1, we have easily

$$\psi(I,-\tfrac{1}{2}X_2X_1^{-1},0,\ldots,0)\psi(X_1,\ldots,X_m)\psi(I,-\tfrac{1}{2}X_1^{-1}X_2,0,\ldots,0)$$
$$= \psi(X_1,0,Y_3,\ldots,Y_m),$$

where Y_i's are still real symmetric or real skew-symmetric. Once again,

$$\psi\big(I,0,-\tfrac{1}{2}Y_3X_1^{-1},0,\ldots,0\big)\psi(X_1,0,Y_3,\ldots,Y_m)\psi\big(I,0,-\tfrac{1}{2}X_1^{-1}Y_3,0,\ldots,0\big)$$
$$= \psi(X_1,0,0,Z_4,\ldots,Z_m).$$

Each time we annihilate one matrix. Repeat this process till we get $\psi(X_1,0,\ldots,0)$. The proof is then completed by taking a nonsingular matrix P such that $P_1X_1P_1'$ is equal to the X in the assertion of the lemma.

Lemma 4.3. *Assume that $X_1,\ldots,X_m$ are all symmetric or all skew-symmetric and X_1 is nonsingular. Then there are $P_1,\ldots,P_m$, P_1 nonsingular, such that*

$$\psi(P_1,\ldots,P_m)\psi(X_1,\ldots,X_m)\psi(P_1',\ldots,P_m') = \psi(X,0,\ldots,0),$$

[3] Lemmas 4.2 and 4.3 are known.

where

$$X = I \qquad \textit{if} \quad \textit{Xi's are symmetric,}$$
$$X = F \qquad \textit{if} \quad X_i\textit{'s are skew-symmetric.}$$

Proof. The proof of Lemma 4.2 applies here. In the last step, X can be taken as I in the case of symmetric X_i's, since we are not restricted to the real number field.

One more notation: We use Γ_s to denote the set of square matrices

$$\begin{pmatrix} Y & Z \\ s\overline{Z} & -s\overline{Y} \end{pmatrix} \qquad (s = \pm 1;\ Y, Z \text{ are square matrices}).$$

We speak of Γ_+ or Γ_- if $s = +1$ or $s = -1$. An easy observation: Γ_s is the set of all solutions X of the equation

$$XF = -sF\overline{X}. \tag{75}$$

Lemma 4.4. *If a nonsingular X belongs to Γ_s, so does X^{-1}.*

Proof. If X is a solution of (75), then X^{-1} is also a solution.

Lemma 4.5. (i) *if $X \in \Gamma_s$, $Y \in \Gamma_s$, then $XY \in \Gamma_s$.* (ii) *if $X \in \Gamma_-$, $Y \in \Gamma_s$, $Z \in \Gamma_-$, then $XYZ \in \Gamma_s$.*

Proof. Direct verification.

Lemma 4.6. *If $X_i \in \Gamma_s$, $X_i' = sX_i$ $(i = 1, \ldots, m)$, and X_1 is nonsingular, then there are matrices $P_2, \ldots, P_m$ in Γ_- such that*

$$\psi(I, P_2, \ldots, P_m)\psi(X_1, \ldots, X_m)\psi(I, P_2', \ldots, P_m') = \psi(X_1, 0, \ldots, 0). \tag{76}$$

Proof. First let $P = -\frac{1}{2}X_2X_1^{-1}$. By Lemma 4.5 (i) and Lemma 4.4, $P \in \Gamma_-$. We have

$$\psi(I, P, 0, \ldots, 0)\psi(X_1, \ldots, X_m)\psi(I, P', 0, \ldots, 0) = \psi(X_1, 0, Y_3, \ldots, Y_m). \tag{77}$$

Next, by Lemma 4.5 (ii), $Y_i \in \Gamma_s$. Hence let $Q = -\frac{1}{2}Y_3X_1^{-1}$, then $Q \in \Gamma_-$ and

$$\psi(I, 0, Q, 0, \ldots, 0)\psi(X_1, 0, Y_3, \ldots, Y_m)\psi(I, 0, Q', 0, \ldots, 0)$$
$$= \psi(X_1, 0, 0, Z_4, \ldots, Z_m).$$

Proceeding this way, we get $\psi(X_1, 0, \ldots, 0)$. Form the product of $\psi(I, P, 0, \ldots, 0)$, $\psi(I, 0, Q, 0, \ldots, 0)$ and so on in the reversed order we get a matrix $\psi(I, P_2, \ldots, P_m)$ such that (76) holds. At the same time $P_i \in \Gamma_-$ by Lemma 4.5 (i). Q.E.D.

Lemma 4.7. *If $H \in \Gamma_s$ is a hermitian matrix of order $2n$:*

$$H = \begin{pmatrix} A & B \\ sB & -s\overline{A} \end{pmatrix}, \qquad H' = \overline{H},$$

then there is a nonsingular $P \in \Gamma_-$ *such that*

$$PH\bar{P}' = D_0 \dotplus (-D_0), \qquad D_0 = I \dotplus 0 \qquad \text{if} \quad s = +1, \tag{78}$$

$$PH\bar{P}' = D \dotplus D, \qquad D = I \dotplus (-I) \dotplus 0 \qquad \text{if} \quad s = -1. \tag{79}$$

Proof. Let $A = [a_{ij}]$, $B = [b_{ij}]$,

$$X_{ij} = \begin{bmatrix} a_{ij} & b_{ij} \\ s\bar{b}_{ij} & -s\bar{a}_{ij} \end{bmatrix}, \qquad X = \begin{bmatrix} X_{11} & \cdots & X_{1n} \\ \cdots & \cdots & \cdots \\ \cdots & \cdots & \cdots \\ X_{n1} & \cdots & X_{nn} \end{bmatrix}.$$

X is nothing but a rearrangement of H. Now we proceed to prove that there is a nonsingular Z such that

$$Z = \begin{bmatrix} Z_{11} & \cdots & Z_{1n} \\ \cdots & \cdots & \cdots \\ \cdots & \cdots & \cdots \\ Z_{n1} & \cdots & Z_{nn} \end{bmatrix}, \qquad Z_{ij} = \begin{pmatrix} u_{ij} & v_{ij} \\ -\bar{v}_{ij} & \bar{u}_{ij} \end{pmatrix}, \tag{80}$$

and such that $ZX\bar{Z}'$ is diagonal. The proof is very similar to the classical diagonalization procedure of Lagrange and we only give a sketch here. We can assume $X \neq 0$. If one of $X_{ii} \neq 0$, we take a rearrangement matrix Q such that the minor $(1,1)$ of $Q \times Q'$ is X_{ii} (notice that Q has the form (80)). If all X_{ii} are zero, then we can assume $X_{12} \neq 0$ for the same reason. Let

$$Y = \begin{pmatrix} a_{12} & b_{12} \\ -\bar{b}_{12} & \bar{a}_{12} \end{pmatrix}, \qquad Z_1 = \begin{pmatrix} I_2 & Y \\ 0 & I_2 \end{pmatrix} \dotplus I_{2n-4}.$$

Z has the form (80) and

$$Z_1 X \bar{Z}_1' = \begin{bmatrix} a & 0 & \cdots \\ 0 & -sa & \cdots \\ \cdots & \cdots & \cdots \\ \cdots & \cdots & \cdots \end{bmatrix}, \qquad a = 2|a_{12}|^2 + 2|b_{12}|^2 \neq 0.$$

We could, therefore, always take $X_{11} \neq 0$, i.e., X_{11} nonsingular. The matrix

$$Z_2 = \begin{bmatrix} I_2 & & & 0 \\ Y_2 & I_2 & & \\ \vdots & \vdots & \ddots & \\ Y_n & 0 & \cdots & I_2 \end{bmatrix}, \qquad Y_i = -X_{i1} X_{11}^{-1}$$

has the form (80) and

$$Z_2 X \bar{Z}_2' = X_{11} \dotplus * .$$

This process can be repeated and X becomes a direct sum of n square matrices of order 2:

$$X_1 \dotplus X_2 \dotplus \cdots \dotplus X_n .$$

It suffices to consider one of these matrices which is not zero. Let it be

$$\begin{pmatrix} a & b \\ s\bar{b} & -s\bar{a} \end{pmatrix} \qquad (a \text{ real}).$$

When $s = -1$, this matrix is just aI_2, since it is hermitian; clearly it can be changed into I_2 or $-I_2$. When $s = +1$, three cases arise:

1. If $b = 0$, $a > 0$, then

$$\begin{pmatrix} a^{-1/2} & 0 \\ 0 & a^{-1/2} \end{pmatrix}\begin{pmatrix} a & 0 \\ 0 & a \end{pmatrix}\begin{pmatrix} a^{-1/2} & 0 \\ 0 & a^{-1/2} \end{pmatrix} = \begin{pmatrix} 1 & 0 \\ 0 & -1 \end{pmatrix}.$$

2. If $b = 0$, $a < 0$, then

$$\begin{bmatrix} 0 & (-a)^{-1/2} \\ -(-a)^{-1/2} & 0 \end{bmatrix}\begin{bmatrix} a & 0 \\ 0 & a \end{bmatrix}\begin{bmatrix} 0 & -(-a)^{-1/2} \\ (-a)^{-1/2} & 0 \end{bmatrix} = \begin{bmatrix} 1 & 0 \\ 0 & 1 \end{bmatrix}.$$

3. If $b \neq 0$, then let $c = a + (a^2 + b\bar{b})^{1/2}$. We have

$$\begin{pmatrix} c & b \\ -\bar{b} & c \end{pmatrix}\begin{pmatrix} a & b \\ \bar{b} & -a \end{pmatrix}\begin{pmatrix} c & -b \\ \bar{b} & c \end{pmatrix} = 2(a^2 + b\bar{b})c\begin{pmatrix} 1 & 0 \\ 0 & -1 \end{pmatrix}.$$

Thus this case can be reduced to Case 1.

We then have

$$ZX\bar{Z}' = \sum{}' \begin{pmatrix} 1 & 0 \\ 0 & -1 \end{pmatrix} \dot{+} 0 \qquad \text{if} \quad s = +1, \tag{81}$$

$$ZX\bar{Z}' = \sum{}' I_2 \dot{+} \sum{}'(-I_2) \dot{+} 0 \qquad \text{if} \quad s = -1, \tag{82}$$

where Z is a nonsingular matrix of the form (80).

If we rearrange the rows and columns of X such that it changes to H, the same rearrangement will change Z into a nonsingular matrix P which belongs to Γ_-, and change the right-hand sides of (81) and (82) into the right-hand sides of (78) and (79). Q.E.D.

Lemma 4.8. *If $W \in \Gamma_s$, $W' = sW$, then there exists a nonsingular $P \in \Gamma_-$ such that*

$$PWP' = \begin{pmatrix} D_0 & 0 \\ 0 & -D_0 \end{pmatrix}, \qquad D_0 = I \dot{+} 0 \qquad \textit{if} \quad s = +1,$$

$$PWP' = \begin{pmatrix} 0 & D \\ -D & 0 \end{pmatrix}, \qquad D = I \dot{+} (-I) \dot{+} 0 \qquad \textit{if} \quad s = -1.$$

Proof. We know from the properties of W that

$$W\begin{pmatrix} 0 & -I \\ I & 0 \end{pmatrix}$$

is hermitian and belongs to Γ_s. According to Lemma 4.7, there is a nonsingular matrix

$$\begin{pmatrix} Q & R \\ -\bar{R} & \bar{Q} \end{pmatrix}$$

such that

$$\begin{pmatrix} Q & R \\ -\bar{R} & \bar{Q} \end{pmatrix} W \begin{pmatrix} 0 & I \\ I & 0 \end{pmatrix} \begin{pmatrix} \bar{Q}' & -R' \\ \bar{R}' & Q' \end{pmatrix} = \begin{cases} D_0 \dotplus (-D_0) & \text{if } s = +1, \\ D \dotplus D & \text{if } s = -1. \end{cases}$$

Multiplying the first equation by $2^{-1/2}\left(\begin{smallmatrix} I & I \\ -I & I \end{smallmatrix}\right)$ from the left and from the right, and multiplying the second equation by $\left(\begin{smallmatrix} 0 & I \\ -I & 0 \end{smallmatrix}\right)$ from the left and from the right, we obtain

$$P_1 W P_1' = D_0 \dotplus (-D_0) \qquad \text{when} \quad s = +1,$$

$$P_2 W P_2' = \begin{pmatrix} 0 & D \\ -D & 0 \end{pmatrix} \qquad \text{when} \quad s = -1,$$

where both

$$P_1 = \begin{pmatrix} Q - \bar{R} & \bar{Q} + R \\ -Q - \bar{R} & \bar{Q} - R \end{pmatrix} \quad \text{and} \quad P_2 = \begin{pmatrix} Q & R \\ -\bar{R} & \bar{Q} \end{pmatrix}$$

belong to Γ_-. Q.E.D.

Lemma 4.9. *Assume that* $X_i \in \Gamma_s$, $X_i' = sX_i$ $(i = 1, \dots, m)$ *and* X_1 *is nonsingular. There are* $P_1, \dots, P_m$, $P_i \in \Gamma_-$, P_1 *nonsingular, such that*

$$\psi(P_1, \dots, P_m)\psi(X_1, \dots, X_m)\psi(P_1', \dots, P_m') = \psi(X_0, 0, \dots, 0),$$

where

$$X_0 = I \dotplus (-I) \qquad \textit{when} \quad s = +1 \qquad (\textit{two } I\textit{'s have the same size}),$$

$$X_0 = \begin{pmatrix} 0 & D \\ -D & 0 \end{pmatrix}, \qquad D = \operatorname{diag}(\pm 1, \dots, \pm 1) \qquad \textit{when} \quad s = -1.$$

Proof. Using Lemma 4.6, first we annihilate $X_2, \dots, X_m$. By Lemma 4.8 and the nonsingularity of X_1, there is a nonsingular $P \in \Gamma_-$ such that PX_1P' equals the X_0 in the assertion of the present lemma. Hence

$$\psi(P, 0, \dots, 0)\psi(X_1, 0, \dots, 0)\psi(P', 0, \dots, 0) = \psi(X_0, 0, \dots, 0).$$

Q.E.D.

Now we are ready to construct the canonical form of the ϕ_s-pair in Lemma 3.3. We will state the results through a series of lemmas.

Lemma 4.10$_+$. *The* ϕ_+*-pair* (*i*) *in Lemma* 3.3 *is* $\mathcal{C}$ *congruent to a direct sum of pairs of the form*

$$\epsilon a M_m N_m, \epsilon N_m \qquad (\epsilon = \pm 1). \tag{83}$$

Proof. It is already known that the matrix C in the ϕ_+-pair (i) can be written in the form (43) (see Section 3, (I)). In other words, $C = \psi(X_1, \ldots, X_m)N_{mn}$, X_i's are real. Since C is symmetric and nonsingular, all of X_i's are real symmetric and X_i is nonsingular. Hence, by Lemma 4.2, there is a real nonsingular matrix $P = \psi(P_1, \ldots, P_m)$ such that

$$P\psi(X_1, \ldots, X_m)\psi(P_1', \ldots, P_m') = \psi(D, 0, \ldots, 0), \tag{84}$$

$$D = \operatorname{diag}(\pm 1, \ldots, \pm 1). \tag{85}$$

Now transforming the ϕ_+-pair (i) by P, we obtain

$$PCP' = P\psi(X_1, \ldots, X_m)\psi(P_1', \ldots, P_m')N_{mn},$$

$$P(aM_{mn}C)P' = aM_{mn}PCP'.$$

By (84),

$$P(aM_{mn}C)P' = aM_{mn}\psi(D, 0, \ldots, 0)N_{mn}, \tag{86}$$

$$PCP' = \psi(D, 0, \ldots, 0)N_{mn}. \tag{87}$$

It is easily seen that for D in (85), after a symmetric rearrangement the right-hand side of (86) becomes a direct sum of n matrices $\pm aM_mN_m$. At the same time the right-hand side of (87) becomes a direct sum of n matrices $\pm N_m$. Q.E.D.

Lemma 4.10. *The ϕ-pair (i) in Lemma 3.3 is $\mathfrak{C}$ congruent to a direct sum of $\frac{1}{2}n$ pairs of the form*

$$-a^2M_mN_m \dotplus N_m, F_m. \tag{88}$$

Proof. It is already known (see Section 3 (I)) that C in the ϕ_--pair can be written in the form of i times (43). In other words, $C = i\psi(X_1, \ldots, X_m)N_{mn}$, all of the X_i's real. Since C is skew-symmetric and nonsingular, all X_i's are skew-symmetric and X_1 is nonsingular. Hence, by Lemma 4.2, there is a real nonsingular $P = \psi(P_1, \ldots, P_m)$ such that

$$P\psi(X_1, \ldots, X_m)\psi(P_1', \ldots, P_m') = \psi(F, 0, \ldots, 0).$$

Now transform the ϕ_--pair (i) by P. After the transformation we obtain the matrix pair

$$P(aM_{mn}C)P' = a_iM_{mn}\psi(F, 0, \ldots, 0)N_{mn}, \tag{89}$$

$$P\bar{C}P' = -i\psi(F, 0, \ldots, 0)N_{mn}; \tag{90}$$

n is even, for the order of F is n. The right-hand side of (89) and (90) is divided into $m \times m$ blocks. Each block being subdivided into 4 blocks of the same size, the matrix itself is divided into $2m \times 2m$ blocks. Now rearranging these blocks by the order $(1, 3, 5, \ldots, 2, 4, \ldots, 2m)$, we obtain

$$ai\begin{pmatrix} 0 & M_{mn'}N_{mn'} \\ -M_{mn'}N_{mn'} & 0 \end{pmatrix}, \begin{pmatrix} 0 & N_{mn'} \\ -N_{mn'} & 0 \end{pmatrix} \quad \left(n' = \frac{1}{2}n\right). \tag{91}$$

Transforming the pair (91) by the matrix

$$2^{-1/2}\begin{pmatrix} -I & iI \\ -N_{mn'} & -iN_{mn'} \end{pmatrix}$$

we get

$$a(-M_{mn'}N_{mn'} \dotplus N_{mn'}M_{mn'}), F. \tag{92}$$

(92) can be made into a direct sum of n identical pairs

$$a(-M_mN_m \dotplus N_mM_m), F_m. \tag{93}$$

Consider the real symmetric pair $S_1 = aM_mN_m$, $S_2 = -(1/a)M_m^{-1}N_m^{-1} = -(1/a)M_m^{-1}N_m$. Because the elementary divisor of $S_1 - \lambda S_2$ is $(\lambda + a^2)^m$, there is a real nonsingular matrix Q such that $QS_1Q' = \pm a^2M_mN_m$, $QS_2Q' = \mp N_m$. Transforming (93) by $Q \dotplus Q'^{-1}$, we have

$$\mp a^2M_mN_m \dotplus (\pm N_m), F_m \tag{94}$$

We want to prove that we can always take the upper sign in (94), i.e., we can always change (94) into (88). In fact, let $f(\lambda)$ be the sum of the first m terms in the Taylor expansion of $(1+\lambda)^{1/2}$. Then $f^2(\lambda) = 1 + \lambda \pmod{\lambda^m}$. Hence $f^2(J_m) = I + J_m = M_m$. The matrix $S = f(J_m)N_m$ is real symmetric, and $SN_mS = M_mN_m$. Consequently $S^{-1}M_mN_mS^{-1} = N_m$. Thus, if we transform the pair (94) by

$$\begin{pmatrix} 0 & aS \\ -a^{-1}S^{-1} & 0 \end{pmatrix},$$

then the first matrix changes sign and the second remains unchanged. Q.E.D.

Lemma 4.11$_+$. *The ϕ_+-pair (ii) in Lemma 3.3 is $\mathcal{C}$ congruent to the direct sum of n identical pairs*

$$bi\begin{pmatrix} 0 & M_mN_m \\ -M_mN_m & 0 \end{pmatrix}, \begin{pmatrix} N_m & 0 \\ 0 & N_m \end{pmatrix}. \tag{95}$$

Proof. We already know that the general form of C in the ϕ_+-pair (ii) is

$$C = \begin{pmatrix} Y & Z \\ \bar{Z} & -\bar{Y} \end{pmatrix},$$

where Y, Z have the form (43) (see Section 3 (II)). Using new notation, we have

$$Y = \psi(Y_1, \ldots, Y_m)N_{mn}, \qquad Z = \psi(Z_1, \ldots, Z_m)N_{mn}. \tag{96}$$

Two matrices in (ii) are divided into $2m \times 2m$ blocks. If we rearrange these blocks in the order $(1, 3, 5, \ldots, 2, 4, \ldots, 2m)$, (ii) becomes the pair

$$\left.\begin{aligned} &b\psi(F_n, F_n, 0, \ldots, 0)\psi(\bar{X}_1, \ldots, \bar{X}_m)N_{m,2n}, \\ &\psi(X_1, \ldots, X_m)N_{m,2n}, \end{aligned}\right\} \tag{97}$$

where

$$X_i = \begin{bmatrix} \bar{Y}_i & Z_i \\ \bar{Z}_i & -\bar{Y}_i \end{bmatrix}.$$

Hence $X_i \in \Gamma_+$. Since C is symmetric and nonsingular, all X_i's are symmetric and X_1 is nonsingular. All conditions in Lemma 4.9 (when $s = +1$) are satisfied. There exists a nonsingular $P = \psi(P_1, \ldots, P_m)$, $P_i \in \Gamma_-$, such that

$$P\psi(X_1, \ldots, X_m)\psi(P_1', \ldots, P_m') = \psi(X_0, 0, \ldots, 0), \tag{98}$$
$$X_0 = I_n \dotplus (-I_n).$$

Because P_i satisfies (75) ($s = -1$), it is easily seen that

$$P\psi(F_n, F_n, 0, \ldots, 0) = \psi(F_n, F_n, 0, \ldots, 0)\psi(\bar{P}_1, \ldots, \bar{P}_m). \tag{99}$$

Transform (97) by P. Using (98) and (99), we find the pair after transformation

$$\left.\begin{array}{l} b\psi(F_n, F_n, 0, \ldots, 0)\psi(X_0, 0, \ldots, 0)N_{m,2n}, \\ \psi(X_0, 0, \ldots, 0)N_{m,2n}, \end{array}\right\} \tag{100}$$

or

$$\left.\begin{array}{l} -b\psi\left(\begin{pmatrix} 0 & I \\ I & 0 \end{pmatrix}, \begin{pmatrix} 0 & I \\ I & 0 \end{pmatrix}, 0, \ldots, 0\right)N_{m,2n}, \\ \psi\left(\begin{pmatrix} I & 0 \\ 0 & I \end{pmatrix}, 0, \ldots, 0\right)N_{m,2n} \end{array}\right\}. \tag{101}$$

Rearranging the $2m \times 2m$ blocks of the two matrices in (101) by the order $(1, 3, 5, \ldots, 2, 4, \ldots, 2m)$, we obtain

$$-b\begin{pmatrix} 0 & M_{mn}N_{mn} \\ M_{mn}N_{mn} & 0 \end{pmatrix}, \begin{pmatrix} N_{mn} & 0 \\ 0 & N_{mn} \end{pmatrix}.$$

This pair, after an obvious rearrangement, becomes the direct sum of n identical pairs

$$-b\begin{pmatrix} 0 & M_mN_m \\ M_mN_m & 0 \end{pmatrix}, \begin{pmatrix} N_m & 0 \\ 0 & N_m \end{pmatrix} \tag{102}$$

Transforming (102) by $-I_m \dotplus (-iI_m)$, we have (95). Q.E.D.

Lemma 4.11$_-$. *The ϕ_- pair (ii) in Lemma 3.3 is $\mathcal{C}$ congruent to the direct sum of n identical pairs*

$$\epsilon(b^2M_mN_m \dotplus N_m), F_m \qquad (\epsilon = \pm 1). \tag{103}$$

Proof. The proof is the same as that of Lemma 4.11$_+$ up to (98), if only we regard X_0 as

$$X_0 = \begin{pmatrix} 0 & D \\ -D & 0 \end{pmatrix}; \qquad D = \operatorname{diag}(\pm 1, \ldots, \pm 1).$$

(100) becomes

$$\left.\begin{array}{l} -b\psi(D \dotplus D, D \dotplus D, 0, \ldots, 0)N_{m,2n}, \\ \psi\left(\begin{pmatrix} 0 & D \\ -D & 0 \end{pmatrix}, 0, \ldots, 0\right)N_{m,2n}. \end{array}\right\} \tag{104}$$

Now, rearranging the $2m\times 2m$ blocks in (104) by the order $(1.3,5,\ldots,2,4,\ldots,2m)$, we have

$$-b\begin{pmatrix}\psi(D,D,0,\ldots,0)N_{mn} & 0\\ 0 & \psi(D,D,0,\ldots,0)\end{pmatrix},$$

$$\times\begin{pmatrix}0 & \psi(D,0,\ldots,0)N_{mn}\\ -\psi(D,0,\ldots,0)N_{mn} & 0\end{pmatrix}.$$

This pair, after an obvious rearrangement, becomes the direct sum of the n identical pairs

$$-\eta b\begin{pmatrix}M_mN_m & 0\\ 0 & M_mN_m\end{pmatrix},\eta\begin{pmatrix}0 & N_m\\ -N_m & 0\end{pmatrix}\qquad(\eta=\pm 1). \tag{105}$$

Transforming (105) by $I\dotplus \eta N_m$:

$$-b\eta M_mN_m\dotplus(-b\eta)N_mM_m, F_m. \tag{106}$$

Consider the real symmetric pair $S_1=-b\eta M_mN_m$, $S_2=-b^{-1}\eta M_m^{-1}N_m$. Since the elementary divisor of $S_1-\lambda S_2$ is $(\lambda-b^2)^m$, there is a real nonsingular matrix P such that $PS_1P'=\epsilon b^2M_mN_m$, $PS_2P'=\epsilon N_m$ $(\epsilon=\pm 1)$. Transforming (106) by $P\dotplus P'^{-1}$, we obtain (103). Q.E.D.

Lemma 4.12$_+$. *The ϕ_+-pair (iii) in Lemma 3.3 is $\mathcal{C}$ congruent to the direct sum of the n identical pairs*

$$\begin{pmatrix}0 & \alpha M_mN_m\\ \bar{\alpha}M_mN_m & 0\end{pmatrix},\begin{pmatrix}N_m & 0\\ 0 & N_m\end{pmatrix}. \tag{107}$$

Proof. C in the ϕ_+-pair (iii) has the general form

$$C=\psi(\bar{X}_1,\ldots,\bar{X}_m)N_{mn}\dotplus\psi(X_1,\ldots,X_m)N_{mn} \tag{108}$$

(see Section 3 (III)). Because C is symmetric and nonsingular, all X_i's are symmetric and X_1 is nonsingular. Therefore, by Lemma 4.3, there is a nonsingular $P=\psi(P_1,\ldots,P_m)$ such that

$$P\psi(X_1,\ldots,X_m)\psi(P_1',\ldots,P_m')=\psi(I,0,\ldots,0)=I. \tag{109}$$

Transform (iii) $(s=+1)$ by $P\dotplus\bar{P}$. Using (108) and (109) after the transformation, we find the pair

$$\begin{pmatrix}0 & \alpha M_{mn}N_{mn}\\ \bar{\alpha}M_{mn}N_{mn} & 0\end{pmatrix},\begin{pmatrix}N_{mn} & 0\\ 0 & N_{mn}\end{pmatrix}.$$

This pair, after an obvious rearrangement, becomes the direct sum of n identical pairs (107). Q.E.D.

Lemma 4.12$_-$. *The ϕ_--pair (iii) in Lemma 3.3 is $\mathcal{C}$ congruent to the direct sum of $\frac{1}{2}n$ (n even) identical pairs*

$$\begin{bmatrix} 0 & \beta M_m & 0 & 0 \\ \beta' M_m' & 0 & 0 & 0 \\ 0 & 0 & 0 & I \\ 0 & 0 & I & 0 \end{bmatrix}, F_{2m}, \qquad (\beta \text{ has a positive imaginary part}). \tag{110}$$

Proof. The general form of C in the ϕ_--pair (iii) is

$$C = \psi(\bar{X}_1, \ldots, \bar{X}_m)N_{mn} \dotplus (-\psi(X_1, \ldots, X_m)N_{mn}) \tag{111}$$

(see Section 3 (III)). Because C is skew-symmetric and nonsingular, all X_i's are skew-symmetric and X_1 is nonsingular. By Lemma 4.3, there is a nonsingular $P = \psi(P_1, \ldots, P_m)$ such that

$$P\psi(X_1, \ldots, X_m)\psi(P_1', \ldots, P_m') = \psi(F, 0, \ldots, 0). \tag{112}$$

Transform (iii) ($s = -1$) by $P \dotplus \bar{P}$. Using (111) and (112) after the transformation, we get the pair

$$\begin{pmatrix} 0 & -\alpha\psi(F, F, 0, \ldots, 0)N_{mn} \\ \bar{\alpha}\psi(F, F, 0, \ldots, 0)N_{mn} & 0 \end{pmatrix},$$
$$\times \begin{pmatrix} \psi(F, 0, \ldots, 0)N_{mn} & 0 \\ 0 & -\psi(F, 0, \ldots, 0)N_{mn} \end{pmatrix} \tag{113}$$

Notice that here the order n of F is even. Dividing the two matrices in (113) into $4m \times 4m$ blocks of the same size, then rearranging them in the order $(1, 3, \ldots, 4m-1, 2, 4, \ldots, 4m)$, we obtain

$$\begin{bmatrix} 0 & 0 & 0 & -\alpha MN \\ 0 & 0 & \bar{\alpha}MN & 0 \\ 0 & \alpha MN & 0 & 0 \\ -\bar{\alpha}MN & 0 & 0 & 0 \end{bmatrix}, \begin{bmatrix} 0 & 0 & N & 0 \\ 0 & 0 & 0 & -N \\ -N & 0 & 0 & 0 \\ 0 & N & 0 & 0 \end{bmatrix}.$$
$$\left(M = M_{m, \frac{n}{2}}, \qquad N = N_{m, \frac{n}{2}}\right).$$

Obviously this pair can be rearranged into the direct sum of $n/2$ identical pairs

$$\begin{bmatrix} 0 & 0 & 0 & -\alpha M_m N_m \\ 0 & 0 & \bar{\alpha}M_m N_m & 0 \\ 0 & \alpha M_m N_m & 0 & 0 \\ -\bar{\alpha}M_m N_m & 0 & 0 & 0 \end{bmatrix}, \begin{bmatrix} 0 & 0 & N_m & 0 \\ 0 & 0 & 0 & -N_m \\ -N_m & 0 & 0 & 0 \\ 0 & N_m & 0 & 0 \end{bmatrix}. \tag{114}$$

Now take a real nonsingular G such that

$$GM_m^2G^{-1} = M_m'.$$

This is apparently possible. Transforming (114) by

$$2^{-1/2}\begin{bmatrix} 0 & -\bar{\alpha}G'^{-1}N_mM_m & 0 & \bar{\alpha}G'^{-1}N_mM_m \\ G & 0 & G & 0 \\ 0 & \bar{\alpha}^{-1}GM_m^{-1} & 0 & \bar{\alpha}^{-1}GM_m^{-1} \\ -G'^{-1}N_m & 0 & G'^{-1}N_m & 0 \end{bmatrix},$$

we get (110) with $\beta = -\bar{\alpha}^2$. The number β has positive imaginary part, for the real and imaginary parts of α both are positive. Q.E.D.

Lemma 4.13$_+$. *The ϕ_+-pair (iv) in Lemma 3.3 is $\mathcal{C}$ congruent to a direct sum of n identical pairs*

$$\epsilon J_mN_m, \epsilon N_m \qquad (\epsilon = \pm 1) \tag{115}$$

when m is even; it is $\mathcal{C}$ congruent to the direct sum of n identical pairs

$$J_mN_m, N_m \tag{116}$$

when m is odd.

Lemma 4.13$_-$. *The ϕ_--pair (iv) in Lemma 3.3 is $\mathcal{C}$ congruent to a direct sum of the 2n identical pairs*

$$\epsilon(N_{m/2} \dotplus J_{m/2}N_{m/2}), F_{m/2} \qquad (\epsilon = \pm 1) \tag{117}$$

when m is even; it is $\mathcal{C}$ congruent to the direct sum of n/2 (n even) identical pairs

$$J_mN_m \dotplus N_mJ_m, F_m \tag{118}$$

when m is odd.

Lemma 4.14$_+$. *The ϕ_+-pair (v) in Lemma 3.3 is $\mathcal{C}$ congruent to the direct sum of n identical pairs*

$$N_m, J_mN_m \tag{119}$$

if m is even; it is $\mathcal{C}$ congruent to the direct sum of n identical pairs

$$\epsilon N_m, \epsilon J_mN_m \qquad (\epsilon = \pm 1) \tag{120}$$

if m is odd.

Lemma 4.14$_-$. *The ϕ_--pair (v) in Lemma 3.3 is $\mathcal{C}$ congruent to the direct sum of n/2 (n even) identical pairs*

$$\begin{pmatrix} 0 & I_m \\ I_m & 0 \end{pmatrix}, \begin{pmatrix} 0 & J_m \\ -J_m' & 0 \end{pmatrix} \tag{121}$$

if m is even; it is $\mathcal{C}$ *congruent to the direct sum of n identical pairs*

$$\epsilon N_m, \quad \begin{bmatrix} 0 & & & & & 1 & 0 \\ & & & & \cdot^{\cdot^{\cdot}} & & \\ & & & 1 & & & \\ & & -1 & 0 & & & \\ & \cdot^{\cdot^{\cdot}} & & & & & \\ -1 & 0 & & & & 0 & \\ 0 & & & & & & \end{bmatrix} \quad (\epsilon = \pm 1) \tag{122}$$

if m is odd.

In order to prove Lemmas $4.13_{\pm}$ and $4.14_{\pm}$, we need a new notation and a lemma.

We will use $\chi_s\,(X_1, \ldots, X_m)$ to denote the square matrix

$$\begin{bmatrix} X_1 & X_2 & \cdots & \cdots & \cdots & X_{m-1} & X_m \\ & s\overline{X}_1 & s\overline{X}_2 & \ddots & & & s\overline{X}_{m-1} \\ & & \ddots & \ddots & \ddots & & \vdots \\ & & & \ddots & \ddots & & \vdots \\ 0 & & & & & \ddots & \vdots \end{bmatrix} \quad (s = \pm 1)$$

where each X_i is an $n \times n$ matrix. χ_s is χ_+ or χ_- if $s = +1$ or $s = -1$. The following relation is easily verified:

$$\chi_s(X_1, \ldots, X_m)J_{mn} = sJ_{mn}\chi_s(\overline{X}_1, \ldots, \overline{X}_m) \tag{123}$$

Lemma 4.15. *Let* $C = \chi_s(X_1, \ldots, X_m)$ $(s = \pm 1)$, X_1 *be nonsingular.*

(α) *If* $C' = sC$ $(s = \pm 1)$, *then there is a nonsingular* $P = \chi_+(P_1, \ldots, P_m)$ *such that*

$$PC\overline{P}' = \chi_s(X_0, 0, \ldots, 0)N_{mn}, \tag{124}$$

where

$$X_0 = \begin{cases} D = \operatorname{diag}(\pm 1, \ldots, \pm 1) & \textit{if m is even,} \\ I & \textit{if m is odd and } s = +1, \\ F & \textit{if m is odd and } s = -1. \end{cases}$$

(β) *If* $C' = \overline{C}$, *there is a nonsingular* $P = \chi_+(P_1, \ldots, P_m)$, *such that*

$$PC\overline{P}' = \chi_s(X_0, 0, \ldots, 0)N_{mn} \tag{125}$$

where

$$X_0 = \begin{cases} I & \textit{if m is even and } s = +1, \\ F & \textit{if m is even and } s = -1, \\ D = \operatorname{diag}(\pm 1, \ldots, \pm 1) & \textit{if m is odd.} \end{cases}$$

Proof. We first prove the following proposition. There is a matrix $Q = \chi_+(I, Q_2, \ldots, Q_m)$ such that

$$QCQ' = \chi_s(X_1, 0, \ldots, 0)N_{mn} \qquad \text{if} \quad C' = sC,$$

$$QC\overline{Q}' = \chi_s(X_1, 0, \ldots, 0)N_{mn} \qquad \text{if} \quad C' = \overline{C}.$$

The proposition is proved by induction on m. It is clearly true for $m = 1$. Consider $m > 1$ and assume the proposition true for $m - 1$.

We can write

$$C = \begin{pmatrix} * & Y \\ Z & 0 \end{pmatrix}.$$

Here $Y = \chi_s(X_1, \ldots, X_{m-1})N_{m-1,n}$, and Z is X_1 or $s\overline{X}_1$. By inspection we have $Y' = \overline{Y}$ or $Y' = sY$ if $C' = sC$ or $C' = \overline{C}$. Hence by the induction hypothesis there exists a $R = \chi_+(I, R_2, \ldots, R_m)$ such that

$$RYR' = \chi_s\Big(X_1, \overbrace{0, \ldots, 0}^{m-2}\Big)N_{m-1,n} \qquad \text{if} \quad C' = sC, \tag{126}$$

$$RY\overline{R}' = \chi_s\Big(X_1, \overbrace{0, \ldots, 0}^{m-2}\Big)N_{m-1,n} \qquad \text{if} \quad C' = \overline{C}. \tag{127}$$

Let

$$V = \chi_+(I, R_2, \ldots, R_{m-1}, 0).$$

Dividing the matrix into blocks, we have

$$V = \begin{pmatrix} R & * \\ 0 & I \end{pmatrix}, \qquad V = \begin{pmatrix} I & * \\ 0 & \overline{R} \end{pmatrix}.$$

Thus

$$VCV' = \begin{pmatrix} R & * \\ 0 & I \end{pmatrix}\begin{pmatrix} * & Y \\ Z & 0 \end{pmatrix}\begin{pmatrix} I & 0 \\ * & \overline{R}' \end{pmatrix} = \begin{pmatrix} 0 & RY\overline{R}' \\ Z & 0 \end{pmatrix}, \tag{128}$$

$$VC\overline{V}' = \begin{pmatrix} R & * \\ 0 & I \end{pmatrix}\begin{pmatrix} * & Y \\ Z & 0 \end{pmatrix}\begin{pmatrix} I & 0 \\ * & R' \end{pmatrix} = \begin{pmatrix} 0 & RYR' \\ Z & 0 \end{pmatrix} \tag{129}$$

if $C' = sC$ (128) is symmetric or skew-symmetric, whereas $RY\overline{R}'$ has the special form (126). It follows that

$$VCV' = \chi_s(X_1, 0, \ldots, 0, W)N_{mn} \qquad \text{if} \quad C' = sC. \tag{130}$$

Similarly, we have

$$VC\overline{V}' = \chi_s(X_1, 0, \ldots, 0, W)N_{mn} \qquad \text{if} \quad C' = \overline{C}. \tag{131}$$

Corresponding to (130) and (131), we let, respectively,

$$Q = \chi_+\Big(I, 0, \ldots, 0, -\frac{s}{2}WX_1'^{-1}\Big)V,$$

$$Q = \chi_+\Big(I, 0, \ldots, 0, -\frac{1}{2}W\overline{X}_1'^{-1}\Big)V.$$

Then Q has the form described in the proposition. Moreover, we have

$$QCQ' = \chi_s(X_1, 0, \ldots, 0)N_{mn} \qquad \text{if} \quad C' = sC,$$

$$QC\overline{Q}' = \chi_s(X_1, 0, \ldots, 0)N_{mn} \qquad \text{if} \quad C' = \overline{C}.$$

The proposition is proved.

Now we return to Lemma 4.15. First of all, we choose Q according to the preceding proposition. Observe the following facts.

(α) If $C' = sC$, then

$$X_1' = \overline{X}_1 \qquad \text{if } m \text{ is even},$$
$$X_1' = sX_1 \qquad \text{if } m \text{ is odd}.$$

(β) If $C' = sC$, then

$$X_1' = sX_1 \qquad \text{if } m \text{ is even},$$

$$X_1' = \overline{X}_1 \qquad \text{if } m \text{ is odd}.$$

Take a nonsingular P such that $PX_1P' = I$ (if X_1 is symmetric) or $PX_1P' = F$ (if X_1 is skew-symmetric) or $PX_1\overline{P}' = D = \operatorname{diag}(\pm 1, \ldots, \pm 1)$ (if X_1 is hermitian). Let $P = \chi_+(P_1, 0, \ldots, 0)Q$, then P has the form required in Lemma 4.15 and we can check that (124) and (125) hold in each case. Lemma 4.15 is proved.

Proof of Lemma 4.13 *and* 4.14. In Section 3 (iv) and (v) we got the general form of C in the ϕ_s-pair (iv) and (v). Now this formula can be written as

$$C = \chi_s(\overline{X}_1, \ldots, \overline{X}_m)N_{mn}, \qquad C \text{ nonsingular}.$$

In (iv), $C' = sC$. By virtue of Lemma 4.15, there is a nonsingular $P = \chi_+(P_1, \ldots, P_m)$ such that

$$P\overline{C}P' = \chi_s(X_0, 0, \ldots, 0)N_{mn}.$$

In (v), $C' = \overline{C}$. By the same lemma, there is a nonsingular $P = \chi_+(P_1, \ldots, P_m)$ such that

$$P\overline{C}\overline{P}' = \chi_s(X_0, 0, \ldots, 0)N_{mn}.$$

In both cases (iv) and (v), we have $PJ_{mn} = J_{mn}\overline{P}$ by (123). Now transforming the pairs (iv) and (v) by P, the two matrices $J_{mn}C, \overline{C}$ of (iv) become

$$\left.\begin{aligned} PJ_{mn}C\overline{P}' &= J_{mn}\overline{P}C\overline{P}' = J_{mn}\chi_s(X_0, 0, \ldots, 0)N_{mn}, \\ P\overline{C}\overline{P}' &= \chi_s(X_0, 0, \ldots, 0)N_{mn}. \end{aligned}\right\} \tag{132}$$

The two matrices in (v) become

$$\left.\begin{aligned} P\overline{C}\overline{P}' &= \chi_s(X_0, 0, \ldots, 0)N_{mn}, \\ PJ_{mn}CP' &= J_{mn}\overline{P}CP' = J_{mn}\chi_s(X_0, 0, \ldots, 0)N_{mn}. \end{aligned}\right\} \tag{133}$$

The pair (132) has the following cases.

S	m	X_0	The pair (132)
+	even	$D = \operatorname{diag}(\pm 1, \ldots, \pm 1)$	$J_{mn}\chi_+(D,0,\ldots,0)N_{mn}, \chi_+(D,0,\ldots,0)N_{mn}$
+	odd	I	$J_{mn}N_{mn}, N_{mn}$
−	even	$D = \operatorname{diag}(\pm 1, \ldots, \pm 1)$	$J_{mn}\chi_-(D,0,\ldots,0)N_{mn}\chi_-(D,0,\ldots,0)N_{mn}$
−	odd	F	$J_{mn}\chi_-(F,0,\ldots,0)N_{mn}, \chi_-(F,0,\ldots,0)N_{mn}$

(134)

The pair (133) has the following cases.

S	m	X_0	The pair (132)
+	even	I	$N_{mn}, J_{mn}N_{mn}$
+	odd	$D = \operatorname{diag}(\pm 1, \ldots, \pm 1)$	$\chi_+(D,0,\ldots,0)N_{mn}, J_{mn}\chi_+(D,0,\ldots,0)N_{mn}$
−	even	F	$\chi_-(F,0,\ldots,0)N_{mn}, J_{mn}\chi_-(F,0,\ldots,0)N_{mn}$
−	odd	$D = \operatorname{diag}(\pm 1, \ldots, \pm 1)$	$\chi_-(D,0,\ldots,0)N_{mn}, J_{mn}\chi_-(D,0,\ldots,0)N_{mn}$

(135)

Lemma 4.13_+ is a direct consequence of the first two lines in (134). Lemma 4.14_+ is a direct consequence of the first two lines in (135). Thus the two lemmas are proved.

After rearrangement, the matrix pair of the third line in (134) becomes the direct sum of n identical pairs

$$\epsilon \operatorname{diag}(1,-1,\ldots,1,-1)J_m N_m, \epsilon \operatorname{diag}(1,-1,\ldots,1,-1)N_m$$

Rearranging their rows and columns by the order $(1,3,\ldots,m-1,2,4,\ldots,2m)$, we obtain

$$\epsilon\begin{pmatrix} N_{m/2} & 0 \\ 0 & -J_{m/2}N_{m/2} \end{pmatrix}, \epsilon\begin{pmatrix} 0 & N_{m/2} \\ -N_{m/2} & 0 \end{pmatrix}.$$

Transforming these matrices by $N_{m/2} \dotplus \epsilon I_{m/2}$, we have

$$\epsilon N_{m/2} \dotplus \epsilon(-J_{m/2}N_{m/2}), F_{m/2}.$$

Transforming by $\operatorname{diag}(1,-1,\ldots,1,-1)$, we have $(-\epsilon(N_{m/2} \dotplus J_{m/2}N_{m/2}), F)$ (if $m/2$ is even) or $(\epsilon(N_{m/2} \dotplus J_{m/2}N_{m/2}), -F)$ (if $m/2$ is odd). The former already has the form (117). In the latter case, transforming by $I_{m/2} \dotplus (-I_{m/2})$, we get (117). This proves the case m even of Lemma 4.13_-.

The fourth line in (134) after rearrangement becomes the direct sum of the following $n/2$ identical pairs:

$$\operatorname{diag}(1,-1,-1,1,1,-1,\ldots,1,-1)J_{2m}^2 N_{2m},$$

$$\operatorname{diag}(1,-1,-1,1,1,-1,\ldots,1,-1)N_{2m}.$$

Rearranging them by the order $(1,3,\ldots,2m-1,2,4,\ldots,2m)$, we obtain

$$\operatorname{diag}(1,-1,1,-1,\ldots,1,-1)\begin{pmatrix} 0 & J_mN_m \\ J_mN_m & 0 \end{pmatrix}, \quad \operatorname{diag}(1,-1,1,-1,\ldots,1,-1)N_{2m}.$$

Transforming by $\operatorname{diag}(\overbrace{1,-1,1,-1,\ldots}^{m}) \dotplus I_m$, we have

$$\begin{pmatrix} 0 & J_mN_m \\ J_mN_m & 0 \end{pmatrix}, \begin{pmatrix} 0 & N_m \\ -N_m & 0 \end{pmatrix}.$$

Transforming by $2^{-1/2}\left(\begin{smallmatrix} I_m & I_m \\ -N_m & N_m \end{smallmatrix}\right)$, we have

$$J_mN_m \dotplus (-N_mJ_m), F. \tag{136}$$

Let

$$\Delta_m = \operatorname{diag}(\overbrace{1,-1,-1,1,1,-1,-1,1,\ldots,1,-1,-1,1}^{4k},1) \qquad \text{if} \quad m = 4k+1,$$

$$\Delta_m = \operatorname{diag}(\overbrace{1,1,-1,-1,\ldots,1,1,-1,-1}^{4k},1,1-1) \qquad \text{if} \quad m = 4k+3.$$

We can prove that

$$\Delta_mJ_mN_m\Delta_m = J_mN_m, \qquad \Delta_mN_mJ_m\Delta_m = -J_mN_m,$$

m being any even integer. Thus, transforming (136) by $\Delta_m \dotplus \Delta_m$, we get (138). This proves the case m odd of Lemma 4.13_ .

The third line in (135) after rearrangement becomes the direct sum of the $n/2$ identical pairs:

$$\operatorname{diag}(1,-1,-1,1,1,-1,\ldots,-1,1)N_{2m},$$

$$\operatorname{diag}(1,-1,-1,1,1,-1,\ldots,-1,1)J_{2m}^2N_{2m},$$

Rearranging them according to the order $(1,3,\ldots,2m-1,2,4,\ldots,2m)$ we obtain

$$\operatorname{diag}(1,-1,1,-1,\ldots,1,-1)\begin{pmatrix} 1 & N_m \\ -N_m & 0 \end{pmatrix},$$

$$\operatorname{diag}(1,-1,1,-1,\ldots,1,-1)\begin{pmatrix} 0 & J_mN_m \\ -j_mN_m & 0 \end{pmatrix}.$$

Transforming by $I_m \dotplus N_m$, we get (121). This proves the case $m =$ even of Lemma 4.14_ .

The matrix pair of the fourth line in (135) after rearrangement becomes the direct sum of n identical pairs,

$$\epsilon\operatorname{diag}(1,-1,1,-1,\ldots,1,-1,1)N_m, \qquad \epsilon\operatorname{diag}(1,-1,1,-1,\ldots,1,-1,1)J_mN_m. \tag{137}$$

If $m = 4k+1$, transform (137) by

$$\operatorname{diag}(\overbrace{1,-1,\ldots,1,-1}^{2k}) \dotplus I_{2k+1};$$

if $m = 4k+3$, transform (137) by

$$\operatorname{diag}(\overbrace{-1,1,\ldots,-1,1}^{2k+2}) \dotplus I_{2k+1}.$$

In either case we get (122). This proves the case $m =$ odd of Lemma 4.14$_-$.

5. Canonical Forms of a General ϕ_+-pair

In this section we will find canonical forms for general ϕ_+-pairs and solve the uniqueness problem in this case. The same questions for ϕ_--pairs will be answered in the next section. From now on a ϕ_+-pair will be denoted by (H,S) (H hermitian and S symmetric).

Combining Lemmas 3.3, 4.10, 4.11, 4.12, 4.13, and 4.14 and noticing at the same time that the pair (95) in Lemma 4.11 and the pair (107) in Lemma 4.12 are the same thing ((107) implies (95) as a special case if the condition on α in (107) is lessened as the real part of $\alpha \geqslant 0$, the imaginary part > 0), we obtain:

Theorem 5.1. *Any ϕ_+-pair (H,S) is $\mathcal{C}$ congruent to a matrix pair of the form*

$$(H,S) \approx \Big(\sum_\nu{}' H_\nu \dotplus 0, \sum_\nu{}' S_\nu \dotplus 0\Big), \tag{138}$$

where each summand (H_ν, S_ν) takes one of the following seven forms.

(1°) $\epsilon a M_m N_m, \epsilon N_m \quad (\epsilon = \pm 1, a > 0)$

(2°) $\begin{pmatrix} 0 & \alpha M_m N_m \\ \bar{\alpha} M_m N_m & 0 \end{pmatrix}, N_m \dotplus N_m \quad (\operatorname{Re}\alpha \geqslant 0, \operatorname{Im}\alpha > 0)$

(3°) $\epsilon J_{2m} N_{2m}, \epsilon N_{2m} \quad (\epsilon = \pm 1)$

(4°) $J_{2m+1} N_{2m+1}, N_{2m+1}$

(5°) $N_{2m}, J_{2m} N_{2m}$

(6°) $\epsilon N_{2m+1}, \epsilon J_{2m+1} N_{2m+1} \quad (\epsilon = \pm 1)$

(7°) $\begin{pmatrix} 0 & K_m \\ K'_m & 0 \end{pmatrix}, \begin{pmatrix} 0 & L_m \\ L'_m & 0 \end{pmatrix}.$

The pair on the right-hand side of (138) will be called the $\mathcal{C}$ canonical form of the pair (H,S). In order to prove the uniqueness, we need some knowledge of invariants of hermitian matrix pencils under conjunctive transformations. The knowledge about this is already available. For details, the reader is referred to [3]. For convenience, we give a brief account of the relevant results.

A matrix pencil $\lambda H_1 + \mu H_2$ is hermitian if H_1, H_2 are two square hermitian matrices of the same size. A transformation of the form

$$H_1 \to P H_1 \bar{P}', \qquad H_2 \to P H_2 \bar{P}' \qquad (P \text{ nonsingular}) \tag{139}$$

is called a conjunctive transformation of the matrix pencil $\lambda H_1 + \mu H_2$. If two matrix pencils can be taken into each other by a conjunctive transformation, then we say they are conjunctive. When we choose a fixed P and make the transformation (139) we say: transform the pencil $\lambda H_1 + \mu H_2$ by P.

The set of the minimal indices for the rows of a hermitian pencil is the same as that for the columns. The collection of these common integers is called the minimal indices of the hermitian pencil.

The complex elementary divisors of a hermitian pencil $\lambda H_1 + \mu H_2$ can be grouped into pairs of the form: $(\lambda\alpha + \mu)^m, (\lambda\bar{\alpha} + \mu)^m$ (α is a complex number).

The following theorem is known.

Theorem I. *If an elementary divisor e of the hermitian pencil $\lambda H_1 + \mu H_2$ appears exactly n times, where e is $(\lambda a + \mu)^m$ (a real) or μ^m, or λ^m, then there is a conjunctive transformation which takes the pencil $\lambda H_1 + \mu H_2$ into a direct sum such that one of its summands is*

$$\sum_{i=1}^{n}{}' \epsilon_i(\lambda a M_m N_m + \mu N_m), \quad \text{or} \quad \sum_{i=1}^{n}{}' \epsilon_i(\lambda J_m N_m + \mu N_m), \quad \text{or} \quad \sum_{i=1}^{n} \epsilon_i(\lambda N_m + \mu J_m N_m).$$

Here $\epsilon_i = \pm(i = 1, \ldots, n)$ and the integer $\epsilon_1 + \cdots + \epsilon_n$ is uniquely determined by the pencil $\lambda H_1 + \mu H_2$.

The integer $\epsilon_1 + \cdots + \epsilon_n$ in Theorem I will be called the signature associated with the elementary divisor e. Thus each distinct real elementary divisor of a hermitian pencil has a signature. The collection of these signatures is called the signature set of the pencil $\lambda H_1 + \mu H_2$ (notice that complex elementary divisors have no signatures).

Theorem II (known). *The minimal indices, the elementary divisors and the signatures form the complete invariants of a hermitian pencil.*

The invariants in Theorem II are called classical invariants of the hermitian pencil.

To sum up the above known results, we are able to prove the uniqueness of the $\mathcal{C}$ canonical form (138) for a ϕ_+-pair.

Given a ϕ_+-pair (H, S), the hermitian pencil of matrices

$$\begin{pmatrix} \lambda H & \mu S \\ \mu\bar{S} & \lambda\bar{H} \end{pmatrix} \tag{140}$$

is called the pencil associated with the pair (H, S). If (H, S) and (H^*, S^*) are $\mathcal{C}$ congruent, then their corresponding associated pencils are conjunctive, because transforming the pair (H, S) by P corresponds to transforming the pencil (140) by $P \dotplus \bar{P}$.

Now we are going to find the classical invariants of the pencil (140). By the foregoing discussion it is enough to consider the $\mathcal{C}$ canonical form of (H, S). After substitution we obtain a pencil

$$\sum_{\nu}{}' \begin{pmatrix} \lambda H_\nu & \mu S_\nu \\ \mu \bar{S}_\nu & \lambda \bar{H}_\nu \end{pmatrix} \dotplus 0, \tag{141}$$

since the direct sum passes through the matrix notation. In (141) each pair takes one of the forms (1°)–(7°). Thus the problem boils down to the associated pencil of each pair (1°)–(7°).

First consider the associated pencil of (7°). It is easily seen that this pencil becomes

$$\begin{pmatrix} \lambda K_m & \mu L_m \\ \mu L_m & \lambda K_m \end{pmatrix} \dotplus \begin{pmatrix} \lambda K'_m & \mu L'_m \\ \mu L'_m & \lambda K'_m \end{pmatrix}$$

after a (not necessarily symmetric) rearrangement of its rows and columns. We have already known[4] that the above pencil has no elementary divisors and contributes only m, m to the minimal indices. Moreover, the order of the summand 0 in (141) is twice that of the summand 0 in (138); therefore, each minimal index of the pencil (140) appears an even number of times.

Next consider the associated pencil of (2°). It is easily seen that the elementary divisors for this pencil are

$$(\lambda\alpha + \mu)^m, (-\lambda\alpha + \mu)^m, (\lambda\bar{\alpha} + \mu)^m, (-\lambda\bar{\alpha} + \mu)^m.$$

Consider the associated pencils of (1°), (3°), (4°), (5°), (6°). All of these pencils have the form

$$\begin{pmatrix} \lambda S_1 & \mu S_2 \\ \mu S_2 & \lambda S_1 \end{pmatrix},$$

where S_1, S_2 are real symmetric. Transforming by $2^{-1/2}\left(\begin{smallmatrix} I & I \\ -I & I \end{smallmatrix}\right)$, we have

$$(\lambda S_1 + \mu S_2) \dotplus (\lambda S_1 - \mu S_2). \tag{142}$$

In case (1°), $S_1 = \epsilon a M_m N_m$, $S_2 = \epsilon N_m$. Substituting in (142), we get

$$\epsilon(\lambda a M_m N_m + \mu N_m) \dotplus (-\epsilon)(-\lambda a M_m N_m + \mu N_m).$$

This pencil contributes elementary divisors $(\lambda a + \mu)^m, (-\lambda a + \mu)^m$ and the associated signatures $\epsilon, -\epsilon$.

In cases (3°), (4°), (5°), and (6°), (142) takes the form

$$\epsilon(\lambda J_{2m} N_{2m} + \mu N_{2m}) \dotplus \epsilon(\lambda J_{2m} N_{2m} - \mu N_{2m}),$$
$$(\lambda J_{2m+1} N_{2m+1} + \mu N_{2m+1}) \dotplus (\lambda J_{2m+1} N_{2m+1} - \mu N_{2m+1}),$$
$$(\lambda N_{2m} + \mu J_{2m} N_{2m}) \dotplus (\lambda N_{2m} - \mu J_{2m} N_{2m}),$$
$$\epsilon(\lambda N_{2m+1} + \mu J_{2m+1} N_{2m+1}) \dotplus \epsilon(\lambda N_{2m+1} - \mu J_{2m+1} N_{2m+1}),$$

respectively. Representing the second summands of these pencils by diag(1, −1,

[4]See [2], p. 13.

$1, -1, \ldots)$ we get

$$\epsilon(\lambda J_{2m}N_{2m} + \mu N_{2m}) \dotplus \epsilon(\lambda J_{2m}N_{2m} + \mu N_{2m}),$$
$$(\lambda J_{2m+1}N_{2m+1} + \mu N_{2m+1}) \dotplus (-1)(\lambda J_{2m+1}N_{2m+1} + \mu N_{2m+1}),$$
$$(\lambda N_{2m} + \mu J_{2m}N_{2m}) \dotplus (-1)(\lambda N_{2m} + \mu J_{2m}N_{2m}),$$
$$\epsilon(\lambda N_{2m+1} + \mu J_{2m+1}N_{2m+1}) + \epsilon(\lambda N_{2m+1} \dotplus \mu J_{2m+1}N_{2m+1}).$$

The contributions of these four pencils are

elementary divisors	μ^{2m}	twice with associated signature	2ϵ
elementary divisors	μ^{2m+1}	twice with associated signature	0
elementary divisors	λ^{2m}	twice with associated signature	0
elementary divisors	λ^{2m+1}	twice with associated signature	2ϵ

Theorem 5.2 follows easily from the results just obtained and is stated here without proof.

Theorem 5.2. *The necessary and sufficient condition for a hermitian pencil to be conjunctive to a pair of the form*

$$\begin{pmatrix} \lambda H & \mu S \\ \mu \bar{S} & \lambda \bar{H} \end{pmatrix}$$

is that its classical invariants satisfy the following conditions.

(α) *Each minimal index appears an even number of times.*
(β) *The set of complex elementary divisors can be grouped into "quadruples":*

$$(\lambda\alpha + \mu)^m, (-\lambda\alpha + \mu)^m, (\lambda\bar{\alpha} + \mu)^m, (-\lambda\bar{\alpha} + \mu)^m. \tag{143}$$

(γ) *The set of elementary divisors of the form $(\lambda a + \mu)^m$ ($a \neq 0$ real) can be grouped into pairs: $(\lambda a + \mu)^m, (-\lambda a + \mu)^m$. The signatures of these two elementary divisors differ only in sign.*
(δ) *Each elementary divisor of the form μ^{2m} and λ^{2m+1} appears an even number of times; the associated signatures and the number of times of appearance are congruent mod 4.*
(ϵ) *Each elementary divisor of the form μ^{2m+1} and λ^{2m} appears an even number of times; the associated signatures are zero.*

Theorem 5.3. *A ϕ_+-pair (H, S) uniquely determines its $\mathcal{C}$ canonical form* (138) *through the following properties of the pair.*

(i) *If the zero summand has order g, then there are exactly $2g$ zeros in the set of the minimal indices of the associated pencil.*
(ii) *If for some m, the pair* (7°) *appears exactly n times, then m appears exactly $2n$ times in the set of the minimal indices of the associated pencil.*
(iii) *If for some m and complex α the pair* (2°) *appears exactly n times, then the quadruple* (143) *appears exactly n times in the set of elementary divisors of the associated pencil.*

(iv) *If for some m and $a > 0$ the pair (1°) appears exactly n times with signs $\epsilon_1, \ldots, \epsilon_n$, then the pair $(\lambda a + \mu)^m, (-\lambda a + \mu)^m$ appears exactly n times in the set of elementary divisors of the associated pencil and the signature of the elementary divisor $(\lambda a + \mu)^m$ is $\epsilon_1 + \cdots + \epsilon_n$.*

(v) *If for some m the pair (3°) (or (6°)) appears exactly n times with signs $\epsilon_1, \ldots, \epsilon_n$, then the elementary divisor μ^m (or λ^{2m+1}) appears exactly $2n$ times with signature $2(\epsilon_1 + \cdots + \epsilon_n)$.*

(vi) *If for some m the pair (4°) (or (5°)) appears n times, then the elementary divisor μ^{2m+1} (or λ^{2m}) appears exactly $2n$ times.*

(vii) *The associated pencil has no other minimal indices except for those in (i) and (ii); it has no other elementary divisors except for those in (iv), (v), and (vi).*

Theorem 5.4. *The necessary and sufficient condition for two ϕ_+-pairs to be $\mathcal{C}$ congruent is that their associated matrix pencils are conjunctive.*

Remark. As a special case of Theorem 5.1, we consider a pair (H, S) where both H and S are nonsingular. The zero summand does not appear at all in the canonical form (138) and only the summands (1°) and (2°) possibly appear. Hence the $\mathcal{C}$ canonical form of (H, S) is a direct sum of certain matrix pairs of type (1°) and (2°). This canonical form was obtained by Hua in his paper [4]. In his results, however, the summand (2°) has a plus or minus sign (when α is not purely imaginary). It can be seen directly that this sign can be dispensed with by transforming the pair (2°) by $iI_m \dotplus (-iI_m)$, because this transformation changes the sign of (2°).

6. Canonical Forms for a General ϕ_--pair.

We use (H, T) to denote a ϕ_--pair, H hermitian, T skew-symmetric. In combining Lemma 3.3, 4.10, 4.11, 4.12, 4.13, and 4.14, we obtain

Theorem 6.1. *Each ϕ_--pair (H, T) is $\mathcal{C}$ congruent to a direct sum of pairs*

$$(H,T) \approx \Big(\sum_{\nu}' H_\nu \dotplus 0, \sum_{\nu}' T_\nu \dotplus 0 \Big). \tag{144}$$

Here each summand (H_ν, T_ν) takes one of the following eight forms.

(1°) $-a^2 M_m N_m \dotplus N_m, F_m \quad (a > 0)$.

(2°) $\epsilon(b^2 M_m N_m \dotplus N_m), F_m \quad (\epsilon = \pm 1, b > 0)$.

(3°)

$$\begin{pmatrix} 0 & \beta M_m \\ \bar{\beta} M_m' & 0 \end{pmatrix} \dotplus \begin{pmatrix} 0 & I_m \\ I_m & 0 \end{pmatrix}, F_{2m} \quad \text{(the imaginary part of } \beta > 0).$$

(4°) $\epsilon(N_m \dotplus J_m N_m), F_m \quad (\epsilon = \pm 1)$.

(5°) $J_{2m+1} N_{2m+1} \dotplus N_{2m+1} J_{2m+1}, F_{2m+1}$.

(6°)

$$\begin{pmatrix} 0 & I_{2m} \\ I_{2m} & 0 \end{pmatrix}, \begin{pmatrix} 0 & J_{2m} \\ -J'_{2m} & 0 \end{pmatrix}.$$

(7°)

$$\epsilon N_{2m+1}, \epsilon \begin{bmatrix} 0 & & & & & 1 & 0 \\ & & & & \cdot^{\cdot^{\cdot}} & & \\ & & & 1 & & & \\ & & -1 & 0 & & & \\ & \cdot^{\cdot^{\cdot}} & & & & & \\ -1 & & & & & & \\ 0 & & & & & & 0 \end{bmatrix}.$$

(8°)

$$\begin{pmatrix} 0 & K_m \\ K'_m & 0 \end{pmatrix}, \begin{pmatrix} 0 & L_m \\ -L'_m & 0 \end{pmatrix}.$$

The right-hand side of (144) is called the $\mathcal{C}$ canonical form of the ϕ_--pair (H, T). In order to prove the uniqueness of the $\mathcal{C}$ canonical form, we introduce a hermitian pencil

$$\begin{pmatrix} \lambda H & \mu T \\ -\mu \overline{T} & \lambda \overline{H} \end{pmatrix}, \tag{145}$$

i.e., the associated pencil of the ϕ_--pair (H, T). As in Section 5, if two ϕ_--pairs are $\mathcal{C}$ congruent, then their associated pencils are conjunctive. Hence, to find the classical invariants of the pencil we could use the $\mathcal{C}$ canonical form (144) of (H, T). After substitution, we get the pencil

$$\sum_{\nu}{}' \begin{pmatrix} \lambda H_\nu & \mu T_\nu \\ -\mu \overline{T}_\nu & \lambda \overline{H}_\nu \end{pmatrix} \dot{+} 0, \tag{146}$$

since the direct sum passes through the matrix notation. Except for the zero summand, each summand of this pencil is the associated pencil of one of the pairs in (1°) to (8°).

As in section 5, the zero summand in (144) contributes $2g$ zeros in the minimal indices of (146) (assuming the zero summand has order g), and (8°) contributes the minimal indices m, m. Therefore, each minimal index of the associated pencil appears an even number of times.

It is easily seen by direct computation that the associated pencil of (1°) has exactly the following elementary divisors.

$$(\lambda ai + \mu)^m, (\lambda ai + \mu)^m, (-\lambda ai + \mu)^m, (-\lambda ai + \mu)^m.$$

The associated pencil of (3°) has the following elementary divisors

$$(\lambda \alpha + \mu)^m, (-\lambda \alpha + \mu)^m, (\lambda \overline{\alpha} + \mu)^m, (-\lambda \overline{\alpha} + \mu)^m,$$

each term appearing twice; here $\alpha = \sqrt{\beta}$ (either root is all right, since the real and imaginary parts of α are not zero).

Now we consider the associated pencils of (2°), (4°), (5°), (6°), and (7°).

The associated pencil of (2°) splits into a direct sum:

$$\begin{pmatrix} \lambda\epsilon b^2 M_m N_m & \mu I_m \\ \mu I_m & \lambda\epsilon N_m \end{pmatrix} \dotplus \begin{pmatrix} \lambda\epsilon N_m & -\mu I_m \\ -\mu I_m & \lambda\epsilon b^2 M_m I_m \end{pmatrix}. \tag{147}$$

Transforming the second summand by F_m, we obtain the first; it is then enough to consider the first summand. Transforming the first summand by $-b^{-1/2}I + b^{1/2}N_m$, we get

$$\begin{pmatrix} \lambda\epsilon b M_m N_m & \mu N_m \\ \mu N_m & \lambda\epsilon b N_m \end{pmatrix}. \tag{148}$$

Obviously, this pencil has elementary divisors $(\lambda b + \mu)^m$ and $(-\lambda b + \mu)^m$. According to Theorem I of Section 5, it is conjunctive to

$$\eta_1(\lambda b M_m N_m + \mu N_m) \dotplus \eta_2(-\lambda b M_m N_m + \mu N_m), \tag{149}$$

where $\eta_1, \eta_2 = \pm 1$. η_1, η_2 are determined as follows. Since (148) and (149) are conjunctive, for any real λ and μ, we have[5]

$$\operatorname{sig}\begin{pmatrix} \lambda\epsilon b M_m N_m & \mu N_m \\ \mu N_m & \lambda\epsilon b N_m \end{pmatrix} = \eta_1 \operatorname{sig}(\lambda b M_m N_m + \mu N_m) + \eta_2 \operatorname{sig}(-\lambda b M_m N_m + \mu N_m). \tag{150}$$

Taking $\lambda = 1$ and $\mu = \pm b$,

$$\operatorname{sig}\begin{pmatrix} \epsilon M_m N_m & \pm N_m \\ \pm N_m & \epsilon N_m \end{pmatrix} = \eta_1 \operatorname{sig}(M_m N_m \pm N_m) + \eta_2 \operatorname{sig}(-M_m N_m \pm N_m). \tag{151}$$

But the left-hand side is equal to $\epsilon \operatorname{sig}(J_m N_m \dotplus N_m)$, because

$$\begin{pmatrix} I & \mp\epsilon I \\ 0 & I \end{pmatrix}\begin{pmatrix} \epsilon M_m N_m & \pm N_m \\ \pm N_m & \epsilon N_m \end{pmatrix}\begin{pmatrix} I & 0 \\ \mp\epsilon I & I \end{pmatrix} = \epsilon(J_m N_m \dotplus N_m).$$

Clearly one of $\operatorname{sig} J_m N_m$ and $\operatorname{sig} N_m$ is zero and the other is one; hence the left-hand side of (151) is ϵ. Taking the different signs in (151) and adding the two equations, we obtain

$$2\epsilon = (\eta_1 - \eta_2)\operatorname{sig}((2I + J_m)N_m \dotplus J_m N_m). \tag{152}$$

Take a polynomial $f(\lambda)$ such that $f^2(\lambda) \equiv 2 + \lambda (\operatorname{mod} \lambda^m)$. Let $Q = f(J_m)$. Then $Q^2 = 2I + J_m$ and $QN_m = N_m Q'$. Consequently $(2I_m + J_m)N_m = Q^2 N_m = QN_m Q'$. It follows that $\operatorname{sig}((2I_m + J_m)N_m \dotplus J_m N_m) = \operatorname{sig}(N_m + J_m N_m) = 1$. From (152) we conclude that $2\epsilon = \eta_1 - \eta_2$, that is, $\eta_1 = \epsilon$, $\eta_2 = -\epsilon$.

We proved then that the two summands in (147) are conjunctive to

$$\epsilon(\lambda b M_m N_m + \mu N_m) \dotplus (-\epsilon)(-\lambda b M_m N_m + \mu N_m).$$

[5] $\operatorname{sig} A$ denotes the signature of a hermitian matrix A.

In other words, the elementary divisors of the associated pencil of (2°) are precisely

$$(\lambda b + \mu)^m, (\lambda b + \mu)^m, (-\lambda b + \mu)^m, (-\lambda b + \mu)^m$$

and the signatures of $(\lambda b + \mu)^m$ and $(-\lambda b + \mu)^m$ are 2ϵ and -2ϵ, respectively.

The associated pencil of (4°) splits into a direct sum

$$\begin{pmatrix} \lambda\epsilon N_m & \mu I \\ \mu I & \lambda\epsilon J_m N_m \end{pmatrix} \dot{+} \begin{pmatrix} \lambda\epsilon J_m N_m & -\mu I \\ -\mu I & \lambda\epsilon N_m \end{pmatrix}.$$

Transforming by $(N_m \dot{+} \epsilon I_m) \dot{+} (N_m \dot{+} \epsilon I_m)F_m$, we have

$$\epsilon\begin{pmatrix} \lambda N_m & \mu N_m \\ \mu N_m & \lambda J_m N_m \end{pmatrix} \dot{+} \epsilon\begin{pmatrix} \lambda N_m & \mu N_m \\ \mu N_m & \lambda J_m N_m \end{pmatrix}. \tag{153}$$

Let l_i denote the ith column of I_{2m}, $P = [l_1, l_3, \ldots, l_{2m-1}, l_2, l_4, \ldots, l_{2m}]$. Transforming (152) by $P \dot{+} P$, the resulting matrix pencil is easily found to be

$$\epsilon(\lambda J_{2m} N_{2m} + \mu N_{2m}) \dot{+} \epsilon(\lambda J_{2m} N_{2m} + \mu N_{2m}).$$

It follows that the associated pencil of (4°) has the two identical elementary divisors μ^{2m} with signature 2ϵ.

The pencil associated with (5°) splits into a direct sum

$$\begin{pmatrix} \lambda J_{2m+1} N_{2m+1} & \mu I \\ \mu I & \lambda N_{2m+1} J_{2m+1} \end{pmatrix} \dot{+} \begin{pmatrix} \lambda N_{2m+1} J_{2m+1} & -\mu I \\ -\mu I & \lambda J_{2m+1} N_{2m+1} \end{pmatrix}.$$

Transforming by

$$2^{-1/2}\begin{pmatrix} I & N_{2m+1} \\ -I & N_{2m+1} \end{pmatrix} \dot{+} 2^{-1/2}\begin{pmatrix} N_{2m+1} & -I \\ N_{2m+1} & I \end{pmatrix},$$

the result is a direct sum of four summands: two of them are $\lambda J_{2m+1} N_{2m+1} + \mu N_{2m+1}$, and the other two are $\lambda J_{2m+1} N_{2m+1} - \mu N_{2m+1}$. The latter becomes $-(\lambda J_{2m+1} N_{2m+1} + \mu N_{2m+1})$ after transforming by $\operatorname{diag}(1, -1, 1, -1, \ldots, 1, -1, 1)$. Hence the associated pencil of (5°) has four identical elementary divisors μ^{2m+1} with the signature zero.

The associated pencil of (6°) is

$$\begin{bmatrix} 0 & \lambda I_{2m} & 0 & \mu J_{2m} \\ \lambda I_{2m} & 0 & -\mu J'_{2m} & 0 \\ 0 & -\mu J_{2m} & 0 & \lambda I_{2m} \\ \mu J'_{2m} & 0 & \lambda I_{2m} & 0 \end{bmatrix}.$$

Transforming by

$$\begin{bmatrix} I_{2m} & -N_{2m} & 0 & 0 \\ 0 & 0 & -I_{2m} & N_{2m} \\ 0 & 0 & I_{2m} & N_{2m} \\ I_{2m} & -N_{2m} & 0 & 0 \end{bmatrix},$$

we obtain a direct sum of the two identical pencils:

$$\begin{pmatrix} \lambda N_{2m} & \mu J_{2m} N_{2m+1} \\ \mu J_{2m} N_{2m} & -\lambda N_{2m} \end{pmatrix}. \tag{154}$$

Pencil (154) is equal to the product

$$\frac{1}{2}\begin{pmatrix} -I & I \\ iI & iI \end{pmatrix}\begin{pmatrix} \lambda I + \mu i J_{2m} & 0 \\ 0 & \lambda I - \mu i J_{2m} \end{pmatrix}\begin{pmatrix} -N_{2m} & iN_{2m} \\ N_{2m} & iN_{2m} \end{pmatrix},$$

therefore the pencil (154) has two identical elementary divisors λ^{2m}. By Theorem I of Section 5, the pencil (154) is conjunctive to $\eta_1(\lambda N_{2m} + \mu J_{2m} N_{2m}) \dotplus \eta_2(\lambda N_{2m} + \mu J_{2m} N_{2m})$. On comparing the two hermitian matrices multiplied by μ we obtain

$$\operatorname{sig}\begin{pmatrix} 0 & J_{2m} N_{2m} \\ J_{2m} N_{2m} & 0 \end{pmatrix} = (\eta_1 + \eta_2)\operatorname{sig}(J_{2m} N_{2m})$$

The left-hand side is clearly zero, thus $\eta_1 + \eta_2 = 0$. Therefore, the associated pencil of (6°) has the four identical elementary divisors λ^{2m} with signature zero.

The associated pencil of (7°) is

$$\epsilon\begin{pmatrix} \lambda N_{2m+1} & \mu R \\ -\mu R & \lambda N_{2m+1} \end{pmatrix}. \tag{155}$$

Here R denotes the second matrix in (7°). Choose a sequence of signs $\epsilon_1, \dots, \epsilon_m$ ($\epsilon_i = \pm 1$) such that

$$D_1 = \operatorname{diag}(\epsilon_1, \epsilon_2, \dots, \epsilon_{m-1}, \epsilon_m, -\epsilon_{m-1}, \epsilon_m, \epsilon_{m-1}, \dots, \epsilon_2, \epsilon_1)$$

$$D_2 = \operatorname{diag}(-\epsilon_2, \epsilon_1, \dots, \epsilon_{m-2}, \epsilon_{m-1}, \epsilon_m, \epsilon_{m-1}, \epsilon_{m-2}, \dots, \epsilon_1, -\epsilon_2)$$

Then we have $D_i N_{2m+1} D_i = N_{2m+1} (i = 1, 2)$. $D_1 R D_2 = J_{2m+1} N_{2m+1}$. Transforming (155) by $D_1 \dotplus D_2$, we get

$$\epsilon\begin{pmatrix} \lambda N_{2m+1} & \mu J_{2m+1} N_{2m+1} \\ \mu J_{2m+1} N_{2m+1} & \lambda N_{2m+1} \end{pmatrix}.$$

Transforming this by $2^{-1/2}\left(\begin{smallmatrix} I & I \\ -I & I \end{smallmatrix}\right)$, we get $\epsilon(\lambda N_{2m+1} + \mu J_{2m+1} N_{2m+1}) \dotplus \epsilon(\lambda N_{2m-1} - \mu J_{2m+1} N_{2m+1})$. Then, transforming the second summand by $\operatorname{diag}(1, -1, 1, -1, \dots, 1, -1, 1)$, we get $\epsilon(\lambda N_{2m+1} + \mu J_{2m+1} N_{2m+1}) \dotplus \epsilon(\lambda N_{2m+1} + \mu J_{2m+1} N_{2m+1})$. It follows that the associated pencil of (7°) has two identical elementary divisors λ^{2m+1} with a signature 2ϵ.

We have now completed the computation of the classical invariants of the associated pencils. The following theorem is obvious from what we obtained above and will be stated without proof.

Theorem 6.2. *The necessary and sufficient condition for a hermitian pencil to be conjunctive to the pencil*

$$\begin{pmatrix} \lambda H & \mu T \\ -\mu \overline{T} & \lambda \overline{H} \end{pmatrix} \qquad (H \text{ hermitian, } T \text{ skew-symmetric})$$

is that the following conditions are satisfied.

(α) *Each minimal index appears an even number of times.*

(β) *Each complex elementary divisor appears an even number of times; the purely imaginary elementary divisors can be grouped into the pairs* $(\lambda ai + \mu)^m, (-\lambda ai + \mu)^m$; *(a nonzero real); the nonpurely imaginary elementary divisors can be grouped into quadruples of the form:*

$$\lambda(\alpha + \mu)^m, (-\lambda\alpha + \mu)^m, (\lambda\bar{\alpha} + \mu)^m, (-\lambda\bar{\alpha} + \mu)^m.$$

(γ) *Each elementary divisor of the form* $(\lambda b + \mu)^m$ *(b nonzero real) appears an even number of times, and the associated signature is congruent to the number of times of appearence mod 4. These elementary divisors can be grouped into pairs of the form* $(\lambda b + \mu)^m, (-\lambda b + \mu)^m$; *the signature of* $(\lambda b + \mu)^m$ *and that of* $(-\lambda b + \mu)^m$ *differ only in sign.*

(δ) *Each elementary divisor of the form* μ^{2m} *or* λ^{2m+1} *appears an even number of times; the associated signature is congruent to the number of times of appearance mod 4.*

(ϵ) *Each elementary divisor of the form* μ^{2m+1} *or* λ^{2m} *appears 0 (mod 4) times with the signature zero.*

Theorem 6.3. *The* $\mathcal{C}$ *canonical form* (144) *of a* ϕ_-*-pair* (H, T) *is determined uniquely by the following properties of* (H, T).

(i) *If the zero summand has order g, then the minimal index set of the associated pencil has 2g zeros.*

(ii) *If for some m, the summand* (8°) *appears exactly n times, then m appears 2n times in the minimal index set of the associated pencil.*

(iii) *If for some m and positive a the summand* (1°) *appears exactly n times, then the pair* $(\lambda ai + \mu)^m, (-\lambda ai + \mu)^m$ *appears 2n times in the elementary divisors of the associated pencil.*

(iv) *If for some m and complex number* β *(the imaginary part of* $\beta > 0$*) the summand* (3°) *appears n times, then the pair* $(\lambda\alpha + \mu)^m, (-\lambda\alpha + \mu)^m$ *appears 2n times in the elementary divisors of the associated pencil* ($\alpha^2 = \beta$).

(v) *If for some m and positive b the summand* (2°) *appears exactly n times with signs* $\epsilon_1, \ldots, \epsilon_n$ ($\epsilon_i = \pm 1$), *then* $(\lambda b + \mu)^m, (-\lambda b + \mu)^m$ *appears 2n times in the elementary divisors of the associated pencil m with signature* $2(\epsilon_1 + \cdots + \epsilon_n)$ *associated to the former.*

(vi) *If for some m the summand* (4°) [*or* (7°)] *appears exactly n times with signs* $\epsilon_1, \ldots, \epsilon_n$ ($\epsilon_i = \pm 1$), *then* μ^{2m} *(or* λ^{2m+1}*) appears 2n times in the elementary divisors of the associated pencil with the signature* $2(\epsilon_1 + \cdots + \epsilon_n)$.

(vii) *If for some m the summand* (5°) [*or* (6°)] *appears exactly n times, then* μ^{2m+1} *(or* λ^{2m}*) appears 4n times in the elementary divisors of the associated pencil.*

(viii) *Except for those in (i) and (ii), the associated pencil has no other minimal indices; except for those in (iii)–(vii), the associated pencil has no other elementary divisors.*

Theorem 6.4. *The necessary and sufficient condition of two* ϕ_-*-pairs to be* $\mathcal{C}$ *congruent is that their associated pencils are conjunctive.*

Remark. Let us consider specially the case when T is nonsingular in the pair (H, T). Clearly in this case the canonical form (144) has no zero summand, and only summands (1°)–(5°) possibly appear. We can write, therefore,

$$(H,T)\approx(\tilde{H},\tilde{T}),$$

where

$$\tilde{H}=\sum_i{}'\left(-a_i^2M_{r_i}N_{r_i}+N_{r_i}\right)\dotplus\sum_j{}'\epsilon_j\left(b_j^2M_{s_j}N_{s_j}\dotplus N_{s_j}\right)$$
$$\dotplus\sum_k{}'\begin{bmatrix}0 & \beta_kM_{t_k}\\ \bar{\beta}_kM'_{t_k} & 0\end{bmatrix}\dotplus\sum_k{}'\begin{bmatrix}0 & I_{t_k}\\ I_{t_k} & 0\end{bmatrix}$$
$$\dotplus\sum_l{}'\eta_l(N_{p_l}\dotplus J_{p_l}N_{p_l})\dotplus\sum_h{}'(J_{2q_h+1}N_{2q_h+1}\dotplus N_{2q_h+1}J_{2q_h+1}),$$
$$\tilde{T}=\sum_i{}'F_{r_i}\dotplus\sum_j{}'F_{s_i}+\sum_k{}'F_{t_k}+\sum_k{}'F_{t_k}+\sum_l{}'F_{p_l}+\sum_h F_{2q_h+1}.$$

After an obvious rearrangement (taking $\tilde{T}$ into F), we obtain

$$(H,T)\approx(H_1\dotplus H_2,F), \tag{156}$$

where

$$H_1=\sum_i{}'\left(-a_i^2M_{r_i}N_{r_i}\right)\dotplus\sum_j{}'\epsilon_jb_j^2M_{s_j}N_{s_j}\dotplus\sum_k{}'\begin{bmatrix}0 & \beta_kM_{t_k}\\ \bar{\beta}_kM'_{t_k} & 0\end{bmatrix}$$
$$+\sum_l\eta_lN_{p_l}\dotplus\sum_h{}'J_{2q_h+1}N_{2q_h+1} \tag{157}$$
$$H_2=\sum_i{}'N_{r_i}\dotplus\sum_j{}'\epsilon_jN_{\bar{s}_j}\dotplus\sum_k{}'\begin{pmatrix}0 & I_{t_k}\\ I_{t_k} & 0\end{pmatrix}$$
$$\dotplus\sum_l{}'\eta_lJ_{p_l}N_{p_l}\dotplus\sum_h{}'N_{2q_h+1}J_{2q_h+1} \tag{158}$$

$$\left(\epsilon_j=\pm1,\quad \eta_l=\pm1,\quad a_i^2>0,\quad b_i^2>0,\quad \beta_k \text{ has a positive real part}\right).$$

In (156) let $T=F$. If H is hermitian of order 2ν, then there is a matrix P such that

$$PF_\nu P'=F_\nu, \tag{159}$$

$$PH\bar{P}'=H_1\dotplus H_2. \tag{160}$$

A matrix which satisfies (159) is called symplectic. If P is symplectic, then the transformation which takes P into $PH\bar{P}'$ is called a symplectic transformation, H being a hermitian matrix of even order. If two such hermitian matrices can be taken into each other by a symplectic transformation, then we say they are symplectically conjunctive. The theory developed in this section subsumes the whole theory of symplectic transformations. From (160) we conclude that the canonical form $H_1\dotplus H_2$ of a hermitian matrix of even order under symplectic transformations is given by

(157) and (158). Theorem 6.4 says in this special case that two hermitian matrices H and H^* of order 2ν are symplectically conjunctive if and only if the associated pencils of (H, F_ν) and (H^*, F_ν) are conjunctive. Theorem 6.2 gives the detail of determining the canonical form $H_1 \dotplus H_2$ from the classical invariants of the associated pencil of (H, F_ν).

In this paper [5], Hua claimed that every hermitian matrix of order 2ν is symplectically conjunctive to a direct sum $H_1^* \dotplus H_2^*$ (H_1^*, H_2^* are of order ν) of the form

$$H_1^* = A_1 \dotplus 0, \qquad H_2^* = A_2 \dotplus 0, \tag{161}$$

where A_1 is nonsingular, A_2 has the same order as A_1. He called such a direct sum $H_1^* \dotplus H_2^*$ "binomial". However, the statement "Every hermitian matrix of even order is symplectically conjunctive to a 'binomial' direct sum" is not correct, and it should be replaced by the following theorem.

A hermitian matrix of order 2ν is symplectically conjunctive to a "binomial" direct sum if and only if the associated pencil of (H, F_ν) has no elementary divisors of the form μ^{2m+1} except for μ.

Proof (necessity). If H is symplectically conjunctive to $H_1^* + H_2^*$, then the associated pencil of (H, F_ν) is conjunctive to that of $(H_1^* + H_2^*, F_\nu)$, which is decomposed into a direct sum

$$\begin{pmatrix} \lambda H_1^* & \mu I \\ \mu I & \lambda \overline{H}_2^* \end{pmatrix} \dotplus \begin{pmatrix} \lambda H_2^* & -\mu I \\ -\mu I & \lambda H_1^* \end{pmatrix}. \tag{162}$$

Substituting the expression (161) for H_1^* and H_2^* in (162), the direct sum splits further:

$$\begin{pmatrix} \lambda A_1 & \mu I \\ \mu I & \lambda \overline{A}_2 \end{pmatrix} \dotplus \begin{pmatrix} \lambda A_2 & -\mu I \\ -\mu I & \lambda A_1 \end{pmatrix} \dotplus \begin{pmatrix} 0 & \mu I \\ \mu I & 0 \end{pmatrix} \dotplus \begin{pmatrix} 0 & -\mu I \\ -\mu I & 0 \end{pmatrix}. \tag{163}$$

Now we compute the elementary divisor in the first summand of (163) (put $\lambda = 1$). Since A_1 is nonsingular, we have

$$\begin{pmatrix} I & 0 \\ -\mu A_1^{-1} & I \end{pmatrix} \begin{pmatrix} A_1 & \mu I \\ \mu I & \overline{A}_2 \end{pmatrix} \begin{pmatrix} I & -\mu A_1^{-1} \\ 0 & I \end{pmatrix} = A_1 \dotplus (\overline{A}_2 - \mu^2 A^{-1}).$$

Hence the first summand in (163) cannot contribute an elementary divisor μ^{2m+1}. Neither can the second summand for the same reason. Clearly the elementary divisor of the third summand and the fourth summand are μ's. Thus the necessity is proved.

(Sufficiency). Suppose the associated pencil of (H, F_ν) has no elementary divisors μ^{2m+1} except for μ. According to Theorem 6.3, in the last two summations $\sum'$ of (157) and (158) for H_1 and H_2 in the canonical form (156), each q_r has to be zero. Hence $H_1 \dotplus H_2$ is itself a "binomial" direct sum. This proves the sufficiency.

References

1. Hsu, Pao-lu, *Acta Math. Sinica* **5** (1955), 333–346 (in Chinese).
2. Hsu, Pao-lu, *Bull. Peking Univ.* (Natural science edition) **1** (1955), 1–16 (in Chinese).
3. Turnbull, H. W., *Proc. Lond. Math. Soc.* **39** (1935), 232–248.
4. Hua, Loo-keng, *Trans. Amer. Math. Soc.* **59** (1946), 508–523.
5. Hua, Loo-keng, *Amer. J. Math.* **66** (1944), 531–563.

Comments by Duan Xue-fu

With the impetus of certain geometrical studies, especially the theory of symmetric spaces, by the late forties and fifties the classification problem of matrices had been enriched in its contents and further developed.

This paper discusses the classification over the complex field of matric pairs (A, B) under the transformations $A \to PA\bar{P}'$, $B \to PBP'$, where A is a hermitian matrix and B a symmetric or skew-symmetric matrix. Here P is assumed to be non-singular, whereas either A or B may be singular. In the paper all the possible canonical forms are determined and the classification is thereby completed. Here also a complete system of invariants for the matric pencil $\lambda A + \mu B$ is given. When B is non-singular, P can be taken as an orthogonal or a symmetric matrix (with respect to the complex field), and the problem considered here is just the classification of hermitian matrices under the transformation group of matrices of either kind. This is what Loo-keng Hua did in his work on the geometry of matrices. However, when B is skew-symmetric, the present paper gives an important supplment to Hua's result on the symplectic congruence of a hermitian matrix of an even degree with a "binomial" direct sum (*Amer. J. Math* **66** (1944), 531–563) by supplying a necessary and sufficient condition. Use is made here of results in the two previous papers.*

*"On a kind of transformations of matrices" and "On a kind of transformations of matrix pairs".

Reprinted from
Acta Sci. Natur. Univ. Pekinensis
4 (1958), 145–150.

The Absolute Continuity of the Distribution Functions in the Class L*

The theory of the distribution functions in the Class L can be found in reference [1], Chapter 6, §§29, 30. Here, we only state the results which we shall need in the following.

A necessary and sufficient condition for a distribution function $F(x)$ belonging to the Class L is that its characteristic function $f(t)$ can be expressed as

$$\ln f(t) = i\gamma t - \tfrac{1}{2}\sigma^2 t^2 + \int_{-\infty}^{-0}\left(e^{itu} - 1 - \frac{itu}{1+u^2}\right) dM(u)$$
$$+ \int_{+0}^{+\infty}\left(e^{itu} - 1 - \frac{itu}{1+u^2}\right) dN(u), \tag{1}$$

in which $M(u)$ and $N(u)$ satisfy the following conditions:

(i) $M(u)$ is a non-decreasing function in $(-\infty, 0)$ and $N(u)$ is a non-decreasing function in $(0, +\infty)$;
(ii) $M(-\infty) = 0$, $N(+\infty) = 0$;
(iii) $\int_{-\infty}^{-0}(u^2/1+u^2)\,dM(u) < \infty$, $\int_{+0}^{+\infty}(u^2/1+u^2)\,dN(u) < \infty$;
(iv) If we write

$$T(v) = -M(-e^v), \qquad S(v) = N(e^v), \qquad -\infty < v < +\infty \tag{2}$$

then, for any $a < b$ and $h > 0$,

$$T(a+h) - T(a) \geqslant T(b+h) - T(b),$$
$$S(a+h) - S(a) \geqslant S(b+h) - S(b).$$

It was proved in [1], p. 151, that $-S(v)$ is continuous and convex. (In the original text, it was stated that $S(v)$ is convex. For consistency with the convention used in [2] here we say that $-S(v)$ is convex.) According to [2], p. 69, Theorem 4.141, $-S(v)$ can be expressed as

$$-S(v) = -S(\alpha) - \int_{\alpha}^{v} \psi(t)\,dt, \tag{3}$$

in which $\psi(v)$ is a non-decreasing function. Since $S(v)$ is non-decreasing, $\psi(v) \geqslant 0$. Condition (ii) implies $S(+\infty) = 0$. Therefore, letting $v \to \infty$ in (3), we have

$$0 = S(\alpha) + \int_{\alpha}^{+\infty} \psi(t)\,dt. \tag{4}$$

*Received January 24, 1958. Translated by Tze-chien Sun, Wayne State University.

Adding (3) and (4), we obtain

$$S(v) = -\int_{v}^{+\infty} \psi(t)\,dt. \tag{5}$$

It follows from (2) and (5) that

$$N(u) = S(\ln u) = -\int_{\ln u}^{+\infty} \psi(t)\,dt = -\int_{u}^{+\infty} \frac{n(v)}{v}\,dv, \qquad u > 0, \tag{6}$$

where $n(u) = \psi(\ln u)$ is a non-increasing and non-negative function in $(0, \infty)$.

Similarly, we can derive

$$M(u) = \int_{-\infty}^{u} \frac{m(v)}{v}\,dv, \qquad u > 0, \tag{7}$$

where $m(u)$ is a non-decreasing and non-negative function in $(-\infty, 0)$.

From (6), (7), (iii), we obtain

$$\int_{-\infty}^{0} \frac{u \cdot m(u)}{1+u^2}\,du < \infty, \qquad \int_{0}^{+\infty} \frac{u \cdot n(u)}{1+u^2}\,du < \infty. \tag{8}$$

Substituting (7) and (6) into (1), we have

$$\ln f(t) = i\gamma t - \tfrac{1}{2}\sigma^2 t^2 + \int_{-\infty}^{0}\left(e^{itu} - 1 - \frac{itu}{1+u^2}\right)\frac{m(u)}{u}\,du$$

$$+ \int_{0}^{+\infty}\left(e^{itu} - 1 - \frac{itu}{1+u^2}\right)\frac{n(u)}{u}\,du. \tag{9}$$

Thus, we have derived the canonical expression (9) for the characteristic function $f(t)$ of a distribution function $F(x)$ in the Class L, in which $m(u)$ is a non-decreasing and non-negative function in $(-\infty, 0)$ and $n(u)$ is a non-increasing and non-negative function in $(0, +\infty)$, both satisfying condition (8).

In the following we shall prove that every distribution function $F(x)$ in the Class L is absolutely continuous. Clearly, we only have to consider the situation

$$\ln f(t) = \int_{-\infty}^{0}\left(e^{itu} - 1 - \frac{itu}{1+u^2}\right)\frac{m(u)}{u}\,du + \int_{0}^{+\infty}\left(e^{itu} - 1 - \frac{itu}{1+u^2}\right)\frac{n(u)}{u}\,du. \tag{10}$$

The proof is divided into two cases.

(I) $m(-0) + n(+0) = +\infty$. From (10), we have

$$\ln|f(t)| = -\int_{0}^{+\infty} \frac{1-\cos tu}{u}\,\phi_1(u)\,du, \tag{11}$$

where $\phi_1(u) = m(-u) + n(u)$.

Since $\phi_1(+0) = +\infty$, we can choose $a > 0$ so that $\phi_1(a) \geqslant 2$. Since $\phi_1(u)$ is non-increasing and non-negative,

$$\ln|f(t)| \leqslant -\int_{0}^{a} \frac{1-\cos tu}{u}\,\phi_1(u)\,du \leqslant -2\int_{0}^{a} \frac{1-\cos tu}{u}\,du$$

$$= -2\int_{0}^{a|t|} \frac{1-\cos u}{u}\,du.$$

Therefore,

$$\ln|f(t)| \leqslant -2\int_1^{a|t|}\frac{1-\cos u}{u}\,du = -2\ln a|t| + 2\int_1^{a|t|}\frac{\cos u}{u}\,du, \qquad \text{for} \quad |t| > \frac{1}{a}.$$

Applying the second mean-value theorem, we have

$$\left|\int_1^{a|t|}\frac{\cos u}{u}\,du\right| \leqslant \left|\int_1^{\xi}\cos u\,du\right| + \left|\int_{\xi}^{a|t|}\cos u\,du\right| \leqslant 4$$

So,

$$\ln|f(t)| \leqslant 8 - \ln a^2t^2,$$

or

$$|f(t)| \leqslant \frac{e^8}{a^2t^2} \qquad \text{for} \quad |t| > \frac{1}{a}.$$

Thus, $f(t)$ is integrable. It follows that $F(x)$ is absolute continuous.

(II) $m(-0) + n(+0) < \infty$. We shall write (10) as

$$\ln f(t) = -\int_0^{+\infty}\frac{1-\cos tu}{u}\phi_1(u)\,du + i\int_0^{+\infty}\left(\frac{\sin tu}{u} - \frac{t}{1+u^2}\right)\phi_2(u)\,du, \quad (12)$$

where $\phi_1(u)$ is defined in (11) and $\phi_2(u) = n(u) - m(-u)$. Since $\phi_1(+0) < +\infty$, $\phi_1(u)$ is bounded. It follows from $|\phi_2(u)| \leqslant \phi_1(u)$ that $\phi_2(u)$ is also bounded. Hence,

$$\int_0^{+\infty}\frac{|\phi_2(u)|}{1+u^2}\,du < \infty.$$

Therefore, instead of (12), we only have to consider

$$\ln f(t) = -\int_0^{+\infty}\frac{1-\cos tu}{u}\phi_1(u)\,du + i\int_0^{+\infty}\frac{\sin tu}{u}\phi_2(u)\,du. \quad (13)$$

If $\phi_1(u) \equiv 0$ then $\phi_2(u) \equiv 0$, so $F(x)$ is a degenerated distribution. Of course, we shall not consider this case. Thus, we can choose $a > 0$ such that $\phi_1(a) = b > 0$. Hence

$$\ln|f(t)| \leqslant -\int_0^{a}\frac{1-\cos tu}{u}\phi_1(u)\,du \leqslant -b\int_0^{a}\frac{1-\cos tu}{u}\,du$$

By a similar argument as above, we can obtain

$$|f(t)| \leqslant \frac{e^{4b}}{a^b|t|^b}, \qquad \text{for} \quad |t| > \frac{1}{a}. \quad (14)$$

Put

$$g(u) = \int_0^{u}\frac{1-\cos v}{v}\,dv, \qquad h(u) = \int_0^{u}\frac{\sin v}{v}\,v. \quad (15)$$

When $t \neq 0$, (13) can be written

$$\ln f(t) = -\int_0^{+\infty}\phi_1(u)\,dg(|t|u) + i\,\mathrm{sgn}\,t\int_0^{\infty}\phi_2(u)\,dh(|t|u).$$

Using the method of integration by parts, we have

$$\ln f(t) = -\phi_1(u)\, g(|t|u)|_{+0}^{+\infty} + i \operatorname{sgn} t \phi_2(u) h(|t|u)|_{+0}^{+\infty}$$
$$+ \int_{+0}^{+\infty} g(|t|u)\, d\phi_1(u) - i \operatorname{sgn} t \int_{+0}^{+\infty} h(|t|u)\, d\phi_2(u). \tag{16}$$

The first two terms are all zero. In fact $g(+0) = h(+0) = 0$ and ϕ_1 and ϕ_2 are bounded. Besides, $\lim_{u\to\infty} \phi_2(u)h(|t|u) = 0$, because h is bounded and $\phi_2(+\infty) = 0$. We shall next show

$$\lim_{u\to\infty} \phi_1(u) g(|t|u) = 0. \tag{17}$$

From (8), we see that $\int^{+\infty}(\phi_1(u)/u)\, du < \infty$, it follows that $\phi_1(u) = o(1/\ln u)$. From (15), clearly $g(u) = O(\ln u)$. This proves (17). Hence, (16) becomes

$$\ln f(t) = \int_{+0}^{+\infty} g(|t|u)\, d\phi_1(u) - i \operatorname{sgn} t \int_{+0}^{+\infty} h(|t|u)\, d\phi_2(u).$$

Clearly, it is legitimate to differentiate with respect to t under the integral sign on the right side of the above expression. Doing so, we obtain

$$f'(t) = \frac{f(t)}{t}\left(\int_{+0}^{+\infty} (1 - \cos tu)\, d\phi_1(u) - i \int_{+0}^{+\infty} \sin tu\, d\phi_2(u)\right), \qquad t \neq 0;$$

thus,

$$|f'(t)| \leqslant \frac{A|f(t)|}{t}, \qquad A \text{ is a constant}, \qquad t \neq 0. \tag{18}$$

It follows from (14) and (18) that in order to show that $F(x)$ is absolutely continuous, we only have to prove the following lemma.

Lemma. *Suppose $F(x)$ is a distribution function and $f(t)$ is its characteristic function. Suppose $f'(t)$ exists at all $t \neq 0$ and there exist positive numbers T, δ, C_1, C_2 such that*

$$|f(t)| \leqslant \frac{C_1}{|t|^\delta}, \qquad |t| > T, \tag{19}$$

$$|f'(t)| \leqslant \frac{C_2|f(t)|}{|t|}, \qquad t \neq 0. \tag{20}$$

Then $F(x)$ is absolutely continuous.

Proof. Clearly $F(x)$ is continuous. By the inversion formula, ([1], p. 48)

$$F(x) - F(0) = \frac{1}{2\pi} \int_{-\infty}^{+\infty} \frac{e^{-itx} - 1}{-it} f(t)\, dt.$$

When $x \neq 0$, applying the method of integration by parts, we have

$$F(x) - F(0) = \frac{1}{2\pi i x} \int_{-\infty}^{+\infty} \frac{f(t)}{it}\, d(e^{-itx} - 1 + itx)$$
$$= \frac{1}{2\pi x} \int_{-\infty}^{+\infty} (e^{-itx} - 1 + itx)\left(\frac{f'(t)}{t} - \frac{f(t)}{t^2}\right) dt.$$

Let us write

$$F(x) - F(0) = \frac{H(x)}{2\pi x}, \qquad x \neq 0, \tag{21}$$

where $H(x)$ is the integral on the right side in the above expression. It is easy to see that $H(x)$ has derivatives at all x. And it is legitimate to differentiate under the integral sign. Thus,

$$H'(x) = \int_{-\infty}^{+\infty} it(1 - e^{-itx})\left(\frac{f'(t)}{t} - \frac{f(t)}{t^2}\right) dt,$$

$$|H'(x)| \leqslant \int_{-\infty}^{+\infty} |1 - e^{-itx}|\left(|f'(t)| + \frac{|f(t)|}{|t|}\right) dt.$$

Hence, by (19) and (20),

$$\begin{aligned} |H'(x)| &\leqslant (C_2 + 1)\int_{-\infty}^{+\infty} |1 - e^{-itx}| \frac{|f(t)|}{|t|}\, dt \\ &\leqslant 2(C_2 + 1)C_1 \int_{|t|>T} \frac{dt}{|t|^{1+\sigma}} + (C_2 + 1)|x| \int_{|t|<T} |f(t)|\, dt \\ &\leqslant A_1|x| + A_2, \end{aligned}$$

where A_1 and A_2 are constants. So, within each bounded interval, $H(x)$ has bounded derivatives. It follows that $H(x)$ is absolutely continuous. From (21), we can see that $F(x)$ is absolutely continuous outside every neighborhood of 0. Since $F(x)$ is also continuous at zero, so it is absolutely continuous at all x.

References

1. B. V. Gnedenko and A. N. Kolmogorov, *Limit Distributions for Sums of Independent Random Variables* (translated by K. L. Chung). Addison-Wesley, 1954. (In the original printing, only the Russian edition was cited.)
2. A. Zygmund, *Trigonometrical Series*. Dover, 1955.

Comments by Tze-chien Sun

In this paper of 1958, Professor Hsu proved that *a distribution function in the class L is absolutely continuous*. Apparently this result was not known to people outside China for quite some time. In 1962 Fisz proved in [3] that a distribution function in the class L is always continuous, using a result obtained in [1]. Of course, this is just a special case of Hsu's result. In the next year Fisz and Varadarajan proved in [4] the absolute continuity of distribution functions in the class L for the first time outside China. The main result obtained in their paper is more general than Hsu's, but they did not refer to Hsu's result in 1958, apparently not knowing its existence. Tucker obtained the same result independently in [9].

A problem related to this result of Hsu's is the unimodality of the distribution functions in the class L. The solution of the unimodality problem has an interesting history. First, in their book of 1949 ([5]), Gnedenko and Kolmogorov gave a proof of the theorem: *The distribution functions in the class L are all unimodal.* In their proof, they used a result by Lapin which stated that the convolution of two unimodal distribution functions is again unimodal. However in 1953 Chung gave a counterexample ([2]) to show that Lapin's result is false, and therefore Gnedenko and Kolmogorov's proof is not valid. In 1957, Ibragimov ([7]) gave a class of distribution functions in the class L which, he claimed, were not unimodal. It seemed for a while that the problem was settled on the negative side. But ten years later Sun showed ([8]) that Ibragimov's class of distribution functions were actually all unimodal. The problem was finally settled positively in 1978 when Yamazato successively proved ([11]) that all the distribution functions in the class L are unimodal. Professor Hsu must have learned of Chung's counterexample or Ibragimov's paper before he wrote this paper, for otherwise he would only have to show that a distribution function in the class L is continuous at its mode, because the unimodality property implies that its distribution function is absolutely continuous except, possibly, at its mode. It makes one wonder: had Hsu decided to attack the unimodality problem, given his masterful skill in the method of characteristic functions, could he have solved this problem some twenty years earlier?

Another related problem which should be mentioned here is the continuity of infinitely divisible distribution functions. The same necessary and sufficient condition for the continuity of infinitely divisible distribution functions was obtained independently by Hartman and Wintner ([6]) and Blum and Rosenblatt ([1]). A necessary and sufficient condition for the absolute continuity of infinitely divisible distribution functions was obtained by Tucker ([10]) in 1965.

References

1. J. R. Blum and M. Rosenblatt, On the structure of the infinitely divisible distributions. *Pacific J. Math.* **9** (1959), 1–7.
2. K. L. Chung, Sur les lois de probabilités unimodales. *C. R. Acad. Sci. Paris* **236** (1953), 583–584.
3. M. Fisz, On the continuity of the L-distribution functions. *Zeit. Wahr.* **1** (1962), 25–27.
4. M. Fisz and V. S. Varadarajan, A condition for absolute continuity of infinitely divisible distribution functions. *Zeit. Wahr.* **1** (1963), 335–339.
5. B. V. Gnedenko and A. N. Kolmogorov, *Limit Distribution Functions for Sums of Independent Random Variables*. Addison-Wesley, Reading, MA, 1954 (translated by K. L. Chung).
6. P. Hartman and A. Wintner, On the infinitesimal generators of integral convolutions. *Am. J. Math.* **64** (1942), 273–298.
7. I. A. Ibragimov, A remark on probability distributions of class L. *Theory Prob. Appl.* **2** (1957), 117–119.

8. T. C. Sun, A note on the unimodality of distribution functions of class *L*. *Ann. Math. Statistics* **38** (1967), 1296–1299.
9. H. G. Tucker, Absolute continuity of infinitely divisible distributions. *Pacific J. Math.* **12** (1962), 1125–1129.
10. H. G. Tucker, On a necessary and sufficient condition that a infinitely divisible distribution be absolutely continuous, *Trans. AMS* **118** (1965), 316–330.
11. M. Yamazato, Unimodality of infinitely divisible distribution functions of class *L*. *Ann. Prob.* **6** (1978), 523–531.

Reprinted from
Acta Sci. Natur. Univ. Pekinensis
4 (1958), 257–270.

The Differentiability of the Probability Transition Function of a Purely Discontinuous Stationary Markov Process on the Euclidean Space*†

§1

The probability transition function of a purely discontinuous stationary Markov process on a general space is defined as follows.

Suppose $\{X, \mathfrak{T}\}$ is a measurable space—i.e., X is an arbitrary set, and $\mathfrak{T}$ is a σ-field of subsets of X including X itself as a member. We assume that every set consisting of a single point of X is in $\mathfrak{T}$. Then the probability transition function of a purely discontinuous stationary Markov process on $\{X, \mathfrak{T}\}$ is a function $P(t, x, E)$ of three variables satisfying the following six conditions:

(I) The domain of t is $[0, \infty)$; the domain of x is Y, where Y is a subset in $\mathfrak{T}$; the domain of E is $\mathfrak{T}$. $P(t, x, E)$ is real-valued and $0 \leqslant P(t, x, E) \leqslant 1$.

(II) For each fixed t and E, $P(t, x, E)$, as a function of x, is $\mathfrak{T}$-measurable; and for each fixed t and x, $P(t, x, E)$, as a function of E, is a probability measure on $\mathfrak{T}$.

(III) $P(t, x, Y) = 1$.

(IV) $P(s + t, x, E) = \int_Y P(s, x, dy) P(t, y, E)$.‡

(V) $P(0, x, E) = \chi_E(x)$. Here (and also in the whole paper) χ_E denotes the characteristic function of E.

(VI) $\lim_{t \downarrow 0} P(t, x, \{x\}) = 1$. Here (and also in the whole paper) $\{x\}$ denotes a set with a single element x.

Definition. A *probability transition function* $P(t, x, E)$ is a function satisfying the properties (I)–(VI).

When, in particular, X is a Euclidean space and $\mathfrak{T}$ is the class of all Borel subsets of X, a function $P(t, x, E)$ as defined above is called a probability transition function on a Euclidean space. Furthermore, when X is a real line and Y is the set of all positive integers, we have the probability transition function of a stationary Markov process with a denumerable number of states. In this case, the variable x takes only the positive integer values; we shall rewrite it as i. According to the

*Received January 24, 1958.

† Translated by Tze-chien Sun, Wayne State University.

‡ Clearly, we can use X for Y as the range of integration, because although the integrand is not defined on $X - Y$, the set $X - Y$ has a measure of zero with respect to the measure used in the integration.

condition (III), $P(t,i,E)$ is uniquely determined by $P(t,i,\{j\})$ for all i,j. We usually rewrite $P(t,i,\{j\})$ as $p_{ij}(t)$ and call it the probability transition function. The conditions satisfied by $p_{ij}(t)$ can be obtained from (I) to (VI) correspondingly, i.e.,

(i) $0 \leqslant p_{ij}(t) \leqslant 1$,
(iii) $\sum_{j=1}^{\infty} p_{ij}(t) = 1$,
(iv) $p_{ij}(s+t) = \sum_{k=1}^{\infty} p_{ik}(s)p_{kj}(t)$,
(v) $p_{ij}(0) = \delta_{ij}$, where δ_{ij} is the Kronecker delta,
(vi) $\lim_{t\downarrow 0} p_{ij}(t) = \delta_{ij}$.

We are particularly interested in the analytic property of $P(t,x,E)$ as a function of t. In the special case $p_{ij}(t)$, this property was studied in detail by Kolmogorov (see [1]). He was the first to prove the existence of the following limits:

$$\lim_{t\downarrow 0} \frac{1-p_{ii}(t)}{t} = q_i \leqslant \infty, \qquad \lim_{t\downarrow 0} \frac{p_{ij}(t)}{t} = q_{ij} < \infty \qquad (i \neq j). \tag{1}$$

Kendall (see [2]) generalized the above results to probability transition functions on a general space.

Not long ago, Austin [3] (using purely analytic methods) and K. L. Chung [4] (using probability methods) proved that, in the case of $p_{ij}(t)$, if for a certain i, $q_i < \infty$ in (1), then $p_{ij}'(t)$ exists almost everywhere. They also derived several properties satisfied by $p_{ij}'(t)$. In this paper, the results in [3] and [4] will be generalized to probability transition probabilities on an Euclidean space. The method we used here is more elementary than those in [3] and [4] and the results we obtained are sharper.

§2

In the later part, we shall need the following simple lemma.

Lemma 1. *Suppose a real-valued function $p(t)$, $t \in [0,\infty)$ and a constant $q > 0$ satisfy the following conditions*:

$$0 \leqslant p(t) \leqslant 1, \tag{2}$$

$$p(s+t) \geqslant e^{-qs}p(t), \qquad 1 - p(s+t) \geqslant e^{-qs}(1-p(t)). \tag{3}$$

Then, $p(t)$ can be expressed as

$$p(t) = q\int_0^t e^{-q(t-s)}r(s)\,ds + p(0)e^{-qt} \tag{4}$$

in which $r(t)$ is measurable on $[0,\infty)$ and $0 \leqslant r(t) \leqslant 1$.

Proof. From the first part of (3), we have

$$p(s+t) - p(t) \geqslant (e^{-qs}-1)p(t) \geqslant e^{-qs} - 1.$$

Similarly, from the second part of (3), we have

$$p(s+t)-p(t)\leqslant 1-e^{-qs}.$$

Hence

$$|p(s+t)-p(t)|\leqslant 1-e^{-qs}\leqslant qs.$$

It follows that $p(t)$ is absolutely continuous in t. Also, from the first part of (3), we have

$$e^{q(s+t)}p(s+t)\geqslant e^{qt}p(t).$$

So $e^{qt}p(t)$ is a nondecreasing function. Similarly, $e^{qt}(1-p(t))$ is also a nondecreasing function. Let

$$r(t)=\frac{1}{q}e^{-qt}\frac{d}{dt}\left(e^{qt}p(t)\right)\qquad \text{a.e.}$$

Then $r(t)\geqslant 0$ a.e.
Since

$$1-r(t)=\frac{1}{q}e^{-qt}\frac{d}{dt}\left[e^{qt}(1-p(t))\right]\qquad \text{a.e.,}$$

we have $r(t)\leqslant 1$ a.e.
From the definition $r(t)$, we have, by integration,

$$e^{qt}p(t)=q\int_0^t e^{qs}r(s)\,ds+p(0). \tag{5}$$

Certainly, we can choose $r(t)$ so that $0\leqslant r(t)\leqslant 1$ for all t. Since (5) is just (4), the lemma is proved.

Now, let us consider a probability transition function $P(t,x,E)$ on a general space. From (IV), we have

$$P(s+t,x,E)\geqslant\int_{\{x\}}P(s,x,dy)P(t,y,E)=P(s,x,\{x\})P(t,x,E). \tag{6}$$

In the particular case when $E=\{x\}$, we have

$$P(s+t,x,\{x\})\geqslant P(s,x,\{x\})P(t,x,\{x\}).$$

It follows that

$$-\log P(s+t,x,\{x\})\leqslant -\log P(s,x,\{x\})-\log P(t,x,\{x\}). \tag{7}$$

Also, from (VI), we have

$$\lim_{t\downarrow 0}\left[-\log P(t,x,\{x\})\right]=0. \tag{8}$$

A function $f(t)$, $0\leqslant t<\infty$, satisfying the inequality

$$f(s+t)\leqslant f(s)+f(t),$$

is called a sub-additive function. If a sub-additive function satisfies the condition

$$\lim_{t\downarrow 0}f(t)=0 \tag{9}$$

then

$$\lim_{t\downarrow 0} \frac{f(t)}{t} \quad \text{exists and is equal to} \quad \sup_{t>0} \frac{f(t)}{t}.^*$$

From (7) and (8), $-\log P(t, x, \{x\})$ is sub-additive in t and satisfies (9). Hence

$$\lim_{t\downarrow 0} \frac{-\log P(t, x, \{x\})}{t} = q(x), \tag{10}$$

$$\lim_{t>0} \frac{-\log P(t, x, \{x\})}{t} = q(x). \tag{11}$$

From (10) we have

$$\log P(t, x, \{x\}) = -q(x)t + o(t) \qquad \text{as} \quad t\downarrow 0,$$

and, hence,

$$P(t, x, \{x\}) = e^{-q(x)t} + o(t).$$

It follows that

$$\lim_{t\downarrow 0} \frac{1 - P(t, x, \{x\})}{t} = q(x). \tag{12}$$

From (11), we have

$$P(t, x, \{x\}) \geqslant e^{-q(x)t}. \tag{13}$$

The $q(x)$ in (12) is just the q_i for the particular case $p_{ij}(t)$ in (1). The result in (12) is already known in reference [2].

Again, from (6) and (13), we have

$$P(s + t, x, E) \geqslant e^{-q(x)s} P(t, x, E). \tag{14}$$

Replacing E by $X - E$ in (14), we have

$$1 - P(s + t, x, E) \geqslant e^{-q(x)s}(1 - P(t, x, E)). \tag{15}$$

When $q(x) = 0$, (14) and (15) imply that $P(t, x, E)$ is independent of t, so $P(t, x, E) = P(0, x, E) = \chi_E(x)$.† Suppose $q(x) > 0$. Equations (14) and (15) together imply that $P(t, x, E)$ and $q(x)$ satisfy the conditions in Lemma 1 for $p(t)$ and q. Using the initial condition $P(0, x, E) = \chi_E(x)$, we obtain

$$P(t, x, E) = q(x) \int_0^t e^{-q(x)(t-s)} r(s, x, E)\, ds + e^{-q(x)t} \chi_E(x), \tag{16}$$

in which $\gamma(t, x, E)$ is a measurable function in t and

$$0 \leqslant r(t, x, E) \leqslant 1. \tag{17}$$

Putting $E = X$ in (16), we have

$$1 - e^{-q(x)t} = q(x) \int_0^t e^{-q(x)(t-s)} r(s, x, X)\, ds,$$

*See [5], p. 180, Theorem 6.11.1.

†This result is already known.

or

$$\int_0^t e^{-q(x)(t-s)}(1 - r(s,x,X))\,ds = 0.$$

Therefore

$$r(t,x,X) = 1 \qquad \text{for almost all } t. \tag{18}$$

Suppose $\{E_n\}$ is a sequence of disjoint sets in $\mathfrak{T}$. We have, from (16),

$$\int_0^t e^{q(x)s} \sum_n r(s,x,E_n)\,ds = \int_0^t e^{q(x)s} r\Big(s,x,\bigcup_n E_n\Big)\,ds.$$

Hence

$$\sum_n r(t,x,E_n) = r\Big(t,x,\bigcup_n E_n\Big) \qquad \text{for almost all } t. \tag{19}$$

We have thus proved the following theorem.

Theorem 1. *Suppose $P(t,x,E)$ is probability transition function and let*

$$q(x) = \lim_{t \downarrow 0} \frac{1 - P(t,x,\{x\})}{t}.$$

Suppose for a certain x, $q(x) < \infty$. If $q(x) = 0$, then

$$P(t,x,E) = \chi_E(x).$$

If $q(x) > 0$, then $P(t,x,E)$ has the expression (16), *in which $r(t,x,E)$ is a measurable function in t satisfying the conditions* (17), (18), *and* (19).

§3

In this section we shall always assume that X is a Euclidean space and $\mathfrak{T}$ is the class of all Borel subsets of X.

Let us consider a probability transition function $P(t,x,E)$. Let

$$q(x) = \lim_{t \downarrow 0} \frac{1 - P(t,x,\{x\})}{t}.$$

Suppose, for a certain x, $q(x) < \infty$. (In this section, x is always fixed, so $P(t,x,E)$ is a function of t and E only. This is true for other functions also.) If $q(x) = 0$, then we already know

$$P(t,x,E) = \chi_E(x). \tag{20}$$

If $q(x) > 0$, according to Theorem 1, we have the expression (16) in which $r(t,x,E)$ satisfies (17), (18) and (19). Notice that here E varies within the class of Borel sets. For each function $r(t,x,E)$ satisfying (17), (18) and (19), we can construct a function $r_1(t,x,E)$ such that

1° for each fixed E, $r_1(t,x,E) = r(t,x,E)$ for almost all t,
2° for each fixed t, $r_1(t,x,E)$ is a probability measure on $\mathfrak{T}$.*

---

*See [6], pp. 29–30.

From (16) and 1°, we have

$$P(t,x,E)=q(x)\int_0^t e^{-q(x)(t-s)}r_1(s,x,E)\,ds+e^{-q(x)t}\chi_E(x). \tag{21}$$

From (21) and 2°, we have, for each bounded measurable function $\varphi(y)$ on Y,

$$\int_Y \varphi(y)P(t,x,dy)=q(x)\int_0^t e^{-q(x)(t-s)}\,ds\int_Y\varphi(y)r_1(s,x,dy)+e^{-q(x)t}\varphi(x). \tag{22}$$

In particular*, putting $\varphi(y)=P(t',y,E)$, we have, according to (IV),

$$P(t+t',x,E)=q(x)\int_0^t e^{-q(x)(t-s)}\,ds\int_Y P(t',y,E)r_1(s,x,dy)$$
$$+e^{-q(x)t}P(t',s,E).$$

Using (21) again, we have

$$P(t+t',x,E)=q(x)\int_0^t e^{-q(x)(t-s)}\,ds\int_Y P(t',y,E)r_1(t,x,dy)$$
$$+q(x)e^{-q(x)t}\int_0^{t'}e^{-q(x)(t'-s)}r_1(s,x,E)\,ds$$
$$+e^{-q(x)(t+t')}\chi_E(x). \tag{23}$$

On the other hand, from (21), we also have

$$P(t+t',x,E)=q(x)\int_0^{t+t'}e^{-q(x)(t+t'-s)}r_1(s,x,E)\,ds+e^{-q(x)(t+t')}\chi_E(x). \tag{24}$$

Combining (23) and (24), we obtain

$$\int_0^t e^{q(x)s}\,ds\int_Y P(t',y,E)r_1(s,x,dy)=e^{-q(x)t'}\int_{t'}^{t+t'}e^{q(x)s}r_1(s,x,E)\,ds$$
$$=\int_0^t e^{q(x)s}r_1(s+t',x,E)\,ds. \tag{25}$$

Therefore, for each fixed t', there exists a null set $N_{t'}$ such that

$$r_1(t+t',x,E)=\int_Y P(t',y,E)r_1(t,x,dy)\qquad\text{for each}\quad t\notin N_{t'}. \tag{26}$$

(Of course, $N_{t'}$ depends also on E. However, since E is assumed to be fixed, the dependence on E is ignored here.) We claim that both sides of (26) are jointly measurable functions of (t,t'). Clearly the left side is. The right side is also, because for each fixed t' it is measurable in t and for each fixed t, it is continuous in t'. Therefore, the set

$$S=\left\{(t,t'):t>0,\ t'>0,\ r_1(t+t',x,E)\neq\int_Y P(t',y,E)r_1(t,x,dy)\right\}$$

is a two-dimensional measurable set. By (26) and the Fubini Theorem, the measure of the set S is zero, i.e.,

$$\iint_S dt\,dt'=0.$$

*The method used in here is similar to that used in [4], pp. 199–200.

Using the transformation $t = u$, $t + t' = v$, we have

$$\iint_M du\, dv = 0,$$

where

$$M = \left\{(u,v) : u > 0, v > u, \gamma_1(v,x,E) \neq \int_Y P(v-u, y, E) r_1(u,x,dy)\right\}.$$

According to the Fubini Theorem, there exists a null set H on the positive u-axis such that for each $u \notin H$ there exists a null set H_u such that

$$r_1(v,x,E) = \int_Y P(v-u, y, E) r_1(u,x,dy) \tag{27}$$

whenever $u \notin H$, $v \notin H_u$ and $v > u$.

Now, for each $v > 0$ we construct a function $R(v,x,E)$ as follows. Choose a u' with $0 < u' < v$ and $u' \notin H$, then let

$$R(v,x,E) = \int_Y P(v-u', y, E) r_1(u',x,dy). \tag{28}$$

We should point out here that (28) is independent of the choice of u'. In fact, suppose $v_0 > 0$ is given, and suppose $0 < u' < v_0$, $0 < u'' < v_0$ and $u', u'' \notin H$. Then, when $v \notin H_{u'} \cup H_{u''}$ and $v > \max(u', u'')$, we have

$$\int_Y P(v-u', y, E) r_1(u',x,dy) = \int_Y P(v-u'', y, E) r_1(u'',x,dy), \tag{29}$$

because, according to (27), both sides of (29) are equal to $r_1(v,x,E)$. In the meantime, both sides of (29) are continuous functions of v, so (29) holds for all $v > \max(u', u'')$, in particular, for $v = v_0$. This shows that the values of $R(v_0,x,E)$ are the same for u' and u''.

The constructed function $R(v,x,E)$ has the following properties:

(α) For each fixed v, $R(v,x,E)$ is a probability measure on $\mathcal{T}$.
(β) For each fixed E, $R(v,x,E)$ is a continuous function of $v > 0$.
(γ) For each fixed E, $R(v,x,E) = r_1(v,x,E)$ for almost all v.

(α) follows from (28) directly. The proof of (β) and (γ) is given as follows: for any given $a > 0$, choose any u_0 with $0 < u_0 < a$, $u_0 \notin H$. Then, we have, for each $v > a$,

$$R(v,x,E) = \int_Y P(v-u_0, y, E) r_1(u_0,x,dy), \tag{30}$$

and hence, $R(v,x,E)$ is a continuous function on (a, ∞). Because a is arbitrary, $R(v,x,E)$ is a continuous function on $(0, \infty)$. In the meantime, from (28) and (27), we have

$$R(v,x,E) = r_1(v,x,E) \qquad \text{whenever} \quad v > a, \quad v \notin H_{u_0}.$$

In other words, the above equality holds for almost all $v > a$. Because a is arbitrary, this equality holds for almost all $v > 0$. Therefore (β) and (γ) are proved.

From (21) and (γ) we have

$$P(t,x,E) = q(x)\int_0^t e^{-q(x)(t-s)} R(s,x,E)\, ds + e^{-q(x)t}\chi_E(x). \tag{31}$$

Since from (21) we can derive (26), so from (31), we can conclude that if we replace $r_1(s,x,E)$ by $R(s,x,E)$ in (26), it still holds. But after the replacement, both sides are continuous functions of t, so equality almost everywhere becomes identity. Therefore

$$R(s+t,x,E)=\int_Y R(s,x,dy)P(t,y,E). \tag{32}$$

In particular, putting $E=X-Y$, we have, according to (III), $R(t,x,X-Y)=0$. Hence, according to (α),

$$R(t,x,Y)=1.$$

Finally, we want to point out that $R(t,x,E)$ is the unique function which satisfies (31) and is continuous in t.

Summarizing the above results, we obtain

Theorem 2. *Suppose $P(t,x,E)$ is a probability transition function on a Euclidean space with*

$$q(x)=\lim_{t\downarrow 0}\frac{1-P(t,x,\{x\})}{t}.$$

Suppose for a certain x, $0<q(x)<\infty$. Then there exists a unique function $R(t,x,E)$, $t>0$, $E\in\mathfrak{T}$, satisfying the following properties:

(α) *For each fixed t, it is a probability measure on $\mathfrak{T}$.*
(β) *For each fixed E, it is a continuous function of t,*

such that

$$P(t,x,E)=q(x)\int_0^t e^{-q(x)(t-s)}R(s,x,E)\,ds+e^{-q(x)t}\chi_E(x). \tag{31}$$

Moreover,

$$R(t,x,Y)=1, \tag{33}$$

$$R(s+t,x,E)=\int_Y R(s,x,dy)P(t,y,E). \tag{32}$$

We also have the following theorem.

Theorem 3. *$P(t,x,E)$ and $q(x)$ are the same as in Theorem 2. Suppose for a certain x, $q(x)<\infty$. Then $P(t,x,E)$ has a derivative $P'(t,x,E)$ for all $t>0$. If $q(x)=0$, then $P'(t,x,E)=0$. If $q(x)>0$, then*

$$P'(t,x,E)=q(x)(R(t,x,E)-P(t,x,E)). \tag{34}$$

$P'(t,x,E)$ is a continuous function in t and is a σ-additive function of bounded variation in E with a total variation of no more than $q(x)$. Moreover,

$$P'(s+t,x,E)=\int_Y P'(s,x,dy)P(t,y,E). \tag{35}$$

Proof. When $q(x)=0$, the conclusion is clearly correct. When $q(x)>0$, from (31) and the continuity property of $R(t,x,E)$ in t, we obtain the differentiability of

$P(t,x,E)$ immediately and

$$\frac{d}{dt}e^{q(x)t}P(t,x,E) = q(x)e^{q(x)t}R(t,x,E).$$

This is (34). The continuity (in t) and the σ-additivity (in E) of $P(t,x,E)$ and its bounded variation all follow from (34). (35) follows directly from (33), (34) and (IV). Q.E.D.

Theorem 4. *Under the conditions in Theorem* 2, *we have*

$$\lim_{t\downarrow 0} R(t,x,\{x\}) = 0, \tag{36}$$

$$\lim_{t\downarrow 0} P'(t,x,\{x\}) = -q(x).^* \tag{37}$$

Proof. Putting $E = \{x\}$ in (33), we have

$$\begin{aligned} R(s+t,x,\{x\}) &\geqslant \int_{\{x\}} R(s,x,dy)P(t,y,\{x\}) \\ &= R(s,x,\{x\})P(t,x,\{x\}). \end{aligned}$$

Letting $s\downarrow 0$, we obtain

$$R(t,x,\{x\}) \geqslant P(t,x,\{x\}) \limsup_{s\downarrow 0} R(s,x,\{x\}).$$

Then, letting $t\downarrow 0$, we obtain, according to (VI),

$$\limsup_{t\downarrow 0} R(t,x,\{x\}) \geqslant \limsup_{s\downarrow 0} R(s,x,\{x\}).$$

Therefore, $\lim_{t\downarrow 0} R(t,x,\{x\})$ exists. Putting $E = \{x\}$ in (31), we have

$$\frac{e^{q(x)t}p(t,x,\{x\}) - 1}{q(x)t} = \frac{1}{t}\int_0^t e^{q(x)s}R(s,x,\{x\})\,ds.$$

Let $t\downarrow 0$. According to (12), the limit of the left-hand side is zero, while the limit of the right side is $\lim_{t\downarrow 0} R(t,x,\{x\})$. Hence, (36) is proved. (37) follows from (36), (34) and (IV).

§4

In this section, we still assume X is a Euclidean space and $\mathfrak{T}$ is in the class of Borel subsets of X.

The results we have obtained in the last section are all based on the assumption that at a certain x, $q(x) < \infty$. Now, we shall derive more results by strengthening this condition.

Kendall proved in [2] that, corresponding to a probability transition function

---

*Of course, this equality is also true when $q(x) = 0$.

$P(t,x,E)$ on a general space, if $E \subset Y$ is a set (in $\mathfrak{T}$) such that

$$\lim_{t\downarrow 0} P(t,y,\{y\}) = 1 \qquad \text{uniformly for all} \quad y \in E,$$

then the limit

$$\lim_{t\downarrow 0} \frac{P(t,x,E)}{t} = q(x,E) \tag{38}$$

exists whenever $x \notin E$, and $q(x,E) < \infty$, $q(x,E) \leqslant q(x)$. $q(x,E)$ is a generalization of q_{ij} in (1). Now, we are going to show:

Theorem 5. *$P(t,x,E)$ and $q(x)$ are the same as in Theorem* 2. *Suppose a point x and a set E ($E \in \mathfrak{T}, E \subset Y$) satisfy the following conditions*:

1. $x \notin E$,
2. $0 < q(x) < \infty$,
3. *$\lim_{t\downarrow 0} p(t,y,\{y\}) = 1$, uniformly for all $y \in E$.*

Then

$$\lim_{t\downarrow 0} R(t,x,E) = \frac{q(x,E)}{q(x)}, \tag{39}$$

$$\lim_{t\downarrow 0} P'(t,x,E) = q(x,E).^* \tag{40}$$

Proof. For each given $\epsilon > 0$, by (3) there exists a $\delta > 0$ such that

$$\inf_{y\in E} P(t,y,\{y\}) > 1-\epsilon \qquad \text{whenever} \quad 0 < t < \delta.$$

From (33), we have

$$R(s+t,x,E) \geqslant \int_E R(s,x,dy) P(t,y,E).$$

However, $P(t,y,E) \geqslant P(t,y,\{y\}) > 1-\epsilon$ for $0 < t < \delta$ and $y \in E$. Hence

$$R(s+t,x,E) \geqslant (1-\epsilon) R(s,x,E), \qquad 0 < t < \delta.$$

Letting $s\downarrow 0$, we have

$$R(t,x,E) \geqslant (1-\epsilon) \limsup_{s\downarrow 0} R(s,x,E), \qquad 0 < t < \delta.$$

So,

$$\liminf_{t\downarrow 0} R(t,x,E) \geqslant (1-\epsilon) \limsup_{s\downarrow 0} R(s,x,E).$$

Since ϵ is arbitrary, the above inequality remains true when $\epsilon = 0$. This proves the existence of $\lim_{t\downarrow 0} R(t,x,E)$.

From (31), we have (since $\chi_E(x) = 0$)

$$\frac{e^{q(x)t} P(t,x,E)}{q(x)t} = \frac{1}{t}\int_0^t e^{q(x)s} R(s,x,E)\, ds.$$

*Of course this equality still holds when $q(x) = 0$.

Let $t \downarrow 0$. By (38) the limit of the left side is $q(x,E)/q(x)$. The limit of the right side is clearly $\lim_{t\downarrow 0} R(t,x,E)$. Hence (39) is proved. (40) follows from (39), (34) and (VI). Q.E.D.

A set E ($E \in \mathfrak{T}, E \subset Y$) is called q-bounded if $q(y)$ is not only finite but also bounded on E.

Theorem 6. *$P(t,x,E)$ and $q(x)$ are the same as in Theorem 2. If $0 < q(x) < \infty$ and E is q-bounded, then $R(t,x,E)$ and $P'(t,x,E)$ are of bounded variation in t.*

Proof. Suppose

$$\sup_{y \in E} q(y) = c < \infty. \tag{41}$$

From (IV) and (33), we have

$$P(s+t,x,E) \geqslant \int_E P(s,x,dy)P(t,y,E),$$

$$R(s+t,x,E) \geqslant \int_E R(s,x,dy)P(t,y,E).$$

For each $y \in E$, from (13) and (41), we have

$$P(t,y,E) \geqslant P(t,y,\{y\}) \geqslant e^{-q(y)t} \geqslant e^{-ct}.$$

Hence

$$P(s+t,x,E) \geqslant e^{-ct}P(s,x,E),$$

$$R(s+t,x,E) \geqslant e^{-ct}R(s,x,E).$$

So

$$e^{c(s+t)}P(s+t,x,E) \geqslant e^{cs}P(s,x,E),$$

$$e^{c(s+t)}R(s+t,x,E) \geqslant e^{cs}R(s,x,E).$$

It follows that both $e^{ct}P(t,x,E)$ and $e^{ct}R(t,x,E)$ are non-decreasing functions in t. Therefore $P(t,x,E)$ and $R(t,x,E)$ are of bounded variation in t. From (34), we see that $P'(t,x,E)$ is also of bounded variation in t. Q.E.D.

Remark. We can see here, under the conditions in Theorem 6, the second derivative in t of $P(t,x,E)$ exists for almost all t. This is a proper generalization of Theorem 3 in K. L. Chung's paper [4], p. 202. That theorem says if $q_i < \infty$ (with no additional conditions), then $p''_{ij}(t)$ exists almost everywhere. But its proof is not correct.*

If $q(x)$ is finite for all $x \in Y$, then $P'(t,x,E)$ is a function of three variables ($t > 0$, $x \in Y$, $E \in \mathfrak{T}$). By (II), it is a measurable function of x.

---

* *Note by K. L. Chung*: If j is stable as well as i, then p''_{ij} exists a.e. See my book *Markov Chains with Stationary Transition Probabilities* (1967), Theorem II.15.7 (read p''_{ij} for p''_{jj} in line 18 there).

Theorem 7. *$P(t,x,E)$ and $q(x)$ are the same as in Theorem 2. Suppose $q(x)$ is finite for all $x \in Y$. If E is q-bounded, then*

$$P'(s+t,x,E) = \int_Y P(s,x,dy)P'(t,y,E). \tag{42}$$

Proof. We shall still use (41). In Theorem 6, we have already showed that $e^{ct}P(t,x,E)$ is a nondecreasing function of t. Hence,

$$U(t,x,E) = e^{ct}\frac{d}{dt}e^{ct}P(t,x,E) = P'(t,x,E) + cP(t,x,E) \geqslant 0. \tag{43}$$

Since $|P'(t,x,E)| \leqslant q(x)$ (see Theorem 3), we have

$$U(t,x,E) \leqslant q(x) + c. \tag{44}$$

According to the definition (43) of $U(t,x,E)$, we have

$$P(t,x,E) = \int_0^t e^{-c(t-s)}U(s,x,E)\,ds + e^{-ct}\chi_E(x). \tag{45}$$

From (45) and (IV), we have

$$\begin{aligned} P(t+t',x,E) &= \int_Y P(t',x,dy)\left[\int_0^t e^{-c(t-s)}U(s,y,E)\,ds + e^{-ct}\chi_E(y)\right] \\ &= \int_0^t e^{-c(t-s)}\,ds\int_Y U(s,y,E)P(t',x,dy) + e^{-ct}P(t',x,E). \end{aligned}$$

Again using (45), we have

$$\begin{aligned} P(t+t',x,E) &= \int_0^t e^{-c(t-s)}\,ds\int_Y U(s,y,E)P'(t,x,dy) \\ &\quad + e^{-ct}\int_0^{t'} e^{-c(t'-s)}U(s,x,E)\,ds + e^{-c(t+t')}\chi_E(x). \end{aligned} \tag{46}$$

On the other hand, from (45) directly, we have

$$P(t+t',x,E) = \int_0^{t+t'} e^{-c(t+t''s)}U(s,x,E)\,ds + e^{-c(t+t')}\chi_E(x). \tag{47}$$

Equating (46) and (47), we can obtain

$$\begin{aligned} \int_0^t e^{cs}\,ds\int_Y U(s,y,E)P(t',x,dy) &= e^{-ct'}\int_{t'}^{t+t'} e^{cs}U(s,x,E)\,ds \\ &= \int_0^t e^{cs}U(s+t',x,E)\,ds. \end{aligned}$$

Hence, for each $t' > 0$, there exists a null set $N_{t'}$ such that

$$U(t+t',x,E) = \int_Y U(t,y,E)P(t',x,dy) \qquad \text{for} \quad t \notin N_{t'}. \tag{48}$$

Let

$$Y_n = \{y = q(y) \leqslant n\}.$$

Then (48) implies

$$U(t+t',x,E) \geqslant \int_{Y_n} U(t,y,E)P(t',x,dy) \qquad \text{for} \quad t \notin N_{t'}. \tag{49}$$

Since $U(t, y, E) \leqslant q(y) + c \leqslant n + c$ for $y \in Y_n$ (see (44)), the integrand in the right side of (49) is bounded on Y_n. Also, since $U(t, y, E)$ is a continuous function of t (see (43)) the right side of (49) is a continuous function of t. Of course the left side of (49) is also a continuous function of t. Hence (49) holds for all t. Therefore, we obtain, by letting $n \to \infty$.

$$U(t + t', x, E) \geqslant \int_Y U(t, y, E) P(t', x, dy) \tag{50}$$

for all t, t', x, E. In the meantime, the right side of (49) is also a continuous function of t' (in fact, it is continuous in (t, t')), so it is jointly measurable in (t, t'). The right side of (48), being the limit of the right side of (49) as $n \to \infty$, is jointly measurable in (t, t'). Of course the left side of (48) is also jointly measurable in (t, t'). Hence, by the Fubini Theorem, there exists a null set H on the positive t-axis such that for each $t \notin H$, there exists a null set H_t such that

$$U(t + t', x, E) = \int_Y U(t, y, E) P(t', x, dy) \qquad \text{for} \quad t \notin H, \quad t' \in H_t. \tag{51}$$

Now suppose there exists a $t_0 \in H$ and a $t_0' \in H_{t_0}$ such that (51) does not hold. Then, according to (50),

$$U(t_0 + t_0', x, E) > \int_Y U(t_0, y, e) P(t_0', x, dy). \tag{52}$$

Thus, for each $t' > t_0'$, we have

$$\begin{aligned}
\int_Y U(t_0, y, E) P(t', x, dy) &= \int_Y U(t_0, y, E) \int_Y P(t' - t_0', x, dz) P(t_0', z, dy) \\
&= \int_Y P(t' - t_0', x, dz) \int_Y U(t_0, y, E) P(t_0', z, dy) \\
&= \left(\int_{\{x\}} + \int_{Y - \{x\}} \right) P(t' - t_0', x, dz) \int_Y U(t_0, y, E) P(t_0', z, dy) \\
&= P(t' - t_0', x, \{x\}) \int_Y U(t_0, y, E) P(t_0', x, dy) \\
&\quad + \int_{Y - \{x\}} P(t' - t_0', x, dz) \int_Y U(t_0, y, E) P(t_0', z, dy).
\end{aligned}$$

Since $P(t' - t_0', x, \{x\}) \geqslant e^{-q(x)(t' - t_0')} > 0$, by (52) and (50), we have

$$\begin{aligned}
\int_Y U(t_0, y, E) P(t', x, dy) &< P(t' - t_0', x, \{x\}) \cup (t_0 + t_0', x, E) \\
&\quad + \int_{Y - \{x\}} P(t' - t_0', x, dz) \cup (t_0 + t_0', z, E) \\
&= \int_Y P(t' - t_0', x, dz) \cup (t_0 + t_0', z, E).
\end{aligned}$$

Again, by (50), we obtain

$$\int_Y U(t_0, y, E) P(t', x, dy) < U(t_0 + t', x, E).$$

The above inequality holds for every $t' > t_0$. It contradicts (51). Therefore, the H_t's in (51) are all empty sets. Thus (51) becomes

$$U(t + t', x, E) = \int_Y U(t, y, E) P(t', x, dy) \qquad \text{whenever} \quad t \notin H. \tag{53}$$

For each given $t > 0$, choose a t_1 such that $0 < t_1 < t$, $t_1 \notin H$. Then, according to (53), we have

$$\begin{aligned} U(t+t',x,E) &= U(t_1+(t+t'-t_1),x,E) \\ &= \int_Y U(t_1,y,E)P(t+t'-t_1,x,dy) \\ &= \int_Y U(t_1,y,E)\int_Y P(t',x,dz)P(t-t_1,z,dy) \\ &= \int_Y P(t',x,dz)\int_Y U(t_1,y,E)P(t-t_1,z,dy). \end{aligned}$$

Using (50) again, we have

$$U(t+t',x,E) \leqslant \int_Y P(t',x,dz)U(t,z,E).$$

This inequality, together with (50), implies

$$U(t+t',x,E) = \int_Y U(t,y,E)P(t',x,dy). \tag{54}$$

We can have (42) by substituting (43) into (54). Q.E.D.

References

1. A. N. Kolmogorov, *Uch. Zap. MGU* (Matom) **4** (1951), 53–60.
2. D. G. Kendall, *Trans. Amer. Math. Soc.* **78** (1955), 509–510.
3. D. G. Austin, *Proc. Nat. Acad. Sci. USA* **41** (1955), 224–226.
4. K. L. Chung, *Trans. Amer. Math. Soc.* **81** (1956), 195–210.
5. E. Hille, *Functional Analysis and Semigroups, III*. 1951.
6. J. L. Doob, *Stochastic Processes*, New York, 1953.

Reprinted from
Prog. Math.
7, 240–281 (1964).

AN ASSOCIATION SCHEME $M_3(6)$ WHICH IS NOT AN L_3-SCHEME

XU BAO-LU (HSÜ PAO-LU)
(*Peking University*)

Given s^2 objects, an $M_3(s)$ association scheme is a scheme for associating each object with each other object in either of two ways, such that the following conditions are satisfied.

1) Any two distinct objects are associated with each other either in the first way or else in the second way. This association is symmetric; i. e., for any two objects x, y, if x is associated with y in the ith way ($i = 1, 2$), then y is associated with x in the ith way.

2) Each object is associated in the first way with $3(s-1)$ other objects.

3) Given any two objects x, y which are associated with each other in the first way, there exist exactly s objects each of which is associated in the first way with x and also with y.

4) Given any two objects z and w which are associated with each other in the second way, there exist exactly 6 objects each of which is associated in the first way with z and also with w.

We now give the definition of an L_3-scheme as follows. An $s \times s$ square array which is formed from s distinct elements in such a way that each element appears exactly once in each column and in each row is called an s-order "Latin square". Let us construct an arbitrary s-order Latin square $A(s)$ and let the s^2 objects discussed above be placed in an arbitrary one-to-one way in the s^2 positions of $A(s)$. Then let any two distinct objects be associated with each other in the first way if they are in the same column of $A(s)$ or in the same row or if they are in positions occupied by the same element of $A(s)$, and otherwise let the two objects be associated with each other in the second way. Then this association is clearly an $M_3(s)$ association scheme, and an $L_3(s)$ association scheme is defined as an $M_3(s)$ scheme which can be put in correspondence in this way with a Latin square $A(s)$.

Thus every $L_3(s)$ scheme is an $M_3(s)$ scheme. But the question arises whether conversely every $M_3(s)$ scheme is an $L_3(s)$ scheme. In the paper [1] the following theorem is given.

Theorem 1. *A necessary and sufficient condition for an $M_3(s)$ scheme to be an $L_3(s)$ scheme is: the $3(s-1)$ objects which are associated in the first way with any given object can always be divided into three sets $\{y_1, y_2, \cdots, y_{s-1}\}$, $\{z_1, z_2, \cdots, z_{s-1}\}$, $\{w_1, w_2, \cdots, w_{s-1}\}$ such that any two objects within the same set are associated with each other in the first way.*

In the paper [1], the question whether every $M_3(s)$ scheme is an $L_3(s)$ scheme was answered for $s \leq 5$ or $s \geq 24$.

The following table gives an $M_3(6)$ association scheme which is not an L_3 scheme.

Received July 17, 1963.

objects	objects which are associated in the first way with the objects listed on the left													
1	2	3	4	12	13	17	20	21	22	24	29	30	31	33	36
2	1	3	4	12	14	16	18	19	24	26	27	31	33	34	35
3	1	2	4	12	15	16	17	20	23	25	26	27	28	32	36
4	1	2	3	18	19	21	22	23	25	28	29	30	32	34	35
5	6	7	8	12	13	24	25	26	27	28	29	30	34	35	36
6	5	7	8	12	14	17	18	19	20	21	22	23	24	25	26
7	5	6	8	12	15	16	21	22	23	31	32	33	34	35	36
8	5	6	7	16	17	18	19	20	27	28	29	30	31	32	33
9	10	11	12	13	14	15	16	17	18	21	22	27	28	29	34
10	9	11	13	14	16	19	20	21	23	25	27	30	31	35	36
11	9	10	13	15	17	18	23	24	25	26	30	31	32	33	34
12	1	2	3	5	6	7	9	13	16	17	21	23	24	27	34
13	1	5	9	10	11	12	17	19	20	24	29	32	34	35	36
14	2	6	9	10	15	19	20	21	22	24	26	27	28	33	35
15	3	7	9	11	14	16	22	23	24	26	28	29	32	33	36
16	2	3	7	8	9	10	12	15	18	19	23	27	29	31	36
17	1	3	6	8	9	11	12	13	18	20	22	25	27	32	33
18	2	4	6	8	9	11	16	17	19	22	25	26	29	31	34
19	2	4	6	8	10	13	14	16	18	20	23	24	29	32	35
20	1	3	6	8	10	13	14	17	19	21	26	28	31	32	36
21	1	4	6	7	9	10	12	14	20	22	23	28	30	31	34
22	1	4	6	7	9	14	15	17	18	21	25	29	33	35	36
23	3	4	6	7	10	11	12	15	16	19	21	24	25	30	32
24	1	2	5	6	11	12	13	14	15	19	23	26	29	30	33
25	3	4	5	6	10	11	17	18	22	23	26	27	30	35	36
26	2	3	5	6	11	14	15	18	20	24	25	28	31	34	36
27	2	3	5	8	9	10	12	14	16	17	25	28	30	33	35
28	3	4	5	8	9	14	15	20	21	26	27	29	30	32	34
29	1	4	5	8	9	13	15	16	18	19	22	24	28	30	36
30	1	4	5	8	10	11	21	23	24	25	27	28	29	31	33
31	1	2	7	8	10	11	16	18	20	21	26	30	33	34	36
32	3	4	7	8	11	13	15	17	19	20	23	28	33	34	35
33	1	2	7	8	11	14	15	17	22	24	27	30	31	32	35
34	2	4	5	7	9	11	12	13	18	21	26	28	31	32	35
35	2	4	5	7	10	13	14	19	22	25	27	32	33	34	36
36	1	3	5	7	10	13	15	16	20	22	25	26	29	31	35

BIBLIOGRAPHY

[1] Liu Wan-ru (Liu Wan-ju), *On the question whether an M_3 association scheme can or can not be put in a Latin square*, Peking University Report (Natural Science), 1962, No. 2, pp. 105–116. (Chinese)

Translated by:
T. C. T. Ting

Comments by Ching-shui Cheng

Shrikhande (1959) showed that an L_2 scheme with n^2 objects is uniquely determined by its parameters unless $n = 4$; if $n = 4$, then there is one exception. This paper of Hsu's is related to Shrikhande's work. He gave an association scheme with 36 objects which has the same parameters as an L_3 scheme but is not an L_3 scheme. This in fact is only one of many such schemes; see, e.g., Seidel (1979).

References

Seidel, J. J. Strongly regular graphs. *Proc. 7th British Combin. Confer.*, Cambridge, 1979.

Shrikhande, S. S. The uniqueness of the L_2 association scheme. *Ann. Math. Statist.* **30** (1959), 781–798.

ACTA MATHEMATICA SINICA
Volume Fourteen (1964), No. 5, pp. 694–714

THE LIMITING DISTRIBUTIONS OF ORDER STATISTICS

BAN CHENG (PAN CH'ENG) 1)

(*Peking University, Department of Mathematical Mechanics*)

§1. Introduction

Let $x_1, x_2, \cdots, x_n$ be mutually independent random variables, all of which have the same distribution $F(x)$. Denote the order statistics of $x_1, x_2, \cdots, x_n$ by

$$\xi_1^{(n)}, \xi_2^{(n)}, \cdots, \xi_n^{(n)}, \quad (\xi_1^{(n)} \leqslant \xi_2^{(n)} \leqslant \cdots \leqslant \xi_n^{(n)}).$$

The distribution of $\xi_k^{(n)}$ is

$$\Phi_{k_n}(x) = n\binom{n-1}{k-1}\int_0^{F(x)} t^{k-1}(1-t)^{n-k}\,dt.$$

The ratio k/n is called here the relative rank of $\xi_k^{(n)}$. We assume that the relative rank of $\xi_k^{(n)}$ tends to a constant λ $(0 \le \lambda \le 1)$ as n tends to infinity. Because the case $\lambda = 1$ can be transformed into the case $\lambda = 0$, we always assume hereafter that $0 \le \lambda < 1$.

N. V. Smirnov [1] has made a detailed study of the case $\sqrt{n}\,(k_n/n - \lambda) \longrightarrow t$, where $0 \le \lambda < 1$, t is finite and k_n are constants, and he also found out all possible limiting laws and the domains of attraction of all types of law under these conditions.

But this paper is based on a different point of view from that of Smirnov. We do not make restrictions on the relative rank of the order statistics; instead we add certain restrictions on the order statistics themselves. In §3, we study some types of the limiting laws and their domains of attraction of the order statistics under these restrictions. In §4, we give examples concerning the distributions of some types of the limiting laws in their domains of attraction. Also in §2, we discuss the solution of a class of functional equations; we often meet these equations when seeking the limiting distribution of the order statistics.

We do not explain the symbols used in this section, since they will be defined again in the following sections.

§2. Solutions for a class of functional equations

Lemma 2.1. *If $u(x)$ is monotone increasing, continuous from the left, $u(-\infty) = -\infty$, $u(+\infty) = +\infty$, and nondegenerate (i.e. takes on all values except $\pm\infty$), and if for two real numbers $d_1 \neq 0$, $d_2 \neq 0$, with irrational quotient d_1/d_2 there exist real numbers $a_1 > 0$, b_1; $a_2 > 0$, b_2 such that*

$$\begin{cases} u(x) + d_1 = u(a_1 x + b_1), \\ u(x) + d_2 = u(a_2 x + b_2), \end{cases} \tag{1}$$

Received May 25, 1963.

1) Some results were reached in the discussion of a group directed by Xu Bao-lu (Hsü Pao-lu) in June, 1962, with notes by Sun Shan-ze (Sun Shan-tse).

then $u(x)$ *must be one of the several following functions:*

$$u_1(x) = \alpha x + \beta, \ \alpha > 0, \quad -\infty < x < +\infty.$$

$$u_2(x) = \begin{cases} -\infty, & x \leqslant c, \\ \alpha \log (x - c) + \beta, \ \alpha > 0, & x > c, \end{cases}$$

where $c = b_1/(1 - a_1) = b_2/(1 - a_2)$;

$$u_3(x) = \begin{cases} +\infty, & x \geqslant c, \\ -\alpha \log |x - c| + \beta, \alpha > 0, & x < c, \end{cases}$$

where $c = b_1/(1 - a_1) = b_2/(1 - a_2)$.

We now prove a complementary theorem to be used in the proof of Lemma 2.1.

Complementary Theorem. *If* $u(x)$ *is monotone increasing, continuous from the left,* $u(-\infty) = -\infty$, $u(+\infty) = +\infty$, *and for a set of real numbers* $d \neq 0$, $a > 0$, *and* b *satisfies*

$$u(x) + d = u(ax + b),$$

then $a = 1$ *is the necessary and sufficient condition for* $u(x)$ *not to assume the values* $\pm\infty$ *in* $(-\infty, +\infty)$.

Proof. Necessity. We prove the theorem by contradiction. If $a \neq 1$ and we set

$$\bar{u}(x) = u\left(x + \frac{b}{1 - a}\right),$$

then $\bar{u}(x)$ is monotone increasing, continuous from the left, $\bar{u}(-\infty) = -\infty$, $\bar{u}(+\infty) = +\infty$ and

$$\bar{u}(x) + d = \bar{u}(ax).$$

If $d > 0$, $ax \geq x$ is true for all x by the monotoneness of $\bar{u}(x)$, that is, for all x $(a - 1)x \geq 0$; but we assumed $(a - 1) \neq 0$, and thus it is impossible. If $d < 0$, we have $ax \leq x$ for all x, but this is also impossible if $(a - 1) \neq 0$. The contradiction leads to $a = 1$.

Sufficiency. Suppose $a = 1$, then we have

$$u(x) + d = u(x + b).$$

Let there be a point $x_0 \in (-\infty, +\infty)$, such that $u(x_0) = +\infty$. From $d \neq 0$, we have $b \neq 0$. We can assume $b > 0$ in the proof. When $k \longrightarrow +\infty$, we have $kb \longrightarrow +\infty$. Thus for any $x \in (-\infty, +\infty)$, there is a positive integer k_0 such that

$$x + k_0 b > x_0,$$

and therefore

$$\bar{u}(x) = \bar{u}(x + k_0 b) - k_0 d \geq \bar{u}(x_0) - k_0 d = +\infty,$$

and we get $\bar{u}(x) = +\infty$. But this contradicts the fact that $\bar{u}(-\infty) = -\infty$, and thus $\bar{u}(x)$ can not take on the value $+\infty$ in $(-\infty, +\infty)$. Similarly we can prove that $\bar{u}(x) \neq -\infty$ in $(-\infty, +\infty)$.

Now we return to the proof of Lemma 2.1. All we have to do is to solve the system of equations (1) under all situations.

i) If $u(x)$ does not take on values $\pm\infty$. The complementary Theorem assures $a_1 = a_2 = 1$. The system of equations (1) becomes

$$\begin{cases} u(x) + d_1 = u(x + b_1), \\ u(x) + d_2 = u(x + b_2). \end{cases} \tag{1.1}$$

From (1.1) we derive that for integers k_1, k_2

$$u(x + k_1b_1 + k_2b_2) = u(x) + k_1d_1 + k_2d_2.$$

But we know that the set of points

$$\{k_1d_1 + k_2d_2\colon\ d_1/d_2 \text{ irrational},\ k_1, k_2 = 0, \pm 1, \pm 2, \cdots\}$$

is dense in $(-\infty, +\infty)$. Because the values of $u(x)$ are dense in $(-\infty, +\infty)$, and $u(x)$ is not equal to $\pm\infty$, and also is monotone increasing, hence $u(x)$ is continuous in $(-\infty, +\infty)$.

Since d_1/d_2 is irrational, it follows that b_1/b_2 is also irrational. In fact, if b_1/b_2 were rational, there would exist two integers M, N such that

$$Nb_1 + Mb_2 = 0,$$

and therefore

$$\begin{aligned} u(0) &= u(Nb_1 + Mb_2) \\ &= u(0) + Nd_1 + Md_2. \end{aligned}$$

This must imply

$$Nd_1 + Md_2 = 0.$$

So it contradicts the fact that d_1/d_2 is irrational, and thus b_1/b_2 is irrational. Thus $\{k_1b_1 + k_2b_2\colon k_1, k_2 = 0, \pm 1, \pm 2, \cdots\}$ is dense in $(-\infty, +\infty)$. Hence for any arbitrary $x, y \in (-\infty, +\infty)$, there is a sequence $k_1^{(n)}b_1 + k_2^{(n)}b_2$ monotonely increasing with y for an upper bound. By the monotoneness of $u(x)$ it is easy to see that

$$k_1^{(n)}d_1 + k_2^{(n)}d_2 = u(k_1^{(n)}b_1 + k_2^{(n)}b_2) - u(0)$$

is also monotone increasing, and has upper bound $u(y) - u(0)$. Thus $\lim_{n\to\infty}(k_1^{(n)}d_1 + k_2^{(n)}d_2)$ exists. But $u(x)$ is continuous, and therefore

$$\begin{aligned} u(y) &= \lim_{n\to\infty} u(k_1^{(n)}b_1 + k_2^{(n)}b_2) \\ &= \lim_{n\to\infty} [u(0) + k_1^{(n)}d_1 + k_2^{(n)}d_2] \\ &= u(0) + \lim_{n\to\infty} [k_1^{(n)}d_1 + k_2^{(n)}d_2], \\ u(x+y) &= \lim_{n\to\infty} u(x + k_1^{(n)}b_1 + k_2^{(n)}b_2) \\ &= u(x) + \lim_{n\to\infty} [k_1^{(n)}d_1 + k_2^{(n)}d_2] \\ &= u(x) + u(y) - u(0). \end{aligned}$$

This result and the continuity and monotone increasing properties of $u(x)$ imply

$$u(x) = \alpha x + u(0),\ \alpha > 0.$$

That is, $u(x)$ is a function of the $u_1(x)$ type in (1).

ii) If there exists $x_0 \in (-\infty, +\infty)$ such that $u(x_0) = -\infty$. Here the complementary theorem gives $a_1 = 1$, $a_2 = 1$. We can also prove $b_1/(1 - a_1) = b_2/(1 - a_2)$. If $b_1/(1 - a_1) \neq b_2/(1 - a_2)$, we can assume $b_1/(1 - a_1) < b_2/(1 - a_2)$. Now we assert that there must be a point x_0' such that $u(x_0') = -\infty$ and $x_0' > b_1/(1 - a_1)$. When $x_0 > b_1/(1 - a_1)$, we pick $x_0' = x_0$ to meet the requirement. When $x_0 \leq b_1/(1 - a_1)$, then $x_0 < b_2/(1 - a_2)$. Let $u'(x) = u(x + b_2/(1 - a_2))$. Then (1) implies

$$u'(x) + d_2 = u'(a_2x).$$

Since $a_2 \neq 1$, hence for any $x < 0$ there exists an integer n such that

$$a_2^n x < x_0 - \frac{b_2}{1 - a_2},$$

and therefore

$$\begin{aligned} u'(x) &= u'(a_2^n x) - nd_2 \\ &\leqslant u'\left(x_0 - \frac{b_2}{1 - a_2}\right) - nd_2 \\ &= u(x_0) - nd_2 \\ &= -\infty. \end{aligned}$$

This leads to $u'(x) = -\infty$ for all $x < 0$, that is $u(x) = -\infty$ for all $x < b_2/(1 - a_2)$. Hence we can find a point x_0' such that $x_0' > b_1/(1 - a_1)$ and $u(x_0') = -\infty$, and thus our assertion is proved.

But the existence of a point x_0' such that $u(x_0') = -\infty$ and $x_0' > b_1/(1 - a_1)$ is impossible. For if we make a transformation $u''(x) = u(x + b_1/(1 - a_1))$, then from (1) we have

$$u''(x) + d_1 = u''(a_1 x).$$

Since $a_1 \neq 1$, hence for any $x > 0$, there exists an integer n such that

$$a_1^n x < x_0' - \frac{b_1}{1 - a_1},$$

and so

$$\begin{aligned} u''(x) &= u''(a_1^n x) - nd_1 \\ &\leqslant u''\left(x_0' - \frac{b_1}{1 - a_1}\right) - nd_1 \\ &= u(x_0') - nd_1 \\ &= -\infty. \end{aligned}$$

This implies that we always have $u(x) = -\infty$ for any $x > 0$, which contradicts the fact $u(+\infty) = +\infty$. Therefore such a point x_0' does not exist, so that we must have $b_1/(1 - a_1) = b_2/(1 - a_2) = c$.

From the above discussion, we can also point out that there must be an $x_0 \leq c$ and for any $x \leq c$, we always have $u(x) = -\infty$, but $u(x)$ can not be equal to $\pm\infty$, since otherwise $u(x)$ would be degenerate. Hence we are finished if we can find an expression for $u(x)$ when $x > c$.

Now let $\bar{u}(x) = u(x + c)$. The function $\bar{u}(x)$ remains monotone increasing, continuous from the left, with $\bar{u}(-\infty) = -\infty$, $\bar{u}(+\infty) = +\infty$, and nondegenerate. Also

$$\begin{cases} \bar{u}(x) + d_1 = \bar{u}(a_1 x), \\ \bar{u}(x) + d_2 = \bar{u}(a_2 x). \end{cases} \tag{1.2}$$

From (1.2) we get, for any two integers k_1, k_2

$$\bar{u}(a_1^{k_1} a_2^{k_2} x) = \bar{u}(x) + k_1 d_1 + k_2 d_2.$$

We now show that $\log a_1/\log a_2$ is irrational, since d_1/d_2 is irrational. For otherwise, there would exist two integers N, M, such that

$$N \log a_1 + M \log a_2 = 0,$$

or

$$a_1^N a_2^M = 1,$$

and therefore

$$\bar{u}(1) = \bar{u}(a_1^N a_2^M) = \bar{u}(1) + Nd_1 + Md_2,$$

and we must have

$$Nd_1 + Md_2 = 0,$$

which contradicts the fact that d_1/d_2 is irrational, and therefore $\log a_1/\log a_2$ is irrational. Hence $\{k_1 \log a_1 + k_2 \log a_2\colon\ k_1, k_2 = 0, \pm 1, \pm 2, \cdots\}$ is dense in $(-\infty, +\infty)$, and so $\{a_1^{k_1} a_2^{k_2}\colon\ k_1, k_2 = 0, \pm 1, \pm 2, \cdots\}$ is dense in $(0, +\infty)$.

That $u(x) \neq \pm\infty$ when $x > c$ implies that $\bar{u}(x) \neq \pm\infty$ when $x > 0$. Also, since the range of $\bar{u}(x)$ is dense in $(-\infty, +\infty)$ and $\bar{u}(x)$ is monotone increasing, we can conclude that $\bar{u}(x)$ is continuous in $(0, +\infty)$.

Now for arbitrary $x > 0$, $y > 0$, there is a sequence $a_1^{k_1^{(n)}} a_2^{k_2^{(n)}}$ which is monotone increasing and tends to y. Also $k_1^{(n)} d_1 + k_2^{(n)} d_2$ is monotone increasing and has a limit. Hence,

$$\bar{u}(y) = \bar{u}(1) + \lim_{n\to\infty} (k_1^{(n)} d_1 + k_2^{(n)} d_2),$$

$$\begin{aligned}\bar{u}(xy) &= \lim_{n\to\infty} \bar{u}(a_1^{k_1^{(n)}} a_2^{k_2^{(n)}} x)\\ &= \bar{u}(x) + \lim_{n\to\infty} (k_1^{(n)} d_1 + k_2^{(n)} d_2)\\ &= \bar{u}(x) + \bar{u}(y) - \bar{u}(1).\end{aligned}$$

This result and the properties of continuity and monotone increase of $\bar{u}(x)$ imply

$$\bar{u}(x) = \alpha \log x + \bar{u}(1), \quad \alpha > 0,\ x > 0,$$

and therefore

$$\bar{u}(x) = \begin{cases} -\infty, & x \leqslant 0,\\ \alpha \log x + \bar{u}(1), \quad \alpha > 0, & x > 0.\end{cases}$$

$$u(x) = \begin{cases} -\infty, & x \leqslant c,\\ \alpha \log (x - c) + \bar{u}(1), \quad \alpha > 0, & x > c.\end{cases}$$

This is the $u_2(x)$ type of function in (1).

iii) If there exists $x_0 \in (-\infty, +\infty)$ such that $u(x_0) = +\infty$. We also have $a_1 \neq 1$, $a_2 \neq 1$, and $b_1/(1 - a_1) = b_2/(1 - a_2) = c$. Again we make a transformation

$$\bar{u}(x) = -u(-x-c).$$

Then $\bar{u}(x)$ remains monotone increasing, with $\bar{u}(-\infty) = -\infty$, $\bar{u}(+\infty) = +\infty$, and nondegenerate. Also

$$\begin{cases} \bar{u}(x) - d_1 = \bar{u}(a_1 x),\\ \bar{u}(x) - d_2 = \bar{u}(a_2 x).\end{cases}$$

From the result of ii) we have

$$\bar{u}(x) = \begin{cases} -\infty, & x \leqslant 0,\\ \alpha \log x + \beta, \quad \alpha > 0, & x > 0.\end{cases}$$

Hence

$$u(x) = \begin{cases} +\infty, & x \geqslant c,\\ -\alpha \log |x - c| + \beta, \quad \alpha > 0, & x < c.\end{cases}$$

This is the $u_3(x)$ type of function in (1).

We have thus finished the discussion of the system of equations (1) under all situations, and so the proof of Lemma 2.1 is complete.

Lemma 2.2. *If $u(x)$ is monotone increasing, continuous from the left, $u(-\infty) = -\infty$, $u(+\infty) = +\infty$, nondegenerate, and for real $\lambda \neq 1$ (> 0), $d \neq 0$, there exist $a_1 > 0$, b_1, b_2 such that*

$$\begin{cases} \lambda u(x) = u(a_1 x + b_1), \\ u(x) + d = u(x + b_2), \end{cases} \tag{2}$$

then $u(x) = \alpha x + \beta$, $\alpha > 0$, $-\infty < x < +\infty$.

Proof. $\lambda \neq 1$ implies that $a_1 \neq 1$. In fact, if $a_1 = 1$, we must have $b_1 = 0$, otherwise, $u(x)$ never changes signs or is identically equal to 0 for all x; but this contradicts either $u(-\infty) = -\infty$ or $u(+\infty) = +\infty$. Hence

$$\lambda u(x) = u(x).$$

When $\lambda \neq 1$, the nondegenerate $u(x)$ that satisfies this equation must be

$$u(x) = \begin{cases} +\infty, & \text{if } x > A, \\ 0, & A \geqslant x > B, \\ -\infty, & x \leqslant B, \end{cases}$$

where A, B are fixed constants. But obviously this solution does not satisfy

$$u(x) + d = u(x + b_2);$$

thus $a_1 \neq 1$ is true. Now we make a transformation

$$\bar{u}(x) = u\left(x + \frac{x}{1 - a_1}\right),$$

then $\bar{u}(x)$ remains monotone increasing, continuous from the left, $\bar{u}(-\infty) = -\infty$, $\bar{u}(+\infty) = +\infty$, nondegenerate, and

$$\begin{cases} \lambda \bar{u}(x) = \bar{u}(a_1 x), \\ \bar{u}(x) + d = \bar{u}(x + b_2), \end{cases} \tag{2.1}$$

where $\lambda \neq 1$ (> 0), $d \neq 0$. By use of the preceding complementary theorem, from the equation we know that $\bar{u}(x)$ will not take on the values $\pm\infty$. For simplicity, let $\lambda > 1$, $d > 0$, and the monotone increasing property of $\bar{u}(x)$ assures us that $a_1 > 0$, $b_2 > 0$. From (2.1) we have for arbitrary integers n, m

$$\bar{u}(a_1^n) = \lambda^n \bar{u}(1), \quad \bar{u}(0) = 0,$$
$$\bar{u}(m b_2) = md.$$

For arbitrary n, there exist integers $k(n)$ such that

$$k(n) b_2 \leqslant a_1^n < (k(n) + 1) b_2.$$

As $n \longrightarrow \infty$,

$$\left| 1 - \frac{k(n) b_2}{a_1^n} \right| < \left| \frac{(k(n) + 1) b_2}{a_1^n} - \frac{k(n) b_2}{a_1^n} \right| = \left| \frac{b_2}{a_1^n} \right| \longrightarrow 0,$$

hence

$$\frac{k(n)}{a_1^n} \to \frac{1}{b_2}.$$

But $\bar{u}(x)$ is monotone increasing, hence

$$\bar{u}(k(n)b_2) \leqslant \bar{u}(a_1^n) \leqslant \bar{u}((k(n)+1)b_2),$$

or

$$k(n)d \leqslant \lambda^n \bar{u}(1) \leqslant (k(n)+1)d,$$

so that

$$\frac{k(n)d}{a_1^n} \leqslant \left(\frac{\lambda}{a_1}\right)^n \bar{u}(1) \leqslant \frac{k(n)d}{a_1^n} + \frac{d}{a_1^n}.$$

Thus we get

$$\overline{\lim_{n\to\infty}} \left(\frac{\lambda}{a_1}\right)^n \bar{u}(1) \leqslant \frac{d}{b_2},$$

$$\underline{\lim_{n\to\infty}} \left(\frac{\lambda}{a_1}\right)^n \bar{u}(1) \geqslant \frac{d}{b_2},$$

therefore

$$\lim_{n\to\infty} \left(\frac{\lambda}{a_1}\right)^n \bar{u}(1) = \frac{d}{b_2}.$$

The necessary and sufficient conditions for this equation to be true are: $a_1 = \lambda$, $d = \bar{u}(1)b_2$. And so

$$\bar{u}(a_1^n) = a_1^n \bar{u}(1),$$
$$\bar{u}(mb_2) = \bar{u}(1)mb_2,$$
$$\bar{u}\left(\frac{mb_2}{a_1^n}\right) = \bar{u}(1)\frac{mb_2}{a_1^n}.$$

But $\{mb_2/a_1^n\colon\ a_1 > 1,\ b_2 > 0,\ m, n \text{ integers}\}$ is dense in $(-\infty, +\infty)$. Thus for any $x \in (-\infty, +\infty)$, there is a subsequence $\{m_k b_2/a_1^{n_k}\}$ that is monotone increasing toward x $(k \to \infty)$. Then

$$\begin{aligned}\bar{u}(x) &= \lim_{k\to\infty} \bar{u}\left(\frac{m_k b_2}{a_1^{n_k}}\right)\\ &= \lim_{k\to\infty} \bar{u}(1)\frac{m_k b_2}{a_1^{n_k}}\\ &= \bar{u}(1)x.\end{aligned}$$

It is not hard to see that $\bar{u}(1)$ is greater than zero. Hence

$$u(x) = \alpha\left(x - \frac{b_1}{1-a_1}\right) = \alpha x + \beta,\ \alpha > 0,\ -\infty < x < +\infty,$$

where $\alpha = \bar{u}(1)$, $\beta = \alpha b_1/(1-a_1)$.

Remark I. N. V. Smirnov [1] used the following functional equation: if $u(x)$ is monotone increasing, continuous from the left, $u(-\infty) = -\infty$, $u(+\infty) = +\infty$, and $u(x)$ is nondegenerate, then for every integer $\lambda > 1$ there exist $a_\lambda > 1$, b_λ such that

$$u(x) = \sqrt{\lambda}\, u(a_\lambda x + b_\lambda).$$

In fact, if for two real numbers λ_1, λ_2 which satisfy the following conditions: $\lambda_1 > 0$, $\lambda_2 > 0$, $\log\lambda_1/\log\lambda_2$ irrational, there exist $a_1 > 0$, b_1; $a_2 > 0$, b_2 such that:

$$\begin{cases} \lambda_1 u(x) = u(a_1 x + b_1), \\ \lambda_2 u(x) = u(a_2 x + b_2), \end{cases} \tag{3}$$

then we can solve for $u(x)$. The method is similar to that of solving (1); there are four types of functions in the solution of (3), but we omit them here since they have already been given in [1].

Remark II. Whenever (d_1/d_2)[or $\log\lambda_1/\log\lambda_2$] is rational in the system of equations (1) [or (3)], the system (1) [or (3)] has no definite solution. Hence we have added the condition that (d_1/d_2) [or $\log\lambda_1/\log\lambda_2$] is irrational.

Remark III. If we have obtained solutions of the systems (1), (2), (3) we are in fact able to solve the following system of equations (4).

If $u(x)$ is monotone increasing, continuous from the left, $u(-\infty) = -\infty$, $u(+\infty) = +\infty$, and $u(x)$ is nondegenerate, then for real numbers $\lambda_i > 0$, $d_i \neq 0$ $(i = 1, 2)$, there exist real numbers $a_i > 0$, b_i $(i = 1, 2)$ such that

$$\begin{cases} \lambda_1 u(x) + d_1 = u(a_1 x + b_1), \\ \lambda_2 u(x) + d_2 = u(a_2 x + b_2). \end{cases} \tag{4}$$

This is because (4) can always be reduced to the forms (1), (2), or (3). But now we shall give the complete calculation as follows:

i) If $\lambda_1 = \lambda_2 = 1$ in (4), then (4) is (1).

ii) In (4), if λ_1, λ_2 are not both 1, we can let $\lambda_1 \neq 1$, and make a transformation

$$v(x) = u(x) + \frac{d_1}{\lambda_1 - 1}.$$

The function $v(x)$ remains monotone increasing, is continuous from the left, and $v(-\infty) = -\infty$, $v(+\infty) = +\infty$; $v(x)$ is nondegenerate, and satisfies

$$\begin{cases} \lambda_1 v(x) = v(a_1 x + b_1), \\ \lambda_2 v(x) + \left(\dfrac{1 - \lambda_2}{\lambda_1 - 1} d_1 + d_2\right) = v(a_2 x + b_2), \end{cases} \tag{5}$$

when $((1 - \lambda_2)/(\lambda_1 - 1))d_1 + d_2 = 0$, then (5) is (3). The case $((1 - \lambda_2)/(\lambda_1 - 1))d_1 + d_2 \neq 0$ is as follows.

iii) In the case $((1 - \lambda_2)/(\lambda_1 - 1))d_1 + d_2 \neq 0$, (5) has the form

$$\begin{cases} \lambda_1 v(x) = v(a_1 x + b_1), \\ \lambda_2 v(x) + c = v(a_2 x + b_2), \end{cases}$$

where $\lambda_1 \neq 1$, $c \neq 0$. Now we make another transformation

$$w(x) = v(a_1 a_2 x) - c.$$

The function $w(x)$ remains monotone increasing, is continuous from the left, and $w(-\infty) = -\infty$, $w(+\infty) = +\infty$; $w(x)$ is nondegenerate, and satisfies

$$\begin{cases} \lambda_1 w(x) = w\left(a_1 x + \dfrac{a_2 b_1 - a_1 b_2 + b_2}{a_1 a_2}\right), \\ w(x) + (\lambda_1 - 1)c = w\left(x + \dfrac{(a_1 - 1)b_2 + (1 - a_2)b_1}{a_1 a_2}\right). \end{cases} \tag{6}$$

This is in the form of the system (2).

Consequently, the problem of solving the system of equations (4) can always be reduced to the problem of solving (1), (2) or (3). Thus we solve (4) whenever we solve the others.

§3. The limiting distributions of order statistics

Definition 3.1. If for a distribution $F(x)$ there is a sequence of positive integers $\{k_n\}$, with $k_n \to \infty$, $k_n/n \to \lambda$, $0 \le \lambda < 1$, when $n \to \infty$; if there are $a_n > 0$, b_n such that

$$\Phi_{k_n n}(a_n x + b_n) \to \Phi(x),$$

then we say that $F(x)$ *belongs to the* λ*-domain of attraction of* $\Phi(x)$. Also we say that $\Phi(x)$ is the *limiting distribution law of* $F(x)$.

Since N. V. Smirnov [1] has already studied the case when $\Phi(x)$ is a degenerate distribution law, we assume in the following that $\Phi(x)$ is always nondegenerate. We quote from [1]

Lemma 3.1. *If* $k_n \to \infty$, $k_n/n \to \lambda$, $0 \le \lambda < 1$, *as* $n \to \infty$. *Then the existence of* $a_n > 0$, b_n *such that*

$$\Phi_{k_n n}(a_n x + b_n) \to \Phi(x)$$

is a necessary and sufficient condition for

$$\frac{n}{\sqrt{k_n}}\left[F(a_n x + b_n) - \frac{k_n}{n}\right] \to \sqrt{1-\lambda}\, u(x),$$

where $u(x)$ *is monotone increasing, continuous from the left,* $u(-\infty) = -\infty$, $u(+\infty) = +\infty$, *and* $u(x)$ *is nondegenerate.*

If the condition is satisfied, then

$$\Phi(x) = \frac{1}{\sqrt{2\pi}}\int_{-\infty}^{u(x)} e^{-\frac{t^2}{2}}\,dt.$$

Hereafter, when $F(x)$ belongs to the λ domain of attraction of $\Phi(x)$ we shall also say that $F(x)$ belongs to the λ domain of attraction of the $u(x)$ to which $\Phi(x)$ corresponds.

Theorem 3.1. *Suppose there exist three sequences of positive integers* $\{k_n^{(i)}\}$, $i = 1, 2, 3$, *which satisfy:* $k_n^{(i)} \to \infty$, $k_n^{(i)}/n \to \infty$, $i = 1, 2, 3$, $0 \le \lambda < 1$, $(k_n^{(2)} - k_n^{(1)})/\sqrt{k_n^{(1)}} \to d_1$, $(k_n^{(3)} - k_n^{(1)})/\sqrt{k_n^{(1)}} \to d_2$, d_1/d_2 *irrational, and also that with respect to these three sequences, there exist* $a_n^{(i)} > 0$, $b_n^{(i)}$, $i = 1, 2, 3$, *such that*

$$\Phi_{k_n^{(i)} n}(a_n^{(i)} x + b_n^{(i)}) \to \Phi(x), \quad i = 1, 2, 3,$$

then $\Phi(x)$ *can only be one of the following types of laws:* 1)

$$\Phi_1(x) = \frac{1}{\sqrt{2\pi}}\int_{-\infty}^{x} e^{-\frac{t^2}{2}}\,dt, \quad -\infty < x < +\infty,$$

$$\Phi_2(x) = \begin{cases} 0, & x \leqslant 0, \\ \dfrac{1}{\sqrt{2\pi}}\displaystyle\int_{-\infty}^{\alpha \log x + \beta} e^{-\frac{t^2}{2}}\,dt, \ \alpha > 0, & x > 0, \end{cases}$$

$$\Phi_3(x) = \begin{cases} \dfrac{1}{\sqrt{2\pi}}\displaystyle\int_{-\infty}^{-\alpha \log |x| + \beta} e^{-\frac{t^2}{2}}\,dt, \ \alpha > 0, & x < 0, \\ 1, & x \geqslant 0. \end{cases}$$

---

1) In this section, β is considered as a fixed constant, since a different β generates the same type of law.

Proof. We have from Lemma 3.1

$$\frac{n}{\sqrt{k_n^{(1)}}}\left[F(a_n^{(1)}x+b_n^{(1)})-\frac{k_n^{(1)}}{n}\right]\to\sqrt{1-\lambda}\,u(x),$$

$$\frac{n}{\sqrt{k_n^{(2)}}}\left[F(a_n^{(2)}x+b_n^{(2)})-\frac{k_n^{(2)}}{n}\right]\to\sqrt{1-\lambda}\,u(x),$$

$$\frac{n}{\sqrt{k_n^{(3)}}}\left[F(a_n^{(3)}x+b_n^{(3)})-\frac{k_n^{(3)}}{n}\right]\to\sqrt{1-\lambda}\,u(x),$$

where $u(x)$ is determined by $\Phi(x)=(1/\sqrt{2\pi})\int_{-\infty}^{u(x)}e^{-t^2/2}dt$.

By $(k_n^{(2)}-k_n^{(1)})/\sqrt{k_n^{(1)}}\to d_1$, we get

$$k_n^{(2)}=k_n^{(1)}+d_1\sqrt{k_n^{(1)}}+o(\sqrt{k_n^{(1)}}).$$

Thus

$$\frac{n}{\sqrt{k_n^{(2)}}}\left[F(a_n^{(1)}x+b_n^{(1)})-\frac{k_n^{(2)}}{n}\right]$$

$$=\frac{n}{\sqrt{k_n^{(2)}}}\left[F(a_n^{(1)}x+b_n^{(1)})-\frac{k_n^{(1)}+d_1\sqrt{k_n^{(1)}}+o(\sqrt{k_n^{(1)}})}{n}\right]$$

$$=\sqrt{\frac{k_n^{(1)}}{k_n^{(2)}}}\frac{n}{\sqrt{k_n^{(1)}}}\left[F(a_n^{(1)}x+b_n^{(1)})-\frac{k_n^{(1)}}{n}\right]-d_1\sqrt{\frac{k_n^{(1)}}{k_n^{(2)}}}-\frac{o(\sqrt{k_n^{(1)}})}{\sqrt{k_n^{(2)}}}.$$

Since $\sqrt{k_n^{(1)}/k_n^{(2)}}\to 1$, hence

$$\frac{n}{\sqrt{k_n^{(2)}}}\left[F(a_n^{(1)}x+b_n^{(1)})-\frac{k_n^{(2)}}{n}\right]\to\sqrt{1-\lambda}\,u(x)-d_1,$$

but the theorem of A. Hinčin assures us there exist $\alpha_1>0$, β_1 such that

$$u(\alpha_1x+\beta_1)=u(x)+\frac{d_1}{\sqrt{1-\lambda}}.$$

Similarly, we get the existence of $\alpha_2>0$, β_2 such that

$$u(\alpha_2x+\beta_2)=u(x)+\frac{d_2}{\sqrt{1-\lambda}}.$$

The solution of the system generated by these two equations is guaranteed by Lemma 2.1. Thus the theorem is true.

Theorem 3.2. *Suppose there exist three sequences of positive integers* $\{k_n^{(i)}\}$, $i=1,2,3$, *which satisfy* $k_n^{(i)}\to\infty$, $k_n^{(i)}/n\to\lambda$, $i=1,2,3$, $0\le\lambda<1$, $(k_n^{(2)}-k_n^{(1)})/\sqrt{k_n^{(1)}}\to d$, $(\nu k_n^{(1)}-k_{[\nu n]}^{(3)})/\sqrt{k_n^{(1)}}\to 0$, *where* $\nu\neq 1$ (>0), $d\neq 0$, *and* $[\nu n]$ *denotes the largest integer of* νn. *If with respect to these three sequences, there exist* $a_n^{(i)}>0$, $b_n^{(i)}$, $i=1,2,3$, *such that*

$$\Phi_{k_n^{(i)},n}(a_n^{(i)}x+b_n^{(i)})\to\Phi(x),\quad i=1,2,3,$$

then $\Phi(x)$ *can only be a normal distribution law.*

Proof. By Lemma 3.1 we get

93

$$\frac{n}{\sqrt{k_n^{(1)}}}\left[F(a_n^{(1)}x + b_n^{(1)}) - \frac{k_n^{(1)}}{n}\right] \to \sqrt{1-\lambda}\, u(x),$$

$$\frac{n}{\sqrt{k_n^{(2)}}}\left[F(a_n^{(2)}x + b_n^{(2)}) - \frac{k_n^{(2)}}{n}\right] \to \sqrt{1-\lambda}\, u(x),$$

$$\frac{n}{\sqrt{k_n^{(3)}}}\left[F(a_n^{(3)}x + b_n^{(3)}) - \frac{k_n^{(3)}}{n}\right] \to \sqrt{1-\lambda}\, u(x),$$

where $u(x)$ is determined by $\Phi(x) = (1/\sqrt{2\pi})\int_{-\infty}^{u(x)} e^{-t^2/2}dt$.

From the proof of Theorem 3.1 we have: there exist $\alpha_1 > 0$, β_1 such that

$$u(\alpha_1 x + \beta_1) = u(x) + d.$$

Also, from $(\nu k_n^{(1)} - k_{[\nu n]}^{(3)})/\sqrt{k_n^{(1)}} \to 0$ we get

$$k_{[\nu n]}^{(3)} = \nu k_n^{(1)} + o(\sqrt{k_n^{(1)}}).$$

Therefore

$$\frac{[\nu n]}{\sqrt{k_{[\nu n]}^{(3)}}}\left[F(a_n^{(1)}x + b_n^{(1)}) - \frac{k_{[\nu n]}^{(3)}}{[\nu n]}\right]$$

$$= \frac{[\nu n]}{\sqrt{k_{[\nu n]}^{(3)}}}\left[F(a_n^{(1)}x + b_n^{(1)}) - \frac{\nu k_n^{(1)} + o(\sqrt{k_n^{(1)}})}{[\nu n]}\right]$$

$$= \frac{[\nu n]}{\nu n}\sqrt{\frac{\nu k_n^{(1)}}{k_{[\nu n]}^{(3)}}}\frac{\sqrt{\nu}\, n}{\sqrt{k_n^{(1)}}}\left[F(a_n^{(1)}x + b_n^{(1)}) - \frac{k_n^{(1)}}{n}\right]$$

$$+ \frac{k_n^{(1)}([\nu n] - \nu n)}{n\sqrt{k_{[\nu n]}^{(3)}}} + \frac{o(\sqrt{k_n^{(1)}})}{\sqrt{k_{[\nu n]}^{(3)}}}.$$

Since $\sqrt{k_{[\nu n]}^{(3)}} \to \infty$, $\nu k_n^{(1)}/k_{[\nu n]}^{(3)} \to 1$, $[\nu n]/\nu n \to 1$, therefore

$$\frac{[\nu n]}{\sqrt{k_{[\nu n]}^{(3)}}}\left[F(a_n^{(1)}x + b_n^{(1)}) - \frac{k_{[\nu n]}^{(3)}}{[\nu n]}\right] \to \sqrt{\nu(1-\lambda)}\, u(x).$$

Also, the A. Hinčin theorem assures us of the existence of $\alpha_2 > 0$, β_2 such that

$$\sqrt{\nu}\, u(x) = u(\alpha_2 x + \beta_2),$$

and then $u(x)$ satisfies

$$\begin{cases} u(x) + d = u(\alpha_1 x + \beta_1), \\ \sqrt{\nu}\, u(x) = u(\alpha_2 x + \beta_2). \end{cases}$$

We now prove that there we must have $\alpha_1 = 1$ in the preceding system of equations. Use the method of contradiction. Let $\alpha_1 \neq 1$; by the method used to prove Lemma 2.1 in §2, we get $\beta_1/(1-\alpha_1) = \beta_2/(1-\alpha_2)$. Now make a transformation

$$\bar{u}(x) = u\left(x + \frac{\beta_1}{1-\alpha_1}\right);$$

then $\bar{u}(x)$ satisfies

$$\begin{cases} \bar{u}(x) + d = \bar{u}(\alpha_1 x), \\ \sqrt{v}\, \bar{u}(x) = \bar{u}(\alpha_2 x), \end{cases}$$

and since $\alpha_1 \neq 1$, by the complementary theorem in §2, we know that there is a point x_0 such that $u(x_0) = +\infty$ or $-\infty$. We can let $x_0 \leq \beta_1/(1 - \alpha_1)$, $\lambda > 1$, $d > 0$. Hence, when $x < 0$, we always have $\bar{u}(x) = -\infty$, $\alpha_2 > 1$, $\alpha_1 > 1$.

Since $\bar{u}(x)$ satisfies

$$\bar{u}(x) + d = \bar{u}(\alpha_1 x),$$

we get

$$\bar{u}\left(\frac{1}{\alpha_1^n}\right) = \bar{u}(1) - nd,$$

$$\lim_{n \to \infty} \bar{u}\left(\frac{1}{\alpha_1^n}\right) = -\infty.$$

Thus

$$\lim_{x \to +0} \bar{u}(x) = -\infty.$$

Also since $\bar{u}(x)$ satisfies

$$\sqrt{\lambda}\, \bar{u}(x) = \bar{u}(\alpha_2 x),$$

we get

$$\bar{u}\left(\frac{1}{\alpha_2^n}\right) = \left(\frac{1}{\lambda}\right)^{n/2} \bar{u}(1),$$

$$\lim_{n \to \infty} \bar{u}\left(\frac{1}{\alpha_2^n}\right) = 0.$$

Thus

$$\lim_{x \to +0} \bar{u}(x) = 0.$$

But these two results contradict each other; hence $\alpha_1 \neq 1$ can not be true and so we must have $\alpha_1 = 1$. This leads to the conclusion that the system of equations which $u(x)$ satisfies is

$$\begin{cases} u(x) + d = u(x + \beta_1), \\ \sqrt{v}\, u(x) = u(\alpha_2 x + \beta_2). \end{cases}$$

The solution can be reached by Lemma 2.2, so that $\Phi(x)$ must have a normal distribution law.

Theorem 3.3. *If $u(x)$ is strictly monotone increasing in an interval (a, b), then a necessary and sufficient condition for $F(x)$ to belong to the λ domain of attraction of $u(x)$ is that there exist $k_n \to \infty$, $k_n/n \to \lambda$, $0 \leq \lambda < 1$, with*

$$a_n' = \frac{1}{x_2 - x_1}\left[\bar{a}_{\left(\frac{k_n}{n} + \frac{\sqrt{(1-\lambda)k_n}}{n} u(x_2)\right)} - \bar{a}_{\left(\frac{k_n}{n} + \frac{\sqrt{(1-\lambda)k_n}}{n} u(x_1)\right)}\right],$$

$$b_n' = \frac{1}{x_2 - x_1}\left[x_2 \bar{a}_{\left(\frac{k_n}{n} + \frac{\sqrt{(1-\lambda)k_n}}{n} u(x_1)\right)} - x_1 \bar{a}_{\left(\frac{k_n}{n} + \frac{\sqrt{(1-\lambda)k_n}}{n} u(x_2)\right)}\right],$$

such that

$$\frac{n}{\sqrt{k_n}}\left[F(a_n' x + b_n') - \frac{k_n}{n}\right] \to \sqrt{(1 - \lambda)}\, u(x),$$

where $\bar{a}_M = \inf\{x: F(x) > M\}$. *Here* $x_2 > x_1$ *are two points in the inteval* (a, b).

Proof. The sufficiency is obvious.

Necessity. By Lemma 3.1, if $F(x)$ belongs to the λ domain of attraction, then there exist $k_n \to \infty$, $k_n/n \to \lambda$, and we have $a_n > 0$, b_n such that

$$\frac{n}{\sqrt{k_n}}\left[F(a_n x + b_n) - \frac{k_n}{n}\right] \to \sqrt{1-\lambda}\, u(x).$$

In particular, for two points x_1, x_2 inside the interval (a, b) we have

$$\frac{n}{\sqrt{k_n}}\left[F(a_n x_1 + b_n) - \frac{k_n}{n}\right] \to \sqrt{1-\lambda}\, u(x_1),$$

$$\frac{n}{\sqrt{k_n}}\left[F(a_n x_2 + b_n) - \frac{k_n}{n}\right] \to \sqrt{1-\lambda}\, u(x_2).$$

Since $u(x)$ is strictly monotone increasing in the interval (a, b), it follows that with sufficiently large n and arbitrary $\epsilon > 0$ we have

$$\frac{n}{\sqrt{k_n}}\left[F(a_n(x_1 + \varepsilon) + b_n) - \frac{k_n}{n}\right] > \sqrt{1-\lambda}\, u(x_1),$$

$$\frac{n}{\sqrt{k_n}}\left[F(a_n(x_1 - \varepsilon) + b_n) - \frac{k_n}{n}\right] < \sqrt{1-\lambda}\, u(x_1),$$

$$\frac{n}{\sqrt{k_n}}\left[F(a_n(x_2 + \varepsilon) + b_n) - \frac{k_n}{n}\right] > \sqrt{1-\lambda}\, u(x_2),$$

$$\frac{n}{\sqrt{k_n}}\left[F(a_n(x_2 - \varepsilon) + b_n) - \frac{k_n}{n}\right] < \sqrt{1-\lambda}\, u(x_2).$$

This gives us

$$\begin{aligned} a_n(x_1 - \varepsilon) + b_n &\leqslant \bar{a}_{\left(\frac{k_n}{n} + \frac{\sqrt{(1-\lambda)k_n}}{n} u(x_1)\right)} \\ &\leqslant a_n(x_1 + \varepsilon) + b_n, \\ a_n(x_2 - \varepsilon) + b_n &\leqslant \bar{a}_{\left(\frac{k_n}{n} + \frac{\sqrt{(1-\lambda)k_n}}{n} u(x_2)\right)} \\ &\leqslant a_n(x_2 + \varepsilon) + b_n. \end{aligned}$$

Therefore if we let

$$a_n' = \frac{1}{x_2 - x_1}\left[\bar{a}_{\left(\frac{k_n}{n} + \frac{\sqrt{(1-\lambda)k_n}}{n} u(x_2)\right)} - \bar{a}_{\left(\frac{k_n}{n} + \frac{\sqrt{(1-\lambda)k_n}}{n} u(x_1)\right)}\right],$$

$$b_n' = \frac{1}{x_2 - x_1}\left[x_2\bar{a}_{\left(\frac{k_n}{n} + \frac{\sqrt{(1-\lambda)k_n}}{n} u(x_1)\right)} - x_1\bar{a}_{\left(\frac{k_n}{n} + \frac{\sqrt{(1-\lambda)k_n}}{n} u(x_2)\right)}\right],$$

we have when $n \to \infty$

$$\frac{a_n'}{a_n} \to 1, \quad \frac{b_n - b_n'}{a_n} \to 0.$$

By the A. Hinčin theorem on the types of law, we know it is also true for a_n', b_n' that

$$\frac{n}{\sqrt{k_n}}\left[F(a_n' x + b_n') - \frac{k_n}{n}\right] \to \sqrt{1-\lambda}\, u(x).$$

96

The theorem is thus proved.

Corollary 1. *A necessary and sufficient condition for $F(x)$ to belong to the λ domain of attraction of $u(x) = x$ is that there exist $k_n \to \infty$, $k_n/n \to \lambda$, $0 \le \lambda < 1$, and*

$$a'_n = \bar{a}_{\left(\frac{k_n}{n} + \frac{\sqrt{(1-\lambda)k_n}}{n}\right)} - \bar{a}_{\frac{k_n}{n}},$$

$$b'_n = \bar{a}_{\frac{k_n}{n}}$$

such that

$$\frac{n}{\sqrt{k_n}}\left[F(a'_n x + b'_n) - \frac{k_n}{n}\right] \to \sqrt{1-\lambda}\, x.$$

Proof. In Theorem 3.3 take $u(x) = x$, $x_1 = 0$, $x_2 = 1$.

Corollary 2. *A necessary and sufficient condition for $F(x)$ to belong to the λ domain of attraction of $u(x)$*

$$u(x) = \begin{cases} -\infty, & x \leqslant 0, \\ \alpha \log x + \beta,\ \alpha > 0, & x > 0 \end{cases}$$

is that there exist $k_n \to \infty$, $k_n/n \to \lambda$, $0 \le \lambda < 1$, and

$$a'_n = e^{\beta/\alpha}(e^{1/\alpha} - 1)^{-1}\left(\bar{a}_{\left(\frac{k_n}{n} + \frac{\sqrt{(1-\lambda)k_n}}{n}\right)} - \bar{a}_{\frac{k_n}{n}}\right),$$

$$b'_n = (e^{1/\alpha} - 1)^{-1}\left(e^{1/\alpha}\, \bar{a}_{\frac{k_n}{n}} - \bar{a}_{\left(\frac{k_n}{n} + \frac{\sqrt{(1-\lambda)k_n}}{n}\right)}\right),$$

such that

$$\frac{n}{\sqrt{k_n}}\left[F(a'_n x + b'_n) - \frac{k_n}{n}\right] \to \begin{cases} -\infty, & x \leqslant 0, \\ \alpha \log x + \beta, & x > 0. \end{cases}$$

Proof. In Theorem 3.3, take

$$u(x) = \begin{cases} -\infty, & x \leqslant 0, \\ \alpha \log x + \beta,\ \alpha > 0, & x > 0, \end{cases}$$

$x_1 = e^{-\beta/\alpha}$, $x_2 = e^{(1-\beta)/\alpha}$, from which the desired result follows.

Corollary 3. *A necessary and sufficient condition for $F(x)$ to belong to the λ domain of attraction of $u(x)$,*

$$u(x) = \begin{cases} -\alpha \log |x| + \beta,\ \alpha > 0, & x < 0, \\ +\infty, & x \geqslant 0 \end{cases}$$

is that there exist $k_n \to \infty$, $k_n/n \to \lambda$, $0 \le \lambda < 1$, and

$$a'_n = e^{-\beta/\alpha}(1 - e^{-1/\alpha})^{-1}\left(\bar{a}_{\left(\frac{k_n}{\alpha} + \frac{\sqrt{(1-\lambda)k_n}}{n}\right)} - \bar{a}_{\frac{k_n}{n}}\right),$$

$$b'_n = (1 - e^{-1/\alpha})^{-1}\left(\bar{a}_{\left(\frac{k_n}{n} + \frac{\sqrt{(1-\lambda)k_n}}{n}\right)} - e^{-1/\alpha}\, \bar{a}_{\frac{k_n}{n}}\right),$$

such that

$$\frac{n}{\sqrt{k_n}}\left[F(a'_n x + b'_n) - \frac{k_n}{n}\right] \to \begin{cases} -\alpha \log |x| + \beta, & x < 0, \\ +\infty, & x \geqslant 0. \end{cases}$$

Proof. In Theorem 3.3 take

$$u(x) = \begin{cases} -\alpha \log |x| + \beta, \ \alpha > 0, & x < 0, \\ +\infty, & x \geq 0, \end{cases}$$

$x_1 = -e^{\beta/\alpha}$, $x_2 = -e^{(\beta-1)/\alpha}$, from which the result follows.

Theorem 3.4. *Let $u(x)$ have jumps at the points x_1, x_2 $(x_1 < x_2)$. Then a necessary and sufficient condition for $F(x)$ to belong to the λ domain of attraction of $u(x)$ is that there exist $k_n \to \infty$, $k_n/n \to \lambda$, $0 \leq \lambda < 1$, with*

$$a_n' = \frac{1}{x_2 - x_1}\left[\bar{a}_{\left(\frac{k_n}{n}+\frac{\sqrt{(1-\lambda)k_n}}{n}u_2\right)} - \bar{a}_{\left(\frac{k_n}{n}+\frac{\sqrt{(1-\lambda)k_n}}{n}u_1\right)}\right],$$

$$b_n' = \frac{1}{x_2 - x_1}\left[x_2\bar{a}_{\left(\frac{k_n}{n}+\frac{\sqrt{(1-\lambda)k_n}}{n}u_1\right)} - x_1\bar{a}_{\left(\frac{k_n}{n}+\frac{\sqrt{(1-\lambda)k_n}}{n}u_2\right)}\right],$$

such that

$$\frac{n}{\sqrt{k_n}}\left[F(a_n'x + b_n') - \frac{k_n}{n}\right] \to \sqrt{1-\lambda}\, u(x),$$

where u_1, u_2 are two fixed constants satisfying

$$u(x_i - 0) < u_i < u(x_i + 0), \quad i = 1, 2.$$

Proof. The proof is similar to that of Theorem 3.3.

Theorem 3.5. *Suppose $F(x_0) = \lambda$, $0 \leq \lambda < 1$, and inside the interval $(x_0 - \delta, x_0 + \delta)$, $F'(x) > 0$, where $\delta > 0$ is a constant, and also $F'(x)$ is continuous at x_0. Then for arbitrary $k_n \to \infty$, $k_n/n \to \lambda$, there exist $a_n > 0$, b_n such that*

$$\Phi_{k_n n}(a_n x + b_n) \to \frac{1}{\sqrt{2\pi}}\int_{-\infty}^{x} e^{-t^2/2}\,dt, \quad -\infty < x < +\infty.$$

Proof. Note that in the interval $(x_0 - \delta, x_0 + \delta)$, $F(x)$ is strictly monotone increasing and continuous, and also $F(x_0) = \lambda$. Also, for $k_n/n \to \lambda$, with n sufficiently large, we would have $F(x_0 - \delta + 0) < k_n/n < F(x_0 + \delta)$. Hence there is a point $\bar{a}_{k_n/n}$ inside the interval $(x_0 - \delta, x_0 + \delta)$ such that $F(\bar{a}_{k_n/n}) = k_n/n$. Now we take

$$b_n = \bar{a}_{\frac{k_n}{n}}, \quad a_n = \frac{\sqrt{k_n}}{nF'(b_n)} \cdot \sqrt{1-\lambda};$$

then $b_n \to x_0$, $a_n \to 0$ when $n \to \infty$, and therefore for any x, and sufficiently large n, we have $x_0 - \delta < a_n x + b_n < x_0 + \delta$. Thus

$$\begin{aligned} &\frac{n}{\sqrt{k_n}}\left[F(a_n x + b_n) - \frac{k_n}{n}\right] \\ &= \frac{n}{\sqrt{k_n}}\left[F(b_n) + a_n x F'(b_n + a_n\theta_n x) - \frac{k_n}{n}\right] \\ &= \frac{F'(b_n + a_n\theta_n x)}{F'(b_n)} x \cdot \sqrt{1-\lambda}, \quad |\theta_n| \leq 1. \end{aligned}$$

Since $b_n \to x_0$, $b_n + a_n\theta_n x \to x_0$ when $n \to \infty$, and $F'(x)$ is continuous at the point x_0, hence

$$\frac{F'(b_n + a_n\theta_n x)}{F'(b_n)} \to 1.$$

Consequently,

$$\frac{n}{\sqrt{k_n}}\left[F(a_nx+b_n)-\frac{k_n}{n}\right]\to\sqrt{1-\lambda}\,x.$$

By Lemma 3.1 we have

$$\Phi_{k_nn}(a_nx+b_n)\to\frac{1}{\sqrt{2\pi}}\int_{-\infty}^{x}e^{-t^2/2}\,dt.$$

Theorem 3.6. *Suppose $F(x_0+0)=0$, and in the interval $(x_0, x_0+\delta)$, $F'(x)>0$, also $\lim_{x\to x_0+0}F'(x)=c>0$. Then for arbitrary $k_n\to\infty$, $k_n/n\to 0$, there exist $a_n>0$, b_n such that*

$$\Phi_{k_nn}(a_nx+b_n)\to\frac{1}{\sqrt{2\pi}}\int_{-\infty}^{x}e^{-t^2/2}\,dt.$$

Proof. The proof is similar to that of Theorem 3.5. We note here that we take

$$b_n=\bar{a}_{\frac{k_n}{n}},\quad a_n=\frac{\sqrt{k_n}}{nF'(b_n)}.$$

Now we have that b_n is greater than x_0 and tends to x_0. But since $(k_n/n)/(b_n-x_0)=[F(b_n)-F(x_0)]/(b_n-x_0)$ is bounded, therefore $a_n/(b_n-x_0)$ tends to 0. Thus for any x, and sufficiently large n, we have $x_0<a_nx+b_n<x_0+\delta$. Then we can make a Taylor expansion.

Theorem 3.7. *Suppose $F'(x)>0$, when $-\infty<x<c$, and also in this interval $(-\infty, c)$*

$$\left|\frac{F(x)}{x^pF'(x)}\right|<M,\quad\left|\frac{F(x)F''(x+o(x^p))}{[F'(x)]^2}\right|<M,$$

where $c>-\infty$, $M>0$, $p\le 1$ are all given constants. Then for arbitrary $k_n\to\infty$, $k_n/n\to 0$, there exist $a_n>0$, b_n such that

$$\Phi_{k_nn}(a_nx+b_n)\to\frac{1}{\sqrt{2\pi}}\int_{-\infty}^{x}e^{-t^2/2}\,dt.$$

Proof. In $(-\infty, c)$ the function $F(x)$ is greater than 0, strictly monotone increasing and continuous; hence when n is sufficiently large, there is a point $\bar{a}_{k_n/n}$ such that $F(\bar{a}_{k_n/n})=k_n/n$. Take

$$b_n=\bar{a}_{\frac{k_n}{n}},\quad a_n=\frac{\sqrt{k_n}}{n}\frac{1}{F'(b_n)}.$$

Thus we have

$$b_n\to-\infty,\ \frac{a_n}{b_n^p}=\frac{1}{\sqrt{k_n}}\frac{F(b_n)}{(b_n)^pF'(b_n)}\to 0.$$

Also $p\le 1$; hence for any x, we have $a_nx+b_n\to-\infty$, for n sufficiently large, $a_nx+b_n<c$. Therefore

$$\begin{aligned}&F(a_nx+b_n)\\&\quad=F(b_n)+a_nxF'(b_n)+\frac{1}{2}(a_nx)^2F''(b_n+a_n\theta_nx),\quad|\theta_n|\leqslant 1.\end{aligned}$$

Since $a_n\theta_nx/(b_n)^p\to 0$, we have

$$\frac{n}{\sqrt{k_n}}(a_n x)^2 F''(b_n + a_n\theta_n x)$$
$$= \frac{x^2}{\sqrt{k_n}} \frac{F(b_n)F''(b_n + a_n\theta_n x)}{(F'(b_n))^2} \to 0.$$

But $F(b_n) = k_n/n$, $(n/\sqrt{k_n})(a_n x)F'(b_n) = x$. Consequently

$$\frac{n}{\sqrt{k_n}}\left[F(a_n x + b_n) - \frac{k_n}{n}\right] \to x.$$

Then by Lemma 3.1, we get

$$\Phi_{k_n n}(a_n x + b_n) \to \frac{1}{\sqrt{2\pi}}\int_{-\infty}^{x} e^{-t^2/2}\,dt.$$

§4. Examples

Example 1.

$$F_1(x) = \begin{cases} 0, & x \leqslant 0, \\ x, & 0 < x \leqslant 1, \\ 1, & x > 1. \end{cases}$$

By Theorems 3.5 and 3.6, we have that for arbitrary $k_n \to \infty$, $k_n/n \to \lambda$, there exist $a_n > 0$, b_n such that

$$\Phi_{k_n n}(a_n x + b_n) \to \frac{1}{\sqrt{2\pi}}\int_{-\infty}^{x} e^{-t^2/2}\,dt.$$

Example 2.

$$F_2(x) = \begin{cases} 0, & x \leqslant 0, \\ 1 - e^{-x}, & x > 0. \end{cases}$$

By Theorems 3.5 and 3.6, we have that for arbitrary $k_n \to \infty$, $k_n/n \to \lambda$, there exist $a_n > 0$, b_n such that

$$\Phi_{k_n n}(a_n x + b_n) \to \frac{1}{\sqrt{2\pi}}\int_{-\infty}^{x} e^{-t^2/2}\,dt.$$

Example 3.

$$F_3(x) = \begin{cases} 0, & x \leqslant 0, \\ \lambda - \dfrac{1}{\alpha \log x}, & 0 < x \leqslant e^{\frac{-1}{\alpha(1-\lambda)}}, \\ 1, & x > e^{\frac{-1}{\alpha(1-\lambda)}}. \end{cases}$$

i) When $k_n \to \infty$, $k_n/n \to \lambda$, and $(n/\sqrt{k_n})(k_n/n - \lambda)^2 \to \infty$, by Corollary 1 of Theorem 3.3, we can assert that there exist $a_n > 0$, b_n such that

$$\Phi_{k_n n}(a_n x + b_n) \to \frac{1}{\sqrt{2\pi}}\int_{-\infty}^{x} e^{-t^2/2}\,dt.$$

In fact,

$$\bar{a}_{\frac{k_n}{n}} = \overline{e^{\alpha(\lambda n - k_n)}}^{\,n},$$

$$\bar{a}_{\left(\frac{k_n}{n}+\frac{\sqrt{(1-\lambda)k_n}}{n}\right)} = \overline{e^{\alpha(\lambda n - k_n - \sqrt{(1-\lambda)k_n})}}^{\,n},$$

$$a'_n = \overline{e^{\alpha(\lambda n - k_n)}}^{\,n}\,[\overline{e^{\alpha(\lambda n - k_n - \sqrt{(1-\lambda)k_n})}}^{\,n} - 1],$$

$$b'_n = \overline{e^{\alpha(\lambda n - k_n)}}^{\,n}.$$

By use of the A. Hinčin theorem on the types of law, we can simplify a'_n, b'_n to

$$a_n = \frac{n\sqrt{(1-\lambda)k_n}}{\alpha(\lambda n - k_n)^2}\,\overline{e^{\alpha(\lambda n - k_n)}}^{\,n},$$

$$b_n = \overline{e^{\alpha(\lambda n - k_n)}}^{\,n}.$$

Since

$$b_n \to +0, \quad \frac{a_n}{b_n} \to 0,$$

therefore for any x, and sufficiently large n, we have $0 < a_n x + b_n < e^{-1/\alpha(1-\lambda)}$. Thus

$$\frac{n}{\sqrt{k_n}}\left[F_3(a_n x + b_n) - \frac{k_n}{n}\right]$$

$$= \frac{n}{\sqrt{k_n}}\left[\lambda - \frac{1}{\dfrac{n}{(\lambda n - k_n)} + \alpha \log\left(1 + \dfrac{n\sqrt{(1-\lambda)k_n}}{\alpha(\lambda n - k_n)^2}x\right)} - \frac{k_n}{n}\right]$$

$$= \frac{n}{\sqrt{k_n}}\left[\lambda - \frac{1}{\dfrac{n}{(\lambda n - k_n)} + \alpha\left[\dfrac{n\sqrt{(1-\lambda)k_n}}{\alpha(\lambda n - k_n)^2}x + O\left(\dfrac{n^2 k_n}{(\lambda n - k_n)^4}\right)\right]} - \frac{k_n}{n}\right]$$

$$= \frac{n}{\sqrt{k_n}}\left[\lambda - \left(\frac{n}{\lambda n - k_n}\right)^{-1} + \left(\frac{n}{\lambda n - k_n}\right)^{-2}\frac{n\sqrt{(1-\lambda)k_n}}{(\lambda n - k_n)^2}x \right.$$
$$\left. + O\left(\frac{n^2}{k_n}\left(\frac{k_n}{n} - \lambda\right)^2\right)^{-1} - \frac{k_n}{n}\right]$$

$$= \frac{n}{\sqrt{k_n}}\left[\frac{\sqrt{k_n}}{n}\sqrt{1-\lambda}\,x + O\left(\frac{n^2}{k_n}\left(\frac{k_n}{n} - \lambda\right)\right)^{-1}\right].$$

When $n \to \infty$, we have

$$\frac{n}{\sqrt{k_n}}\left[F_3(a_n x + b_n) - \frac{k_n}{n}\right] \to \sqrt{1-\lambda}\,x.$$

ii) When $k_n \to \infty$, $k_n/n \to \lambda$, and $(n/\sqrt{k_n})(k_n/n - \lambda)^2 \to \sqrt{1-\lambda}$, by Corollary 2 of Theorem 3.3, we can assert that there are $a_n > 0$, b_n such that

$$\Phi_{k_n n}(a_n x + b_n) \to \frac{1}{\sqrt{2\pi}}\int_{-\infty}^{u(x)} e^{-t^2/2}\,dt.$$

where

$$u(x) = \begin{cases} -\infty, & x \leqslant 0, \\ \alpha \log x + \beta, \quad \alpha > 0, & x > 0. \end{cases}$$

In fact,

$$\bar{a}_{\frac{k_n}{n}} = e^{\frac{n}{\alpha(\lambda n - k_n)}},$$

$$\bar{a}_{\left(\frac{k_n}{n} + \frac{\sqrt{(1-\lambda)k_n}}{n}\right)} = e^{\frac{n}{\alpha(\lambda n - k_n - \sqrt{(1-\lambda)k_n})}},$$

$$a_n' = e^{\beta/\alpha}(e^{1/\alpha} - 1)^{-1} e^{\frac{n}{\alpha(\lambda n - k_n)}} \left[e^{\frac{n\sqrt{(1-\lambda)k_n}}{\alpha(\lambda n - k_n)(\lambda n - k_n - \sqrt{(1-\lambda)k_n})}} - 1 \right],$$

$$b_n' = (e^{1/\alpha} - 1)^{-1} e^{\frac{n}{\alpha(\lambda n - k_n)}} \left[e^{1/\alpha} - e^{\frac{n\sqrt{(1-\lambda)k_n}}{\alpha(\lambda n - k_n)(\lambda n - k_n - \sqrt{(1-\lambda)k_n})}} \right].$$

By use of the A. Hinčin theorem on the types of law, we can simplify a_n', b_n' to

$$a_n = e^{\beta/\alpha} \cdot e^{\frac{n}{\alpha(\lambda n - k_n)}}.$$

$$b_n = 0.$$

Since $a_n \longrightarrow +0$, hence for any $x > 0$, and sufficiently large n, we have $0 < a_n x + b_n < e^{-1/\alpha(1-\lambda)}$. Therefore

$$\frac{n}{\sqrt{k_n}} \left[F_3(a_n x + b_n) - \frac{k_n}{n} \right]$$

$$= \frac{n}{\sqrt{k_n}} \left[\lambda - \frac{1}{\frac{n}{(\lambda n - k_n)} + (\alpha \log x + \beta)} - \frac{k_n}{n} \right]$$

$$= \frac{n}{\sqrt{k_n}} \left[\lambda - \left(\frac{n}{\lambda n - k_n} \right)^{-1} + \left(\frac{n}{\lambda n - k_n} \right)^{-2} (\alpha \log x + \beta) \right.$$

$$\left. + o\left(\frac{\lambda n - k_n}{n} \right)^3 - \frac{k_n}{n} \right]$$

$$= \frac{n}{\sqrt{k_n}} \left(\frac{k_n}{n} - \lambda \right)^2 (\alpha \log x + \beta) + o\left(\frac{k_n}{n} - \lambda \right),$$

and thus when $n \longrightarrow \infty$

$$\frac{n}{\sqrt{k_n}} \left[F_3(a_n x + b_n) - \frac{k_n}{n} \right] \rightarrow \sqrt{1 - \lambda}(\alpha \log x + \beta).$$

Also for any $x < 0$, we have $a_n x + b_n < 0$, and hence

$$\frac{n}{\sqrt{k_n}} \left[F_3(a_n x + b_n) - \frac{k_n}{n} \right] = -\sqrt{k_n},$$

so that when $n \longrightarrow \infty$,

$$\frac{n}{\sqrt{k_n}} \left[F_3(a_n x + b_n) - \frac{k_n}{n} \right] \rightarrow -\infty.$$

Example 4.

$$F_4(x) = \frac{a}{\pi}\int_{-\infty}^{x} \frac{1}{a^2+t^2}\,dt, \quad a>0, \quad -\infty < x < +\infty.$$

We have

$$F_4'(x) = \frac{a}{\pi(a^2+x^2)} > 0, \quad -\infty < x < +\infty.$$

Also by use of the L'Hopital rule we get

$$\begin{aligned}
&\lim_{x\to-\infty} \frac{F_4(x)}{xF_4'(x)} \\
&\quad = \lim_{x\to-\infty} \frac{F_4'(x)}{F_4'(x) + xF_4''(x)} \\
&\quad = \lim_{x\to-\infty} \left(\frac{a}{\pi(a^2+x^2)}\right)\left(\frac{a(a^2-x^2)}{\pi(a^2+x^2)^2}\right)^{-1} \\
&\quad = -1, \\
&\lim_{x\to-\infty} \frac{xF_4''(x+o(x))}{F_4'(x)} = -2a.
\end{aligned}$$

Hence by the case $p = 1$ in Theorem 3.7 we have: for arbitrary $k_n \to \infty$, $k_n/n \to 0$ there exist $a_n > 0$, b_n such that

$$\Phi_{k_n n}(a_n x + b_n) \to \frac{1}{\sqrt{2\pi}}\int_{-\infty}^{x} e^{-t^2/2}\,dt.$$

Example 5.

$$F_5(x) = \frac{a}{2}\int_{-\infty}^{x} e^{-a|t|}\,dt, \quad a>0, \quad -\infty < x < +\infty.$$

We have

$$F_5'(x) = \frac{a}{2}e^{-a|x|} > 0, \quad -\infty < x < +\infty,$$

$$\lim_{x\to-\infty} \frac{F_5(x)}{F_5'(x)} = 1,$$

$$\lim_{x\to-\infty} \frac{F_5''(x+o(1))}{F_5'(x)} = 1.$$

Thus by the case $p = 0$ in Theorem 3.7 we have: for arbitrary $k_n \to \infty$, $k_n/n \to 0$ there exist $a_n > 0$, b_n such that

$$\Phi_{k_n n}(a_n x + b_n) \to \frac{1}{\sqrt{2\pi}}\int_{-\infty}^{x} e^{-t^2/2}\,dt.$$

Example 6.

$$F_6(x) = \frac{1}{\sqrt{2\pi}}\int_{-\infty}^{x} e^{-t^2/2}\,dt, \quad -\infty < x < +\infty.$$

We have

$$F_6'(x) = \frac{1}{\sqrt{2\pi}} e^{-x^2/2} > 0, \quad -\infty < x < +\infty,$$

$$\lim_{x \to -\infty} \frac{x F_6(x)}{F_6'(x)} = -1,$$

$$\lim_{x \to -\infty} \frac{F_6''\left(x + o\left(\frac{1}{x}\right)\right)}{x F_6'(x)} = -1.$$

Therefore we use the case $p = -1$ in the Theorem 3.7 and get: for arbitrary $k_n \to \infty$, $k_n/n \to 0$, there are $a_n > 0$, b_n such that

$$\Phi_{k_n n}(a_n x + b_n) \to \frac{1}{\sqrt{2\pi}} \int_{-\infty}^{x} e^{-t^2/2} dt.$$

The study of the limiting distributions of order statistics in this article is not at all definitive. A great many problems that must be cleared up are still left open. For example, if we relax the restrictions on the order statistics in our article, what results would we get? For the three types of limiting distributions that were given in this article can we find necessary and sufficient conditions for a distribution function $F(x)$ to belong to their λ domain of attraction? To solve these problems we must have deeper research on this subject. On the other hand, this article does in fact partially solve some problems left by N. V. Smirnov [1]: the limiting distribution of sequences of extreme terms ($k_n \to \infty$, $k_n/n \to 0$) and that of sequences of central terms ($k_n/n \to \lambda$, $0 < \lambda < 1$, $\sqrt{n}\,(k_n/n - \lambda) \to \infty$). Obviously these two cases satisfy the conditions $k_n \to \infty$, $k_n/n \to \lambda$, $0 \leq \lambda < 1$, in our article.

BIBLIOGRAPHY

[1] N. V. Smirnov, *The limit distributions for the terms of a variational series*, Trudy Mat. Inst. Steklov. 25 (1949); English transl., Translation No. 67, Amer. Math. Soc., Providence, R. I., 1952; reprint, Amer. Math. Soc. Transl. (1) 11 (1962), 82–143; Chinese transl. by Wang Shou-ren (Wang Shou-jen), 1954. MR 11, 605; MR 13, 853.

[2] B. Gnedenko, *Sur la distribution limite du terme maximum d'une série aléatoire*, Ann. of Math. 44 (1943), 423–453. MR 5, 41.

Translated by:
Monbill Tong

Reprinted from
Prog. Math. 7 (1964), 240–281.

Partially Balanced Incomplete Block Designs*†

BAN CHENG‡

Foreword

For simplicity, generally we will not repeat the proof of any theorem already proven in the literature. If a method can be used to treat different problems, detailed discussions will be given only when it first appears.

§1. Definitions, Simple Properties of Parameters

Definition 1.1. Suppose that there are v treatments and m relations. We use $\alpha, \beta, \ldots$ to denote the treatments and $(\alpha, \beta) = i$ to denote that treatment β is an ith associate of α (α and β have the ith relation). If the following conditions are satisfied, then we say that the v treatments form an association scheme with m associate classes, or an m-association scheme in brief, under the m relations:

1. For any two different treatments α, β, there is a unique i, $1 \leqslant i \leqslant m$, such that $(\alpha, \beta) = i$. Furthermore $(\alpha, \beta) = i$ implies $(\beta, \alpha) = i$.
2. Any treatment α has exactly n_i ith associates and the number n_i is independent of α, $1 \leqslant i \leqslant m$.
3. For any triple (i, j, k), $1 \leqslant i, j, k \leqslant m$ if $(\alpha, \beta) = i$, then the number of treatments that are both jth associates of α and kth associates of β is p^i_{jk}, and is independent of α and β.

The numbers v, n_i, p^i_{jk} $(i, j, k = 1, 2, \ldots, m)$ are called the parameters of the association scheme.

Definition 1.2. For a given m-association scheme with parameters v, n_i, p^i_{jk} $(i, j, k = 1, 2, \ldots, m)$, if the v treatments can be arranged in b blocks so that

1. each treatment appears in r blocks;
2. each block size is k;
3. if $(\alpha, \beta) = i$, then α and β appear together in λ_i blocks,

---

*Received March 13, 1963.

† Translated by Ching-shui Cheng.

‡ This paper was the collective work of a seminar led by Professor Xu Baolu (Pao-Lu Hsu), written by Zhang Yaoting, and published here under the pseudonym "Ban Cheng".

then this arrangement is called an m-associate class, partially balanced incomplete block design (PBIB design).* The numbers $v, b, r, k, \lambda_i, n_i, p_{jk}^i$ $(i, j, k = 1, 2, \ldots, m)$ are called the parameters of the design. If $m = 1$, then the design is called a balanced incomplete block design (BIB design).

For the convenience of later discussions, we shall introduce matrix notations of association schemes and designs.

Suppose that an association scheme with parameters v, n_i, p_{jk}^i, $i, j, k = 1, 2, \ldots, m$, exists. Let

$$b_{\alpha i}^{\beta} = \begin{cases} 1, & \text{if } (\alpha, \beta) = i, \\ 0, & \text{if } (\alpha, \beta) \neq i, \end{cases} \qquad \alpha, \beta = 1, 2, \ldots, v, \quad i = 1, 2, \ldots, m,$$

$$B_i = (b_{\alpha i}^{\beta}) \qquad i = 1, 2, \ldots, m.$$

For convenience, each treatment is considered to be the 0th associate of itself. Let

$$b_{\alpha 0}^{\beta} = \begin{cases} 1, & \text{if } \alpha = \beta, \\ 0, & \text{if } \alpha \neq \beta, \end{cases} \qquad \alpha, \beta = 1, 2, \ldots, v,$$

$$B_0 = (b_{\alpha 0}^{\beta}),$$

$$n_0 = 1, \qquad \begin{aligned} p_{jk}^0 &= n_j \delta_{jk}, & j, k &= 1, 2, \ldots, m, \\ p_{j0}^k &= \delta_{jk}, & j, k &= 1, 2, \ldots, m. \end{aligned}$$

Then we obtain $m + 1$ matrices $B_0, B_1, \ldots, B_m$ with 0, 1 entries such that

$$B_0 = I, \qquad B_i' = B_i, \qquad i = 1, 2, \ldots, m, \tag{A.1}$$

$$\sum_{i=0}^{m} B_i = E, \tag{A.2}$$

$$B_j B_k = \sum_{i=0}^{m} p_{jk}^i B_i, \qquad j, k = 0, 1, \ldots, m, \tag{A.3}$$

where I is the identity matrix, and E is the matrix with all entries equal to 1. Conversely, if there exist $m + 1$ $v \times v$ (0, 1)-matrices $B_0, B_1, \ldots, B_m$ such that (A.1)–(A.3) are satisfied, then it is not difficult to show that an association scheme with parameters v, p_{jk}^i, n_i $(= p_{ii}^0)$ exists (see [1]).

Let there be a PBIB design with parameters $v, b, r, k, \lambda_i, n_i, p_{jk}^i$ $(i, j, k = 1, 2, \ldots, m)$. Then the corresponding association scheme can be represented by matrices $B_0, B_1, \ldots, B_m$, which satisfy (A.1) – (A.3).

Let

$$n_{\alpha j} = \begin{cases} 1, & \text{if treatment } \alpha \text{ appears in the } j\text{th block}, \\ 0, & \text{if treatment } \alpha \text{ does not appear in the } j\text{th block}, \end{cases}$$

$$\alpha = 1, 2, \ldots, v, \qquad j = 1, 2, \ldots, b,$$

$$N = (n_{\alpha j}).$$

* *Translator's note*: More precisely, the design has to be binary, i.e., each treatment appears in each block at most once.

Then N is a $v \times b$ (0, 1)-matrix such that

$$N 1_b = r 1_v, \tag{D.1}$$

$$N' 1_v = k 1_b, \tag{D.2}$$

$$NN' = rB_0 + \lambda_1 B_1 + \cdots + \lambda_m B_m, \tag{D.3}$$

where **1** is a column vector of ones whose dimension can be determined from context and is sometimes explicitly displayed as a subscript. Conversely, for a given association scheme $B_0, B_1, \ldots, B_m$ with parameters v, n_i, p_{jk}^i, if there is a (0, 1)-matrix N satisfying (D.1)–(D.3), then it is not hard to show (see [1]) that N defines a PBIB design with parameters $v, b, r, k, \lambda_i, n_i, p_{jk}^i$.

Definition 1.3. Suppose $\{A_0, \ldots, A_m\}$, $\{B_0, \ldots, B_m\}$ are two association schemes with the same parameters. If there is a permutation matrix T such that $TA_iT' = B_i$, $i = 1, \ldots, m$, then schemes $\{A_0, \ldots, A_m\}$ and $\{B_0, \ldots, B_m\}$ are said to be equivalent.

It is easy to see that two schemes are equivalent if and only if one scheme can be obtained from the other by appropriately changing the labels of the treatments. Therefore we will not distinguish between equivalent schemes.

Calculating the row sums of the matrices in (A.1)–(A.3) and (D.1)–(D.3), we immediately obtain the following relations among the parameters of a PBIB design and the corresponding association scheme:

$$B_i 1 = n_i 1, \qquad i = 1, 2, \ldots, m, \tag{1.1}$$

$$v = \sum_{i=0}^{m} n_i, \tag{1.2}$$

$$\begin{aligned} n_i &= 1 + \sum_{k=1}^{m} p_{ik}^i = \sum_{k=0}^{m} p_{ik}^i \qquad & i = 0, 1, \ldots, m, \\ n_i &= \sum_{k=1}^{m} p_{ik}^j, & i \neq j, \quad i, j = 1, 2, \ldots, m. \end{aligned} \tag{1.3}$$

$$bk = rv, \tag{1.4}$$

$$r(k-1) = \sum_{i=1}^{m} n_i \lambda_i. \tag{1.5}$$

§2. Properties of the Parameters of Association Schemes

Throughout this section, we assume that an association scheme with parameters v, n_i, p_{jk}^i, $i, j, k = 1, 2, \ldots, m$, exists. Let the corresponding association matrices be $B_0, \ldots, B_m$.

Theorem 2.1. *Let* $\mathfrak{B} = \{B : B = \sum_{i=0}^{m} c_i B_i,\ c_i$ *is a real number*, $i = 0, 1, \ldots, m\}$. *Then* $\mathfrak{B}$ *is an* $(m+1)$*-dimensional linear associative algebra over the real field under* *matrix addition, matrix multiplication, and scalar multiplication. The matrices* B_0, $B_1, \ldots, B_m$ *constitute a basis and the constants of multiplication corresponding to this basis are* $\{p_{jk}^i\}$.

Proof. From (A.2), $\sum_{i=0}^{m} c_i B_i = 0$ if and only if $c_i = 0$, $i = 0, 1, \ldots, m$. Hence $\mathfrak{B}$ is a linear associative algebra over the real field with basis $B_0, \ldots, B_m$. By (A.3), the constants of multiplication corresponding to $B_0, \ldots, B_m$ are $\{p_{jk}^i\}$.

Consider the matrices of constants of multiplication $P_i = (p_{si}^t)$, $i = 0, 1, 2, \ldots, m$. It is easy to see that P_i has the following properties.

$$P_0 = I, \tag{2.1}$$

$$P_j P_k = \sum_{t=0}^{m} p_{jk}^t P_t, \qquad j, k = 0, 1, \ldots, m, \tag{2.2}$$

$$p_{jk}^i = p_{kj}^i, \qquad i, j, k = 0, 1, \ldots, m, \tag{2.3}$$

$$P_j P_k = P_k P_j, \qquad j, k = 0, 1, \ldots, m, \tag{2.4}$$

$$\sum_{i=0}^{m} P_i = \begin{bmatrix} 1 & 1 & \cdots & 1 \\ n_1 & n_1 & \cdots & n_1 \\ \vdots & \vdots & \ddots & \vdots \\ n_m & n_m & \cdots & n_m \end{bmatrix}. \tag{2.5}$$

Theorem 2.2. *Let* $\mathfrak{P} = \{P : P = \sum_{i=0}^{m} c_i P_i,\ c_i$ *is a real number*, $i = 0, 1, \ldots, m\}$. *Then* $\mathfrak{P}$ *is an* $(m + 1)$*-dimensional linear associative algebra over the real field under matrix addition, matrix multiplication, and scalar multiplication. The matrices* $P_0, P_1, \ldots, P_m$ *constitute a basis and the constants of multiplication corresponding to this basis are* $\{p_{jk}^i\}$.

Proof. Notice that the first row of P_i is $(\overbrace{0 \cdots 0}^{i-1}10 \cdots 0)$. Therefore, $\sum_{i=0}^{m} c_i P_i = 0$ if and only if $c_i = 0$, $i = 0, 1, \ldots, m$. Hence $\mathfrak{P}$ is an $(m + 1)$-dimensional linear associative algebra over the real field with basis $P_0, P_1, \ldots, P_m$. By (2.2), the constants of multiplication corresponding to $P_0, \ldots, P_m$ are $\{p_{jk}^i\}$.

It is easy to see that the function from $\mathfrak{B}$ to $\mathfrak{P}$ defined by

$$\phi : B = \sum_{i=0}^{m} c_i B_i \rightarrow P = \sum_{i=0}^{m} c_i P_i$$

is an isomorphism. Therefore we have

Theorem 2.3. B *and* $\phi(B)$ *have the same minimal polynomial and the same distinct characteristic roots.*

Since B is a real symmetric matrix, all the characteristic roots of B are real. Therefore we have

Theorem 2.4. *All the characteristic roots of each* P_i *are real.*

Since P is a square matrix of order $m + 1$, it has at most $m + 1$ distinct characteristic roots. Therefore one concludes

Theorem 2.5. *B has at most $m + 1$ distinct characteristic roots.*

By (2.4), $\mathcal{P}$ is commutative under matrix multiplication. Using Theorem 16.1 of [5], we obtain

Theorem 2.6. *For a suitable ordering of the characteristic roots $\{z_{ui} : u = 0, 1, \ldots, m\}$ of each P_i, any matrix $P = \sum_{i=0}^{m} c_i P_i \in \mathcal{P}$ has characteristic roots $\sum_{i=0}^{m} c_i z_{ui}$, $u = 0, 1, \ldots, m$.*

Hereafter we shall use $\{z_{ui}\}$ to denote the characteristic roots of P_i ordered as specified in Theorem 2.6. Let $Z = (z_{ui})$.

Theorem 2.7. $|Z| \neq 0$.

Proof. Suppose $c = (c_0, c_1, \ldots, c_m)$ is a solution of $Zx = 0$. Let $P = \sum_{i=0}^{m} c_i P_i$. From Theorem 2.6, all the characteristic roots of P are zero. By Theorem 2.3, all the characteristic roots of $B = \sum_{i=0}^{m} c_i B_i$ are also zero. Since B is symmetric, $B = 0$. Hence $c_i = 0$, $i = 0, 1, \ldots, m$. This shows that $Zx = 0$ only has a trivial solution. Therefore $|Z| \neq 0$.

Theorem 2.8. *There is a matrix Q in $\mathcal{P}$ such that all the characteristic roots of Q are simple.*

Proof. Let $\theta_0, \theta_1, \ldots, \theta_m$ be $m + 1$ arbitrary distinct real numbers and $\theta' = (\theta_0 \theta_1 \ldots \theta_m)$. Then by Theorem 2.7, equation $Zx = \theta$ has a unique solution. Suppose the solution is $c' = (c_0 c_1 \ldots c_m)$. Then the characteristic roots of the matrix $\sum_{i=0}^{m} c_i P_i$ are $\theta_0, \theta_1, \ldots, \theta_m$, which are distinct. Let $Q = \sum_{i=0}^{m} c_i P_i$. Then the theorem is proved.

Theorem 2.9. *Each $P \in \mathcal{P}$ only has linear elementary divisors.*

Proof. Let the characteristic roots of the matrix Q in Theorem 2.8 be $\theta_0, \theta_1, \ldots, \theta_m$. Let

$$D_\theta = \begin{bmatrix} \theta_0 & & & 0 \\ & \theta_1 & & \\ & & \ddots & \\ 0 & & & \theta_m \end{bmatrix}.$$

Then there is a matrix G, $|G| \neq 0$, such that $Q = G D_\theta G^{-1}$. Obviously $Q^k = G D_\theta^k G^{-1}$. Since the $m + 1$ characteristic roots of Q are all distinct, the degree of its minimal polynomial must be $m + 1$. It follows that $Q^0, Q^1, \ldots, Q^m$ are linearly independent and form a basis of $\mathcal{P}$. For each $P \in \mathcal{P}$, there exist $c_0, c_1, \ldots, c_m$ such that $P = \sum_{i=0}^{m} c_i Q^i$. Then

$$P = \sum_{i=0}^{m} c_i Q^i = \sum_{i=0}^{m} c_i G D_\theta^i G^{-1} = G \begin{bmatrix} \sum_{i=0}^{m} c_i \theta_0^i & & 0 \\ & \ddots & \\ 0 & & \sum_{i=0}^{m} c_i \theta_m^i \end{bmatrix} G^{-1}.$$

This proves Theorem 2.9.

Theorem 2.9 can be rephrased as

Theorem 2.10. *If λ is a characteristic root of $P \in \mathcal{P}$, then the rank of $\lambda I - P$ is $m + 1 - l$, where l is the multiplicity of λ.*

Theorem 2.11. *There exist $m + 1$ integers $\alpha_0, \alpha_1, \ldots, \alpha_m$, such that for any $B \in \mathcal{B}$, as long as B has $m + 1$ distinct characteristic roots, the multiplicities of these characteristic roots are $\alpha_0, \alpha_1, \ldots, \alpha_m$.*

Proof. Let $A = \phi^{-1}(Q)$ be the matrix in $\mathcal{B}$ corresponding to the matrix Q in Theorem 2.8. Then $A^0, A^1, \ldots, A^m$ are a basis of $\mathcal{B}$. Matrix A has the same distinct characteristic roots as Q, i.e., $\theta_0, \theta_1, \ldots, \theta_m$. Let their multiplicities be $\alpha_0, \alpha_1, \ldots, \alpha_m$, respectively. Suppose $B = \sum_{i=0}^{m} c_i A^i$ has $m + 1$ distinct characteristic roots. Since $B = \sum_{i=0}^{m} c_i A^i$, the characteristic roots of B are

$$\underbrace{\sum_{i=0}^{m} c_i \theta_0^i, \ldots, \sum_{i=0}^{m} c_i \theta_0^i}_{\alpha_0}, \ldots, \underbrace{\sum_{i=0}^{m} c_i \theta_m^i, \ldots, \sum_{i=0}^{m} c_i \theta_m^i}_{\alpha_m}.$$

Matrix B has $m + 1$ distinct characteristic roots only when $\sum_{i=0}^{m} c_i \theta_u^i$ are all different, $u = 0, 1, \ldots, m$. Therefore the characteristic roots of B are $\sum_{i=0}^{m} c_i \theta_u^i$, $u = 0, 1, \ldots, m$, with multiplicities $\alpha_0, \ldots, \alpha_m$.

Theorem 2.12. *The α_0 in Theorem 2.11 can be chosen to be 1.*

Proof. Consider the A, θ_i, α_i, $i = 0, 1, \ldots, m$ in the proof of Theorem 2.11. Since $E \in \mathcal{B}$, there are $c_0, c_1, \ldots, c_m$ such that $E = \sum_{i=0}^{m} c_i A^i$. Now the characteristic roots of E are $v, \underbrace{0, \ldots, 0}_{v-1}$; on the other hand, they are

$$\sum_{i=0}^{m} c_i \theta_0^i, \underbrace{\ldots,}_{\alpha_0} \sum_{i=0}^{m} c_i \theta_0^i, \ldots, \sum_{i=0}^{m} c_i \theta_m^i, \underbrace{\ldots,}_{\alpha_m} \sum_{i=0}^{m} c_i \theta_m^i.$$

So one of $\sum_{i=0}^{m} c_i \theta_u^i$, where $u = 0, 1, \ldots, m$, is v. Let θ_0 correspond to v. Then $\alpha_0 = 1$.

Theorem 2.13. *Suppose the distinct characteristic roots of B_i are $\theta_1, \ldots, \theta_k$ with multiplicities $r_1, \ldots, r_k$, respectively. If $r_1 = r_2 = \cdots = r_j$, $r_{j+i} \neq r_1$, $i = 1, 2, \ldots, k-j$, then $\theta_1, \ldots, \theta_j$ are the roots of a monic polynomial with degree j and integral coefficients.*

In order to prove Theorem 2.13, we shall list a sequence of results on polynomials. The proofs can be found in §38 of [6].

1. Every algebraic number has a unique minimal polynomial which is irreducible and has no multiple roots.
2. If $P(x)$ is the minimal polynomial of an algebraic number, then $P(x)$ is the minimal polynomial of any root of $P(x)$.
3. If $f(x)$ is a polynomial such that $f(a) = 0$ for an algebraic number a, then $f(x)$ is divisible by the minimal polynomial of a.
4. **Gauss's Lemma.** *If $f(x)$ is a monic polynomial with integral coefficients, $g(x)$ and $h(x)$ are monic polynomials with rational coefficients, and $f(x) = g(x)h(x)$, then $g(x)$ and $h(x)$ have integral coefficients.*

Theorem 2.14. *Suppose $f(x)$ is a monic polynomial with integral coefficients. The roots of $f(x)$ are $\theta_1, \ldots, \theta_k$ with multiplicities $r_1, \ldots, r_k$, respectively. If $r_1 = \cdots = r_j$, $r_{j+i} \neq r_1$, $i = 1, \ldots, k-j$, then $\theta_1, \ldots, \theta_j$ are the roots of a monic polynomial with integral coefficients and degree j.*

Proof. Let $g_1(x)$ be the minimal polynomial of θ_1. It follows from 3 that the roots of $g_1(x)$ are among $\theta_1, \ldots, \theta_k$. We will show that the roots of $g_1(x)$ are among $\theta_1, \ldots, \theta_j$. If not, we may assume that θ_k is a root of $g_1(x)$.

(i) If $r_k < r_1$, then $[g_1(x)]^{r_k}$ divides $f(x)$. Let $f(x) = g(x)[g_1(x)]^{r_k}$. Obviously $g(\theta_1) = 0$, but $g_1(x)$ does not divide $g(x)$. This contradicts 3.
(ii) If $r_k > r_1$, then $[g_1(x)]^{r_1}$ divides $f(x)$. Let $f(x) = h(x)[g_1(x)]^{r_1}$. Obviously $h(\theta_k) = 0$, but $g_1(x)$ does not divide $h(x)$. This again contradicts 3.

If the degree of $g_1(x)$ is j, then the theorem follows from the Gauss lemma. If the degree of $g_1(x)$ is less than j, let $\theta_1, \ldots, \theta_{l_1}$, $l_1 < j$, be all the roots of $g_1(x)$. Consider the minimal polynomial $g_2(x)$ of θ_{l_1+1}. By 2, $\theta_1, \ldots, \theta_{l_1}$ are not roots of $g_2(x)$. Repeating the above discussion, we know that the roots of $g_2(x)$ are among $\theta_{l_1+1}, \ldots, \theta_j$. Let the degree of $g_2(x)$ be l_2. If $l_1 + l_2 = j$, then the theorem is proved. If $l_1 + l_2 < j$, then we can repeat the above procedure. Since j is finite, there must exist $g_1(x), g_2(x), \ldots, g_t(x)$ with degrees $l_1, \ldots, l_t$, respectively, such that $l_1 + l_2 + \cdots + l_t = j$. Then $g_0(x) = g_1(x)g_2(x) \ldots g_t(x)$ is the desired polynomial in the theorem.

Theorem 2.13 follows immediately from Theorem 2.14.
From Theorem 2.13 we conclude:

Theorem 2.15. *If all the multiplicities of the distinct characteristic roots of B_i are different, then the characteristic roots of B_i are integers.*

§3. Association Schemes and PBIB Designs with Two Associate Classes

There are seven parameters in an association scheme with two associate classes. From (1.2)–(1.4), it is not hard to see that among the seven parameters, only three are independent. For convenience, we select four parameters and give them special notations. These four parameters are v, n_1, $a = p_{11}^1$, and $c = p_{11}^2$.

An association scheme with two associate classes is defined by three matrices B_0, B_1, B_2. Since $B_0 = I$, $B_2 = E - I - B_1$, it is enough to determine B_1. Therefore the problem of association schemes can be stated as follows:

1. Given parameters v, n_1, a, c, is there a matrix B satisfying the following conditions?
   (i) B is a $v \times v$ (0, 1)-matrix with zero diagonal elements; (3.1)
   (ii) $B' = B$; (3.2)
   (iii) $B1 = n_1 1$; (3.3)
   (iv) $B^2 = n_1 I + aB + c(E - I - B)$. (3.4)
2. If problem (1) has a solution, how many nonequivalent solutions are there?

Definition 3.1. If there is a matrix B satisfying (3.1)–(3.4), then $B^* = E - I - B$ also satisfies (3.1)–(3.4). Let v^*, n_1^*, a^*, c^* be the corresponding parameters. Then

$$v^* = v, \qquad n_1^* = v - n_1 - 1, \qquad a^* = v - 2n_1 + c - 2, \qquad c^* = v - 2n_1 + a. \tag{3.5}$$

We say that B^* is the complement scheme of B.

Obviously B exists if and only if B^* exists. Furthermore, $(B^*)^* = B$.

I. Properties of Parameters

From §2, we know that B has at most three distinct characteristic roots which are the solutions of the equation $|P_1 - \lambda I| = 0$. Notice that when $m = 2$,

$$P_1 = \begin{bmatrix} 0 & 1 & 0 \\ n_1 & a & c \\ 0 & n_1 - a - 1 & n_1 - c \end{bmatrix}.$$

Thus $|\lambda I - P_1| = (\lambda - n_1)[\lambda^2 - (a - c)\lambda + (c - n_1)]$.

Let $\gamma = a - c$, $\Delta = (a - c)^2 + 4(n_1 - c)$. Then the solutions of $|\lambda I - P_1| = 0$ are

$$n_1, \qquad \theta_1 = \tfrac{1}{2}(\gamma + \sqrt{\Delta}), \qquad \theta_2 = \tfrac{1}{2}(\gamma - \sqrt{\Delta}).$$

Theorem 3.1. $\theta_1 \neq \theta_2$.

Proof. If this is not true, then $\Delta = 0$, $a = c$, and $n_1 = c$. Taking the row sums on both sides of (3.4), we get

$$n_1^2 = n_1 + an_1 + c(v - n_1 - 1). \tag{3.6}$$

Substituting $a = c = n_1$ into (3.6), we get $n_1(v - n_1) = 0$ which is not possible.

Therefore there are two possibilities for the characteristic roots of P_1:

1. n_1 is a multiple root of P_1; *
2. n_1 is a simple root of P_1.

Theorem 3.2. *If there is a matrix B satisfying* (3.1)–(3.4), *then the following are equivalent.*

(i) *n_1 is a multiple root of P_1;*
(ii) $c = 0$;
(iii) $a = n_1 - 1$;

(iv) $$B = I \times (E - I)_{n_1+1} = \begin{bmatrix} E-I & & & 0 \\ & E-I & & \\ & & \ddots & \\ 0 & & & E-I \end{bmatrix}.$$

Proof. (i)$\Rightarrow$(ii). Since n_1 is a multiple root of P_1, it is also a multiple root of B and is the maximum characteristic root of B. According to Theorem 2 in §2, Chapter 13 of [7], B is of the form $\begin{pmatrix} D_1 & 0 \\ 0 & D_2 \end{pmatrix}$. Substituting this into (3.4) and comparing the entries on both sides we obtain $c = 0$.

(ii)$\Rightarrow$(iii). Substituting $c = 0$ into (3.6), we immediately obtain $a = n_1 - 1$.

(iii)$\Rightarrow$(iv). By the definition of a and B.

(iv)$\Rightarrow$(i). The parameters corresponding to B are $v = t(n_1 + 1)$, n_1, $a = n_1 - 1$, $c = 0$. Substituting these into the formulas of the characteristic roots of P_1, we get $\theta_1 = n_1$.

Theorem 3.2 provides the solution for case (1). It is not hard to see that the solution is unique (up to equivalence).

The following corollaries can also be obtained from Theorem 3.2.

Corollary 1. *n_1 is a multiple root of B if and only if it is a multiple root of P.*

Corollary 2. "*$c = n_1$ and B exists*" *if and only if $B^* = I \times (E - I)$.*

Corollary 3. *If $n_1 = v - 2$, then $c = n_1$.*

Corollary 4. *If $n_1 = 1$, then $c^* = n_1^*$.*

Thus, hereafter, it suffices to discuss case (2) under the following restrictions on the parameters:

$$\begin{cases} 0 \leqslant a \leqslant n_1 - 2 & (n \geqslant 2), \\ 1 \leqslant c \leqslant n_1 - 1, & \\ 2 \leqslant n_1 \leqslant v - 3 & (v \geqslant 5), \\ a^* \geqslant 0. & \end{cases} \tag{3.7}$$

* For convenience, hereafter, if θ is a characteristic root of matrix A with multiplicity p, then we say that θ is a p-multiple root of A.

In case (2), n_1, θ_1, θ_2 are all different. By Corollary 1 of Theorem 3.2, the multiplicity of n_1 is 1. Let the multiplicities of θ_1 and θ_2 be α and $v - \alpha - 1$, respectively. Then

$$n_1 \cdot 1 + \theta_1 \cdot \alpha + \theta_2 \cdot (v - \alpha - 1) = \operatorname{tr} B = 0.$$

Solving for α, we get

$$\alpha = \frac{v-1}{2} + \frac{(c-a)(v-1) - 2n_1}{2\sqrt{\Delta}},$$

$$v - \alpha - 1 = \frac{v-1}{2} - \frac{(c-a)(v-1) - 2n_1}{2\sqrt{\Delta}}.$$

Theorem 3.3. *If B exists and the multiplicities of θ_1 and θ_2 are the same, then the parameters of B must satisfy the following conditions.*

$$v = 4a + 5, \qquad n_1 = 2a + 2, \qquad c = a + 1. \tag{3.8}$$

Proof. If $\alpha = v - \alpha - 1$, then $(c - a)(v - 1) = 2n_1$. By (3.6),

$$c = a + 1, \qquad n_1 = 2a + 2, \qquad v = 4a + 5.$$

Theorem 3.4. *If B exists and the multiplicities of θ_1, θ_2 are different, then Δ must be the square of an integer.*

Proof. From Theorem 2.15, we know that if the multiplicities of θ_1 and θ_2 are different, then θ_1 and θ_2 must be integers. Therefore $\sqrt{\Delta}$ must be an integer.

Hereafter, we will only consider the case where the multiplicities of θ_1 and θ_2 are different.

Since $\Delta > (a-c)^2$, $\theta_2 = \frac{1}{2}((a-c) - \sqrt{\Delta}) < 0$. Let $\nu = -\theta_2 > 0$.

II. A Complete Solution for $\nu = 1, 2$

When $\nu = 1$, $4(a - c) + 4 = 4(n_1 - c)$, we have $a = n_1 - 1$. By Theorem 3.2, the scheme exists and is unique.

When $\nu \geqslant 2$, besides condition (3.7), we also have

$$\alpha = \frac{\nu(v-1) - n_1}{a - c + 2\nu} \quad \text{which is an integer and} \quad n_1 = c + (a-c)\nu + \nu^2. \tag{3.9}$$

Since B is a symmetric matrix, and the characteristic vector corresponding to n_1 is $(1/\sqrt{v})1$, there is an orthogonal matrix Γ:

$$\Gamma = \left(\frac{1}{\sqrt{v}} 1 \Gamma_1 \Gamma_2 \right),$$

such that

$$B = \Gamma \left[\begin{array}{ccccccc} n_1 & & & & & & 0 \\ & \theta_1 & & \alpha & & & \\ & & \ddots & & & & \\ & & & \theta_1 & & & \\ & & & & \theta_2 & & v-\alpha-1 \\ & & & & & \ddots & \\ 0 & & & & & & \theta_2 \end{array} \right] \Gamma',$$

where Γ_1, Γ_2 are semi-orthogonal matrices (i.e., $\Gamma_i'\Gamma_i = I$). Thus

$$B - \theta_2 I = \frac{n - \theta_2}{v} E + (\theta_1 - \theta_2)\Gamma_1\Gamma_1',$$

$$\theta_1 I - B = \frac{\theta_1 - n_1}{v} E + (\theta_1 - \theta_2)\Gamma_2\Gamma_2'.$$

Let

$$T = B - \theta_2 I - \frac{n_1 - \theta_2}{v} E.$$

Theorem 3.5. *If B exists, then T is semi-positive definite, and the rank of T is α.*

Proof. Since $T = (\theta_1 - \theta_2)\Gamma_1\Gamma_1'$ and $\theta_1 - \theta_2 > 0$, T is semi-positive definite. Furthermore, the rank of Γ_1 is α, so the rank of T is α.

The case $\nu > 2$ is more complicated. Theorem 3.5 can be used to obtain satisfactory results for the case $\nu = 2$. From now on, we will always assume $\nu = 2$ unless declared otherwise.

When $\nu = 2$, the corresponding parameters are

$$\left\{ \begin{array}{l} n_1 = 2a - c + 4, \\ \theta_1 = a - c + 2, \qquad \theta_2 = -2, \\ v = \dfrac{1}{c}(2a - c + 6)(a + 2), \\ \alpha = \dfrac{(2a - c + 6)(2a - c + 4)}{c(a - c + 4)}. \end{array} \right. \tag{3.10}$$

If $\theta_1 = 0$, then $c = a + 2$. Therefore $v = a + 4$ and $n_1 = a + 2$. Considering the parameters of the complement scheme, we obtain $n_1^* = a + 1$ and $a^* = c^* = a$. Therefore $2a + 4 = v = v^* = (1/a)n_1^*(n_1^* - 1) + 1 = a + 2$, and hence $a = -2$, which is not possible. Without loss of generality, hereafter we may assume that $\theta_1 \geqslant 1$ (i.e., $a \geqslant c - 1$). Then the inequality $a^* \geqslant 0$ in (3.7) can be modified to $a \geqslant \frac{1}{2}(3c - 6)$.

Lemma 3.1. *If B exists and*

$$F = \begin{bmatrix} 0 & 1 & 1 & 1 \\ 1 & 0 & 0 & 0 \\ 1 & 0 & 0 & 0 \\ 1 & 0 & 0 & 0 \end{bmatrix}$$

is not a principal submatrix of B, then

$$\begin{pmatrix} \overbrace{E - I}^{n_1 - a} & e_1 \\ e_1' & 0 \end{pmatrix}$$

must be a principal submatrix of B, where $e_1' = (100 \ldots 0)$.

Proof. Without loss of generality, assume that the upper-left corner of B is

$$\begin{array}{r|c|c|c} & 0 & \overbrace{1\ 1 \cdots 1}^{n_1 - a} & \overbrace{1\ 1 \cdots 1}^{a} \\ n_1 - a \left\{ \begin{array}{c} \\ \\ \\ \\ \end{array} \right. & \begin{array}{c} 1 \\ 1 \\ \vdots \\ 1 \end{array} & \begin{array}{c} 0\ 0 \cdots 0 \\ 0 \qquad \\ \vdots \quad C \\ 0 \qquad \end{array} & \begin{array}{c} 1\ 1 \cdots 1 \\ \\ * \\ \\ \end{array} \\ a \left\{ \begin{array}{c} \\ \\ \\ \end{array} \right. & \begin{array}{c} 1 \\ \vdots \\ 1 \end{array} & \begin{array}{c} 1 \qquad \\ \vdots \quad * \\ 1 \qquad \end{array} & * \end{array}$$

If C has a zero off-diagonal element, then F is a principal submatrix of B. This contradicts the assumption. Thus $C = E - I$. After appropriate permutations of rows and columns, the upper-left corner of B becomes

$$\begin{pmatrix} \overbrace{E - I}^{n_1 - a} & e_1 \\ e_1' & 0 \end{pmatrix}.$$

Lemma 3.2. *Suppose B exists. If*

$$\begin{pmatrix} \overbrace{E - I}^{n_1 - a} & e_1 \\ e_1' & 0 \end{pmatrix}$$

is a principal submatrix of B, then $n_1 - a \leqslant \alpha$.

Proof. Consider T. By assumption, T has an $(n_1 - a) \times (n_1 - a)$ principal submatrix

$$T_1 = E - I + 2I - \frac{n_1 + 2}{v} E.$$

Then

$$|T_1| = 1 + \frac{v-(n_1+2)}{v}(n_1-a) \geqslant 1.$$

Therefore T_1 is a full-rank submatrix of T, and n_1-a, the order of T_1, cannot be bigger than α, the rank of T.

Lemma 3.3. *If B exists and F is a principal submatrix of B, then $a = 2c-2$.*

Proof. If F is a principal submatrix of B, then T has a principal submatrix T_1:

$$T_1 = F + 2I - \frac{n_1+2}{v}E,$$

$$(a+2)T_1 = \begin{bmatrix} 2a-c+4 & a-c+2 & a-c+2 & a-c+2 \\ a-c+2 & 2a-c+4 & -c & -c \\ a-c+2 & -c & 2a-c+4 & -c \\ a-c+2 & -c & -c & 2a-c+4 \end{bmatrix}.$$

Write $(a+2)T_1$ as $\binom{T_{11}\ T_{12}}{T_{21}\ T_{22}}$. By the semi-positive definiteness of $T_1, T_2 = T_{22} - T_{21}T_{11}^{-1}T_{12}$ is semi-positive definite. Take $T_{11} = 2a-c+4$; then

$$T_2 = (2a-c+4)I - c(E-I) - \frac{(a-c+2)^2}{2a-c+4}E$$

is semi-positive definite. Therefore $1'T_2 1 \geqslant 0$, i.e., $a \geqslant 2c-2$.

Furthermore, since F is a principal submatrix of B, without loss of generality, we may assume that the first four rows of B are

$$\begin{bmatrix} 0 & 1 & 1 & 1 & \overbrace{1 \quad 1 \quad \cdots \quad 1}^{n_1-3} & \overbrace{0 \quad 0 \quad \cdots \quad 0}^{n_2} \\ 1 & 0 & 0 & 0 & & \\ 1 & 0 & 0 & 0 & Y' & Z' \\ 1 & 0 & 0 & 0 & & \end{bmatrix}.$$

Substituting into (3.4) and comparing the entries on the two sides, we have

$$Y'Y + Z'Z = (n_1-c)I + (c-1)E = (n_1-1)I + (c-1)(E-I).$$

Therefore each entry of $Y'Y$ is less than or equal to the corresponding entry on the right side. Let m_k be the number of column vectors y of Y' such that $y'1 = k$. Then $0 \leqslant k \leqslant 3$, and

$$m_3 + m_2 + m_1 + m_0 = n_1 - 3 = 2a - c + 1; \tag{3.11}$$

$$3m_3 + 2m_2 + m_1 = 3a. \tag{3.12}$$

Let

$$Y'Y = \begin{pmatrix} a & x & y \\ x & a & z \\ y & z & a \end{pmatrix}.$$

Then x, y, z are all less than or equal to $c-1$. Notice that

$$1'Y' = (\underbrace{3 \cdots 3}_{m_3} \quad \underbrace{2 \cdots 2}_{m_2} \quad \underbrace{1 \cdots 1}_{m_1} \quad \underbrace{0 \cdots 0}_{m_0}).$$

Thus

$$1'Y'Y1 = 9m_3 + 4m_2 + m_1. \tag{3.13}$$

On the other hand

$$1'Y'Y1 = 3a + 2(x+y+z) \leqslant 3a + 6(c-1).$$

From (3.13) we obtain

$$9m_3 + 4m_2 + m_1 \leqslant 3a + 6(c-1). \tag{3.14}$$

Subtracting (3.12) from (3.14), we get

$$3m_3 + m_2 \leqslant 3(c-1).$$

Subtracting the above from (3.12), we get

$$m_2 + m_1 \geqslant 3a - 3c + 3.$$

Subtracting the above from (3.11), we get

$$m_3 + m_0 \leqslant 2c - a - 2.$$

Therefore

$$2c - 2 \geqslant a.$$

According to Lemma 3.3, we divide the association schemes with $\nu = 2$ into two cases: $a = 2c-2$, and $a \neq 2c-2$. Eventually all the schemes with $\nu = 2$ will be obtained; see Table 3.3 for detailed results.

If $a = 2c-2$, by (3.10), the parameters are

$$v = 6c + 4, \qquad n_1 = 3c, \qquad \alpha = 9 - \frac{12}{c+2}.$$

Therefore c can only be 1, 2, 4, or 10. All the possible parameters are listed in Table 3.1.

It is easy to show that scheme (I.1) exists and is unique. Its association matrix will be given in Appendix II.

It is also easy to see that in case (I.2), schemes exist but are not unique. There are two solutions and their association matrices are presented in Appendix II.

L. C. Chang (Zhang Li-qien) [3] pointed out the nonuniqueness of schemes in case (I.3). There are four solutions which can be found in [3]; they will not be discussed in the present paper.

Table 3.1

No.	c	v	n_1	a	α	Number of solutions
(I.1)	1	10	3	0	5	1 (see Appendix II)
(I.2)	2	16	6	2	6	2 (see Appendix II)
(I.3)	4	28	12	6	7	4 (see [3])
(I.4)	10	64	30	18	8	0

So what remains to be done is to show that scheme (I.4) does not exist.

If a scheme in case (I.4) exists, without loss of generality, we may assume that the first three rows of its association matrix B are

$$\begin{array}{cccccccc} 0&1&1& \overbrace{1\ \cdots\ 1}^{17} & \overbrace{1\ \cdots\ 1}^{11} & \overbrace{0\ \cdots\ 0}^{11} & \overbrace{0\ \cdots\ 0}^{22} \\ 1&0&1& 1\ \cdots\ 1 & 0\ \cdots\ 0 & 1\ \cdots\ 1 & 0\ \cdots\ 0 \\ 1&1&0& *\ \cdots\ * & & & \end{array}.$$

Obviously, $*$ cannot all be zero. Therefore

$$S_0 = \begin{bmatrix} 0&1&1&1\\ 1&0&1&1\\ 1&1&0&1\\ 1&1&1&0 \end{bmatrix}$$

is a principal submatrix of B. Let the first four rows of B be $(S_0\ X')$. Then we have

Proposition 3.1. *Among the columns of X', a column with three* 1*'s and one* 0 *or three* 0*'s and one* 1 *can appear at most three times.*

Proof. Suppose this is not the case. Assume that $(1\ 1\ 1\ 0)'$ appears at least four times as columns of X'; then

$$\begin{bmatrix} \begin{matrix} & & & \\ & S_0 & & \\ & & & \\ & & & \end{matrix} & \begin{matrix} 1&1&1&1\\ 1&1&1&1\\ 1&1&1&1\\ 0&0&0&0 \end{matrix} \\ \begin{matrix} 1&1&1&0\\ 1&1&1&0\\ 1&1&1&0\\ 1&1&1&0 \end{matrix} & \begin{matrix} 0& & *' & \\ & 0 & & \\ *' & & 0 & \\ & & & 0 \end{matrix} \end{bmatrix}$$

is a principal submatrix of B. The corresponding matrix in T is

$$T_0 = \begin{bmatrix} 3&1&1&1&1&1&1&1\\ 1&3&1&1&1&1&1&1\\ 1&1&3&1&1&1&1&1\\ 1&1&1&3&-1&-1&-1&-1\\ 1&1&1&-1&3&*&*&*\\ 1&1&1&-1&*&3&*&*\\ 1&1&1&-1&*&*&3&*\\ 1&1&1&-1&*&*&*&3 \end{bmatrix},$$

where $*$ is $+1$ or -1. By the semi-positive definiteness of T_0, the matrix

$$\begin{bmatrix} 3&*&*&*\\ *&3&*&*\\ *&*&3&*\\ *&*&*&3 \end{bmatrix} - \begin{bmatrix}1\\1\\1\\1\end{bmatrix}(1\ \ 1\ \ 1\ -1)\left(\tfrac{1}{2}I - \tfrac{1}{12}11'\right)\begin{bmatrix}1\\1\\1\\-1\end{bmatrix}(1\ \ 1\ \ 1\ \ 1)$$

is semi-positive definite, i.e.,

$$\begin{bmatrix} 4 & 3\epsilon_1 - 5 & 3\epsilon_2 - 5 & 3\epsilon_3 - 5 \\ 3\epsilon_1 - 5 & 4 & 3\epsilon_4 - 5 & 3\epsilon_5 - 5 \\ 3\epsilon_2 - 5 & 3\epsilon_4 - 5 & 4 & 3\epsilon_6 - 5 \\ 3\epsilon_3 - 5 & 3\epsilon_5 - 5 & 3\epsilon_6 - 5 & 4 \end{bmatrix}$$

is semi-positive definite, where $\epsilon_i = \pm 1$. Obviously, all the ϵ_i must be 1. After substitution, we conclude that

$$\begin{bmatrix} 2 & -1 & -1 & -1 \\ -1 & 2 & -1 & -1 \\ -1 & -1 & 2 & -1 \\ -1 & -1 & -1 & 2 \end{bmatrix}$$

is semi-positive definite. But it cannot be semi-positive definite, since -1 is one of its characteristic roots. This shows that $(1\ \ 1\ \ 1\ \ 0)'$ cannot appear more than 3 times in X'. Similarly, one can show that $(0\ \ 0\ \ 0\ \ 1)'$ cannot appear more than 3 times in X'.

Proposition 3.2. *Among the columns of X', a column with two 1's and two 0's can appear at most twice.*

Proof. Similar to the proof of Proposition 3.1.

Let m_i, $i = 0, 1, 2, 3$, be the number of columns in X' in which 1 appears i times. By Propositions 3.1 and 3.2,

$$m_3 \leqslant 12, \qquad m_1 \leqslant 12, \qquad m_2 \leqslant 12.$$

Furthermore, from $B^2 = 20I + 8B + 10E$, we obtain

$$\begin{bmatrix} 3 & 2 & 2 & 2 \\ 2 & 3 & 2 & 2 \\ 2 & 2 & 3 & 2 \\ 2 & 2 & 2 & 3 \end{bmatrix} + X'X = \begin{bmatrix} 30 & 18 & 18 & 18 \\ 18 & 30 & 18 & 18 \\ 18 & 18 & 30 & 18 \\ 18 & 18 & 18 & 30 \end{bmatrix}.$$

Therefore

$$X'X = \begin{bmatrix} 27 & 16 & 16 & 16 \\ 16 & 27 & 16 & 16 \\ 16 & 16 & 27 & 16 \\ 16 & 16 & 16 & 27 \end{bmatrix}.$$

Thus

$$\begin{aligned} m_4 + m_3 + m_2 + m_1 + m_0 &= 60, \\ 4m_4 + 3m_3 + 2m_2 + m_1 &= 108, \\ 6m_4 + 3m_3 + m_2 &= 96. \end{aligned}$$

Let $m_2 = 12 - 3k$, $k \geqslant 0$, $m_3 = 12 - l$, $l \geqslant 0$. Substituting into the above expression, we get

$$2m_4 = k + l + 16.$$

Therefore $m_1 = 16 + l + 4k > 12$, which is a contradiction. This shows the nonexistence of a scheme in case (I.4).

If $a \neq 2c - 2$, then by Lemmas 3.1–3.3, $n_1 - a \leqslant \alpha$. We have assumed $\theta_1 \geqslant 1$ (i.e., $a \geqslant c - 1$) and $a^* \geqslant 0$ (i.e., $a \geqslant \frac{1}{2}(3c - 6)$), therefore

$$a \geqslant c - 1, \qquad a \geqslant \tfrac{1}{2}(3c - 6), \qquad a \neq 2c - 2, \tag{3.15}$$

$$(c - 4)a^2 - 2(c^2 - 6c + 10)a + (c - 2)(c - 3)(c - 4) \leqslant 0. \tag{3.16}$$

It follows from (3.15) and (3.16) that when $a \neq 2c - 2$, the only possible values of c are $1 \leqslant c \leqslant 10$. For convenience of discussion, we will prove several useful lemmas before we proceed.

Lemma 3.4. *Suppose B has a principal submatrix $E - I$. If $\nu \geqslant 2$ and $n_1 + \nu < v$, then the principal submatrix T_0 of T corresponding to $E - I$ has full rank.*

Proof. We have

$$T_0 = E - I + \nu I - \frac{n_1 + \nu}{v} E = (\nu - 1)I + \left(1 - \frac{n_1 + \nu}{v}\right)E;$$

thus

$$|T_0| = (\nu - 1)^{s-1}\left(\nu - 1 + \left(1 - \frac{n_1 + \nu}{v}\right)s\right),$$

where s is the order of the principal submatrix $E - I$. By assumption $|T_0| \neq 0$, thus T_0 has full rank.

Since, when $\nu = 2$, all the parameters can be expressed in terms of a and c, we have

Corollary 1. *If $\nu = 2$ and B has a principal submatrix $E - I$, then the principal submatrix T_0 of T corresponding to $E - I$ has full rank.*

Lemma 3.5. *Let A be a symmetric matrix of the form*

$$\begin{pmatrix} U & X \\ X' & V \end{pmatrix},$$

and $\mathrm{rk}(U) = \mathrm{rk}(A)$, $|U| \neq 0$. *Then the entries of V can be expressed in terms of the entries of U and X.*

Proof. Let $X = (x_1 x_2 \ldots x_t)$ and $V = (v_{ij})_{t \times t}$, where x_i's are column vectors. For any entry v_{ij} of V,

$$\begin{vmatrix} U & x_j \\ x_i' & v_{ij} \end{vmatrix} = 0.$$

Therefore,

$$v_{ij} = x_i' U^{-1} x_j .$$

For convenience, hereafter we call X the side matrix of U. Lemma 3.5 pointed out that if a symmetric matrix A has a full-rank principal submatrix which has the same rank as A, then this submatrix and its side matrix determine A.

Lemma 3.6. *Suppose $\nu \geqslant 2$. If the matrix T corresponding to B is of the form*

$$\begin{pmatrix} T_0 & Y \\ Y' & D \end{pmatrix},$$

the order of T_0 is α, and $|T_0| \neq 0$, then all the columns of Y are different.

Proof. Suppose Y has two identical columns. Without loss of generality, assume $Y = (\gamma_1 \gamma_2 \ldots \gamma_t)$, where γ_1 is the same as γ_2. Then

$$\begin{vmatrix} T_0 & y_1 \\ y_1' & d_{11} \end{vmatrix} = 0, \qquad \begin{vmatrix} T_0 & y_1 \\ y_2' & d_{21} \end{vmatrix} = 0 \qquad (d_{ij} \text{ is an entry of } D).$$

Since $y_1 = y_2$, $d_{11} = y_1' T_0^{-1} y_1 = y_2' T_0^{-1} y_1 = d_{21}$. By the definition of T,

$$d_{11} = \nu - \frac{n_1 + \nu}{v}, \qquad d_{21} = b_{\alpha+2\alpha+1} - \frac{n_1 + \nu}{v}.$$

Therefore $b_{\alpha+2\alpha+1} = \nu \geqslant 2$. But since B is a (0, 1)-matrix, $b_{\alpha+2\alpha+1}$ cannot be bigger than 1. This is a contradiction and Lemma 3.6 is proved.

Lemma 3.7. *Suppose $\nu \geqslant 2$, B exists and has the form*

$$\begin{pmatrix} S_0 & X \\ X' & C \end{pmatrix},$$

where the order of S_0 is α, and the principal submatrix T_0 of T corresponding to S_0 has full rank. Then all the columns of X are different.

Proof. By Lemma 3.6.

Lemma 3.8. *Suppose the association matrix B has an $\alpha \times \alpha$ principal submatrix $E - I$ with side matrix X. Then the number of 1's in each column of X, say l, satisfies the following quadratic equation:*

$$Rv^2 l^2 - [2RSv - Sv^2 + R\alpha l^2] l + S(\alpha S + R)(R + S) - \alpha S(v - R)^2 = 0, \qquad (3.17)$$

where $R = v - n_1 - \nu$, $S = (\nu - 1)v$.

Proof. By Corollary 1 of Lemma 3.4, the principal submatrix T_0 of T corresponding to the principal submatrix $E - I$ of B has full rank. Since the order of $E - I$ is α, $\text{rk}(T_0) = \text{rk}(T)$. For any given column x of X, let T_1 be the submatrix of T corresponding to

$$\begin{pmatrix} E - I & x \\ x' & 0 \end{pmatrix}.$$

Then $|T_1| = 0$, and equation (3.17) follows.

Now we shall proceed with the possible values of c.

(II.1) $c = 1$.
In this case, the parameters are

$$v = (2a+5)(a+2), \qquad n_1 = 2a+3, \qquad a \geqslant 1, \qquad \alpha = \frac{(2a+3)(2a+5)}{a+3}.$$

Since α is not an integer when $a \geqslant 1$, there is no such scheme.

(II.2) $c = 2$.
In this case, the parameters are

$$v = (a+2)^2, \qquad n_1 = 2(a+1), \qquad a \geqslant 1, \qquad a \neq 2, \qquad \alpha = 2a+3.$$

Obviously, these are the parameters of the L_2 schemes. By Lemmas 3.1 and 3.3, B has an $(a+2) \times (a+2)$ principal submatrix $E - I$. It is easy to see that the side matrix of $E - I$ is $I_{a+2} \times 1'_{a+1}$. Suppose

$$B = \begin{pmatrix} \overbrace{E-I}^{a+2} & X \\ X' & A \end{pmatrix}, \qquad X = I_{a+2} \times 1'_{a+1},$$

$$A = \begin{bmatrix} A_{11} & \cdots & A_{1\,a+2} \\ \vdots & \ddots & \vdots \\ A_{a+2\,1} & \cdots & A_{a+2\,a+2} \end{bmatrix}.$$

Then A_{11} has no zero off-diagonal entries; otherwise, by appropriately permuting the columns and rows, B would have the following principal submatrix:

$$a+2\left\{\begin{bmatrix} \overbrace{\qquad E-I \qquad}^{a+2} & \begin{matrix} 1 & 1 \\ 0 & 0 \\ \vdots & \vdots \end{matrix} \\ \begin{matrix} 1 & 0 & \cdots & 0 \\ 1 & 0 & \cdots & 0 \end{matrix} & \begin{matrix} 0 & 0 \\ 0 & 0 \end{matrix} \end{bmatrix}\right..$$

Then the corresponding principal submatrix T_1 of T would have a negative determinant. This contradicts the semi-positive definiteness of T. Therefore, $A_{11} = E - I$, and $A_{1j} = 0$ for $j \geqslant 2$. Similarly, one can show that $A_{ii} = E - I$, and $A_{ij} = 0$ for $i \neq j$. Thus in case (II.2), an association scheme either does not exist, or is unique. Now we already know that the parameters of (II.2) are those of L_2 schemes. Therefore scheme (II.2) exists and is unique. This also shows the uniqueness of an L_2 scheme.

(II.3) $c = 3$.
In this case, the parameters are

$$v = \frac{(2a+3)(a+2)}{3}, \quad n_1 = 2a+1, \quad a \geqslant 2, \quad a \neq 4, \quad \alpha = \frac{(2a+3)(2a+1)}{3(a+1)}.$$

Since α is not an integer when $a \geqslant 2$, there is no such scheme.

(II.4) $c = 4$.

In this case, the parameters are

$$v = \tfrac{1}{2}(a+1)(a+2), \qquad n_1 = 2a, \qquad a \geqslant 3, \qquad a \neq 6, \qquad \alpha = a+1.$$

Obviously, these are parameters of triangular schemes. By Lemmas 3.3 and 3.1, B has the principal submatrix

$$S_0 = \begin{pmatrix} E - I & e_1 \\ e_1' & 0 \end{pmatrix},$$

where the order of S_0 is $n_1 - a + 1 = a + 1 = \alpha$. The principal submatrix T_0 of T corresponding to S_0 is $S_0 + 2I - [4/(a+2)]E$. It is not hard to show that $|T_0| \neq 0$. Thus T_0 is a full-rank principal submatrix of T and has the same rank as T. Let the side matrix of S_0 be $X = (x_1, \ldots, x_{v-a-1})$. For an arbitrary x_i, let l be the number of 1's in the first a elements of x_i, and x be the last element of x_i. If T_1 is the principal submatrix of T corresponding to

$$\begin{pmatrix} S_0 & x_i \\ x_i' & 0 \end{pmatrix},$$

then

$$(a+2)T_1 = \begin{bmatrix} & & & a-2 & y_1 \\ & & & -4 & y_2 \\ (a+2) & (E+I) & -4E & \vdots & \vdots \\ & & & -4 & y_a \\ a-2 & -4 \quad \cdots & -4 & 2a & y \\ y_1 & y_2 \quad \cdots & y_a & y & 2a \end{bmatrix},$$

where $y_1, \ldots, y_a$ are those entries of T corresponding to the first a elements of x_i, and y is the entry of T corresponding to x. Since the rank of T is α and the order of T_1 is $\alpha + 1$, $|T_1| = 0$. Therefore, we have the following equation.

$$\begin{aligned} &(a-6)(a+2)^2 l^2 - \left[(a+2)^3(a-6) + 2(a+2)^2(y - y_1)\right] l \\ &+ 2a(a-6)(a+2)^2 + 8a(a+2)(y - y_1) \\ &- (a^2 - a + 2)(y - y_1)^2 = 0. \end{aligned} \tag{3.18}$$

Now B is a (0, 1)-matrix, so y_1 and y only have two possible values. It follows from equation (3.18) that the columns of X must be of the following four kinds:

$$\begin{array}{ccccc} & \text{(a)} & \text{(b)} & \text{(c)} & \text{(d)} \\ & 1 & 1 & 0 & 0 \\ a-1 \left\{ \begin{matrix} \\ \\ \\ \\ \end{matrix} \right. & \begin{matrix} 1 \\ 1 \\ \vdots \\ 1 \end{matrix} & \left. \begin{matrix} * \\ \vdots \\ \vdots \\ * \end{matrix} \right\} \text{one } 1 & \left. \begin{matrix} * \\ \vdots \\ \vdots \\ * \end{matrix} \right\} \text{two } 1\text{'s} & \left. \begin{matrix} * \\ \vdots \\ \vdots \\ * \end{matrix} \right\} \text{one } 1\,. \\ & 1 & 1 & 0 & 1 \end{array}$$

By Lemma 3.7, all the columns of X are different. Therefore, it has at most one column as (a), at most $a-1$ columns as (b), at most $(a-1)(a-2)/2$ columns as (c), and at most $a-1$ columns as (d). But

$$1+2(a-1)+\tfrac{1}{2}(a-1)(a-2)=\frac{a}{2}(a+1)=v-a-1;$$

therefore X is completely determined. By Lemma 3.5, B is uniquely determined. This shows that scheme (II.4) either does not exist or is unique. Now we already know that the parameters of (II.4) are those of triangular schemes. Accordingly, scheme (II.4) exists and is unique. This also shows the uniqueness of a triangular scheme.

(II.5) $c=5$.
In this case, the parameters are

$$v=\tfrac{1}{5}(2a+1)(a+2),\quad n_1=2a-1,\quad a\geqslant 4,\quad a\neq 8,\quad \alpha=\frac{(2a+1)(a-1)}{5(a-1)}.$$

Since α is not an integer when $a\geqslant 4$, there is no such scheme.

(II.6) $c=6$.
In this case, the parameters are

$$v=\frac{2a(a+2)}{6},\qquad n_1=2a-2,\qquad a\geqslant 5,\qquad a\neq 10,\qquad \alpha=\frac{2a(a-1)}{3(a-2)}.$$

Since $\alpha=(a/3)[2+2/(a-2)]$, the only possible value of $a\geqslant 5$ for which α is an integer is $a=6$. Therefore in the present case, the only possible parameter values for which a scheme may exist are

$$v=16,\qquad n_1=10,\qquad a=6,\qquad c=6,\qquad \alpha=5.$$

Now we shall show that such a scheme exists and is unique.

By Lemmas 3.3 and 3.1, B must have a principal submatrix $E-I$ whose order is $n_1-a=4$. Using the same method as in (I.4), we can show that B must have a 5×5 principal submatrix $E-I$. By Lemma 3.8, the number of 1's in each column of the side matrix X of $(E-I)_{5\times 5}$ satisfies equation (3.17). The solutions are $l=3,0$. Furthermore, it follows from Lemmas 3.4 and 3.7 that all the columns of the side matrix of $E-I$ are different. Now $\binom{5}{3}+\binom{5}{0}=10+1=11=v-\alpha$; therefore X is completely determined. By Lemma 3.5, B is unique if it exists. It is easy to show that B indeed exists. This association matrix will be presented in Appendix II.

(II.7) $c=7$.
In this case, the parameters are

$$v=\tfrac{1}{7}(2a-1)(a+2),\qquad n_1=2a-3,\qquad a=8\quad\text{or}\quad 9.$$

Since v is not an integer for $a=8$ or 9, such a scheme does not exist.

(II.8) $c=8$.
In this case, the parameters are

$$v=\tfrac{1}{8}(2a-2)(a+2),\qquad n_1=2a-4,\qquad a=10.$$

Therefore, in the present case, the only possible parameter values for which a scheme may exist are

$$v = 27, \qquad n_1 = 16, \qquad a = 10, \qquad c = 8, \qquad \alpha = 6.$$

By Lemmas 3.3 and 3.1, B must have a principal submatrix $(E - I)_{6\times 6}$. Furthermore, by Lemmas 3.4 and 3.7, all the columns of the side matrix X of $E - I$ are different. It follows from Lemma 3.8 that the number of 1's in each column of the side matrix X satisfies equation (3.17). The solutions are $l = 1,4$. Now $\binom{6}{1} + \binom{6}{4} = 6 + 15 = 21 =$ the number of columns of X; therefore X is uniquely determined. By Lemma 3.5, B is unique if it exists. It is easy to show that B indeed exists. This association matrix will be given in Appendix II.

(II.9) $c = 9$.

In this case, the parameters are

$$v = \tfrac{1}{9}(2a - 3)(a + 2), \qquad n_1 = 2a - 5, \qquad a = 12.$$

Since v is not an integer when $a = 12$, such a scheme does not exist.

(II.10) $c = 10$.

In this case, the parameters are

$$v = \tfrac{1}{10}(2a - 4)(a + 2), \qquad n_1 = 2a - 6, \qquad a = 12.$$

Thus the only possible parameter values for which a scheme may exist are

$$v = 28, \qquad n_1 = 18, \qquad a = 12, \qquad c = 10, \qquad \alpha = 6.$$

By Lemmas 3.3 and 3.1, if B exists, then B must have a principal submatrix

$$\begin{pmatrix} E - I & e_1 \\ e_1' & 0 \end{pmatrix},$$

where $E - I$ is 6×6. It is easy to show that the corresponding principal submatrix of T has a nonzero determinant. Therefore the rank of T is not less than 7, which is contradictory to $\alpha = 6$. It follows that scheme (II.10) does not exist.

Summarizing the above discussion for the case $a \neq 2c - 2$, we have Table 3.2.

Combining all of the above discussions regarding the case $\nu = 2$, we obtain all the association schemes with $\nu = 2$ (see Table 3.3).

Now consider the complement schemes. We know that the complement schemes of (II.1), (II.3), (II.5), (II.7), (II.9), (II.10), (I.4) do not exist, while the complement schemes of (I.1), (I.2), (I.3), (II.2), (II.4), (II.6), (II.8) exist and we know all the solutions. It is interesting that, except (I.1), the parameter ν^* of the complement scheme is not less than 3. The readers can easily calculate the parameters of these schemes which will not be listed here.

III. Using BIB Designs with $\lambda = 1$ to Construct Association Schemes and PBIB Designs with Two Associate Classes

Hereafter we use v', b', r', k', λ' to denote the parameters of a BIB design, and use $v, b, r, k, \lambda_i, n_i, p_{jk}^i$ to denote the parameters of an association scheme and PBIB

Table 3.2

No.	c	v	n_1	a	α	Number of solutions
(II.1)	1	$(2a+5)(a+2)$	$2a+3$	$\geqslant 1$	$\dfrac{(2a+3)(2a+5)}{a+3}$	zero
(II.2)	2	$(a+2)^2$	$2(a+1)$	$\geqslant 1$, $\neq 2$	$2a+3$	1 for each a (L_2)
(II.3)	3	$\dfrac{(2a+3)(a+2)}{3}$	$2a+1$	$\geqslant 2$, $\neq 4$	$\dfrac{(2a+3)(2a+1)}{3(a+1)}$	zero
(II.4)	4	$\dfrac{(a+1)(a+2)}{2}$	$2a$	$\geqslant 3$, $\neq 6$	$a+1$	1 for each a (triangular)
(II.5)	5	$\dfrac{(2a+1)(a+2)}{5}$	$2a-1$	$\geqslant 4$, $\neq 8$	$\dfrac{(2a+1)(2a-1)}{5(a-1)}$	zero
(II.6)	6	16	10	6	5	one (Appendix II)
(II.7)	7	$\dfrac{(2a-1)(a+2)}{7}$	$2a-3$	8 or 9		zero
(II.8)	8	27	16	10	6	one (Appendix II)
(II.9)	9	$\dfrac{(2a-3)(a+2)}{9}$	$2a-5$	12		zero
(11.10)	10	28	18	12	6	zero

Table 3.3

No.	c	v	n_1	a	α	Number of solutions
(I.1)	1	10	3	0	5	one (Appendix II)
(I.2)	2	16	6	2	6	two (Appendix II)
(I.3)	4	28	12	6	7	four (see [3])
(II.2)	2	$(a+2)^2$	$2(a+1)$	≥ 1, $\neq 2$	$2a+3$	one for each a (L_2)
(II.4)	4	$\frac{(a+1)(a+2)}{2}$	$2a$	≥ 3, $\neq 6$	$a+1$	one for each a (triangular)
(II.6)	6	16	10	6	5	one (Appendix II)
(II.8)	8	27	16	10	6	one (Appendix II)

design with two associate classes. Furthermore, p_{11}^1 and p_{11}^2 are denoted by a and c, respectively. From [13, 14], we know:

Theorem 3.6. *If a BIB design with parameters $v', b', r', k', \lambda' = 1$ exists, then there is an association scheme with parameters*

$$v = b', \qquad n_1 = k'(r'-1), \qquad a = r' - 2 + (k'-1)^2 \tag{3.19}$$

and a corresponding PBIB design with parameters

$$k = r', \qquad r = k', \qquad \lambda_1 = 1, \qquad \lambda_2 = 0, \qquad b = v' \tag{3.20}$$

exists.

Proof. Let M be the design matrix with parameters $v', b', r', k', \lambda' = 1$. Since $\lambda' = 1$, any two columns of M have at most one common treatment. Therefore

$$M'M = k'I + 1 \cdot D + 0 \cdot (E - I - D), \tag{3.21}$$

where D is a symmetric (0, 1)-matrix. Notice that M is a BIB design matrix. Hence

$$M1_{b'} = r'1_{v'}, \qquad M'1_{v'} = k'1_{b'}, \qquad MM' = r'I + (E - I). \tag{3.22}$$

Taking the row sums of (3.21), we get

$$D1 = k'(r'-1)1. \tag{3.23}$$

Also,

$$\begin{aligned} D^2 &= (M'M - k'I)(M'M - k'I) = (r' - 2k' - 1)M'M + k'^2E + k'^2I \\ &= k'(r'-1)I + \big((k'-1)^2 + r' - 2\big)D + k'^2(E - D - I). \end{aligned}$$

It follows from (3.23), (3.21), and the above expression that D satisfies (3.1)–(3.4) with respect to the parameters in (3.19). This shows that D is an association matrix with the parameters in (3.19). Furthermore, by (3.22) and (3.21), it is easy to see that M' is a PBIB design matrix with the parameters in (3.19) and (3.20).

Through simple calculation, we can show that if an association scheme with the parameters in (3.19) exists, then all the other parameters can be expressed in terms of v', b', r', k'. The formulas are

$$\begin{cases} n_2 = b' - k'(r'-1), & c = k'^2, \\ \Delta = (r'-1)^2, \quad \nu = k', & \alpha = r'(k'-1). \end{cases} \tag{3.24}$$

Using Theorem 3.6, we can obtain a series of association schemes with two associate classes. It is known [8,9] that, when $k' = 2, 3, 4$ all the BIB designs with $\lambda' = 1$ have been found. Therefore the association schemes with the following parameters exist:

No.	$k' = \nu$	v	n_1	a	c	α		Remark
(III.1)	2	$\frac{v'}{2}(v'-1)$	$2(v'-2)$	$v'-2$	4	$v'-1$	$v' \geqslant 3$	(triangular schemes)
(III.2)	3	$(2t+1)(3t+1)$	$9t$	$3(t+1)$	9	$6t+2$	$t \geqslant 1$	
(III.3)	3	$t(6t+1)$	$3(3t-1)$	$3t+2$	9	$6t$	$t \geqslant 1$	
(III.4)	4	$(3t+1)(4t+1)$	$16t$	$4(t+2)$	16	$3(4t+1)$	$t \geqslant 1$,	
(III.5)	4	$t(12t+1)$	$4(4t-1)$	$4t+5$	16	$12t$	$t \geqslant 1$.	

From [9], it is also known that when $k' = 5$, except $B(141, 5)$* (whose existence is still unknown), all the other designs with $\lambda' = 1$ have been found. Thus the association schemes with the following parameters exist:

(III.6) 5 $(4t+1)(5t+1)$ $25t$ $5(t+3)$ 24 $4(5t+1)$ $t \geqslant 1$,

(III.7) 5 $t(20t+1)$ $5(5t-1)$ $5t+14$ 25 $20t$ $t \geqslant 1, \quad t \neq 7$.

Furthermore, if it can be shown that an association scheme with the following parameters does not exist, then $B(141, 5)$ also does not exist:

(III.8) $v = 987, \quad n_1 = 170, \quad a = 49, \quad c = 25, \quad \alpha = 140, \quad \nu = 5.$

For the same reason, we know that the two BIB designs in Yates' Table which have not been solved, i.e., those with parameters

(24) $b' = 69, \quad v' = 46, \quad k' = 6, \quad r' = 9, \quad \lambda' = 1,$

(31) $b' = 85, \quad v' = 51, \quad k' = 6, \quad r' = 10, \quad \lambda' = 1$

are related to the following two schemes.

(III.9) $v = 69, \quad n_1 = 48, \quad a = 32, \quad c = 36, \quad \alpha = 45, \quad \nu = 6,$

(III.10) $v = 85, \quad n_1 = 54, \quad a = 33, \quad c = 36, \quad \alpha = 50, \quad \nu = 6.$

* $B(v', k')$ denotes a BIB design of block size k' with v' treatments and $\lambda' = 1$.

If we can show that schemes (III.9) and (III.10) do not exist, then designs (24) and (31) in Yates' table also do not exist.

From [10], we know that for $k' \geqslant 7$, as long as k' is a prime power, $B(k'^l, k')$ exists for all integers $l \geqslant 2$. Therefore the following scheme exists:

$$\text{(III.11)} \qquad v = k'^{l-1}\left(\frac{k'^l - 1}{k' - 1}\right), \qquad n_1 = k'\left(\frac{k'^l - 1}{k' - 1} - 1\right),$$
$$a = \frac{k'^l - 1}{k' - 1} - 2 + (k' - 1)^2, \qquad c = k'^2,$$
$$\nu = k', \qquad \alpha = k'^l - k' + 1,$$

where k' is a prime power.

From schemes (III.1)–(III.7) and (III.11), we can construct their complement schemes. The parameters of these complement schemes can easily be determined from (3.5).

Theorem 3.6 also tells us that the association schemes (III.1)–(III.7), and (III.11) are also PBIB designs with the parameters in (3.20). It is natural to ask under what conditions the PBIB designs with two associate classes obtained from $B(v', k')$ are group-divisible designs. Through simple calculation, we obtain:

Theorem 3.7. *If $B(v', k')$ exists, then the PBIB design obtained by applying Theorem* 3.6 *is a group-divisible design if and only if the parameters of $B(v', k')$ satisfy one of the following conditions.*

$$(1) \quad r' = k';$$
$$(2) \quad b' = (k' - 1)(2r' - k' - 1) + r'.$$

Corollary 1. *If $r' = k' + 1$, then the PBIB design obtained by applying Theorem* 3.6 *to $B(v', k')$ is group-divisible (i.e., the PBIB design obtained from $B(k'^2, k')$ is group-divisible).*

The complement of the scheme (and design) obtained by applying Theorem 3.6 to $B(k'^2, k')$ has parameters

$$\begin{cases} v^* = k'(k' + 1), & n_1^* = k' - 1, \quad n_2^* = k'^2, \\ \lambda_1^* = 0, \quad \lambda_2^* = 1, & r^* = k', \quad k^* = r' + 1. \end{cases} \tag{3.25}$$

From [11], we know that a group-divisible design with the parameters in (3.25) exists if and only if there are $k' - 1$ mutually orthogonal Latin squares of order k'. We already know that if there are $k' - 2$ mutually orthogonal Latin squares of order k', then a $B(k'^2, k')$ exists. Therefore, the maximum number of mutually orthogonal Latin squares of order k' cannot be $k' - 2$. This is an interesting corollary of Theorem 3.7.

§4. Properties of the Parameters of Association Schemes with Three Associate Classes and Construction of Some Schemes

We rewrite (1.2), (1.3), (2.3)–(2.5) in §1 and §2 for association schemes with three associate classes:

$$1 + n_1 + n_2 + n_3 = v, \tag{4.1}$$

$$\left.\begin{matrix} n_1 p_{12}^1 = n_2 p_{11}^2, & n_1 p_{13}^1 = n_3 p_{11}^3; & n_1 p_{33}^1 = n_3 p_{31}^3, & n_2 p_{13}^2 = n_3 p_{12}^3 \\ n_1 p_{22}^1 = n_2 p_{21}^2, & n_1 p_{23}^1 = n_3 p_{21}^3; & n_2 p_{23}^2 = n_3 p_{22}^3, & n_2 p_{33}^2 = n_3 p_{32}^3 \end{matrix}\right\}, \tag{4.2}$$

$$\begin{gathered} \sum_{j=0}^{3} p_{jk}^1 = n_k, \quad k = 1,2,3; \qquad \sum_{j=0}^{3} p_{jk}^2 = n_k, \quad k = 1,2,3; \\ \sum_{j=0}^{3} p_{jk}^3 = n_k, \quad k = 1,2,3, \end{gathered} \tag{4.3}$$

$$\sum_{u=0}^{3} p_{11}^u p_{u2}^2 = \sum_{u=0}^{3} p_{12}^u p_{u1}^2. \tag{4.4}$$

Imitating the case of two associate classes, we will first try to find the characteristic roots of P_1. Since the characteristic roots of P_1 are related to a cubic equation, there is no explicit formula. But the problem is simpler if not all the characteristic roots of P_1 are simple. Now let us discuss this case. We have

$$P_1 - \theta I = \begin{bmatrix} -\theta & 1 & 0 & 0 \\ n_1 & p_{11}^1 - \theta & p_{11}^2 & p_{11}^3 \\ 0 & p_{21}^1 & p_{21}^2 - \theta & p_{21}^3 \\ 0 & p_{31}^1 & p_{31}^2 & p_{31}^3 - \theta \end{bmatrix}. \tag{4.5}$$

Let e_i $(i = 1,2,3,4)$ be the ith column of $P_1 - \theta I$.

Theorem 4.1. *$|P_1 - \theta I| = 0$ has no 4-multiple root.*

Proof. Use the fact that $\operatorname{rk}(P_1 - \theta I) \geqslant 1$ and Theorem 2.10.

Theorem 4.2. *If $|P_1 - \theta I| = 0$ has a 3-multiple root, then this root must be n_1 and the parameters of the association scheme are*

$$P_0 = I, \qquad P_1 = \begin{bmatrix} 0 & 1 & 0 & 0 \\ n_1 & n_1 - 1 & 0 & 0 \\ 0 & 0 & n_1 & 0 \\ 0 & 0 & 0 & n_1 \end{bmatrix}, \qquad P_2 = \begin{bmatrix} 0 & 0 & 1 & 0 \\ 0 & 0 & n_1 & 0 \\ n_2 & n_2 & p_{22}^2 & p_{22}^3 \\ 0 & 0 & p_{32}^2 & p_{32}^3 \end{bmatrix},$$

$$P_3 = \begin{bmatrix} 0 & 0 & 0 & 1 \\ 0 & 0 & 0 & n_1 \\ 0 & 0 & p_{23}^2 & p_{23}^3 \\ n_3 & n_3 & p_{33}^2 & p_{33}^3 \end{bmatrix}. \tag{4.6}$$

Proof. Since θ is a 3-multiple root of $P_1 - \theta I$, it follows from Theorem 2.10 that the rank of $P_1 - \theta I$ is 1. Therefore, $\theta = n_1$ and the parameters are given by (4.6).

Now we will determine the association matrix with the parameters in (4.6). Since $B_1^2 = n_1 I + (n_1 - 1)B_1$, without loss of generality, we may assume that

$$B_1 = I_t \times (E - I)_{n_1+1} = \begin{bmatrix} E-I & 0 & \cdots & 0 \\ 0 & E-I & \cdots & 0 \\ \vdots & \vdots & \ddots & \vdots \\ 0 & 0 & \cdots & E-I \end{bmatrix}, \qquad t = \frac{v}{1+n_1}.$$

We may also make a corresponding partition of B_2. Let

$$B_2 = \begin{bmatrix} c_{11} & c_{12} & \cdots & c_{1t} \\ \vdots & \vdots & \ddots & \vdots \\ c_{t1} & c_{t2} & \cdots & c_{tt} \end{bmatrix}.$$

Since $\sum_{i=0}^{3} B_i = E$, $c_{ii} = 0$ for $i = 1, 2, \ldots, t$. Furthermore, since $B_2 B_1 = B_1 B_2 = n_1 B_2$, we have

$$Ec_{ij} = c_{ij}E = (n_1 + 1)c_{ij}, \qquad i \neq j, \qquad i, j = 1, 2, \ldots, t.$$

Thus for $i \neq j$, $c_{ij} = 0$ or $c_{ij} = E$. Therefore $B_2 = \tilde{B} \times E$, and obviously $B_3 = (E - I - \tilde{B}) \times E$. It is easy to see that $\tilde{B}$ has the following properties:

$$\tilde{B} \text{ is a } t \times t \text{ (0,1)-matrix with zero diagonal elements,} \tag{4.7}$$

$$\tilde{B}' = \tilde{B}, \tag{4.8}$$

$$\tilde{B}1 = \frac{n_2}{1+n_1} 1, \tag{4.9}$$

$$\tilde{B}^2 = \frac{n_2}{1+n_1} I + \frac{p_{22}^2}{1+n_1} \tilde{B} + \frac{p_{22}^3}{1+n_1} (E - I - \tilde{B}). \tag{4.10}$$

Comparing (3.1)–(3.4) with (4.7)–(4.10), we conclude that $\tilde{B}$ is an association scheme with two associate classes. Therefore, we have:

Theorem 4.3. *An association scheme with parameters* (4.6) *exists if and only if there is a 2-association scheme with parameters*

$$v' = \frac{v}{1+n_1}, \qquad n_1' = \frac{n_2}{1+n_1}, \qquad a = \frac{p_{22}^2}{1+n_1}, \qquad c = \frac{p_{22}^3}{1+n_1}. \tag{4.11}$$

Now let us consider the case where P_1 has a 2-multiple root but no 3-multiple root.

If θ is a 2-multiple root of P_1, then, by Theorem 2.10, the rank of $P_1 - \theta I$ is 2, i.e., any three of e_1, e_2, e_3, e_4 are linearly dependent, and at least two of them are linearly independent. Obviously, $e_3 = e_4 = 0$ is not possible. In what follows, several cases will be separately discussed.

1. $e_3 = 0$, $e_4 \neq 0$ (or $e_4 = 0$, $e_3 \neq 0$). In this case, $\theta = n_1$, i.e., n_1 is a 2-multiple root of P_1. Then the parameters p^i_{jk} and the characteristic roots of P_i are as follows.

$$P_0 = I, \qquad P_1 = \begin{bmatrix} 0 & 1 & 0 & 0 \\ n_1 & p^1_{11} & p^2_{11} & 0 \\ 0 & p^1_{21} & p^2_{21} & 0 \\ 0 & 0 & 0 & n_1 \end{bmatrix}, \qquad P_2 = \begin{bmatrix} 0 & 0 & 1 & 0 \\ 0 & p^1_{12} & p^2_{12} & 0 \\ n_2 & p^1_{22} & p^2_{22} & 0 \\ 0 & 0 & 0 & n_2 \end{bmatrix},$$

$$P_3 = \begin{bmatrix} 0 & 0 & 0 & 1 \\ 0 & 0 & 0 & n_1 \\ 0 & 0 & 0 & n_2 \\ n_3 & n_3 & n_3 & n_3 - (1 + n_1 + n_2) \end{bmatrix}, \tag{4.12}$$

$$Z = \begin{bmatrix} 1 & n_1 & n_2 & n_3 \\ 1 & n_1 & n_2 & -(n_1 + n_2 + 1) \\ 1 & \dfrac{p^1_{11} - p^2_{11} - \sqrt{\Delta}}{2} & \dfrac{p^2_{22} - p^1_{22} + \sqrt{\Delta}}{2} & 0 \\ 1 & \dfrac{p^1_{11} - p^2_{11} + \sqrt{\Delta}}{2} & \dfrac{p^2_{22} - p^1_{22} - \sqrt{\Delta}}{2} & 0 \end{bmatrix}, \tag{4.13}$$

where $\Delta = (p^1_{11} - p^2_{11})^2 + 4(n_1 - p^2_{11}) = (p^2_{22} - p^1_{22})^2 + 4(n_2 - p^1_{22})$. It is easy to see that $\Delta \neq 0$. Therefore, the other two characteristic roots of P_1 are simple. Now we will determine the association matrix with parameters (4.12).

Since $B_1 + B_2$ has a multiple root $n_1 + n_2$, it is decomposable. Furthermore, $(B_1 + B_2)^2 = (n_1 + n_2)I + (n_1 + n_2 - 1)(B_1 + B_2)$. Without loss of generality, we may assume that

$$B_1 + B_2 = I_t \times (E - I)_{(1+n_1+n_2)}, \qquad t = \frac{v}{1 + n_1 + n_2}.$$

Then $B_3 = (E - I)_t \times E_{(1+n_1+n_2)}$.
Let

$$B_1 = \begin{bmatrix} B_{11} & 0 & \cdots & 0 \\ 0 & B_{21} & \cdots & 0 \\ \vdots & \vdots & \ddots & \vdots \\ 0 & 0 & \cdots & B_{t1} \end{bmatrix}.$$

Then

$$B_2 = \begin{bmatrix} E - I - B_{11} & 0 & \cdots & 0 \\ 0 & E - I - B_{21} & \cdots & 0 \\ \vdots & \vdots & & \\ 0 & 0 & \cdots & E - I - B_{t1} \end{bmatrix}.$$

Since $B_1^2 = n_1 I + p^1_{11} B_1 + p^2_{11} B_2$, we have

$$B_{j1}^2 = n_1 I + p^1_{11} B_{j1} + p^2_{11}(E - I - B_{j1}), \qquad j = 1, 2, \ldots, t. \tag{4.14}$$

It is easy to see that $B_{11}, B_{21}, \ldots, B_{t1}$ have the following properties.

$$B_{j1} \text{ is a } (1+n_1+n_2)\times(1+n_1+n_2)\ (0,1)\text{-matrix with zero diagonal elements} \tag{4.15}$$

$$B_{j1}' = B_{j1}, \tag{4.16}$$

$$B_{j1}1 = n_1 1. \tag{4.17}$$

Comparing (4.14)–(4.17) with (3.1)–(3.4), we conclude that B_{j1} is an association matrix with two associate classes. Therefore, we have:

Theorem 4.4. *An association scheme with parameters* (4.12) *exists if and only if there is a 2-association scheme with parameters*

$$v' = 1 + n_1 + n_2, \qquad n_1' = n_1, \qquad a = p_{11}^1, \qquad c = p_{11}^2. \tag{4.18}$$

2. $e_3 \neq 0, e_4 \neq 0.$

Theorem 4.5. *If* $e_3 \neq 0$, $e_4 \neq 0$, *then* $P_1 - \theta I$ *has a multiple root if and only if*

$$p_{11}^2 = p_{11}^3. \tag{4.19}$$

Proof. It is easy to see that if θ is a multiple root of $P_1 - \theta I$, then e_3, e_4 must be linearly dependent. Otherwise, e_2, e_3, e_4 would be linearly independent. Then $\text{rk}(P_1 - \theta I) \geqslant 3$, contradicting Theorem 2.10. Therefore, there is a λ such that $e_3 = \lambda e_4$. Taking the column sums of both sides, we obtain $(n_1 - \theta) = \lambda(n_1 - \theta)$. If $\theta = n_1$, then $p_{11}^2 = 0$, $p_{11}^3 = 0$; if $\theta \neq n_1$, then $\lambda = 1$. Thus $p_{11}^2 = p_{11}^3$, which shows the necessity.

If $p_{11}^2 = p_{11}^3 = 0$, then the characteristic roots of P_1 are $n_1, n_1, -1, -1$ which are obviously multiple. If $p_{11}^2 = p_{11}^3$, consider the 3×3 principal submatrix on the upper-left corner of (4.5):

$$P_0 = \begin{bmatrix} -\theta & 1 & 0 \\ n_1 & p_{11}^1 - \theta & p_{11}^2 \\ 0 & p_{21}^1 & p_{21}^2 - \theta \end{bmatrix}.$$

By (4.4), when $p_{11}^2 = p_{11}^3$, we have

$$G = \frac{n_2 + n_3}{n_3}\left(p_{31}^2\right)^2 - \left(1 + n_1 + \frac{v - n_1 - 1}{n_1}\,p_{11}^2\right)p_{31}^2 - \frac{n_3}{n_1}\left(p_{11}^2\right)^2 + n_3 p_{11}^2 = 0,$$

and when $\theta = p_{21}^2 - p_{21}^3$,

$$|P_0| = \frac{n_2}{n_3}\left(\frac{n_2 + n_3}{n_3}\,p_{31}^2 + p_{11}^2\right)G = 0.$$

Therefore the rank of $P_1 - (p_{21}^2 - p_{21}^3)I$ cannot be bigger than 2, while $p_{21}^2 - p_{21}^3$ is a characteristic root of P_1, This shows that $|P_1 - \theta I| = 0$ has a multiple root.

When $p_{11}^2 = p_{11}^3$, we can write down the characteristic roots of P_i and their multiplicities:

$$Z=\begin{pmatrix} 1 & n_1 & n_2 & n_3 \\ 1 & p_{11}^1+p_{31}^2+p_{21}^3-n_1 & p_{12}^1-p_{12}^3 & p_{13}^1-p_{13}^2 \\ 1 & p_{21}^2-p_{21}^3 & \dfrac{p_{22}^2-p_{22}^3+\sqrt{\Delta}}{2} & \dfrac{p_{33}^3-p_{33}^2-\sqrt{\Delta}}{2} \\ 1 & p_{21}^2-p_{21}^3 & \dfrac{p_{22}^2-p_{22}^3-\sqrt{\Delta}}{2} & \dfrac{p_{33}^3-p_{33}^2+\sqrt{\Delta}}{2} \end{pmatrix}, \tag{4.20}$$

where

$$\Delta=\left(p_{22}^2-p_{22}^3\right)^2+4\left[\left(p_{22}^1-p_{22}^3\right)\left(p_{12}^2-p_{12}^3\right)+\left(n_2-p_{22}^3\right)\right],$$

$$\begin{cases} \alpha_1=\dfrac{n_1+(v-1)z_{21}}{z_{21}-z_{11}}, \\ \alpha_2=\dfrac{n_2+\alpha_1 z_{12}+(v-1-\alpha_1)z_{32}}{z_{32}-z_{22}}, \\ \alpha_3=\dfrac{n_2+\alpha_1 z_{12}+(v-1-\alpha_1)z_{22}}{z_{22}-z_{32}}. \end{cases} \tag{4.21}$$

Theorem 4.6. *If $p_{11}^2=p_{11}^3$ and $\alpha_2\neq\alpha_3$, then Δ is the square of an integer and $z_{22}, z_{32}, z_{23}, z_{33}$ are all integers.*

Proof. By Theorem 2.15 of §2.

There are two subcases in case 2:

2a. n_1 is a 2-multiple root of P_1. In this case, $p_{11}^2=p_{11}^3=0$, and the other characteristic roots of P_1 are -1, -1.

2b. n_1 is a simple root of P_1. In this case $p_{21}^2-p_{21}^3$ is a 2-multiple root of P_1.

We cannot find all the schemes in 2a and 2b. Only some special cases are solved.

Theorem 4.7. *In case 2a, the parameters of the association scheme are*

$$P_0=I,\quad P_1=\begin{pmatrix} 0 & 1 & 0 & 0 \\ n_1 & n_1-1 & 0 & 0 \\ 0 & 0 & d & n_1-e \\ 0 & 0 & n_1-d & e \end{pmatrix},\quad P_2=\begin{pmatrix} 0 & 0 & 1 & 0 \\ 0 & 0 & d & n_1-e \\ n_2 & \dfrac{n_2}{n_1}d & p_{22}^2 & p_{22}^3 \\ 0 & n_2-d\dfrac{n_2}{n_1} & p_{32}^2 & p_{32}^3 \end{pmatrix},$$

$$P_3=\begin{pmatrix} 0 & 0 & 0 & 0 \\ 0 & 0 & n_1-d & e \\ 0 & n_3-e\dfrac{n_3}{n_1} & p_{23}^2 & p_{23}^3 \\ n_3 & \dfrac{n_3}{n_1}e & p_{33}^2 & p_{33}^3 \end{pmatrix}, \tag{4.22}$$

where

$$d = \frac{n_1 n_2 - n_3}{n_2 + n_3}, \qquad e = \frac{n_1 n_3 - n_2}{n_2 + n_3}.$$

Proof. By (4.4) and the fact that in case 2a, $p_{11}^2 = p_{11}^3 = 0$.

It is easy to see that in (4.22), if $p_{21}^2 = 0$, i.e., $n_1 n_2 = n_3$, then the rectangular association schemes are included as special cases.

A rectangular association scheme is obtained by arranging v letters in an $(n_1 + 1) \times (n_2 + 1)$ rectangular array. The elements in the same column are first associates of one another, and elements in the same row are second associates of one another, and two elements are third associates of each other if they are neither in the same column nor in the same row. Obviously, this is an association scheme with three associate classes. The parameters are

$$P_0 = I, \quad P_1 = \begin{bmatrix} 0 & 1 & 0 & 0 \\ n_1 & n_1 - 1 & 0 & 0 \\ 0 & 0 & 0 & 1 \\ 0 & 0 & n_1 & n_1 - 1 \end{bmatrix}, \quad P_2 = \begin{bmatrix} 0 & 0 & 1 & 0 \\ 0 & 0 & 0 & 1 \\ n_2 & 0 & n_2 - 1 & 0 \\ 0 & n_2 & 0 & n_2 - 1 \end{bmatrix},$$
$$P_3 = \begin{bmatrix} 0 & 0 & 0 & 1 \\ 0 & 0 & n_1 & n_1 - 1 \\ 0 & n_2 & 0 & n_2 - 1 \\ n_1 n_2 & n_2(n_1 - 1) & n_1(n_2 - 1) & (n_1 - 1)(n_2 - 1) \end{bmatrix}. \tag{4.23}$$

It is clear that in (4.22), if $p_{21}^2 = 0$ and $p_{22}^2 = n_2 - 1$, then the parameters in (4.22) are those of a rectangular scheme. Therefore, we have:

Theorem 4.8. *If* $p_{21}^2 = 0$, $p_{22}^2 = n_2 - 1$, *then an association scheme with the parameters in* (4.22) *exists*.

If $p_{21}^2 = 0$ and $n_1 = 1$, then (4.22) becomes

$$P_0 = I, \quad P_1 = \begin{bmatrix} 0 & 1 & 0 & 0 \\ 1 & 0 & 0 & 0 \\ 0 & 0 & 0 & 1 \\ 0 & 0 & 1 & 0 \end{bmatrix}, \quad P_2 = \begin{bmatrix} 0 & 0 & 1 & 0 \\ 0 & 0 & 0 & 1 \\ n_2 & 0 & p_{22}^2 & p_{22}^3 \\ 0 & n_2 & p_{32}^2 & p_{32}^3 \end{bmatrix},$$
$$P_3 = \begin{bmatrix} 0 & 0 & 0 & 1 \\ 0 & 0 & 1 & 0 \\ 0 & n_2 & p_{23}^2 & p_{23}^3 \\ n_2 & 0 & p_{33}^2 & p_{33}^3 \end{bmatrix}. \tag{4.24}$$

Using a method similar to the proof of Theorem 4.4, we obtain:

Theorem 4.9. *Suppose* $n_2 = v' - 1$ *and* $p_{22}^2 = n_1' + n_1'/(n_2' - n_1)$. *If a 2-association scheme* $\tilde{B}$ *with parameters*

$$v', n_1', c = \frac{n_1'}{2} - \frac{n_1'}{2(n_2' - n_1')} \tag{4.25}$$

exists, then an association scheme with the parameters in (4.24) *exists and the corresponding association matrices are*

$$B_0 = I, \qquad B_1 = (E - I)_2 \times I,$$
$$B_2 = I_2 \times \tilde{B} + (E - I)_2 \times (E - \tilde{B} - I),$$
$$B_3 = (E - 1)_2 \times \tilde{B} + I_2 \times (E - I - \tilde{B}).$$

Theorem 4.10. *If* $p_{22}^2 = 0$, *then an association scheme with the parameters in* (4.24) *exists. The association matrices are*

$$B_0 = I, \quad B_1 = (E - I)_2 \times I, \quad B_2 = (E - I)_2 \times (E - I), \quad B_3 = I_2 \times (E - I).$$

Theorem 4.11. *If* $n_2 = 4n + 1$, *then an association scheme with the parameters in* (4.24) *exists if and only if there is a 2-association scheme* $\tilde{B}$ *with parameters*

$$\begin{aligned} v' &= 4n + 1, & n_1' &= 2n - 1, \\ c &= n, & a &= n - 1 \end{aligned} \tag{4.26}$$

Furthermore, we have the following relationship between the association matrices:

$$B_2 = U \times I_2 + V \times (E - I)_2,$$
$$B_3 = U \times (E - I)_2 + V \times I_2,$$

$$U = \begin{bmatrix} 0 & 1 & 1 & \cdots & 1 \\ 1 & & & & \\ 1 & & & & \\ \vdots & & \tilde{B} & & \\ 1 & & & & \end{bmatrix}, \qquad V = \begin{bmatrix} 0 & 0 & \cdots & 0 \\ 0 & & & \\ \vdots & & E - I - \tilde{B} & \\ 0 & & & \end{bmatrix}.$$

In (4.22), if $p_{21}^2 = n_2 - 1$, then the parameter matrices are

$$P_0 = I, \quad P_1 = \begin{bmatrix} 0 & 1 & 0 & 0 \\ n_1 & n_1 - 1 & 0 & 0 \\ 0 & 0 & n_1 - n_3 & n_2 \\ 0 & 0 & n_3 & n_1 - n_2 \end{bmatrix}, \quad P_2 = \begin{bmatrix} 0 & 0 & 1 & 0 \\ 0 & 0 & n_2 - 1 & n_2 \\ n_2 & n_2 - d & 0 & 0 \\ 0 & d & 0 & 0 \end{bmatrix},$$

$$P_3 = \begin{bmatrix} 0 & 0 & 0 & 1 \\ 0 & 0 & n_3 & n_3 - 1 \\ 0 & d & 0 & 0 \\ n_3 & n_3 - d & 0 & 0 \end{bmatrix}, \tag{4.27}$$

where $d = n_2 n_3 / n_1$ and $n_1 = n_2 + n_3 - 1$.

By a method similar to what was used above, we obtain:

Theorem 4.12. *An association scheme with the parameters in* (4.27) *exists if and only if there is a symmetric BIB design M with parameters*

$$v' = n_2 + n_3, \qquad r' = n_2, \qquad \lambda' = n_2 - d. \tag{4.28}$$

Furthermore the association matrices and the design matrix M have the following relations.

$$B_2 = \begin{pmatrix} 0 & M \\ M' & 0 \end{pmatrix}, \qquad B_3 = \begin{pmatrix} 0 & E - M \\ E - M' & 0 \end{pmatrix}.$$

Since there is a symmetric BIBD with $\lambda = 1$, $r = p^l$, where p is a prime number, this shows that if n_2 is a prime power, and $n_2 - d = 1$, then an association scheme with the parameters in (4.27) exists.

Case 2b is more complicated. Generally speaking, it is more difficult. It is easy to see that when $p_{11}^2 = p_{11}^3 = n_1$, P_2 (respectively, P_3) has a 2-multiple root n_2 (respectively, n_3) and the other two roots are different. We may consider P_2 as P_1; then the problem is solved by using Theorem 4.4.

Summarizing the above discussion, we obtain Table 4.1. As to the case where P_2 (or P_3) has a multiple characteristic root, we only need to consider P_2 (or P_3) as P_1; then results similar to the above can be obtained, which we will not repeat. Now we shall discuss the case where the number of multiplicities α_1, α_2 and α_3 are equal.

Theorem 4.13. *If P_1, P_2 or P_3 has at least a multiple characteristic root, and $\alpha_1 = \alpha_2 = \alpha_3 = n$, then the parameters of the association scheme are*

$$\begin{cases} v = 1 + 3n, \quad n_1 = n_2 = n_3 = n, \quad x = n^2 - 2na, \\ a_{12} = a_{21} = a_{23} = a_{32} = a_{13} = a_{31} = na, \end{cases} \tag{4.29}$$

where $a_{ij} = n_i p_{ij}^i$, $x = n_1 p_{23}^1 = n_2 p_{31}^2 = n_3 p_{12}^3$, $a = \frac{1}{9}(3n - 1 \pm \sqrt{3n+1})$.

Proof. When $\alpha_2 = \alpha_3$, from (4.23), we obtain $2(n_2 + \alpha_1 z_{12}) + (v - 1 - \alpha_1)(p_{22}^2 - p_{22}^3) = 0$. Substituting $\alpha_1 = n$, we get $n_2/n + z_{12} + (p_{22}^2 - p_{22}^3) = 0$. Therefore $n_2 = k_2 n$, where k_2 is an integer. For the same reason, $n_3 = k_3 n$, where k_3 is an integer. Since $\alpha_0 + \alpha_1 + \alpha_2 + \alpha_3 = v$, $v = 3n + 1$. Therefore $n_1 + (k_2 + k_3)n = 3n$, which implies that $k_2 = k_3 = 1$, $n_1 = n_2 = n_3 = n$.

By expression (4.23) of α_1 and $\alpha_1 = n$, we get $2a_{12} + x = n^2$. Furthermore, $a_{12} + a_{21} + x = n_1 n_2 = n^2$, thus $a_{12} = a_{21} = (n^2 - x)/2$. By the same method, we can show that $a_{12} = a_{21} = a_{23} = a_{32} = a_{13} = a_{31} = na$. Substituting this into (4.4), we find that a satisfies the equation

$$9a^2 - 6na + 2a + n^2 - n = 0.$$

Therefore $a = \frac{1}{9}(3n - 1 \pm \sqrt{3n+1})$.

A scheme with $n = 5$, $v = 16$, $n_1 = n_2 = n_3 = 5$, $a = 2$ is given in Appendix II.

It is easy to show that if P_i has at least one multiple characteristic root, then a scheme with parameters (4.29) must satisfy $\alpha_1 = \alpha_2 = \alpha_3 = n$. Therefore we call a scheme with parameters (4.29) an association scheme of equi-multiplicity type.

§5. Construction of Association Schemes and PBIB Designs with m Associate Classes

In this section, we present two methods for the construction of association schemes and PBIB designs with m associate classes.

1. Lattice Schemes and Designs

Suppose $m \geqslant 2$ and R_m is the m-dimensional Euclidean space. Consider the following point set in R_m:

$$V = \{(x_1, \ldots, x_m) : 0 \leqslant x_i \leqslant t, x_i \text{ is an integer}, i = 1, 2, \ldots, m\}. \tag{5.1}$$

Definition 5.1. If two points P, Q in V have exactly $m - i$ common coordinates, then we say that P, Q are ith associates of each other. This relation is denoted by $(P, Q) = i$.

We now prove that the points in V constitute an association scheme with respect to the relations in Definition 5.1. Some corresponding PBIB designs are also obtained. It is easy to see that either one of the following two transformations (T and O) maps V into V and does not change the relation between the points of V:

$$T(x_i) = \begin{cases} x_i, & i \neq k, \quad i = 1, 2, \ldots, m, \\ x_k, & x_k \neq r, s, \quad 0 \leqslant r < s \leqslant t, \\ r, & x_k = s, \\ s, & x_k = r, \end{cases} \tag{5.2}$$

where k is a preassigned integer, $1 \leqslant k \leqslant m$;

$$O(x_i) = \begin{cases} x_i, & i \neq k, \quad l, i = 1, 2, \ldots, m, \\ x_k, & i = l, \\ x_l, & i = k, \end{cases} \tag{5.3}$$

where k, l are two preassigned integers, $1 \leqslant k, l \leqslant m$.

Lemma 5.1. *For any point P in V, the number of points in V which are the ith associates of P is $\binom{m}{i} t^i$, and is independent of P.*

Proof. Any given point P can be transformed into the origin by transformations T. The number of points which have exactly $m - i$ common coordinates with the origin is $\binom{m}{i} t^i$. This proves Lemma 5.1.

For any two different points P, Q in V, there is i, $1 \leqslant i \leqslant m$, such that $(P, Q) = i$. The two points P, Q may be transformed into the following by transformations T and O:

$$P_0 = (\overbrace{0 \ \ 0 \cdots 0}^{m-i} \ \overbrace{0 \ \ 0 \cdots 0}^{i}),$$

$$Q_0 = (\overbrace{0 \ \ 0 \cdots 0}^{m-i} \ \overbrace{1 \ \ 1 \cdots 1}^{i}).$$

Table 4.1. $m = 3, P_1$ has a multiple characteristic root

<table>
<tr><td>(1)</td><td>(2)</td><td colspan="10">(3)</td></tr>
<tr><td rowspan="6">P_1 has a 4-multiple root (Theorem 4.1) closed</td><td rowspan="6">P_1 has a 3-multiple root (Theorems 4.2 and 4.3) closed</td><td colspan="10">P_1 has a 2-multiple root</td></tr>
<tr><td colspan="8">n_1 is a 2-multiple root</td><td colspan="2">n_1 is a simple root</td></tr>
<tr><td rowspan="4">other roots not equal (Theorem 4.4) closed</td><td colspan="7">other roots equal $(n_1, n_1, -1, -1)$</td><td rowspan="4">$p_{11}^2 = p_{11}^3 = n_1$ (Theorem 4.4) closed</td><td rowspan="4">others open</td></tr>
<tr><td colspan="5">$p_{21}^2 = 0$</td><td rowspan="3">$p_{21}^2 = n_2 - 1$ (Theorem 4.12) closed</td><td rowspan="3">others open</td></tr>
<tr><td rowspan="2">$p_{22}^2 = n_2 - 1$ (Theorem 4.8) closed</td><td colspan="3">$n_1 = 1$</td><td rowspan="2">others open</td></tr>
<tr><td>$n_2 = n_3 = v' - 1$
$p_{22}^2 = n_1' + \frac{n_1'}{n_2' - n_1'}$
(Theorem 4.9)</td><td>$p_{22}^2 = 0$ (Theorem 4.10) closed</td><td>$n_3 = 4n + 1$ (Theorem 4.11) closed</td></tr>
</table>

Then the number of points in the point set

$$V_{jk}(P, Q) = \{R : R \in V, (P,R) = j, (Q,R) = k\}$$

is the same as that of $V_{jk}(P_0, Q_0)$.

If $R \in V_{jk}(P_0, Q_0)$, then R has $m - j$ common coordinates with P_0. Suppose there are α zeros among the first $m - i$ coordinates and $m - j - \alpha$ zeros among the last i coordinates. Since $(Q_0, R) = k$, there must be $m - k - \alpha$ ones among the last i coordinates of R. Conversely, if a point R in V has the above property, then R must be in $V_{jk}(P_0, Q_0)$.

Let $p^i_{jk}(t)$ be the number of points in $V_{jk}(P_0, Q_0)$. Then

$$p^i_{jk}(1) = \begin{cases} \begin{pmatrix} m-i \\ m-\frac{s}{2} \end{pmatrix} \begin{pmatrix} i \\ \frac{s}{2}-j \end{pmatrix}, & \begin{array}{l} 0 \leqslant m - \frac{s}{2} \leqslant m - i, \\ 0 \leqslant \frac{s}{2} - j \leqslant i, \ \frac{s}{2} \text{ is an integer}, \end{array} \\ 0 & \text{otherwise}, \end{cases} \tag{5.4}$$

where $s = i + j + k$.

When $t \geqslant 2$,

$$p^i_{jk}(t) = \sum_{\max\left(0, m-\frac{s}{2}\right) \leqslant \alpha \leqslant m-i} \binom{m-i}{\alpha} \frac{i!\,t^{\alpha}(t-1)^{s+2(\alpha-m)}}{(m-j-\alpha)!\,(m-k-\alpha)!\,(s+2\alpha-2m)!}, \qquad s = i+j+k. \tag{5.5}$$

Therefore, we have

Lemma 5.2. *For any two different points P, Q in V such that $(P, Q) = i$, the number of points in V which are both jth associates of P and kth associates of Q is $p^i_{jk}(t)$, which only depends on i, j, k, and is independent of P, Q.*

By Lemmas 5.1 and 5.2, we obtain

Theorem 5.1. *There exists an m-association scheme with the following parameters:*

$$\begin{cases} v = (t+1)^m, & \\ n_i = \binom{m}{i} t^i & i = 1, \ldots, m, \\ p^i_{jk} = p^i_{jk}(t) & i, j, k = 1, \ldots, m, \end{cases} \tag{5.6}$$

where $t \geqslant 1$.

The readers can easily see that this is a generalization of an L_2 scheme. We call the scheme with parameters (5.6) constructed from points of V a lattice scheme. It is interesting to consider the converse: Must a scheme with parameters (5.6) be a lattice scheme?

For any $l \geqslant 1$, consider the hyperplane or straight line

$$\pi_l : x_{ik} = c_k, \qquad k = 1, 2, \ldots, l,$$
$$1 \leqslant i_1 < i_2 < \cdots < i_l \leqslant m,$$

where c_k is a fixed integer and $0 \leqslant c_k \leqslant t$.

For fixed l, there are $\binom{m}{l}(t+1)^l$ such π_l's. Each π_l has exactly $(t+1)^{m-l}$ points in V. For any two different points P, Q in V, if $(P, Q) = i$, then P, Q meet in $\binom{m-i}{l}$ such π_l's (when $m - i < l$, $\binom{m-i}{l}$ is defined to be zero). Thus we have:

Theorem 5.2. *There exists a PBIB design with parameters* (5.6) *and*

$$\begin{cases} b = \binom{m}{l}(t+1)^l, & k = (t+1)^{m-l}, \quad r = \binom{m}{l}, \\ \lambda_i = \binom{m-i}{l}, & i = 1, 2, \dots, m, \end{cases} \tag{5.7}$$

where l is a fixed number and $1 \leqslant l \leqslant m-1$.

2. Geometric Schemes and Designs

Suppose G is a finite field and G_n is an n-dimensional linear space over G. Let the subspaces of G_n be denoted by $A, B, \dots,$ and let the dimension of A be denoted by $\dim A$. In the subsequent discussion, we will always assume that G has q elements.

For any integer $m \geqslant 1$, $n \geqslant 2m$, consider the totality of the m-dimensional subspaces of G_n. We use $\mathfrak{M}(A, B)$ to denote the subspace spanned by A and B. For any two different m-dimensional subspaces A, B of G_n, there is i, $1 \leqslant i \leqslant m$, such that $\dim \mathfrak{M}(A, B) = m + i$. Thus we have the following definition.

Definition 5.2. Suppose $\dim A = m$, $\dim B = m$. If $\dim \mathfrak{M}(A, B) = m + i$, then we say that A and B are the ith associates of each other.

It is easy to see that for any two m-dimensional subspaces A, B of G_n, there must be an i, $1 \leqslant i \leqslant m$, such that $(A, B) = i$. Furthermore, $(A, B) = i$ implies $(B, A) = i$. For any m-dimensional subspace A of G_n, by Theorem 2.4′ of Appendix I, the cardinality of the set $\mathfrak{N}_i(A) = \{B : (A, B) = i\}$ is $\mu(\mathfrak{D}_{nmi})$, which is independent of A. For any two different m-dimensional subspaces A and B, if $(A, B) = i$, then by Theorem 2.5′ of Appendix I, the cardinality of the set $\mathfrak{N}^i_{jk}(A, B) = \{C : (A, C) = j, (B, C) = k\}$ is $\mu(\mathfrak{E}_{nmijk})$, which is independent of A, B. Therefore all the m-dimensional subspaces of G_n constitute an association scheme with m associate classes under the m relations in Definition 5.2. Thus we have:

Theorem 5.3. *Suppose p is a prime number, l is a natural number, and $q = p^l$, $n \geqslant 2m$. Then there exists an m-association scheme with the following parameters.*

$$\begin{cases} v = \mu(\mathfrak{B}_{nm}) = \prod_{0 \leqslant \alpha \leqslant m-1} \dfrac{(q^{n-\alpha}-1)}{(q^{m-\alpha}-1)}, \\ n_i = \mu(\mathfrak{D}_{nmi}) = q^{i^2} \prod_{0 \leqslant \alpha \leqslant m-i-1} \dfrac{(q^{m-\alpha}-1)}{(q^{m-i-\alpha}-1)} \prod_{0 \leqslant \alpha \leqslant i-1} \dfrac{(q^{n-m-\alpha}-1)}{(q^{i-\alpha}-1)}, \\ \qquad\qquad i = 1, \dots, m, \\ p^i_{jk} = \mu(\mathfrak{E}_{nmijk}), \qquad i, j, k = 1, \dots, m. \end{cases} \tag{5.8}$$

Proof. From the previous discussion, we know that all the m-dimensional subspaces of G_n constitute an association scheme under the relations in Definition 5.2. The parameters in (5.8) are obtained from Theorems 2.2′, 2.4′, and 2.5′ in Appendix I.

When $m = 2$, Theorem 5.3 gives a series of two-association schemes. For $n \geqslant 4$, their parameters are

$$\begin{cases} v = \dfrac{(q^n - 1)(q^{n-1} - 1)}{(q^2 - 1)(q - 1)}, \\ n_1 = q(q+1)\dfrac{(q^{n-2} - 1)}{(q - 1)}, \\ a = p_{11}^1 = \dfrac{q^2(q^{n-3} - 1)}{q - 1} + q^2 + q - 1, \\ c = p_{11}^2 = (q+1)^2. \end{cases} \tag{5.9}$$

Notice that for this scheme, $\nu = q + 1$.

When $m = 3$, Theorem 5.3 also gives a series of association schemes. For $n \geqslant 6$, their parameters are

$$\begin{cases} v = \dfrac{(q^n - 1)(q^{n-1} - 1)(q^{n-2} - 1)}{(q^3 - 1)(q^2 - 1)(q - 1)}, \\ n_1 = q^2(1 + q + q^2)\dfrac{(q^{n-3} - 1)}{(q - 1)}, \\ n_2 = q^4(1 + q + q^2)\dfrac{(q^{n-3} - 1)(q^{n-4} - 1)}{(q^2 - 1)(q - 1)}, \\ p_{11}^1 = q^2\dfrac{(q^{n-4} - 1)}{(q - 1)} + q^3 + q^2 + q - 1, \\ p_{12}^1 = q^3(q+1)\dfrac{q^{n-4} - 1}{q - 1}, \\ p_{11}^2 = (q+1)^2, \\ p_{12}^2 = q^3(q+1)\dfrac{q^{n-5} - 1}{q - 1} + q^4 + 2q^3 + q^2 - q - 1. \end{cases} \tag{5.10}$$

It is interesting that for $m \geqslant 2$, we always have $p_{11}^2 = (q+1)^2$.

Suppose $n > 2m$, $t \geqslant 2m$. Consider all the m-dimensional subspaces of G_n as treatments and consider all the t-dimensional subspaces of G_n as blocks. Then we have

Theorem 5.4. *There exists a PBIB design with parameters* (5.8) *and*

$$\begin{cases} b = \mu(\mathcal{B}_{nt}), \\ k = \mu(\mathcal{B}_{tm}), \\ r = \mu(\mathcal{B}_{nt})\mu(\mathcal{B}_{tm})/\mu(\mathcal{B}_{nm}), \\ \lambda_i = \mu(\mathcal{B}_{nt})\mu(\mathcal{B}_{t\,m+i})/\mu(\mathcal{B}_{n\,m+i}), \qquad i = 1, 2, \ldots, m. \end{cases} \tag{5.11}$$

Proof. The existence of an association scheme with parameters (5.8) follows from Theorem 5.3. By Theorem 2.2′ of Appendix I, the number of blocks b is $\mu(\mathcal{B}_{nt})$ and the block size k is $\mu(\mathcal{B}_{tm})$. Furthermore, by Theorem 2.6 of Appendix I, each m-dimensional subspace is contained in exactly $\mu(\mathcal{B}_{nt})\mu(\mathcal{B}_{tm})/\mu(\mathcal{B}_{nm})$ distinct t-dimensional subspaces. This obtains r and λ_i in (5.11).

Appendix I. Some Combinatorial Properties of Linear Space over Finite Fields

§1. Canonical Forms of Full-Rank Column Matrices under Right Similar Transformations

Suppose G is an arbitrarily given field. Consider matrices consisting of elements of G.

Definition 1.1. Suppose A is an $n \times m$ matrix. If $n \geqslant m$, then A is called a column matrix.

Definition 1.2. Suppose A is an $n \times m$ column matrix. If $\mathrm{rk}(A) = m$, then we say that A is a column matrix of full rank.

Definition 1.3. Suppose A is a matrix. If a full-rank square matrix is multiplied to the right of A, then we say that a right transformation is applied to A. If A can be transformed to B by a right transformation, then we say that A and B are right similar, denoted by $A \overset{r}{\sim} B$.

Obviously, $A \overset{r}{\sim} B$ is an equivalence relation.

Definition 1.4. Suppose A is an $n \times m$ full-rank column matrix and the $m \times m$ submatrices of A are arranged lexicographically. If the first full-rank submatrix is I, then we say that A is in r.c. form (right canonical form).

It is not hard to show that:

Lemma 1.1. *If A is a full-rank column matrix, then* $\mathrm{rk}(AB) = \mathrm{rk}(B)$.

Lemma 1.2. *If A is a full-rank column matrix, and $AB = AC$, then $B = C$.*

Theorem 1.1. *If A, B are two column matrices in r.c. forms and $A \overset{r}{\sim} B$, then $A = B$.*

Proof. Let $A(\alpha_1 \ldots \alpha_m)$ be the submatrix consisting of the α_1th, ..., α_mth rows of A. Suppose A and B are $n \times m$, and their first full-rank submatrices are $A(\alpha_1 \ldots \alpha_m)$ and $B(\beta_1 \ldots \beta_m)$, respectively. Since $A \overset{r}{\sim} B$, there exists T, $|T| \neq 0$, such that $AT = B$. Therefore,

$$A(\alpha_1 \ldots \alpha_m)T = B(\alpha_1 \ldots \alpha_m). \tag{1.1$'$}$$

By Lemma 1.1, $B(\alpha_1 \ldots \alpha_m)$ is a full-rank submatrix of B. For the same reason, $A(\beta_1 \ldots \beta_m)$ is a full-rank submatrix of A. Hence $\alpha_i = \beta_i$, $i = 1, 2, \ldots, m$. Notice that A, B are column matrices in r.c. forms. Thus (1.1) becomes $IT = I$, i.e., $T = I$. Therefore, $A = B$.

Theorem 1.2. *For any full-rank column matrix A, there is a unique column matrix B in r.c. form which is right similar to A.*

Proof. Suppose A is $n \times m$ and its first full-rank $m \times m$ submatrix is T. Then AT^{-1} is a column matrix in r.c. form. Obviously, $A \overset{r}{\sim} AT^{-1}$. If B is in r.c. form, and $A \overset{r}{\sim} B$, then $AT^{-1} \sim B$. By Theorem 1.1, $B = AT^{-1}$.

Theorem 1.3. *For any matrix A, there is a unique column matrix B in r.c. form such that $A \overset{r}{\sim} (0, B)$.*

Proof. Suppose the rank of A is m. Permute the columns of A so that the m rightmost columns are linearly independent. Through a further elementary transformation, we immediately obtain

$$A \overset{r}{\sim} (0 \quad C),$$

where C is a full-rank column matrix. By Theorem 1.2, there is a column matrix B in r.c. form such that $C \overset{r}{\sim} B$. Therefore, $A \overset{r}{\sim} (0 \quad B)$.

Let B_1 and B_2 be two column matrices in r.c. forms such that $A \overset{r}{\sim} (0 \quad B_1)$ and $A \overset{r}{\sim} (0, B_2)$. Then $\text{rk}(A) = \text{rk}(B_1) = \text{rk}(B_2)$. Therefore B_1 and B_2 are of the same size. Furthermore, since $(0 \quad B_1) \overset{r}{\sim} (0 \quad B_2)$, there exists $T = \left(\begin{smallmatrix} P & Q \\ R & S \end{smallmatrix}\right)$, $|T| \neq 0$, such that $(0 \quad B_1) = (0 \quad B_2)T$. Then $(0 \quad B_1) = (B_2R \quad B_2S)$. Comparing the two sides, we obtain $R = 0$. Thus S is of full rank. By $B_1 = B_2S$ and Theorem 1.1, $B_1 = B_2$. This shows the uniqueness.

For the convenience of the subsequent discussion, we shall introduce some notations. Suppose A is an $n \times m$ full-rank column matrix with the first $m \times m$ full-rank submatrix $A(\alpha_1 \ldots \alpha_m)$. Let the rows of A not in $A(\alpha_1 \ldots \alpha_m)$ be the β_1th, ..., and β_{n-m}th rows. Let $T(A)$ denote the following matrix.

(a) $T(A)$ is $n \times (n - m)$;

(b) The submatrix of $T(A)$ consisting of the β_1th, ..., and β_{n-m}th rows is I, and all the other entries are zero.

It is easy to see that $T(A)$ has the following properties.

$$\mathrm{rk}(AT(A)) = n, \tag{1.2$'$}$$

$$\mathrm{rk}(T(A)) = n - m. \tag{1.3$'$}$$

Theorem 1.4. *Suppose A is an $n \times m$ full-rank column matrix. Then any matrix M with the same number of rows as A can be uniquely expressed as $AX' + T(A)Y'$.*

Proof. It is easy to see that $(AT(A))$ is a full-rank square matrix. Therefore,

$$M = (AT(A))(AT(A))^{-1}M = AX' + T(A)Y'. \tag{1.4$'$}$$

If $M = AX' + T(A)Y' = AU' + T(A)V'$, then

$$(AT(A))(XY)' = (AT(A))(UV)'. \tag{1.5$'$}$$

Multiplying $(AT(A))^{-1}$ to the left of both sides of (1.5), we obtain $(XY)' = (UV)'$. This shows the uniqueness of the representation.

Definition 1.5. Let A be a full-rank column matrix. If

$$A = \begin{array}{c} \begin{matrix} t_1 & t_2 & \cdots & t_\alpha \end{matrix} \\ \begin{bmatrix} B_1 & T(B_1)C_{11} & \cdots & T(B_1)C_{1k} \\ 0 & B_2 & \cdots & T(B_2)C_{2k} \\ \vdots & \vdots & \ddots & \vdots \\ 0 & 0 & \cdots & B_i \end{bmatrix} \end{array} \begin{matrix} l_1 \\ l_2 \\ \vdots \\ l_k \end{matrix} \qquad \begin{matrix} l_i > 0, \\ l_2 = 1, 2, \ldots, k, \end{matrix}$$

where B_i is a column matrix in r.c. form, $T(B_i)C_{ij} = C_{ij}$ when $t_i = 0$, and $T(B_i)C_{ij} = 0$ when $t_i = l_j$, then we say that A is in k-block r.c. form of size $(l_1, \ldots, l_k)$.

Theorem 1.5. *If A and B are two column matrices in k-block* r.c. *forms of size $(l_1, \ldots, l_k)$ and $A \overset{r}{\sim} B$, then $A = B$.*

Proof. The case $k = 1$ follows from Theorem 1.1. Suppose the present theorem is true for any natural number less than k. Now we shall show that it also holds for k. Let

$$A = \begin{bmatrix} A_1 & T(A_1)Y_{11} & \cdots & T(A_1)Y_{1k} \\ 0 & A_2 & \cdots & T(A_2)Y_{2k} \\ \vdots & \vdots & \ddots & \vdots \\ 0 & 0 & \cdots & A_k \end{bmatrix} = \begin{pmatrix} A^* & S(A^*) \\ 0 & A_k \end{pmatrix},$$

$$B = \begin{bmatrix} B_1 & T(B_1)Z_{11} & \cdots & T(B_1)Z_{1k} \\ 0 & B_2 & \cdots & T(B_2)Z_{2k} \\ \vdots & \vdots & \ddots & \vdots \\ 0 & 0 & \cdots & B_k \end{bmatrix} = \begin{pmatrix} B^* & S(B^*) \\ 0 & B_k \end{pmatrix},$$

By assumption, there exists T, $|T| \neq 0$, such that

$$A = BT. \tag{1.6)'$$

Let $T = \begin{pmatrix} P & Q \\ U & V \end{pmatrix}$, where V is an $l_k \times l_k$ square matrix. By (1.6) we have

$$(0 \quad A_k) = (0 \quad B_k)T = (0 \quad B_k)\begin{pmatrix} P & Q \\ U & V \end{pmatrix}.$$

By Theorem 1.3, $A_k = B_k$. Therefore $V = I$, $U = 0$, and hence $|P| \neq 0$. By (1.6),

$$A^* = B^* P,$$

so $A^* \overset{r}{\sim} B^*$. Since the theorem is true for any natural number less than $k, A^* = B^*$. Therefore $P = I$. Using (1.6) again, we obtain $Q = 0$. This shows that the theorem also holds for k. By induction, the theorem is proved.

Theorem 1.6. *For any full rank column matrix A ($n \times m$) and integers $l_1, \ldots, l_k$, $l_i > 0$ and $\sum_{i=1}^{k} l_i = n$, there is a unique column matrix in k-block r.c. form of size $(l_1, \ldots, l_k)$ which is right similar to A.*

Proof. For $k = 1$, the theorem reduces to Theorem 1.2. Suppose the present theorem is true for any natural number less than k. Now we shall show that it also holds for k. Let

$$A = \begin{bmatrix} D_1 \\ D_2 \\ \vdots \\ D_k \end{bmatrix}, \qquad D_i \text{ has } l_i \text{ rows}, \qquad i = 1, 2, \ldots, k.$$

By Theorem 1.3, there exists T, $|T| \neq 0$ such that $D_k T = (0 \quad A_k)$, where A_k is a column matrix in r.c. form. Therefore,

$$AT = \begin{pmatrix} A^* & S(A^*) \\ 0 & A_k \end{pmatrix}.$$

Since the theorem holds for $k - 1$, there exists P, $|P| \neq 0$, such that $A^* P$ is a column matrix in $(k-1)$-block r.c. form of size $(l_1, \ldots, l_{k-1})$. Without loss of generality, assume

$$A^* P = \begin{bmatrix} A_1 & T(A_1)C_{11} & \cdots & T(A_1)C_{1\,k-1} \\ 0 & A_2 & \cdots & T(A_2)C_{2\,k-1} \\ \vdots & \vdots & \ddots & \vdots \\ 0 & 0 & \cdots & A_{k-1} \end{bmatrix},$$

where $A_1, \ldots, A_{k-1}$ are column matrices in r.c. forms with $l_1, \ldots, l_{k-1}$ rows, respectively. Then

$$AT\begin{pmatrix} P & 0 \\ 0 & I \end{pmatrix} = \begin{pmatrix} A^*P & X \\ 0 & A_k \end{pmatrix}.$$

It is easy to see that through some elementary transformations, $\begin{pmatrix} A^*P & X \\ 0 & A_k \end{pmatrix}$ can be transformed into a column matrix in k-block r.c. form of size $l_1, \ldots, l_k$. This shows that the theorem also holds for k. By induction, for any $(l_1, \ldots, l_k)$, as long as $l_i > 0$ and $\sum_{i=1}^{k} l_i = n$, there is a B in k-block r.c. form of size $(l_1, \ldots, l_k)$ which is right similar to A. The uniqueness of B follows immediately from Theorem 1.5.

§2. Some Combinatorial Properties of Linear Spaces over Finite Fields

Let G be a field. Consider an n-dimensional linear space G_n over G. We shall use lowercase Latin letters $a, b, \ldots,$ to denote the vectors in G_n. The subspace spanned by the vectors $a_1, \ldots, a_k$ is denoted by $\mathfrak{M}(a_1, \ldots, a_k)$, and the subspace generated by the column vectors of matrix A is denoted by $\mathfrak{M}(A)$.

Definition 2.1. Let A, B be two $n \times m$ matrices. If $\mathfrak{M}(A) = \mathfrak{M}(B)$, then we say that A and B are equivalent.

Lemma 2.1. *Let A, B be two full-rank column matrices. Then A and B are equivalent if and only if $A \overset{r}{\sim} B$.*

Proof. Suppose A and B are equivalent. Let $A = (a_1, \ldots, a_m)$, $B = (b_1 \ldots b_m)$. Then

$$a_i = \sum_{\alpha=1}^{m} t_{\alpha i} b_\alpha, \qquad i = 1, 2, \ldots, m.$$

Let $T = (T_{\alpha i})$. Then $|T| \neq 0$ and $A = BT$, so $A \overset{r}{\sim} B$. Notice that the above argument can be reversed. The lemma is proved.

In the subsequent discussion, we shall assume that G is a finite field with q elements. The subspace generated by the column vectors of matrices A and B is denoted by $\mathfrak{M}(A, B)$ and the cardinality of a set $\mathcal{C}$ is denoted by $\mu(\mathcal{C})$.

Let

$$\mathcal{C}_{nm} = \{A : A \text{ is an } n \times m \text{ full-rank column matrix}\},$$
$$\mathcal{B}_{nm} = \{B : B \text{ is an } n \times m \text{ column matrix in r.c. form}\}.$$

Theorem 2.1.*

$$\mu(\mathcal{C}_{nm}) = \begin{cases} \prod_{0 \leqslant i \leqslant m-1} (q^m - q^i), & m \geqslant 1, \\ 1, & m = 0. \end{cases}$$

Proof. (Omitted)

*If Γ is the empty set, then $\prod_{\alpha \in \Gamma}$ is defined to be 1.

Theorem 2.2.

$$\mu(\mathcal{B}_{nm}) = \prod_{0 \leqslant \alpha \leqslant m-1} \frac{(q^{n-\alpha} - 1)}{(q^{m-\alpha} - 1)}.$$

Proof. From Theorem 1.2, we know that when $A \in \mathcal{C}_{nm}$, there exists a unique $B \in \mathcal{B}_{nm}$ such that $A = BT$, $|T| \neq 0$. So $\mu(\mathcal{C}_{nm}) = \mu(\mathcal{B}_{nm})\mu(\mathcal{C}_{mm})$. The expression of $\mu(\mathcal{B}_{nm})$ follows immediately from Theorem 2.1.

Let $\mathcal{C}_{nmx} = \{A : A \text{ is } n \times m \text{ and of rank } x\}$.

Theorem 2.3.

$$\mu(\mathcal{C}_{nmx}) = q^{x(x-1)/2} \prod_{0 \leqslant \alpha \leqslant x-1} \frac{(q^{n-\alpha} - 1)(q^{m-\alpha} - 1)}{(q^{x-\alpha} - 1)}, \qquad x \geqslant 1;$$

and when $x = 0$, $\mu(\mathcal{C}_{nm0}) = 1$.

Proof. When $x = 0$, C_{nm0} only contains the zero matrix. So $\mu(\mathcal{C}_{nm0}) = 1$. When $x \geqslant 1$, from Theorem 1.3, we obtain

$$\mu(\mathcal{C}_{nmx}) = \mu(\mathcal{C}_{mx})\,\mu(\mathcal{B}_{nx}).$$

Theorem 2.3 then follows from Theorems 2.1 and 2.2.

Let

$$\mathcal{D}_{nmi} = \left\{ \begin{pmatrix} A \\ B \end{pmatrix}^{\!m}_{\!n-m}\!\!\!\!\!\!\!\!\!\overset{m}{} : \begin{pmatrix} A \\ B \end{pmatrix} \text{ is in r.c. form, } \operatorname{rk}\begin{pmatrix} A \\ B \end{pmatrix} = m,\ \operatorname{rk}(B) = i \right\}.$$

Theorem 2.4.

$$\mu(\mathcal{D}_{nmi}) = q^{i^2} \prod_{0 \leqslant \alpha \leqslant m-i-1} \frac{(q^{m-\alpha} - 1)}{(q^{m-i-\alpha} - 1)} \prod_{0 \leqslant \alpha \leqslant i-1} \frac{(q^{n-m-\alpha} - 1)}{(q^{i-\alpha} - 1)}.$$

Proof. Consider the two-block r.c. forms of matrices in $\mathcal{D}_{nmi}$,

$$\begin{array}{cc} & \begin{array}{cc} \beta_1 & \beta_2 \end{array} \\ \begin{array}{c} m \\ n-m \end{array} & \begin{pmatrix} B_1 & T(B_1)C \\ 0 & B_2 \end{pmatrix}. \end{array}$$

Then $\beta_2 = i$, $\beta_1 = m - i$. Therefore,

$$\mu(\mathcal{D}_{n\,mi}) = \mu(\mathcal{B}_{m\,m-i})\,\mu(\mathcal{B}_{n-m\,i})q^{i^2}.$$

Theorem 2.4 then follows from Theorem 2.2.

Let

$$\mathcal{E}_{nmijk} = \left\{ \overset{m}{\begin{pmatrix} A_1 \\ A_2 \\ A_3 \\ A_4 \end{pmatrix}} \begin{matrix} m-i \\ i \\ i \\ n-m-i \end{matrix} : \begin{pmatrix} A_1 \\ A_2 \\ A_3 \\ A_4 \end{pmatrix} \text{ is in r.c. form, } \operatorname{rk}\begin{pmatrix} A_3 \\ A_4 \end{pmatrix} = j,\ \operatorname{rk}\begin{pmatrix} A_2 \\ A_4 \end{pmatrix} = k \right\}.$$

Theorem 2.5.

$$\mu(\mathcal{E}_{nmijk}) = \sum_{\substack{\alpha+\beta=m-j \\ \gamma+\delta=j}} q^{(m-i-\alpha)(\beta+j)+(2i-\beta-\gamma)\delta} \mu(\mathcal{C}_{i-\beta\gamma k-\beta-\delta}) \mu(\mathcal{B}_{m-i\,\alpha})$$
$$\cdot \mu(\mathcal{B}_{i\beta}) \mu(\mathcal{B}_{i\gamma}) \mu(\mathcal{B}_{n-m-i\,\delta}).$$

Proof. Consider the four-block r.c. forms of matrices in $\mathcal{E}_{nmijk}$,

$$\begin{matrix} \alpha & \beta & \gamma & \delta & \\ \end{matrix}$$
$$\begin{bmatrix} B_1 & T(B_1)C_{12} & T(B_1)C_{13} & T(B_1)C_{14} \\ 0 & B_2 & T(B_2)C_{23} & T(B_2)C_{24} \\ 0 & 0 & B_3 & T(B_3)C_{34} \\ 0 & 0 & 0 & B_4 \end{bmatrix} \begin{matrix} m-i \\ i \\ i \\ n-m-i \end{matrix}.$$

Then

$$\alpha+\beta+\gamma+\delta = m, \qquad j = \operatorname{rk}\begin{pmatrix} A_3 \\ A_4 \end{pmatrix} = \gamma+\delta,$$

$$k = \operatorname{rk}(B_2 T(B_2) C_{23}) + \delta = \beta + \delta + \operatorname{rk}(C_{23}).$$

Therefore $\alpha+\beta = m-j$, $\gamma+\delta = j$, $\operatorname{rk}(C_{23}) = k-\beta-\delta$, and the other C_{ij}'s are free. Thus

$$\mu(\mathcal{E}_{nmijk}) = \sum_{\substack{\alpha+\beta=m-j \\ \gamma+\delta=j}} q^{(m-i-\alpha)(\beta+j)+(2i-\beta-\gamma)\delta} \mu(\mathcal{C}_{i-\beta\gamma\,k-\beta-\delta}) \mu(\mathcal{B}_{m-i\,\alpha})$$
$$\cdot \mu(\mathcal{B}_{i\beta}) \mu(\mathcal{B}_{i\gamma}) \mu(\mathcal{B}_{n-m-i\,\delta}).$$

Noticing Lemma 2.1, we can state Theorems 2.2–2.5 in the language of subspaces and obtain:

Theorem 2.2′. *If* $1 \leqslant m \leqslant n$, *then* G_n *has* $\mu(\mathcal{B}_{nm})$ *different m-dimensional subspaces.*

For convenience, we use $A, B, \ldots,$ to denote subspaces of G_n. The dimension of A is denoted by $\dim A$ and the subspace spanned by subspaces A and B is denoted by $\mathcal{M}(A, B)$. Then we have:

Theorem 2.4′. *If $1 \leqslant m < n/2$, A is an m-dimensional subspace of G_n, and*

$$\mathfrak{N}_i(A) = \{B : \dim B = m,\ \dim \mathfrak{M}(A,B) = m + i\},$$

then $\mu(\mathfrak{N}_i(A)) = \mu(\mathfrak{D}_{nmi})$.

Theorem 2.5′. *If $1 \leqslant m < n/2$, A, B are two different m-dimensional subspaces of G_n, $\dim \mathfrak{M}(A,B) = m + i$, and $\mathfrak{N}^i_{jk}(A,B) = \{C : \dim C = m,\ \dim \mathfrak{M}(A,C) = m + j,\ \dim \mathfrak{M}(B,C) = m + k\}$, then $\mu(\mathfrak{N}^i_{jk}(A,B)) = \mu(\mathfrak{E}_{nmijk})$.*

If $n > t \geqslant m$, and A is an m-dimensional subspace of G_n, then $\lambda(n,t,m)$, the number of t-dimensional subspaces of G_n containing A, is independent of A and only depends on t, n, m. Thus $\lambda(n,t,m)\mu(\mathfrak{B}_{nm}) = \mu(\mathfrak{B}_{nt})\mu(\mathfrak{B}_{tm})$. Therefore we have:

Theorem 2.6. $\lambda(n,t,m) = \mu(\mathfrak{B}_{nt})\mu(\mathfrak{B}_{tm})/\mu(\mathfrak{B}_{nm})$.

Appendix II

1. The association matrix B of scheme (II.1), $v = 10$, $n_1 = 3$, $a = 0$, $c = 1$.

$$\begin{bmatrix}
0 & 1 & 1 & 1 & 0 & 0 & 0 & 0 & 0 & 0\\
1 & 0 & 0 & 0 & 1 & 1 & 0 & 0 & 0 & 0\\
1 & 0 & 0 & 0 & 0 & 0 & 1 & 1 & 0 & 0\\
1 & 0 & 0 & 0 & 0 & 0 & 0 & 0 & 1 & 1\\
0 & 1 & 0 & 0 & 0 & 0 & 1 & 0 & 1 & 0\\
0 & 1 & 0 & 0 & 0 & 0 & 0 & 1 & 0 & 1\\
0 & 0 & 1 & 0 & 1 & 0 & 0 & 0 & 0 & 1\\
0 & 0 & 1 & 0 & 0 & 1 & 0 & 0 & 1 & 0\\
0 & 0 & 0 & 1 & 1 & 0 & 0 & 1 & 0 & 0\\
0 & 0 & 0 & 1 & 0 & 1 & 1 & 0 & 0 & 0
\end{bmatrix}.$$

2. The association matrix B of scheme (II.2), $v = 16$, $n_1 = 6$, $a = 2$, $c = 2$.

$$(A)\quad \begin{matrix} 1 & 2 & 3 & 4\\ 5 & 8 & 11 & 14\\ 6 & 9 & 12 & 15\\ 7 & 10 & 13 & 16 \end{matrix} \qquad
\begin{bmatrix}
0 & 1 & 1 & 1 & 1 & 1 & 1 & 0 & 0 & 0 & 0 & 0 & 0 & 0 & 0 & 0\\
1 & 0 & 1 & 1 & 0 & 0 & 0 & 1 & 1 & 1 & 0 & 0 & 0 & 0 & 0 & 0\\
1 & 1 & 0 & 1 & 0 & 0 & 0 & 0 & 0 & 0 & 1 & 1 & 1 & 0 & 0 & 0\\
1 & 1 & 1 & 0 & 0 & 0 & 0 & 0 & 0 & 0 & 0 & 0 & 0 & 1 & 1 & 1\\
1 & 0 & 0 & 0 & 0 & 1 & 1 & 1 & 0 & 0 & 1 & 0 & 0 & 1 & 0 & 0\\
1 & 0 & 0 & 0 & 1 & 0 & 1 & 0 & 1 & 0 & 0 & 1 & 0 & 0 & 1 & 0\\
1 & 0 & 0 & 0 & 1 & 1 & 0 & 0 & 0 & 1 & 0 & 0 & 1 & 0 & 0 & 1\\
0 & 1 & 0 & 0 & 1 & 0 & 0 & 0 & 1 & 1 & 1 & 0 & 0 & 1 & 0 & 0\\
0 & 1 & 0 & 0 & 0 & 1 & 0 & 1 & 0 & 1 & 0 & 1 & 0 & 0 & 1 & 0\\
0 & 1 & 0 & 0 & 0 & 0 & 1 & 1 & 1 & 0 & 0 & 0 & 1 & 0 & 0 & 1\\
0 & 0 & 1 & 0 & 1 & 0 & 0 & 1 & 0 & 0 & 0 & 1 & 1 & 1 & 0 & 0\\
0 & 0 & 1 & 0 & 0 & 1 & 0 & 0 & 1 & 0 & 1 & 0 & 1 & 0 & 1 & 0\\
0 & 0 & 1 & 0 & 0 & 0 & 1 & 0 & 0 & 1 & 1 & 1 & 0 & 0 & 0 & 1\\
0 & 0 & 0 & 1 & 1 & 0 & 0 & 1 & 0 & 0 & 1 & 0 & 0 & 0 & 1 & 1\\
0 & 0 & 0 & 1 & 0 & 1 & 0 & 0 & 1 & 0 & 0 & 1 & 0 & 1 & 0 & 1\\
0 & 0 & 0 & 1 & 0 & 0 & 1 & 0 & 0 & 1 & 0 & 0 & 1 & 1 & 1 & 0
\end{bmatrix},$$

$$
(B)\begin{pmatrix}
0&1&1&1&1&1&1&0&0&0&0&0&0&0&0&0\\
1&0&0&0&1&1&0&1&1&1&0&0&0&0&0&0\\
1&0&0&0&1&0&1&0&0&0&1&1&1&0&0&0\\
1&0&0&0&0&1&1&0&0&0&0&0&0&1&1&1\\
1&1&1&0&0&0&0&1&0&0&1&0&0&1&0&0\\
1&1&0&1&0&0&0&0&1&0&0&1&0&0&1&0\\
1&0&1&1&0&0&0&0&0&1&0&0&1&0&0&1\\
0&1&0&0&1&0&0&0&0&1&0&0&1&1&1&0\\
0&1&0&0&0&1&0&0&0&1&1&1&0&0&0&1\\
0&1&0&0&0&0&1&1&1&0&0&0&1&0&0&1\\
0&0&1&0&1&0&0&0&1&0&0&1&0&1&0&1\\
0&0&1&0&0&1&0&0&1&0&1&0&1&0&1&0\\
0&0&1&0&0&0&1&1&0&1&0&1&0&0&1&0\\
0&0&0&1&1&0&0&1&0&0&1&0&0&0&1&1\\
0&0&0&1&0&1&0&1&0&0&0&1&1&1&0&0\\
0&0&0&1&0&0&1&0&1&1&1&0&0&1&0&0
\end{pmatrix}.
$$

3. The association matrix B of Scheme (II.6), $v = 16$, $n_1 = 10$, $a = 6$, $c = 6$.

$$
\begin{pmatrix}
0&1&1&1&1&1&1&1&1&1&1&0&0&0&0&0\\
1&0&1&1&1&1&1&1&0&0&0&1&1&1&0&0\\
1&1&0&1&1&1&0&0&1&1&0&1&1&0&1&0\\
1&1&1&0&1&0&1&0&1&0&1&1&0&1&1&0\\
1&1&1&1&0&0&0&1&0&1&1&0&1&1&1&0\\
1&1&1&0&0&0&1&1&1&1&0&1&1&0&0&1\\
1&1&0&1&0&1&0&1&1&0&1&1&0&1&0&1\\
1&1&0&0&1&1&1&0&0&1&1&0&1&1&0&1\\
1&0&1&1&0&1&1&0&0&1&1&1&0&0&1&1\\
1&0&1&0&1&1&0&1&1&0&1&0&1&0&1&1\\
1&0&0&1&1&0&1&1&1&1&0&0&0&1&1&1\\
0&1&1&1&0&1&1&0&1&0&0&0&1&1&1&1\\
0&1&1&0&1&1&0&1&1&0&0&1&0&1&1&1\\
0&1&0&1&1&0&1&1&0&0&1&1&1&0&1&1\\
0&0&1&1&1&0&0&0&1&1&1&1&1&1&0&1\\
0&0&0&0&0&1&1&1&1&1&1&1&1&1&1&0
\end{pmatrix}.
$$

4. The association matrix B of scheme (II.8), $v = 27$, $n_1 = 16$, $a = 10$, $c = 8$.

$$
\begin{bmatrix}
0&1&1&1&1&1&1&1&1&1&1&1&1&1&1&1&1&0&0&0&0&0&0&0&0&0&0\\
1&0&1&1&1&1&1&1&1&1&1&1&0&0&0&0&0&1&1&1&1&1&0&0&0&0&0\\
1&1&0&1&1&1&1&1&1&0&0&0&1&1&1&0&0&1&1&1&0&0&1&1&0&0&0\\
1&1&1&0&1&1&1&0&0&1&1&0&1&1&0&1&0&1&1&0&1&0&1&0&1&0&0\\
1&1&1&1&0&1&0&1&0&1&0&1&1&0&1&1&0&1&0&1&1&0&1&0&0&1&0\\
1&1&1&1&1&0&0&0&1&0&1&1&0&1&1&1&0&0&1&1&1&0&1&0&0&0&1\\
1&1&1&1&0&0&0&1&1&1&1&0&1&1&0&0&1&1&1&0&0&1&0&1&1&0&0\\
1&1&1&0&1&0&1&0&1&1&0&1&1&0&1&0&1&1&0&1&0&1&0&1&0&1&0\\
1&1&1&0&0&1&1&1&0&0&1&1&0&1&1&0&1&0&1&1&0&1&0&1&0&0&1\\
1&1&0&1&1&0&1&1&0&0&1&1&1&0&0&1&1&1&0&0&1&1&0&0&1&1&0\\
1&1&0&1&0&1&1&0&1&1&0&1&0&1&0&1&1&0&1&0&1&1&0&0&1&0&1\\
1&1&0&0&1&1&0&1&1&1&1&0&0&0&1&1&1&0&0&1&1&1&0&0&0&1&1\\
1&0&1&1&1&0&1&1&0&1&0&0&0&1&1&1&1&1&0&0&0&0&1&1&1&1&0\\
1&0&1&1&0&1&0&1&1&0&0&1&1&0&1&1&1&0&1&0&0&0&1&1&1&0&1\\
1&0&1&0&1&1&0&1&1&0&0&1&1&1&0&1&1&0&0&1&0&0&1&1&0&1&1\\
1&0&0&1&1&1&0&0&0&1&1&1&1&1&1&0&1&0&0&0&1&0&1&0&1&1&1\\
1&0&0&0&0&0&1&1&1&1&1&1&1&1&1&1&0&0&0&0&0&1&0&1&1&1&1\\
0&1&1&1&1&0&1&1&0&1&0&0&1&0&0&0&0&0&1&1&1&1&1&1&1&1&0\\
0&1&1&1&0&1&1&0&1&0&1&0&0&1&0&0&0&1&0&1&1&1&1&1&1&0&1\\
0&1&1&0&1&1&0&1&1&0&0&1&0&0&1&0&0&1&1&0&1&1&1&1&0&1&1\\
0&1&0&1&1&1&0&0&0&1&1&1&0&0&0&1&0&1&1&1&0&1&1&0&1&1&1\\
0&1&0&0&0&0&1&1&1&1&1&1&0&0&0&0&1&1&1&1&1&0&0&1&1&1&1\\
0&0&1&1&1&1&0&0&0&0&0&0&1&1&1&1&0&1&1&1&1&0&0&1&1&1&1\\
0&0&1&0&0&0&1&1&1&0&0&0&1&1&1&0&1&1&1&1&0&1&1&0&1&1&1\\
0&0&0&1&0&0&1&0&0&1&1&0&1&1&0&1&1&1&1&0&1&1&1&1&0&1&1\\
0&0&0&0&1&0&0&1&0&1&0&1&1&0&1&1&1&1&0&1&1&1&1&1&1&0&1\\
0&0&0&0&0&1&0&0&1&0&1&1&0&1&1&1&1&0&1&1&1&1&1&1&1&1&0
\end{bmatrix}
$$

5. The association matrices for $v = 16$, $n_1 = n_2 = n_3 = 5$, $n = 5$, $a = 2$. In this case,

$$
P_0 = \begin{bmatrix} 1&0&0&0\\ 0&1&0&0\\ 0&0&1&0\\ 0&0&0&1 \end{bmatrix}, \qquad
P_1 = \begin{bmatrix} 0&1&0&0\\ 5&0&2&2\\ 0&2&2&1\\ 0&2&1&2 \end{bmatrix},
$$

$$
P_2 = \begin{bmatrix} 0&0&1&0\\ 0&2&2&1\\ 5&2&0&2\\ 0&1&2&2 \end{bmatrix}, \qquad
P_3 = \begin{bmatrix} 0&0&0&1\\ 0&2&1&2\\ 0&1&2&2\\ 5&2&2&0 \end{bmatrix}.
$$

$B_0 = I_{16}$. Let

$$
C_1 = \begin{bmatrix} 0&1&1&0&0\\ 1&0&0&1&0\\ 1&0&0&0&1\\ 0&1&0&0&1\\ 0&0&1&1&0 \end{bmatrix}, \qquad
C_2 = \begin{bmatrix} 0&0&0&1&1\\ 0&0&1&0&1\\ 0&1&0&1&0\\ 1&0&1&0&0\\ 1&1&0&0&0 \end{bmatrix}.
$$

Then

$$B_1 = \begin{pmatrix} 0 & 1\ 1\ 1\ 1\ 1 & 0\ 0\ 0\ 0\ 0 & 0\ 0\ 0\ 0\ 0 \\ \begin{matrix}1\\1\\1\\1\\1\end{matrix} & 0 & C_1 & C_2 \\ \begin{matrix}0\\0\\0\\0\\0\end{matrix} & C_1 & C_1 & I \\ \begin{matrix}0\\0\\0\\0\\0\end{matrix} & C_2 & I & C_2 \end{pmatrix},$$

$$B_2 = \begin{pmatrix} 0 & 0\ 0\ 0\ 0\ 0 & 1\ 1\ 1\ 1\ 1 & 0\ 0\ 0\ 0\ 0 \\ \begin{matrix}0\\0\\0\\0\\0\end{matrix} & C_2 & C_2 & I \\ \begin{matrix}1\\1\\1\\1\\1\end{matrix} & C_2 & 0 & C_1 \\ \begin{matrix}0\\0\\0\\0\\0\end{matrix} & I & C_1 & C_1 \end{pmatrix},$$

$$B_3 = \begin{bmatrix} 0 & 0\;0\;0\;0\;0 & 0\;0\;0\;0\;0 & 1\;1\;1\;1\;1 \\ \begin{matrix}0\\0\\0\\0\\0\end{matrix} & C_1 & I & C_1 \\ \begin{matrix}0\\0\\0\\0\\0\end{matrix} & I & C_2 & C_2 \\ \begin{matrix}1\\1\\1\\1\\1\end{matrix} & C_1 & C_2 & 0 \end{bmatrix}.$$

Appendix III

In this paper, Theorem 2.9 of §2, some parts of (II) and (III) in §3, and some material in §4, §5, and Appendix I are new. It is not hard to see that some familiar theorems can be obtained by applying the results in Appendix I to the construction of BIB designs.

References

1. R. C. Bose and Dale M. Mesner, On linear associative algebras corresponding to association schemes of partially balanced designs. *Ann. Math. Stat.* **30** (1959), 21–38.
2. W. S. Connor and W. H. Clatworthy, Some theorems for partially balanced designs. *Ann. Math. Stat.* **25** (1954), 100–112.
3. L. C. Chang (L. Q. Zhang), The uniqueness and nonuniqueness of the triangular association schemes, *Science Record* **3:12** (1959), 485–492; Association schemes of partially balanced designs with parameters $v = 28$, $n_1 = 12$, $n_2 = 15$, and $p_{11}^2 = 4$, *Science Record* **4:1** (1960), 9–14.
4. S. S. Shrikhande, The uniqueness of the L_2 association scheme, *Ann. Math. Stat.* **30** (1959), 781–789.
5. C. C. MacDuffee, *The Theory of Matrices*. Chelsea, 1946.
6. C. C. MacDuffee, *An Introduction to Abstract Algebra*. Wiley, New York, 1940.
7. Φ. P. Гантмахер, *Matrix Theory*, 1953.
8. M. Hall, Jr., *A Survey of Combinatorial Analysis, Some Aspects of Analysis and Probability*. Wiley, New York, 1958, pp. 76–104.

9. Haim Hanani, The existence and construction of balanced incomplete block designs. *Ann. Math. Stat.* **32** (1961), 361–386.
10. R. C. Bose and S. S. Shrikhande, On the composition of B.I.B. designs. *Can. J. Math.* **12** (1960), 177–188.
11. R. C. Bose and K. N. Bhattacharya, On the construction of group divisible incomplete block designs. *Ann. Math. Stat.* **24** (1953), 167–195.
12. M. L. Yang, C. X. Xu, and Y. Y. Lu, Properties of the parameters of association schemes with three associate classes. Unpublished.
13. S. S. Shrikhande, On the dual of some balanced incomplete block designs. *Biometrics* **8** (1952), 66–71.
14. P. V. Rao, The dual of a balanced incomplete block design. *Ann. Math. Stat.* **31** (1960), 779–785.

Reprinted from
Acta. Sci. Natur. Univ. Pekinensis
1 (1980), 21–47.

On the Coincidence Property of Stochastic Matrices†

BY XU BAO-LU, CHEN JIA-DING, AND ZHENG ZHONG-GUO

Suppose that E is a countable set, and $P = (p_{ij})$ is a stochastic matrix on E (i.e., $p_{ij} \geqslant 0$ for all i, j in E and $\sum_{j \in E} p_{ij} = 1$ for all i in E). We shall call a temporally homogeneous Markov chain $(x_n,\ n \geqslant 0)$ with E as its state space and P as its transition matrix a P-chain. Following [1] we have:

Definition. A stochastic matrix P on E is said to have the coincidence property (to be abbreviated as c.p.) if for any i, j in E and two independent P-chains $(x_n,\ n \geqslant 0)$ and $(y_n,\ n \geqslant 0)$ [defined of course on a common probability space $(\Omega, \mathfrak{F}, \mathbb{P})$], with $\mathbb{P}(x_0 = i) = \mathbb{P}(y_0 = j) = 1$, we have

$$\mathbb{P}(x_n = y_n \text{ for some } n \geqslant 1) = 1.$$

In other words, if for any i, j in E the probability is 1 that two independent P-chains, starting at i, j, respectively, meet (coincide) at some time $n \geqslant 1$, then we say that P has the c.p. Here, independence of $(x_n,\ n \geqslant 0)$ and $(y_n,\ n \geqslant 0)$ means that any $A \in \mathfrak{F}(x_n,\ n \geqslant 0)$ and $B \in \mathfrak{F}(y_n,\ n \geqslant 0)$ are independent. We use $\mathfrak{F}(\xi_\lambda, \lambda \in \Lambda)$ to denote the minimal σ-algebra of subsets of Ω containing $\{\xi_\lambda = i\}$ for all $i \in E$, $\lambda \in \Lambda$.

Under what conditions does P have the c.p.? It is the problem to be investigated in this paper. We have found some sufficient conditions for P to have the c.p., and have studied especially in detail what we call "three-ray matrices" and spatially homogeneous matrices (i.e., those of random walks) on the m-dimensional integer lattice, and obtained necessary and sufficient conditions for a spatially homogeneous matrix to have the c.p. The main results are Corollary 1.2 and Theorems 1.2, 3.1, 4.1, 4.2, and 5.2.

The paper is divided into five sections: §1, first criterion; §2, second criterion, §3, a theorem about three-ray matrices; §4, the coincidence property of three-ray matrices; §5, the coincidence property of spatially homogeneous matrices.

We follow [2] for various notations about a Markov chain, except when specifically mentioned otherwise. For instance, f_{ij}^* denotes the probability of visiting j (at some time $n \geqslant 1$) when starting at i, $p_{ij}^{(n)}$ the probability of being at j at time n when starting at i, etc. Say that i leads to j if $p_{ij}^{(n)} > 0$ for some $n \geqslant 1$. The states of a

†Translated by Shih Chung-Tuo, University of Michigan at Ann Arbor. Most of the results in this paper were obtained in a seminar on Markov processes conducted by the late Professor Xu Bao-lu from the winter of 1964 to the spring of 1965; the rest have been obtained in recent years.

P-chain are simply called the states of P. If all states of P are recurrent (respectively, positive recurrent, null recurrent, nonrecurrent), we say that P is recurrent (respectively, positive recurrent, null recurrent, nonrecurrent). If all states of P have period d, we say that P has period d.

For convenience, we use the convention that when $\mathbb{P}(\Lambda_1) = 0$, $\mathbb{P}(\Lambda_2 \mid \Lambda_1) = 0$.

§1. First Criterion

Let $P = (p_{ij})$ be a stochastic matrix on E. In this section we study first necessary conditions for the c.p. of P, then its sufficient conditions.

Let $\tilde{E} = E \times E$, and

$$\tilde{p}_{(i,j)(k,l)} = p_{ik}p_{jl}. \tag{1.1}$$

Then $\tilde{P} = (\tilde{p}_{(i,j)(k,l)})$ is a stochastic matrix on $\tilde{E}$, called the direct product of P and P, and denoted $\tilde{P} = P \times P$.

Lemma 1.1. *Suppose that $(x_n, n \geqslant 0)$ and $(y_n, n \geqslant 0)$ are two independent P-chains. Then $((x_n, y_n), n \geqslant 0)$ is a $\tilde{P}$-chain.*

Proof. For given (i_m, j_m), $m = 0, 1, \ldots, n+1$, $\mathbb{P}((x_m, y_m) = (i_m, j_m), 0 \leqslant m \leqslant n+1) = \mathbb{P}(x_m = i_m, 0 \leqslant m \leqslant n+1)\mathbb{P}(y_m = j_m, 0 \leqslant m \leqslant n+1) = \mathbb{P}(x_m = i_m, 0 \leqslant m \leqslant n)p_{i_n i_{n+1}}\mathbb{P}(y_m = j_m, 0 \leqslant m \leqslant n)p_{j_n j_{n+1}} = \mathbb{P}((x_m, y_m) = (i_m, j_m), 0 \leqslant m \leqslant n)\tilde{p}_{(i_n, j_n)(i_{n+1}, j_{n+1})}$. This proves the lemma.

It is easy to see that n-step transition probabilities of a $\tilde{P}$-chain

$$\tilde{p}^{(n)}_{(i,j)(k,l)} = p^{(n)}_{ij}p^{(n)}_{kl}. \tag{1.2}$$

Definition 1.1. A nonnegative real-valued function h on E is called a P-harmonic function (or a harmonic function of P) if for all $i \in E$

$$\sum_{j \in E} p_{ij}h(j) = h(i). \tag{1.3}$$

Definition 1.2. A state i is an essential state of P (see [2]) if, for any state j, j leads to i whenever i leads to j.

Theorem 1.1. *Suppose that a stochastic matrix P has the c.p. Then*

(i) *any bounded P-harmonic function is constant;*
(ii) *any essential state of P (if existing) has period* 1.

Proof. Let P have the c.p. First prove (i). Suppose that $(x_n, n \geqslant 0)$, $(y_n, n \geqslant 0)$ are independent P-chains with $\mathbb{P}(x_0 = i) = \mathbb{P}(y_0 = j) = 1$. Define $\tau = \inf\{n : n \geqslant 1, x_n = y_n\}$, with the convention that $\inf\{n : n \in A\} = \infty$ when A is the empty set. Since P has the c.p., $\mathbb{P}(\tau < \infty) = 1$. Let f be a bounded $\tilde{P}$-harmonic function. Since $((x_n, y_n), n \geqslant 0)$ is a $\tilde{P}$-chain, $(f(x_n, y_n), n \geqslant 0)$ is a martingale. Because τ is a finite

Markov time (stopping time), we have $Ef(x_0, y_0) = Ef(x_\tau, y_\tau)$. But $Ef(x_0, y_0) = f(i, j)$. So $f(i, j) = \sum_{k \in E} \mathbb{P}(x_\tau = k, y_\tau = k) f(k, k)$. Consider now a bounded P-harmonic function h, and let $f(u, v) = h(u)$, $g(u, v) = h(v)$ for all u, v in E. f and g are $\tilde{P}$-harmonic functions. Thus $h(i) = f(i, j) = \sum_{k \in E} \mathbb{P}(x_\tau = k, y_\tau = k) h(k) = g(i, j) = h(j)$. It follows that h is constant.

(ii) will be proved by contradiction. So suppose that i_0 is an essential state of P having period $d \geqslant 2$. Let $E_1 = \{i : p_{i_0 i}^{(n)} > 0 \text{ for some } n \geqslant 0\}$. From [2] we know that E_1 is an irreducible closed set, and $E_1 = \bigcup_{r=1}^{d} D_r$ where $D_1, D_2, \ldots, D_d$ are pairwise disjoint, and from a state in D_r, $l \leqslant r < d$, a one-step transition can only be made to states in D_{r+1}, and from a state in D_d only to states in D_1. Let $i \in D_1$, $j \in D_2$. Then two independent chains starting from i, j, respectively, will not meet with probability 1 (in fact, they will meet with probability 0), contradicting the assumption that P has the c.p. The proof is complete.

Lemma 1.2. *Suppose that C is a minimal closed set of P (i.e., C is a closed set and any two states in C lead to each other) and has period 1. Then $C \times C$ is a minimal closed set of $\tilde{P} = P \times P$ and has period 1.*

Proof. For $(i, j) \in C \times C$,

$$\sum_{(k,l) \in C \times C} \tilde{p}_{(i,j)(k,l)} = \sum_{k \in C} p_{ik} \sum_{l \in C} p_{jl} = 1.$$

So $C \times C$ is closed. Fix $(i, j), (k, l)$ in $C \times C$. Since C is minimal closed and has period 1, there exist $n_1 > 0$ and $n_2 > 0$ such that $p_{ik}^{(n)} > 0$ for all $n \geqslant n_1$ and $p_{jl}^{(n)} > 0$ for all $n \geqslant n_2$. Let $n = n_1 + n_2$; then

$$\tilde{p}_{(i,j)(k,l)}^{(n)} = p_{ik}^{(n)} p_{jl}^{(n)} > 0.$$

Thus we see that $C \times C$ is a minimal closed set. The equality $p_{(i,i)(i,i)}^{(n)} = [p_{ii}^{(n)}]^2$ shows that (i, i) has period 1 if i has period 1. It follows that $C \times C$ has period 1, completing the proof.

Let us introduce the concept of a "sure-exit set", which will be used many times.

Definition 1.3. Let P be a stochastic matrix on E. A subset F of E is called a sure-exit set if for any $i \in F$ and a P-chain $(x_n, n \geqslant 0)$ with $\mathbb{P}(x_0 = i) = 1$, $\mathbb{P}(x_n \in F \text{ for all } n \geqslant 1) = 0$.

We know from the theory of Markov chains that the state space $E = (\bigcup_r C_r) \cup E_1$ where E_1 consists of all nonrecurrent states and the C_r's are pairwise disjoint recurrent classes. Of course, some of these sets may be empty.

Lemma 1.3. *Suppose that there exists a recurrent state and that every bounded P-harmonic function is constant. Then there is only one recurrent class, and all nonrecurrent states, if any, form a sure-exit set.*

Proof. Under the assumption of the lemma, it is obvious that there is only one recurrent class. To prove the second assertion, let E_1 be the set of all nonrecurrent states and consider a P-chain $(x_n, n \geqslant 0)$ satisfying $\mathbb{P}(x_0 = i) > 0$ for all $i \in E$. Define $h(i) = \mathbb{P}(x_n \in E_1$ for all $n \geqslant 1 \mid x_0 = i)$. Since $h(i) = \sum_{j \in E_1} p_{ij} \mathbb{P}(x_n \in E_1$ for all $n \geqslant 2 \mid x_1 = j) = \sum_{j \in E_1} p_{ij} h(j) = \sum_{j \in E} p_{ij} h(j)$ [note that $h(j) = 0$ if $j \notin E_1$], h is a bounded P-harmonic function. Therefore, by assumption h is constant. Since there exists a recurrent state i_0 and $h(i_0) = 0$, $h(i) \equiv 0$. This proves that E_1 is a sure-exit set.

We point out in passing that one can prove that the converse of Lemma 1.3 is also valid.

Theorem 1.2 (First criterion). *Suppose that P satisfies*:

(i) *every bounded P-harmonic function is constant,*
(ii) *there exists i_0 with period 1 and with*

$$\sum_{n=1}^{\infty} \left[p_{i_0\, i_0}^{(n)} \right]^2 = \infty. \tag{1.4}$$

Then P has the c.p.

Proof. Since $p_{ii}^{(n)} \geqslant [p_{ii}^{(n)}]^2$, from (1.4) $\sum_{n=1}^{\infty} p_{i_0\, i_0}^{(n)} = \infty$, and so i_0 is recurrent. From condition (i) and Lemma 1.3, there is a unique recurrent class C, and all nonrecurrent states (if any) form a sure-exit set E_1, i.e., $E = C \cup E_1$ where $C \neq \phi$ but E_1 may be empty. Of course, C is a minimal closed set. Since $i_0 \in C$ and has period 1, Lemma 1.2 implies that $C \times C$ is a minimal closed set of $\tilde{P}$. Since $\sum_{n=1}^{\infty} \tilde{p}_{(i_0,i_0)(i_0,i_0)}^{(n)} = \sum_{n=1}^{\infty} [p_{i_0\, i_0}^{(n)}]^2$, it follows from (1.4) that (i_0, i_0) is a recurrent state of $\tilde{P}$, so that $C \times C$ is a recurrent class of $\tilde{P}$.

For fixed i, j, in E, suppose that $(x_n, n \geqslant 0)$ and $(y_n, n \geqslant 0)$ are independent P-chains with $\mathbb{P}(x_0 = i) = \mathbb{P}(y_0 = j) = 1$, and denote $z_n = (x_n, y_n)$. From Lemma 1.1 $(z_n, n \geqslant 0)$ is a $\tilde{P}$-chain. Let $\tau_1 = \inf\{n : n \geqslant 1$ and $x_n \in C\}$, $\tau_2 = \inf\{n : n \geqslant 1$ and $y_n \in C\}$. τ_1, τ_2 are Markov times relative to $(x_n, n \geqslant 0)$ and $(y_n, n \geqslant 0)$, respectively. Let $\tau = \max(\tau_1, \tau_2)$; then τ is a Markov time relative to $(z_n, n \geqslant 0)$. Let us show $\mathbb{P}(\tau < \infty) = 1$. Indeed, if $i \in C$ then $\mathbb{P}(\tau_1 = 1) = \mathbb{P}(x_1 \in C) = 1$ (since C is closed), and if $i \notin C$ then, since $E - C$ is a sure-exit set, $\mathbb{P}(x_n \in C$ for some $n \geqslant 1) = 1$. Thus $\mathbb{P}(\tau_1 < \infty) = 1$. Similarly $\mathbb{P}(\tau_2 < \infty) = 1$. It follows that $\mathbb{P}(\tau < \infty) = 1$. Now define

$$\Lambda_1 = \{x_n \in C \text{ for all } n \geqslant \tau_1\}, \qquad \Lambda_2 = \{y_n \in C \text{ for all } n \geqslant \tau_2\}.$$

Then $\mathbb{P}(\Lambda_1) = \sum_{m=1}^{\infty} \mathbb{P}(\tau_1 = m, x_n \in C$ for all $n \geqslant m) = \sum_{m=1}^{\infty} \sum_{k \in C} \mathbb{P}(\tau_1 = m, x_m = k) \mathbb{P}(x_n \in C$ for all $n \geqslant m+1 \mid x_m = k)$. Since C is closed, $\mathbb{P}(x_n \in C$ for all $n \geqslant m+1 \mid x_m = k) = 1$ if $k \in C$ and $\mathbb{P}(x_m = k) > 0$. So $\mathbb{P}(\Lambda_1) = \sum_{m=1}^{\infty} \sum_{k \in C} \mathbb{P}(\tau_1 = m, x_m = k) = \mathbb{P}(\tau_1 < \infty)$. Similarly, $\mathbb{P}(\Lambda_2) = 1$; so $\mathbb{P}(\Lambda_1 \Lambda_2) = 1$.

Obviously, if $\omega \in \Lambda_1 \cap \Lambda_2 \cap \{\tau < \infty\}$, then $(x_\tau(\omega), y_\tau(\omega)) \in C \times C$. Thus $\mathbb{P}(z_\tau \in C \times C) = 1$. For a given $k \in C$, $\mathbb{P}(z_n = (k,k)$ i.o.$) = \sum_{u,v \in C} \mathbb{P}(z_\tau = (u,v); z_n = (k,k)$ i.o.$) = \sum_{u,v \in C} \mathbb{P}(z_\tau (u,v); z_{\tau+n} = (k,k)$ i.o.$) = \sum_{u,v \in C} \sum_{m=1}^{\infty} \mathbb{P}(z_\tau = (u,v), \tau = m)$ $\mathbb{P}(z_{m+n} = (k,k)$ i.o.$\mid z_m = (u,v))$. (Of course, $\{x_n = (k,k)$ i.o. (infinitely often)$\}$ $= \{x_n = (k,k)$ for infinitely many $n\} = \overline{\lim}_{n \to \infty} \{x_n = (k,k)\}$.) Since $C \times C$ is a recurrent class of $\tilde{P}$, $P(z_{m+n} = (k,k)$ i.o.$\mid z_m = (u,v)) = 1$ if $(u,v) \in C \times C$ and

$\mathbb{P}(z_m = (u, v)) > 0$. It follows that $\mathbb{P}(z_n = (k, k) \text{ i.o.}) = \sum_{u,v \in C} \sum_{m=1}^{\infty} \mathbb{P}(z_\tau = (u, v), \tau = m) = \mathbb{P}(z_\tau \in C \times C) = 1$. Therefore $\mathbb{P}(x_n = y_n \text{ for some } n \geqslant 1) = 1$, proving that P has the c.p. The proof of Theorem 1.2 is complete.

Corollary 1.1. *Suppose that P is irreducible and has period 1, and that for some i*

$$\sum_{n=1}^{\infty} \left[p_{ii}^{(n)} \right]^2 = \infty. \tag{1.5}$$

Then P has the c.p.

Proof. Under condition (1.5) and irreducibility P is recurrent. That P is irreducible then also implies that every bounded harmonic function is constant. Now the conclusion follows from Theorem 1.2.

Theorem 1.3. *Suppose that P has period 1 and contains a positive recurrent state. Then in order for P to have the c.p., it is necessary and sufficient that every bounded P-harmonic function be constant.*

Proof. The necessity follows from Theorem 1.1, so we need only to prove the sufficiency. Assume that i_0 is a positive recurrent state with period 1 and that every bounded P-harmonic function is constant. Since $\lim_{n\to\infty} P_{i_0\, i_0}^{(n)} > 0$, we have $\sum_{n=1}^{\infty} [P_{i_0\, i_0}^{(n)}]^2 = \infty$. It follows from Theorem 1.2 that P has the c.p.

Corollary 1.2. *Suppose that P is irreducible, has period 1, and is positive recurrent. Then P has the c.p.*

Proof. This is because the assumption of Corollary 1.1 holds.

Corollary 1.3. *Suppose that the state space E of P is finite. Then in order for P to have the c.p. it is necessary and sufficient that*

(i) *every bounded P-harmonic function be constant,*
(ii) *all recurrent states have period 1.*

Proof. Since E is finite P must have positive recurrent states. Thus the corollary follows immediately from Theorem 1.3.

Example 1. Let $E = \{0, 1, 2, \ldots\}$ and P be the following matrix on E (which we will call a "three-ray matrix").

$$p = \begin{bmatrix} r_0 & p_0 & & & & 0 \\ q_1 & r_1 & p_1 & & & \\ & q_2 & r_2 & p2 & & \\ & & q_3 & r_3 & p_3 & \\ & & & \cdot & \cdot & \cdot \\ & 0 & & & \cdot & \cdot & \cdot \end{bmatrix}, \tag{1.6}$$

where $p_i > 0$, $r_i \geqslant 0$ for $i \geqslant 0$, $q_i > 0$ for $i \geqslant 1$, $r_0 + p_0 = 1$, $p_i + r_i + q_i = 1$ for $i \geqslant 1$.

The following facts about the three-ray matrix P are well-known.

(i) P is irreducible.

(ii) In order for P to have period 1 it is necessary and sufficient that $r_i > 0$ for some i.

(iii) In order for P to be recurrent it is necessary and sufficient that $\sum_{k=1}^{\infty} q_1 \cdots q_k / p_1 \cdots p_k = \infty$.

(iv) In order for P to be positive recurrent it is necessary and sufficient that

$$\sum_{k=1}^{\infty} \frac{q_1 \cdots q_k}{p_1 \cdots p_k} = \infty, \qquad \sum_{k=1}^{\infty} \frac{p_0 \cdots p_{k-1}}{q_1 \cdots q_i} < \infty.$$

(v) Every P-harmonic function is constant.

From Corollary 1.2 we know that if $r_i > 0$ for some i and

$$\sum_{k=1}^{\infty} \frac{q_1 \cdots q_k}{p_1 \cdots p_k} = \infty, \qquad \sum_{k=1}^{\infty} \frac{p_0 \cdots p_{k-1}}{q_0 \cdots q_k} < \infty,$$

then P in (1.6) has the c.p.

We will see in §4 that there exist (nonrecurrent) three-ray matrices with period 1 that do not have the c.p. This shows that conditions (i) and (ii) of Theorem 1.1 are not sufficient for a stochastic matrix P to have the c.p.

Example 2. Let $E = \{0, 1, 2, \ldots\}$ and

$$P = (p_{ij}) = \begin{bmatrix} f_1 & f_2 & f_3 & f_4 & \cdots \\ 1 & 0 & 0 & 0 & \cdots \\ 0 & 1 & 0 & 0 & \cdots \\ 0 & 0 & 1 & 0 & \cdots \\ \cdot & \cdot & \cdot & \cdot & \cdots \\ \cdot & \cdot & \cdot & \cdot & \cdots \\ \cdot & \cdot & \cdot & \cdot & \cdots \end{bmatrix}, \tag{1.7}$$

where $p_{0j} = f_{j+1} > 0$ for $j \geqslant 0$, $\sum_{j=1}^{\infty} f_j = 1$, $p_{i,i-1} = 1$ for $i \geqslant 1$, and all other $p_{ij} = 0$.

(A) Clearly P is irreducible, has period 1, and is recurrent, and for P to be null recurrent it is necessary and sufficient that

$$\sum_{n=1}^{\infty} n f_n = \infty. \tag{1.8}$$

(B) We will show that a necessary and sufficient condition for P to have the c.p. is

$$\sum_{n=1}^{\infty} \left[P_{00}^{(n)} \right]^2 = \infty. \tag{1.9}$$

Only the necessity needs to be proved (the sufficiency follows from Corollary 1.1).

Suppose that P has the c.p. We will prove that $(0,0)$ is a recurrent state of $\tilde{P} = P \times P$. Let $(x_n,\ n \geqslant 0)$ and $(y_n,\ n \geqslant 0)$ be two independent P-chains with $\mathbb{P}(x_0 = 0) = \mathbb{P}(y_0 = 0) = 1$. Then $\mathbb{P}(x_n = y_n$ for some $n \geqslant 1) = 1$. Define $\tau = \inf\{n\colon n \geqslant 1,\ x_n = y_n\}$; then $\mathbb{P}(\tau < \infty) = 1$. Let $\Lambda_1 = \{\tau < \infty,\ x_{\tau + x_\tau} = 0\}$. Then

$$P(\Lambda_1) = \sum_{m=1}^{\infty} \sum_{i \in E} \mathbb{P}(\tau = m,\ x_\tau = i,\ x_{\tau + x_\tau} = 0)$$

$$= \sum_{m=1}^{\infty} \sum_{i \in E} \mathbb{P}(\tau = m,\ x_m = i)\mathbb{P}(x_{m+i} = 0 \mid x_m = i).$$

When $i = 0$, $\mathbb{P}(x_{m+i} = 0 \mid x_m = i) = 1$; when $i > 0$ and $\mathbb{P}(x_m = i) > 0$, $\mathbb{P}(x_{m+i} = 0 \mid x_m = i) = p_{i,i-1}p_{i-1,i-2} \cdots p_{10} = 1$. Thus $\mathbb{P}(\Lambda_1) = \sum_{m=1}^{\infty}\sum_{i \in E}\mathbb{P}(\tau = m, x_m = i) = 1$. Similarly let $\Lambda_2 = \{\tau < \infty,\ y_{\tau + y_\tau} = 0\}$; then $\mathbb{P}(\Lambda_2) = 1$. Now if $\omega \in \Lambda_1\Lambda_2$ then $x_{\tau + x_\tau}(\omega) = 0 = y_{\tau + x_\tau}(\omega)$. Hence $\Lambda_1\Lambda_2 \subset \{(x_n, y_n) = (0,0)$ for some $n \geqslant 1\}$, and so $\mathbb{P}((x_n, y_n) = (0,0)$ for some $n \geqslant 1) = 1$. Therefore $(0,0)$ is $\tilde{P}$-recurrent, and it follows that $\sum_{n=1}^{\infty}[P_{00}^{(n)}]^2 = \sum_{n=1}^{\infty}\tilde{P}_{(0,0)(0,0)}^{(n)} = \infty$, proving the necessity of (1.9).

(C) Let $F(x) = \sum_{n=1}^{\infty} f_n x^n$, $|x| < 1$. Then in order for P to have the c.p. it is necessary and sufficient that

$$\sum_{n=1}^{\infty} \left[\frac{1}{n!} \left(\frac{1}{1 - F(x)} \right)_0^{(n)} \right]^2 = \infty. \tag{1.10}$$

Indeed, let $(x_n,\ n \geqslant 0)$ be a P-chain with $\mathbb{P}(x_0 = 0) = 1$; then $f_n = \mathbb{P}(x_1 \neq 0, \ldots, x_{n-1} \neq 0,\ x_n = 0) = f_{00}^{(n)}$. So $F(x) = \sum_{n=1}^{\infty} f_{00}^{(n)} x^n$. Let $U(x) = \sum_{n=0}^{\infty} p_{00}^{(n)} x^n$; then $U(x) = 1/(1 - F(x))$. Thus $p_{00}^{(n)} = (1/n!)U^{(n)}(0) = (1/n!)(1/(1 - F(x)))_0^{(n)}$, proving the equivalence of (1.10) and (1.9). So the necessity and sufficiency of (1.10) follows from (B).

(D) In (1.7), let $f_n = (-1)^{n+1}\binom{\delta}{n}$, $n \geqslant 1$, where $0 < \delta < 1$. Here

$$\binom{\delta}{n} = \frac{\delta(\delta - 1) \cdots (\delta - n + 1)}{n!}.$$

We show that (i) P is null recurrent; (ii) when $1/2 \leqslant \delta < 1$, P has the c.p.; and (iii) when $0 < \delta < 1/2$, P does not have the c.p.

$F(x) = \sum_{n=1}^{\infty} f_n x^n = 1 - (1 - x)^\delta$, $|x| < 1$ [which incidentally verifies $\sum_{n=1}^{\infty} f_n = \lim_{x \to 1-} F(x) = 1$]. Since $F'(x) = \delta(1 - x)^{\delta - 1}$, $\sum_{n=1}^{\infty} n f_n = \lim_{x \to 1-} F'(x) = \infty$. From (A) above we know that P is null recurrent.

Since $1/(1 - F(x)) = (1 - x)^{-\delta}$, $(1/(1 - F(x)))_0^{(n)} = (-\delta)(-\delta - 1) \cdots (-\delta - n + 1)(-1)^n$. Hence

$$\sum_{n=1}^{\infty} \left[\frac{1}{n!} \left(\frac{1}{1 - F(x)} \right)_0^{(n)} \right]^2 = \sum_{n=1}^{\infty} \left[(-1)^n \binom{-\delta}{n} \right]^2 = \sum_{n=1}^{\infty} \binom{-\delta}{n}^2.$$

But

$$\binom{-\delta}{n}^2 \Big/ \binom{-\delta}{n+1} = \left(\frac{n+1}{\delta+n}\right)^2 = \left(1+\frac{1-\delta}{n+\delta}\right)^2$$

$$= 1 + \frac{2(1-\delta)}{n+\delta} + \left(\frac{1-\delta}{n+\delta}\right)^2$$

$$= 1 + \frac{2(1-\delta)}{n} + \frac{(1-\delta)^2}{(n+\delta)^2} - \left[\frac{2(1-\delta)}{n} - \frac{2(1-\delta)}{n+\delta}\right]$$

$$= 1 + \frac{2(1-\delta)}{n} + \frac{(1-\delta)^2}{(n+\delta)^2} - \frac{2\delta(1-\delta)}{n(n+\delta)}$$

$$= 1 + \frac{2(1-\delta)}{n} + O\left(\frac{1}{n^2}\right).$$

From the Gauss criterion, when $2(1-\delta) > 1$, $\sum_{n=1}^{\infty}\binom{-\delta}{n}^2 < \infty$; when $2(1-\delta) \leqslant 1$, $\sum_{n=1}^{\infty}\binom{-\delta}{n} = \infty$. Thus by (C), if $0 < \delta < 1/2$, P does not have the c.p. and if $1/2 \leqslant \delta < 1$, P has the c.p.

§2. Second Criterion

Although Theorem 1.2 gives a useful criterion when a stochastic matrix has the c.p., its application is narrow because it is inapplicable to a nonrecurrent matrix. In this section we use the concept of a sure-exit set to study more general sufficient conditions.

Lemma 2.1. *Suppose that F is a sure-exit set of a stochastic matrix P and $(x_n, n \geqslant 0)$ is a P-chain. Then*

$$\mathbb{P}(x_n \in F \textit{ for all sufficiently large } n) = 0. \tag{2.1}$$

Proof. Since F is a sure-exit set, for any $j \in F$, $\mathbb{P}(x_n \in F$ for all $n \geqslant m+1 \mid x_m = j) = 0$. So $\mathbb{P}(x_n \in F$ for all $n \geqslant m) = \sum_{j\in F}\mathbb{P}(x_m = j)\mathbb{P}(x_n \in F$ for all $n \geqslant m+1 \mid x_m = j) = 0$. Thus $\mathbb{P}(x_n \in F$ for all sufficiently large $n) \leqslant \sum_{m=0}^{\infty}\mathbb{P}(x_n \in F$ for all $n \geqslant m) = 0$, proving (2.1).

Let $(x_n, n \geqslant 0)$ be a P-chain with $\mathbb{P}(x_0 = i) = 1$. For $A \subset E$ denote $f_{iA}^* = \mathbb{P}(x_n \in A$ for some $n \geqslant 1)$, $g_{iA} = \mathbb{P}(x_n \in A$ for infinitely many $n)$.

Lemma 2.2. *Let P be a stochastic matrix on E, and $E = E_1 \cup E_2 \cup E_3$ where E_1, E_2, E_3 are pairwise disjoint and satisfy*:

(i) *E_2 is a sure-exit set,*
(ii) $\inf_{j\in E_1} f_{jE_3}^* > 0$.

Then for all $i \in E$, $g_{iE_3} = 0$.

Proof.† Suppose $(x_n, n \geqslant 0)$ is a P-chain with $\mathbb{P}(x_0 = i) = 1$. From Lemma 2.1 $\mathbb{P}(x_n \notin E_2 \text{ i.o.}) = 1$. So $g_{iE_3} = \mathbb{P}(x_n \in E_3 \text{ i.o.}) = 1 - \mathbb{P}(x_n \in E_1 \cup E_2$ for all sufficiently large $n) = 1 - \mathbb{P}(x_n \in E_1 \text{ i.o.}, x_n \in E_1 \cup E_2$ for all sufficiently large n). Thus $g_{iE_3} = 1$ if the last probability is 0, which will hold if we can show $\mathbb{P}(x_n \in E_1$ i.o., $x_n \notin E_3$ for all $n \geqslant m) = 0$ for any $m \geqslant 0$. For any $M > m$ define $\tau_M = \inf\{n : n \geqslant M, x_n \in E_1\}$. Then

$$\begin{aligned}
&\mathbb{P}(\tau_M < \infty, x_n \notin E_3 \text{ for all } n \geqslant m)\\
&\quad = \sum_{N=M}^{\infty} \mathbb{P}(\tau_M = N, x_n \notin E_3 \text{ for all } n \geqslant m)\\
&\quad = \sum_{N=M}^{\infty} \sum_{j \in E_1} \mathbb{P}(\tau_m = N, x_N = j, x_n \notin E_3 \text{ for } m \leqslant n \leqslant N)\\
&\qquad \times \mathbb{P}(x_n \notin E_3 \text{ for all } n > N \mid x_N = j)\\
&\quad = \sum_{N=M}^{\infty} \sum_{j \in E_1} \mathbb{P}(\tau_M = N, x_N = j, x_n \notin E_3 \text{ for } m \leqslant n \leqslant N)(1 - f^*_{jE_3})\\
&\quad \leqslant \mathbb{P}(\tau_M < \infty, x_n \notin E_3 \text{ for } m \leqslant n \leqslant \tau_M)\Big(1 - \inf_{j \in E_1} f^*_{jE_3}\Big).
\end{aligned}$$

Let a denote $\mathbb{P}(x_n \in E_1$ i.o., $x_n \notin E_3$ for all $n \geqslant m$). As $M \to \infty$ the first and last expressions of the above display tend, respectively, to a and $a(1 - \inf_{j \in E_1} f^*_{jE_3})$.‡ So we have $a \leqslant a(1 - \inf_{j \in E_1} f^*_{jE_3})$, which forces $a = 0$. The proof is complete.

Suppose that $E = E_1 \cup E_2 \cup E_3$ where E_1, E_2, E_3 are pairwise disjoint and the corresponding partition of a stochastic matrix P on E is

$$P = \begin{array}{c} \\ \end{array}\begin{array}{ccc} E_1 & E_2 & E_3 \\ \end{array}$$

$$P = \begin{bmatrix} P_{11} & P_{12} & P_{13} \\ P_{21} & P_{22} & P_{23} \\ P_{31} & P_{32} & P_{33} \end{bmatrix} \begin{matrix} E_1 \\ E_2. \\ E_3 \end{matrix} \tag{2.2}$$

Denote $G_1 = \sum_{n=0}^{\infty} P_{11}^n$, where P_{11}^n is the usual nth power of P_{11} (P_{11}^0 is the identity matrix on E_1). Also, if F_1, F_2 are countable sets and $A = (a_{ij})$ is a (real) matrix on $F_1 \times F_2$ (i varies in F_1 and j in F_2), we will use $e_i'A$ to denote the ith row and Ae_j the jth column of A. Here e_i' can be seen as a row vector on F_1 whose ith component is 1 and other components are 0, so that $e_i'A$ is understood as the usual product of e_i' and A, and similarly for e_j (a column vector) and Ae_j. Furthermore, a column vector with all components equal to 1 will be denoted also by 1; thus $A1$ is the column vector obtained by summing the rows of A.

† *Translator's note.* The original proof contains a minor error that has now been corrected.

‡ *Translator's note.* $\Lambda_M = \{\tau_M < \infty\} \downarrow \Lambda = \{x_n = E_1 \text{ i.o.}\}$ and $\Gamma_M = \{x_n \notin E_3 \text{ for } m \leqslant n \leqslant \tau_M\} \uparrow \{x_n \notin E_3 \text{ for all } n \geqslant m\}$ as $M \to \infty$. So $|\mathbb{P}(\Lambda_M\Gamma_M) - \mathbb{P}(\Lambda\Gamma)| \leqslant \mathbb{P}(\Lambda_M\Gamma_M - \Lambda\Gamma) + \mathbb{P}(\Lambda\Gamma - \Lambda_M\Gamma_M) \leqslant \mathbb{P}(\Lambda_M - \Lambda) + \mathbb{P}(\Gamma - \Gamma_M) \to 0$ as $M \to \infty$.

Lemma 2.3. *Suppose that the stochastic matrix P in* (2.2) *satisfies*

$$\inf_{i\in E_1} e_i' G_1 P_{13} 1 > 0. \tag{2.3}$$

Then $\inf_{i\in E_1} f^*_{iE_3} > 0$.

Proof. Let $(x_n, n > 0)$ be a P-chain with $\mathbb{P}(x_0 = i) = 1$. Then $f^*_{iE_3} = \mathbb{P}(x_n \in E_3$ for some $n \geqslant 1) \geqslant \sum_{n=1}^{\infty}\mathbb{P}(x_1 \in E_1, \ldots, x_{n-1} \in E_1, x_n \in E_3) = \sum_{n=1}^{\infty} e_i' P_{11}^{n-1} P_{13} 1 = e_i'(\sum_{n=1}^{\infty} P_{11}^{n-1}) P_{13} 1 = e_i' G_1 P_{13} 1$. The lemma follows.

Applying the above results to $\tilde{P} = P \times P$, we obtain a criterion for P to have the c.p.

Theorem 2.1 (Second criterion). *Suppose that P is a stochastic matrix on E, $\tilde{E} = E \times E = \tilde{E}_1 \cup \tilde{E}_2 \cup \tilde{E}_3$ where $\tilde{E}_1, \tilde{E}_2, \tilde{E}_3$ are pairwise disjoint, and the corresponding partition of $\tilde{P} = P \times P$ is*

$$P = \begin{array}{c} \begin{matrix} \tilde{E}_1 & \tilde{E}_2 & \tilde{E}_3 \end{matrix} \\ \begin{bmatrix} \tilde{P}_{11} & \tilde{P}_{12} & \tilde{P}_{13} \\ \tilde{P}_{21} & \tilde{P}_{22} & \tilde{P}_{23} \\ \tilde{P}_{31} & \tilde{P}_{32} & \tilde{P}_{33} \end{bmatrix} \end{array} \begin{matrix} \tilde{E}_1 \\ \tilde{E}_2. \\ \tilde{E}_3 \end{matrix}$$

If (i) $\tilde{E}_3 = \{(i, i) : i \in E\}$, (ii) $\tilde{E}_2$ *is a sure-exit set of* $\tilde{P}$, *and* (iii) $\inf_{\lambda\in E_1} f^*_{\lambda\tilde{E}_3} > 0$, *then P has the c.p.*

Proof. This follows directly from Lemma 2.2 and the definition of $g_{\lambda\tilde{E}_3}$.

Theorem 2.2. *If in Theorem 2.1 condition* (iii) *of the assumption is changed to* (iii′) $\inf_{\lambda\in\tilde{E}_1} e_\lambda' \tilde{G}_1 u > 0$, *where* $\tilde{G}_1 = \sum_{n=0}^{\infty} \tilde{P}_{11}^n$, $u = \tilde{P}_{13} 1$, *while the rest of the assumption is kept unchanged, P has the c.p.*

Proof. From Lemma 2.1, condition (iii′) implies condition (iii); so the theorem follows from Theorem 2.1.

Corollary 2.1. *If, in Theorem 2.1, condition (iii) of the assumption is changed to* $\inf e_\lambda' \tilde{P}_{11}^n u > 0$ *for some* $n \geqslant 0$ (*where* $u = \tilde{P}_{13} 1$), *while the rest of the assumption is kept unchanged, then P has the c.p.*

Proof. This follows immediately from Theorem 2.2 since $\tilde{G}_1 \geqslant \tilde{P}_{11}^n$.

Although Theorems 2.1, 2.2 and Corollary 2.1 obtain sufficient conditions for the c.p. of a stochastic matrix on P, their actual application may still present a problem, because it involves finding an appropriate sure-exit set and computing $\tilde{G}_1$, u, etc. Fortunately in the case of a three-ray matrix these are not too difficult. In order to find an appropriate sure-exit set for a three-ray matrix, we establish an important theorem in §3 about such a matrix, while in §4 the computations of $\tilde{P}_{11}$, u, etc. are carried out.

§3. A Theorem about Three-Ray Matrices

In this section, $P = (p_{ij})$ is a "three-ray matrix" mentioned in §1, i.e., P is the following matrix on $E = \{0, 1, 2, \dots\}$.

$$P = \begin{bmatrix} r_0 & p_0 & & & & \\ q_1 & r_1 & p_1 & & 0 & \\ & q_2 & r_2 & p_2 & & \\ & & \cdot & \cdot & \cdot & \\ 0 & & & \cdot & \cdot & \cdot \\ & & & & \cdot & \cdot & \cdot \end{bmatrix}, \tag{3.1}$$

where $p_i = p_{i,i+1} > 0$, $r_i = p_{ii} \geqslant 0$ for $i \geqslant 0$, $q_i = p_{i,i-1} > 0$ for $i \geqslant 1$, $p_0 + r_0 = 1$, and $p_i + r_i + q_i = 1$ for $i \geqslant 1$.

Fix i_0, j_0 in E, and let $(x_n,\ n \geqslant 0)$, $(y_n,\ n \geqslant 0)$ be independent P-chains with $\mathbb{P}(x_0 = i_0) = \mathbb{P}(y_0 = j_0) = 1$. For any $k \in E$, define

$$\tau_k = \inf\{n : n \geqslant 1, x_n = k\}, \tag{3.2}$$

$$\tau'_k = \inf\{n : n \geqslant 1, y_n = k\}. \tag{3.3}$$

The principal result of this section is

Theorem 3.1. *Suppose that P in* (3.1) *satisfies*

$$\prod_{k=0}^{\infty} p_k = 0. \tag{3.4}$$

Then

$$\mathbb{P}(\tau_k \leqslant \tau'_k \text{ for infinitely many } k) = 1. \tag{3.5}$$

The proof is lengthy; so we divide it into a series of lemmas.

Lemma 3.1. *If $i < j$, $f^*_{ij} = 1$.*

Proof. For a fixed $j \geqslant 1$ define $u_i = f^*_{ij}$ when $i \neq j$; $= 1$ when $i = j$. Then $u_i = \sum_{l \in E} P_{il} u_l$ for all $i \neq j$. So

$$\begin{aligned} u_0 &= r_0 u_0 + p_0 u_1 \\ u_1 &= q_1 u_0 + r_1 u_1 + p_1 u_2 \\ &\vdots \\ u_i &= q_i u_{i-1} + r_i u_i + p_i u_{i+1} \qquad (i < j). \end{aligned}$$

It follows that $u_0 = u_1$, $u_{i+1} - u_i = (q_i/p_i)(u_i - u_{i-1})$ for $1 \leqslant i < j$. Thus $u_0 = u_1 = \cdots = u_j$. Since $u_j = 1$, we have $u_i = 1$ for all $i < j$, proving the lemma.

Lemma 3.2. *For any $k > \max(i_0, j_0)$, $\mathbb{P}(\tau_k < \infty) = \mathbb{P}(\tau'_k < \infty) = 1$.*

Proof. This follows from Lemma 3.1 since $\mathbb{P}(\tau_k < \infty) = f^*_{i_0 k}$, $\mathbb{P}(\tau'_k < \infty) = f^*_{j_0 k}$.

Lemma 3.3. *If $k_2 > k_1 > \max(i_0, j_0)$, then $\mathbb{P}(\tau_{k_1} < \tau_{k_2}) = \mathbb{P}(\tau'_{k_1} < \tau'_{k_2}) = 1$.*

Proof[†]. This follows easily from the fact $p_{ij} = 0$ if $j > i + 1$. Thus, for example, if $\tau_{k_2} = n$ and $x_0 = i_0,\ x_1 = i_1, \ldots,\ x_n = i_n = k_2$ [from Lemma 3.2 $\mathbb{P}(\tau_{k_2} < \infty) = 1$], then we may assume $i_{m+1} \leqslant i_m + 1$ for $0 \leqslant m < n$; since, as one can easily see, $i_{m_1} = k_1$ where $m_1 = \max\{m : 0 \leqslant m < n,\ i_m \leqslant k_1\}$, we have $\tau_{k_1} < \tau_{k_2}$.

Because of Lemmas 3.2 and 3.3, we may assume below that for all $k_2 > k_1 > \max(i_0, j_0)$, $\tau_{k_1}(\omega) < \tau_{k_2}(\omega) < \infty$ and $\tau'_{k_1}(\omega) < \tau'_{k_2}(\omega) < \infty$ for all $\omega \in \Omega$.

Lemma 3.4. *Let* $l > \max(i_0, j_0)$; *then*

$$\tau_l\,, \tau_{l+1} - \tau_l\,, \ldots, \tau_{l+k} - \tau_{l+k-1}, \ldots \tag{3.6}$$

and

$$\tau'_l\,, \tau'_{l+1} - \tau'_l\,, \ldots, \tau'_{l+k} - \tau_{l+k-1}, \ldots \tag{3.7}$$

are series of independent random variables; moreover, the two series are independent of each other.

Proof. Since the random variables in (3.6) are measurable with respect to the σ-algebra $\mathfrak{F}(x_n, n \geqslant 0)$ and those in (3.7) measurable with respect to $\mathfrak{F}(y_n, n \geqslant 0)$, and since $(x_n, n \geqslant 0), (y_n, n \geqslant 0)$ are independent, the two series are independent. It remains to show that (3.6) is an independent series [the proof of (3.7) is similar]. Since τ_j is a Markov time of $(x_n, n \geqslant 0)$, $\tau_1, \tau_2, \ldots, \tau_k$ are measurable with respect to $\mathfrak{F}_{\tau_k} = \{\Lambda : \Lambda \in \mathfrak{F};$ for all $n \geqslant 0$, $\Lambda \cap \{\tau_k \leqslant n\} \in \mathfrak{F}(x_0, x_1, \ldots, x_n)\}$. Now $\tau_{k+1} - \tau_k$ is measurable with respect to $\mathfrak{F}(x_{\tau_k + n}, n \geqslant 0)$, being the time of first visit to $k + 1$ for the P-chain $(x_{\tau_k + n},\ n \geqslant 0)$. Since $\mathbb{P}(x_{\tau_k} = k) = 1$, it follows from the general theory of Markov chains that the σ-algebras $\mathfrak{F}_{\tau_k}$ and $\mathfrak{F}(x_{\tau_k + n},\ n \geqslant 0)$ are independent. In particular, $\tau_l, \ldots, \tau_k$ (for any $k > l$) are independent of $\tau_{k+1} - \tau_k$. This implies the series (3.6) is independent.

Lemma 3.5. $\tau_{k+1} - \tau_k$ *and* $\tau'_{k+1} - \tau'_k$ *have the same distribution, where* $k > \max(i_0, j_0)$.

Proof. This is because $(x_{\tau_k + n},\ n \geqslant 0), (y_{\tau'_k + n},\ n \geqslant 0)$ are both P-chains starting at k.

In the next lemmas, let $\eta_k = (\tau'_{k+1} - \tau'_k) - (\tau_{k+1} - \tau_k)$, where $k > \max(i_0, j_0)$.

Lemma 3.6. *Suppose* $\prod_{k=0}^{\infty} p_k = 0$; *then*

$$\mathbb{P}(\eta_k \neq 0 \textit{ for infinitely many } k) = 1. \tag{3.8}$$

Proof. For any $n \geqslant 0$,

$$\begin{aligned}\mathbb{P}(\tau_{k+1} - \tau_k = n) &\leqslant \mathbb{P}\big(x_{\tau_k + n - 1} = k,\ x_{\tau_k + n} = k + 1\big)\\ &= \mathbb{P}\big(x_{\tau_k + n - 1} = k\big)\mathbb{P}\big(x_{\tau_k + n} = k + 1 \,|\, x_{\tau_k + n - 1} = k\big) \leqslant p_k\,.\end{aligned}$$

[†] *Translator's note.* We have omitted the detail in the original proof.

Thus

$$\mathbb{P}(\eta_k = 0) = \sum_{n=1}^{\infty} \mathbb{P}(\tau'_{k+1} - \tau'_k = n)\mathbb{P}(\tau_{k+1} - \tau_k = n)$$

$$\leqslant \sum_{n=1}^{\infty} \mathbb{P}(\tau'_{k+1} - \tau'_k = n)p_k = p_k .$$

It follows that $\sum_{k=l}\mathbb{P}(\eta_k \neq 0) \geqslant \sum_{k=l}^{\infty}(1 - p_k)$ (where $l > \max(i_0, j_0)$). By assumption $\prod_{k=0}^{\infty} p_k = 0$; so $\sum_{k=0}^{\infty}(1 - p_k) = \infty$ and we have $\sum_{k=l}^{\infty}\mathbb{P}(\eta_k \neq 0) = \infty$. Since $\eta_l, \eta_{l+1}, \ldots$ are independent, from the Borel–Cantelli Lemma we obtain $\mathbb{P}(\eta_k \neq 0$ for infinitely many $k) = 1$.

We need the following concept that has been used by Doob and others: A series $\sum_{k=1}^{\infty}\zeta_k$ of random variables is said to converge essentially if there exist constants $c_l, c_{l+1}, \ldots$ such that $\sum_{k=l}^{\infty}(\zeta_k - c_k)$ converges with probability 1.

Lemma 3.7. *Suppose* $\prod_{k=0}^{\infty} p_k = 0$, $l > \max(i_0, j_0)$; *then* $\sum_{k=l}^{\infty}\eta_k$ *does not converge essentially.*

Proof. We prove it by contradiction. Suppose that $\sum_{k=l}^{\infty}\eta_k$ converges essentially; then for some constants $c_l, c_{l+1}, \ldots,$ $\sum_{k=l}^{\infty}(\eta_k - c_k)$ converges a.s. Since the η_k have symmetric distributions by Lemma 3.5, it follows from Theorem 2.6 of [3] that $\sum_{k=l}^{\infty}\eta_k$ converges a.s. But this is impossible by Lemma 3.6, since the η_k are integer valued.

Lemma 3.8. *Suppose that* $\prod_{k=0}^{\infty} p_k = 0$, $l > \max(i_0, j_0)$, *and* $\xi_n = \sum_{k=l}^{n}\eta_k$ *for* $n > l$; *then* $\mathbb{P}(\overline{\lim}_{n\to\infty}\xi_n = \infty) = 1$.

Proof. From Lemma 3.7 and Theorem 2.9 of [3] we know that for all $m > 0$, $\lim_{n\to\infty}\mathbb{P}(|\xi_n| > m) = 1$. Since the ξ_n have symmetric distributions $\lim_{n\to\infty}\mathbb{P}(\xi_n > m) = \frac{1}{2}$, and so $\mathbb{P}(\overline{\lim}_{n\to\infty}\{\xi_n > m\}) \geqslant \overline{\lim}_{n\to\infty}\mathbb{P}(\xi_n > m) = \frac{1}{2}$. But

$$\overline{\lim}_{n\to\infty}\{\xi_n > m\}\downarrow\{\overline{\lim}_{n\to\infty}\xi_n = \infty\} \qquad \text{as} \quad m\to\infty;$$

it follows that

$$\mathbb{P}(\overline{\lim}_{n\to\infty}\xi_n = \infty) \geqslant \tfrac{1}{2}. \tag{3.9}$$

Since $\{\overline{\lim}_{n\to\infty}\xi_n = \infty\}$ is a tail event of the sequence of independent random variables $\eta_l, \eta_{l+1}, \ldots$ (see Lemma 3.4), by the Kolmogorov 0–1 law $\mathbb{P}(\overline{\lim}_{n\to\infty}\xi_n = \infty) = 0$ or 1. The lemma now follows from (3.9).

Lemma 3.9. *Suppose* $\prod_{k=0}^{\infty} p_k = 0$; *then for all* $l \geqslant 1$

$$\mathbb{P}(\tau_k \leqslant \tau'_k \text{ for some } k > l) = 1. \tag{3.10}$$

Proof. We may assume $l > \max(i_0, j_0)$. Let us prove

$$\mathbb{P}(\tau'_k < \tau_k \text{ for all } k > l) = 0. \tag{3.11}$$

Indeed, $\{\tau'_k < \tau_k$ for all $k > l\} = \{\sum_{j=l}^{k-1}(\tau'_{j+1} - \tau'_j) + \tau'_l < \sum_{j=l}^{k-1}(\tau_{j+1} - \tau_j) + \tau_l$ for all $k > l\} = \{\xi_{k-1} < \tau_l - \tau'_l$ for all $k > l\}$ (the ξ_k are as defined in Lemma

3.8) $\subset \{\overline{\lim}_{k\to\infty}\xi_k \leqslant \tau_l - \tau_l'\}$. So $\mathbb{P}(\tau_k' < \tau_k$ for all $k > l) \leqslant \mathbb{P}(\overline{\lim}_{k\to\infty}\xi_k \leqslant \tau_l - \tau_l') \leqslant \mathbb{P}(\overline{\lim}_{k\to\infty}\xi_k < \infty)$, since τ_l and τ_l' are a.s. finite. (3.11) now follows from Lemma 3.8, and (3.10) from (3.11).

From Lemma 3.9 we have $\mathbb{P}(\tau_k \leqslant \tau_k'$ for infinitely many $k) = 1$. This is (3.5), and so the proof of Theorem 3.1 is complete.

Using Theorem 3.1 we can obtain appropriate sure-exit sets of $\tilde{P} = P \times P$. Let

$$A = \{(i,j) : 0 \leqslant i < j\}, \qquad B = \{(i,j) : i > j \geqslant 0\}.$$

Theorem 3.2. *If P in* (3.1) *satisfies $\prod_{k=1}^{\infty} p_k = 0$, then A, B are sure-exit sets of $\tilde{P}$.*

Proof. We only prove that A is a sure-exit set; the proof for B is similar. Fix $(i_0, j_0) \in A$ and let $((x_n, y_n), n \geqslant 0)$ be a $\tilde{P}$-chain with $\mathbb{P}((x_0, y_0) = (i_0, j_0)) = 1$. It is easy to see that $(x_n, n \geqslant 0)$ and $(y_n, n \geqslant 0)$ are independent P-chains. From Theorem 3.1 we have [where τ_k, τ_k' are as defined in (3.2) and (3.3)]

$$\mathbb{P}(\tau_k \leqslant \tau_k' \text{ for some } k > j_0) = 1. \tag{3.12}$$

Let $\hat{\Omega} = \{\tau_k' < \infty$ for all $k > j_0\}$; then $\mathbb{P}(\hat{\Omega}) = 1$ by Lemma 3.2. We point out that

$$\hat{\Omega} \cap \{\tau_k \leqslant \tau_k' \text{ for some } k > j_0\} \subset \{(x_{\tau_k}, y_{\tau_k}) \notin A \text{ for some } k > j_0\}. \tag{3.13}$$

Indeed, let $\omega \in \hat{\Omega} \cap \{\tau_k \leqslant \tau_k'$ for some $k > j_0\}$; then for some $k > j_0$, $\tau_k(\omega) \leqslant \tau_k'(\omega) < \infty$, and so $x_{\tau_k}(\omega) = k$, $y_{\tau_k}(\omega) \leqslant k$, and it follows that $(x_{\tau_k}(\omega), y_{\tau_k}(\omega)) \notin A$. This proves (3.13). From (3.12) and (3.13) we have $\mathbb{P}((x_n, y_n) \in A$ for all $n \geqslant 1) = 0$. Since this is true for all $(i_0, j_0) \in A$, A is a sure-exit set for $\tilde{P}$.

Theorem 3.3. *Under the assumption of Theorem* 3.2, *the set $\tilde{E}_2 = \{(i,j) : i \geqslant 0, j \geqslant 0, |i-j| \geqslant 2\}$ is a sure-exit set of $\tilde{P}$.*

Proof. $C_1 = A \cap \tilde{E}_2$ and $C_2 = B \cap \tilde{E}_2$ are sure-exit sets of $\tilde{P}$ by Theorem 3.2, where A, B are as above. Since from a state of C_1 a one-step transition does not enter C_2 and vice versa, $\tilde{E}_2 = C_1 \cup C_2$ is a sure-exit set.

The existence of the sure-exit sets $A, B, \tilde{E}_2$ is the building block which enables us to study the c.p. for three-ray matrices, based on Theorems 2.1 and 2.2.

Remark. Theorems 3.1 and 3.2 can be generalized to certain more general stochastic matrices than the three-ray matrices. Suppose $P = (p_{ij})$ is a stochastic matrix on $E = \{0, 1, 2, \ldots\}$ with $p_{ij} = 0$ for $j \geqslant i + 2$. Then if $P_{i,i+1} > 0$ for all $i \geqslant 0$ and $\prod_{i=0}^{\infty} p_{i,i+1} = 0$, using the same proofs as in this section one can obtain the same conclusions as Theorems 3.1 and 3.2 for such a matrix.

§4. Coincidence Property of Three-Ray Matrices

Here notations are as above: $E = \{0, 1, 2, \ldots\}$, $\tilde{E} = E \times E$, P is a three-ray matrix displayed in (3.1), $\tilde{P} = P \times P$, $\tilde{E}_2 = \{(i,j) : i, j \in E, |i-j| \geqslant 2\}$. Let $\tilde{E}_3 = \{(i, i : i \in E\}$, $\tilde{E}_1 = \{(i,j) : i, j \in E, |i-j| = 1\}$. Then $\tilde{E} = \tilde{E}_1 \cup \tilde{E}_2 \cup \tilde{E}_3$ and $\tilde{E}_1$,

$\tilde{E}_2, \tilde{E}_3$ are pairwise disjoint. In order to apply the results of §2 to P, we need to compute $\tilde{P}_{11}$, $u = \tilde{P}_{13}1 = 1 - \tilde{P}_{11}1 - \tilde{P}_{12}1$ and $\tilde{G}_1 = \sum_{n=0}^{\infty} \tilde{P}_{11}^n$ in Theorems 2.1 and 2.2.

Let $A_1 = \{(i, i+1) : i \geqslant 0\}$, $A_2 = \{(i+1, i) : i \geqslant 0\}$. Then $\tilde{E}_1 = A_1 \cup A_2$.

(A) *Computation of u.* Denote

$$u = \binom{v}{w} \qquad \text{with} \quad v = \begin{bmatrix} v_{01} \\ v_{12} \\ v_{23} \\ \vdots \end{bmatrix}, \quad w = \begin{bmatrix} w_{10} \\ w_{21} \\ w_{32} \\ \vdots \end{bmatrix}.^{\dagger}$$

Since $u = 1 - \tilde{P}_{11}1 - \tilde{P}_{12}1 = \tilde{P}_{13}1$, we know that $v_{i,i+1} = w_{i+1,i} = p_i r_{i+1} + r_i q_{i+1}$, $i \geqslant 0$. So

$$u = \binom{v}{v} \qquad \text{where} \quad v = (p_i r_{i+1} + r_i q_{i+1}). \tag{4.1}$$

(B) *Computations of $\tilde{P}_{11}u$ and $\tilde{G}_1 u$.* Denote

$$\tilde{P}_{11} = \begin{array}{c} \begin{array}{cc} A_1 & A_2 \end{array} \\ \begin{bmatrix} Q_1 & Q_2 \\ Q_3 & Q_4 \end{bmatrix} \end{array} \begin{array}{c} \\ A_1 \\ A_2 \end{array}.$$

Since

$$\begin{aligned} \tilde{P}_{(i,i+1)(j,j+1)} &= p_i p_{i+1} && \text{when} \quad j = i+1, \quad i \geqslant 0 \\ &= r_i r_{i+1} && \text{when} \quad j = i, \quad i \geqslant 0 \\ &= q_i q_{i+1} && \text{when} \quad j = i-1, \quad i \geqslant 1 \\ &= 0 && \text{otherwise,} \end{aligned}$$

we have

$$Q_1 = \begin{bmatrix} r_0 r_1 & p_0 p_1 & & & 0 \\ q_1 q_2 & r_1 r_2 & p_1 p_2 & & \\ & q_2 q_3 & r_2 r_3 & p_2 p_3 & \\ 0 & & \cdot & \vdots & \\ & & & & \ddots \end{bmatrix}. \tag{4.2}$$

Since

$$\begin{aligned} \tilde{P}_{(i,i+1)(j+1,j)} &= p_i q_{i+1} && \text{when} \quad j = i, \quad i \geqslant 0, \\ &= 0 && \text{otherwise,} \end{aligned}$$

we have

$$Q_2 = \begin{bmatrix} p_0 q_1 & & 0 & \\ & p_1 q_2 & & \\ & & p_2 q_3 & \\ 0 & & & \ddots \end{bmatrix}. \tag{4.3}$$

[†] *Translator's note.* v, w are column vectors on A_1, A_2, respectively, whose ith components are $(i, i+1)$, $(i+1, i)$, respectively. Here and below, one should think of A_1, A_2 as $E = \{i : i \geqslant 0\}$.

It is easily verified that $Q_3 = Q_2$, $Q_4 = Q_1$. Hence

$$\tilde{P}_{11}u = \begin{bmatrix} (Q_1 + Q_2)v \\ (Q_1 + Q_2)v \end{bmatrix} = \begin{bmatrix} Qv \\ Qv \end{bmatrix},$$

where

$$Q = \begin{bmatrix} r_0r_1 + p_0q_1 & p_0p_1 & & & \\ q_1q_2 & r_1r_2 + p_1q_2 & p_1p_2 & & 0 \\ & q_2q_3 & r_2r_3 + p_2q_3 & p_2p_3 & \\ & & & \ddots & \ddots \\ & 0 & & & \ddots \end{bmatrix}. \tag{4.4}$$

In general,

$$\tilde{P}_{11}^n u = \begin{bmatrix} Q^n v \\ Q^n v \end{bmatrix}$$

for all $n \geqslant 0$. Therefore

$$\tilde{G}_1 u = \begin{bmatrix} \left(\sum_{n=0}^{\infty} Q^n \right) v \\ \left(\sum_{n=0}^{\infty} Q^n \right) v \end{bmatrix}. \tag{4.5}$$

Theorem 4.1. *Suppose that P in* (3.1) *satisfies*

$$\inf_{i \geqslant 0} e_i' \left(\sum_{n=0}^{\infty} Q^n \right) v > 0, \tag{4.6}$$

where v is given in (4.1), *Q in* (4.4). *Then P has the c.p.*

We first prove a lemma.

Lemma 4.1. *Under the assumption of Theorem* 4.1, $\prod_{k=0}^{\infty} p_k = 0$.

Proof. From (4.5), (4.6) we have $\inf_{\lambda \in \tilde{E}_1} e_\lambda' \tilde{G}_1 u > 0$; then from Lemma 2.3 $\inf_{\lambda \in \tilde{E}_1} f_{\lambda \tilde{E}_3}^* = \epsilon > 0$. Let $((x_n, y_n), n \geqslant 0)$ be a $\tilde{P}$-chain with $\mathbb{P}((x_0, y_0) = (0, 1)) = 1$. Then for any $\lambda \in \tilde{E}_1$,

$$\mathbb{P}((x_m, y_m) \in \tilde{E}_1 \text{ for all } m > n \mid (x_n, y_n) = \lambda) \leqslant 1 - \epsilon. \tag{4.7}$$

[*Note*: When $\mathbb{P}((x_n, y_n) = \lambda) = 0$ the left-hand side of (4.7) is understood to be 0.] Sc for all $n \geqslant 1$,

$$\begin{aligned} &\mathbb{P}((x_m, y_m) \in \tilde{E}_1 \text{ for all } m \geqslant 1) \\ &\quad = \sum_{\lambda \in \tilde{E}_1} \mathbb{P}((x_m, y_m) \in \tilde{E}_1 \text{ for } 1 \leqslant m < n, (x_n, y_n) = \lambda) \\ &\qquad \times \mathbb{P}((x_m, y_m) \in \tilde{E}_1 \text{ for all } m > n \mid (x_n, y_n) = \lambda) \\ &\quad \leqslant \mathbb{P}((x_m, y_m) \in \tilde{E}_1 \text{ for } 1 \leqslant m \leqslant n)(1 - \epsilon). \end{aligned}$$

As $n \to \infty$, we obtain

$$\mathbb{P}((x_m, y_m) \in \tilde{E}_1 \text{ for all } m \geqslant 1) \leqslant \mathbb{P}((x_m, y_m) \in \tilde{E}_1 \text{ for all } m \geqslant 1)(1-\epsilon),$$

which forces the left-hand side of the above to equal 0. Now

$$\begin{aligned}\prod_{k=0}^{\infty} p_k \prod_{k=1}^{\infty} p_k &= \mathbb{P}(x_1 = 1, x_2 = 2, \ldots)\mathbb{P}(y_1 = 2, y_2 = 3, \ldots) \\ &= \mathbb{P}((x_1, y_1) = (1,2), (x_2, y_2) = (2,3), \ldots, (x_n, y_n) = (n, n+1)) \\ &\leqslant \mathbb{P}((x_n, y_n) \in \tilde{E}_1 \text{ for all } n \geqslant 1) = 0.\end{aligned}$$

It follows that $\prod_{k=0}^{\infty} p_k = 0$; Lemma 4.1 is proved.

Proof of Theorem 4.1. Under condition (4.6) $\tilde{E}_2$ is a sure-exit set of $\tilde{P}$ by Lemma 4.1 and Theorem 3.3. Since $\inf_{\lambda \in \tilde{E}_1} e'_\lambda \tilde{G}_1 u = \inf_{i \geqslant 0} e'_i (\sum_{n=0}^{\infty} Q^n) v > 0$, it follows from Theorem 2.2 that P has the c.p.

Corollary 4.1. *Suppose P in* (3.1) *satisfies*

$$\inf_{i \geqslant 0} (p_i r_{i+1} + r_i q_{i+1}) > 0. \tag{4.8}$$

Then P has the c.p.

Proof. Indeed, since $e'_i(\sum_{n=0}^{\infty} Q^n) v \geqslant e'_i v = p_i r_{i+1} + r_i q_{i+1}$, (4.8) implies (4.6) and thus that P has the c.p.

Corollary 4.2. *Suppose that P in* (3.1) *satisfies*

$$\begin{aligned}\inf_{i \geqslant 0} \big(& q_i q_{i+2} p_{i-1} r_i + q_i^2 q_{i+2} r_{i-1} + p_i r_i r_{i+1}^2 + q_{i+1} r_i^2 r_{i+1} + p_i^2 p_{i+1} r_{i+1} \\ & + p_i p_{i+1} q_{i+1} r_i + p_i p_{i+1}^2 r_{i+2} + p_i p_{i+1} q_{i+2} r_{i+1} \big) > 0.\end{aligned} \tag{4.9}$$

Then P has the c.p.

Proof. Since $e'_i \sum_{n=0}^{\infty} Q^n v \geqslant e'_i Q v$ and since a simple computation gives equality of $e'_i Q v$ and the expression inside the parentheses in (4.9), (4.9) implies (4.6) and that P has the c.p.

Example 1.

$$P = \begin{bmatrix} r_0 & p_0 & & & & & \\ q & r & p & & & 0 & \\ & q & r & p & & & \\ & & q & r & p & & \\ & 0 & & \cdot & \cdot & \cdot & \\ & & & & \cdot & \cdot & \cdot \\ & & & & \cdot & \cdot & \cdot \end{bmatrix} \tag{4.10}$$

where $r_0 \geqslant 0$, $p_0 > 0$, $q > 0$, $r > 0$, $p > 0$, $r_0 + p_0 = q + r + p = 1$.

Here (4.8) holds. So the three-ray matrix P in (4.10) has the c.p. It is worth noting that when $p < q$, the matrix P is positive recurrent; when $p = q$, P is null recurrent; and when $p > q$, it is nonrecurrent. But P always has the c.p.

Example 2.

$$P = \begin{bmatrix} r_0 & p_0 & & & & & & & & \\ q & 0 & p & & & & & & & \\ & q & 0 & p & & & & 0 & & \\ & & q' & r & p' & & & & & \\ & & & q & 0 & p & & & & \\ & & & & q & 0 & p & & & \\ & 0 & & & & q' & r & p' & & \\ & & & & & & \cdot & \cdot & \cdot & \\ & & & & & & & \cdot & \cdot & \cdot \\ & & & & & & & & \cdot & \cdot \end{bmatrix} \tag{4.11}$$

where $p_0, r_0, p, q, r, p', q'$ are all positive, i.e., P is the matrix in (3.1) with $P_{3k} = p'$, $q_{3k} = q'$, $r_{3k} = r$ for $k \geqslant 1$, $p_{3k+1} = P_{3k+2} = p$, $q_{3k+1} = q_{3k+2} = q$, $r_{3k+1} = r_{3k+2} = 0$ for $k \geqslant 1$.

One verifies that (4.9) holds. So the matrix P in (4.11) has the c.p. Note that P does not satisfy (4.8).

We now establish another criterion that parallels Theorem 4.1.

Theorem 4.2. *Suppose that P in* (3.1) *satisfies*

(i) $\prod_{k=0}^{\infty} p_k = 0,$ (4.12)

(ii) $\inf_{i>0}(p_i r_{i+1} + r_i q_{i+1})/p_i q_{i+1} > 0.$ (4.13)

Then P has the c.p.

Proof. We will utilize Theorem 2.1. Still denote $\tilde{E}_1 = \{(i,j) : |i-j| = 1,\ i, j \geqslant 0\}$, $\tilde{E}_2 = \{(i,j) : |i-j| \geqslant 2,\ i, j \geqslant 0\}$, $\tilde{E}_3 = \{(i,i) : i \geqslant 0\}$, $A = \{(i,j) : 0 \leqslant i < j\}$, $B = \{(i,j) : i > j \geqslant 0\}$. Let ϵ be the infimum in (4.13); so we have for $i \geqslant 0$

$$p_i r_{i+1} r_i q_{i+1} \geqslant \epsilon p_i q_{i+1}. \tag{4.14}$$

We will show that

$$\inf_{\lambda \in \tilde{E}_1} f^{*}_{\lambda \tilde{E}_3} \geqslant \epsilon/(1+\epsilon). \tag{4.15}$$

Let $(z_n,\ n \geqslant 0)$ be a $\tilde{P}$-chain with $\mathbb{P}(z_0 = (i, i+1)) = 1$. Let $\tau = \inf\{n : n \geqslant 1, z_n \notin A\}$. Since, by Theorem 3.1, A is a sure-exit set, $\mathbb{P}(\tau < \infty) = 1$. So

$$\mathbb{P}(z_\tau \in \tilde{E}_3) + \mathbb{P}(z_\tau \in B) = 1. \tag{4.16}$$

But

$$\mathbb{P}(z_\tau \in \tilde{E}_3) \geqslant \sum_{k=0}^{\infty} \mathbb{P}(z_\tau \in \tilde{E}_3, z_{\tau-1} = (k, k+1)), \tag{4.17}$$

$$\mathbb{P}(z_\tau \in B) = \sum_{k=0}^{\infty} \mathbb{P}(z_\tau \in B, z_{\tau-1} = (k, k+1)\}. \tag{4.18}$$

Now

$$\begin{aligned}\mathbb{P}(z_\tau \in \tilde{E}_3, z_{\tau-1} = (k,k+1)) &= \sum_{n=1}^{\infty} \mathbb{P}(z_1 \in A, \ldots, z_{n-2} \in A, z_{n-1} = (k,k+1)) \\ &\quad \times \mathbb{P}(z_n \in \tilde{E}_3 \mid z_{n-1} = (k,k+1)) \\ &= \sum_{n=1}^{\infty} \mathbb{P}(z_1 \in A, \ldots, z_{n-2} \in A, z_{n-1} = (k,k+1)) \\ &\quad \times (p_k r_{k+1} + r_k q_{k+1}),\end{aligned}$$

and from (4.14) this is

$$\begin{aligned}&\geqslant \sum_{n=1}^{\infty} \mathbb{P}(z_1 \in A, \ldots, z_{n-2} \in A, z_{n-1} = (k,k+1)) \cdot \epsilon p_k q_{k+1} \\ &= \epsilon \sum_{n=1}^{\infty} \mathbb{P}(z_1 \in A, \ldots, z_{n-2} \in A, z_{n-1} = (k,k+1)) \mathbb{P}(z_n \in B \mid z_{n-1} = (k,k+1)) \\ &= \epsilon \mathbb{P}(z_\tau \in B, z_{\tau-1} = (k,k+1)).\end{aligned}$$

Therefore, from (4.17) and (4.18), we obtain

$$\mathbb{P}(z_\tau \in \tilde{E}_3) \geqslant \epsilon \mathbb{P}(z_\tau \in B). \tag{4.19}$$

Substituting (4.19) into (4.16), we have $\mathbb{P}(z_\tau \in \tilde{E}_3) \geqslant \epsilon/(1+\epsilon)$. Now $f^*_{(i,i+1)\tilde{E}_3} \geqslant \mathbb{P}(z_\tau \in \tilde{E}_3) \geqslant \epsilon/(1+\epsilon)$. Similarly, $f^*_{(i+1,i)\tilde{E}_3} \geqslant \epsilon/(1+\epsilon)$. Thus (4.15) is proved. Finally, by Theorem 2.1, P has the c.p. The theorem is proved.

Corollary 4.3. *Suppose that P in (3.1) satisfies one of the following conditions.*

(i) $\inf_{i \geqslant 0} r_{2i+1} > 0,$ (4.20)
(ii) $\inf_{i \geqslant 0} r_{2i} > 0.$ (4.21)

Then P has the c.p.

Proof. (4.2) implies (4.12) and (4.13). Indeed, let $\epsilon = \inf_{i \geqslant 0} r_{2i+1}$; then $p_{2i+1} \leqslant 1-\epsilon$ and thus $\prod_{k=0}^{\infty} p_k \leqslant \prod_{i=0}^{\infty} p_{2i+1} \leqslant \lim_{n\to\infty}(1-\epsilon)^n = 0$, proving (4.12). On the other hand, $(p_i r_{i+1} + r_i q_{i+1})/p_i q_{i+1} \geqslant r_{i+1} + r_i \geqslant \epsilon$, proving (4.13). By Theorem 4.2, (4.20) implies that P has the c.p. Similarly, (4.21) implies that P has the c.p.

Example 3. Suppose that the matrix P of (3.1), $p_0 = 1/2$, $q_1 = 1/6$, $r_k = 1/2$ for $k \geqslant 0$, $p_{2k-1} = q_{2k} = 1/3k$, $p_{2k} = q_{2k-1} = 1 - 1/2 - 1/3k$ for $k \geqslant 1$. It follows from Corollary 4.3 that P has the c.p. It is worth remarking that one learns after some close study that condition (4.6) is not satisfied.

Example 4. Consider the three-ray matrix

$$P = (p_{ij}) = \begin{bmatrix} a & b & & & & & & 0 \\ b & 0 & a & & & & & \\ & a & 0 & b & & & & \\ & & b & 0 & a & & & \\ & & & a & 0 & b & & \\ 0 & & & & \cdot & \cdot & \cdot & \\ & & & & & \cdot & \cdot & \cdot \\ & & & & & & \cdot & \cdot & \cdot \end{bmatrix}, \tag{4.22}$$

where $a > 0$, $b > 0$, $a + b = 1$, i.e., $p_{00} = p_{2i,2i-1} = p_{2i-1,2i} = a$ for $i \geqslant 1$, $p_{2i,2i+1} = p_{2i+1,2i} = b$ for $i \geqslant 0$, $p_{ij} = 0$ otherwise. It is easy to see that P is irreducible, has period 1, and is null recurrent. It does not satisfy condition (4.13), and close study also shows it does not satisfy condition (4.6) of Theorem 4.1. But we can prove that P has the c.p. Its proof is somewhat long. We sketch it as follows.

(A) Let $Q = (q_{ij}) = P^2$. Then $q_{ii} = a^2 + b^2$ for $i \geqslant 0$, $q_{01} = q_{10} = ab$, $q_{i,i+2} = q_{i+2,i} = ab$ for $i \geqslant 0$, $q_{ij} = 0$ otherwise. Denote $Q_n = (q_{ij}^{(n)})$. Then $q_{ij}^{(n)} = p_{ij}^{(2n)}$. Using induction, one can show that

$$q_{0,2k-1}^{(n)} = q_{0,2k}^{(n)}, \qquad k \geqslant 1; \qquad q_{0j}^{(n)} = 0, \qquad j \geqslant 2n+1. \tag{4.23}$$

(B) Let $Z = \{\ldots, -1, 0, 1, \ldots\}$, i.e., the set of all integers. Let $\hat{Q} = (\hat{q}_{ij})$ be the stochastic matrix on Z with $\hat{q}_{ij} = a^2 + b^2$, $\hat{q}_{i,i+2} = \hat{q}_{i+2,i} = ab$, $q_{ij} = 0$ otherwise. Denote $\hat{Q}^n = (\hat{q}_{ij}^{(n)})$. Using (4.23), one can show that

$$q_{0k}^{(n)} \geqslant \hat{q}_{0k}^{(n)}, \qquad k \geqslant 0. \tag{4.24}$$

(C) We show that

$$\sum_{n=1}^{\infty} \left[\hat{q}_{00}^{(n)}\right]^2 = \infty. \tag{4.25}$$

Let $\zeta_1, \zeta_2, \ldots, \zeta_n, \ldots$ be independent and identically distributed random variables assuming only values $0, 2, -2$ and satisfying $\mathbb{P}(\zeta_n = 0) = a^2 + b^2$, $\mathbb{P}(\zeta_n = -2) = \mathbb{P}(\zeta_n = 2) = ab$. Let $x_0 = y_0 = 0$; $x_n = \sum_{m=1}^{n} \zeta_{2m-1}$, $y_n = \sum_{m=1}^{n} \zeta_{2m}$ for $n \geqslant 1$. Then $(x_n, n \geqslant 0)$, $(y_n, n \geqslant 0)$ are independent $\hat{Q}$-chains, and $((x_n, y_n), n \geqslant 0)$ is a random walk on the two-dimensional integer lattice. Since $E(x_1^2 + y_1^2) < \infty$, $Ex_1 = Ey_1 = 0$, it can be proved that $((x_n, y_n), n \geqslant 0)$ is recurrent (see [5]). It follows that $\sum_{n=1}^{\infty} \mathbb{P}((x_n, y_n) = (0,0)) = \infty$ (note that $(x_0, y_0) = (0,0)$). But $\mathbb{P}((x_n, y_n) = (0,0)) = \mathbb{P}(x_n = 0)\mathbb{P}(y_n = 0) = [\hat{q}_{00}^{(n)}]^2$, proving (4.25).

(D) *The three-ray matrix P of* (4.22) *always has the c.p.*

Indeed, since $P_{00}^{(2n)} = q_{00}^{(n)}$, from (4.24) and (4.25) we have $\sum_{n=1}^{\infty} [P_{00}^{(n)}]^2 = \infty$. Now it follows from Corollary 1.1 that P has the c.p.

Is it true that every irreducible† null-recurrent three-ray with period 1 has the c.p.? We have not resolved this question. But we can prove that there are plenty of three-ray matrices that do not have the c.p.

Theorem 4.3. *Suppose that the three matrix P of* (3.1) *is nonrecurrent* (*i.e., satisfies* $\sum_{k=1}^{\infty} (q_1 \ldots q_k)/(p_1 \ldots p_k) < \infty$), *and there are only finitely many r_k not equal to* 0. *Then P does not have the c.p.*

Proof. If $r_k = 0$ for all k, then P has period 2, and so it does not have the c.p. We assume $k \geqslant 1$ satisfying $r_{k-1} > 0$, $r_i = 0$ for all $i \geqslant k$. Consider independent

† *Translator's note.* A three-ray matrix P in (3.1) is always irreducible, as the paper noted earlier, because $p_i > 0$, $q_i > 0$ for all i.

P-chains $(x_n, n \geqslant 0)$, $(y_n, n \geqslant 0)$ satisfying $\mathbb{P}(x_0 = k-1, y_0 = k+1) = 1$. Let

$$\Lambda_1 = \{x_0 = k-1, x_1 = k-1, x_2 = k, x_n \neq k-1 \text{ for all } n > 2\},$$
$$\Lambda_2 = \{y_0 = k+1, y_1 = k, y_2 = k+1, y_n \neq k-1 \text{ for all } n > 2\}.$$

Then

$$\begin{aligned}\mathbb{P}(\Lambda_1\Lambda_2) &= \mathbb{P}(\Lambda_1)\mathbb{P}(\Lambda_2)\\ &= r_{k-1}p_{k-1}\mathbb{P}(x_n \neq k-1 \text{ for all } n > 2 \mid x_2 = k)\\ &\quad \times q_{k+1}p_k\mathbb{P}(y_n \neq k-1 \text{ for all } n > 2 \mid y_2 = k+1)\\ &= r_{k-1}p_{k-1}(1 - f^*_{k,k-1})q_{k+1}p_k(1 - f^*_{k+1,k-1}).\end{aligned}$$

Since $f^*_{k,k-1}f^*_{k-1,k} \leqslant f^*_{kk} < 1$ (by assumption P is nonrecurrent) and if $f^*_{k-1,k} = 1$ (see Lemma 3.1), we have $f^*_{k,k-1} < 1$. Similarly, $f^*_{k+1,k-1} \leqslant f^*_{k+1,k-1}f^*_{k-1,k+1} \leqslant f^*_{k+1,k+1} < 1$. So $\mathbb{P}(\Lambda_1\Lambda_2) > 0$. Since $\mathbb{P}(x_{n+1} = x_n, x_n \geqslant k) = \sum_{j=k}^{\infty}\mathbb{P}(x_n = j)r_j = 0$, and similarly $\mathbb{P}(y_{n+1} = y_n, y_n \geqslant k) = 0$, we may assume that for $\omega \in \Lambda_1\Lambda_2$, $n \geqslant 2$, $|x_{n+1}(\omega) - x_n(\omega)| = |y_{n+1}(\omega) - y_n(\omega)| = 1$. Since $x_2 = k$, $y_2 = k+1$ on $\Lambda_1\Lambda_2$, we have $\mathbb{P}(x_n \neq y_n \text{ for all } n) \geqslant \mathbb{P}(\Lambda_1\Lambda_2) > 0$, proving that P does not have the c.p.

§5. Coincidence Property of Spatially Homogeneous Matrices

Denote $Z^m = \{(z_1, \ldots, z_m) : z_1, \ldots, z_m \text{ are integers}\}$ where $m \geqslant 1$; i.e., Z^m is the m-dimensional integer lattice.

Definition 5.1. A stochastic matrix $P = (p_{xy})$ on Z^m is said to be spatially homogeneous if for all x, y in Z^m, $p_{xy} = p_{0,y-x}$.

Definition 5.2. If P is a spatially homogeneous (stochastic) matrix on Z^m, any P-chain is called a random walk on Z^m.

As is well known, a sequence $(s_n, n \geqslant 0)$ of Z^m-valued random variables is a random walk iff $s_0, s_1 - s_0, s_2 - s_1, \ldots, s_n - s_{n-1}, \ldots$ are independent and $s_1 - s_0$, $s_2 - s_1, \ldots, s_n - s_{n-1}, \ldots$ are identically distributed. (The interested reader is referred to Chap. 2 of [4] or [5] for random walks.)

In this section we will obtain necessary and sufficient conditions for spatially homogeneous matrices to have the c.p.

Lemma 5.1. *Suppose that $P = (p_{xy})$ is a spatially homogeneous matrix on Z^m and $(x_n, n \geqslant 0)$, $(y_n, n \geqslant 0)$ are two independent P-chains. Let*

$$q_{xy} = \sum_{t \in Z^m} p_{xt}p_{yt}, \qquad x \in Z^m, \qquad y \in Z^m; \qquad Q = (q_{xy}); \tag{5.1}$$

and $s_n = x_n - y_n$. Then Q is a spatially homogeneous matrix on Z^m, and $(s_n, n \geqslant 0)$ is a Q-chain.

Proof. The spatial homogeneity of Q is easily verified as follows.

$$q_{0,y-x}=\sum_{y\in Z^m} p_{0t}p_{y-x,t}=\sum_{t\in Z^m} p_{x,x+t}p_{y,x+t}=\sum_{z\in Z^m} p_{xz}p_{yz}=q_{xy}.$$

Next we show that $(s_n,\ n\geqslant 0)$ is a Q-chain. Let $\xi_n=x_n-x_{n-1}$, $\eta_n=y_n-y_{n-1}$, and $\sigma_n=s_n-s_{n-1}$ for $n\geqslant 1$. Then, since ξ_n,η_n are independent and

$$\mathbb{P}(\xi_n=t)=\mathbb{P}(\eta_n=t)=p_{0t},$$

$$\mathbb{P}(\sigma_n=z)=\mathbb{P}(\xi_n-\eta_n=z)=\sum_{x\in Z^m}\mathbb{P}(\xi_n=x)\mathbb{P}(\eta_n=x-z)$$

$$=\sum_{x\in Z^m} p_{0x}p_{0,x-z}=\sum_{x\in Z^m} p_{0x}p_{zx}=q_{0z}$$

(which, incidentally, shows that Q is a stochastic matrix, since it is spatially homogeneous). Now, since $\sigma_n=s_n-s_{n-1}$ is independent of $s_0,s_1,\ldots,s_{n-1}$ for any $n\geqslant 1$ and $z_0,\ldots,z_{n-1},z_n$ in Z^m,

$$\mathbb{P}(s_0=z_0,\ldots,s_{n-1}=z_{n-1},s_n=z_n)$$
$$=\mathbb{P}(s_0=z_0,\ldots,s_{n-1}=z_{n-1})\mathbb{P}(\sigma_n=z_n-z_{n-1})$$
$$=\mathbb{P}(s_0=z_0,\ldots,s_{n-1}=z_{n-1})q_{0,z_n-z_{n-1}}=\mathbb{P}(s_0=z_0,\ldots,s_{n-1}=z_{n-1})q_{z_{n-1}z_n}.$$

This shows that $(s_n,\ n\geqslant 0)$ is a Q-chain.

We denote by 0 the zero $(0,0,\ldots,0)$ in Z^m, and use $f^*_{z0}(Q)$ to stand for the probability that a Q-chain starting at z will visit 0 (at some time $n\geqslant 1$).

Lemma 5.2. *In order for a spatially homogeneous matrix P on Z^m to have the c.p., it is necessary and sufficient that for all $z\in Z^m$*

$$f^*_{z0}(Q)=1,$$

where $Q=(q_{xy})$ is given by (5.1).

Proof. Let $(x_n,\ n\geqslant 0)$, $(y_n,\ n\geqslant 0)$ be independent P-chains with $\mathbb{P}(x_0=x)=\mathbb{P}(y_0=y)=1$ for some fixed x, y in Z^m. Let $s_n=x_n-y_n$. Then $\mathbb{P}(x_n=y_n$ for some $n\geqslant 1)=\mathbb{P}(s_n=0$ for some $n\geqslant 1)=f^*_{x-y,0}(Q)$ by Lemma 5.1. The lemma clearly follows.

Theorem 5.1. *In order for a spatially homogeneous matrix P to have the c.p., it is necessary and sufficient that Q defined in* (5.1) *be irreducible and recurrent.*

Proof. This is immediate from Lemma 5.2, since Q is irreducible and recurrent iff $f^*_{z0}(Q)=1$ for all z.

For a fixed spatially homogeneous matrix $P=(p_{xy})$ on Z^m, let

$$D(P)=\{x:x\in Z^m, p_{0x}>0\},$$

$$H_+(P)=\{x:x\in Z^m, p^{(n)}_{0x}>0 \text{ for some } n\geqslant 0\},$$

$$H(P)=\{x:x\in Z^m, x=y-z \text{ for some } y,z \text{ in } H_+(P)\}.$$

Clearly, $H_+(P)$ is closed with respect to addition, and $H(P)$ is a group with respect to addition.

Lemma 5.3. *Let P be a spatially homogeneous matrix on Z^m and $Q = (q_{xy})$ be as defined in* (5.1). *Then for Q to be irreducible, it is necessary and sufficient that $H(Q) = Z^m$.*

Proof. Necessity: Suppose Q is irreducible. Then in a Q-chain, 0 leads to any other state; so $H_+(Q) = Z^m$. Therefore, $H(Q) = Z^m$.

Sufficiency: Suppose $H(Z) = Z^m$. We first prove $H(Q) = H_+(Q)$. Since $D(Q) = \{z : q_{0z} > 0\}$, the following shows $z \in D(Q)$ iff $-z \in D(Q)$.

$$q_{0z} = \sum_{t \in Z^m} P_{0t}P_{zt} = \sum_{y \in Z^m} P_{-z,t-z}P_{0,t-z} = \sum_{y \in Z^m} P_{-z,y}P_{0y} = q_{0,-z}\,.$$

Now $z \in H_+(Q)$ iff $z = 0$ or there exist $z_1, \ldots, z_n$ in $D(Q)$ such that $z = z_1 + \cdots + z_n$. Hence $z \in H_+(Q)$ iff $-z \in H_+(Q)$. Since $H_+(Q)$ is closed with respect to addition, $H_+(Q) = H(Q)$. It follows that $H_+(Q) = Z^m$. Now let $x, y \in Z^m$ with $x \neq y$. Since $y - x \in H_+(Q)$ and $y - x \neq 0$, there exists $n \geqslant 1$ such that $q^{(n)}_{0,y-x} > 0$ and $q^{(n)}_{xy} > 0$, so that x leads to y. Therefore, Q is irreducible.

Theorem 5.2. *Let P be a spatially homogeneous matrix on Z^m. Then for P to have the c.p., it is necessary and sufficient that*

(i) $|J(u)| \neq 1$ *for all* $u \neq 2\pi v$, $v \in Z^m$;

(ii) $$\int_{\pi}^{\pi} \cdots \int_{\pi}^{\pi} \frac{1}{1 - |J(u)|^2}\, du = \infty. \tag{5.2}$$

Here $J(u) = \sum_{x \in Z^m} p_{0x}e^{i(u,x)}$, $u = (u_1, \ldots, u_m) \in \mathbb{R}^m$, where (u, x) denotes the inner product $u \cdot x$.

Proof. From Theorem 5.1 and Lemma 5.3, for P to have the c.p. it is necessary and sufficient that Q be recurrent and $H(Q) = Z^m$. By a theorem in Chap. 2 of [4], $H(Q) = Z^m$ iff $\tilde{J}(u) \neq 1$ for all $u \neq 2\pi v$, $v \in Z^m$, where $\tilde{J}(u) = \sum_{x \in Z^m} q_{0x}e^{i(u,x)}$. It is easy to verify that $\tilde{J}(u) = |J(u)|^2$. So $H(Q) = Z^m$ iff (i) holds. By the same theorem of [4], Q is recurrent iff

$$\lim_{a \to 1^-} \int_{\pi}^{\pi} \cdots \int_{\pi}^{\pi} \operatorname{Re} \frac{1}{1 - a\tilde{J}(u)}\, du = \infty, \tag{5.3}$$

where Re stands for "the real part of". Since $\tilde{J}(u) = |J(u)|^2$, (5.3) is equivalent to (5.2). So Q is recurrent iff (ii) holds. The proof is complete.

Corollary 5.1. *If $m \geqslant 3$, a spatially homogeneous matrix P on Z^m does not have the c.p.*

Proof. When $m \geqslant 3$, conditions (i) and (ii) of Theorem 5.2 cannot hold simultaneously. Indeed, if (i) holds then $h(Q) = Z^m$; since $m \geqslant 3$, this implies Q is nonrecurrent, and so (ii) fails.

In the following we obtain, for the cases $m = 1$ and 2, sufficient conditions for P to have the c.p. that are suitable for application.

Theorem 5.3. *Suppose that $P = (p_{xy})$ is a spatially homogeneous matrix on $Z = Z^1$ satisfying*

(i) $\text{g.c.d.}\{x : x = y - z \text{ for some } y, z \text{ in } Z \text{ with } p_{0y} > 0, p_{0z} > 0\} = 1,$ (5.4)

(ii) $\sum_{x \in Z} |x| p_{0x} < \infty.$ (5.5)

Then P has the c.p. (Here g.c.d. *stands for the greatest common divisor.)*

Proof. Still let $Q = (q_{xy})$ be as in (5.1). First we show that (5.4) implies Q is irreducible. Indeed, let A denote the set appearing in (5.4); then $x \in A$ iff $q_{0x} = \sum_{y \in Z} p_{0y} p_{xy} = \sum_{y \in Z} p_{0y} p_{0,x-y} > 0$, i.e., $A = D(Q)$. Thus $A \subset H(Q)$ and so $\text{g.c.d.} A \geqslant \text{g.c.d.} H(Q)$. From (5.4) we have $\text{g.c.d.} H(Q) = 1$. But $H(Q)$ is a subgroup of Z; so $H(Q) = Z$. By Lemma 5.3, Q is irreducible. Next we show that (5.5) implies Q is recurrent. Let $(x_n, n \geqslant 0)$, $(y_n, n \geqslant 0)$ be independent P-chains with $\mathbb{P}(x_0 = 0) = \mathbb{P}(y_0 = 0) = 1$. Let $s_n = x_n - y_n$. Then $(s_n, n \geqslant 0)$ is a Q-chain with $\mathbb{P}(s_0 = 0) = 1$. Since $E|s_1| \leqslant E|x_1| + E|y_1| = 2\sum_{x \in Z} |x| p_{0x} < \infty$, $Es_1 = Ex_1 - E_{y_1} = \sum_{x \in Z} x p_{0x} - \sum_{x \in Z} x p_{0x} = 0$, it follows from a criterion for the recurrence of random walks that Q is recurrent. Since Q is irreducible and recurrent, Theorem 5.1 implies that P has the c.p.

We remark that in Theorem 5.3 (5.4) is a necessary condition for P to have the c.p. This follows from Theorem 5.1, Lemma 5.3, and the fact that $H(Q)$ is generated by $A = D(Q)$.

Theorem 5.4. *Suppose that $P = (p_{xy})$ is a spatially homogeneous matrix on Z^2 satisfying*

(i) $p_{00} > 0$ *and* $H(P) = Z^2$,

(ii) $\sum_{x \in Z^2} |x|^2 p_{0x} < \infty.$

Then P has the c.p.

Proof.† Suppose that conditions (i) and (ii) hold. Let $J(u)$ be as defined in Theorem 5.2. We show that if $|J(u)| = 1$ then $u = 2\pi v$ for some $v \in Z^2$. Indeed, if $|J(u)| = 1$ then $J(u) = e^{i\alpha}$ for some real α and so $\sum_{x \in Z^2} p_{0x} e^{i[(u,x) - \alpha]} = 1$. This implies that $(u, x) - \alpha$ is an integral multiple of 2π for all x with $p_{0x} > 0$. Since $p_{00} > 0$, $\alpha = 2n\pi$ for some n. So (u, x) is an integral multiple of 2π if $p_{0x} > 0$. The condition $H(P) = Z^2$ now implies that (u, x) is an integral multiple of 2π for all $x \in Z^2$. From this it follows that $u = 2\pi v$ for some $v \in Z^2$. Therefore, condition (i) of Theorem 5.2 holds and, as seen in the proof of that theorem, Q is irreducible. Next we prove that Q is recurrent. Let $(x_n, n \geqslant 0)$, $(y_n, n \geqslant 0)$ be independent

† *Translator's note.* That Q is irreducible is more easily proved as follows. Since $p_{00} > 0$, $q_{0x} \geqslant p_{0x} p_{xx} = p_{0x} p_{00} > 0$ if $p_{0x} > 0$, it follows that $H(P) \subset H(Q)$. Since $H(P) = Z^2$, we have $H(Q) = Z^2$, so that Q is irreducible (by Lemma 5.3).

P-chains with $\mathbb{P}(x_0 = 0) = \mathbb{P}(y_0 = 0) = 1$ and let $s_n = x_n - y_n$. $(s_n, n \geqslant 0)$ is a Q-chain. Since Q is irreducible, $H(Q) = Z^2$. But $E|s_1|^2 \leqslant 2(E|x_1|^2 + E|y_1|^2) \leqslant 4\sum_{x \in Z^2}|x|^2 p_{0x} < \infty$ and $Es_1 = 0$. By a criterion for recurrence of random walks Q is then recurrent. Finally, from Theorem 5.1, P has the c.p.

Example. Let $P = (p_{xy})$ be the matrix on Z^2 satisfying $p_{xy} = 1/2$ when $y - x = (1,0)$ or $(0,1)$, $p_{xy} = 0$ otherwise. Then P does not have the c.p. Indeed, here $q_{0z} = \frac{1}{2}p_{z,(1,0)} + \frac{1}{2}p_{z,(0,1)}$ and it is easy to see that $H(Q) = \{k(1,-1) : k \in Z\} \neq Z^2$. Thus from Theorem 5.1 and Lemma 5.3, P does not have the c.p.

References

1. J. Lamperti and J. L. Snell, Martin boundaries for certain Markov chains. *J. Math. Soc. Japan* **15**(2) (1963), 113–128.
2. K. L. Chung, *Markov Chains with Stationary Transition Probabilities*. Springer-Verlag, Berlin, 1967.
3. J. L. Doob, *Stochastic Processes*. Wiley, New York, 1953.
4. I. I. Gihman and A. V. Skorohod, *The Theory of Stochastic Processes*, Vol. I (English translation). Springer-Verlag, Berlin, 1974.
5. F. Spitzer, *Principles of Random Walk*. Van Nostrand, Princeton, NJ, 1964.

BIB Matrices, Simple Codes and Orthogonal Codes*

0. Balanced Incomplete Block (BIB) Matrices

Explanation: The number of 1's in any row (or column) of a $(0,1)$-matrix is equal to the sum of entries in that row (or column). Hereafter, for simplicity, the number of 1's in a row (or column) is called a row sum (or column sum). Notice that if 0 and 1 are the elements of the field $\{0,1\}$, then $1+1=0$, and one cannot replace "the number of 1's" by "sum". We shall also call the inner product of two rows their intersection number. For a $(0,1)$-matrix $\mathbf{T}$, let $\mathbf{T}'$ be its transpose, then the diagonal elements of $\mathbf{TT}'$ are the row sums of $\mathbf{T}$ and each off-diagonal element is the intersection number of the two corresponding rows. These facts will be used frequently and will not be mentioned again.

Definition 0.1. A BIB matrix is a $(0,1)$-matrix satisfying the following conditions:

1. All the column sums are equal.
2. The intersection number of any two rows is a constant.

Corollary. *All the row sums of a BIB matrix are also equal.*

Proof. Let the number of rows, common column sum, and common intersection number of this matrix be v, k, and λ, respectively. Let the row sum of the first row be r. Then the first row has r 1's, below each of which there are $k-1$ 1's. The total number is $r(k-1)$, which of course is the sum of the intersection numbers of the first row and the other $v-1$ rows, i.e., $(v-1)\lambda$. Therefore, $r(k-1)=(v-1)\lambda$, and $r=(v-1)\lambda/(k-1)$. Since any row can be considered as the first row, all the row sums are equal to $(v-1)\lambda/(k-1)$.

The following two formulas summarize the properties of a BIB matrix $\mathbf{T}$.

$$\mathbf{TT}' = \begin{bmatrix} r & \lambda & \lambda & \cdots \\ \lambda & r & \lambda & \cdots \\ \lambda & \lambda & r & \cdots \\ \vdots & \vdots & \vdots & \ddots \end{bmatrix},$$

$$\mathbf{T1} = r\mathbf{1}, \qquad \mathbf{1}'\mathbf{T} = k\mathbf{1}',$$

*Translated by Ching-shui Cheng, University of California at Berkeley.

where $\mathbf{1}$ is the column vector with all entries equal to 1 and $\mathbf{1}'$ is a row vector.

There are five parameters in a BIB matrix:

the number of rows v,
the number of columns b,
the row sum r,
the column sum k,
the intersection number λ.

A BIB matrix with the above parameters will be called a BIB (v, b, r, k, λ) matrix. These five parameters are not independent and satisfy the following two constraints:

First constraint: $vr = bk$ (since the number of 1's in the whole matrix is equal to vr, and also equal to bk).
Second constraint: $r(k-1) = (v-1)\lambda$. (This has been obtained in the proof of the corollary.)

For the need of Section 1, we shall state two theorems:

Theorem 0.1. *Suppose a* $(0, 1)$*-matrix* $\mathbf{T}$ *satisfies the following conditions.*

(1°)

$$\mathbf{T}\mathbf{T}' = \begin{bmatrix} r & \lambda & \lambda & \cdots \\ \lambda & r & \lambda & \cdots \\ \lambda & \lambda & r & \cdots \\ \vdots & \vdots & \vdots & \ddots \end{bmatrix}, \qquad \textit{the number of rows} = m;$$

(2°) *each column sum of* $\mathbf{T}$ *is* k *or* $k-1$;
(3°) $r(k-1) = m\lambda$.

Then a BIB $(m+1, *, r, k, \lambda)$ *matrix can be obtained by adding a row to the bottom of* $\mathbf{T}$ *in the following manner. The new entry is* 1 *if the corresponding column sum in* $\mathbf{T}$ *is* $k-1$ *and is* 0 *otherwise.*

Proof. After the addition of such a row, all the column sums of the new matrix are equal to k. Therefore, it is enough to show that the intersection number of the new row and any old row is λ. Without loss of generality, we may just consider the intersection number of the first row and the new row. In the new matrix, there are $k-1$ 1's under each 1 in the first row. The total number of these 1's is $r(k-1)$ and of course is the sum of the intersection numbers of the first row and the other rows (including the new row). But the intersection number of the first row and any of the $(m-1)$ old rows is λ. Therefore, the intersection number of the first row and the new row is $r(k-1) - (m-1)\lambda$, which is λ according to (3°). Q.E.D.

Counterexample. The matrix

$$\mathbf{T} = \begin{bmatrix} 1 & 1 & 1 & 0 & 0 & 0 & 1 & 1 & 0 & 0 & 0 & 0 & 0 & 0 \\ 1 & 0 & 0 & 1 & 1 & 0 & 0 & 0 & 1 & 1 & 0 & 0 & 0 & 0 \\ 0 & 1 & 0 & 1 & 0 & 1 & 0 & 0 & 0 & 0 & 1 & 1 & 0 & 0 \\ 0 & 0 & 1 & 0 & 1 & 1 & 0 & 0 & 0 & 0 & 0 & 0 & 1 & 1 \end{bmatrix}$$

satisfies conditions (1°) and (2°) ($m = 4$, $r = 5$, $\lambda = 1$, $k = 2$), but does not satisfy (3°). The addition of a new row (0 0 0 0 0 0 1 1 1 1 1 1 1 1) does not result in a BIB matrix.

Theorem 0.2. *If* $\mathbf{T}_1, \ldots, \mathbf{T}_g$ *are* g *BIB* (v,b,r,k,λ) *matrices (which can be the same), then the superposition*

$$(\mathbf{T}_1, \ldots, \mathbf{T}_g)$$

is a BIB $(v, gb, gr, k, g\lambda)$ *matrix.*

This theorem is straightforward.

0′. Cyclic BIB Matrices

In this section, we shall introduce cyclic BIB (v,v,k,k,λ) matrices.
We shall use the following notation:

$$\mathbf{e}_i' = (0 \ldots 0\ 1\ 0 \ldots 0) \qquad (1 \times v) \qquad (i = 1, \ldots, v),$$

where the unique 1 is at the ith position. Let

$$\pi = \begin{bmatrix} e_2' \\ \vdots \\ e_v' \\ e_1' \end{bmatrix} \qquad (v \times v).$$

Then π is called a cyclic translation matrix. We have the following facts about π:

1°. For any matrix with v rows,

$$\mathbf{A} = \begin{bmatrix} \mathbf{a}_1' \\ \mathbf{a}_2' \\ \vdots \\ \mathbf{a}_v' \end{bmatrix},$$

one has

$$\pi\mathbf{A} = \begin{bmatrix} \mathbf{a}_2' \\ \vdots \\ \mathbf{a}_v' \\ \mathbf{a}_1' \end{bmatrix}.$$

In particular,

$$\boldsymbol{\pi}^2 = \boldsymbol{\pi}\boldsymbol{\pi} = \begin{bmatrix} \mathbf{e}_3' \\ \vdots \\ \mathbf{e}_v' \\ \mathbf{e}_1' \\ \mathbf{e}_2' \end{bmatrix},$$

$$\vdots$$

$$\boldsymbol{\pi}^i = \boldsymbol{\pi}\boldsymbol{\pi}^{i-1} = \begin{bmatrix} \mathbf{e}_{i+1}' \\ \vdots \\ \mathbf{e}_v' \\ \mathbf{e}_1' \\ \vdots \\ \mathbf{e}_i' \end{bmatrix},$$

$$\vdots$$

$$\boldsymbol{\pi}^v = \begin{bmatrix} \mathbf{e}_1' \\ \vdots \\ \mathbf{e}_v' \end{bmatrix} = \mathbf{I} = \boldsymbol{\pi}^0.$$

Therefore,

$$\boldsymbol{\pi}^0 + \boldsymbol{\pi}^1 + \cdots + \boldsymbol{\pi}^{v-1} = \mathbf{1}\mathbf{1}'. \tag{1}$$

2°. The matrix

$$c_0\boldsymbol{\pi}^0 + c_1\boldsymbol{\pi}^1 + \cdots + c_{v-1}\boldsymbol{\pi}^{v-1} \qquad (v \times v) \tag{2}$$

can be described as follows. Its first row is $(c_0, c_1, \ldots, c_{v-1})$, and the other rows are obtained from the first row by successive cyclic translations. For example, the second row is $(c_{v-1}, c_0, c_1, \ldots, c_{v-2})$, the third row is $(c_{v-2}, c_{v-1}, c_0, c_1, \ldots, c_{v-3})$, etc. Such a matrix is called a cyclic matrix. In particular, if $c_0, \ldots, c_{v-1}$ are 0 or 1, then (2) is a (0, 1)-matrix. In other words, the sum of any different powers of $\boldsymbol{\pi}$ (with exponents not exceeding v) is a cyclic (0, 1)-matrix. Notice that if matrix (2) is zero, then all the c_i's must be 0. Therefore, $\boldsymbol{\pi}^0, \boldsymbol{\pi}^1, \ldots, \boldsymbol{\pi}^{v-1}$ are linearly independent.

3°. $\boldsymbol{\pi}\boldsymbol{\pi}' = \mathbf{I}$, therefore $\boldsymbol{\pi}' = \boldsymbol{\pi}^{-1}$.

The above facts are obvious, so the proofs are omitted. In Sections 1 and 2, we shall see that simple codes and orthogonal codes can be constructed through BIB matrices. The corresponding code is easier to generate if we use a cyclic BIB matrix, because in this case we only need to generate or store one code word and obtain the other code words by translation. Since cyclic BIB matrices have this advantage in practical applications, we shall study their existence and construction.

Example.

$$\pi^0 + \pi^1 + \pi^3 = \begin{bmatrix} 1 & 1 & 0 & 1 & 0 & 0 & 0 \\ 0 & 1 & 1 & 0 & 1 & 0 & 0 \\ 0 & 0 & 1 & 1 & 0 & 1 & 0 \\ 0 & 0 & 0 & 1 & 1 & 0 & 1 \\ 1 & 0 & 0 & 0 & 1 & 1 & 0 \\ 0 & 1 & 0 & 0 & 0 & 1 & 1 \\ 1 & 0 & 1 & 0 & 0 & 0 & 1 \end{bmatrix}$$

is a cyclic BIB $(7, 7, 3, 3, 1)$ matrix.

Theorem 0′.1. *The cyclic* $(0, 1)$-*matrix*

$$\mathbf{T} = \pi^{d_1} + \cdots + \pi^{d_k} \qquad (v \times v) \tag{3}$$

is a BIB (v, v, k, k, λ) *matrix, where* $d_1, \ldots, d_k$ *belong to* $\{0, 1, \ldots, v-1\}$, *if and only if among the* $k(k-1)$ *differences* $d_i - d_j$ (*mod* v), $i \neq j$, *each of* $1, \ldots, v-1$ *appears exactly* λ *times.*

Proof. $\mathbf{T}$ is a BIB (v, v, k, k, λ) matrix

$$\Leftrightarrow \mathbf{TT}' = \begin{bmatrix} k & \lambda & \lambda & \cdots \\ \lambda & k & \lambda & \cdots \\ \lambda & \lambda & k & \cdots \\ \vdots & \vdots & \vdots & \ddots \end{bmatrix} = (k-\lambda)\mathbf{I} + \lambda \mathbf{1}\mathbf{1}'$$

$$= (k-\lambda)\mathbf{I} + \lambda(\pi^0 + \pi^1 + \cdots + \pi^{v-1})$$

$$= k\mathbf{I} + \lambda(\pi^1 + \cdots + \pi^{v-1}).$$

But

$$\mathbf{TT}' = \left(\sum_{i=1}^{k} \pi^{d_i}\right)\left(\sum_{i=1}^{k} \pi^{-d_i}\right) = k\pi^0 + \sum_{i \neq j} \pi^{d_i - d_j}.$$

Combine the terms in $\sum_{i \neq j} \pi^{d_i - d_j}$ mod v, and let

$$\sum_{i \neq j} \pi^{d_i - d_j} = \lambda_1 \pi^1 + \cdots + \lambda_{v-1} \pi^{v-1};$$

then $\mathbf{T}$ is a BIB (v, v, k, k, λ) matrix

$$\Leftrightarrow k\mathbf{I} + \lambda_1 \pi^1 + \cdots + \lambda_{v-1}\pi^{v-1} = k\mathbf{I} + \lambda(\pi^1 + \cdots + \pi^{v-1})$$

$$\Leftrightarrow \lambda_1 = \cdots = \lambda_{v-1} = \lambda \quad \text{(by linear independence)},$$

which is the property stated in the theorem. Q.E.D.

In view of Theorem 0′.1, we propose the following definition.

Definition 0′.1. If a subset $\{d_1, \ldots, d_k\}$ of $\{1, \ldots, v\}$ satisfies the properties in Theorem 0′.1, then it is called a difference set.

There are three parameters in a difference set: v, k, λ. They must satisfy the relation $k(k-1) = (v-1)\lambda$. This is because both sides are the total number of the differences $d_i - d_j$ $(i \neq j)$. This in fact is the second constraint of a BIB (v, v, k, k, λ) matrix. A difference set with parameters v, k, λ will be called a difference set (v, k, λ).

Theorem 0′.1 can be stated as follows. Whether there is a cyclic BIB (v, v, k, k, λ) matrix depends on whether there is a difference set (v, k, λ). A cyclic BIB (v, v, k, k, λ) matrix (3) generates a difference set $(v, k, \lambda)\{d_1, \ldots, d_k\}$; on the other hand, the latter generates the former, i.e., (3).

According to Theorem 0′.1, difference sets are indispensable tools for constructing cyclic BIB matrices. As to the existence and construction of difference sets, we shall only give the definition of a multiplier, state the extended multiplier theorem (without proof), and give two examples for illustration.

As usual, we shall not care about the ordering of the elements in a set. Any subset of $\{1, 2, \ldots, v-1\}$ (mod v) will be called a mod v set. For any mod v set D and integers t, s, the meanings of the symbols tD and $D + s$ are self-evident. The set $D + s$ is said to be equivalent to D. If D_1 and D_2 are two equivalent mod v sets, then we write $D_1 \sim D_2$. Of course, a mod v set equivalent to a difference set is also a difference set.

Definition 0′.2. If integer t $(\neq 1)$ and mod v set D satisfy the relation $tD \sim D$, then we say that t is a multiplier of D or D has t as a multiplier.

Notice that if v is a prime, then for any mod v set with t as a multiplier, there must be an equivalent mod v set D such that $tD = D$. This can be proved as follows. Suppose t is a multiplier of the mod v set D_1, i.e., $tD_1 = D_1 + s$ for some s. Then $t(D_1 + m) = tD_1 + tm = D_1 + s + tm = D_1 + m + s + (t-1)m$ for any m. Since v is a prime, there is a solution, say m_0, to $(t-1)m + s \equiv 0 \pmod{v}$. Then $t(D_1 + m_0) = D_1 + m_0$. Therefore, $D = D_1 + m_0$ satisfies $tD = D$. The proof is complete. Furthermore, suppose D has k elements, one of which is $d \neq 0$. Then td also belongs to D, hence $t^2 d$ is in D. Repeating this argument, we conclude that all the $t^j d$'s $(0 \leqslant j \leqslant k)$ belong to D. Since there are $k+1$ such elements, there must be repetitions, i.e., $t^{j_1} d = t^{j_2} d$ for some $0 \leqslant j_1 < j_2 \leqslant k$. Therefore $(t^{j_2 - j_1} - 1)t^{j_1} d \equiv 0 \pmod{v}$. If v is a prime, then $t^{j_2 - j_1} - 1 \equiv 0 \pmod{v}$, i.e., there is a power t^l $(1 \leqslant l \leqslant k)$ of t such that $t^l - 1 \equiv 0 \pmod{v}$.

Theorem 0′.2 (Extended multiplier theorem). *Suppose integers v, k, λ, l satisfy the following conditions.*

(1) *l is a divisor of $k - \lambda$.*
(2) $l > \lambda$.
(3) $(l, v) = 1$.
(4) *For all the distinct prime divisors $p_1, p_2, \ldots,$ of l, there exist $j_1, j_2, \ldots$ such that*

$$p_1^{j_1} \equiv p_2^{j_2} \equiv \cdots \pmod{v}.$$

Let t be the common value of $p_i^{j_i}$ (mod v) in (4); then any difference set (v,k,λ) has t as a multiplier.

The use of this theorem can be understood through the following examples.

Example 1. Let $v=41$, $k=16$, $\lambda=6$, and $l=10$. Then conditions (1), (2) and (3) are satisfied. The prime divisors of 10 are 2 and 5, and

$$2 \equiv 5^3 (\equiv 2) \qquad (\text{mod } 41).$$

Therefore, the common residue 2 is a multiplier of any difference set (41, 16, 6). Now v is a prime; from the last sentence before Theorem 0′.2, if there is a difference set (41, 16, 6), then at least one of $2^1, \ldots, 2^{16}$ must be $\equiv 1$ (mod 41). But these 16 residues are

$$2,4,8,16,32,23,5,10,20,40,39,37,33,25,9,18.$$

Therefore there is no difference set (41, 16, 6), nor does a cyclic BIB (41, 41, 16, 16, 6) exist.

Example 2. Let $v=23$, $k=11$, $\lambda=5$, and $l=6$. Then conditions (1), (2) and (3) are satisfied. The prime divisors of 6 are 2 and 3, and

$$2^5 \equiv 3^2 (\equiv 9) \qquad (\text{mod } 23).$$

The common residue is $t=9$. Now v is a prime. Using the result in the paragraph before Theorem 0′.2 we conclude that if there is a difference set (23, 11, 5), then there must be an equivalent D such that $9D=D$. If $d \neq 0$ belongs to D, then all of $9^0 d, \ldots, 9^{11} d$ must also be in D and there must be repetitions. Now $9^0, \ldots, 9^{11}$ (mod v) are

$$1,9,12,16,6,8,3,4,13,2,18,1.$$

There are indeed repetitions, so the possibility of existence cannot be ruled out. Since there is no repetition among the first 11 numbers, up to equivalence, a difference set (23, 11, 5) must be of the form

$$\{d,9d,12d,16d,6d,8d,3d,4d,13d,2d,18d\}.$$

Let $d=1$; after rearranging the order, we obtain the set

$$\{1,2,3,4,6,8,9,12,13,16,18\}.$$

When we choose other values of d, we do not have to choose any element already in the above set, since otherwise the same set would be generated. Let $d=5$; after rearrangement we obtain

$$\{5,7,10,11,14,15,17,19,20,21,22\}.$$

Then all the nonzero residues are exhausted. Therefore, the above two sets are the only possible nonequivalent difference sets (23, 11, 5). Whether they are difference sets needs to be verified. Since one can be obtained by subtracting the other from 23, either both are difference sets or both are not. So it is enough to verify the first one. By direct verification we conclude that it is indeed a difference set.

Remark. From the complexity viewpoint, the problem of verification may be important and needs to be explored further.

A. Simple Codes

Suppose that the transmitter and the receiver agree to make use of an $n \times \omega(1, -1)$-matrix $\mathbf{M}$ such that the transmitter will send only rows of $\mathbf{M}$ which of course has no repetitive rows. Then $\mathbf{M}$ will be called a code.

If the receiver receives a wrong row, can he discover it? There are two possibilities.

First, if the wrong row received does not belong to $\mathbf{M}$, then the receiver will discover the mistake immediately.

Second, if the wrong row received is in $\mathbf{M}$, then the receiver will not be able to discover the mistake. Therefore, a criterion for choosing $\mathbf{M}$ is to make the occurrence of this second possibility as rare as possible. An occurrence of the second possibility means that the original row and the wrong row both belong to $\mathbf{M}$ but are different at $e \geqslant 1$ positions. e is called the magnitude of error.

We have

the inner product of the ith row and the jth row $(i \neq j)$ of $\mathbf{M}$

$= (\mathbf{MM}')_{ij}$

= the number of positions in which the two rows have the same sign minus the number of positions in which the two rows have opposite signs

$= \omega - 2$ (the number of positions in which the two rows have different signs).

Therefore,

the number of positions in which the ith row and the jth row have opposite signs

$$= \frac{\omega - (\mathbf{MM}')_{ij}}{2}.$$

Thus

the minimum number of positions in which two rows of $\mathbf{M}$ have opposite signs

$$= \tfrac{1}{2}(\omega - \max_{i \neq j}(\mathbf{MM}')_{ij}). \tag{1}$$

If the minimum number of positions in which two rows of $\mathbf{M}$ have opposite signs is $\geqslant d$ and the magnitude of error is $< d$, then the wrong row received is not in $\mathbf{M}$ and the mistake will be discovered by the receiver. In other words, a code with the minimum number of positions in which two rows have opposite signs $\geqslant d$ can detect all the errors with magnitude less than d. Thus the minimum number of positions in which two rows of $\mathbf{M}$ have opposite signs should be as large as possible. In view of (1), this is the same as saying that $\max_{i \neq j}(\mathbf{MM}')_{ij}$ should be as small as possible. This is the background for simple codes discussed in this section.

Theorem A.1.

$$\max_{i\neq j}(\mathbf{MM}')_{ij} \geqslant \frac{1}{n(n-1)}\sum_{i\neq j}(\mathbf{MM}')_{ij}\begin{cases} \geqslant -\dfrac{\omega}{n-1} & \textit{when } n \textit{ is even,} \\ \geqslant -\dfrac{\omega}{n} & \textit{when } n \textit{ is odd.}\end{cases}$$

Proof. The first inequality is obvious. We have

$$\frac{1}{n(n-1)}\sum_{i\neq j}(\mathbf{MM}')_{ij} = \frac{\mathbf{1}'\mathbf{MM}'\mathbf{1} - \mathrm{tr}\mathbf{MM}'}{n(n-1)} = \frac{\mathbf{1}'\mathbf{MM}'\mathbf{1}}{n(n-1)} - \frac{\omega}{n-1}.$$

Since $\mathbf{1}'\mathbf{MM}'\mathbf{1} \geqslant 0$, when n is even, the second inequality is true in any case. When n is odd, $\mathbf{1}'\mathbf{M}$ has ω odd integers each of which has a square $\geqslant 1$. Therefore, their sum of squares is $\geqslant \omega$, i.e., $\mathbf{1}'\mathbf{MM}'\mathbf{1} \geqslant \omega$. This shows the second inequality where n is odd. Q.E.D.

Notice that equality holds in the second inequality only if: When n is even, all the column sums of $\mathbf{M}$ are zero, i.e., each column of $\mathbf{M}$ has exactly $n/2$ 1's; when n is odd, all the column sums of $\mathbf{M}$ are ± 1, i.e., each column of $\mathbf{M}$ has exactly $(n+1)/2$ 1's or exactly $(n-1)/2$ 1's.

Definition A.1. If

$$\max_{i\neq j}(\mathbf{MM}')_{ij} = \begin{cases} -\dfrac{\omega}{n-1} & \text{when } n \text{ is even,} \\ -\dfrac{\omega}{n} & \text{when } n \text{ is odd,}\end{cases}$$

then $\mathbf{M}$ is called a simple code.

Theorem A.2. $\mathbf{M}$ *is a simple code if and only if*

$$(\alpha) \qquad \mathbf{MM}' = \begin{bmatrix} \omega & -\dfrac{\omega}{n-1} & -\dfrac{\omega}{n-1} & \cdots \\ -\dfrac{\omega}{n-1} & \omega & -\dfrac{\omega}{n-1} & \cdots \\ -\dfrac{\omega}{n-1} & -\dfrac{\omega}{n-1} & \omega & \cdots \\ \vdots & \vdots & \vdots & \ddots \end{bmatrix} \qquad \textit{when } n \textit{ is even;}$$

$$\mathbf{MM}' = \begin{bmatrix} \omega & -\dfrac{\omega}{n} & -\dfrac{\omega}{n} & \cdots \\ -\dfrac{\omega}{n} & \omega & -\dfrac{\omega}{n} & \cdots \\ -\dfrac{\omega}{n} & -\dfrac{\omega}{n} & \omega & \cdots \\ \vdots & \vdots & \vdots & \ddots \end{bmatrix} \qquad \textit{when } n \textit{ is odd.}$$

(β) *When n is even, each column of* $\mathbf{M}$ *has exactly $n/2$ 1's, and when n is odd, each column of* $\mathbf{M}$ *has exactly $(n+1)/2$ 1's or exactly $(n-1)/2$ 1's.*

Proof. Since condition (α) alone is enough to guarantee that **M** is a simple code, sufficiency is not a problem. Now we shall prove the necessity. If **M** is a simple code, then, by definition, the first and third terms in Theorem A.1 are equal. Therefore, the middle term is also equal to the first term. This obviously implies that all the entries $(\mathbf{MM}')_{ij}$ are equal to the first term. Thus (α) holds. The necessity of (β) has been demonstrated in the remark before Definition A.1.

Now we shall investigate the existence and construction of simple codes. The simplicity of a code is invariant under the following transformations:

1. row permutation;
2. column permutation;
3. changing signs of all the entries in a column.

By using the third transformation, we may assume that the first row of **M** is $\mathbf{1}'$:

$$\mathbf{M} = \begin{matrix} & \omega \\ 1 & \\ n-1 & \end{matrix} \begin{pmatrix} \mathbf{1}' \\ \mathbf{F} \end{pmatrix}.$$

Thus

$$\mathbf{MM}' = \begin{pmatrix} \omega & \mathbf{1}'\mathbf{F}' \\ \mathbf{F1} & \mathbf{FF}' \end{pmatrix}.$$

From Theorem A.2, we obtain:
when n is even,

$$\left.\begin{array}{l} \mathbf{F1} = -\dfrac{\omega}{n-1}\mathbf{1}, \qquad \mathbf{FF}' = \begin{bmatrix} \omega & -\dfrac{\omega}{n-1} & \cdots \\ -\dfrac{\omega}{n-1} & \omega & \cdots \\ \vdots & \vdots & \ddots \end{bmatrix} \\ \text{and each column of } \mathbf{F} \text{ has exactly } n/2 - 1 \text{ 1's;} \end{array}\right\} \quad (2\text{ even})$$

when n is odd,

$$\left.\begin{array}{l} \mathbf{F1} = -\dfrac{\omega}{n}\mathbf{1}, \qquad \mathbf{FF}' = \begin{bmatrix} \omega & -\dfrac{\omega}{n} & \cdots \\ -\dfrac{\omega}{n} & \omega & \cdots \\ \vdots & \vdots & \ddots \end{bmatrix} \\ \text{and each column of } \mathbf{F} \text{ has exactly } (n-1)/2 \text{ 1's,} \\ \text{or exactly } (n-3)/2 \text{ 1's.} \end{array}\right\} \quad (2\text{ odd})$$

Our fundamental method is to consider the matrix $(\mathbf{F} + \mathbf{11}')/2$ obtained by changing the -1's of **F** to 0's while keeping the 1's unchanged.

From (2 even) and (2 odd), we obtain:

when n is even,

$$\left.\begin{aligned}\left(\frac{\mathbf{F}+\mathbf{11}'}{2}\right)\left(\frac{\mathbf{F}'+\mathbf{11}'}{2}\right) &= \tfrac{1}{4}\left(\mathbf{FF}' - \frac{2\omega}{n-1}\mathbf{11}' + \omega\mathbf{11}'\right)\\ &= \frac{1}{4}\begin{bmatrix} 2\omega - \dfrac{2\omega}{n-1} & \omega - \dfrac{3\omega}{n-1} & \cdots \\ \omega - \dfrac{3\omega}{n-1} & 2\omega - \dfrac{2\omega}{n-1} & \cdots \\ \vdots & \vdots & \ddots \end{bmatrix}\\ &\text{and all the column sums of } (\mathbf{F}+\mathbf{11}')/2 \text{ are } n/2-1;\end{aligned}\right\} \quad (3\text{ even})$$

when n is odd,

$$\left.\begin{aligned}\left(\frac{\mathbf{F}+\mathbf{11}'}{2}\right)\left(\frac{\mathbf{F}'+\mathbf{11}'}{2}\right) &= \tfrac{1}{4}\left(\mathbf{FF}' - \frac{2\omega}{n}\mathbf{11}' + \omega\mathbf{11}'\right)\\ &= \frac{1}{4}\begin{bmatrix} 2\omega - \dfrac{2\omega}{n} & \omega - \dfrac{3\omega}{n} & \cdots \\ \omega - \dfrac{3\omega}{n} & 2\omega - \dfrac{2\omega}{n} & \cdots \\ \vdots & \vdots & \ddots \end{bmatrix}\\ &\text{and all the column sums of } (\mathbf{F}+\mathbf{11}')/2 \text{ are } (n-1)/2 \text{ or } (n-3)/2.\end{aligned}\right\} \quad (3\text{ odd})$$

Now we shall divide the discussion into four cases.

(I) $n = 4t$.
From (3 even) we obtain

$$\left.\begin{aligned}\left(\frac{\mathbf{F}+\mathbf{11}'}{2}\right)\left(\frac{\mathbf{F}'+\mathbf{11}'}{2}\right) &= \begin{bmatrix} \dfrac{2t-1}{4t-1}\omega & \dfrac{t-1}{4t-1}\omega & \cdots \\ \dfrac{t-1}{4t-1}\omega & \dfrac{2t-1}{4t-1}\omega & \cdots \\ \vdots & \vdots & \ddots \end{bmatrix}\\ &\text{and all the column sums of } (\mathbf{F}+\mathbf{11}')/2 \text{ are } 2t-1.\end{aligned}\right\}$$

Since $\omega(2t-1)/(4t-1)$ is an integer and $2t-1$ is relatively prime to $4t-1$, ω is a multiple of $4t-1$.

Let $\omega = g(4t-1)$. Then

$$\left.\begin{aligned}\left(\frac{\mathbf{F}+\mathbf{11}'}{2}\right)\left(\frac{\mathbf{F}'+\mathbf{11}'}{2}\right) &= \begin{bmatrix} g(2t-1) & g(t-1) & \cdots \\ g(t-1) & g(2t-1) & \cdots \\ \vdots & \vdots & \ddots \end{bmatrix}\\ &\text{and all the column sums of } \frac{\mathbf{F}+\mathbf{11}'}{2} \text{ are } 2t-1.\end{aligned}\right\}$$

Therefore, $(\mathbf{F}+\mathbf{11}')/2$ is a BIB $(4t-1,\ g(4t-1),\ g(2t-1), 2t-1,\ g(t-1))$ matrix.

Conversely, if the above BIB matrix exists, then a simple code can be obtained by changing the 0's of this matrix to -1's, and then adding a row of 1's on top of it.

(II) $n = 4t+2 \quad (t \geqslant 1)$.
By (3 even),

$$\left.\begin{array}{l}\left(\dfrac{\mathbf{F}+\mathbf{11}'}{2}\right)\left(\dfrac{\mathbf{F}'+\mathbf{11}'}{2}\right) = \begin{bmatrix} \dfrac{2t\omega}{4t+1} & \dfrac{(2t-1)\omega}{2(4t+1)} & \cdots \\ \dfrac{(2t-1)\omega}{2(4t+1)} & \dfrac{2t\omega}{4t+1} & \cdots \\ \vdots & \vdots & \ddots \end{bmatrix} \\ \text{and all the column sums of } (\mathbf{F}+\mathbf{11}')/2 \text{ are equal to } 2t. \end{array}\right\}$$

Since ω is a multiple of $4t+1$ (by examination of the diagonal elements) and is even (by examination of the off-diagonal elements), it must be a multiple of $2(4t+1)$.

Let $\omega = 2g(4t+1)$. Then,

$$\left.\begin{array}{l}\left(\dfrac{\mathbf{F}+\mathbf{11}'}{2}\right)\left(\dfrac{\mathbf{F}'+\mathbf{11}'}{2}\right) = \begin{bmatrix} 4gt & g(2t-1) & \cdots \\ g(2t-1) & 4gt & \cdots \\ \vdots & \vdots & \ddots \end{bmatrix} \\ \text{and all the column sums of } (\mathbf{F}+\mathbf{11}')/2 \text{ are equal to } 2t. \end{array}\right\}$$

Therefore, $(\mathbf{F}+\mathbf{11}')/2$ is a BIB $(4t+1,\ 2g(4t+1),\ 4gt,\ 2t,\ g(2t-1))$ matrix.

Conversely, if the above BIB matrix exists, then a simple code can be obtained by changing the 0's of this matrix to -1's and then adding a row of 1's on top of it.

(III) $n = 4t+1 \quad (t \geqslant 1)$.
By (3 odd),

$$\left.\begin{array}{l}\left(\dfrac{\mathbf{F}+\mathbf{11}'}{2}\right)\left(\dfrac{\mathbf{F}'+\mathbf{11}'}{2}\right) = \begin{bmatrix} \dfrac{2t\omega}{4t+1} & \dfrac{(2t-1)\omega}{2(4t+1)} & \cdots \\ \dfrac{(2t-1)\omega}{2(4t+1)} & \dfrac{2t\omega}{4t+1} & \cdots \\ \vdots & \vdots & \ddots \end{bmatrix} \\ \text{and all the column sums of } (\mathbf{F}+\mathbf{11}')/2 \text{ are } 2t \text{ or } 2t-1. \end{array}\right\}$$

Similar to (II), $\omega = 2g(4t+1)$, and therefore

$$\left(\frac{\mathbf{F}+\mathbf{11}'}{2}\right)\left(\frac{\mathbf{F}'+\mathbf{11}'}{2}\right) = \begin{bmatrix} 4gt & g(2t-1) & \cdots \\ g(2t-1) & 4gt & \cdots \\ \vdots & \vdots & \ddots \end{bmatrix}.$$

Since $(\mathbf{F}+\mathbf{11}')/2$ satisfies all the conditions in Theorem 0.1 (with $m = 4t$, $r = 4gt$, $\lambda = g(2t-1)$, $k = 2t$), a BIB $(4t+1, 2g(4t+1), 4gt, 2t, g(2t-1))$ matrix can be obtained by adding a row to $(\mathbf{F}+\mathbf{11}')/2$ as described in the theorem. Conversely, if such a BIB matrix exists, then a simple code can be obtained by changing all the 0's to -1's.

(IV) $n = 4t-1$.
By (3 odd),

$$\left.\begin{aligned} &\left(\frac{\mathbf{F}+\mathbf{11}'}{2}\right)\left(\frac{\mathbf{F}'+\mathbf{11}'}{2}\right) = \begin{bmatrix} \frac{(2t-1)\omega}{4t-1} & \frac{(t-1)\omega}{4t-1} & \cdots \\ \frac{(t-1)\omega}{4t-1} & \frac{(2t-1)\omega}{4t-1} & \cdots \\ \vdots & \vdots & \ddots \end{bmatrix} \\ &\text{and all the column sums of } (\mathbf{F}+\mathbf{11}')/2 \text{ are } 2t-1 \text{ or } 2t-2. \end{aligned}\right\}$$

Similar to (I), $\omega = g(4t-1)$, and therefore

$$\left(\frac{\mathbf{F}+\mathbf{11}'}{2}\right)\left(\frac{\mathbf{F}'+\mathbf{11}'}{2}\right) = \begin{bmatrix} g(2t-1) & g(t-1) & \cdots \\ g(t-1) & g(2t-1) & \cdots \\ \vdots & \vdots & \ddots \end{bmatrix}.$$

Since $(\mathbf{F}+\mathbf{11}')/2$ satisfies all the conditions in Theorem 0.1 (with $m = 4t-2$, $r = g(2t-1)$, $\lambda = g(t-1)$, $k = 2t-1$), a BIB $(4t-1, g(4t-1), g(2t-1), 2t-1, g(t-1))$ matrix can be obtained by adding a row to $(\mathbf{F}+\mathbf{11}')/2$ as described in the theorem. Conversely, if such a BIB matrix exists, then a simple code can be obtained by changing all the 0's to -1's.

We now summarize the above results in the following theorem.

Theorem A.3.

(i) *There are only four possibilities for the dimension of a simple code* $\mathbf{M}$:

(I) $4t \times g(4t-1)$,
(II) $(4t+2) \times 2g(4t+1)$,
(III) $(4t+1) \times 2g(4t+1)$,
(IV) $(4t-1) \times g(4t-1)$.

(ii) *Whether a simple code of type* (I) *or* (IV) *exists depends on whether a BIB* $(4t-1, g(4t-1), g(2t-1), 2t-1, g(t-1))$ *matrix exists. Let* $\mathbf{M}$ *be transformed*

into the form $\binom{\mathbf{1}'}{\mathbf{F}}$ *and all the* -1*'s of* **F** *changed to* 0*'s. If* **M** *is of type* (I), *then the transformed* **F** *is a BIB* $(4t-1, g(4t-1), g(2t-1), 2t-1, g(t-1))$ *matrix. If* **M** *is of type* (IV), *then a BIB* $(4t-1, g(4t-1), g(2t-1), 2t-1, g(t-1))$ *matrix can be obtained by adding a row to the transformed* **F** *according to the rule of Theorem* 0.1. *Conversely, if a BIB* $(4t-1, g(4t-1), g(2t-1), 2t-1, g(t-1))$ *matrix exists, then changing all the* 0*'s in this matrix to* -1*'s results in a simple code of type* (IV); *supplementing one more row of* 1*'s to this simple code then gives a simple code of type* (I).

A simple code of type (II) *or* (III) *exists if and only if a BIB* $(4t+1, 2g(4t+1), 4gt, 2t, g(2t-1))$ *matrix exists. Let* **M** *be transformed into the form* $\binom{\mathbf{1}'}{\mathbf{F}}$ *and all the* -1*'s of* **F** *changed to* 0*'s. If* **M** *is of type* (II), *then the transformed* **F** *is a BIB* $(4t+1, 2g(4t+1), 4gt, 2t, g(2t-1))$ *matrix. If* **M** *is a type* (III), *then a BIB* $(4t+1, 2g(4t+1), 4gt, 2t, g(2t-1))$ *matrix can be obtained by adding a row to the transformed* **F** *according to the rule of Theorem* 0.1. *Conversely, if such a BIB* $(4t+1, 2g(4t+1), 4gt, 2t, g(2t-1))$ *matrix exists, then a simple code of type* (III) *can be obtained by changing the* 0*'s in the BIB matrix to* -1*'s; supplementing one more row of* 1*'s to this simple code then gives a simple code of type* (II).

Remarks.

(1) We only gave proofs on constructing BIB matrices from simple codes. The construction of simple codes from BIB matrices is left to the readers.

(2) From Definition A.1 and the numbers of rows and columns of a simple code, we see immediately that the inner product of a simple code of type (I) or (IV) is $-g$, and the inner product of a simple code of type (II) or (III) is $-2g$. Therefore (see (1)), a simple code of type (I) or (IV) can detect all the errors with magnitude not equal to $2gt$, and a simple code of type (II) or (III) can detect all the errors with magnitude not equal to $2g(2t+1)$.

(3) The reason we presented Theorem 0.2 in Section 0 is that the parameters of all the BIB matrices in this section are of the form $(*, g*, g*, *, g*)$. Note that Theorem 0.2 states that if there is such a BIB matrix with $g = 1$, then a BIB matrix with an arbitrary g also exists.

(4) The number of positions in which two rows have opposite signs is usually called the Hamming distance of the two rows. The three conditions of a distance are satisfied.

B. Orthogonal Codes

Definition B.1. If an $n \times \omega$ code **M** is an orthogonal matrix (i.e., $\mathbf{MM}' = \omega\mathbf{I}$), then **M** is called an orthogonal code.

The orthogonality of a code is invariant under the following transformations: row permutation, column permutation, changing signs of all the entries in a row, changing signs of all the entries in a column.

Using these transformations, we may assume that

$$\mathbf{M} = \begin{array}{c} \\ 1 \\ n-1 \end{array} \!\!\begin{array}{c} \begin{array}{cc} 1 & \omega - 1 \end{array} \\ \begin{pmatrix} 1 & \mathbf{1}' \\ \mathbf{1} & \mathbf{F} \end{pmatrix} \end{array}. \tag{1}$$

(1) is called the standard form of **M**. The orthogonality of **M** is equivalent to

$$\mathbf{F1} = -\mathbf{1}, \qquad \mathbf{FF}' = \begin{bmatrix} \omega - 1 & -1 & -1 & \cdots \\ -1 & \omega - 1 & -1 & \cdots \\ -1 & -1 & \omega - 1 & \cdots \\ \vdots & \vdots & \vdots & \ddots \end{bmatrix}. \tag{2}$$

(2) is also equivalent to

$$\left(\frac{\mathbf{F} + \mathbf{11}'}{2} \right)\left(\frac{\mathbf{F}' + \mathbf{11}'}{2} \right) = \frac{1}{4} \begin{bmatrix} 2\omega - 4 & \omega - 4 & \cdots \\ \omega - 4 & 2\omega - 4 & \cdots \\ \vdots & \vdots & \ddots \end{bmatrix}. \tag{3}$$

If $n > 2$, then (3) indeed has an off-diagonal element $\omega - 4$. Therefore, $\omega = 4t$. Thus

$$\left(\frac{\mathbf{F} + \mathbf{11}'}{2} \right)\left(\frac{\mathbf{F}' + \mathbf{11}'}{2} \right) = \begin{bmatrix} 2t - 1 & t - 1 & \cdots \\ t - 1 & 2t - 1 & \cdots \\ \vdots & \vdots & \ddots \end{bmatrix}. \tag{4}$$

Of course, n is always $\leqslant \omega$. When $n = \omega$, $(1/\sqrt{\omega}\,)\mathbf{M}$ is a standardized orthogonal matrix, thus any two columns of **M** are also orthogonal to each other. In this case, by (1), all the column sums of **F** are equal to -1, i.e., each column of **F** has exactly $\omega/2 - 1$ 1's. In other words, all the column sums of $(\mathbf{F} + \mathbf{11}')/2$ are equal to $\omega/2 - 1$. Combining this fact with (4), we conclude that $(\mathbf{F} + \mathbf{11}')/2$ is a BIB $(4t - 1, 4t - 1, 2t - 1, 2t - 1, t - 1)$ matrix.

We summarize the above result as:

Theorem B.1. *For any $n \times \omega$ orthogonal code* **M**,

(i) *except for $n = 1, 2$, $n \leqslant \omega = 4t$;*
(ii) *under the standard form*

$$\mathbf{M} = \begin{array}{c} \\ 1 \\ n-1 \end{array} \!\!\begin{array}{c} \begin{array}{cc} 1 & 4t - 1 \end{array} \\ \begin{pmatrix} 1 & \mathbf{1}' \\ \mathbf{1} & \mathbf{F} \end{pmatrix} \end{array},$$

the orthogonality of **M** *is equivalent to*

$$\left(\frac{\mathbf{F} + \mathbf{11}'}{2} \right)\left(\frac{\mathbf{F}' + \mathbf{11}'}{2} \right) = \begin{bmatrix} 2t - 1 & t - 1 & \cdots \\ t - 1 & 2t - 1 & \cdots \\ \vdots & \vdots & \ddots \end{bmatrix}.$$

If $n = 4t$, then the orthogonality is equivalent to $(\mathbf{F} + \mathbf{11}')/2$ *being a BIB* $(4t - 1, 4t - 1, 2t - 1, 2t - 1, t - 1)$ *matrix.*

Remarks.

(1) The number of positions in which any two rows of an $n \times 4t$ orthogonal code have opposite signs is $2t$, therefore it can detect all the errors with magnitude not equal to $2t$.

(2) A $4t \times 4t$ orthogonal code is usually called a Hadamard matrix in literature.

(3) Not all of the $n \times 4t$ orthogonal codes with $n < 4t$ can be extended to a $4t \times 4t$ orthogonal code. The following are two counterexamples.

Example 1.

$$\frac{\mathbf{F}+\mathbf{1}\mathbf{1}'}{2} = \begin{bmatrix} 1 & 1 & 1 & 1 & 1 & 0 & 0 & 0 & 0 & 0 & 0 \\ 1 & 1 & 0 & 0 & 0 & 1 & 1 & 1 & 0 & 0 & 0 \\ 1 & 1 & 0 & 0 & 0 & 0 & 0 & 0 & 1 & 1 & 1 \end{bmatrix},$$

$$\mathbf{F} = \begin{bmatrix} 1 & 1 & 1 & 1 & 1 & -1 & -1 & -1 & -1 & -1 & -1 \\ 1 & 1 & -1 & -1 & -1 & 1 & 1 & 1 & -1 & -1 & -1 \\ 1 & 1 & -1 & -1 & -1 & -1 & -1 & -1 & 1 & 1 & 1 \end{bmatrix},$$

$$\mathbf{M} = \begin{bmatrix} 1 & 1 & 1 & 1 & 1 & 1 & 1 & 1 & 1 & 1 & 1 & 1 \\ 1 & 1 & 1 & 1 & 1 & 1 & -1 & -1 & -1 & -1 & -1 & -1 \\ 1 & 1 & 1 & -1 & -1 & -1 & 1 & 1 & 1 & -1 & -1 & -1 \\ 1 & 1 & 1 & -1 & -1 & -1 & -1 & -1 & -1 & 1 & 1 & 1 \end{bmatrix}.$$

Example 2.

$$\frac{\mathbf{F}+\mathbf{1}\mathbf{1}'}{2} = \begin{bmatrix} 1 & 1 & 1 & 1 & 1 & 1 & 1 & 1 & 1 & 0 & 0 & 0 & 0 & 0 & 0 & 0 & 0 & 0 & 0 \\ 1 & 1 & 1 & 0 & 1 & 0 & 0 & 0 & 0 & 1 & 1 & 1 & 1 & 1 & 0 & 0 & 0 & 0 & 0 \\ 1 & 1 & 1 & 0 & 0 & 1 & 0 & 0 & 0 & 1 & 0 & 0 & 0 & 0 & 1 & 1 & 1 & 1 & 0 \\ 1 & 1 & 0 & 1 & 0 & 0 & 1 & 0 & 0 & 0 & 1 & 1 & 0 & 0 & 1 & 1 & 0 & 0 & 1 \\ 1 & 1 & 0 & 1 & 0 & 0 & 0 & 1 & 0 & 0 & 0 & 0 & 1 & 1 & 0 & 0 & 1 & 1 & 1 \\ 1 & 0 & 0 & 0 & 1 & 1 & 0 & 0 & 1 & 0 & 1 & 0 & 1 & 0 & 1 & 0 & 1 & 0 & 1 \\ 1 & 0 & 0 & 0 & 1 & 0 & 1 & 0 & 1 & 1 & 0 & 0 & 0 & 1 & 0 & 1 & 0 & 1 & 1 \\ 0 & 1 & 0 & 0 & 0 & 1 & 0 & 1 & 1 & 1 & 1 & 1 & 0 & 0 & 0 & 0 & 0 & 1 & 1 \\ 0 & 1 & 0 & 0 & 1 & 0 & 0 & 1 & 1 & 0 & 0 & 1 & 0 & 1 & 1 & 1 & 1 & 0 & 0 \\ 0 & 0 & 1 & 1 & 1 & 1 & 0 & 0 & 0 & 0 & 0 & 1 & 0 & 1 & 1 & 0 & 0 & 1 & 1 \\ 0 & 0 & 1 & 1 & 1 & 0 & 0 & 1 & 0 & 1 & 1 & 0 & 0 & 0 & 0 & 1 & 1 & 0 & 1 \end{bmatrix},$$

$$\mathbf{F} = \begin{bmatrix} 1 & 1 & 1 & 1 & 1 & 1 & 1 & 1 & 1 & -1 & -1 & -1 & -1 & -1 & -1 & -1 & -1 & -1 & -1 \\ 1 & 1 & 1 & -1 & 1 & -1 & -1 & -1 & -1 & 1 & 1 & 1 & 1 & 1 & -1 & -1 & -1 & -1 & -1 \\ 1 & 1 & 1 & -1 & -1 & 1 & -1 & -1 & -1 & 1 & -1 & -1 & -1 & -1 & 1 & 1 & 1 & 1 & -1 \\ 1 & 1 & -1 & 1 & -1 & -1 & 1 & -1 & -1 & -1 & 1 & 1 & -1 & -1 & 1 & 1 & -1 & -1 & 1 \\ 1 & 1 & -1 & 1 & -1 & -1 & -1 & 1 & -1 & -1 & -1 & -1 & 1 & 1 & -1 & -1 & 1 & 1 & 1 \\ 1 & -1 & -1 & -1 & 1 & 1 & -1 & -1 & 1 & -1 & 1 & -1 & 1 & -1 & 1 & -1 & 1 & -1 & 1 \\ 1 & -1 & -1 & -1 & 1 & -1 & 1 & -1 & 1 & 1 & -1 & -1 & -1 & 1 & -1 & 1 & -1 & 1 & 1 \\ -1 & 1 & -1 & -1 & -1 & 1 & -1 & 1 & 1 & 1 & 1 & 1 & -1 & -1 & -1 & -1 & -1 & 1 & 1 \\ -1 & 1 & -1 & -1 & 1 & -1 & -1 & 1 & 1 & -1 & -1 & 1 & -1 & 1 & 1 & 1 & 1 & -1 & -1 \\ -1 & -1 & 1 & 1 & 1 & 1 & -1 & -1 & -1 & -1 & -1 & 1 & -1 & 1 & 1 & -1 & -1 & 1 & 1 \\ -1 & -1 & 1 & 1 & 1 & -1 & -1 & 1 & -1 & 1 & 1 & -1 & -1 & -1 & -1 & 1 & 1 & -1 & 1 \end{bmatrix},$$

$$
\mathbf{M} = \left[\begin{array}{ccccccccc}
1 & 1 & 1 & 1 & 1 & 1 & 1 & 1 & 1 \\
1 & 1 & 1 & 1 & 1 & 1 & 1 & 1 & 1 \\
1 & 1 & 1 & 1 & -1 & 1 & -1 & -1 & -1 \\
1 & 1 & 1 & 1 & -1 & -1 & 1 & -1 & -1 \\
1 & 1 & 1 & -1 & 1 & -1 & -1 & 1 & -1 \\
1 & 1 & 1 & -1 & 1 & -1 & -1 & -1 & 1 \\
1 & 1 & -1 & -1 & -1 & 1 & 1 & -1 & -1 \\
1 & 1 & -1 & -1 & -1 & 1 & -1 & 1 & -1 \\
1 & -1 & 1 & -1 & -1 & -1 & 1 & -1 & 1 \\
1 & -1 & 1 & -1 & -1 & 1 & -1 & -1 & 1 \\
1 & -1 & -1 & 1 & 1 & 1 & 1 & -1 & -1 \\
1 & -1 & -1 & 1 & 1 & 1 & -1 & -1 & 1
\end{array}\right.
$$

$$
\left.\begin{array}{ccccccccccc}
1 & 1 & 1 & 1 & 1 & 1 & 1 & 1 & 1 & 1 & 1 \\
1 & -1 & -1 & -1 & -1 & -1 & -1 & -1 & -1 & -1 & -1 \\
-1 & 1 & 1 & 1 & 1 & 1 & -1 & -1 & -1 & -1 & -1 \\
-1 & 1 & -1 & -1 & -1 & -1 & 1 & 1 & 1 & 1 & -1 \\
-1 & -1 & 1 & 1 & -1 & -1 & 1 & 1 & -1 & -1 & 1 \\
-1 & -1 & -1 & -1 & 1 & 1 & -1 & -1 & 1 & 1 & 1 \\
1 & -1 & 1 & -1 & 1 & -1 & 1 & -1 & 1 & -1 & 1 \\
1 & 1 & -1 & -1 & -1 & 1 & -1 & 1 & -1 & 1 & 1 \\
1 & 1 & 1 & 1 & -1 & -1 & -1 & -1 & -1 & 1 & 1 \\
1 & -1 & -1 & 1 & -1 & 1 & 1 & 1 & 1 & -1 & -1 \\
-1 & -1 & -1 & 1 & -1 & 1 & 1 & -1 & -1 & 1 & 1 \\
-1 & 1 & 1 & -1 & -1 & -1 & -1 & 1 & 1 & -1 & 1
\end{array}\right].
$$

The above two orthogonal codes cannot be extended, i.e., it is not possible to add a row of $+1$'s and -1's to keep the orthogonality.

Comments by Zhang Yao-ting

This paper was Professor Pao-lu Hsu's last article, completed in October 1970. He died in December of the same year. The Cultural Revolution lasting already four years at the time had not yet ended. He had suffered tremendously in this calamity, and physically he was paralyzed. This paper was completed when he was bedridden. The only journal he could have access to was the *Annals of Mathematical Statistics*. It was said that when he gave this paper to Mr. H. F. Tuan, being no longer able to speak clearly, he was using his hands to express himself.

The material discussed in this article is closely related to the contents covered in his early 1966 seminars on combinatorial analysis. During the later years of his life, he was devoted to using matrices to describe and prove results in combinatorial analysis. This article is a typical representative of this idea.

Reprinted from
Biometrika
38, 345–367 (1951).

Appendix:

THE JACOBIANS OF CERTAIN MATRIX TRANSFORMATIONS USEFUL IN MULTIVARIATE ANALYSIS

BASED ON LECTURES OF P. L. HSU AT THE UNIVERSITY OF NORTH CAROLINA, 1947

BY WALTER L. DEEMER AND INGRAM OLKIN, *University of North Carolina*

Editorial Note. The following paper was submitted by Prof. Hotelling in the summer of 1950 for publication in *Biometrika* with the accompanying Note from Col. Deemer and Mr Olkin:

'In 1947 Prof. P. L. Hsu gave courses in multivariate analysis at the University of North Carolina in which he developed new techniques for finding Jacobians of certain matrix transformations. Hsu returned to China at the end of that academic year, leaving as a record of this material only the notes of students in his classes.

'We (Deemer and Olkin) became interested in these matrix transformations in the course of our work in multivariate analysis. Since we did not take Hsu's courses, we used Ralph Bradley's notes as a basis for our studies.

'In the spring of 1948 there was a seminar in multivariate analysis under the direction of Prof. Harold Hotelling. At his request we prepared some lectures on matrix transformations. At the completion of the seminar, Prof. Hotelling and Prof. R. C. Bose suggested that in view of the importance of these techniques and their non-availability, we should prepare an expository paper giving a systematic development with all proofs given in detail.

'All the new ideas of importance in this paper are due to Hsu. Our contribution has been to organize the material in logical form, making all proofs complete with the necessary lemmas explicitly stated and proved.

'Efforts were made to communicate with Prof. Hsu in order that he could review this material before it was circulated. To date such efforts have failed.'

Since this contribution was received contact has been made with Prof. Hsu and the paper is now published with his approval. A suggestion which he made for improvement has been added as a Note on p. 361. However, in view of the liberty they have taken in the exposition of Hsu's ideas, the American authors would like it to be clear that they are to be held responsible for any errors.

FOREWORD

BY HAROLD HOTELLING

We are apparently at the beginning of a major development in the use of statistical procedures for joint treatment of a multiplicity of correlated variates. Many of the new methods depend ultimately on the distribution of the roots of certain determinantal equations. This distribution was published simultaneously in 1939 by P. L. Hsu (1939) and R. A. Fisher (1939) in the *Annals of Eugenics*, and, excepting for a constant multiplier, by S. N. Roy (1938–40) in *Sankhyā*. Hsu's derivation, which, like Roy's, is otherwise complete, demonstrates the correctness of his formula for this constant multiplier only for the case of three variates. Proof that the formula is correct for the general case has turned out to be unexpectedly difficult to develop. In seeking to close this gap while lecturing on advanced multivariate analysis at Chapel Hill, Hsu developed what appears to be an extraordinarily powerful method of dealing with a wide class of transformations of the elements of one matrix into those of another induced by a functional relation between the matrices. This technique not only leads to a solution of the original problem but appears to offer the possibility of a fresh attack on other problems of multivariate distribution theory as well. Heretofore it has been recorded only in lecture notes made by students in the spring of 1947. Col. Deemer and Mr Olkin have not merely reproduced lecture notes but have reworked the material and improved it in certain essential respects.

1. Summary and introduction

In finding the distribution of multivariate statistics it is often desirable to make certain transformations. However, the calculation of the Jacobian is frequently rather difficult. We have therefore assembled in this paper the Jacobians of many of the most used transformations.

Certain general methods are applied in the calculation of the Jacobians; these methods will be found useful in the calculation of other Jacobians not given here. Three techniques of particular value for non-linear transformations are:

(*a*) Taking differentials to get a linear equation in the differentials, e.g. Theorem 4·1.

(*b*) Transformations to new variables so that the Jacobian can be evaluated as the product of other Jacobians which are more easily calculated, e.g. Theorem 4·2.

(*c*) Introduction of new matrices such that the Jacobian of the original transformation is the discriminant of the bilinear form in the variables of the new matrices, e.g. Lemma 4·3 and Theorem 4·3.

2. Notation

Because the shape and symmetry of matrices are important in finding Jacobians, we shall consistently use the following notation:

(2·1) Lower-case letters will denote scalars.

(2·2) As is customary, a prime will denote the transpose of a matrix.

(2·3) Lower-case letters with a superior bar will denote column vectors, e.g. $\bar{u}' = (u_1, \ldots, u_p)$.

(2·4) Capital letters will denote matrices. A system of superior symbols will be used as follows:

(*a*) $\bar{A}$ is a symmetric matrix of elements a_{ij}.

(*b*) $\hat{A}$ is a skew-symmetric matrix of elements a_{ij} above the main diagonal and $a_{ji} = -a_{ij}$ below the main diagonal and zeros in the main diagonal.

(*c*) $\tilde{A}$ is a triangular matrix with zeros above the main diagonal, e.g.

$$\begin{bmatrix} a_{11} & \cdots & \cdots & \cdot \\ a_{12} & a_{22} & \cdots & \cdot \\ \cdots & \cdots & \cdots & \cdot \\ a_{1p} & a_{2p} & \cdots & a_{pp} \end{bmatrix}.$$

(*d*) A matrix not in classes (*a*), (*b*), (*c*) will be denoted by a capital letter with no superior symbol.

(*e*) (dX) is the matrix of the same shape as X, whose elements are the differentials of the elements of X. This will be denoted by X^*.

(2·5) Diagonal matrices will be denoted by D with a subscript showing the elements, e.g. D_r is the diagonal matrix with elements $r_1, r_2, \ldots, r_p$.

(2·6) Orthogonal matrices will be denoted by Δ and Γ.

(2·7) $A: p \times q$ means the matrix A has p rows and q columns; the same notation will be used for matrices with superior symbols.

(2·8) Since it is the absolute value of the Jacobian that is wanted, it is to be understood that all determinants are evaluated up to the sign only.

(2·9) Let X, Y be two matrices which satisfy an equation $Y = f(X)$. Let X and Y have the same number n of independent elements, and denote them by $x_1, \ldots, x_n$ and $y_1, \ldots, y_n$, respectively. Then the Jacobian, $\dfrac{\partial(y_1, \ldots, y_n)}{\partial(x_1, \ldots, x_n)}$, will be denoted by $D(X;\ Y)$.

For purposes of calculation it is sometimes desirable to express this in terms of the elements of X and Y, for example, $D(X;\ Y)$ might be written $D(x_{ii}, x_{ij};\ y_{ii}, y_{ij})$.

More generally, if $X_1, \ldots, X_k$, $Y_1, \ldots, Y_m$ satisfy the equations $Y_i = f_i(x_1, \ldots, x_k)$ $(i = 1, \ldots, m)$ and $(X_1, \ldots, X_k)$ and $(Y_1, \ldots, Y_m)$ have the independent elements $x_1, \ldots, x_n$ and $y_1, \ldots, y_n$, respectively, then $D(X_1, \ldots, X_k;\ Y_1, \ldots, Y_m)$ denotes

$$\frac{\partial(y_1, \ldots, y_n)}{\partial(x_1, \ldots, x_n)}.$$

(2·10) The following abbreviations will be used: p.d. for positive definite; n.s. for non-singular; J for Jacobian; tr for trace; I_p for the identity matrix of order p.

3. THE JACOBIANS OF LINEAR TRANSFORMATIONS

We state here a number of theorems on the J of linear transformations, some without proof. When the J does not immediately follow from previous theorems we will give the proof.

THEOREM 3·1 *a*. The J of the transformation $Y = aX$ $(Y, X\colon p \times q)$, where a is a constant scalar $\neq 0$, is $D(X;\ Y) = a^{pq}$.

Proof. $y_{ij} = ax_{ij}$ and $\partial y_{ij}/\partial x_{ij} = a$. Therefore the scheme of coefficients is diagonal with pq a's in the main diagonal. Q.E.D.

THEOREM 3·1 *b*. The J of the transformation $\overline{Y} = a\overline{X}$ $(\overline{Y}, \overline{X}\colon p \times p)$, where a is a constant scalar $\neq 0$, is $D(\overline{X};\ \overline{Y}) = a^{\frac{1}{2}p(p+1)}$.

THEOREM 3·2. The J of the linear transformation $\overline{y} = A\overline{x}$ $(\overline{y}, \overline{x}\colon p \times 1;\ A\colon p \times p)$ is $D(\overline{x};\ \overline{y}) = |A|$.

Proof. $y_i = \sum_{k=1}^{p} a_{ik}x_k$ and $\partial y_i/\partial x_j = a_{ij}$, therefore $J = D(\overline{x};\ \overline{y}) = |A|$. Q.E.D.

COROLLARY 3·2. The J of the transformation $\overline{x} = A^{-1}\overline{y}$ $(\overline{x}, \overline{y}\colon p \times 1;\ A\colon p \times p)$ is $D(\overline{y};\ \overline{x}) = |A|^{-1}$.

This follows from Theorem 3·2, since $|A^{-1}| = |A|^{-1}$.

THEOREM 3·3. The J of the transformation

$$\overline{y}' = \overline{x}'A'(\overline{y}, \overline{x}\colon p \times 1;\ A\colon p \times p) \quad \text{is} \quad D(\overline{x};\ \overline{y}) = |A|.$$

THEOREM 3·4. The J of the transformation

$$Y = AX(Y, X\colon p \times q;\ A\colon p \times p) \quad \text{is} \quad D(X;\ Y) = |A|^q.$$

Proof. This follows from Theorem 3·2 and 5 B. 4†, since the transformation of each column of Y is independent of the others and there are q such columns of Y, the J of each column transformation being $|A|$.

THEOREM 3·5. The J of $Y = XA$ $(Y, X\colon p \times q;\ A\colon q \times q)$ is $D(X;\ Y) = |A|^p$.

Proof. Write $Y' = A'X'$ and the result follows directly from Theorem 3·4.

† References of this form are to the appendix to the present paper.

THEOREM 3·6. The J of the transformation

$$Y = AXB(Y, X\colon p\times q;\ A\colon p\times p;\ B\colon q\times q) \quad \text{is} \quad D(X;\ Y) = |A|^q |B|^p.$$

Proof. Let $Z = AX$ and then $Y = ZB$. $D(X;\ Z) = |A|^q$ and $D(Z;\ Y) = |B|^p$. Using 5 B. 2 the theorem follows.

LEMMA 3·7. Let
$$Y = A_n A_{n-1} \ldots A_1 X A_1' \ldots A_{n-1}' A_n',$$
then
$$D(X;\ Y) = D(X;\ Y_1)\, D(Y_1;\ Y_2) \ldots D(Y_{n-1};\ Y),$$
where
$$Y_i = A_i Y_{i-1} A_i' \quad (i = 1, \ldots, n); \qquad Y_0 \equiv X, \quad Y_n \equiv Y.$$

Proof. Follows directly from 5 B. 2.

THEOREM 3·7. The J of the transformation

$$\bar{Y} = B\bar{X}B'(\bar{Y}, B, \bar{X}\colon p\times p) \quad \text{is} \quad D(\bar{X};\ \bar{Y}) = |B|^{p+1}.$$

Proof. By 5 A. 5 we can write $B = F_m F_{m-1} \ldots F_1$, where some of the F's are of the form 5 A. 1 and the rest are of the form 5 A. 2. Then

$$\bar{Y} = F_m \ldots F_1 \bar{X} F_1' \ldots F_m'. \tag{3·7·1}$$

By Lemma 3·7:
$$D(\bar{X};\ \bar{Y}) = D(\bar{X};\ \bar{Y}_1)\, D(\bar{Y}_1;\ \bar{Y}_2) \ldots D(\bar{Y}_{m-1};\ \bar{Y}_m),$$
where
$$\bar{Y}_\nu = F_\nu \bar{Y}_{\nu-1} F_\nu' \quad (\nu = 1, \ldots, m); \qquad \bar{Y}_0 \equiv \bar{X}, \quad \bar{Y}_m \equiv \bar{Y}.$$
We show that
$$D(\bar{Y}_{\nu-1};\ \bar{Y}_\nu) = |F_\nu|^{p+1} \quad (\nu = 1, \ldots, m),$$
and hence it will follow that

$$D(\bar{X};\ \bar{Y}) = |F_m|^{p+1} \ldots |F_1|^{p+1} = |F_m \ldots F_1|^{p+1} = |B|^{p+1}.$$

(1) Let G be any of the F's which are of the form 5 A. 1; the transformation $\bar{Y}_\nu = G\bar{Y}_{\nu-1}G'$ implies
$$y_{ii} = a^2 x_{ii}; \qquad y_{ij} = a x_{ij} \quad (i \neq j); \qquad y_{jk} = x_{jk} \quad (j, k \neq i).$$
The determinant of the partial derivatives is therefore diagonal with $p-1$ elements a, one element a^2, and the remaining elements 1. Therefore $D(\bar{Y}_{\nu-1};\ \bar{Y}_\nu) = a^{p+1} = |G|^{p+1}$.

(2) Let H be any of the F's which are of the form 5 A. 2; the transformation $\bar{Y}_\nu = H\bar{Y}_{\nu-1}H'$ implies
$$y_{ii} = x_{ii} + 2a x_{ij} + a^2 x_{jj},$$
$$y_{ki} = y_{ik} = x_{ik} + a x_{jk} \quad (k \neq i),$$
$$y_{jk} = x_{jk} \quad (j, k \neq i).$$
The determinant of partial derivatives will have 1's in the main diagonal and 0's below the main diagonal. Hence $D(\bar{Y}_{\nu-1};\ \bar{Y}_\nu) = 1 = |H|^{p+1}$. Q.E.D.

COROLLARY 3·7. The J of the transformation

$$\bar{Y} = A^{-1}\bar{X}A'^{-1} \ (\bar{Y}, \bar{X}, A\colon p\times p) \quad \text{is} \quad D(\bar{X};\ \bar{Y}) = |A|^{-(p+1)}.$$

THEOREM 3·8. The J of the transformation $\tilde{Y} = \tilde{A}\tilde{X}\ (\tilde{Y}, \tilde{A}, \tilde{X}\colon p\times p)$ is $D(\tilde{X};\ \tilde{Y}) = \prod_{i=1}^{p} a_{ii}^{i}$.

Proof.
$$y_{ij} = \sum_{\nu=i}^{j} a_{\nu j} x_{i\nu} \quad (i \leqslant j).$$

The elements on one side of the main diagonal of the determinant of partial derivatives are zero. The main diagonal contains a_{11} from $\partial y_{11}/\partial x_{11}$; it contains a_{22} twice: once from $\partial y_{12}/\partial x_{12}$, and once from $\partial y_{22}/\partial x_{22}$; it contains a_{ii} i times: once from each of $\partial y_{ji}/\partial x_{ji}$ $(j = 1, \ldots, i)$. Therefore the determinant is $\prod_{i=1}^{p} a_{ii}^{i}$. Q.E.D.

THEOREM 3·9. The J of the transformation

$$\bar{Y} = \tilde{A}\tilde{X}' + \tilde{X}\tilde{A}' \; (\bar{Y}, \tilde{A}, \tilde{X}: p \times p) \quad \text{is} \quad D(\tilde{X};\, \bar{Y}) = 2^p \prod_{i=1}^{p} a_{ii}^{p-i+1}.$$

Proof. Pre- and post-multiply by $\tilde{A}^{-1}$ and $\tilde{A}'^{-1}$ respectively:

$$\tilde{A}^{-1}\bar{Y}\tilde{A}'^{-1} = \tilde{X}'\tilde{A}'^{-1} + \tilde{A}^{-1}\tilde{X}. \tag{3·9·1}$$

Let $\bar{Z} = \tilde{A}^{-1}\bar{Y}\tilde{A}'^{-1}$ and $\tilde{U} = \tilde{A}^{-1}\tilde{X}$. Then

$$\bar{Z} = \tilde{U}' + \tilde{U}, \tag{3·9·2}$$

$$D(\tilde{X};\, \bar{Y}) = D(\tilde{X};\, \tilde{U})\, D(\tilde{U};\, \bar{Z})\, D(\bar{Z};\, \bar{Y}) \tag{3·9·3}$$

by 5 B. 2;

$$D(\tilde{X};\, \tilde{U}) = \prod_{i=1}^{p} a_{ii}^{-i} \tag{3·9·4}$$

by Theorem 3·8;

$$D(\bar{Z};\, \bar{Y}) = |\,\tilde{A}\,|^{p+1} = \prod_{i=1}^{p} a_{ii}^{p+1} \tag{3·9·5}$$

by Corollary 3·7 and 5 B. 1;

$$z_{ii} = 2u_{ii} \quad \text{and} \quad z_{ij} = u_{ij} \tag{3·9·6}$$

from 3·9·2. The determinant of partial derivatives is therefore diagonal. It contains p elements equal to 2 and the remainder 1's. Therefore

$$D(\tilde{U};\, \bar{Z}) = 2^p. \tag{3·9·7}$$

Substituting in (3·9·3) $\quad D(\tilde{X};\, \bar{Y}) = 2^p \prod_{i=1}^{p} a_{ii}^{p-i+1}.$ Q.E.D.

THEOREM 3·10. The J of the transformation

$$\hat{Y} = G\hat{X}G' \; (\hat{Y}, \hat{X}, G: p \times p) \quad \text{is} \quad D(\hat{X};\, \hat{Y}) = |\,G\,|^{p-1}.$$

Proof. The proof follows along the same lines as the proof of Theorem 3·7 with the necessary changes due to $\hat{X}$ and $\hat{Y}$ being skew symmetric.

4. NON-LINEAR TRANSFORMATIONS

The general method for finding the J of a non-linear transformation is to take the differential of both sides and then to find the J of the resulting linear transformation in the differentials (see 5 B. 3).

THEOREM 4·1. The J of the transformation $\bar{V} = \tilde{T}\tilde{T}'$ is $D(\tilde{T};\, \bar{V}) = 2^p \prod_{i=1}^{p} t_{ii}^{p-i+1}$.

Proof. Taking differentials of $\bar{V} = \tilde{T}\tilde{T}'$,

$$d\bar{V} = \tilde{T}(d\tilde{T})' + (d\tilde{T})\,\tilde{T}'. \tag{4·1·1}$$

Let $\bar{V}^* = d\bar{V}$; $\tilde{T}^* = d\tilde{T}$, then

$$\bar{V}^* = \tilde{T}\tilde{T}^{*\prime} + \tilde{T}^*\tilde{T}'. \tag{4·1·2}$$

This is the transformation of Theorem 3·9, and hence the theorem follows. Q.E.D.

LEMMA 4·2. Let $\bar{A}$, $\bar{B}$ be p.d., $p \times p$ matrices. Let $|\,\bar{A} - \phi\bar{B}\,| = 0$ have p distinct roots $\phi_1 > \phi_2 > \ldots > \phi_p > 0$, and let D_ϕ be the diagonal matrix with elements $\phi_1, \ldots, \phi_p$. Then there exists a unique matrix $W = (w_{ij})$: $p \times p$, where $w_{1i} > 0$ $(i = 1, \ldots, p)$ such that $\bar{A} = WD_\phi W'$ and $\bar{B} = WW'$.

23-2

THEOREM 4·2.† The J of the transformation $\bar{A} = WD_\phi W'$; $\bar{B} = WW'$ (all matrices $p \times p$) is

$$D(W, \phi; \bar{A}, \bar{B}) = | W |^{p+2} 2^p \prod_{i<j}^{p} (\phi_i - \phi_j).$$

Proof. Taking differentials:

$$d\bar{A} = (dW) D_\phi W' + W(dD_\phi) W' + WD_\phi (dW)', \tag{4·2·1}$$

$$d\bar{B} = W(dW)' + (dW) W'. \tag{4·2·2}$$

Let $\bar{A}^* = d\bar{A}$; $\bar{B}^* = d\bar{B}$; $W^* = dW$; $D_\eta = dD_\phi$,

then

$$\bar{A}^* = W^* D_\phi W' + WD_\eta W' + WD_\phi W^{*\prime}, \tag{4·2·3}$$

$$\bar{B}^* = WW^{*\prime} + W^{*\prime} W'. \tag{4·2·4}$$

By 5 B. 3: $J = D(W^*, \eta; \bar{A}^*, \bar{B}^*)$. Pre- and post-multiply (4·2·3) and (4·2·4) by W^{-1} and W'^{-1} respectively:

$$W^{-1}\bar{A}^* W'^{-1} = W^{-1} W^* D_\phi + D_\eta + D_\phi W^{*\prime} W'^{-1}, \tag{4·2·5}$$

$$W^{-1}\bar{B}^* W'^{-1} = W^{*\prime} W'^{-1} + W^{-1} W^*. \tag{4·2·6}$$

Let

$$\bar{U} = W^{-1}\bar{A}^* W'^{-1}, \tag{4·2·7}$$

and hence $\bar{A}^* = W\bar{U}W'$;

$$\bar{V} = W^{-1}\bar{B}^* W'^{-1}, \tag{4·2·8}$$

and hence $\bar{B}^* = W\bar{V}W'$;

$$R = W^{-1} W^*. \tag{4·2·9}$$

Substituting in (4·2·5) and (4·2·6):

$$\bar{U} = RD_\phi + D_\phi R' + D_\eta, \tag{4·2·10}$$

$$\bar{V} = R + R'; \tag{4·2·11}$$

$$D(W^*, \eta; \bar{A}^*, \bar{B}^*) = D(W^*, \eta; R, \eta)\, D(R, \eta; \bar{U}, \bar{V})\, D(\bar{U}, \bar{V}; \bar{A}^*, \bar{B}^*) \tag{4·2·12}$$

using 5 B. 2;

$$D(W^*, \eta; R, \eta) = | W |^{-p} \tag{4·2·13}$$

from (4·2·9) using Theorem 3·4 and 5 B. 4;

$$D(\bar{U}, \bar{V}; \bar{A}^*, \bar{B}^*) = D(\bar{U}; \bar{A}^*)\, D(\bar{V}; \bar{B}^*) \tag{4·2·14}$$

by 5 B. 4;

$$D(\bar{U}; \bar{A}^*) = D(\bar{V}; \bar{B}^*) = | W |^{p+1} \tag{4·2·15}$$

by Theorem 3·7. Hence

$$D(\bar{U}, \bar{V}; \bar{A}^*, \bar{B}^*) = | W |^{2(p+1)}. \tag{4·2·16}$$

To find $D(R, \eta; \bar{U}, \bar{V})$ we use (4·2·10) and (4·2·11):

$$u_{ii} = 2r_{ii}\phi_i + \eta_i, \quad v_{ii} = 2r_{ii},$$

$$u_{ij} = r_{ij}\phi_j + r_{ji}\phi_i, \quad v_{ij} = r_{ij} + r_{ji} \quad (i \neq j).$$

† For an application see Hsu (1939).

The scheme of partial derivatives will then be as follows:

	η_i $(i = 1, \ldots, p)$	r_{ii} $(i = 1, \ldots, p)$	r_{ij} $(i>j)$	r_{ij} $(i<j)$
u_{ii} $(i = 1, \ldots, p)$	I	L	0	0
v_{ii} $(i = 1, \ldots, p)$	0	$2I$	0	0
u_{ij} $(i<j)$	0	0	D	C
v_{ij} $(i<j)$	0	0	I	I

where C is the diagonal matrix with elements $\phi_2, \phi_3, \ldots, \phi_p; \phi_3, \ldots, \phi_p; \ldots; \phi_{p-1}, \phi_p; \phi_p$, and D is the diagonal matrix with elements $\phi_1, \ldots, \phi_1$ $(p-1$ terms), $\phi_2, \ldots, \phi_2$ $(p-2$ terms), $\ldots$, ϕ_{p-1}.

The matrix L does not affect the value of the determinant, which is equal to

$$2^p \,|D-C| = 2^p \prod_{i<j}^{p} (\phi_i - \phi_j)$$

by applying 5 A. 7 twice. Hence

$$J = 2^p \,|\, W \,|^{p+2} \prod_{i<j}^{p} (\phi_i - \phi_j). \quad \text{Q.E.D.}$$

LEMMA 4·3. Let $X_1 = (x_{ij}^{(1)})$, $(p_1 \times q_1)$ and $X_2 = (x_{ij}^{(2)})$, $(p_2 \times q_1)$ be two matrices containing a total of N independent variables $(N \leqslant p_1 q_1 + p_2 q_1)$ and $Y_1 = (y_{ij}^{(1)})$, $(p_1 \times q_1)$, $Y_2 = (y_{ij}^{(2)})$, $(p_2 \times q_1)$ be two other matrices containing the same number of variables.

Let $\Xi: N \times 1$ be the vector formed from all N x's by using the elements in the first column of X_1 as the first p_1 elements of Ξ, the p_1 elements in the second column of X_1 as the next p_1 elements of Ξ, etc., through the columns of X_2.

Let $\mathrm{H}: N \times 1$ be the vector formed from the columns of Y_1 and Y_2 in the same way that Ξ was formed from X_1 and X_2. Thus

$$\Xi' = (x_{11}^{(1)}\, x_{21}^{(1)} \ldots x_{p_1 q_1}^{(1)} x_{11}^{(2)}\, x_{21}^{(2)} \ldots x_{p_2 q_1}^{(2)}),$$

$$\mathrm{H}' = (y_{11}^{(1)}\, y_{21}^{(1)} \ldots y_{p_1 q_1}^{(1)} y_{11}^{(2)}\, y_{21}^{(2)} \ldots y_{p_2 q_1}^{(2)}).$$

Let $\Xi = A\mathrm{H}$ $(A: N \times N)$, where A is a matrix of constants, be the connecting equations between the variables $x_{ij}^{(\nu)}$ and $y_{ij}^{(\nu)}$ $(\nu = 1, 2)$.

Let $R_1 = (r_{ij}^{(1)})$, $(q_1 \times p_1)$ and $R_2 = (r_{ij}^{(2)})$, $(q_1 \times p_2)$ be two matrices with a total of N independent variables. Then $D(Y_1, Y_2; X_1, X_2)$ is equal to the discriminant of $\operatorname{tr}(R_1 X_1 + R_2 X_2)$, where $\operatorname{tr}(R_1 X_1 + R_2 X_2)$ is to be considered as a bilinear form in the variables of R_1, R_2 and Y_1, Y_2.

Proof. Let $\mu(N \times 1)$ be the vector formed from the elements R_1 and R_2 by taking the rows of R_1 in turn and then the rows of R_2 in turn:

$$\mu' = (r_{11}^{(1)}\, r_{12}^{(1)} \ldots r_{p_1 q_1}^{(1)} r_{11}^{(2)}\, r_{12}^{(2)} \ldots r_{p_2 q_1}^{(2)}),$$

$$\operatorname{tr}(R_1 X_1 + R_2 X_2) = \sum_{\nu=1}^{2} \sum_{i=1}^{q_1} \sum_{\sigma=1}^{p_\nu} r_{i\sigma}^{(\nu)} x_{\sigma i}^{(\nu)} = \mu' \Xi = \mu' A \mathrm{H}.$$

The discriminant of this bilinear form is $|\,A\,|$.

The J, $D(Y_1, Y_2; X_1, X_2)$ of the original transformation $\Xi = A\mathrm{H}$ is $|\,A\,|$ (by Theorem 3·2). Q.E.D.

Introduction to Theorem 4·3

Let $\bar{V} = \begin{bmatrix} \bar{V}_{11} & V_{12} \\ V_{21} & \bar{V}_{22} \end{bmatrix}$ be p.d. ($\bar{V}_{11}: p\times p$; $V_{12}: p\times q$; $\bar{V}_{22}: q\times q$) ($p \leqslant q$). There exists a n.s. matrix G of the form $\begin{bmatrix} G_1 & 0 \\ 0 & G_2 \end{bmatrix}$ ($G_1: p\times p$; $G_2: q\times q$) such that

$$(4·3·2) \qquad G\bar{V}G' = \begin{bmatrix} I_p & D_r & 0 \\ D_r & & \\ 0 & & I_q \end{bmatrix},$$

where D_r: $p\times p$ has elements $r_1, \ldots, r_p$; $r_1^2 \ldots r_p^2$ are the roots of the determinantal equation $|V_{12}\bar{V}_{22}^{-1}V_{21} - \theta\bar{V}_{11}| = 0$ (see 5 A. 21).

$$(4·3·3) \qquad \bar{V} = G^{-1}\begin{bmatrix} I_p & D_r & 0 \\ D_r & & \\ 0 & & I_q \end{bmatrix}G'^{-1}$$

from (4·3·2). Let

$$G^{-1} = \begin{bmatrix} P & 0 \\ 0 & Q \end{bmatrix},$$

then

$$(4·3·4) \qquad \bar{V} = \begin{bmatrix} P & 0 \\ 0 & Q \end{bmatrix}\begin{bmatrix} I_p & D_r & 0 \\ D_r & & \\ 0 & & I_q \end{bmatrix}\begin{bmatrix} P' & 0 \\ 0 & Q' \end{bmatrix}.$$

This is a transformation from the v's to the p's, q's and r's. $\bar{V}$ contains $\frac{1}{2}(p+q)(p+q+1)$ variables, and the right side of (4·3·4) contains p^2+q^2+p variables. Let $\delta = q-p$, then $[(p^2+q^2+p) - \frac{1}{2}(p+q)(p+q+1)] = \frac{1}{2}\delta(\delta-1)$, and the number of variables on each side will be equal if $p = q$ or $p = q-1$.

We now partition Q and rewrite (4·3·4):

$$(4·3·5) \qquad \bar{V} = \begin{bmatrix} P & 0 & 0 \\ 0 & Q_{11} & Q_{12} \\ 0 & Q_{21} & Q_{22} \end{bmatrix}\begin{bmatrix} I_p & D_r & 0 \\ D_r & I_p & 0 \\ 0 & 0 & I_\delta \end{bmatrix}\begin{bmatrix} P' & 0 & 0 \\ 0 & Q'_{11} & Q'_{21} \\ 0 & Q'_{12} & Q'_{22} \end{bmatrix},$$

$$(Q_{11}: p\times p;\ Q_{12}: p\times\delta;\ Q_{22}: \delta\times\delta).$$

Post-multiplying Q_{12} and Q_{22} by an orthogonal matrix Δ will not change the value of this product, i.e.

$$\bar{V} = \begin{bmatrix} P & 0 & 0 \\ 0 & Q_{11} & Q_{12}\Delta \\ 0 & Q_{21} & Q_{22}\Delta \end{bmatrix}\begin{bmatrix} I_p & D_r & 0 \\ D_r & I_p & 0 \\ 0 & 0 & I_\delta \end{bmatrix}\begin{bmatrix} P' & 0 & 0 \\ 0 & Q'_{11} & Q'_{21} \\ 0 & \Delta'Q'_{12} & \Delta'Q'_{22} \end{bmatrix},$$

as may be verified by multiplying out this expression. By 5 A. 10 we can choose this Δ so that $Q_{22}\Delta$ is triangular with 0's below the main diagonal. When we do this we reduce the number of variables on the right by $\frac{1}{2}\delta(\delta-1)$ and thus have the same number on each side. The transformation is unique if $r_1 > \ldots > r_p > 0$; $p_{1j} > 0\,(j = 1, \ldots, p)$. We now find its J after the number of variables is equalized.

Hsu has used this transformation in finding the distribution of the canonical correlation coefficient (see Hotelling, 1936) when the population values are not zero. Hsu's work on this has apparently not been published.

THEOREM 4·3. Let
$$Q = \begin{bmatrix} Q_{11} & Q_{12} \\ Q_{21} & \tilde{T}' \end{bmatrix}$$

$$(Q_{11}: p\times p;\ Q_{12}: p\times\delta;\ \tilde{T}: \delta\times\delta)\ (p \leqslant q), \qquad \delta = q-p;\ t_{ii} > 0 \quad (i = 1, \ldots, \delta).$$

Let
$$\Phi = \begin{bmatrix} I_p & D_r & 0 \\ D_r & & \\ 0 & & I_q \end{bmatrix},$$
(D_r: $p \times p$). Then the J of the transformation

(4·3·6)
$$\bar{V} = \begin{bmatrix} P & 0 \\ 0 & Q \end{bmatrix} \Phi \begin{bmatrix} P' & 0 \\ 0 & Q' \end{bmatrix}$$

is
$$D(P, Q, r; \bar{V}) = 2^{p+q} |P|^{q+1} |Q|^{p+1} \left(\prod_{i=1}^{\delta} t_{ii}^{i-1}\right)\left(\prod_{i=1}^{p} r_i^{\delta}\right)\left(\prod_{i<j}^{p} (r_i^2 - r_j^2)\right).$$

When $q = p$ or $q = p+1$, the terms in t do not appear.

Proof. Take differentials of (4·3·6) and let $\bar{V}^* = d\bar{V}$, $P^* = dP$, $Q_{11}^* = dQ_{11}$, $Q_{12}^* = dQ_{12}$, $Q_{21}^* = dQ_{21}$, $\tilde{T}^* = d\tilde{T}$, $D_\eta = dD_r$, and let
$$Q^* = dQ = \begin{bmatrix} Q_{11}^* & Q_{12}^* \\ Q_{21}^* & \tilde{T}^{*\prime} \end{bmatrix};$$
then

(4·3·7)
$$\bar{V}^* = \begin{bmatrix} P^* & 0 \\ 0 & Q^* \end{bmatrix} \Phi \begin{bmatrix} P' & 0 \\ 0 & Q' \end{bmatrix} + \begin{bmatrix} P & 0 \\ 0 & Q \end{bmatrix} \Phi \begin{bmatrix} P^* & 0 \\ 0 & Q^* \end{bmatrix}' + \begin{bmatrix} P & 0 \\ 0 & Q \end{bmatrix} \begin{bmatrix} 0 & D_\eta & 0 \\ D_\eta & 0 & 0 \\ 0 & 0 & 0 \end{bmatrix} \begin{bmatrix} P' & 0 \\ 0 & Q' \end{bmatrix}.$$

$J = D(P^*, Q^*, \eta; \bar{V}^*)$ by 5 B. 3. Pre-multiply (4·3·7) by $\begin{bmatrix} P^{-1} & 0 \\ 0 & Q^{-1} \end{bmatrix}$ and post-multiply by $\begin{bmatrix} P^{-1} & 0 \\ 0 & Q^{-1} \end{bmatrix}'$ and introduce new variables:

(4·3·8)
$$\bar{X} = \begin{bmatrix} P^{-1} & 0 \\ 0 & Q^{-1} \end{bmatrix} \bar{V}^* \begin{bmatrix} P^{-1} & 0 \\ 0 & Q^{-1} \end{bmatrix}',$$

(4·3·9)
$$A = P^{-1}P^*,$$

(4·3·10)
$$\begin{bmatrix} B_{11} \\ B_{21} \end{bmatrix} = Q^{-1} \begin{bmatrix} Q_{11}^* \\ Q_{21}^* \end{bmatrix}.$$

To simplify the notation, we write

(4·3·11)
$$\begin{bmatrix} B_{12} \\ B_{22} \end{bmatrix} = Q^{-1} \begin{bmatrix} Q_{12}^* \\ \tilde{T}^{*\prime} \end{bmatrix},$$

but we do not consider B_{12} and B_{22} as new variables, since $Q^{-1}\tilde{T}^{*\prime}$ is not triangular, and hence the number of variables in B_{22} would be greater than the number of variables in $\tilde{T}^{*\prime}$. We then get from (4·3·7):

(4·3·12)
$$\bar{X} = \begin{bmatrix} A & 0 & 0 \\ 0 & B_{11} & B_{12} \\ 0 & B_{21} & B_{22} \end{bmatrix} \Phi + \Phi \begin{bmatrix} A' & 0 & 0 \\ 0 & B_{11}' & B_{21}' \\ 0 & B_{12}' & B_{22}' \end{bmatrix} + \begin{bmatrix} 0 & D_\eta & 0 \\ D_\eta & 0 & 0 \\ 0 & 0 & 0 \end{bmatrix}$$
$$= \begin{bmatrix} A + A' & AD_r + D_r B_{11}' + D_\eta & D_r B_{21}' \\ (AD_r + D_r B_{11}' + D_\eta)' & B_{11} + B_{11}' & B_{12} + B_{21}' \\ (D_r B_{21}')' & (B_{12} + B_{21}')' & B_{22} + B_{22}' \end{bmatrix}.$$

By 5 B. 2

(A)
$$\begin{aligned} J &= D(P^*, Q^*, \eta; \bar{V}^*) \\ &= D(P^*, Q^*, \eta; A, B_{11}, B_{21}, Q_{12}^*, \tilde{T}^*, \eta)\, D(A, B_{11}, B_{21}, Q_{12}^*, \tilde{T}^*, \eta; \bar{X})\, D(\bar{X}; \bar{V}^*) \\ &= D_1 D_2 D_3. \end{aligned}$$

By 5 B. 4
$$\begin{aligned} D_1 &= D(P^*; A)\, D(Q_{11}^*, Q_{21}^*; B_{11}, B_{21}) \\ &= |P|^{-p} |Q|^{-p} \end{aligned}$$

(from (4·3·9), (4·3·10), using Theorem 3·4). From (4·3·8), using Theorem 3·7,

(B)
$$D_3 = D(\bar{X}; \bar{V}^*) = |P|^{p+q+1} |Q|^{p+q+1}.$$

To compute D_2 we partition $\bar{X}$ and hence can equate the submatrices of $\bar{X}$ to the submatrices on the right of (4·3·12):

$$\bar{X} = \begin{bmatrix} \bar{X}_{11} & X_{12} & X_{13} \\ X'_{12} & \bar{X}_{22} & X_{23} \\ X'_{13} & X'_{23} & \bar{X}_{33} \end{bmatrix} \quad (\bar{X}_{11}, X_{12}, \bar{X}_{22}: p \times p;\; X_{13}, X_{23}: p \times \delta;\; \bar{X}_{33}: \delta \times \delta).$$

We then have

$$\text{(4·3·13)} \qquad \bar{X}_{11} = A + A'; \quad X_{12} = AD_r + D_r B'_{11} + D_\eta; \quad \bar{X}_{22} = B_{11} + B'_{11},$$

$$\text{(4·3·14)} \qquad X_{13} = D_r B'_{21}; \quad X_{23} = B_{12} + B'_{21}; \quad \bar{X}_{33} = B_{22} + B'_{22}.$$

The transformations in (4·3·13) and (4·3·14) are independent, so we can write

$$\text{(C)} \qquad D_2 = D(A, B_{11}, \eta; \bar{X}_{11}, X_{12}, \bar{X}_{22})\, D(B_{21}, Q^*_{12}, \tilde{T}^*; X_{13}, X_{23}, \bar{X}_{33})$$
$$= D_4 D_5.$$

Computation of D_4. For convenience we will relabel the elements of $\bar{X}$:

$$\text{(4·3·15)} \qquad \bar{X}_{11} = (x_{ij}); \quad X_{12} = (y_{ij}); \quad \bar{X}_{22} = (z_{ij}),$$

all $i, j = 1, \ldots, p$, and we denote the elements of the other matrices as follows:

$$A = (a_{ij}); \quad B'_{11} = (b_{ij}) \quad (i, j = 1, \ldots, p); \qquad \eta = (\eta_1 \cdots \eta_p).$$

Using (4·3·13), (4·3·14) and (4·3·15) we get

$$\text{(4·3·16)} \qquad x_{ii} = 2a_{ii}; \quad y_{ii} = r_i a_{ii} + r_i b_{ii} + \eta_i; \quad z_{ii} = 2b_{ii},$$

$$\text{(4·3·17)} \qquad x_{ij} = a_{ij} + a_{ji} \quad (i < j); \qquad y_{ij} = r_j a_{ij} + r_i b_{ij} \quad (i \neq j); \qquad z_{ij} = b_{ij} + b_{ji} \quad (i < j).$$

The transformations in (4·3·16) and (4·3·17) are independent, hence

$$D_4 = D(A, B_{11}, \eta; \bar{X}_{11}, X_{12}, \bar{X}_{22}) = D(a_{ii}, b_{ii}, \eta_i; x_{ii}, y_{ii}, z_{ii})\, D(a_{ij}, b_{ij}; x_{ij}, y_{ij}, z_{ij}).$$

For $D(a_{ii}, b_{ii}, \eta_i; x_{ii}, y_{ii}, z_{ii})$ we have the scheme of coefficients:

$$\begin{array}{c|ccc} & a_{ii} & b_{ii} & \eta_i \\ \hline x_{ii} & 2I & 0 & 0 \\ z_{ii} & 0 & 2I & 0 \\ y_{ii} & L & M & I \end{array} = 2^{2p},$$

the values L and M not affecting the determinant. To compute $D(a_{ij}, b_{ij}; x_{ij}, y_{ij}, z_{ij})$ we rewrite in (4·3·17):

$$y_{ij} = r_j a_{ij} + r_i b_{ij} \quad (i < j = 1, \ldots, p),$$
$$y_{ji} = r_i a_{ji} + r_j b_{ji} \quad (i < j = 1, \ldots, p).$$

We then have the scheme of coefficients ($i < j$ in all cases)

$$\begin{array}{c|cc:cc} & a_{ij} & b_{ji} & a_{ji} & b_{ij} \\ \hline x_{ij} & I & 0 & I & 0 \\ z_{ij} & 0 & I & 0 & I \\ \hdashline y_{ji} & 0 & C & D & 0 \\ y_{ij} & C & 0 & 0 & D \end{array} = \left| \begin{bmatrix} D & 0 \\ 0 & D \end{bmatrix} - \begin{bmatrix} 0 & C \\ C & 0 \end{bmatrix} \right| = |D|\,|D - CD^{-1}C| = |D^2 - C^2|,$$

using 5 A. 6 and 5 A. 7 twice. C is a diagonal matrix with elements

$$r_2, r_3, \ldots, r_p;\; r_3, \ldots, r_p;\; \ldots;\; r_{p-1}, r_p;\; r_p;$$

D is a diagonal matrix with elements

$$r_1, \ldots, r_1\,(p-1 \text{ terms}), r_2, \ldots, r_2\,(p-2 \text{ terms}), \ldots, r_{p-1}.$$

Therefore $$D^2 - C^2 = \prod_{i<j}^{p} (r_i^2 - r_j^2),$$

and hence

(D) $$D_4 = D(A, B_{11}, \eta; \bar{X}_{11}, X_{12}, \bar{X}_{22}) = 2^{2p} \prod_{i<j}^{p} (r_i^2 - r_j^2).$$

Computation of $D(B_{21}, Q_{12}^*, \tilde{T}^*; X_{13}, X_{23}, \bar{X}_{33}) = D_5$. The connecting equations are given in (4·3·14). B_{22} is independent of the other variables, but recall that the original variables are Q_{12}^* and Q_{22}^* and that the relation (4·3·11) does not define B_{12} and B_{22} as new variables.

The scheme of coefficients, from (4·3·14), is

(4·3·18)
$$\begin{array}{c|c:cc} & B_{21} & Q_{12}^* & \tilde{T}^* \\ \hline X_{13} & F & 0 & 0 \\ \hdashline X_{23} & & & \\ \bar{X}_{33} & L & G & \end{array} = |F|\,|G|,$$

so the value of L does not affect the value of the determinant. $|F|$ is the J of the transformation $X_{13} = D_r B_{21}'$ $(X_{13}, B_{21}': p \times \delta;\ D_r: p \times p)$ which by Theorem 3·4 is $|D_r|^\delta = \left(\prod_{i=1}^{p} r_i\right)^\delta$. Since the value of B_{21} does not affect G, we can, for determining G, write $X_{23} = B_{12}$ and $\bar{X}_{33} = B_{22} + B_{22}'$. Therefore

(E) $$D_5 = \left(\prod_{i=1}^{p} r_i\right)^\delta D_6; \quad D_6 = D(Q_{12}^*, \tilde{T}^*; X_{23}, \bar{X}_{33}).$$

Computation of D_6. The connecting equations between $B_{12}, B_{22}; Q_{12}^*, \tilde{T}^*; X_{23}, \bar{X}_{33}$ are

$$\begin{bmatrix} B_{12} \\ B_{22} \end{bmatrix} = Q^{-1} \begin{bmatrix} Q_{12}^* \\ \tilde{T}^{*\prime} \end{bmatrix},$$
$$X_{23} = B_{12},$$
$$\bar{X}_{33} = B_{22} + B_{22}'.$$

As previously stated, B_{12} and B_{22} cannot be considered as new variables. We therefore do not have any direct equations between the variables of $X_{23}, \bar{X}_{33}$ and $Q_{12}^*, \tilde{T}^*$. It is clear, however, that the elements of $X_{23}, \bar{X}_{33}$ are linear functions of the elements of $Q_{12}^*, \tilde{T}^*$, and hence we can apply Lemma 4·3.

Let $W = (w_{ij})$: $(\delta \times p)$ and $\bar{S}$: $(\delta \times \delta)$, where $\bar{S} = (\epsilon_{ij} s_{ij})$ $(i,j = 1, \ldots, \delta)$; $\epsilon_{ij} = 1$ if $i = j$; $\epsilon_{ij} = \frac{1}{2}$ if $i \neq j$.

Let
$$\begin{aligned} \operatorname{tr} H &= \operatorname{tr}(W X_{23} + \bar{S} \bar{X}_{33}) \\ &= \operatorname{tr}(W B_{12} + \bar{S} B_{22} + \bar{S} B_{22}') = \operatorname{tr}(W B_{12} + 2\bar{S} B_{22}) \text{ (by 5 A. 12)}; \\ &= \operatorname{tr}\left((W\ 2\bar{S}) \begin{bmatrix} B_{12} \\ B_{22} \end{bmatrix}\right) \\ &= \operatorname{tr}\left((W\ 2\bar{S})\, K \begin{bmatrix} Q_{12}^* \\ \tilde{T}^{*\prime} \end{bmatrix}\right), \end{aligned}$$

where $K = Q^{-1}$.

Writing $Q_{12}^* = (\beta_{ij})$ $(i = 1, \ldots, p; j = 1, \ldots, \delta)$, $\tilde{T}^* = (\tau_{ij})$ $(i \leqslant j = 1, \ldots, \delta)$:

$$(4\cdot3\cdot19) \qquad \operatorname{tr} H = \operatorname{tr}\begin{bmatrix} w_{11} & \cdots & w_{1p} & 2s_{11} & s_{12} & \cdots & s_{1\delta} \\ \cdots & \cdots & \cdots & s_{12} & 2s_{22} & \cdots & s_{2\delta} \\ \cdots & \cdots & \cdots & \cdots & \cdots & \cdots & \cdots \\ w_{\delta 1} & \cdots & w_{\delta p} & s_{1\delta} & \cdots & \cdots & 2s_{\delta\delta} \end{bmatrix} K \begin{bmatrix} \beta_{11} & \cdots & \beta_{1\delta} \\ \cdots & \cdots & \cdots \\ \beta_{p1} & \cdots & \beta_{p\delta} \\ \tau_{11} & \cdots & \tau_{1\delta} \\ \cdots & \cdots & \cdots \\ 0 & \cdots & \tau_{\delta\delta} \end{bmatrix}$$

H is $\delta \times \delta$ and $\operatorname{tr} H = \sum_{j=1}^{\delta} h_{jj}$. h_{jj} is found by multiplying the jth row of $(W\ 2\bar{S})$ into K and then multiplying the resulting row vector into the jth column of $\begin{bmatrix} Q_{12}^* \\ \tilde{T}^{*\prime} \end{bmatrix}$. Therefore

$$\operatorname{tr} H = \sum_{j=1}^{\delta} (w_{j1} \ldots w_{jp}\, s_{j1} \ldots 2s_{jj} \ldots s_{j\delta})\, K \begin{bmatrix} \beta_{1j} \\ \vdots \\ \beta_{pj} \\ \tau_{1j} \\ \vdots \\ \tau_{jj} \end{bmatrix}$$

We now partition K so that the partition is a function of j:

$$(4\cdot3\cdot20) \qquad Q^{-1} = K = \begin{bmatrix} K_{11}^{(j)} & K_{12}^{(j)} \\ K_{21}^{(j)} & K_{22}^{(j)} \end{bmatrix} \quad (1 \leqslant j \leqslant \delta),$$

$$(K_{11}^{(j)}\colon (p+j) \times (p+j);\ K_{12}^{(j)}\colon (p+j) \times (\delta - j);\ K_{22}^{(j)}\colon (\delta-j) \times (\delta - j)).$$

Now we can write

$$(4\cdot3\cdot21) \qquad \operatorname{tr} H = \sum_{j=1}^{\delta} \left\{ (w_{j1} \ldots w_{jp}\, s_{j1} \ldots 2s_{jj})\, K_{11}^{(j)} \begin{bmatrix} \beta_{1j} \\ \vdots \\ \beta_{pj} \\ \tau_{1j} \\ \vdots \\ \tau_{jj} \end{bmatrix} + (s_{j,j+1} \ldots s_{j\delta})\, K_{21}^{(j)} \begin{bmatrix} \beta_{1j} \\ \vdots \\ \beta_{pj} \\ \tau_{1j} \\ \vdots \\ \tau_{jj} \end{bmatrix} \right\}.$$

By Lemma 4·3, $D(Q_{12}^*, \tilde{T}^*; X_{23}, \bar{X}_{33})$ is the determinant of this bilinear form. Writing the terms for $j = 1, 2, \ldots, \delta$ in order we find that the discriminant has 0's above the main diagonal, and the product of the terms in the main diagonal is

$$(4\cdot3\cdot22) \qquad 2\,|K_{11}^{(1)}|\, 2\,|K_{11}^{(2)}| \ldots 2\,|K_{11}^{(\delta)}| = 2^{\delta} \prod_{j=1}^{\delta} |K_{11}^{(j)}|;$$

$$|K_{11}^{(j)}| = \begin{vmatrix} k_{11} & \cdots & k_{1,p+j} \\ \cdots & \cdots & \cdots \\ k_{p+j,1} & \cdots & k_{p+j,p+j} \end{vmatrix},$$

where the elements k_{ih} are the elements of Q^{-1}. Let A_{ih} be an element of the adjoint of Q. Then $k_{ih} = \frac{1}{|Q|} A_{hi}$. Therefore

$$|K_{11}^{(j)}| = \frac{1}{|Q|^{p+j}} \begin{vmatrix} A_{11} & \cdots & A_{1,p+j} \\ \cdots & \cdots & \cdots \\ A_{p+j,1} & \cdots & A_{p+j,p+j} \end{vmatrix} = \frac{1}{|Q|} \begin{vmatrix} q_{p+j+1,p+j+1} & \cdots & q_{p+j+1,p+\delta} \\ \cdots & \cdots & \cdots \\ q_{p+\delta,p+j+1} & \cdots & q_{p+\delta,p+\delta} \end{vmatrix},$$

by 5 A. 11. But $Q = \begin{bmatrix} Q_{11} & Q_{12} \\ Q_{21} & \tilde{T}' \end{bmatrix}$ $(Q_{11}: p \times p;\ Q_{12}: p \times \delta;\ \tilde{T}': \delta \times \delta)$.

Therefore
$$q_{p+k+1,\, p+h+1} = t_{k+1,\, h+1}.$$
Therefore

(4·3·23)
$$| K_{11}^{(j)} | = \frac{1}{|Q|} \begin{vmatrix} t_{j+1,j+1} & \cdots & t_{j+1,\delta} \\ \cdots & \cdots & \cdots \\ 0 & \cdots & t_{\delta\delta} \end{vmatrix} = \frac{1}{|Q|} t_{j+1,j+1} \cdots t_{\delta\delta}.$$

On substituting this result in (4·3·22) we find

(F)
$$D_6 = D(Q_{12}^*, \tilde{T}^*;\ X_{23}, \bar{X}_{33}) = 2^\delta \prod_{j=1}^{\delta} | K_{11}^{(j)} | = 2^\delta \,| Q |^{-\delta} t_{22} t_{33}^2 \cdots t_{\delta\delta}^{\delta-1}.$$

We now consolidate the results from equations (A) through (F):

$$J = 2^{p+q} \,| P |^{q+1} \,| Q |^{p+1} \left(\prod_{i=2}^{\delta} t_{ii}^{i-1} \right) \left(\prod_{i=1}^{p} r_i \right)^{\delta} \prod_{i<j}^{p} (r_i^2 - r_j^2). \quad \text{Q.E.D.}$$

THEOREM 4·4. The J of the transformation

(4·4·1)
$$\bar{U} = | \bar{V} | \bar{V}^{-1},$$

where $\bar{V}: p \times p$ is p.d., is $D(\bar{V};\ \bar{U}) = (p-1) \,| \bar{V} |^{\frac{1}{2}(p+1)(p-2)}$.

Proof. Take differentials of (4·4·1) and let $\bar{U}^* = d\bar{U}$, $\bar{V}^* = d\bar{V}$; then

(4·4·2)
$$\bar{U}^* = | \bar{V} | \operatorname{tr} (\bar{V}^{-1} \bar{V}^*) \bar{V}^{-1} - | \bar{V} | (\bar{V}^{-1} \bar{V}^* \bar{V}^{-1}),$$

using 5 A. 13, 5 A. 14 and 5 A. 15.

By 5 B. 3, $$J = D(\bar{V}^*;\ \bar{U}^*).$$

$\bar{V}$, and hence $\bar{V}^{-1}$, is p.d., so by 5 A. 16 we can factor $\bar{V}^{-1}$ into $\tilde{T}\tilde{T}'$. Substituting this for $\bar{V}^{-1}$ in (4·4·2) and pre- and post-multiplying the result by $\tilde{T}^{-1}$ and $\tilde{T}'^{-1}$ respectively,

(4·4·3)
$$\tilde{T}^{-1} \bar{U}^* \tilde{T}'^{-1} = | \bar{V} | \operatorname{tr} (\tilde{T}' \bar{V}^* \tilde{T}) I - | \bar{V} | (\tilde{T}' \bar{V}^* \tilde{T}).$$

Let

(4·4·4)
$$\begin{cases} \bar{Z} = \tilde{T}^{-1} \bar{U}^* \tilde{T}'^{-1} & (\bar{Z}: p \times p), \\ \bar{W} = \tilde{T}' \bar{V}^* \tilde{T} & (\bar{W}: p \times p). \end{cases}$$

By Theorem 3·7 and Corollary 3·7

(4·4·5)
$$\begin{cases} D(\bar{U}^*;\ \bar{Z}) = | \tilde{T} |^{-(p+1)}, \\ D(\bar{V}^*;\ \bar{W}) = | \tilde{T} |^{p+1}. \end{cases}$$

Putting (4·4·4) in (4·4·3)

(4·4·6)
$$\bar{Z} = | \bar{V} | (\operatorname{tr} \bar{W}) I - | \bar{V} | \bar{W}.$$

Then
$$\begin{aligned} D(\bar{V}^*;\ \bar{U}^*) &= D(\bar{V}^*;\ \bar{W})\, D(\bar{W};\ \bar{Z})\, D(\bar{Z};\ \bar{U}^*) \quad \text{(by 5 B. 2)} \\ &= | \tilde{T} |^{p+1} D(\bar{W};\ \bar{Z}) \,| \tilde{T} |^{p+1} \\ &= | \bar{V} |^{-(p+1)} D(\bar{W};\ \bar{Z}). \end{aligned}$$

Computation of $D(\bar{W};\ \bar{Z})$. Let

(4·4·7)
$$\bar{S} = | \bar{V} | \bar{W} \quad (\bar{S}: p \times p).$$

By Theorem 3·1 *b*: $$D(\bar{W};\ \bar{S}) = | \bar{V} |^{\frac{1}{2}p(p+1)}.$$

Substitute (4·4·7) in (4·4·6):

(4·4·8)
$$\bar{Z} = (\operatorname{tr} \bar{S}) I - \bar{S},$$

whence we find

(4·4·9)
$$z_{ii} = \sum_{\nu=1}^{p} s_{\nu\nu} - s_{ii} = \sum_{\substack{\nu=1\\ \nu\neq i}}^{p} s_{\nu\nu}.$$

Therefore
$$\frac{\partial z_{ii}}{\partial s_{ii}} = 0; \qquad \frac{\partial z_{ii}}{\partial s_{jj}} = 1 \quad (i \neq j); \qquad \frac{\partial z_{ii}}{\partial s_{ij}} = 0 \quad (i \neq j).$$
$$z_{ij} = -s_{ij} \quad (i<j),$$

and therefore
$$\frac{\partial z_{ij}}{\partial s_{ij}} = -1.$$

The scheme of coefficients is

(4·4·10)

	s_{ii}	s_{ij} $(i<j)$
z_{ii}	H	0
z_{ij} $(i<j)$	0	$-I$

the determinant of which is

(4·4·11)
$$H = \begin{vmatrix} 0 & 1 & 1 & \dots & 1 \\ 1 & 0 & 1 & \dots & 1 \\ \dots & \dots & \dots & \dots & \dots \\ 1 & 1 & 1 & \dots & 0 \end{vmatrix} = D(\bar{S}; \bar{Z}).$$

Let $\bar{h}$ be a column vector, $p \times 1$, consisting entirely of 1's. Then $\bar{h}\bar{h}'$ is a $p \times p$ matrix consisting entirely of 1's, and $D(\bar{S}; \bar{Z}) = |\bar{h}\bar{h}' - I_p|$. Using 5 A. 8 we get
$$D(\bar{S}; \bar{Z}) = |1 - \bar{h}'\bar{h}| = p-1.$$

Consolidating the results
$$\begin{aligned} J = D(\bar{V}^*; \bar{U}^*) &= D(\bar{V}^*; \bar{W})\, D(\bar{W}; \bar{S})\, D(\bar{S}; \bar{Z})\, D(\bar{Z}; \bar{U}^*) \\ &= |\bar{V}|^{-(p+1)} |\bar{V}|^{\frac{1}{2}p(p+1)} (p-1) \\ &= (p-1) |\bar{V}|^{\frac{1}{2}(p+1)(p-2)}. \quad \text{Q.E.D.} \end{aligned}$$

COROLLARY 4·4. The J of the inverse of the transformation of Theorem 4·4 is
$$D(\bar{U}; \bar{V}) = \frac{1}{p-1} |\bar{U}|^{-(p+1)(p-2)/[2(p-1)]}.$$

Proof.
$$|\bar{U}| = |\bar{V}|^{p} |\bar{V}|^{-1} = |\bar{V}|^{p-1},$$

and therefore
$$|\bar{V}| = |\bar{U}|^{1/(p-1)}.$$
$$D(\bar{U}; \bar{V}) = \frac{1}{D(\bar{V}; \bar{U})} = \frac{1}{p-1} |\bar{V}|^{-\frac{1}{2}(p+1)(p-2)}$$
$$= \frac{1}{p-1} |\bar{U}|^{-(p+1)(p-2)/[2(p-1)]}. \quad \text{Q.E.D.}$$

THEOREM 4·5.† The J of the transformation $U = \tilde{T}[2(\hat{X}+I)^{-1} - I]$ is
$$D(\tilde{T}, \hat{X}; U) = 2^{\frac{1}{2}\{p(p-1)\}} \left(\prod_{i=1}^{p-1} t_{ii}^{p-i}\right) |\hat{X}+I|^{-(p-1)} \quad (\hat{X}, \tilde{T}, U: p \times p;\ U \text{ n.s.}).$$

This transformation is unique if $|\tilde{T}^{-1}U + I| \neq 0$ (see 5 A. 18).

† This transformation has been used by P. L. Hsu in unpublished work on the distribution of roots of determinantal equations under certain non-null hypotheses.

Proof. Since $\hat{X}$ is skew symmetric it does not have a characteristic root -1, so $|\hat{X}+I| \neq 0$. By 5 A. 17 and 5 A. 19 we can write $UU' = \tilde{T}\tilde{T}'$. Therefore $\tilde{T}^{-1}UU'\tilde{T}'^{-1} = I$, and hence $\Gamma = \tilde{T}^{-1}U$ is orthogonal. Then

$$U = \tilde{T}\Gamma = \tilde{T}[2(\hat{X}+I)^{-1}-I] \quad \text{(by 5 A. 18)}. \tag{4·5·1}$$

Take differentials of (4.5.1) and put $U^* = dU$; $\tilde{T}^* = d\tilde{T}$; $\hat{X}^* = d\hat{X}$. Then using 5 A. 20 we get

$$U^* = \tilde{T}^*\Gamma - \tfrac{1}{2}\tilde{T}(\Gamma+I)\hat{X}^*(\Gamma+I), \tag{4.5.2}$$

and by 5 B. 3

$$J = D(\tilde{T}^*, \hat{X}^*; U^*). \tag{4·5·3}$$

Let

$$\begin{cases} W = \tilde{T}^{-1}U^*\Gamma', \\ \tilde{R} = \tilde{T}^{-1}\tilde{T}^*, \\ \hat{Y} = (\Gamma+I)\hat{X}^*(\Gamma'+I). \end{cases} \tag{4·5·4}$$

Pre- and post-multiply (4·5·2) by $\tilde{T}^{-1}$ and Γ' respectively and substitute from (4·5·4). Then

$$W = \tilde{R} - \tfrac{1}{2}\hat{Y}. \tag{4·5·5}$$

By 5 B. 2:

$$D(\tilde{T}^*, \hat{X}^*; U^*) = D(\tilde{T}^*, \hat{X}^*; \tilde{R}, \hat{Y})\, D(\tilde{R}, \hat{Y}; W)\, D(W; U^*), \tag{4·5·6}$$

which by 5 B. 4 becomes

$$= D(\tilde{T}^*; \tilde{R})\, D(\hat{X}^*; \hat{Y})\, D(\tilde{R}, \hat{Y}; W)\, D(W; U^*).$$

From (4·5·4), using Theorems 3·6, 3·8 and 3·10,

$$D(W; U^*) = |\tilde{T}|^p \quad (\text{since } |\Gamma| = 1),$$

$$D(\tilde{T}^*; \tilde{R}) = \prod_{i=1}^{p} t_{ii}^{-i},$$

$$D(\hat{X}^*; \hat{Y}) = |\Gamma+I|^{p-1}.$$

From (4·5·5) $\quad w_{ii} = r_{ii}; \quad w_{ij} = -\tfrac{1}{2}y_{ij} \quad (i<j);$

$$w_{ji} = r_{ij} + \tfrac{1}{2}y_{ij} \quad (i<j).$$

The scheme of coefficients is

	r_{ii}	y_{ij} $(i<j)$	r_{ij} $(i<j)$
w_{ii}	I	0	0
w_{ij} $(i<j)$	0	$-\frac{1}{2}I$	0
w_{ji} $(i<j)$	0	$\frac{1}{2}I$	I

Therefore $\quad D(\tilde{R}, \hat{Y}; W) = 2^{-\frac{1}{2}p(p-1)}.$

Hence, consolidating the results,

$$J = 2^{-\frac{1}{2}p(p-1)}\left(\prod_{i=1}^{p-1} t_{ii}^{p-i}\right)|\Gamma+I|^{p-1}.$$

By 5 A. 19, $\quad |\Gamma+I|^{p-1} = 2^{p(p-1)}|\hat{X}+I|^{-(p-1)},$

and we have $\quad J = 2^{\frac{1}{2}p(p-1)}\left(\prod_{i=1}^{p-1} t_{ii}^{p-i}\right)|\hat{X}+I|^{-(p-1)}.$ Q.E.D.

THEOREM 4·6. The J of the transformation

$$\bar{S} = [2(\hat{X}+I)^{-1}-I]\,D_\theta[2(\hat{X}+I)^{-1}-I]'$$

is $$D(\hat{X},\theta;\,\bar{S}) = 2^{\frac{1}{2}p(p-1)}|\,\hat{X}+I\,|^{-(p-1)}\prod_{i<j}^{p}(\theta_i-\theta_j) \quad (\bar{S},\,D_\theta,\,\hat{X}:\,p\times p).$$

Proof. By 5 A. 19 we can write

(4·6·1) $$\bar{S} = \Gamma D_\theta \Gamma'.$$

Let

(4·6·2) $$\begin{cases}\bar{S}^* = d\bar{S},\\ \hat{X}^* = d\hat{X},\\ D_\eta = dD_\theta.\end{cases}$$

Take differentials of (4·6·1) using 5 A. 20 and substitute from (4·6·2):

(4·6·3) $$\bar{S}^* = -\tfrac{1}{2}(\Gamma+I)\,\hat{X}^*(\Gamma+I)\,D_\theta\Gamma'+\tfrac{1}{2}\Gamma D_\theta(\Gamma'+I)\,\hat{X}^*(\Gamma'+I)+\Gamma D_\eta\Gamma'.$$

(Note that $\hat{X}' = -\hat{X}$.) By 5 B. 3

(4·6·4) $$J = D(\hat{X}^*,\eta;\,\bar{S}^*).$$

Let

(4·6·5) $$\begin{cases}\bar{V} = \Gamma'\bar{S}^*\Gamma,\\ \hat{Y} = (\Gamma'+I)\,\hat{X}^*(\Gamma+I).\end{cases}$$

Pre- and post-multiply (4·6·3) by Γ' and Γ respectively and substitute from (4·6·5):

(4·6·6) $$\bar{V} = -\tfrac{1}{2}\hat{Y}D_\theta+\tfrac{1}{2}D_\theta\hat{Y}+D_\eta,$$

whence from 5 B. 4

(4·6·7) $$D(\hat{X}^*,\eta;\,\bar{S}^*) = D(\hat{X}^*,\eta;\,\hat{Y},\eta)\,D(\hat{Y},\eta;\,\bar{V})\,D(\bar{V};\,\bar{S}^*).$$

From (4·6·5), using Theorems 3·7 and 3·10,

(4·6·8) $$\begin{cases}D(\bar{V};\,\bar{S}^*) = 1,\\ D(\hat{X}^*;\,\hat{Y}) = |\,\Gamma+I\,|^{p-1}.\end{cases}$$

From (4·6·6): $\quad v_{ii} = \eta_i;\quad v_{ij} = \tfrac{1}{2}\theta_i y_{ij}-\tfrac{1}{2}\theta_j y_{ij} = \tfrac{1}{2}(\theta_i-\theta_j)\,y_{ij}\quad (i<j).$

The scheme of coefficients is

	η_i	y_{ij}
v_{ii}	I	0
v_{ij}	0	H

where H is diagonal, with elements

$$\tfrac{1}{2}(\theta_1-\theta_2),\ \ldots,\ \tfrac{1}{2}(\theta_1-\theta_p),\ \tfrac{1}{2}(\theta_2-\theta_3),\ \ldots,\ \tfrac{1}{2}(\theta_{p-1}-\theta_p),$$

a total of $\tfrac{1}{2}p(p-1)$ terms. Therefore

(4·6·9) $$D(\hat{Y},\eta;\,\bar{V}) = 2^{-\frac{1}{2}p(p-1)}\prod_{i<j}^{p}(\theta_i-\theta_j),$$

and

(4·6·10) $$J = |\,\Gamma+I\,|^{p-1}\,2^{-\frac{1}{2}p(p-1)}\prod_{i<j}^{p}(\theta_i-\theta_j);$$

whence, using 5 A. 18, $$J = 2^{\frac{1}{2}p(p-1)}|\,\hat{X}+I\,|^{-(p-1)}\prod_{i<j}^{p}(\theta_i-\theta_j).$$ Q.E.D.

Note. The following comments relative to the uniqueness of the transformation of Theorem 4·6 have been made by P. L. Hsu in a communication to the Editor:

'THEOREM. The equation

$$\bar{S} = [2(I+\hat{X})^{-1}-I]D_\theta[2(I+\hat{X})^{-1}-I]'$$

establishes a (1, 1)-correspondence between $\bar{S}$ and $\hat{X}$, provided that

(i) $\bar{S}$ does not belong to a certain set of symmetric matrices which constitutes a set of measure zero (in the $\frac{1}{2}p(p+1)$-dimensional space), and

(ii) the elements (1, 2), (1, 3), ..., (1, p) of the matrix $(I+\hat{X})^{-1}$ are all positive.

Remark on the above theorem. The condition (ii) on $\hat{X}$ may be put in a simpler form as follows: if we write

$$\hat{X} = \begin{pmatrix} 0 & -\bar{x} \\ \bar{x}' & \hat{Y} \end{pmatrix},$$

then the row formed of the elements (1, 2), (1, 3), ..., (1, p) of $(I+\hat{X})^{-1}$ is equal to

$$\frac{\bar{x}(I+\hat{Y})^{-1}}{1+\bar{x}(I+\hat{Y})^{-1}\bar{x}'}.$$

Thus an equivalent to the condition (ii) is the following: $(\text{ii})_1$ every element of the row $\bar{x}(I+\hat{Y})^{-1}$ is positive.

As to the proof of the above theorem, which is not very short, I intend to give it in a paper to be published in one of the journals issued by the Chinese Academy of Sciences.'

5. APPENDIX

A. *Certain matrix theorems*

The following definitions and results from the theory of matrices and determinants are stated for reference purposes.

5 A. 1. Multiplication of the ith $\begin{matrix}\text{row}\\\text{column}\end{matrix}$ of a matrix by a constant $a \neq 0$ may be accomplished by $\begin{matrix}\text{pre-}\\\text{post-}\end{matrix}$multiplication by a diagonal matrix D, with a in the ith position and 1's elsewhere. $|D| = a$; D^{-1} is a matrix of the same type as D (see Frazer, Duncan & Collar (1938), § 3·11).

5 A. 2. To add $(a) \times \begin{matrix}(j\text{th row})\\(i\text{th column})\end{matrix}$ to the $\begin{matrix}i\text{th row}\\j\text{th column}\end{matrix}$, $a \neq 0$, $\begin{matrix}\text{pre-}\\\text{post-}\end{matrix}$multiply by a matrix E containing 1's in the main diagonal, a in the ith row and jth column, and 0's elsewhere. $|E| = 1$; E^{-1} is a matrix of the same type as E (see Frazer *et al.* (1938), § 3·11).

Any combination or succession of a finite number of operations of types 5 A. 1 and 5 A. 2 is called an *elementary transformation.*

5 A. 3. The rank of a matrix is unchanged by an elementary transformation (see Bôcher (1938), § 19).

5 A. 4. *Canonical forms.* A $p \times p$ matrix A of rank r can be reduced by an elementary transformation to a canonical form, i.e. to a diagonal matrix with 1's in the first r positions and 0's elsewhere. The reduction to a diagonal matrix of the characteristic roots can be effected by an orthogonal transformation (see Bôcher (1938), p. 280).

If the matrix A is $p\times q$, $p<q$, of rank r, there exist two orthogonal matrices $\Gamma: p\times p$ and $\Delta: q\times q$ such that $\Gamma' A\Delta = (\Lambda 0)$, where

$$\Lambda = \begin{bmatrix}\lambda_1 & & 0\\ & \ddots & \\ 0 & & \lambda_p\end{bmatrix} \quad (p\times p)$$

and $\lambda_1^2 \ldots \lambda_p^2$ are the characteristic roots of the matrix AA'. Notice that $p-r$ of the λ_i are zero (see Hsu, *Quart. J. Math.*, Oxford Series, **17** (1946), 162).

Proof. AA' is non-negative (see Albert (1946), p. 64), and hence its roots are non-negative and can be written $\lambda_1^2 \ldots \lambda_r^2$. The rank $(A) = r$ and hence there exists an orthogonal matrix Γ such that

$$\Gamma' AA'\Gamma = \begin{bmatrix}\lambda_1^2 & & 0\\ & \ddots & \\ 0 & & \lambda_p^2\end{bmatrix} = \Lambda^2$$

of the theorem (see Cramér, 1946, 11·9). Since we can rearrange the position of the λ's in Λ by an orthogonal transformation we can assume

$$\Lambda^2 = \begin{bmatrix}\lambda_1^2 & 0 & & & \\ & \ddots & & 0 & \\ 0 & \lambda_r^2 & & & \\ & & 0 & & \\ & 0 & & \ddots & \\ & & & & 0\end{bmatrix} = \begin{bmatrix}\Lambda_1^2 & 0\\ 0 & 0\end{bmatrix}$$

and $|\Lambda_1| \neq 0$.

Write $\Gamma = (\Gamma_1\Gamma_2)$, where $\Gamma_1: p\times r$; $\Gamma_2: p\times p-r$. Then

$$\Lambda^2 = \Gamma' AA'\Gamma = \begin{bmatrix}\Gamma_1'\\ \Gamma_2'\end{bmatrix} AA'(\Gamma_1\Gamma_2) = \begin{bmatrix}\Gamma_1' AA'\Gamma_1 & \Gamma_1' AA'\Gamma_2\\ \Gamma_2' AA'\Gamma_1 & \Gamma_2' AA'\Gamma_2\end{bmatrix} = \begin{bmatrix}\Lambda_1^2 & 0\\ 0 & 0\end{bmatrix},$$

(5 A. 4·1) $$\Gamma_1' AA'\Gamma_1 = \Lambda_1^2,$$

(5 A. 4·2) $$\Gamma_2' AA'\Gamma_2 = 0.$$

Pre- and post-multiply (5 A. 4·1) by Λ_1^{-1}:

(5 A. 4·3) $$\Lambda_1^{-1}\Gamma_1' AA'\Gamma_1\Lambda_1^{-1} = I_r.$$

Let $$\Delta_1 = A'\Gamma_1\Lambda_1^{-1} \quad (\Delta_1: q\times r)$$

and then

(5 A. 4·4) $$\Delta_1\Lambda_1 = A'\Gamma_1,$$

(5 A. 4·5) $$\Delta_1'\Delta_1 = I_r \quad \text{(by 5 A. 4·3)}.$$

From (5 A. 4·2) $(\Gamma_2' A)(\Gamma_2' A)' = 0$ and hence $\Gamma_2' A = 0$.

From (5 A. 4·5) we see that the columns of Δ_1 are orthogonal and hence there exists a matrix Δ_2 such that

(5 A. 4·6) $$\Delta = [\Delta_1\Delta_2]$$

is orthogonal. The matrices Γ and Δ satisfy the theorem, since on performing the multiplication we find

$$\Gamma' A = (\Lambda 0)\Delta'. \quad \text{Q.E.D.}$$

5 A. 5. An elementary transformation on a matrix B is equivalent to $E_1 \dots E_n B E_{n+1} \dots E_m$ where some of the E's are of the type 5 A. 1, and the remainder are of type 5 A. 2. Hence if B is square and n.s. we can write (by 5 A. 4)

$$\begin{aligned} B &= E_n^{-1} \dots E_1^{-1} E_m^{-1} \dots E_{n+1}^{-1} \\ &\equiv F_m F_{m-1} \dots F_1, \end{aligned}$$

where the F's are of type 5 A. 1 or 5 A. 2 (see Frazer *et al.* 1938, § 3·12).

5 A. 6. Diagonal matrices are commutative (see R. G. Frazer *et al.* 1938, § 1·6).

5 A. 7. If A and D are square, not necessarily of the same dimensions, and $|D| \neq 0$, then

$$\begin{vmatrix} A & B \\ C & D \end{vmatrix} = |D|\,|A - BD^{-1}C| \quad \text{(see Hsu, 1941–2).}$$

Proof.

$$\begin{vmatrix} A & B \\ C & D \end{vmatrix} = \left| \begin{bmatrix} A & B \\ C & D \end{bmatrix} \begin{bmatrix} I & 0 \\ -D^{-1}C & I \end{bmatrix} \right| = \left| \begin{bmatrix} A - BD^{-1}C & B \\ 0 & D \end{bmatrix} \right|$$

$$= |D|\,|A - BD^{-1}C|. \quad \text{Q.E.D.}$$

5 A. 8. $$|I + \bar{x}\bar{y}'| = 1 + \bar{y}'\bar{x} \quad \text{(see Aitken, 1939, § 37).}$$

5 A. 9. If $\bar{x}$ is $n \times 1$, there exists an orthogonal matrix Γ: $n \times n$ such that $\bar{x}'\Gamma = (0, \dots, 0, a)$, $a \geqslant 0$, with $a = 0$ if and only if $\bar{x}' = (0, \dots, 0)$.

Proof. If $\bar{x}' \neq (0, \dots, 0)$, choose Γ as any orthogonal matrix with elements in the last column

$$x_i \left(\sum_{i=1}^{n} x_i^2 \right)^{-\frac{1}{2}} \quad (i = 1, \dots, n).$$

Then $\bar{x}'\Gamma = \left[0, \dots, 0, \left(\sum_1^n x_i^2 \right)^{\frac{1}{2}} \right]$. If $\bar{x}' = (0, \dots, 0)$, any orthogonal matrix will do, and $a = 0$. Q.E.D.

5 A. 10. If M is any $m \times m$ matrix, there exists an orthogonal matrix Δ: $m \times m$, such that $M\Delta$ is triangular with 0's below the main diagonal, and with diagonal elements $\geqslant 0$.

Proof. By induction. If M is 1×1, then $\Delta = \pm 1$ will do. Assume true for a matrix of dimensions $(m-1) \times (m-1)$. Take Γ so that the product of the last row of M into Γ is $(0, \dots, 0, a)$, $a \geqslant 0$ (by 5 A. 9). Then

$$M\Gamma = \begin{bmatrix} M_1 & b \\ 0 & a \end{bmatrix} \quad (M_1\colon (m-1) \times (m-1);\ b\colon (m-1) \times 1;\ a\colon 1 \times 1).$$

By the hypothesis of the induction there exists an orthogonal matrix Γ_1 such that $M_1\Gamma_1$ is triangular with 0's below the main diagonal. Let

$$\Delta = \Gamma \begin{bmatrix} \Gamma_1 & 0 \\ 0 & 1 \end{bmatrix}.$$

Then $$M\Delta = M\Gamma \begin{bmatrix} \Gamma_1 & 0 \\ 0 & 1 \end{bmatrix} = \begin{bmatrix} M_1 & b \\ 0 & a \end{bmatrix} \begin{bmatrix} \Gamma_1 & 0 \\ 0 & 1 \end{bmatrix} = \begin{bmatrix} M_1\Gamma_1 & b \\ 0 & a \end{bmatrix},$$

which is of the required form. Δ is the product of two orthogonal matrices and hence is orthogonal. Q.E.D.

5 A. 11. If D_1 is the adjoint of any determinant D, and M and M_1 are corresponding m-rowed minors of D and D_1 respectively, then M_1 is equal to the product of D^{m-1} by the algebraic complement of M (see Bôcher, 1938, § 11).

5 A. 12. If A: $m \times n$ and B: $n \times m$, then $\operatorname{tr}(AB) = \operatorname{tr}(BA)$.

Proof.
$$\operatorname{tr}(AB) = \sum_{i=1}^{m} \sum_{j=1}^{n} a_{ij} b_{ji} = \sum_{i=1}^{n} \sum_{j=1}^{m} b_{ij} a_{ji} = \operatorname{tr}(BA). \quad \text{Q.E.D.}$$

DEFINITION 5 A. 13 *a*. If $f(x_1, \ldots, x_n)$ is a differentiable function of $x_1 \ldots x_n$, then
$$df = \sum_{i=1}^{n} \frac{\partial f}{\partial x_i} dx_i$$
is called the differential of f.

DEFINITION 5 A. 13 *b*. If $A = (a_{ij})$ $(i = 1, \ldots, p; j = 1, \ldots, q)$, where the a_{ij} are functions of $(x_1 \ldots x_r) = a_{ij}(x_1 \ldots x_r)$, then
$$dA = (da_{ij}) \quad (i = 1, \ldots, p; j = 1, \ldots, q).$$

5 A. 13.
$$d(AB) = (dA)B + A(dB).$$

Proof.
$$d(AB) = d(\sum_j a_{ij} b_{jk}) = (\sum_j d(a_{ij} b_{jk}))$$
$$= \sum_j [(da_{ij}) b_{jk} + a_{ij}(db_{jk})] = (dA)B + A(dB). \quad \text{Q.E.D.}$$

5 A. 14. If A is square, n.s., $d|A| = |A| \operatorname{tr}[A^{-1}(dA)]$.

Proof. Let
$$A^{-1} = (a^{ij}).$$

(5 A. 14·1)
$$a^{ij} = (\text{cofactor } a_{ji}) |A|^{-1},$$

(5 A. 14·2)
$$d|A| = \sum_i \sum_j \frac{\partial |A|}{\partial a_{ij}} da_{ij} \quad \text{(by Definition 5 A. 13}\,a\text{)}.$$

(5 A. 14·3)
$$\frac{\partial |A|}{\partial a_{ij}} = \text{cofactor } a_{ij} = |A| a^{ji} \quad \text{(by 5 A. 14·1)}.$$

Substitute (5 A. 14·3) in (5 A. 14·2):

(5 A. 14·4)
$$d|A| = |A| \sum_i \sum_j a^{ji} da_{ij}.$$

The (i, i) term of $A^{-1}(dA)$ is $\sum_j a^{ij}(da_{ji})$, and $\operatorname{tr} A^{-1}(dA) = \sum_i \sum_j a^{ij}(da_{ji})$, which when put in (5 A. 14·4) gives the result of the theorem. Q.E.D.

5 A. 15. If A is square, n.s., $(dA^{-1}) = -A^{-1}(dA)A^{-1}$.

Proof.
$$d(AA^{-1}) = (dA)A^{-1} + A(dA^{-1}) = 0. \quad \text{Q.E.D.}$$

5 A. 16. A p.d. matrix $\overline{V}$: $p \times p$ can be factored uniquely into $\overline{V} = \tilde{T}\tilde{T}'$, $t_{ii} > 0$ $(i = 1, \ldots, p)$ (see Mahalanobis, Bose & Roy, 1937).

5 A. 17. If U: $p \times p$ is n.s., then UU' is p.d. (see Birkhoff & MacLane, 1947, p. 246).

5 A. 18. If Γ is orthogonal and $|\Gamma + I| \neq 0$, there is a one-to-one relation between Γ and $\hat{X}$:
$$\Gamma = 2(\hat{X} + I)^{-1} - I,$$
$$\hat{X} = 2(\Gamma + I)^{-1} - I.$$

Proof. Show that $\hat{X}$ is skew symmetric, i.e. $\hat{X} + \hat{X}' = 0$.

5 A. 19. UU' of 5 A. 17 can be expressed uniquely as $\tilde{T}\tilde{T}'$, where $t_{ii} > 0$ $(i = 1, \ldots, p)$. If $|\tilde{T}^{-1}U + I| \neq 0$, U can be expressed uniquely in the form

$$U = \tilde{T}[2(\hat{X} + I)^{-1} - I].$$

Proof. Follows from 5 A. 16, 5 A. 17 and 5 A. 18.

5 A. 20. Given Γ as in 5 A. 18, then

$$d\Gamma = -\tfrac{1}{2}(\Gamma + I)(d\hat{X})(\Gamma + I).$$

5 A. 21. Let
$$\bar{V} = \begin{bmatrix} \bar{V}_{11} & V_{12} \\ V_{21} & \bar{V}_{22} \end{bmatrix}$$

be p.d. ($\bar{V}_{11}$: $p \times p$; V_{12}: $p \times q$; $\bar{V}_{22}$: $q \times q$) $(p \leqslant q)$. Then there exists a matrix G of the form

$$\begin{bmatrix} G_1 & 0 \\ 0 & G_2 \end{bmatrix} \quad (G_1\colon p \times p;\ G_2\colon q \times q)$$

such that
$$G\bar{V}G' = \bar{\Phi} = \begin{bmatrix} I_p & D_r & 0 \\ D_r & & \\ 0 & I_q & \end{bmatrix},$$

where D_r: $p \times p$ has elements $r_1, \ldots, r_p$, and $r_1^2 \ldots r_p^2$ are the roots of the determinantal equation $|V_{12}\bar{V}_{22}^{-1}V_{21} - \theta\bar{V}_{11}| = 0$ (see H. Hotelling, 1936).

Proof. Since $\bar{V}$ is p.d., $\bar{V}_{11}$ and $\bar{V}_{22}$ are of rank p and q respectively, and hence by 5 A. 4 there exist two non-singular matrices F_1 and F_2 such that

$$F_1\bar{V}_{11}F_1' = I_p, \tag{5 A. 21·1}$$

$$F_2\bar{V}_{22}F_2' = I_q. \tag{5 A. 21·2}$$

Let $A = F_1V_{12}F_2'$; then by 5 A. 4 there exist two orthogonal matrices Γ_1 and Γ_2 such that

$$\Gamma_1 A \Gamma_2' = \begin{bmatrix} \lambda_1 & & 0 & \\ & \ddots & & 0 \\ 0 & & \lambda_p & \end{bmatrix}, \tag{5 A. 21·3}$$

where $\lambda_1^2, \ldots, \lambda_p^2$ are the characteristic roots of AA', i.e. roots of $|AA' - \lambda I| = 0$.

Pre- and post-multiply (5 A. 21·1) by Γ_1 and Γ_1' respectively, and pre- and post-multiply (5 A. 21·2) by Γ_2 and Γ_2' respectively. Then

$$\Gamma_1F_1\bar{V}_{11}F_1'\Gamma_1' = \Gamma_1I_p\Gamma_1' = I_p, \tag{5 A. 21·4}$$

$$\Gamma_2F_2\bar{V}_{22}F_2'\Gamma_2' = \Gamma_2I_q\Gamma_2' = I_q. \tag{5 A. 21·5}$$

But $AA' = F_1V_{12}F_2'F_2V_{21}F_1'$, and from (5 A. 21·1) and (5 A. 21·2)

$$\bar{V}_{11} = F_1^{-1}F_1'^{-1} \quad \text{and} \quad \bar{V}_{22}^{-1} = F_2'F_2.$$

Hence

$$\begin{aligned} AA' - \lambda I &= F_1[V_{12}\bar{V}_{22}^{-1}V_{21} - \lambda F_1^{-1}F_1'^{-1}]F_1' \\ &= F_1[V_{12}\bar{V}_{22}^{-1}V_{21} - \lambda\bar{V}_{11}]F_1'. \end{aligned} \tag{5 A. 21·6}$$

F_1 is n.s., so $|AA' - \lambda I| = 0$ is equivalent to $|V_{12}\bar{V}_{22}^{-1}V_{21} - \lambda\bar{V}_{11}| = 0$, and hence $\lambda_1^2, \ldots, \lambda_p^2$ are the roots of this last equation. Hence

$$\begin{bmatrix} \lambda_1 & & 0 & \\ & \ddots & & 0 \\ 0 & & \lambda_p & \end{bmatrix} = (D_r 0). \tag{5 A. 21·7}$$

24-2

Let $G_1 = \Gamma_1 F_1$ and $G_2 = \Gamma_2 F_2$; then G_1 and G_2 are the matrices of the theorem, since by substituting these values in G and multiplying out $G\bar{V}G'$ we get

$$G\bar{V}G' = \begin{bmatrix} \Gamma_1 F_1 \bar{V}_{11} F_1' \Gamma_1' & \Gamma_1 F_1 V_{12} F_2' \Gamma_2' \\ \Gamma_2 F_2 V_{21} F_1' \Gamma_1' & \Gamma_2 F_2 \bar{V}_{22} F_2' \Gamma_2' \end{bmatrix}, \tag{5 A. 21·8}$$

which by (5 A. 21·4), (5 A. 21·5) and (5 A. 21·7) is

$$\begin{bmatrix} I_p & D_r & 0 \\ D_r & & \\ 0 & I_q & \end{bmatrix}. \quad \text{Q.E.D.}$$

B. *General properties of Jacobians*

5 B. 1. $$D(Y;\, X) = \frac{1}{D(X;\, Y)}, \text{ if } D(X;\, Y) \neq 0.$$

5 B. 2. $D(X;\, Y) = D(X;\, U)\,D(U;\, V)\,D(V;\, Y)$, where U and V are any functions of X and Y such that none of the terms on the right vanish. The extension of this to more variables and to the introduction of functions of U and V is immediate.

5 B. 3. Given a matrix transformation, linear or not, $Y = F(X)$. Then the transformation of the differentials, $dY = dF(X)$, is linear in the differentials, and the J of the original transformation is the J of the linear transformation in the differentials.

Proof. $Y = F(X)$ implies a set of transformations in the elements of Y:

$$y_i = F_i(x_1 \dots x_m) \quad (i = 1, \dots, m).$$

Let $$(dy)' = (dy_1 \dots dy_m); \quad (dx)' = (dx_1 \dots dx_m).$$

Then the equation $$dY = dF(X)$$

is of the form $$(dy) = G(dx),$$

where $$G = \left[\frac{\partial f_i}{\partial x_j}\right] \quad (i,j = 1, \dots, m),$$

and G is by definition the J of the original transformation. Q.E.D.

5 B. 4. If $X = F_1(U)$ and $Y = F_2(V)$ are a transformation from variables X and Y to new variables U and V, then

$$J = D(U, V;\, X, Y) = D(U;\, X)\,D(V;\, Y).$$

We wish to express our gratitude to Prof. Harold Hotelling for his help and encouragement in the preparation of this paper.

We are grateful to Dr Ralph A. Bradley for the use of his lecture notes.

REFERENCES

AITKEN, A. C. (1939). *Determinants and Matrices.* Edinburgh: Oliver and Boyd.
ALBERT, A. A. (1946). *Introduction to Algebraic Theories.* Chicago: University of Chicago Press.
BIRKHOFF, G. & MACLANE, S. (1947). *A Survey of Modern Algebra.* New York: Macmillan Co.
BÔCHER, M. (1938). *Introduction to Higher Algebra.* New York: Macmillan Co.
CRAMÉR, H. (1946). *Mathematical Methods of Statistics.* Princeton: Princeton University Press.

FISHER, R. A. (1939). The sampling distribution of some statistics obtained from non-linear equations. *Ann. Eugen., Lond.*, **9**, 238.

FRAZER, R. G., DUNCAN, W. J. & COLLAR, A. R. (1938). *Elementary Matrices and some Applications to Dynamics and Differential Equations.* Cambridge University Press.

HOTELLING, H. (1936). Relations between two sets of variates. *Biometrika*, **28**, 321.

HSU, P. L. (1939). On the distribution of roots of certain determinantal equations. *Ann. Eugen., Lond.*, **9**, 250.

HSU, P. L. (1941–2). On the limiting distribution of the canonical correlations. *Biometrika*, **32**, 38.

MAHALANOBIS, P. C., BOSE, R. C. & ROY, S. N. (1937). Normalization of statistical variates and the use of rectangular coordinates in the theory of sampling distributions. *Sankhyā*, **3**, 1.

ROY, S. N. (1938–40). *p*-Statistics, or some generalizations in analysis of variance appropriate to multivariate problems. *Sankhyā*, **4**, 381.